Springer Series in Brain Dynamics 2

Series Editors:
E. Başar, W. J. Freeman, W.-D. Heiss,
D. Lehmann, F. H. Lopes da Silva,
E.-J. Speckmann

Erol Başar and Theodore H. Bullock (Eds.)

Brain Dynamics

Progress and Perspectives

With 206 Figures, Some in Color

Springer-Verlag Berlin Heidelberg New York
London Paris Tokyo Hong Kong

Professor Dr. EROL BAŞAR
Institute of Physiology
Medical University Lübeck
Ratzeburger Allee 160
D-2400 Lübeck 1, FRG

Professor Dr. THEODORE H. BULLOCK
Dept. of Neurosciences, A-001
University of California, San Diego
La Jolla, CA 92093, USA

ISBN-13:978-3-642-74559-1 e-ISBN-13:978-3-642-74557-7
DOI: 10.1007/978-3-642-74557-7

Library of Congress Cataloging-in-Publication Data. Brain dynamics: progress and perspectives: based on a conference in West Berlin in August 1987/Erol Başar and Theodore H. Bullock (editors) p. cm. – (Springer series in brain dynamics; 2) Based on the Second Brain Dynamics Conference held in Berlin on Aug. 10–14, 1987, as a satellite conference of the Budapest Congress of the International Brain Research Organization. Includes bibliographies ISBN-13:978-3-642-74559-1 (U.S.) 1. Electroencephalography–Congresses. 2. Neural circuitry–Congresses. 3. Brain–Magnetic fields–Congresses. 4. Neurophysiology–Congresses. I. Başar, Erol. II. Bullock, Theodore Holmes. III. Brain Dynamics Conference (2nd: 1987: Berlin, Germany) IV. International Brain Research Organization. Congress (2nd: 1987: Budapest, Hungary) V. Series. [DNLM: 1. Brain–physiology–congresses. 2. Electroencephalography–congresses. 3. Electromagnetics–congresses. 4. Electrophysiology–congresses. 5. Evoked Potentials–congresses. W1 SP65ME v. 2/WL 102 B8124 1987] QP376.5.B73 1989 599′.0188–dc20 DNLM/DLC 89-11362

Softcover reprint of the hardcover 1st edition 1989

2125/3130-543210 – Printed on acid-free paper

A Word about the Series BRAIN DYNAMICS

The *brain* is probably the most complexly structured matter known. For this reason, a deeper understanding of brain function demands a special selection of techniques and procedures without which significant progress cannot be expected. *Dynamics* is a concept that takes into account the changes of state of a system. Different situations are not considered as discrete states existing in isolation; rather, the transitions and mutual interactions between various states are examined.

The aim of the new series BRAIN DYNAMICS is to integrate diverse disciplines of brain research and the central theme of brain dynamics by publishing qualified contributions implementing a broad spectrum of highly sophisticated approaches. The following themes help to characterize the direction of the new series:

- EEG, MEG, evoked and event-related brain responses
- Neuropathology and brain function
- Neural populations and neural networks
- Model epilepsies
- Brain imaging, including computer tomography, PET, and NMR
- Cognitive functions of the brain
- Comparative neurophysiology
- Dynamics of neural populations at the cellular level
- Chaotic dynamics in brain function

The first two books to appear in this series embody the above criteria in that they contain contributions written by recognized neuroscientists presenting a variety of advanced approaches and concepts of great promise for future research. Neuroscientists are encouraged to send proposals for monographs, multiauthor volumes, or conference proceedings on any of the themes outlined above to one of the editors of the series. Suggestions for additional topics are also welcome.

EROL BAŞAR, Coordinating Editor

Preface

This volume is based on contributions to the second Brain Dynamics Conference, held in Berlin on August 10–14, 1987, as a satellite conference of the Budapest Congress of the International Brain Research Organization. Like the volume resulting from the first conference, *Dynamics of Sensory and Cognitive Processing by the Brain,* the present work covers new approaches to brain function, with emphasis on electromagnetic fields, EEG, event-related potentials, connectivistic views, and neural networks. Close attention is also paid to research in the emerging field of deterministic chaos and strange attractors.

The diversity of this collection of papers reflects a multipronged advance in a hitherto relatively neglected domain, i. e., the study of signs of dynamic processes in organized neural tissue in order both to explain them and to exploit them for clues to system function. The need is greater than ever for new windows. This volume reflects a historical moment, the moment when a relatively neglected field of basic research into available signs of dynamic processes ongoing in organized neural tissue is expanding almost explosively to complement other approaches. From the topics treated, this book should appeal, as did its predecessor, to neuroscientists, neurologists, scientists studying complex systems, artificial intelligence, and neural networks, psychobiologists, and all basic and clinical investigators concerned with new techniques of monitoring and analyzing the brain's electromagnetic activity.

The second Brain Dynamics Conference comprised sessions of invited lectures and a closing discussion workshop with nine panelists on the theme "How Brains May Work". In addition to contributions by conference participants, a few authors who did not attend the conference accepted the editors' invitation to write chapters for the second book. The two volumes, which can be considered complementary, are both concerned with presenting concepts and findings that characterize current research. The epilogue explains tentatively how the contributions in different parts of the book, as well as the closing workshop, might serve to widen avenues in brain research which have been relatively neglected in the years since progress in cellular and subcellular research, welcome as it is, has overshadowed the investigation of the organized nervous system.

Since the first Brain Dynamics Conference attracted great interest, one of the editors of this volume (E. B.) was encouraged to organize the second conference only 2 years later. On the basis of their longlasting scientific collaboration and pursuit of common interests, T. H. B., who took on the responsibility of chairing the closing workshop, also accepted E. B.'s proposal to co-edit the present book.

The editors are grateful to the contributors for their readiness and their promptness in making alterations and revisions of their manuscripts.

The present book is divided into five major parts, each introduced by comments from the editors. Part I deals with basic approaches expounding cellular-level and system-theoretical views of various central systems, including systems of glial cells and invertebrate and vertebrate systems, based on EEG and magnetoencephalography. Part II concerns linear and nonlinear theoretical approaches and ends with a discussion covering some aspects of theoretical physics related to artificial intelligence. Since the nonlinear dynamic approach was one of the important threads of the conference, this part of the book covers most of the current lines of work in this area, analyzed by pioneers in the field. Parts III and IV juxtapose human and animal studies of learning, memory, and motor function both at the cellular and at the EEG/ERP level.

Finally, Part V, containing the proceedings of the workshop "How Brains May Work," reflects one of the aims of the conference, to trigger discussion among scientists with different interests and different approaches whose interwoven themes might reinforce useful new trends in brain dynamics research.

E. Başar and T. H. Bullock, Editors

Acknowledgements

The organization of the second Brain Dynamics Conference in Berlin only 2 years after the first one was made possible by the coordinated efforts of members of the Neurophysiology Research Group in Lübeck. I want to express a great vote of thanks to all the group, including my graduate students.

Special thanks are due to my secretary Mrs. Kristine Kay, who coordinated all the correspondence and to Dr. Joachim Röschke, who organized all local events in Berlin with great ability. Our chief engineer Mr. Ferdinand Greitschus, who took the responsibility of technical organization merits additional thanks.

I wish to thank the following institutions for providing financial support: the Berlin Senate; the U.S. Air Force; Bayer AG; Cadwell Europe Ltd; Dantec GmbH; Data Analysis Computer Systeme GmbH; Drägerwerke AG Lübeck; Hoechst AG; Ernst Leitz GmbH; Madaus GmbH; Nicolet Instrument GmbH; Schwarzer Picker International GmbH; Schwind Medizintechnik.

E. BAŞAR, Conference Chairman

Contents

I. Basic Approaches

Introductory Remarks
T. H. BULLOCK 3

The Micro-EEG Represents Varied Degrees of Cooperativity Among Wide-Band Generators: Spatial and Temporal Microstructure of Field Potentials
T. H. BULLOCK 5

Electrogenesis of Evoked Potentials
R. GALAMBOS (With 6 Figures) 13

Cell Membranes, Electromagnetic Fields, and Intercellular Communication
W. R. ADEY (With 3 Figures) 26

The EEG is a Quasi-Deterministic Signal Anticipating Sensory-Cognitive Tasks
E. BAŞAR, C. BAŞAR-EROGLU, J. RÖSCHKE, and A. SCHÜTT (With 17 Figures) 43

Microstates of the Brain in EEG and ERP Mapping Studies
D. LEHMANN (With 2 Figures) 72

Spatiotemporal Aspects of Synergetic Processes in the Auditory Cortex as Revealed by the Magnetoencephalogram
M. HOKE, K. LEHNERTZ, C. PANTEV, and B. LÜTKENHÖNER (With 12 Figures) 84

II. Theoretical Approaches to Brain Function (Linear and Nonlinear)

Introductory Remarks
E. BAŞAR 109

Self-Similarity in Hyperchaotic Data
O. E. RÖSSLER and J. L. HUDSON (With 4 Figures) 113

Estimation of Correlation Dimensions from Single and Multichannel Recordings – A Critical View
A. BABLOYANTZ (With 4 Figures) 122

Correlation Dimensions in Various Parts of Cat and Human Brain in Different States
J. RÖSCHKE and E. BAŞAR (With 14 Figures) 131

Magnetoencephalography and Attractor Dimension: Normal Subjects and Epileptic Patients
K. SAERMARK, J. LEBECH, C. K. BAK, and A. SABERS (With 7 Figures) . . 149

Chaotic Attractors in a Model of Neocortex: Dimensionalities of Olfactory Bulb Surface Potentials Are Spatially Uniform and Event Related
J. E. SKINNER, J. L. MARTIN, C. E. LANDISMAN, M. M. MOMMER, K. FULTON, M. MITRA, W. D. BURTON, and B. SALTZBERG (With 7 Figures) 158

Dimensional Analysis of the Waking EEG
K. E. GRAF and T. ELBERT (With 7 Figures) 174

A Model of the Generation of Electrocortical Rhythms
K. J. BLINOWSKA and P. J. FRANASZCZUK (With 4 Figures) 192

Linearity and Nonlinearity in Electrocortical Waves, and Their Elementary Statistical Dynamics
J. J. WRIGHT (With 4 Figures) 202

Can Artificial Intelligence Help in Finding How Brains May Work?
A. DE CALLATAŸ (With 6 Figures) 214

Discussion Following Remarks by A. DE CALLATAŸ 233

III. Animal Models in Sensory and Cognitive Processes

Introductory Remarks
T. H. BULLOCK 241

Comparisons of Sensory and Cognitive Brain Potentials in the Human and in an Animal Model
J. S. BUCHWALD (With 6 Figures) 242

Evolution of Compound Field Potentials in the Brain
T. H. BULLOCK 258

Discontinuities in Visual Cortex and Possible Functional Implications: Relating Cortical Structure and Function with Multielectrode/ Correlation Techniques
R. ECKHORN, R. BAUER, and H. J. REITBOECK (With 6 Figures) 267

Subcortical Evoked Potential Correlates of Early Information Processing: Mismatch Negativity in Cats
V. CSÉPE, G. KARMOS, and M. MOLNÁR (With 6 Figures) 279

"No-Go Potential" in the Prefrontal Cortex of Monkeys
K. SASAKI and H. GEMBA (With 6 Figures) 290

Epileptic Phenomena in the Neocortex: From Activity of Single Neurons to Field Potentials of Neuronal Pools
H. POCKBERGER, E.-J. SPECKMANN, and J. WALDEN (With 5 Figures) . . . 302

IV. Brain Potentials During Sensory and Cognitive Processes

Introductory Remarks
E. BAŞAR 313

Quantitative Electrophysiological Maps of Mental Activity
E. R. JOHN, L. S. PRICHEP, and R. J. CHABOT (With 8 Figures) 316

Neuroanatomical Contributions to Individual Differences in P300 Morphology
J. POLICH (With 5 Figures) 331

Sensory and Cognitive 40-Hz Event-Related Potentials: Behavioral Correlates, Brain Function, and Clinical Application
D. E. SHEER (With 9 Figures) 339

The CERP: Event-Related Perturbations in Steady-State Responses
S. MAKEIG and R. GALAMBOS (With 15 Figures) 375

Noninvasive Functional Investigations into Human Cortical Motor Physiology, Motor Learning, and Motor Imagery
L. DEECKE, W. LANG, G. GOLDENBERG, and I. PODREKA (With 2 Figures) . 401

Signs of Model Making by the Human Brain
A. S. GEVINS (With 5 Figures) 408

The Clinical Use of P300 Cartography in Diseases with Disturbed Cognitive Processing of the Brain
K. MAURER (With 6 Figures) 420

Influence of Physostigmine on Cognitive Processing of the Brain
R. IHL, L. FRÖLICH, T. DIERKS, and K. MAURER (With 2 Figures) 429

Semantic Distance and the Electrophysiological Priming Effect
T. F. MÜNTE, H. KÜNKEL, and H. J. HEINZE (With 5 Figures) 436

The Search for Face-Specific Evoked Potentials
K. BÖTZEL, O.-J. GRÜSSER, B. HÄUSSLER, and A. NAUMANN (With 13 Figures) 449

Area-Specific Regulation of Slow Cortical Potentials
B. ROCKSTROH (With 4 Figures) 467

V. How Brains May Work

Introductory Remarks
T. H. BULLOCK 481

Panel Discussion: How Brains May Work
Edited by H. STOWELL, T. H. BULLOCK, and E. BAŞAR (With 2 Figures) . 482

Remarks by
R. GALAMBOS: The Glia-Neural Approach 482
O.-J. GRÜSSER: Neurons and Glia in the Retina 484
E. R. JOHN: Comments on Integrative Function in the CNS 487
E. BAŞAR: Brain State Matrix – A Preliminary Proposal to Approach Brain Function Using EEG-EP Feynman Diagrams 491
D. LEHMANN: Temporal Characteristics in the Brain's Electrical Records 497
W. R. ADEY: Bioinstrumentation – Cutting Edge and Limiting Factor in the Future of Brain Research 498
W. J. FREEMAN: Brain Dynamics and Differential Equations 502
L. DEECKE: From Readiness Potential Onwards 505
A. STARR: View of a Neurologist 508

Analysis of Strange Attractors in EEGs with Kinesthetic Experience and 4-D Computer Graphics
W. J. FREEMAN (With 4 Figures) 512

Chaos in Brain Function and the Problem of Nonstationarity: A Commentary
G. J. MPITSOS 521

Epilogue

Signs of Dynamic Processes in Organized Neural Tissue: Extracting Order from Chaotic Data
T. H. BULLOCK 539

Addresses of Participants

ADEY, W.R., Pettis Memorial Veterans Administration Hospital, 11201 Benton Street, Loma Linda, CA 92357, USA

BABLOYANTZ, A., Department of Physical Chemistry II, Free University of Brussels, P.O. Box 231, Boulevard du Triomphe, B-1050 Brussels, Belgium

BAK, C.K., Physics Laboratory I, Technical University of Denmark, DK-2800 Lyngby, Denmark

BAŞAR, E., Institute of Physiology, Medical University Lübeck, Ratzeburger Allee 160, D-2400 Lübeck, FRG

BAŞAR-EROGLU, C., Institute of Physiology, Medical University Lübeck, Ratzeburger Allee 160, D-2400 Lübeck, FRG

BAUER, R., Department of Biophysics, Philipps-University, Renthof 7, D-3550 Marburg, FRG

BLINOWSKA, K.J., Medical Physics Laboratory, Institute of Experimental Physics, Warsaw University, ul. Hoza 69, 00-681 Warsaw, Poland

BÖTZEL, K., Department of Physiology, Free University of Berlin, Arnimallee 22, D-1000 Berlin 33, FRG

BUCHWALD, J.S., Department of Physiology, Brain Research Institute and Mental Retardation Research Center, UCLA School of Medicine, Los Angeles, CA 90024, USA

BULLOCK, T.H., Department of Neurosciences A-001, School of Medicine, University of California, San Diego, La Jolla, CA 92093, USA

BURTON, W.D., Neurosciences Signal Analysis Laboratory, Department of Psychiatry and Behavioral Science, University of Texas Medical School, Houston, TX 77030, USA

CHABOT, R.J., New York University Medical Center, School of Medicine, New York, NY 10016, USA

CSÉPE, V., Institute of Psychology, Hungarian Academy of Sciences, P.O. Box 398, H-1394 Budapest, Hungary

DE CALLATAŸ, A., IBM European System Research Institute, A.K. Watson IEC, Ch. de Bruxelles 135, B-1310 La Hulpe, Belgium

Deecke, L., University Clinic of Neurology, Lazarettgasse 14, A-1097 Vienna, Austria

Dierks, T., Department of Psychiatry, University of Würzburg, Füchsleinstr. 15, D-8700 Würzburg, FRG

Eckhorn, R., Department of Biophysics, Philipps University, Renthof 7, D-3550 Marburg, FRG

Elbert, T., Psychiatry, 116A3, Veterans Administration Medical Center, 3801 Miranda Avenue, Palo Alto, CA 94304, USA

Franaszczuk, P.J., Medical Physics Laboratory, Institute of Experimental Physics, Warsaw University, ul. Hoza 69, 00-681 Warsaw, Poland

Freeman, W.J., Department of Physiology-Anatomy, University of California, Berkeley, CA 94720, USA

Frölich, L., Department of Psychiatry, University of Würzburg, Füchsleinstr. 15, D-8700 Würzburg, FRG

Fulton, K., Neurophysiology Section, Department of Neurology and Neuroscience Program, Baylor College of Medicine, Houston, TX 77030, USA

Galambos, R., Children's Hospital Research Center, 8001 Frost Street, San Diego, CA 92123, USA

Gemba, H., Department of Physiology, Institute for Brain Research, Faculty of Medicine, Kyoto University, 606 Kyoto, Japan

Gevins, A.S., EEG Systems Laboratory, 51 Federal Street, San Francisco, CA 94107, USA

Goldenberg, G., University Clinic of Neurology, Lazarettgasse 14, A-1097 Vienna, Austria

Graf, K.E., Department of Clinical and Physiological Psychology, University of Tübingen, Gartenstr. 29, D-7400 Tübingen, FRG

Grüsser, O.-J., Institute of Physiology, Free University of Berlin, Arnimallee 22, D-1000 Berlin 33, FRG

Häussler, B., Department of Physiology, Free University of Berlin, Arnimallee 22, D-1000 Berlin 33, FRG

Heinze, H.J., Clinic of Neurology and Clinical Neurophysiology, Medical University of Hannover, Konstanty-Gutschow-Str. 8, D-3000 Hannover 61, FRG

Hoke, M., Division of Experimental Audiology, University of Münster, Kardinal-von-Galen-Ring 10, D-4400 Münster, FRG

Hudson, J.L., Department of Chemical Engineering, University of Virginia, Charlottesville, VA 22901, USA

IHL, R., Department of Psychiatry, University of Würzburg, Füchsleinstr. 15, D-8700 Würzburg, FRG

JOHN, E.R., Brain Research Laboratories, New York University Medical Center, 550 First Avenue, New York, NY 10016, USA

KARMOS, G., Institute of Psychology, Hungarian Academy of Sciences, P.O. Box 398, H-1394 Budapest, Hungary

KÜNKEL, H., Clinic of Neurology and Clinical Neurophysiology, Medical University of Hannover, Konstanty-Gutschow-Str. 8, D-3000 Hannover 61, FRG

LANDISMAN, C.E., Neurophysiology Section, Department of Neurology and Neuroscience Program, Baylor College of Medicine, Houston, TX 77030, USA

LANG, W., University Clinic of Neurology, Lazarettgasse 14, A-1097 Vienna, Austria

LEBECH, J., Physics Laboratory I, Technical University of Denmark, DK-2800 Lyngby, Denmark

LEHMANN, D., Department of Neurology, University Hospital, Frauenklinikstr. 26, CH-8091 Zürich, Switzerland

LEHNERTZ, K., Institute of Experimental Audiology, University of Münster, Kardinal-von-Galen-Ring 10, D-4400 Münster, FRG

LÜTKENHÖNER, B., Institute of Experimental Audiology, University of Münster, Kardinal-von-Galen-Ring 10, D-4400 Münster, FRG

MAKEIG, S., Department of Psychiatry, University of California, San Diego, La Jolla, CA 92093, USA

MARTIN, J.L., Neurophysiology Section, Department of Neurology and Neuroscience Program, Baylor College of Medicine, Houston, TX 77030, USA

MAURER, K., Department of Psychiatry, University of Würzburg, Füchsleinstr. 15, D-8700 Würzburg, FRG

MITRA, M., Neurophysiology Section, Department of Neurology and Neuroscience Program, Baylor College of Medicine, Houston, TX 77030, USA

MOLNÁR, M., Institute of Psychology, Hungarian Academy of Sciences, P.O. Box 398, H-1394 Budapest, Hungary

MOMMER, M.M., Neurophysiology Section, Department of Neurology and Neuroscience Program, Baylor College of Medicine, Houston, TX 77030, USA

MPITSOS, G.J., M.O. Hatfield Marine Science Center, Oregon State University, Newport, OR 97365, USA

MÜNTE, T.F., Clinic of Neurology and Clinical Neurophysiology, Medical University of Hannover, Konstanty-Gutschow-Str. 8, D-3000 Hannover 61, FRG

NAUMANN, A., Department of Physiology, Free University of Berlin, Arnimallee 22, D-1000 Berlin 33, FRG

PANTEV, C., Institute of Experimental Audiology, University of Münster, Kardinal-von-Galen-Ring 10, D-4400 Münster, FRG

POCKBERGER, H., Institute of Neurophysiology, University of Vienna, Währinger Str. 17, A-1090 Vienna, Austria

PODREKA, I., University Clinic of Neurology, Lazarettgasse 14, A-1097 Vienna, Austria

POLICH, J., Division of Preclinical Neuroscience and Endocrinology, Research Institute of Scripps Clinic, 10666 North Torrey Pines Road, La Jolla, CA 92037, USA

PRICHEP, L.S., New York University Medical Center, School of Medicine, New York, NY 10016, USA

REITBOECK, H.J., Department of Biophysics, Philipps University, Renthof 7, D-3550 Marburg, FRG

ROCKSTROH, B., Department of Clinical and Physiological Psychology, University of Tübingen, Gartenstr. 29, D-7400 Tübingen, FRG

RÖSCHKE, J., Institute of Physiology, Medical University Lübeck, Ratzeburger Allee 160, D-2400 Lübeck, FRG

RÖSSLER, O.E., Institute of Physical and Theoretical Chemistry, University of Tübingen, Auf der Morgenstelle 8, D-7400 Tübingen, FRG

SABERS, A., Department of Neurology, Hvidovre Hospital, DK-2650 Hvidovre, Denmark

SAERMARK, K., Physics Laboratory I, Technical University of Denmark, DK-2800 Lyngby, Denmark

SALTZBERG, B., Neuroscience Signal Analysis Laboratory, Department of Psychiatry and Behavioral Science, University of Texas Medical School, Houston, TX 77030, USA

SASAKI, K., Department of Physiology, Institute for Brain Research, Faculty of Medicine, Kyoto University, 606 Kyoto, Japan

SCHÜTT, A., Institute of Physiology, Medical University Lübeck, Ratzeburger Allee 160, D-2400 Lübeck, FRG

SHEER, D.E., Department of Psychology, University of Houston, Houston, TX 77004, USA

SKINNER, J.E., Neurophysiology Section, Department of Neurology and Neuroscience Program, Baylor College of Medicine, Houston, TX 77030, USA

SPECKMANN, E.-J., Institute of Physiology, Neurophysiology Section, University of Münster, Robert-Koch-Str. 27a, D-4400 Münster, FRG

STARR, A., Department of Neurology, University of California, Irvine Medical Center, Orange, CA 92668, USA

STOWELL, H., 120 Nature Creek SW, Milledgeville, GA 31061, USA

WALDEN, J., Institute of Physiology, Neurophysiology Section, University of Münster, Robert-Koch-Str. 27a, D-4400 Münster, FRG

WRIGHT, J.J., Department of Psychiatry and Behavioural Science, University of Auckland School of Medicine, Auckland, New Zealand

I. Basic Approaches

Introductory Remarks

T. H. Bullock

The first half dozen chapters are collected together under the somewhat arbitrary rubric Basic Approaches not because many others are less fundamental or because they present the alternative bases for approaching brain dynamics. They are selected as representing so many orthogonal dimensions if we may use this expression, and also for being introductory in character, each urging a point of view that overlaps with but does not conflict with the others.

Bullock wants to look at the compound field potentials from the bottom up, to emphasize the diversity of generators and their orientations, and to raise the question of evaluating degrees of cooperativity among them. Galambos proposes a closer look at the glia, on the model of the retina, as an essential step in the generation of evoked potentials and perhaps ongoing background activity as well.

Adey calls attention to the wide ranging possibilities for causal influence to tissue fields at and below the level of EEG voltage gradiens; field modulations of cell-surface chemical events indicate a major amplification in which calcium plays a key role. Intercellular communication through gap junctions is reported to be similarly sensitive to weak oscillating pericellular electromagnetic fields. Adey hypothesizes that such fields can play a role in promoting cancer by disrupting cell communication, suggesting therapy not by cytotoxic destruction of neoplastic elements but by restoring cell membrane function.

Başar and coworkers present a study which exemplifies the value of a combined analysis of an ensemble of parameters, in this case the human EEG, bandpass filtered EEG, EP, and ERP with quantitative manipulation of the cognitive event. They report that the expectancy or missing stimulus wave amplitude correlates with the strength of the expectation and that about 1 s before the expected stimulus the EEG shows a "coherent, phase-ordered, rhythmic pattern."

Lehmann develops the consequences of the moment-to-moment changes in functional state of the brain, in the sub-second time domain, pointing to evidence that stable microstates can be identified and change in a discontinuous manner, superimposed on the familiar gross brain states. He proposes that the microstates influence higher information processing, provide a wide array of strategies for dealing with information, and reach consciousness only if the duration is beyond a minimum.

Hoke and his coworkers represent for this collection the burgeoning field of magnetoencephalography (MEG); they apply its considerably higher spatial resolution than EEG recording to processes triggered by auditory stimuli and develop a special technique to allow high temporal resolution as well. They believe this method of analysis opens new perspectives in spatiotemporal distribution of

Springer Series in Brain Dynamics 2
Edited by Erol Başar and T. H. Bullock

brain currents and in particular on separating endogenous from exogenous activity.

Further additions to the sampling of basic approaches in this book are found in other sections, especially in Part II, Theoretical Approaches and in several of the contributions in Part V, How Brains May Work. Characteristic of all these chapters is that they look ahead by proposing perspectives for fresh views of the multilevel, multidimensional problem of the brain in action.

The Micro-EEG Represents Varied Degrees of Cooperativity Among Wide-Band Generators:

Spatial and Temporal Microstructure of Field Potentials

T. H. BULLOCK

1 Introduction

The underlying proposition of this paper is that the ongoing, compound field potentials in the brain, including the local activity recorded as the micro-EEG and event-related potentials (ERPs), manifest a certain amount of cooperativity superimposed on the vector sum of stochastic events in great numbers of generators, largely disordered geometrically, and that the degree of cooperativity, which includes synchrony of slower, subthreshold potentials, differs significantly between parts of the brain, species, states, and stages of ontogeny. A corollary of this proposition is that the micro-EEG and ERPs are not merely noise in the system or symptoms of the general health of the brain but information-rich signs to the investigator of moment by moment and locus by locus changes in the information processing status of the neural substratum, even though blurred and smeared in proportion to whether the recording is done grossly or at a distance from the active generators.

2 Descriptive Facts Deserve Priority

Treating compound field potentials as signs means that I give first priority to a better comprehension of the descriptive natural history of a mysterious phenomenon, rather than to the causal analysis, which is certainly the eventual goal. Reductionist analysis aiming at understanding mechanisms is of course the desideratum and is tempting, not to say alluring, single cell technics being so well developed. My claim, however, is that modeling the EEG or ERP in terms of cells and generators is bound to be unrealistically simplified today.

The point I would emphasize is that we badly need a better idea of the details of what can best be called the natural history of brain waves (using that term or the term EEG to embrace the micro-EEG and ongoing electric field potentials as well as ERPs anywhere in the brain) before going too far into theories, just as in the natural history of disease, of behavior, of ecology, and of evolution. In the case of brain waves we need a better picture of the simple descriptive facts. Although we know a great deal about the EEG and ERPs, with the enormous accumulated literature, my position is that we have been extremely limited in the windows available and know relatively little about some fundamental aspects of natural history. These include seven domains that I select for emphasis here.

Springer Series in Brain Dynamics 2
Edited by Erol Başar and T. H. Bullock

1. The *spatial distribution of EEG and ERP activity* is a basic need, recorded on a fractional millimeter or finer scale, in all three dimensions at once. Pioneering studies by such workers as Elul and Freeman and Petsche, among others, usually in one or two dimensions, need to be extended.

2. The *temporal structure* on a scale appropriate to the brain's function, which means fractions of a second, and with phase and sequence information preserved, is a basic need. We cannot be satisfied with spectra or other analyses averaged over many seconds – any more than we could in describing speech.

3. Histological organization among the parts of the brain presents a rich array of *neural differentiation* that we believe to be functionally relevant. Yet we have no systematic body of descriptive data or rules about any array of patterns of activity corresponding to histology, apart from intriguing bits of information such as differences in spectra or coherence between cortical laminae. We still have almost no body of information on the EEG correlates of cell size, cell density, lamination, and other microscopic features, or of brain advancement in differentiation (see the next item). Only a few empirical data are available on EEG differences between such structures as cortex and cerebellum. One major reason is our paucity of suitable descriptors for patterns of activity in a volume of tissue, on the scales of fractions of millimeters and seconds.

4. *Phylogeny* presents us with the outstanding achievement of evolution, the development of higher and higher brains; yet we have no systematic body of descriptive knowledge about any variety of functional patterns corresponding even to the major taxa, such as fish, reptiles, and mammals – again for lack of suitable descriptors.

The next two domains of needed natural history are much more amenable to investigation.

5. Mental activity or *behavioral state* has been extensively studied as a correlate or predictor of EEG and ERPs, but not or only in very limited ways with the fine scale, temporally and spatially, that is needed or with the novel descriptors I am calling for.

6. The relation between *EEG, ERPs, and unit spikes* is one of the oldest questions to be studied, beginning with Adrian, Jasper, Monnier, Gerard, and others. (Since this is a synthetic and interpretive paper, not a review, I will not generally cite the authorities I am leaning on.) Yet we lack, once again, any systematic body of descriptive knowledge about the activity of a fair sample of the neuronal population, recorded at the same time. This is conceptually and technically the simplest of the seven demands I am listing, since it merely calls for multichannel recording on an adequate scale in terms of numbers of units and types of units within a small three-dimensional volume such as a cubic millimeter!

7. Lastly, the relations between *EEG, ERPs and subthreshold potentials,* both of neurons and of glia, represent an even more important research agenda, specifically the relations of a good number of cells in a small volume, all recorded at once. Only when we know the activity of an adequate number of units at once can we estimate how much each form of activity contributes to the gross compound field and how much cooperativity there is among the cells. I do not overlook the formidable problem, given that we had the multichannel recordings from

many units, of what statistical tests would evaluate a mixture of significant cooperativity and the chance level of coincidence.

Here I will not dwell on the formidable problems facing each of these seven pillars of neural wisdom, but instead aim simply to persuade the reader that they are indeed major gaps today and pillars of the future, opportunities for creative innovation, technically and conceptually.

3 Cooperativity in Networks Is Special

Cooperativity being the essential feature behind these aspects of information richness, let us see what it means in this context. It includes the degree of synchrony above the chance level, especially synchrony of neuronal and glial slow potentials. I would exclude from the term cooperativity any consequences of classical synaptic connectivity that depend on stochastic timing of arriving or arising impulses, since including them would mean the term is so broad as to have no distinctive use in the context of connected networks of all-or-none cells. It does include any mechanisms that may produce nonrandom timing of arriving input among parallel or converging channels or lateral interaction among equivalent elements or feedback or feedforward. Thus electrical field effects and the action of transmitters or modulators over micrometers, many of the consequences of so-called local circuits (Rakic 1975), but also other mechanisms (Bullock 1980, 1981, 1983, 1988) should be included. Among these are the patterns in three-dimensional space and time of activity in arrays of terminals of axonal arbors and of arrays of branches of dendritic ramifications wherever geometrically characteristic configurations constrain the distribution of many junctions. Cooperativity is a broad category of phenomena, mostly unfamiliar and little understood. This is the reason for drawing particular attention to them here.

4 The Agenda Recapitulated

Once more, for emphasis, let me repeat the seven *domains of urgently needed knowledge* about ongoing and event-related compound field potentials where, in spite of brilliant pioneering work, our ignorance is seriously limiting reductionist analysis and modeling: (1) spatial distribution on the fractional millimeter scale, (2) temporal structure on the fraction of a second scale, in many loci at once, (3) systematic rules in relation to histological structure, (4) systematic rules in relation to phylogenetic advancement, (5) behavioral and mental correlates of the spatiotemporal pattern on the fractional second and millimeter scales, (6) correlation with spikes of a fair sample of the neighboring units, and (7) correlation with subthreshold neuronal and glial potentials of a fair sample of neighboring cells at once.

One reason I regard this agenda as important is that I sense a certain level of satisfaction or overconfidence in some writings about these signs of brain activity.

Better natural history may provide evidence overturning some of our favorite preconceptions, which are largely based on scalp recording from humans, and substitute less comfortable but more realistic pictures of the nature of electrical activity of organized assemblies of neural tissue. I refer to such notions as synchrony and rhythms, oscillations and resonance, sources and sinks, circuits and local circuits – all of which there are already reasons to question or significantly update. I will come back to some of these below.

5 Micro-EEG and ERPs Are Vector Sums of Diverse Currents from Many Generators

I am not content here merely to pose challenges to prevailing dogma and propose agendas for new research. I would like to point to evidence, meager as it may be, supporting the conclusion that cooperativity of cellular generators varies significantly in degree among well functioning brains. But first I must start with some reminders about what can be asserted from existing knowledge about the *basis of the micro-EEG and ERPs*. It is more than a supposition that the observed compound field potential is the vector sum of a very large number of generators, most of them limited parts of the processes of neurons. The relatively huge voltage differences across cell membranes, between inside and outside, are most likely to be the main driving forces for the extracellular currents whose IR drops (current × resistance) form the field potentials. The important point for our argument is that not simply spikes and not simply synaptic potentials add up to account for the EEG and ERPs, but almost certainly at least half a dozen kinds of cellular events and state changes.

These *events that add up to account for the EEG and ERPs* include synaptic potentials of both polarities and different time courses, as well as local impedance changes when the potential is at equilibrium and does not shift. Nonclassical postsynaptic responses are known rather widely, in which a long-lasting hyperpolarization is accompanied by an impedance increase instead of the decrease typical of inhibitory postsynaptic potentials. Plateau potentials are being found in many places: more or less bistable, triggered states. Pacemaker potentials of several distinct forms are known in favorable preparations; we must assume pacemaker potentials are widespread in the brain. Subthreshold local potentials and long-lasting after-potentials of spikes are doubtless quite general. Potentials caused not by membrane impedance changes but by fluctuating or controlled ion pumps, which are known in some cases to be able to act in the time range as short as 10 ms, may contribute a large share to the field potential, especially the pumps in the membranes of glial cells. This is probably not an exhaustive list of the variety of events that contribute to the compound field.

These events come from a variety of structures, including separate parts of dendrites, synaptic loci, axons and axon terminal arbors, somata, and at least astrocytes if not also other forms of neuroglia. The shapes and durations of these events vary widely, but in general they all have broad power spectra, extending to very low frequencies, essentially DC.

Such a spectrum is just what we see in the usual vertebrate EEG. Whereas an EEG spectrum could be due to the summing of sine waves, there is no need for a rhythm or oscillation to exist at each or any of the frequencies showing energy in the spectrum. An irregular train of pulses, whether wide or narrow, and even a single pulse, has a broad spectrum extending to low frequencies. It is of course heuristic for some purposes to speak of the compound field potentials "as though they were rhythms that arise in oscillators", without asserting that this is the case. To demonstrate that in fact oscillations are the basis of the observed compound field would require evidence (a) that the actual generators are rhythmic at the frequency of the EEG (itself only exceptionally rhythmic, as in alpha and theta states), virtually demanding intracellular evidence, and (b) that such generators are synchronized to a sufficient degree.

Aside from the special cases such as alpha and theta waves and olfactory bulb bursts under particular behavioral conditions and in particular parts of the brain, the observed spectra of EEGs and ERPs are usually broad humped and not marked by narrow peaks. Averaged over 10 or more seconds, they become progressively smoother, without consistent peaks; in epochs of a few seconds or less they are jagged and become indistinguishable from the null hypothesis of a short sample of band-passed white noise.

Reasonable scientific caution dictates that we should speak of synchrony only if we have evidence of more than the chance level of coincidence. Even though some individual neurons are firing spikes more or less periodically, any presumed rhythms in the assembly should be shown to be more than filtered pink noise. Cooperativity should not be assumed but evidenced.

What is the *null hypohesis?* It states that with respect to physiology the generators are independent. Yet we know some of them are not entirely so: postsynaptic potentials follow presynaptic events. If cooperativity meant no more than this, there would be much less room for discussion and research. In our context cooperativity means any kind of working together, including classical synaptic as well as other, unspecified forms of one-way or two-way interaction.

6 Cooperativity Varies

That cooperativity occurs, and in widely differing degrees from nil to extremely weak to substantial, is a conclusion from several lines of evidence, besides the facts of specific connectivity:

1. In some animals, brain regions, or states there is *excess root-mean-square (RMS) voltage* of ongoing brain activity over that in other species, places, or states. We cannot know what RMS amplitude to expect on the null hypothesis of independence of generators in a very anisotropic volume conductor. But there is an apparently generally greater RMS voltage in the mammalian forebrain, even in nonlaminated regions, than in the reptilian, amphibian, fish, and invertebrate, even in well laminated regions (Bullock and Başar 1987). There is increase in amplitude with arousal in fish. The excess in seizures over the resting state in humans and laboratory animals also bespeaks summing by more than chance coincidence

of activity. By sufficient ad hoc hypotheses one might explain such differences in terms of purely classical synaptic connectivity, but it seems more likely that at least some are due to additional mechanisms of interaction. It is worth reminding ourselves that even with classical synaptic connectivity, timing of postsynaptic events is not generally modeled by experiments with spinal motor neurons receiving artificially synchronized input, but more generally involves imperfectly synchronized inputs and frequency modulated outputs.

2. *Estimates of synchrony* in a volume of neural tissue can now be made by the method of plotting steepness of decline of coherence with distance between electrodes, in a given frequency band. The degree of synchrony need not be high. If a computer-generated EEG-like signal is common to two channels and mixed with noise from two independent white noise generators similarly filtered but of 8 times greater amplitude, the coherence is about 0.5. Significant synchrony in excess coherence over randomness can mean a very small percentage of common signal. We find coherence values in the brain from virtually 1.0, the maximum possible, down to the noise level. I would guess that commonly a very limited subset of the cells in a given population is cooperating in the sense of slow potential synchrony.

3. A tight *phase spectrum,* in contrast to the wildly fluctuating phase spectrum of two independent sources, is typical of EEG records, even when the coherence is rather low (0.2–0.3) or the common signal much weaker than the independent noise fraction. By tight I mean the phase relation of any pair of channels is relatively flat and generally <15 degrees over a large part of or all of the spectrum – usually 1–40 or 1–80 Hz.

4. *Current source density* plots across the cortex reveal closely spaced, transient, and shifting sources and sinks of current. These indicate by a physiological measure, namely the ongoing activity in the population, that the assemblage has local sign, spatial differentiation and specialization.

5. Consistent *peaks in the power spectrum* do occur under some circumstances, like the alpha wave in some species when awake with eyes closed, and change in a characteristic way with state.

6. A body of evidence I have summarized elsewhere (Bullock 1980) points to the need for a new imagery and *language beyond circuitry,* even so-called local circuits, to take account of such presumed facts as field effects over at least several micrometers, both chemical and electrical, and the influence of three-dimensional stereogeometry in the known variety of highly characteristic dendritic and axonal terminal arbors.

The first and second of these lines of evidence, RMS voltage and coherence decline with distance, appear to be much weaker in the telencephalon of lower vertebrates and in the lower levels of the brain stem and spinal cord of higher vertebrates, as well as in the highest centers of most invertebrates. These are still only tentative generalizations; doubtless they will have to be modified significantly – as when we found the EEG of octopus is more vertebrate-like than invertebrate-like.

7 Summary

Cooperativity as used here means something more than the vector sum of activity in neurons connected only by impulse traffic and classical synapses. It includes nonclassical interactions in local circuits, the actions of modulators and of transmitters that exert influence over micrometers, the possible influence of field potentials, at least at times and places, and patterns in three-dimensional space of the activity of arrays of terminals in axonal arbors and of arrays of branches of dendritic ramifications. It is the proposition of this paper that such cooperativity is significant in higher levels of the brain in higher vertebrates less so in spinal cord, brain stem, and cerebellum, perhaps also in lower vertebrates, and especially in invertebrates.

A corollary of this proposition is that there is fine structure of the ongoing and evoked compound field potentials of the higher centers of the brain in higher vertebrates that is information-rich about the moment to moment and locus to locus changes in the status of the neural substratum. This in turn leads to the conclusion that we are still profoundly in need of descriptive natural history about the micro-EEG and ERPs, in at least seven areas:

1. Spatial distribution of various parameters of activity on the fractional millimeter scale in all three axes, a problem in current source density distribution in an anisotropic volume packed with microgenerators.
2. Temporal structure on a scale of fractions of a second, preserving the phase and sequence information in the time domain
3. Correlations between electrical pattern, in both of the foregoing domains, with histological differentiation of different parts of the brain
4. Correlations between electrical pattern and phylogeny
5. Correlations between electrical pattern and mental state
6. Relations between micro-EEG, ERPs and unit spikes in a fair sample of the neuronal population recorded at the same time
7. Relations between micro-EEG, ERPs, and a variety of subthreshold, graded, generally slower cell potentials, in many units or loci recorded at the same time

Badly needed are both the extensive application of existing methods of estimating these properties to more taxa, regions of the brain, and developmental and behavioral states, as well as the addition of new methods. What is needed in particular are methods for measuring and characterizing the microstructure and dynamics of wide-band population activity (not just spikes) with fractional millimeter and fractional second resolution.

Two large agendas still ahead are (1) to determine empirically what the microstructure is in different places, states, stages, and species, for ongoing and for event-related brain activity, and (2) to discover channels, mechanisms, and properties of the cooperativity. This paper emphasizes the first of these agendas and the need for new methods of observing and characterizing the micro-EEG and ERPs, particularly wide-band, many-channeled recording, close spacing of the electrodes (<1 mm), and temporal resolution better than 1 s.

References

Bullock TH (1980) Reassessment of neural connectivity and its specification. In: Pinsker HM, Willis WD (eds) Information processing in the nervous system. Raven, New York, pp 199–220

Bullock TH (1981) A comparative neurologist's view of signals and signs in the nervous system. Adv Phys Sci (USSR) 30:1–8

Bullock TH (1983) Electrical signs of activity in assemblies of neurons: compound field potentials as objects of study in their own right. Acta Morphol Acad Sci Hung 31:39–62

Bullock TH (1988) Compound potentials of the brain, ongoing and evoked: perspectives from comparative neurology. In: Basar E (ed) Dynamics of sensory and cognitive processing of the brain. Springer, Berlin Heidelberg New York, pp 3–18

Bullock TH, Başar E (1987) Comparison of ongoing, compound field potentials in the brains of invertebrates and vertebrates. Brain Res Rev 13:57–75

Rakic P (1975) Local circuit neurons. Neurosci Res Program Bull 13(3):289–446, i-ix

Electrogenesis of Evoked Potentials

R. Galambos

1 Introduction

At the first of the Berlin Conferences 2 years ago I pointed out how little we understand about the anatomical and physiological bases of evoked potentials (EPs). What I had in mind is illustrated in Fig. 1, a 15-min experiment during which EPs were recorded through implanted brain electrodes from a rhesus monkey (Galambos 1961 a; see also Galambos and Sheatz 1962). The experimenters outside the experimental room watched the monkey through a window. The time

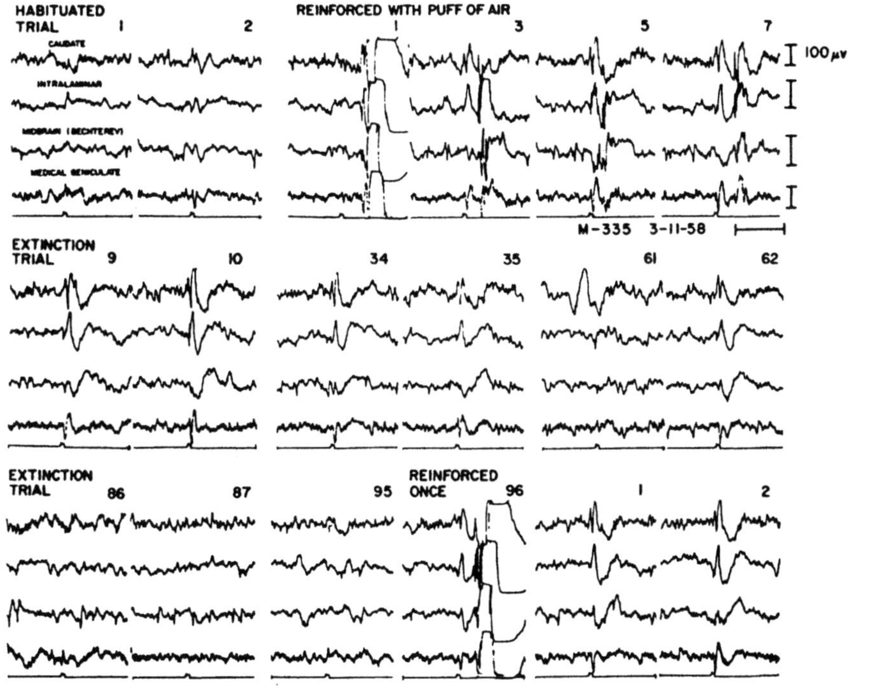

Fig. 1. Evoked potential lability in an unanesthetized monkey undergoing a simple conditioning procedure. Sounds (indicated by the *up-going deflections* in the lowest trace of each block) were occasionally accompanied (*Reinforced*) by a puff of air to the face. The sound produces large brain responses only if the stimuli have recently been presented together

Springer Series in Brain Dynamics 2
Edited by Erol Başar and T. H. Bullock

was BC (Before Computers), so you see here 18 2-s segments of EEG tracings recorded through four bipolar electrodes (located, from above downward, in caudate nucleus, thalamic intralaminar nucleus, midbrain reticular formation, and medial geniculate body). In each segment the pip in the center of the bottom line shows when a fairly loud click was emitted (once every 10 s) through a loudspeaker above the animal's head; lined up above these pips one sees the brain potentials the stimuli evoked.

The purpose of the experiment was to study brain changes during learning, and so from time to time we reinforced the click stimuli with puffs of air into the monkey's face. To deliver these stimuli we simply blew into our end of a long rubber tube, whereupon the monkey received an air puff out of the other end pointed at his face. The loudspeaker had been emitting the clicks day and night for several days when the experiment began (Habituated Trial 1) and the monkey was paying no attention to them; correspondingly, the clicks produced little or no brain response, although they were clearly audible to us through the wall.

The experiment began (at Reinforced with Puff of Air) when we blew into our end of the tube each time we heared the click. The first puff produced a large movement artifact, and every click response thereafter was very large compared to the control. After ten puffs we stopped, and within a few minutes (extinction trial 34) the responses began to decline in size. About 10 min later they had dropped to control size, at which time a single puff immediately restored large amplitudes at all electrode sites.

These data pose the problems this chapter tries to answer: what brain cells are responsible for generating these EPs, where are those cells located, and what rules control when they will and will not generate these bioelectrical currents.

In 1961 I asked the same questions and proposed that neurons and glial cells cooperate in some manner to produce phenomena such as these (Galambos 1961 b). Since that time precise physiological measurements on glia and neurons in the retina have revealed this is exactly the way the electroretinogram (ERG), which is the EP of the retina, is generated. In what follows I will describe these retinal EP experiments and then argue that similar glia–neuronal interactions are probably responsible for EPs recorded in the brain also. Finally I will return to this monkey experiment and use these new concepts to interpret its evidence that dynamic signal processing is under way within the CNS. But first I shall point out some crucial physiological differences between glia and neurons upon which the arguments rest.

2 Some Special Properties of Glial Cells

About 20 years ago Kuffler and his colleagues identified four major differences between neurons and glial cells on the basis of intracellular recordings (Kuffler and Nichols 1966):

1. Glial membranes differ from neuronal membranes in being almost exclusively permeable to K^+; as the concentration of the ion outside the cell, $[K^+]_0$, rises, the glial membrane potential drops linearly.

2. Unlike neurons, glia do not generate action potential spikes.
3. Glia do, however, partially depolarize when action currents pass along nearby axons.
4. Finally, the cytoplasm of one glia cell is frequently coupled to that of another, as demonstrated by the fact that electrical current injected into one cell passes by a low resistance pathway to an electrode located in an adjacent one.

3 Electrogenesis of the Electroretinogram

The fact that the ERG largely reflects the special properties of a glial cell, the retinal Müller cell, is by now a well-known story summarized in several places (e.g., Kline et al. 1978; Roitbak 1983; Ripps and Witkovsky 1985; Newman 1985a, b). When Dowling and his colleagues published the first systematic intracellular measurements on the retinal cells (Miller and Dowling 1970) they uncovered the following facts:

1. The entire population of retinal cells gives only graded responses to light stimuli except for the ganglion cells, which produce true axon spikes that ascend the optic nerve.
2. The transmembrane response of the Müller cell is the only graded depolarization that resembles the ERG b-wave recorded outside the eye.
3. The $[K^+]_0$ increases which light stimuli evoke occur in the plexiform (synaptic) layers of the retina (Kline et al. 1978).

These experiments identified the unique Müller cell contribution to the retinal EP, and demonstrated that synaptic activities in the neuropile (the internal and external plexiform layers) are what initiate the electrical currents visualized at external electrodes.

Since then a third major group of contributors, Newman and his colleagues, have confirmed Dowling's facts and added crucial new information about both retinal and brain glial cells (Newman et al. 1984; Newman 1985a, b).

1. Besides being almost exclusively permeable to K^+ ion, retinal and brain glial cells are *selectively* permeable in the endfoot region. Endfeet contact body fluid reservoirs (blood; cerebrospinal fluid; the vitreous humor in the eye), and something like 95% of the permeability for K^+ ion is to be found in this small region of both brain and retinal astrocytes (Newman 1985b).

   At the present time almost everyone agrees that:

2. K^+ ion enters Müller cells and brain astrocytes wherever neurons produce the ion at a rate that exceeds their ability to reabsorb it, an event initially termed "potassium spatial buffering" by Orkand et al. (1966).
3. K^+ ions move through the glial cytoplasm toward the endfoot region where they are transported through the membrane into blood, CSF, or the vitreous. The fact that glia do this rapidly because K^+ conductivity is concentrated in the endfoot was termed "K^+ siphoning" by Newman et al. (1984).

The following quotation from Newman states the current view of ERG electrogenesis.

K^+ enters glial cells in regions of tissue where $[K^+]_0$ is raised. In order to maintain electrical neutrality, an equal amount of K^+ exits from these cells, or from cells electrically coupled to them, in regions where $[K^+]_0$ is lower. A return current through extracellular space completes the circuit. The net effect of this current flow is to transfer K^+ from regions of tissue where $[K^+]_0$ is high to regions where $[K^+]_0$ is lower. The high conductance of the Müller cell endfoot makes spatial buffering a far more powerful process than previously envisioned.

Because of this endfoot specialization ... almost all of the K^+ current entering Müller cells ... would be shunted out through the endfoot process into the extracellular space at the vitreal surface of the retina. The vitreous would function as an enormous K^+ sink, storing excess K^+ until retinal $[K^+]_0$ levels returned to normal. K^+ could then be transferred back to the retina from the vitreous by a reversal of the K^+ current flow ... (Newman 1985b, p 2237).

4 Electrogenesis of Scalp-Recorded EPs: Astrocytes in Brain May Control $[K^+]_0$ as They Do in Retina

It has occurred to a number of gliologists that K^+ currents might flow through the cytoplasm of brain astrocytes also, in which case an end result could be external field currents recorded from the scalp. First among these theorists were Kuffler and Nicholls, who said in 1966:

The distribution of current flow that results in the cortex from glial depolarization will be ...complex. Massive synchronous activity of groups of neighboring neurons is likely to depolarize glial cells in a circumscribed area... Fused membrane contacts have been found between glial cells in the mammalian cortex, suggesting electrical coupling... Hence glial cells some distance from the depolarized region could act as sources of current and therefore cause changes that could be recorded with external electrodes. No specific analyses of these problems have yet been made in the brain because it has been difficult to identify reliably the glial cells.

Somjen, who has written extensively on the matter (1975, 1979, 1980, 1981a, b, 1984, 1987a, b) is cautious but optimistic in a recent review (1987b, p 18): "It does not seem farfetched to suggest for astrocytes a function similar to that of Müller cells. Astroglia might channel excess K^+ ions from the cerebral interstitium toward the CSF through the pial and ependymal surfaces, or toward the bloodstream by releasing the excess in pericapillary spaces..."

Nearly 10 years ago Dowling suggested "all of the types of generator processes seen in the retina... could also exist in the brain" (Galambos and Hillyard 1981, p 230), and 2 years ago Ripps and Witkovsky (1985, p 205) echoed the view: "A large number of studies ... have led to the conclusion that in the brain and spinal cord, glial cells, responding to changes in the extracellular levels of K^+, are the principal generators of the slow potential shifts recorded from the surface of these structures." Newman (1985b, p 2237) has put it this way:

An important function of glial cells in the central nervous system is to regulate $[K^+]_0$ by the process of spatial buffering... The location of high conductance astrocyte endfeet would determine the dipole orientation of extracellular potentials generated by these cells. The field potentials generated by currents shunted through the endfeet surrounding capillaries would tend to cancel due to their radial orientation. In contrast, the endfeet forming the glia limitans at the surface of the brain are all oriented in the same direction. K^+ shunted through these endfeet would generate negative slow potentials that could be measured by electrodes within active regions of grey matter.

5 A Neuropile-Astrocyte Theory of EP Generation

My current formulation of these ideas is represented in the propositions that follow.

First, the embryological development of brain and retina are similar, and so too are their final structures and functions. Thus:

1. Brain specializes everywhere in neuropile analogues of the retinal inner and outer plexiform layers where nonspiking neurons interact through synapses.
2. Central astrocyte membranes also share with neuronal membranes the extracellular space, which neural activities can enrich with K^+.
3. Brain astrocytes similarly possess endfeet specializations (orthogonal arrays: see, for example, Hatton and Ellisman 1981) in contact with fluid reservoirs (CSF and blood) with which K^+ ion can be exchanged.
4. Gap junctions couple central glia also [but Somjen (1987b) points out that these have not yet been shown to pass K^+ ions in mammals as they do in lower forms].

Second, all brain astrocytes – Müller cells, cerebellar Bergmann glia, and the radial glial cells throughout brain and spinal cord – react with the astrocyte-specific glial fibrillary acidic protein antibody. During embryology one endfoot of every brain astrocyte attaches to the ventricular wall, the other to the pial surface of the brain, and in all forms except birds and mammals this relationship persists in the adult. Each primitive neuron, which originates as an ependymal cell lining the ventricle, crawls toward its final destination along the surface of a particular one of these radial glia cells (Rakic 1982). It is possible that each neuron and its radially oriented glial cell (or the daughter cells derived from them) remain near neighbors and continue to interact in some manner throughout life. However, the functional significances of these morphological relationships, if there are any, remain to be defined.

Of course, important differences between brain and retina exist. For instance, the intrinsic retinal neurons produce no axon spikes, but axons enter and leave all brain regions and their activities produce spikes in abundance. Furthermore, in the cortex the pyramidal cells are believed to inject current directly into the extracellular space, and to the extent this is so the analogy between retinal and cortical electrogenesis is imperfect. The magnitude of these neuronal membrane current contributions to the EP remain, however, to be precisely established.

It should be pointed out in this connection that the highest K^+ flux per unit volume is likely to take place in synaptic regions (the neuropile) because this is where the ratio of membrane to cytoplasm is highest. If this is so the contributions of axon spikes and cell-body activities to brain EPs should be relatively small, but the matter will not as readily be settled in brain as in retina, where there are neither radially oriented pyramidal cells nor axon spikes to complicate the picture.

6 The Model

My next three figures diagram the factors presumed to be relevant to EP electrogenesis in the brain.

Figure 2 is a highly simplistic picture of neuronal and glial membranes separated by extracellular space. It is intended to press home the fact that while neuronal membranes possess a full complement of ion channels, glial membranes specialize in K^+ channels. Glial membranes sense and respond almost exclusively to $[K^+]_0$; that is the message. There is no implication intended, however, that this is the only – or even the most important – function these cells serve in the brain.

Figure 3 shows two systems of glial cells coupled by gap junctions; they are displayed as if removing K^+ ion from the subarachnoid space (arrows entering at the top) and transporting it through cytoplasm into the capillary shown at the bottom. K^+ ions initially enter the cerebrospinal fluid either in the plasma ultrafiltrate produced at the choroid plexus, or through the capillary endothelium, where it oozes across along with the nutrients and other substances that penetrate the blood–brain barrier at this point. The equilibrium reached by the processes diagrammed by the arrows maintains brain $[K^+]_0$ at around 3 mM/liter, a concentration about two-thirds that of the plasma and one that is presumably optimal for the conduct of neuronal activities.

The K^+ current shown flowing into the capillaries contributes to the potential difference of 1–5 mV (cortical surface negative) recorded between the surface of the resting brain and outside structures such as blood and bone (Tschirgi and Taylor 1958).

Figure 4 adds a representative neuropile to Fig. 3. Incoming axons (PRE) are shown contacting postsynaptic elements across diagrammatic synapses; EPSP and IPSP summarize the $[K^+]_0$ changes resulting from activities at these synapses. IPSP covers any event that moves K^+ from the extracellular space into neurons (and hence from the astrocytes into extracellular space), whereas EPSP means those events that transfer K^+ from neurons into extracellular space (and from there into the astrocytes). The K^+ currents so created add algebraically with those shown in Fig. 3 to be steadily transporting K^+ ion into the blood.

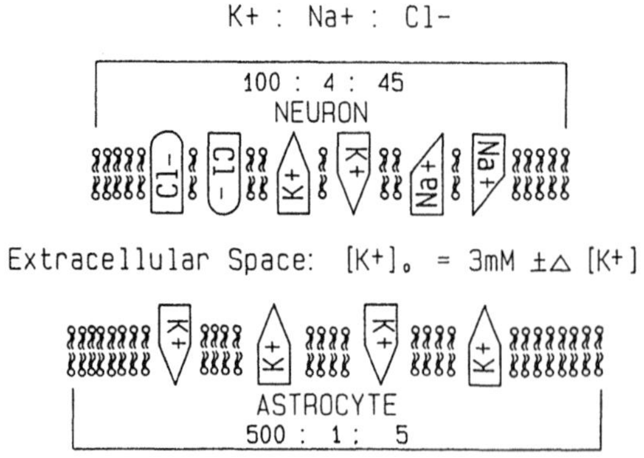

Fig. 2. Diagram of the structures presumably involved in evoked potential electrogenesis: neuronal and glial membranes separated by extracellular space. The ratios of potassium, sodium, and chloride ion channels differ in the two membranes

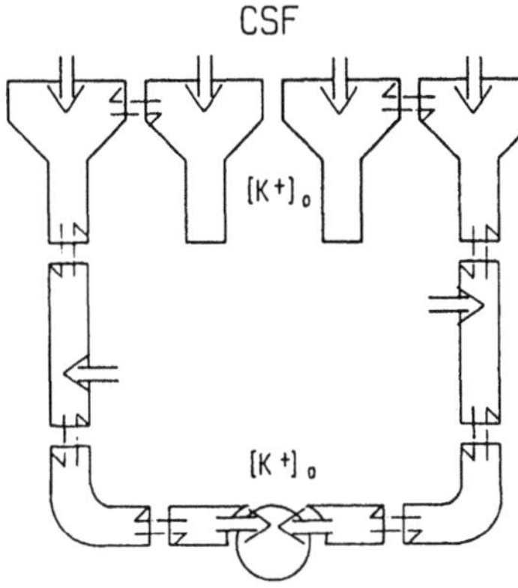

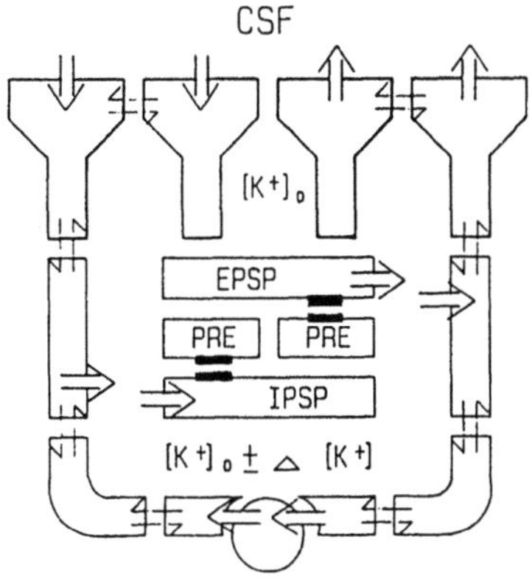

Fig. 3. A diagrammatic neuron-free nervous system in which two groups of glial cells interconnected through gap junctions remove K^+ ions from cerebrospinal fluid and extracellular spaces and transport it through astrocyte cytoplasm to capillaries, an active process that maintains $[K^+]_0$ at about two-thirds that of blood serum

Fig. 4. A diagrammatic synaptic region has been added to the astrocyte collections in Fig. 3. Synaptic activities (*EPSP, IPSP*) constantly alter $[K^+]_0$ as suggested by the *arrows;* whenever the astrocytes counter these changes, K^+ ion current exits or enters through their end-feet at the brain surfaces and through the area surrounding the capillaries.

The above model can account for:

1. The sustained potentials (SPs) measured between two brain regions on the surface or in the depths
2. An increase or decrease in these SPs like those observed during behavior (e. g., the CNV)
3. Both the positive and the negative swings around a baseline that characterize stimulus-evoked potentials

To summarize: EPs appear whenever certain physiological activities go on within a more or less limited brain region in which astrocyte and synaptic membranes are separated by extracellular space. Within this glia–neural unit neuronal activities are continuously changing $[K^+]_0$; astrocytes reciprocate with prompt transfers of K^+ ion into and out of capillaries and the subarachnoid space. The EP is a reflection of this K^+ current which synaptic events initiate, astrocyte cytoplasm transports, and extracellular space returns via other ions.

7 Some Relevant Experimental Evidence

Intracellular recordings from cortical glial cells show that, like the Müller cell, they depolarize during nearby neuronal activities, but they do not follow rapid changes such as axon depolarizations (see Karahashi and Goldring 1966; Krnjevic and Schwartz 1967; and for reviews Varon and Somjen 1979; Somjen 1987 a). Thus, a time base of 5 s is clearly appropriate for the slow depolarizations

Kelly and Van Essen (1974) recorded to moving visual stimuli from the glial ("silent") cells of cat visual cortex. In that study, however, as in most such reports, the animal was anesthetized.

Anesthesia importantly alters stimulus-induced extracellular K^+ ion changes within cortical tissue, according to Molnar and Skinner (1983). They state skin shocks which evoked "a huge" increase in K^+ in the waking cat cortex failed to do so when the animal was anesthetized with Nembutal, and identical electrical shocks to the cortical surface invariably produced smaller K^+ changes in the anesthetized animal.

In normal animals "conspicuously large potassium increases (nearing the 10 mM level) accompanied the 'event-related' potentials elicited by meaningful stimuli... [but] no consistent relationship could be observed between the changes of extracellular field potentials and potassium" (Molnar and Skinner 1983; see also Skinner and Molnar 1983). These authors suggest, nevertheless, that "... both neuronal and glial elements are involved in the genesis of slow evoked potentials."

Further discussion of the published data will not be undertaken here; at the present time the evidence seems, on the whole, to favor some kind of glial contribution to the later EP components but none at poststimulus intervals below perhaps 100 ms.

8 EPs Collected from Animal Brains: Interpretations

8.1 The Puff-Conditioned Monkey

Until I extrapolated the retinal discoveries to brain as outlined above, I was unable to interpret the data of Fig. 1. Now, however, the following propositions seem reasonable to me.

The bipolar electrodes at each of the four brain locations in Fig. 1 monitor the electrical consequences of K^+ current flowing through astrocyte cytoplasm as these cells create and maintain $[K^+]_0$ optimal for the axon and dendrite activities in the neuropile. When the experiment starts, the click, whatever else it may be doing in the rest of the nervous system, creates relatively little synaptic activity in the four neuropile regions the electrodes have isolated for our examination. The first air puff delivered to the face activates one or more distant brain regions that do not belong to the classical auditory pathway described by Cajal. These regions participate in organizing the orienting responses, and their efferent axons project to the synaptic territories under the electrodes. The new input these axons supply initiates many graded membrane changes – depolarizations and hyperpolarizations – that are time-locked to the click; these neuronal events produce corresponding time-locked variations in the value of $[K^+]_0$. *The sudden appearance of evoked potentials reflects this increased K^+ ion flow out of and into neuronal membranes and the reciprocal flow of K^+ ion into and out of astrocyte membranes.* (The number of EP wavelets actually generated in this way is an open question, and in the retina the very earliest events clearly are not. In brain the extracellular currents generated by axon spikes and cortical pyramidal cell activities may well contribute to the scalp-recorded EP, but the retinal observations suggest these contributions will not be responsible for the whole of it.)

To return to the monkey experiment: When the puff reinforcement ends, the axon input from the distant brain region or regions gradually declines, and both the K^+ ion flux and the click-evoked response amplitudes due to that flux drop toward the control size. This process is quickly reversed, however, when a single reinforcement reestablishes the orienting reflex and a new surge of axon discharges enters every neuropile region our electrodes sample.

You will note that this explanation adds to the concepts directly extracted from the retinal model only the idea that axons from a distant place can gain access to a given neuropile and alter the synaptic activities under way there, an idea neither new nor unfamiliar. Now I shall turn to another example for a further test of these ideas.

8.2 The Flicker-Conditioned Cat

To many observers the similarities among the responses in channels 1, 2, and 4 of Fig. 1 imply they share a common generator. A set of records published several years ago by E. Roy John (1967) raises this same problem with unusual clarity. Figure 5, copied without reduction from his book, shows averaged EPs obtained through implanted electrodes from a cat trained to respond to light flickering at 2.5 per second; the top trace was obtained monopolarly from visual cortex, the rest bipolarly from visual cortex, lateral geniculate, and reticular formation. To make records like those in Fig. 5 easier to read, my colleague Scott Makeig has written programs which allow us to redigitize and then manipulate them in our own computer. To redigitize we first prepare a transparency enlarged about 5 times, as in Fig. 6a, and then paste it onto the tube-face of the computer monitor; the cursor is then moved beneath the tracing of interest to enter it into the computer memory.

Figure 6b shows the four traces extracted from our computer memory with all of the 30-ms downward-going peaks aligned. The figure certainly supports the

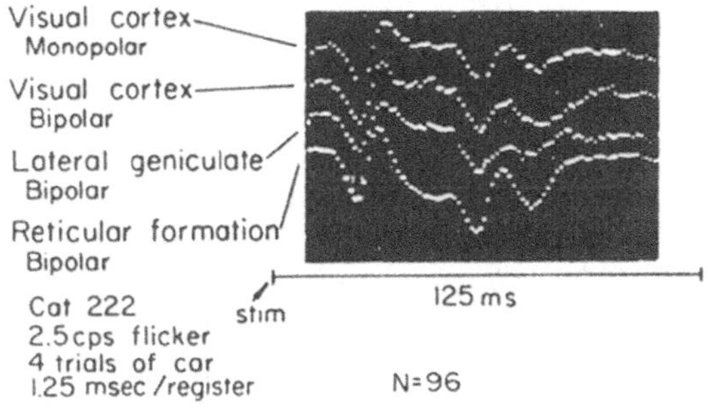

Fig. 5. A cat experiment published in 1967 (see John 1967) is like the monkey experiment of Fig. 1: the evoked responses recorded at widely different brain sites in a conditioned animal resemble each other

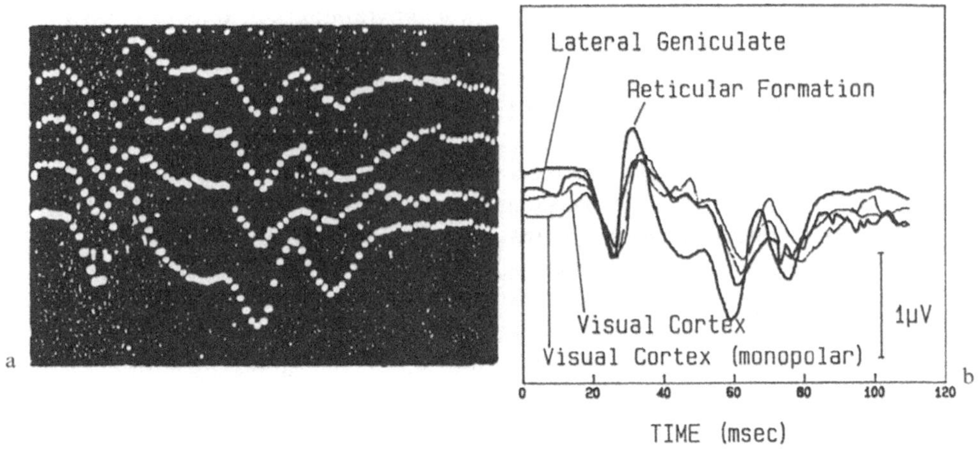

Fig. 6a, b. The data of Fig. 5 after redigitizing as described in the text

claim that EPs from widely different brain locations can resemble each other closely. For instance, several of the peak amplitudes are almost identical, a truly amazing fact. The reticular formation peaks and valleys generally precede their counterparts in the other locations, but the differences are nowhere greater than a few milliseconds.

It is often claimed that recordings like those in Fig. 5 and 6 merely demonstrate that electrical currents due to localized brain events spread readily through brain fluids to distant regions. (Records like these are frequently dismissed with the remark "that's nothing but volume conduction.") Actually, what I propose here is also a volume conduction explanation, but this time the current is assumed to flow partly intracellularly within astrocyte cytoplasm, and partly extracellularly, in the opposite direction, through the extracellular fluids. This, after all, is the way the retina does it, and so why not the brain as well?

Next, it is important to realize that the EP patterns of Figs. 5 and 6 did not exist until after the cat had learned the significance of the flickering light. They developed during training, over time; and furthermore they were unique for the particular behavioral response that animal had been trained to perform. Roy's neurological explanation of the phenomenon is as follows: "Common modes of activity are characteristic of an extensive anatomical system during correct performance to a conditioned stimulus. The neural networks sustaining this common mode activity would seem to correspond to our hypothetical representational system... A change in the time course of discharge in neural units comprising these ensembles must accompany the observed changes in evoked potential shape..." (John 1967, p 342). The *neuronal* activity changes, he says. Alternatively, the striking convergence of EP shape toward a single pattern at all locations might have developed because as the experiment progressed astrocytes progressively interconnected the discharging brain regions in new and different ways. In this alternative explanation *the constellation of glial cells coupled electrically through gap junctions is conceived to change during learning, as a result of which the same (or very similar) patterns of neuronal activity now initiate K^+ ion flow through the cy-*

toplasm of a different, much larger assembly of coupled glial cells. In this view the "new" electrical pattern is in fact merely a somewhat different composite view of the same (or very similar) synaptic events present before the training; thanks to a rearrangement of glial gap junctions, the electrical activities in several once-separated neuropile regions are no longer isolated from each other.

For the cat of Figs. 5 and 6 this learning-created coupling of astrocytes would have to extend from reticular formation through thalamus into the cortex, a distance of several centimeters. If such a coupling actually does take place, all cells involved may well be close relatives of the radial glia that originally spanned this same territory and guided the immature neurons to their final positions in the brain.

9 Summary

1. The electrogenesis of the two major types of potential recorded from the human scalp are here at issue. These are the spontaneous potentials (both brain waves and the standing or DC potentials) and the time-locked potentials (i.e., those shown by averaging procedures to precede or follow external stimuli, voluntary movements, and the like). This chapter discusses their possible generator sources, with emphasis on the possibility that contributions from brain astrocytes (glial cells) have been underestimated.

2. The retina displays both spontaneous and time-locked potentials despite the absence there of two of the three major candidate neuronal generators found in brain: neither axon spikes nor oriented layers of pyramidal cells characterize retinal anatomy and physiology. Furthermore, the Müller cell, a retinal astrocyte, has been securely identified as essential for generation of the electroretinogram, the time-locked potential recorded from outside the eye.

3. Extrapolation of the retinal physiological findings to the brain poses few conceptual problems since the two structures have common embryological origins and similar final structures and functions. The end result is a hypothesis according to which brain astrocytes regulate $[K^+]_0$ (the concentration of potassium ion in extracellular space) both at rest and during synaptic activities; K^+ is transported through astrocyte cytoplasm from regions where $[K^+]_0$ is high to where $[K^+]_0$ is lower, and this intracellular current flow, returned by other ions through extracellular space, is what conventional scalp electrodes detect.

4. While a modest amount of direct experimental evidence supports the idea that K^+ could flow through the cytoplasm of brain astrocytes during neuronal activities, none at all can yet be brought to bear upon two other hypotheses advanced in the text. These propositions are first that certain primitive anatomical and physiological interactions characteristic of embryonic neurons and glia are retained in adulthood, and second that the gap junctions that normally interconnect glial cells increase in number during learning to form large assemblies of coupled cells that interconnect once widely separated regions where synaptic activities alter $[K^+]_0$.

5. These hypotheses are shown to lead to interpretations of published experiments on animals and man which are somewhat at variance with the conventional explanations.

Acknowledgments. Discussions with T. H. Bullock, Glenn Northcutt, and especially Scott Makeig led to significant improvements in the content and organization of this material. Phyllis Galambos prepared the figures and provided moral support throughout. The work was aided in part by NIH grants and financial support from the Children's Hospital Research Center, San Diego (Director: Donald Krebs).

References

Galambos R (1961 a) Processing of auditory information. In: Brazier MAB (ed) Brain and behavior. American Institute of Biological Science, Washington DC, pp 171–203
Galambos R (1961 b) A glia-neural theory of brain function. Proc Natl Acad Sci USA 47:129–136
Galambos R, Hillyard SA (1981) Electrophysiological approaches to human cognitive processing. Neurosci Res Prog Bull 20:141–265
Galambos R, Sheatz GC (1962) An electroencephalograph study of classical conditioning. Am J Physiol 203:173–184
Hatton JD, Ellisman MH (1981) The distribution of orthogonal arrays and their relationship to intercellular junctions in neuroglia of the freeze-fractured hypothalamo-neurohypophysial system. Cell Tissue Res 215:309–323
John ER (1967) Mechanisms of memory. Academic, New York
Karahashi Y, Goldring S (1966) Intracellular potentials from idle cells in cerebral cortex of cat. Electroencephalogr Clin Neurophysiol 20:600–607
Kelly JP, Van Essen DC (1974) Cell structure and function in the visual cortex of the cat. J Physiol (Lond) 238:515–547
Kline RP, Ripps H, Dowling JE (1978) Generation of b-wave currents in the skate retina. Proc Natl Acad Sci USA 75:5727–5731
Krnjevic K, Schwartz S (1967) Some properties of unresponsive cells in the cerebral cortex. Exp Brain Res 3:306–319
Kuffler SW, Nicholls JG (1966) The physiology of neuroglia cells. Ergeb Physiol Biol Chem Exp Pharm 57:1–90
Miller RF, Dowling JE (1970) Intracellular responses of the Müller (glia) cells of mud-puppy retina: their relation to b-wave of the electroretinogram. J Neurophysiol 33:323–341
Molnar M, Skinner JE (1983) Evoked and spontaneous extracellular potassium shifts in the cerebral cortex of unanesthetized cats. Acta Physiol Hung 61(4):265–279
Newman EA (1985 a) Regulation of potassium levels by glial cells in the retina. Trends Neurosci 8(4):156–159
Newman EA (1985 b) Membrane physiology of retinal glial (Müller) cells. J Neurosci 5(8):2225–2239
Newman EA, Frambach DA, Odette LL (1984) Control of extracellular potassium levels by retinal glial cell K^+ siphoning. Science 225:1174–1175
Orkand RK, Nicholls JG, Kuffler SW (1966) Effect of nerve impulses on the membrane potential of glial cells in the central nervous system of amphibia. J Neurophysiol 29:788–806
Rakic P (1982) Early developmental events: cell lineages, acquisition of neuronal positions, and areal and laminar development. Neurosci Res Prog Bull 20:439–451
Ripps H, Witkovsky P (1985) Neuron-glia interaction in the brain and retina. Prog Retinal Res 4:181–219
Roitbak AI (1983) Neuroglia. Fischer, Jena (Brain and behaviour research monograph series, vol 10)

Skinner JE, Molnar M (1983) Event-related extracellular potassium ion activity in frontal cortex of the conscious cat. J Neurophysiol 49:204–215
Somjen GG (1975) Electrophysiology of neuroglia. Annu Rev Physiol 37:163–190
Somjen GG (1980) Influence of potassium and neuroglia in the generation of seizures and their treatment. In: Glaser GH, Penry JK, Woodbury DM (eds) Antiepileptic drugs: mechanisms of action. Raven, New York, pp 155–167
Somjen GG (1981 a) Physiology of glial cells. In: Salanki J (ed) Physiology of non-excitable cells. Akademiai Kiado, Budapest, pp 23–43, Adv Physiol Sci 3
Somjen GG (1981 b) Neuroglia and spinal fluids. J Exp Biol 95:129–133
Somjen GG (1984) Interstitial ion concentration and the role of neuroglia in seizures. In: Wheal HV, Schwartzkroin PA (eds) Electrophysiology of epilepsy. Academic, New York, pp 303–341
Somjen GG (1987 a) Glial cells, functions. In: Adelman G (ed) Encyclopedia of neuroscience. Birkauser, Boston
Somjen GG (1987 b) Functions of glial cells in the cerebral cortex. In: Jones EG, Peters A (eds) Cerebral cortex. Plenum, New York, pp 1–39
Somjen GG, Trachtenberg M (1979) Neuroglia as generator of extracellular current. In: Speckmann EJ, Caspers H (eds) Origin of cerebral field potentials. Thieme, Stuttgart, pp 21–32
Tschirgi RD, Taylor JL (1958) Slowly changing bioelectric potentials associated with the blood–brain barrier. Am J Physiol 195:7–22
Varon SS, Somjen GC (1979) Neuron–glia interactions. Neurosci Res Prog Bull 17:1–239

Cell Membranes, Electromagnetic Fields, and Intercellular Communication

W. R. ADEY

1 Introduction

It has long been assumed that equilibrium models of cellular excitation that focus on depolarization of the membrane potential and associated massive changes in ionic equilibria across the cell membrane also offer an adequate basis for an understanding of the first events in cell membrane transductive coupling of molecular and electrochemical stimuli at the cell surface. For nervous tissue, it has been generally accepted that the Hodgkin–Huxley (1952) model appropriately describes both sequence and energetics of excitatory events. However, this brilliant thesis from relatively limited biological data was originally offered only in the context of a mathematical description of major perturbations in Na^+ and K^+ ionic equilibria that occur at a certain epoch in the course of excitation in giant nerve fibers of the squid.

There has been a philosophical incubus for more than 30 years from the early and seemingly inappropriate extrapolation of this model to threshold phenomena in excitation of nerve cells in the spinal cord (Eccles 1953) and subsequently in cerebral neurons. At that time, excitatory mechanisms in nerve fibers and nerve cells were grouped under a common rubric of ionic equilibrium mechanisms. There was little interest in the possibility that functional organization of cell soma and dendritic membranes might involve threshold sensitivities to both oscillating electric fields and molecular stimuli at energy levels substantially lower than those predicted in Hodgkin–Huxley models, and substantially below typical thresholds seen in nerve fibers. In their electrical nature, these sensitivities appear to involve nonequilibrium and highly cooperative processes that mediate a major amplification of initial weak triggers associated with binding of hormones, antibodies, neurotransmitters, and chemical cancer promoters to their specific cell surface receptor sites (Adey 1983, 1986, 1987a, b; Adey and Lawrence 1984; Lawrence and Adey 1982).

In the interim, there has been an increasing focus on the role of Ca^{2+} ions in excitatory processes. There is clear evidence that Ca^{2+} ions (Adey et al. 1982; Bawin and Adey 1976; Bawin et al. 1978a b; Blackman et al. 1979, 1985a b; Lin-Liu and Adey 1982) and calcium-dependent processes at cell membranes (Dixey and Rein 1982; Kaczmarek and Adey 1974; Lyle et al. 1983) initiate intracellular enzyme responses (Adey 1986; Luben et al. 1982; Luben and Cain 1984; Byus et al. 1987, 1988) that are modulated by extracellular electromagnetic (EM) fields inducing transmembrane currents millions of times weaker than those required for threshold excitation in accordance with a Hodgkin–Huxley model.

Springer Series in Brain Dynamics 2
Edited by Erol Başar and T. H. Bullock

1.1 Pathways for Electric Current Flow in Tissue

The crux of the problem becomes apparent from a consideration of distribution of current flowing in a tissue composed of cells with a high membrane resistance and bathed in a strongly conducting fluid (Cole 1940). Typical cell membrane resistances are in the range 3 000–100 000 ohm cm^{-2}. Extracellular fluid has a specific resistance of only 50 ohm cm^{-1}. Thus, although the extracellular space forms only about 10% of the conducting cross-section of typical tissue, it is clearly a preferred pathway, carrying at least 90% of any imposed or intrinsic induced current.

As a corollary to this scheme of divided current pathways, most current flows *along membrane surfaces*, following narrow intercellular gutters that separate cell membranes. As discussed below, the longitudinal current flow at cell surfaces appears of considerable importance in the action of pericellular fields on Ca^{2+} binding at cell surfaces (Bawin and Adey 1976; Bawin et al. 1978 a; Bawin et al. 1975)

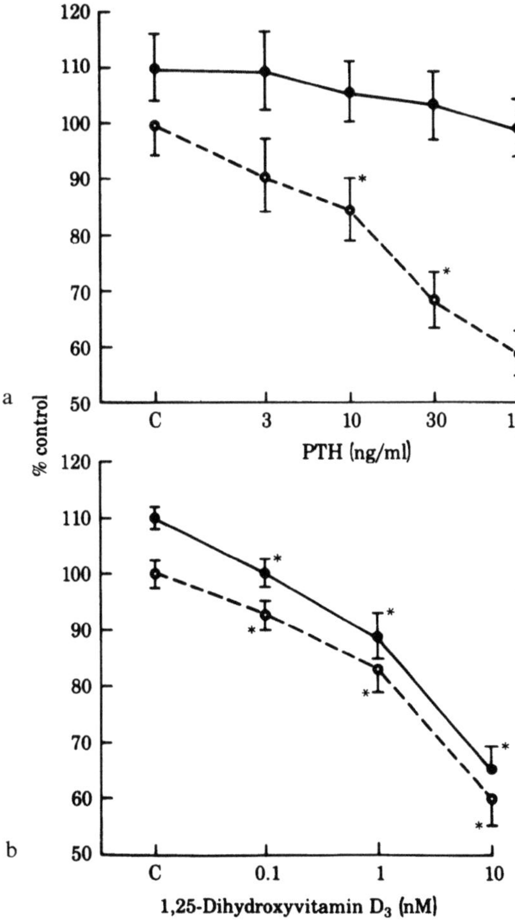

Fig. 1 a, b. Effects of fields on collagen synthesis by bone cells. MMB-1 cells were grown in the presence (*solid circles*) or absence (*open circles*) of a pulsed 72-Hz magnetic field for 12 h. Then various concentration of PTH (a) or 1,25-dihydroxyvitamin D3 (b) were added in fresh culture medium. Culture was continued for 48 h, with the final 4 h being in the presence of ^{3}H-proline at 1 μC/ml. Cells were disrupted and collagen synthesis measured by a collagenase technique. *C*, control; *, significantly different from control, $P < 0.05$. (Luben et al. 1982)

and in concurrent activation along cell membranes of cascades of Ca^{2+}-dependent protein kinases (Nishizuka 1983, 1984).

Hodgkin–Huxley models as originally applied to nerve axons predict a transmembrane current at a threshold of 1.0 mA/cm^2. As just noted, extracellular current densities to achieve this transmembrane level would be 10^2–10^4 times higher. The diverse physiological effects of weak imposed EM fields cited above and described in detail below have shown this to be an inappropriate model. For example, low frequency pulsed magnetic fields effective in therapy of ununited fractures produce tissue gradients of about 3 mV/cm and extracellular current densities around 10^{-6} A/cm^2, or one-millionth of threshold transmembrane currents predicted by Hodgkin–Huxley models (Luben et al. 1982; Bassett et al. 1982). These experimental data (Fig. 1) clearly suggest the need for a radically different approach to understanding these sensitivities of cells to their electrochemical environment.

1.2 Intercellular Signaling Across the Extracellular Space

Furshpan and Potter (1959) reported electrical transmission at the crayfish giant motor synapse. Shortly thereafter, similar cell–cell communication was reported for Mauthner neurons in the goldfish brain (Furshpan and Furakawa 1962) and in mammalian smooth muscle (Dewey and Barr 1962). Contacting cells were found to exchange ions through low resistance pathways connecting adjacent cytoplasms. This electrotonic coupling was bidirectional and occurred with an impedance 3–4 orders of magnitude less than that of the resting transmembrane impedance.

These initial observations on ionic mechanisms were consistent with existing models for excitable tissue (Bennett et al. 1963), but when Kanno and Loewenstein (1966) found that microinjected fluorescein passed between contacting epithelial cells in seconds, it became apparent that the phenomenon of ionic coupling was not confined to excitable tissues nor to ions alone (see Fletcher et al. 1987a for review). This initiated the concept that intercellular communication could have broad significance in cellular regulation (Loewenstein 1968; Bennett and Trinkhaus 1970; Gilula et al. 1972). Added support for this concept has come from evidence that there is exchange of metabolic products between intimately contacting cells in "metabolic cooperation" (Pitts and Finbow 1986). As discussed below, a morphological substrate for this cell–cell communication resides in the narrow but regular intercellular space at the junctional region designated as a *gap junction* (Revel and Karnovsky 1967).

1.3 Imposed Electromagnetic Fields as Tools in Studies of Transmembrane Signaling and Intercellular Communication

Long accepted functional models of the plasma membrane have emphasized the existence in most cells of a major electrical gradient (the membrane potential) across the 40 Å width of the lipid bilayer. In these models, this gradient of 10^5 V/cm has been considered an effective electrical barrier against stimulation by weak EM fields in surrounding fluid. However, much recent research has shown that imposed weak EM fields in the extremely low frequency (ELF) spectrum below 100 Hz at intensities in the pericellular fluid many orders of magnitude weaker than the membrane potential gradient can modulate actions of hormone, antibody, neurotransmitter, and cancer promoting molecules at their cell surface receptor sites. Similar sensitivities have been observed for radiofrequency fields that are amplitude-modulated at ELF frequencies. These modulating actions of EM fields suggest highly cooperative processes in the underlying physical mechanisms. The observed thresholds are as low as 10^{-7} V/cm in the ELF spectrum. Moreover, many of these interactions are *windowed* with respect to field frequency and amplitude, and to duration of field exposure.

Use of imposed EM fields in these studies has allowed identification of some of the sequences and energetics of steps that couple signals from the outside to the inside of the cell, pointing the way to further studies designed to elucidate their cooperative nature. The windowed character of many of these responses points to their nonlinear and nonequilibrium character, and focuses current research on physical substrates for these interactions (Adey 1975, 1977, 1981 a, b, 1983, 1984, 1986; Lawrence and Adey 1982; Maddox 1986), with emphasis on quantum mechanisms and long range ordering. Thus, a carefully designed spectrum of fields at ELF frequencies, or amplitude modulated at ELF frequencies, offers unique tools in evaluating the physical ordering of cell membrane coupling.

2 Cell Membrane Substrates for Transmembrane Signaling and Energy Transfer

With the electron microscope, the cell membrane is defined as a lipid bilayer about 40 Å in width, with polar head groups on external and internal surfaces of the double layer (the plasma membrane). More importantly, these studies of membrane ultrastructure have revealed numerous strands of protein (intramembranous particles, IMPs) inserted into the plasma membrane. These IMPs span the membrane from inside to outside. They have external protrusions into the fluid surrounding the cell, with terminal glycoprotein strands that sense electric fields and form receptor sites for chemical stimuli, including hormones, antibodies, neurotransmitter molecules, and certain cancer promoters. Inside the cell, they make functional contact with key elements in the cell machinery, including enzymes and the numerous fine tubes and filaments of the cytoskeleton. They "float" in the sea of lipid molecules of the plasma membrane, leading to the generally accepted *fluid mosaic* model of the cell membrane (Singer and Nicolson 1972). Thus, these intramembranous protein strands form signaling pathways by which external stimuli are sensed and conveyed to the cell interior.

2.1 Cell Membrane Receptor Proteins for Human Epidermal Growth Factor and Nerve Growth Factor as Models of Membrane Coupling Proteins

Much attention now focuses on membrane receptor proteins for the human epidermal growth factor (EGF) and the nerve growth factor (NGF) as models of coupling proteins in studies of the nature of transmembrane signals.

The entire 1210 amino acid sequence of the EGF receptor protein has been deduced by Ullrich et al. (1985), with striking findings on the sequences that make up the extracellular, intramembranous, and cytoplasmic portions of the chain. Extracellular and intracellular segments are each composed of appoximately 600 hydrophilic amino acids. The molecule appears to cross the membrane only once. The salient and surprising finding is the extremely short length of the intramembranous segment of 23 amino acids, predominantly hydrophobic, and with only a single amino acid with a side chain capable of hydrogen bonding.

Subsequent studies have shown that the NGF receptor protein also has a strikingly similar segment of 23 hydrophobic amino acids within the membrane (Radeke et al. 1987), suggesting that this configuration plays a fundamental role in processes of transmembrane signaling. This view is strengthened by studies with a chimeric protein constructed of the extracellular portion of insulin receptor protein joined to the transmembrane and intracellular domains of the EGF receptor protein (Riedel et al. 1986). In this molecule, the EGF receptor kinase domain of the chimeric protein is activated by insulin binding. The authors conclude that insulin receptors and EGF receptors employ closely related or identical mechanisms for signal transduction across the plasma membrane.

What can be inferred from these studies about the essential nature of the transmembrane signal? Ullrich et al. (1985) point out that this 23 amino acid sequence is probably too short to be involved in conformation changes, and that its hydrophobic character makes unlikely its participation in either ion or proton translocation. As an alternative, they have suggested that an EGF-induced conformation change in the extracellular segment of the receptor protein may be transmitted to the cytoplasmic domain of the protein by movement of this short intramembranous segment, or by receptor aggregation. As a further option, we have hypothesized that this transmembrane signaling may involve nonlinear vibration modes in helical proteins and generation of Davydov–Scott soliton waves (Lawrence and Adey 1982). Although evidence for soliton waves in DNA and helical proteins is not conclusive and has been criticized on theoretical grounds (Lawrence et al. 1987), we have strong supporting evidence for nonlinear, nonequilibrium processes at critical steps in transmembrane coupling, based on windows in EM field frequency, amplitude, and time of exposure that determine stimulus effectiveness.

Despite the hydrophobic character of this short transmembrane coupling segment, addition of EGF to human epidermal cell cultures causes a two- to fourfold increase in cytoplasmic free Ca^{2+} within 30–60 s (Moolenaar et al. 1986). This EGF-induced signal appears to result from Ca^{2+} entry via a voltage-independent channel, since it is not accompanied by a change in membrane potential. It is completely dependent on extracellular Ca^{2+} moving into the cell interior. This action is inhibited by cancer-promoting phorbol esters which have a specific membrane receptor (Ca^{2+}-dependent protein kinase C). EM field interactions with these cancer promoters are discussed below.

2.2 Morphogenetic Concepts Involving Cell Surface Proteins in Brain Tissue

Cell migration in the embryo leading to tissue and organ formation requires varying degrees of cell adhesion that change in the sequence of migratory events. The glycoprotein strands on the cell surface described above in the Singer–Nicolson fluid mosaic membrane model have sialic acid (amino sugar) terminals that are strongly negatively charged, producing a huge polyanionic surface. Edelman (1984) has described patterns in the mosaic of surface glycoproteins in brain tissue as a manifestation of gene expression. These are the *cell adhesion molecules* (N-CAMs in nervous tissue) that determine cell adhesion in embryogenesis. N-CAM levels are high at primary sites near the neural tube, are low during migration, and reappear as they reach their destination and interact with other cells. N-CAMs change form radically in a shift from embryonic (E) to adult (A) forms, with a reduction from 30% to 10% of the total mass of glycoproteins. This ontogenetic change fails to occur in *reeler disease* in mice, manifested as a gross failure in cerebellar development.

These findings are noteworthy in three respects. Until recently, it had been a matter of speculation whether cell surface proteins included enzymes that would participate directly in the initial steps of transduction of electrical and chemical stimuli. Ehrlich et al. (1986) have reported the presence on nerve cell surfaces of an ecto-protein kinase that is selectively stimulated during cell depolarization. It performs the typical functions of protein kinase enzymes in phosphorylating other protein substrates. N-CAMs are a substrate for this ecto-protein kinase system. Also, the histology of cell surface glycoproteins has been reported by Delgado et al. (1982) as abnormal in chick embryos exposed to weak low frequency pulsed magnetic fields; and Byus et al. (1984) have identified a sensitivity to amplitude-modulated radiofrequency fields in a class of protein kinases in human lymphocytes, some of which are known to be membrane related (see below).

2.3 Gap Junction Organization in Intercellular Communication

In high resolution electron microscopy of Mauthner cell lateral dendrites, Robertson (1963) observed specialized regions of contact between membranes of adjacent cells that could serve as the site of electrotonic coupling that Furshpan and Furakawa (1962) had found in the same preparation. These images showed that the membranes were differentiated into a honeycomb of 9-nm facets when viewed parallel to the plane of the membrane, with the subunits spaced 10–11 nm center-to-center and aggregated into a nearly hexagonal lattice (see review by Fletcher et al. 1987 a). Seen in cross section, the specialized junction consisted of plasma membranes in such intimate contact that the intercellular space appeared nonexistent. However, Revel and Karnovsky (1967) subsequently filled the intercellular space of hepatic tissue with the electron dense tracer colloidal lanthanum and showed the junctions described by Robertson in sharp detail. These lanthanum preparations revealed a thin 2- to 3-nm cleft separating adjoining hepatocyte membranes, rather than the virtual fusion noted by Robertson. Revel and Karnovsky assigned the term *gap junction* to this sharply narrowed but regular intercellular space at the junctional region.

From X-ray and electron optical diffraction techniques with image reconstruction (Unwin and Ennis 1984), gap junctions are seen as protein subunits packed as hexamers that form the essential junctional subunit, the *connexon* (Hertzberg 1984). There is a 1.5 nm diameter hydrophilic channel in the center of each connexon that appears to span the entire membrane, so that when connexons of adjacent cell membranes come into register, the cytoplasms of connected cells are effectively in continuity. These connexons thus provide the physical substrate for ionic coupling, fluorescent dye transfer and metabolic cooperation discussed above (Gilula et al. 1972).

Although gap junctions in excitable tissues clearly function as electrical synapses and may mediate ionic and metabolic coupling, much less is known about their role in nonexcitable tissues, where these junctions are most numerous and diverse. They appear to play a critical role in embryogenesis, where monoclonal antibodies to gap junction proteins injected at an early stage lead to arrest of embryogenesis (Warner et al. 1984). In adult tissues where cells are united by gap junctions, there have been implications of involvement in growth control, including tissue repair and neoplastic transformation (Loewenstein 1966, 1979, 1981). As discussed below, initial studies by Loewenstein (1968) indicated a correlation between controlled cell growth and the presence of gap junctions, and uncontrolled growth in their absence, although later studies have shown that this correlation is only partial. Nevertheless, our studies indicate a synergic action of chemical cancer promoters and EM fields at cell membranes in modification of gap junction functions (Fletcher et al. 1986).

3 Modulation of Cell Membrane Transductive Coupling by Pericellular Electromagnetic Fields

Based on the Singer–Nicolson fluid mosaic model discussed above, there is a minimum sequence of three steps in transductive coupling (Adey 1984), and each is calcium dependent: (a) cell surface glycoproteins that are stranded protrusions from intramembranous helical proteins (IMPs) sense the first weak electrochemical events associated with binding of neurohumoral molecules at their receptor sites; (b) transmembrane portions of IMPs signal these events to the cell interior; (c) internally, there is coupling of the signal to intracellular enzyme systems and to the cytoskeleton, and thus to the nucleus and to other organelles. The numerous fixed charges on the terminals of surface glycoprotein strands cause the cell surface to behave as a polyanionic sheet (Elul 1966, 1967). It moves visibly in response to focal current pulses that simulate synaptic action, so that the cell surface may be likened to a field of waving corn, responding to an infinite variety of faint electrochemical breezes that blow along the membrane surface.

The polyanionic surface sheet of stranded glycoproteins attracts a cationic counterion layer, principally of calcium and hydrogen ions, which compete for these anionic binding sites. This ionic bilayer is now recognized as a major element in a membrane surface compartment (Blank 1986). It contributes a negative surface potential of many millivolts that can modulate the action of drugs, such as acetylcholine and curare (Van der Kloot and Cohen 1979). Bass and Moore (1968) predicted a "local alkalosis" in the course of excitation, due to the depletion of the hydrogen ion atmosphere as these ions displace calcium from binding sites on surface proteins. These proteins include calcium-dependent protein kinases (ecto-kinases), activated during cell membrane depolarization by electric fields and necessary for cell adhesion processes in neural development and maturation (Ehrlich et al. 1986).

3.1 Cooperative Modification of Calcium Binding with Amplification of Initial Signals

Initial stimuli associated with weak pericellular EM oscillations and with binding of humoral molecules at their membrane receptor sites elicit a highly cooperative modification of calcium binding to glycoproteins along the membrane surface. This longitudinal spread is consistent with the direction of flow of extracellular currents associated with physiological activity and from imposed EM fields, as noted above. It is also consistent with spreading activation of Ca-dependent protein kinase C, which is also a specific receptor for cancer-promoting phorbol esters (Nishizuka, 1983). This cooperative modification of surface Ca^{2+} binding is an *amplifying stage,* with evidence from concurrent manipulation of these initial events by imposed EM fields that there is a far greater increase or decrease in Ca^{2+} efflux than is accounted for in the events of receptor–ligand binding (Bawin and Adey 1976; Bawin et al.1975; Lin-Liu and Adey 1982). There is further striking evidence for the nonequilibrium character of this modification in Ca^{2+} binding in its occurrence in quite narrow frequency and amplitude *windows* (Adey 1981 a, b; Bawin et al. 1978 b; Blackman et al. 1979).

The pioneering observations of Bawin et al. (Bawin and Adey 1976; Bawin et al. 1975) were the first to show "tuning curves" of altered Ca^{2+} efflux from tissue as a function of low frequencies in imposed EM fields. Most studies have used cerebral tissue, including cerebral cortex in awake cats (Adey et al. 1982), isolated chick cerebral hemispheres (Bawin and Adey 1976; Bawin et al. 1975; Blackman et al. 1982, 1985 a, b), cultured neurons (Dutta et al. 1984), and cerebral synaptosome fractions (Lin-Liu and Adey 1982). Studies in different laboratories are consistent in reporting shifts in Ca^{2+} efflux that depend on low-frequency characteristics and on field intensity.

These effects have been observed with radiofrequency fields amplitude modulated at low frequencies, with low-frequency electric fields, with low-frequency EM fields, and with combined low-frequency EM fields and static magnetic fields. Effects are maximal around 16 Hz and less at higher and lower frequencies. Neither size nor geometry is a determinant of these interactions, which occur over an enormous range of physical dimensions, from intact cerebral cortex to cultured neurons and finally in cerebral synaptosomes with mean diameters around 0.7 μm. The maximum tissue electric gradients induced by these imposed fields are at the levels of the EEG in fluid around brain cells (O.1 V/cm), seen with exposures to radiofrequency fields in the studies of Bawin et al., Blackman et al., and Dutta et al. cited above. At these levels, they are at least 6 orders of magnitude less than the electric barrier of the membrane potential. However, with low-frequency fields (in the spectrum below 1.0 kHz), coupling to tissues and cell cultures is far weaker, inducing electric gradients typically in the range 10^{-7}–10^{-3} V/cm. These fields also modify tissue Ca^{2+} binding (Bawin and Adey 1976) and modulate calcium-dependent cell mechanisms, including neurotransmitter release at gradients of 10^{-4} V/cm (Dixey and Rein 1982) and bone matrix formation at 10^{-7} V/cm (Fitzsimmons et al. 1986). These interactions emphasize the importance of *amplification* in their ultimate effects on intracellular mechanisms discussed below.

3.2 Effects of Combined Static Magnetic and Oscillating ELF Electromagnetic Fields

About 20% of pineal cells in pigeons, guinea pigs, and rats respond to changes in both direction and intensity of the earth's magnetic field (Semm 1983). Experimental inversion of the horizontal component of the earth's magnetic field significantly decreases synthesis and secretion of the pineal peptide hormone melatonin, which powerfully influences circadian rhythms, and also reduces activity in its synthesizing enzymes (Welker et al. 1983). This magnetic sensitivity has been traced to nerve fiber pathways from the retina to the midbrain optic tectum (Semm and Demaine 1986). Units in the basal optic root stimulated while the eyes were illuminated by light of different wavelengths exhibited peaks of magnetic responsiveness at 503 and 582 nm. Cells selectively sensitive to magnetic field direction were found in the midbrain optic tectum. They also showed directional sensitivity to pulsed photic stimuli. Response peaks varied with orientation of the pigeon's head in the horizontal plane.

These results suggest that a substrate for detection and elaboration of information from these very weak environmental magnetic fields exists in the pigeon's visual system. In contrast, no changes in nerve conduction velocity in human ulnar, median, or peroneal nerves were observed before, during, or after exposure to a strong (1 T) static magnetic field, although motor excitability was transiently increased (Hong 1987). Evidence for involvement of Ca^{2+} ions in inhibition of morphine-induced analgesia and behavioral activity in mice by magnetic fields was found in studies using a 60 min exposure to a 0.5 Hz rotating magnetic field at 1.5–90 G (Kavaliers and Ossenkopp 1986). This field exposure significantly reduced day-time analgesia and locomotory effects of morphine. Intraventricular injection of the calcium chelator EGTA blocked these effects, while the calcium ionophore A21387 potentiated them. In a parallel fashion, intraventricular Ca^{2+} reduced these responses in a dose-dependent manner in control mice.

Blackman et al. (1985 b) have shown interactions between the earth's magnetic field and a weak imposed low frequency EM field (40 V peak-to-peak in air, estimated tissue components 10^{-7} V/cm) in determining Ca^{2+} efflux from chick cerebral tissue. For example, halving the local geomagnetic field with Helmholtz coils rendered a previously effective 15 Hz field ineffective, and doubling the geomagnetic field caused an ineffective 30 Hz signal to become effective.

3.3 Intracellular Enzymes as Molecular Markers of Transductive Coupling Through Cell Membranes

We have identified three groups of intracellular enzymes that respond to signals initiated at cell membranes as a response to EM field exposure. These responses occur with or without concurrent interactions with humoral stimuli. The enzymes are: (1) membrane-bound adenylate cyclase involved in activation of protein kinases through conversion of ATP to cAMP, as seen in our studies with bone cells (Luben et al. 1982; Luben and Cain 1984); (2) cAMP-independent protein kinases that perform messenger functions (Byus et al. 1984); (3) ornithine decarboxylase, essential for growth of all cells by its participation in synthesis of polyamines essential for DNA formation (Adey 1986; Byus et al. 1987a, b). All are calcium dependent and their actions have been reviewed in detail elsewhere (Adey 1986; see also Proceedings of Conference I of this series).

4 Intercellular Communication and Cancer Promotion; Enzymatic Markers of Field Interactions with Chemical Cancer Promoters at Cell Membranes

Tumor formation as a manifestation of abnormal control of cell growth is now widely modeled as a multistep process, based on animal tumor models. These models of carcinogenesis envisage initial damage to the DNA genome within the cell nucleus. This stage of *initiation* involves actions of mutagenic substances or agents such as ionizing radiation. Initiated or transformed cells may remain indefinitely in this condition without tumor formation. Tumor formation occurs as a second step requiring *promotion* by agents that are not mutagenic and thus not cancer initiators by an action on DNA in the nucleus.

Tumor promoters include insecticides such as DDT, polychlorbiphenyls (PCBs) used as electrical insulators and coolants, saccharin, and plant lectins now used as cancer promoters in laboratory studies, the phorbol esters. Phorbol esters have a specific receptor at cell membranes. For that reason, we have examined cancer promotion in tissue culture models in studies of joint actions of phorbol esters and nonionizing EM fields, *since both act at cell membranes* (Adey 1986; Byus et al. 1987, 1988). Available evidence indicates that nonionizing EM fields do not function as classical initiators in the etiology of cancer by causing DNA damage and gene mutation. On the other hand, there is increasing evidence implicating environmental EM fields in the home and at the workplace in increased cancer risk; thus there is evidence of increased incidence of childhood leukemia and lymphoma as a result of exposure to high-voltage power line fields (Wertheimer and Leeper 1979; Savitz et al. 1986) and a greater risk of leukemia, lymphoma, and pancreatic cancer among workers in electrical industries (Milham 1985).

The biology of cancer therefore invites consideration of promotional processes occurring at cell membranes, rather than at the cell nucleus. We have observed strong interactions at cell membranes between cancer-promoting phorbol esters and EM fields. We hypothesize that cancer promotion may involve a distorted inward signal stream from cell membranes directed to the nucleus and to other organelles (Adey 1987a, b). At the same time, cell–cell communication through gap junctions discussed above is disrupted by chemical cancer promoters (phorbol esters, DDT, PCBs, and saccharin) and is indicative of a disrupted outward signal stream from cell membranes (Yotti et al. 1979). Moreover, normal cells can exert control over those of their number that have been transformed into cancer cells. For example, differentiation is restored to carcinomatous keratinocytes by contact with normal skin fibroblasts and development of carcinomas is inhibited (Newmark 1987). We shall review evidence implicating modified intercellular communication through distorted inward and outward signal streams at cell membranes in cancer promotion.

4.1 Inward Signals at Cell Membranes from Cancer-Promoting Phorbol Esters; Enzyme Activities of Protein Kinases and Ornithine Decarboxylase

As noted above, an important cAMP-independent protein kinase that functions both as a receptor and as an enzyme occurs widely in cell membranes and is in highest concentrations in brain tissue (Nishizuka 1983, 1984). This enzyme, phosphatidyl serine protein kinase (kinase C), is Ca^{2+}-dependent and normally activated by diacylglycerol formed from inositol phospholipids by the action of cell surface stimuli. Following activation of one molecule of kinase C by a single molecule of diacyl glycerol, Nishizuka has described a spreading domino effect that activates all kinase C molecules around the whole membrane surface. Kinase C is irreversibly activated by phorbol esters. In invertebrate neurons, its activation or injection of the pure enzyme enhances inward Ca^{2+} currents (DeRiemer et al. 1985). Exhaustion of kinase C occurs with continued stimulation by phorbol esters and is associated with loss of cell division. This response is restored by intracellular injection of kinase C (Pasti et al. 1986), thus linking it to paths that signal from cell membranes to nuclear mechanisms mediating cell division. Protein kinase C belongs to a group of cAMP-independent protein kinases which we have identified as sensitive to EM fields (Byus et al. 1984).

We have also shown that induction of ornithine decarboxylase (ODC) follows stimulation of liver and ovary cells with phorbol esters (Fig. 2) and that this response is sharply enhanced by 450 MHz fields (1.0 mW/cm^2), sinusoidally amplitude modulated at 16 Hz (Byus et al. 1987a). Similar enhancement of ODC activity was noted with 60 Hz EM fields at fluid gradients of 0.1–10 mV/cm (Byus et al. 1987b), of the same order as those induced by high-voltage power line exposure. ODC activity in synthesis of polyamines is an essential step in DNA syn-

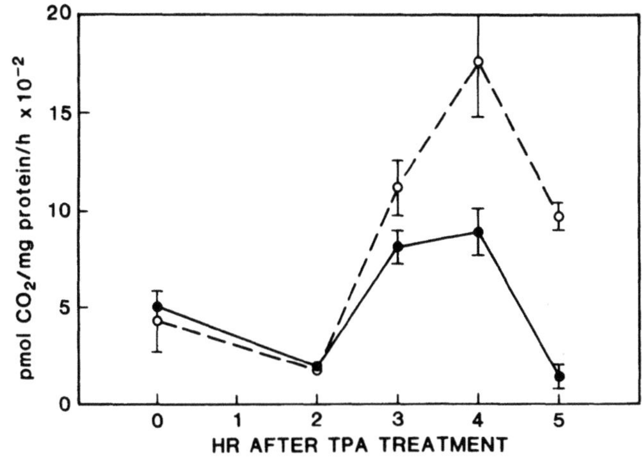

Fig. 2. Ornithine decarboxylase activity in Chinese hamster ovary cells stimulated by a phorbol ester (tetradecanoyl phorbol acetate, *TPA*), with and without concurrent exposure to a 450-Hz field (1.0 mW/cm^2 peak envelope power) amplitude modulated at 16 Hz. *Closed circles*, response to TPA alone; *open circles*, TPA plus field. (Byus et al. 1988)

thesis. Clinically, ODC activity in cultures of suspected cancer cells (for example, human prostatic cancer), has proved a useful index of malignancy. Not all agents that stimulate ODC are cancer promoting, but all cancer promoters stimulate ODC. Although its activation pathways are not well defined, binding of phorbol esters at membrane receptors induces ODC, and ODC activity is increased by 50% in a 3-h test period following a 1-h exposure to the same 450-MHz fields tested above (Byus et al. 1987a). Also, phorbol ester treatment of embryonic fibroblasts previously irradiated with X-rays and microwaves increases transformation frequencies above rates in cells previously irradiated only with X-rays (Balcer-Kubiczek and Harrison 1985). The findings are consistent with persistent membrane effects from prolonged microwave exposure, and these may enhance promoter action.

4.2 Outward Signals Through Gap Junctions; Disruption of Intercellular Communication by Phorbol Esters and Microwaves in Tumor Promotion

We have noted above that phorbol esters and other chemical cancer promoters disrupt transfer of chemical signals between cells (Yotti et al. 1979). Trosko has further hypothesized that two major types of intercellular communication help to maintain "normal orchestration" of proliferation and differentiation during development and between quiescent stem–progenitor and differentiated cells in the adult (Trosko and Chang 1986): one involving transfer of molecular signals from cells of one differentiation or tissue type to another over an extracellular space and distance (for example, in the action of hormones, growth factors, and neurotransmitters), and the other mediated by the transfer of relatively low molecular weight molecules and ions through neighboring cells via gap junctions. In Trosko's model, gap-junctional communication would involve a minimum of four steps: cell recognition or adhesion molecules, as discussed above; functional

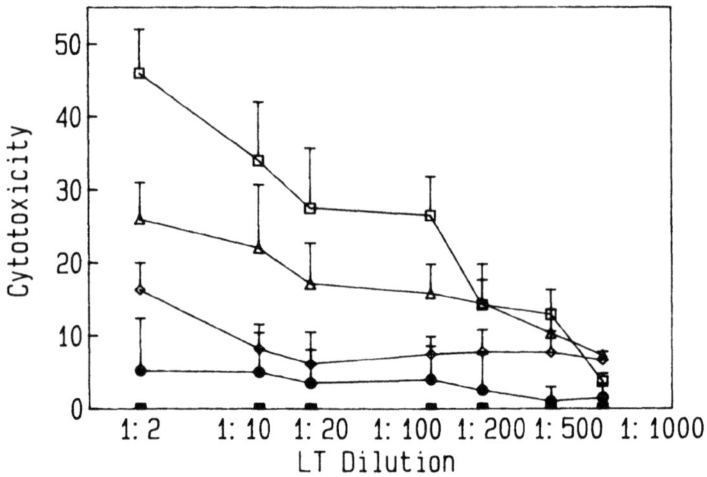

Fig. 3. Effect of the phorbol ester TPA on α-lymphotoxin (*LT*) mediated cytolysis of CHO hamster ovary cells. CHO cells were treated with different doses of TPA in the continuous presence of various concentration of LT. Results are cumulated from two to four experiments ($n = 8$–16). ●, CHO-F + LT only; ◇, CHO-F + 4800 nM TPA + LT; △, CHO-F + 1600 nM TPA + LT; □, CHO-F + 480 nM TPA + LT; ■, CHO-F + 9600 nM TPA + LT. (Fletcher et al. 1987b)

gap junctions; small regulatory or signaling molecules; and transducing protein receptors for these signaling molecules.

We have examined the separate and combined actions of phorbol esters and microwave fields on gap junction communication (Fletcher et al. 1986). In CHO hamster ovary cells, entry of α-lymphotoxin is dependent on their ability to express gap junctions (Fig. 3). This capacity varies widely in different cell strains. Entry is facilitated in strain unable to express gap junctions; and in cells with well developed gap junctions, ability to exclude α-lymphotoxin is disrupted by phorbol esters, an action enhanced by 450-MHz fields (1.0 mW/cm^2, 16 Hz modulation). No effects were noted with 450 MHz fields alone. These findings imply that microwave fields at athermal levels may act synergically with chemical cancer promoters to disrupt normal intercellular communication through gap junctions, leading to autonomous cell growth.

5 Conclusions

Use of weak EM fields at tissue levels of EEG gradients and below (10^{-7}–10^{-1} V/cm) to study the sequence and energetics of events that couple humoral stimuli from surface receptor sites to the cell interior has identified cell membranes as a primary site of these field interactions. Field modulation of cell surface chemical events indicates a major amplification of initial weak triggers associated with binding of hormones, antibodies, and neurotransmitters to their specific receptor sites.

Calcium ions play a key role in this stimulus amplification, probably through highly cooperative alterations in binding to surface glycoproteins, with spreading waves of altered calcium binding across the membrane surface. Protein particles spanning the cell membrane form pathways for signaling and energy transfer to intracellular enzyme systems. Fields millions of times weaker than the membrane potential gradient of 10^5 V/cm modulate cell responses to surface stimulating molecules. The evidence supports nonlinear, nonequilibrium processes at critical steps in transmembrane signal coupling.

Intercellular communication through gap junctions is similarly sensitive to weak oscillating pericellular electromagnetic fields. Cancer-promoting substances that act at cell membranes and disrupt gap junction communication are enhanced in their actions by these fields. We hypothesize that cancer promotion may involve events at cell membranes that are separate steps from cancer initiation occurring as a result of prior damage (mutation) to nuclear DNA. Cancer promotion may involve distortion of inward signals from cell membranes directed to the nucleus and to other organelles. These aberrant signal patterns may arise in loss of normal cell–cell communication through gap junction mechanisms. A possible future focus of cancer therapies may be in the domain of cell membrane dysfunctions, with the ultimate goal of restoration of cell differentiation, rather than in the realm of cytotoxic destruction of all neoplastic elements.

Acknowledgments. Studies in this laboratory have been generously suported by the US Department of Energy, the US Environmental Protection Agency, the FDA Bureau of Radiological Health, the US Office of Naval Research, the Southern California Edison Company, and the US Veterans Administration.

References

Adey WR (1975) Evidence for cooperative mechanisms in the susceptibility of cerebral tissue to environmental and intrinsic electric fields. In: Schmitt FO, Schneider DM, Crothers DM (eds) Functional linkage in biomolecular systems. Raven, New York, p 325

Adey WR (1977) Models of membranes of cerebral cells as substrates for information storage. Biosystems 8:163–178

Adey WR (1981 a) Tissue interactions with nonionizing electromagnetic fields. Physiol Rev 61:435–514

Adey WR (1981 b) Ionic nonequilibrium phenomena in tissue interactions with nonionizing electromagnetic fields. In: Illinger KH (ed) Biological effects of nonionizing radiation. American Chemical Soc, Washington DC, p 271

Adey WR (1983) Molecular aspects of cell membranes as substrates for interactions with electromagnetic fields. In: Basar H, Flohr H, Haken H, Mandell AJ (eds) Synergetics of the brain. Springer, Berlin Heidelberg New York, p 201

Adey WR (1984) Nonlinear, nonequilibrium aspects of electromagnetic field interactions at cell membranes. In: Adey WR, Lawrence AF (eds) Nonlinear electrodynamics in biological systems. Plenum, New York, p 3

Adey WR (1986) The sequence and energetics of cell membrane transductive coupling to intracellular enzyme systems. Bioelectrochem Bioenergetics 15:447–456

Adey WR (1988 a) Cell membranes, the electromagnetic environment and cancer promotion. Neurochem Res 13:671–677

Adey WR (1988 b) Physiological signalling across cell membranes and cooperative influences of extremely low frequency electromagnetic fields. In: Frohlich H (ed) Biological coherence and response to external stimuli. Springer, Berlin Heidelberg New York

Adey WR, Lawrence AF (eds) (1984) Nonlinear electrodynamics in biological systems. Plenum, New York

Adey WR, Bawin SM, Lawrence AF (1982) Effects of weak, amplitude-modulated microwave fields on calcium efflux from awake cat cerebral cortex. Bioelectromagnetics 3:295–307

Balcer-Kubiczek E, Harrison GH (1985) Evidence for microwave carcinogenesis in vitro. Carcinogenesis 6:859–864

Bass L, Moore WJ (1968) A model of nervous excitation based on the Wien dissociation effect. In: Rich A, Davidson CM (eds) Structural chemistry and molecular biology, Freeman, San Francisco, p 356

Bassett CAL, Mitchell N, Gaston SR (1982) Pulsing electromagnetic fields in ununited fractures and failed arthrodeses. J Am Med Assoc 247:623–627

Bawin SM, Adey WR (1976) Sensitivity of calcium binding in cerebral tissue to weak environmental electric fields oscillating at low frequency. Proc Natl Acad Sci USA 73:1999–2003

Bawin SM, Kaczmarek LK, Adey WR (1975) Effects of modulated VHF fields on the central nervous system. Ann NY Acad Sci 247:74–91

Bawin SM, Adey WR, Sabbot IM (1978a) Ionic factors in release of $^{45}Ca^{2+}$ from chick cerebral tissue by electromagnetic fields. Proc Natl Acad Sci USA 75:6314–6318

Bawin SM, Sheppard AR, Adey WR (1978b) Possible mechanisms of weak electromagnetic field coupling in brain tissue. Bioelectrochem Bioenergetics 5:67–76

Bennett MVL, Trinkhaus JP (1970) Electrical coupling between embryonic cells by way of extracellular space and specialized junctions. J Cell Biol 44:592–606

Bennett MVL, Aljure E, Nakajima Y, Pappas GD (1963) Electrotonic junctions between teleost spinal neurons: electrophysiology and ultrastructure. Science 141:262–265

Blackman CF, Elder JA, Weil CM, Benane SG, Eichinger DC, House DE (1979) Induction of calcium ion efflux from brain tissue by radio frequency radiation. Radio Sci. 14:94–98

Blackman CF, Benane SG, Kinney LS, Joines WT, House DE (1982) Effects of ELF fields on calcium-ion efflux from brain tissue in vitro. Radiat Res 92:510–520

Blackman CF, Benane SG, House DE, Joines WT (1985a) Effects of ELF (1–120 Hz) and modulated (50 Hz) RF fields on the efflux of calcium ions from brain tissue in vitro. Bioelectromagnetics 6:1–11

Blackman CF, Benane SG, Rabinowitz JR, House DE, Joines WT (1985b) A role for the magnetic field in the radiation-induced efflux of calcium ions from brain tissue in vitro. Bioelectromagnetics 6:327–338

Blank M (1976) Hemoglobin reactions as interfacial phenomena. J Electrochem Soc 123:1653–1656

Blank M (1986) Electrical double blayers in membrane transport and nerve excitation. Bioelectrochem Soc, First Int School, Pleven, Bulgaria. Proceedings, p 26

Byus CV, Lundak RL, Fletcher RM, Adey WR (1984) Alterations in protein kinase activity following exposure of cultured lymphocytes to modulated microwave fields. Bioelectromagnetics 5:34–51

Byus CV, Pieper S, Adey WR (1987) The effect of environmentally significant low-energy 60 Hz electromagnetic fields upon the cancer-related enzyme ornithine decarboxylase. Carcinogenesis 8:1385–1389

Byus CV, Kartun K, Pieper S, Adey WR (1988) Microwaves act at cell membranes alone or in synergy with cancer-promoting phorbol esters to enhance ornithine decarboxylase activity. Cancer Res 48:4222–4226

Cole KS (1940) Permeability and impermeability of cell membranes for ions. Cold Spring Harbor Symp Quant Biol 8:110–122

Delgado JMR, Leal J, Monteagudo JL, Garcia MG (1982) Embryological changes induced by weak, extremely low frequency electromagnetic fields J Anat 134:533–552

DeRiemer SA, Strong JA, Albert KA, Greengard P, Kaczmarek LK (1985) Enhancement of calcium current in Aplysia neurons by phorbol ester and kinase C. Nature 313:313–316

Dewey MM, Barr L (1962) Intercellular connection between smooth muscle cells: the nexus. Science 137:670–672

Dixey R, Rein G (1982) ^{3}H-Noradrenaline release potentiated in a clonal nerve cell line by low-intensity pulsed magnetic fields. Nature 296:253–255

Dutta SK, Subramaniam A, Ghosh B, Parshad R (1984) Microwave radiation-induced calcium efflux from brain tissue. Bioelectromagnetics 5:71–78

Eccles JC (1953) The neurophysiological basis of mind. Clarendon, Oxford

Edelman GM (1984) Cell adhesion molecules: a molecular basis for animal form. Sci 250(4):118–129

Ehrlich YH, Davis TB, Bock DE, Kornecki E, Lenox R (1986) Exto-protein kinase activity on the external surface of neural cells. Nature 320:67–69

Elul R (1966) Applications of non-uniform electric fields. Part I. Electrophoretic evaluation of absorption. Trans Faraday Soc 62:3484–3492

Elul R (1967) Fixed charge in the cell membrane. J Physiol (Lond) 189:351–365

Fitzsimmons RJ, Farley J, Adey WR, Baylink DJ (1986) Embryonic bone matrix formation is increased after exposure to a low-amplitude capacitively coupled electric field, in vitro. Biochim Biophys Acta 882:51–56

Fletcher WH, Shiu WW, Haviland DL, Ware CF, Adey WR (1986) A modulated-microwave field and tumor promoters similarly enhance the actions of alpha-lymphotoxins. Proceedings of the 8th annual meeting of the Bioelectromagnetics Society, p 12 (abstract)

Fletcher WH, Byus CV, Walsh DA (1987a) Receptor-mediated action without occupancy: a function for cell-cell communication in ovarian follicles. In: Mahesh V (ed) Regulation of ovarian and testicular function, Plenum, New York

Fletcher WH, Shiu WW, Ishida TA, Haviland DL, Ware CF (1987b) Resistance to the cytolytic action of lymphotoxin and tumor necrosis factor coincides with the presence of gap junctions uniting target cells. J Immunol 139(3):1–7

Furshpan EJ, Furakawa T (1962) Intracellular and extracellular responses of several regions of the Mauthner cell in the goldfish. J Neurophysiol 25:732–771

Furshpan EJ, Potter DD (1959) Transmission at the giant motor synapses of the wayfish. J Physiol (Lond) 145:289–325

Gilula NB, Reeves OR, Steinbach A (1972) Metabolic coupling, ionic coupling and cell contacts. Nature 235:262–265

Hertzberg EL (1984) A detergent-independent procedure for the isolation of gap junctions from rat liver. J Biol Chem 259:9936–9943

Hodgkin AL, Huxley AF (1952) A quantitative description of membrane current and its application to conduction and excitation in nerve. J Physiol (Lond) 117:500–517

Hong CZ (1987) Static magnetic field influence on human nerve function. Arch Phys Med Rehabil 68:162–164

Kaczmarek LK, Adey WR (1974) Weak electric gradients change ionic and transmitter fluxes in cortex. Brain Res 66:537–540

Kanno J, Loewenstein WR (1966) Cell-to-cell passage of large molecules. Nature 212:629–631

Kavaliers M, Ossenkopp KP (1986) Magnetic field inhibition of morphine-induced analgesia and behavioral activity in mice: evidence for involvement of calcium ions. Brain Res 379:30–38

Lawrence AF, Adey WR (1982) Nonlinear wave mechanisms in interactions between excitable tissue and electromagnetic fields. Neurol Res 4:115–153

Lawrence AF, McDaniel JC, Chang DB, Birge RR (1987) The nature of phonons and solitary waves in alpha-helical proteins. Biophys J 51:785–793

Lin-Liu S, Adey WR (1982) Low frequency, amplitude-modulated microwave fields change calcium efflux rates from synaptosomes. Bioelectromagnetics 3:309–322

Loewenstein WR (1968) Communication through cell junctions. Implications in growth and differentiation. Dev Biol 2:151–157

Loewenstein WR (1979) Junctional intercellular communication and the control of growth. Biochim Biophys Acta 560:1–65

Loewenstein WR (1981) Junctional intercellular communication: the cell-to-cell membrane channel. Physiol Rev 61:829–913

Loewenstein WR, Kanno Y (1966) Intercellular communication and the control of tissue growth. Nature 209:1248–1250

Luben RA, Cain CD (1984) Use of bone cell hormone responses to investigate bioelectromagnetic effects on membranes in vitro. In: Adey WR, Lawrence AF (eds) Nonlinear electrodynamics in biological systems. Plenum, New York, p 23

Luben RA, Cain CD, Chen MY, Rosen DM, Adey WR (1982) Effects of electromagnetic stimuli on bone and bone cells in vitro: inhibition of responses to parathyroid hormone by low-energy, low-frequency fields. Proc Natl Acad Sci USA 79:4180–4183

Lyle DB, Schechter P, Adey WR, Lundak RL (1983) Suppression of T lymphocyte cytotoxicity following exposure to sinusoidally amplitude-modulated fields. Bioelectromagnetics 4:281–292

Maddox J (1986) Physicists about to hijack DNA? Nature 324:11

Milham S (1985) Mortality in workers exposed to electromagnetic fields. Environ Health Perspect 62:297–300

Moolenaar WH, Aerts RJ, Tertoolen LGJ, DeLast SW (1986) The epidermal growth factor-induced calcium signal in A431 cells. J Biol Chem 261:279–285

Newmark P (1987) Oncogenes and cell growth. Nature 327:101–102

Nishizuka Y (1983) Calcium, phospholipid and transmembrane signalling. Philos Trans R Soc Lond B302:101–112

Nishizuka Y (1984) The role of protein kinase C in cell surface transduction and tumor promotion. Nature 308:693–696

Pasti G, Lacal JC, Warren BS, Aaronson SA, Blumberg PM (1986) Loss of mouse fibroblast response to phorbol esters restored by microinjected protein kinase C. Nature 324:375–377

Pitts JD, Finbow ME (1986) The gap junction. J Cell Sci [Suppl] 4:239–266

Radeke MJ, Misko TP, Hsu C, Herzenberger LA, Shooter EM (1987) Gene transfer and molecular cloning of the rat nerve growth factor receptor. Nature 325:393–397

Revel JP, Karnovsky MJ (1967) Hexagonal arrays of subunits in intercellular junctions of the mouse heart and liver. J Cell Biol 33:C7–C12

Riedel H, Schlessinger J, Ullrich A (1986) A chimeric ligand binding v-erb B/EGF receptor retains transforming potential. Science 236:197–200

Robertson JD (1963) The occurrence of a subunit pattern in the unit membranes of club endings in Mauthner cell synapses in goldfish brains. J Cell Biol 19:201–221

Savitz DA, Wachtel H, Barnes F (1986) National contractor's review. US Department of Energy, Office of Energy Storage and Distribution, Washington DC, and Electric Power Research Institute Health Studies Program Proceedings, Palo Alto, November

Semm P (1983) Neurobiological investigations on the magnetic sensitivity of the pineal gland in rodents and pigeons. Comp Biol Physiol 76A:683–692

Semm P, Demaine C (1986) Neurophysiological properties of magnetic cells in the pigeon's visual system. J Comp Physiol 159:619–625

Singer SJ, Nicolson GL (1972) The fluid mosaic model of the structure of cell membranes. Science 175:720–731

Trosko JE, Chang CC (1986) Oncogene and chemical inhibition of gap-junctional intercellular communication: implications for teratogenesis and carcinogenesis. In: Genetic toxicology of environmental chemicals, part B: Genetic effects and applied mutagenesis. Liss, New York, p 21

Ullrich A, Coussens L, Hayflick JS, Dull TJ, Gray Tam AW, Lee J, Yarden Y, Libermann TA, Schlessinger J, Downard J, Mayes ELV, Whittle N, Waterfield MD, Seeburg PH (1985) Human epidermal growth factor receptor cDNA sequence and aberrant expression of the amplified gene in A431 epidermoid carcinoma cells. Nature 309:428–431

Unwin PMT, Ennis PD (1984) Two configurations of a channel-forming membrane protein. Nature 307:609–613

Van der Kloot WG, Cohen I (1979) Membrane surface potential changes may alter drug interactions: an example, acetyl choline and curare. Science 203:1351–1352

Warner AE, Guthrie SC, Gilula NB (1984) Antibodies to gap-junctional protein selectively disrupt junctional communication in the early amphibian embryo. Nature 311:127–131

Welker HA, Semm P, Willig RP, Wiltschko W, Vollrath L (1983) Effects of an artificial magnetic field on serotonin-*N*-acetyltransferase activity and melatonin content of the rat pineal gland. Ex Brain Res 50:426–531

Wertheimer N, Leeper E (1979) Electrical wiring configurations and childhood cancer. Am, J Epidemiol 109:273–284

Yotti LP, Chang CC, Trosko JE (1979) Elimination of metabolic cooperation in Chinese hamster cells by a tumor promoter. Science 206:1089–1091

The EEG is a Quasi-Deterministic Signal Anticipating Sensory-Cognitive Tasks

E. Başar, C. Başar-Eroglu, J. Röschke, and A. Schütt

1 Introduction

1.1 Remarks About the EEG

One of the main concerns of brain research is to measure the brain's electrical activity and, in this way, to try to detect the coding of behaviorally relevant information in the central nervous system (CNS). Analyses of the EEG, of evoked potentials (EPs), and of endogenous potentials (the P300 family) are among the most fundamental research tools for understanding the sensory and cognitive information processing in the brain. Many reports in this volume show the broad extent of applications of the EEG, of sensory EPs and event-related potentials (ERPs) to the understanding of CNS information processing and behavior.

Since Berger's (1929) first description of the electrical activity of the brain, several approaches have been undertaken in order to correlate the activity at neuronal levels with the electroencephalogram (EEG). Creutzfeldt (1974) pointed out that the spontaneous electrical activity of the CNS and sensory-evoked potentials are sometimes highly correlated to intracellularly measured postsynaptic potentials (EPSPs and IPSPs). Ramos et al. (1976) postulated that it is impossible to specify any general causal or predictable relationship between the waveform of an evoked potential and the firing pattern of a neuron. Some authors take the view that the spontaneous EEG activity is an expression of the incessant, irregular background neural firing (Fessard 1961). Do we have the right to consider the spontaneous activity of the brain as a background noise in the sense of antisignal in ideal communication theory? Or rather, is the EEG a most important fluctuation, which controls the sensory-evoked and event-related potentials? We have written elsewhere that the spontaneous activity plays an active role in the signals transmitted through various structures and recorded at various sites in the brain and that the EEG should not be considered as a noisy signal. Especially, we have assumed that regular patterns of the EEG reflect coherent states of the brain during which cognitive and sensory inputs are processed (Başar 1980, 1983a, b).

1.2 Aim of the Study: A Step to Bridge Sensory and Cognitive Processes with EEG Dynamics

Since the first measurement of the event-related potential (ERP) "P300" by Sutton et al. (1965) several paradigms have been used in order to correlate cognitive

Springer Series in Brain Dynamics 2
Edited by Erol Başar and T. H. Bullock

tasks and behavior with slow waves of the brain. However, there are only a few reports that analyze prestimulus EEG activity during these tasks. Our earlier research which had demonstrated the correlation between EEG and evoked potentials led us to further experiments, in which a strong relation between prestimulus EEG and the P300 wave was also indicated (Başar et al. 1984; Başar and Stampfer 1985; Stampfer and Başar 1985). We observed that during the application of various event-related paradigms the prestimulus EEG tends to attain a phase-ordered pattern prior to expected stimulation. Our preliminary experiments have now been extended to measure the event-related EEG prior to a cognitive task with a chain of paradigms. In the present report we will describe the results indicating the existence of regular, phase-ordered prestimulus EEG rhythms, which tend to show a repeatable pattern formation preceding successful cognitive tasks.

The current major trend in EEG-related investigations on cognitive processes focusses on studies performing defined psychophysiological paradigms. Other groups of scientists working to define sensory processes of the CNS do not usually consider the cognitive loading of subjects under study. The analyzers working with P300 paradigms and the EEG investigators do not tend to combine results and rules of various approaches. Is the family of cognitive potentials, which is called the "P300 family," something completely different from the sensory-evoked potentials? Is the simple EEG, which the evoked-potential investigator usually considers as a background noise, without any importance in defining the cognitive and/or sensory responses?

The present study aims to answer these questions by applying combined analyses of related EEG, sensory-evoked potentials, and event-related potentials.

2 Does the Brain Emit Event-Related Rhythms?

2.1 Inverse Relation Between EEG and EPs

Figure 1 illustrates two EPs recorded during the same experimental session. In this analysis, we selected two different classes of EEG-EP sets (of single trials = epochs = sweeps). The solid curve was obtained by averaging EEG-EP subsets in which the RMS value of the prestimulus EEG was smaller than 10 μV for both the frequency ranges of 8–13 Hz and 4–7 Hz. In other words, for this subset of EEG-EP epochs, the prestimulus EEG should not exceed 10 μV RMS after filtering either with the band-pass filter of 8–13 Hz and with that of 4–7 Hz (two different filters applied!). For the analysis of the 8- to 13-Hz component, a time period of 250 ms prior to auditory stimulation was used. For the analysis of the 4- to 7-Hz frequency component, a duration of 400 ms prior to stimulation was evaluated.

In the second stage of analysis, a subset of EEG-EP epochs that have shown RMS-EEG values larger than 10 μV was averaged (higher EEG activity just prior to stimulation). Both subsets contained an equal number of EEG-EP epochs. Although the two EPs in Fig. 1 have almost similar shapes, the amplitudes of various deflections are different. The dashed curve, the EP with a high-amplitude prestim-

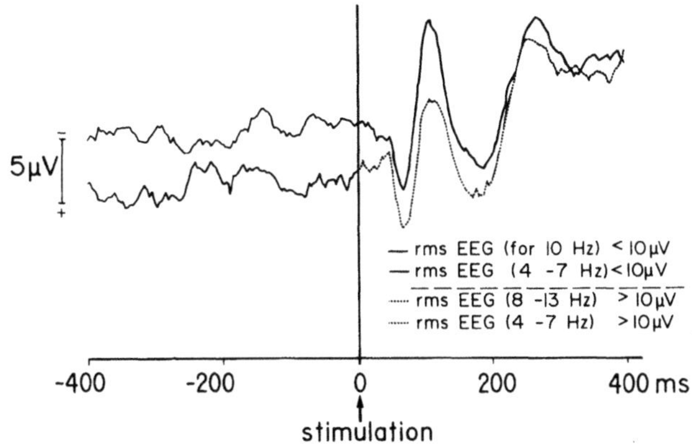

Fig. 1. The AEP (averaged evoked potential) from the human vertex averaged by using two different criteria. *Dashed curve,* the selectively averaged EP using only EEG-EP sets with RMS values of prestimulus EEG greater than 10 μV for filtered EEGs in the 8- to 13-Hz and 4- to 7-Hz frequency ranges. *Solid curve,* the selectively averaged EP using only EEG-EP sets with RMS values of prestimulus EEG less than 10 μV for filtered EEGs in the 8- to 13-Hz and 4- to 7-Hz frequency ranges

ulus EEG, is about 40% smaller than the solid curve containing sweeps with a smaller prestimulus EEG. In other words, there is an inverse relationship between the magnitude of the prestimulus EEG and the amplitudes of the EPs. This example is important for the investigator of EPs who is not yet accustomed to performing an analysis in the frequency or time domain using concepts of alpha or theta responses. This example shows that the conventional EP is a function of the EEG, which should be studied prior to performing a signal averaging of EPs. We have shown here the influence of theta and alpha components in the EEG. The same analysis has been extended to the 40-Hz range, which reveals the same inverse relationship between 40-Hz prestimulus EEG and "middle-latency" responses of the EP (ca. 25 ms latency; Başar et al. 1987).

The 40% reduction in the amplitude of the averaged evoked potential illustrated in Fig. 1 is a typical correction achieved by selective averaging, in which EEG-EP sets with low amplitude EEG preceding the stimulation have been taken into account for evaluation. It is to be emphasized that this reduction can not be considered as a "mean value correction" or a reflection of the statistical interpretation valid for all experiments. We have systematically studied EEG-EP sets of about 20 subjects by performing similar analyses. Although we obtained reductions in AEPs (average evoked potentials) of all subjects, the reduction in the peak-to-peak EP amplitude varied between 20% and 100% depending on subjects, types of experiments, and electrode locations. Moreover, the percentage of EP variabilities are different at the end of the experiments, during which the spontaneous EEG usually gains higher magnitudes in comparison with the beginning. ERPs during cognitive tasks also showed great standard deviations. Accordingly, an extended analysis will soon be published taking into account mean values and standard deviations. However, we doubt that this analysis will make the evaluation of some standard data possible. In order to be able to collect standard data it would be necessary to build several subject groups of alpha and beta types and for example, conduct experiments with "closed eyes" or "open eyes." Therefore, we give only an illustration here in order to point out the importance of this consideration.

A critical objection to this type of analysis is the following: In doing the analysis by selecting the EEG epochs in the same way, we increase the chances that a trigger preceded by *Low RMS* will be followed by spontaneous, background high RMS, since we know that the waxing and waning has at least a strong component in the 3- to 15-Hz frequency range. This effect would certainly create a pseudo-EP with probably different shape, but such pseudo-EPs should also be taken into account and it should be considered whether they significantly contribute to our correction. Therefore, we have performed analyses with EEG sets without stimulation by applying the same selective averaging. In some of the curves there are slight increases in the time-points of hypothetical pseudo-EPs. However, we never recorded changes of 40% or 100% as is the case in the true EPs. This is probably due to two facts: (a) The pseudo-EPs are out of phase so that they are very quickly averaged out. (b) We selected our epochs in two different frequency windows in theta and alpha ranges which do not necessarily have a simultaneous waning and waxing periodicity.

2.2 Internal Evoked Potentials

We consider ten randomly chosen EEG-EP epochs digitally filtered with band-pass filters of 8–13 Hz illustrated in Fig. 2 A (recording from cat reticular forma-

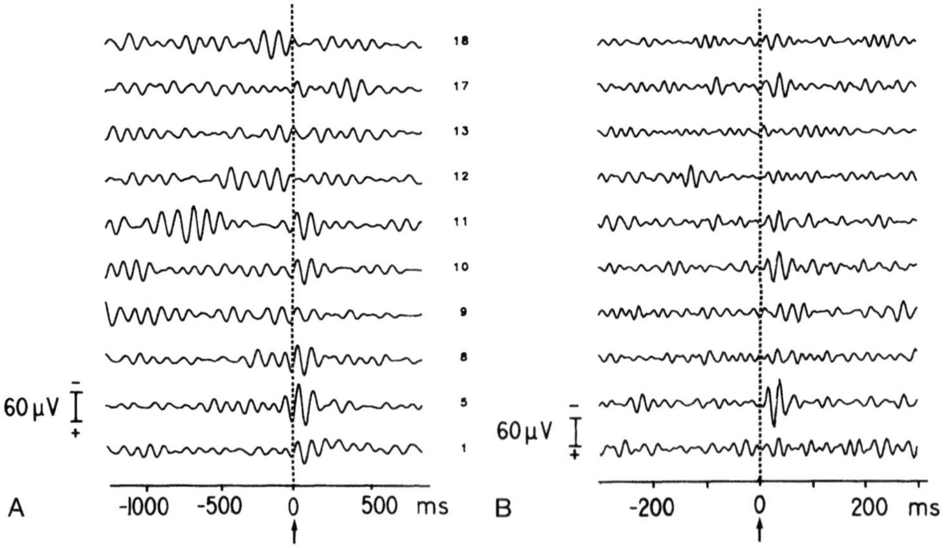

Fig. 2. (A) Filtered EEG-EP epochs in the 8- to 13-Hz frequency range. (B) Filtered EEG-EP epochs in the 30- to 60-Hz frequency range

tion). The 10-Hz activity before stimulation is often regular, the wave packets with large amplitudes almost up to 60 μV. The stimulus-evoked responses are often time-locked (for example, sweep nos. 1, 5, 8, and 10); however, the time-locked patterns do not have shapes different from the elementary waveforms depicted in the spontaneous activity. "Spontaneous patterns which are synchronized without external sensory stimulation" shown in this illustration in the prestimulus activity of sweep nos. 11, 12, and 18 were observed in all relevant frequency ranges of the EEG – for example, 40-Hz activity; several other examples are in Başar (1980).

The differences between the spontaneous EEG and evoked patterns are:

1. The synchronized 10-Hz pattern prior to stimulation occurs randomly, whereas the 10-Hz wave packet is triggered regularly upon external sensory stimulation, provided that there is not ample 10-Hz activity immediately prior to stimulation. In cases where ample ongoing 10-Hz activity precedes the stimulation, there is, as a rule, no time-locked enhancement upon stimulation (sweep nos. 9, 12, and 18).
2. The evoked 10-Hz response usually has a large magnitude in comparison with the magnitudes of the spontaneous 10-Hz wave packets. One of the most important points in this kind of analysis is the fact that in the spontaneous activity (although it has a random distribution), patterns that are comparable with the evoked ones can often be detected. These spontaneous patterns, when they occur, have *comparable* magnitudes, the same frequency, and the same shape as the evoked patterns.

Figure 2 B illustrates the filtered EEG-EP epoch in the 40-Hz frequency range (the filter limit is 30–60 Hz). The 40-Hz spontaneous and evoked wave packets show the same behavior as do the 10-Hz waveforms. The responses usually show time-locked, large-amplitude wave packets, provided that there is no synchronized 40-Hz activity prior to stimulation. In the spontaneous activity, however, randomly occurring 40-Hz waves are often recorded.

As we have seen in the previous examples in Fig. 2, in the spontaneous activity of the brain some patterns can often be detected that are comparable with the EP patterns of the same frequency. The expression "internal evoked potential" is used for the spontaneous patterns, which occur randomly without any external stimulation (originating probably from hidden internal sources) and which have the same frequency and the same shape with comparable amplitudes as the evoked patterns. Based on various experiments using the sets of EEG-EP records, we have tentatively formulated the following working hypothesis, which we call "the excitability rule" (Başar 1980):

> Various brain structures depict spontaneous rhythmic activity in a wide frequency range between 1 Hz and 1000 Hz. Without application of external stimulation, the spontaneous activity of a given brain structure can often show frequency-stable and high-magnitude electrical activity. If regular spontaneous oscillations can be detected in the electrical activity of a defined brain structure during a determined period, it is to be expected that upon external sensory stimulation, this structure will have a response susceptibility in the same frequency channel (for example, 40-Hz activity of the hippocampus and cortex, 250- to 300-Hz activity of the cerebellum, 10-Hz activity of all brain structures, and 10-Hz and 40-Hz activity of human scalp responses). The response susceptibility of a brain structure depends mostly on its susceptibility to its own intrinsic

rhythmic activity. Frequency stabilization, time-locking, and amplification of the spontaneous activity upon stimulation contribute greatly to the genesis of large potential changes which are called evoked potential.

2.3 Why Look for Internal Event-Related Rhythms?

These results now give us roughly the following picture. Various neural populations in the brain can generate coherent states in which a train of 10-Hz activity as well as theta activity are recorded. We do not yet know how such *internally evoked wave trains* can be elicited. Are they elicited by hidden sources? What type of afferent volleys arrive from the periphery to the brain? We only know that we can bring the brain to coherent states by stimulating it also with external sensory stimulation. A light flash can elicit a 10-Hz enhancement in the brain if the brain shows a disordered activity prior to stimulation. We also can evoke a 40-Hz response with sharp onset light or acoustical stimulation and so on. At this point an important question is: Can we find a way to put the brain in such coherent states of EEG activity *without* external sensory stimulation? Can we find a sensory-cognitive task to produce coherent internal evoked potentials, or better, internal event-related potentials?

3 Coherent and Ordered States of EEG Due to Cognitive Tasks

Since in the literature of EEG and ERP the type of experiments, which will be described in this report, are not yet encountered, we prefer to describe the experiments in this study in chronological order, as we did them. In this way the trend can be explained better: According to our earlier findings (Başar et al. 1984) we first tried to evaluate for each subject single EEG sweeps with good congruency by means of superposition. After we had established a good correlation between EEG states and subject's reports, we described the global trend by evaluation of mean value curves from 16 subjects. We then undertook experiments with five subjects who had already been subjects several times. This step thus allowed us the possibility of a reliable statistical interpretation for each subject.

3.1 Preliminary Experiments

3.1.1 Methods

The experiments were carried out with 16 volunteer healthy subjects, mostly students 19–21 years of age. The EEG was recorded in vertex, parietal, and occipital locations against the reference of the ear lobe (Cz, P3, P4, and 01 in the 10–20 system). The EEG signals have been amplified by using a Schwarzer EEG machine. The subjects sat in a soundproof and echo-free room which was dimly illuminated. For stimulus preparation, evaluation of selective averaging procedure and digital filtering, a Hewlett Packard 1000F computer was used. The filtering of EEG and of event-related potentials was carried out by using a digital filtering method (Başar 1980). The digital filters do not create any phase shift. As auditory stimulation 2000 Hz 80 dB tones of 800 ms duration were applied with regular intervals of 2600 ms. Every third or fourth tone was omitted. The subjects were asked to predict and to mark mentally the time of occurrence of the omitted signals. The EEG 1 s prior to omitted stimulation was also recorded with the ERP. The light stimulator was a 20-W fluorescent bulb which was electrically triggered. The duration of the light step was also 800 ms.

3.1.2 Preliminary Results in Special Cases

When subjects had learned and successfully followed the rhythmicity which is contained in the paradigm, they were usually able to increase their attention. Then, rhythmic prestimulus EEG patterns could be observed. Most of the subjects reported that at the beginning of an experimental session with repetitive signals they had difficulty to predict the time of occurrence of the stimulus omission. Usually, in the second half of the experiment they were able to predict the time of the omitted signal. Accordingly, in our signal analyses we applied a selective averaging by grouping approximately the first ten prestimulus sweeps at the beginning of the experiment and the last ten sweeps. Figure 3 illustrates comparatively the averages of the first and the last ten prestimulus EEG epochs (digitally filtered between 1 and 25 Hz) which were recorded at the vertex of a subject who reported that at the beginning of the experiment he felt himself unsure and diffuse. Towards the end of the experimental session he could be more concentrated, and thus performed his task much better. The average of the ten sweeps at the end of the experiment depicted a regular rhythmic behavior with large amplitudes. The first ten sweeps tend also to the same rhythmicity, but the average is less regular and has low amplitudes.

The rhythms, that are in principle similar to those illustrated in Fig. 3, have been observed in all the subjects. The alignment and phase-reordering were not the same in all the subjects. The exact time of regularity and phase-reordering showed fluctuations in a time period 700 to 0 ms prior to the event.

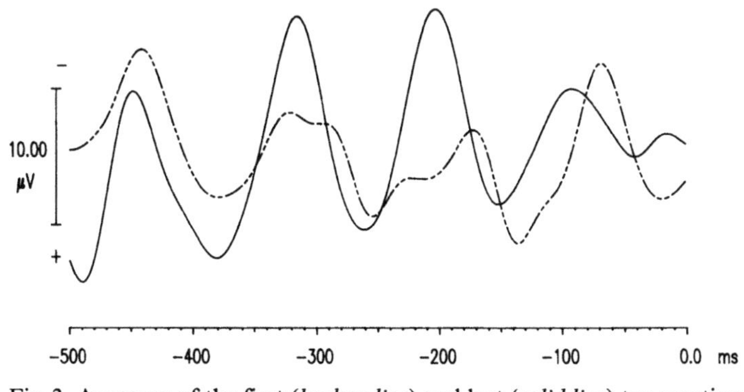

Fig. 3. Averages of the first (*broken line*) and last (*solid line*) ten prestimulus EEG epochs of the experiment, filtered in the 1- to 25 Hz-frequency band

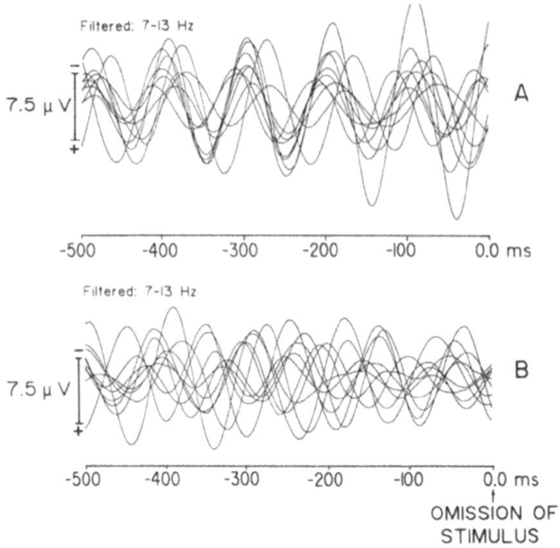

Fig. 4. (A) Approximately ten prestimulus EEG sweeps at the end of the experiment. (B) Approximately ten prestimulus EEG sweeps at the beginning of the experiment

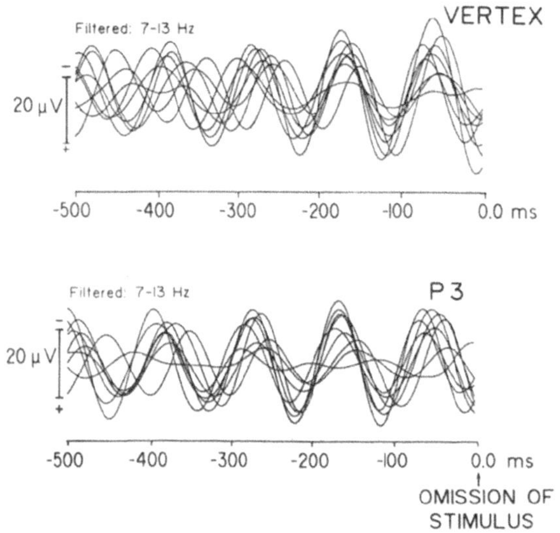

Fig. 5. The last nine sweeps from vertex and parietal (*P3*) leads filtered in the 7- to 13-Hz frequency band

Figure 4 shows approximately ten prestimulus EEG epochs from another subject C at the end (A) and at the beginning (B) of an experiment. Single sweeps here were digitally filtered in the frequency range between 7 and 13 Hz according to the rhythmicity revealed in the wideband curve. It is easy to see that at the end of the experiment repeatable patterns are observed in contrast to the beginning without such patterns.

Figure 5 shows results from another experiment with the third subject. In this case the superposition of the last nine sweeps from vertex and parietal leads is illustrated. The regularity of the single EEG rhythms are easy to observe, because

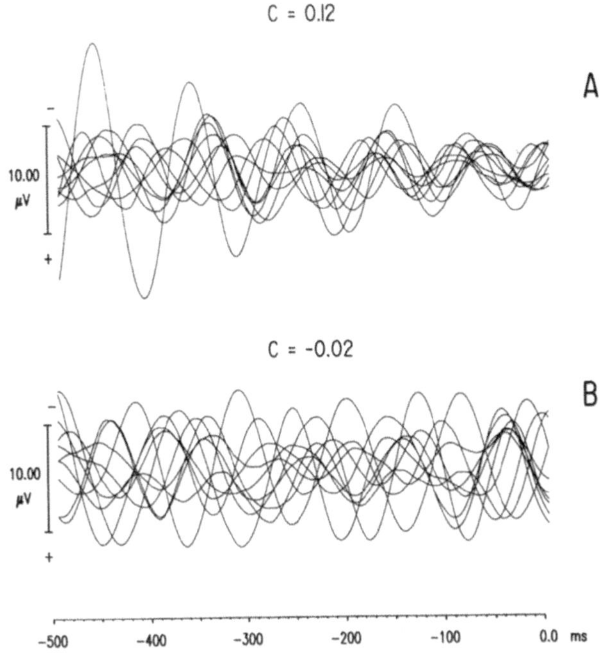

Fig. 6. (A) Single filtered (8–13 Hz) EEG sweeps before the fourth, omitted tone. (B) Single filtered (8–13 Hz) EEG sweeps before the third, omitted tone. Subject C focussed her attention on the fourth, omitted tone

of their template pattern before the stimulation. Although the reactions of various subjects to the sequence of stimuli applied and omitted had in general a common character, there were slight differences among subjects in understanding of the paradigm and in its interpretation. Performance differences were observed, especially among the students. This was also reflected in prestimulus EEG.

Since it is not possible to describe all change of event-related rhythms, we will describe a task-related change of EEG which seemed to reflect a subtile behavior of some subjects. After the experiment most of the subjects ($n=7$) explained that it was usually easier for them to mark the target mentally when the fourth stimulus was omitted. They explained the event in the following manner: "As I heard the third tone I knew that now the fourth signal should be omitted. Then I tried to mark mentally the fourth, omitted signal. I have done this task better by marking it mentally than in the cases of the missed third signal." According to the subjects statements we then performed a selective averaging. Averaging was made with two different categories of EEG signals: (a) EEG prior to the third, omitted signal and (b) EEG prior to the fourth, omitted signal. Figure 6 illustrates the superposition of filtered EEG trials for both groups of data for the subject K. The signals prior to the fourth, omitted tone are ordered, whereas the EEG signals prior to the third, omitted tone show a disordered behavior. Only one of the subjects (subject C) presented the contrary performance. She reported that she had tried to perform the task for the third, omitted signal and mostly had good success. She was usually no longer concentrated or attentive enough to mark the fourth, omitted tone mentally. The EEG curve recorded from the vertex of subject C gave results opposite to those of subject M (not illustrated here).

3.2 Global Trends of Pretarget Event-Related Rhythms. Expectation and Reaction of Subjects

3.2.1 Subject Variability

In order to summarize these preliminary results we can say that subjects often decided on their own target or *mental prediction target*. Some subjects could better predict the omitted signals after the second tones, others after the third tones. Most of the subjects showed better performance at the end of the experiment, but some were able to recognize the time of the occurrence of the target already at the beginning of the experiments. At this stage of the investigation the reliability of the results consists only in comparison of the phase-ordered EEG states with statements of the subjects as to whether or not they were able to mark the target signal mentally. It is well known that during long recording sessions the EEG can show high stationary behavior, random synchronization, and alignments. Our findings showed, however, that these phase-ordered patterns were fairly correlated with the subjects' reports and, as will be shown in Section 4, fluctuations could be recorded in strong and long standard experiments for a number of subjects.

3.2.2 Paradigms with Increasing Occurrence Probability

In order to reduce the possibilities of recording *randomly occurring coherent EEG signals,* we extended our paradigm as follows. In addition to paradigm 1, omitting the third or fourth stimulation (randomly), we extended our measurements to include a more difficult paradigm 2 and thus decrease the probability of the occurrence of the target signals. In paradigm 2 the omitted signal could be changed from the fourth to the seventh one randomly (the occurrence of target signal is 25% when the subject already heard the second tone, see Table 1). During the same recording session the subjects had to perform another experiment, paradigm 3, which offered an easier way to mark the target signal mentally – exactly every fourth signal was omitted. This is the easiest one of our paradigms, since the probability of the target occurrence is 100%, when the subject has already heard the third signal.

The comparison of the experimental results showed that the subjects could emit coherent and phase-ordered pretarget EEG signals almost in all cases of the

Table 1. Comparison of paradigms

Paradigm	Probability of occurrence of target
1. Every 4th or 7th stimulation (randomly) omitted (most difficult paradigm)	25% After 3rd tone
2. Every 3rd to 4th stimulation (randomly) omitted (intermediary paradigm, less difficult)	50% After 2nd tone
3. Every 4th stimulation (no randomness) omitted (easiest paradigm)	100% After 3rd tone

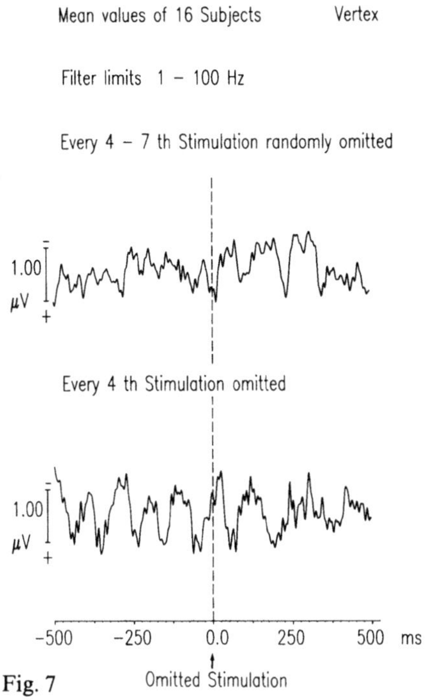

Fig. 7

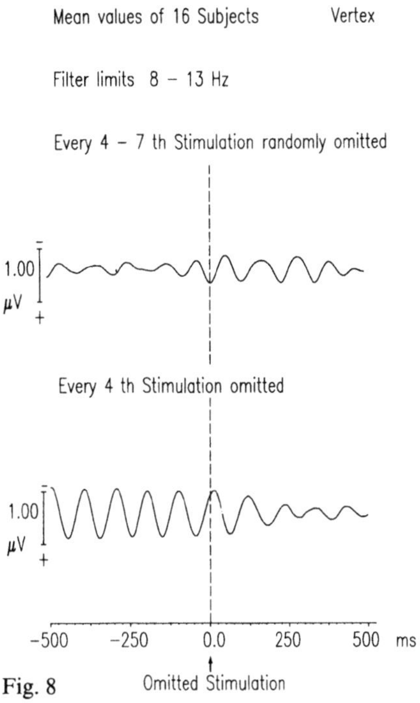

Fig. 8

Fig. 7. Comparison of the most difficult (every 4th to 7th stimulation randomly omitted *top*) and the easiest (every 4th stimulation omitted, *bottom*) paradigms as mean value curve from 16 subjects. Filter limits, broad band 1–100 Hz. Target signal occurs at time "0." Derivation, vertex

Fig. 8. Comparison of the most difficult (every 4th to 7th stimulation randomly omitted *top*) and the easiest (every 4th stimulation omitted, *bottom*) paradigms as mean value curve from 16 subjects. Filter limits, 8–13 Hz (alpha frequency range). Target signal occurs at time "0." Derivation, vertex

easiest paradigm, whereas the same subjects did not show the same good coherent and phase-ordered pretarget EEG during the most difficult paradigm. In this section we describe globally the results with mean value curves from experiments with 16 subjects. In Section 4 we will present descriptions of experiments on single subjects.

Figure 7 illustrates the comparison between the most difficult paradigm and the easiest one. The curves are wide-band filtered (between 1–100 Hz). The results are as follows:

1. Every fourth to seventh signal omitted (the most difficult paradigm):

a) We observed no regular rhythmicity prior to target (or omitted stimulation).

b) We observed an enhancement of the unfiltered EEG following the omitted stimulation (target signal).

We should note here that the occurrence of target is rare. This means that upon the omitted stimulation the surprise effect should be greater. We do not analyze here the variability of the P300 wave among the subjects which showed relevant individual fluctuations and also variabilities at the beginning and the end

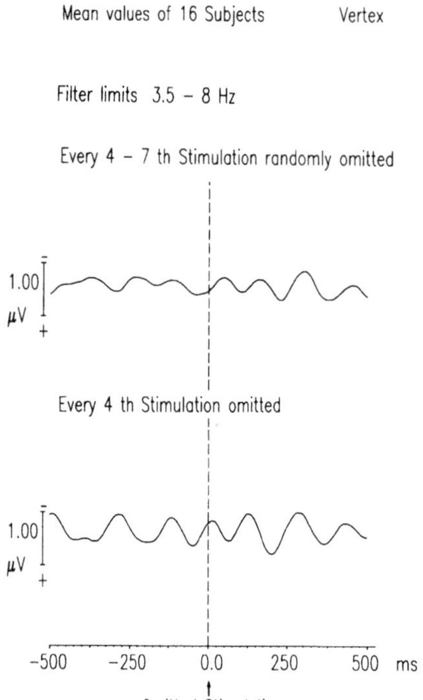

Fig. 9

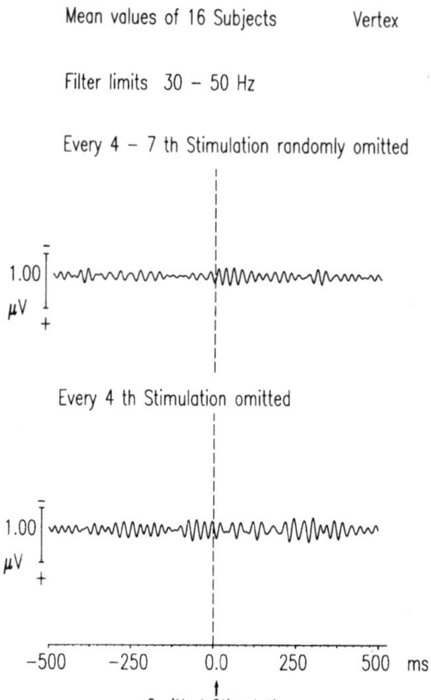

Fig. 10

Fig. 9. Comparison of the most difficult (every 4th to 7th stimulation randomly omitted *top*) and the easiest (every 4th stimulation omitted, *bottom*) paradigms as mean value curve from 16 subjects. Filter limits, 3.5–8 Hz (theta frequency range). Target signal occurs at time "0." Derivation, vertex

Fig. 10. Comparison of the most difficult (every 4th to 7th stimulation randomly omitted *top*) and the easiest (every 4th stimulation omitted, *bottom*) paradigms as mean value curve from 16 subjects. Filter limits, 30–50 Hz (40-Hz frequency range). Target signal occurs at time "0." Derivation, vertex

of the experimental session. The meaning of the mean value curves is as follows: it can be globally stated that mean value curves from 16 subjects showed a slight EEG enhancement with peakings around 300–400 ms. The latency change of waves of the P300 family were large in our measurements. (For the variability of the range of the P300-wave family, see also Galambos and Hillyard (1981).

2. Every fourth signal omitted (the easiest paradigm):

a) The pretarget EEG shows a rhythmicity of around 9.5 Hz.
b) No EEG enhancement is observed after the omitted stimulation. The mean value EEG signal looks like a continuation with a slightly slower rhythm.

Since we have explained in detail the usefulness of digital filtering in earlier studies on event-related potentials and/or event-related rhythms (Başar et al. 1984; Başar and Stampfer 1985; Stampfer and Başar 1985; see also Başar 1980; Bullock and Başar 1988), our results will be analyzed in three different frequency bands, as follows.

8- to 13-Hz Range

During the most difficult paradigm (every fourth to seventh signal omitted) no coherent and ordered 10-Hz activity was observed preceding the target signal (Fig. 8). However, following the omitted stimulation, a 10-Hz enhancement was observed.

During the easiest paradigm (every fourth signal omitted) a coherent 9-Hz rhythmicity preceding the target was observed. On the contrary, after the omitted stimulus a blocking of the regular 10-Hz activity was observed.

3.5- to 8-Hz Range

During the most difficult paradigm, in theta frequency range an enhancement following the omitted stimulation was observed, whereas there is no regular rhythmicity in the pretarget EEG. In other words, by this paradigm there is a reaction due to the surprise (Fig. 9).

During the easiest paradigm we observed a regular rhythmicity before target and also an enhancement after target (Stampfer and Başar 1985).

40-Hz Range

In Fig. 10 the mean value curves of 16 subjects in the 40-Hz frequency range are illustrated.

During the paradigm with every fourth signal omitted we observed an increased regular rhythmicity of 40-Hz activity just prior to stimulation (50 ms prior to omitted stimulation) in the mean value curve. There is a decrease in the 40-Hz activity (or a blocking) following the omitted stimulation.

During the most difficult paradigm (every third to seventh signal omitted) there is no increased regular rhythmicity prior to target but there is an enhancement after the omitted stimulation.

Results of enhancement or blocking of 40 Hz in this global analysis have the same trend as in the 10-Hz activity. Our analysis is not yet sufficient to describe whether the 10-Hz and 40-Hz enhancement (or blocking) occur simultaneously or not.

4 Experiments with Light Stimulation

In the previous sections we described preliminary results from a few subjects by using repetitive acoustical stimuli in order to describe roughly how the experiments have been initiated. Further, we presented as mean value curves results of experiments with 16 subjects. In the following section we will describe some experiments, during which repetitive light stimuli were used, and give more details of the performed analysis. Although a small number of subjects was used, a large number of experiments were carried out for each subject. In these series of experiments we will consider only the 8- to 13-Hz frequency range since a detailed and exact account of all the experiments covering various frequency ranges is not feasible in this study.

Methodological Remarks

We used during these experiments the same experimental setup and procedure as described in Section 3. The light stimulation consisted in light steps of 800 ms duration. The light source was a 20-W fluorescent bulb which can be triggered with a short time constant. The intervals between stimuli were 2600 ms. The evaluation and interpretation of the correlation coefficient are described in the Appendix.

4.1 Examples of Experiments with Varied Probabilities of Stimulus Occurrence

The subject J.K. is a medical student who quickly learned the goal of the experiments and has been very cooperative during the experiments. Figure 11 A, B, and C illustrates samples of the filtered resting EEG as a control before an experiment with the cognitive task. There are three plots of the filtered EEG segments, with ten sweeps in each plot. The three plots present samples during the same recording session. The mean correlation coefficient ($\bar{C}$) of each ensemble of sweeps in a time scale from -500 to 0 ms is also shown in the illustration.

The subject was instructed to be attentive to repetitive light stimuli. Every fourth light stimulation was omitted (the easiest paradigm, cf. Section 3.2.2). Subject J.K. reported that at the beginning of the experiment he could easily mark the target signal; however, after approximately ten omitted signals or after the first 40 sensory stimulations he could not concentrate as well; and toward the end of the measurement he had enormous difficulties in concentrating. Figure 12 A shows the first ten filtered sweeps together with the filtered mean value and wideband mean curves (1–30 Hz). A clear rhythmicity and a good congruency in almost all sweeps are observed. In the following sessions of the experiments (B–C)

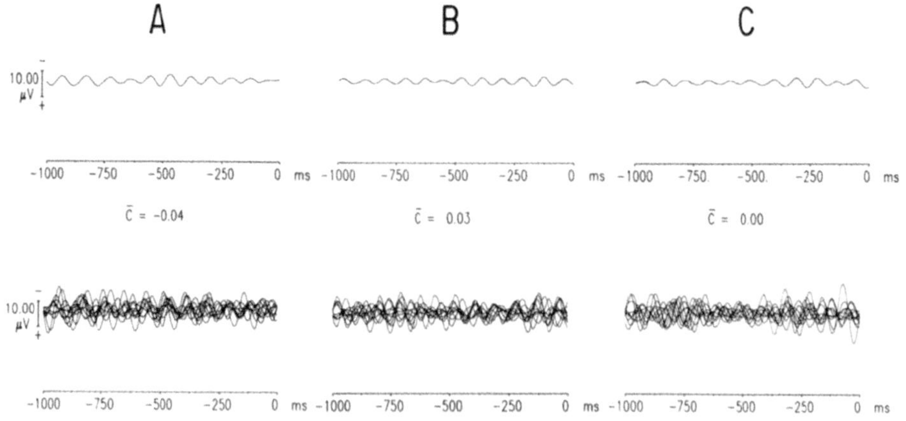

Fig. 11 A–C. Resting EEG of the subject J. K. At the *bottom* ten sweeps of EEG segments which were digitally filtered in the frequency range of 8–13 Hz are shown. Time "0" is arbitrarily chosen. EEG samples were recorded at the beginning (A), middle (B), and end (C) of the recording session. *Top,* mean value results of ten sweeps. Correlation coefficients $\bar{C}$ evaluated from three ensembles of ten sweeps

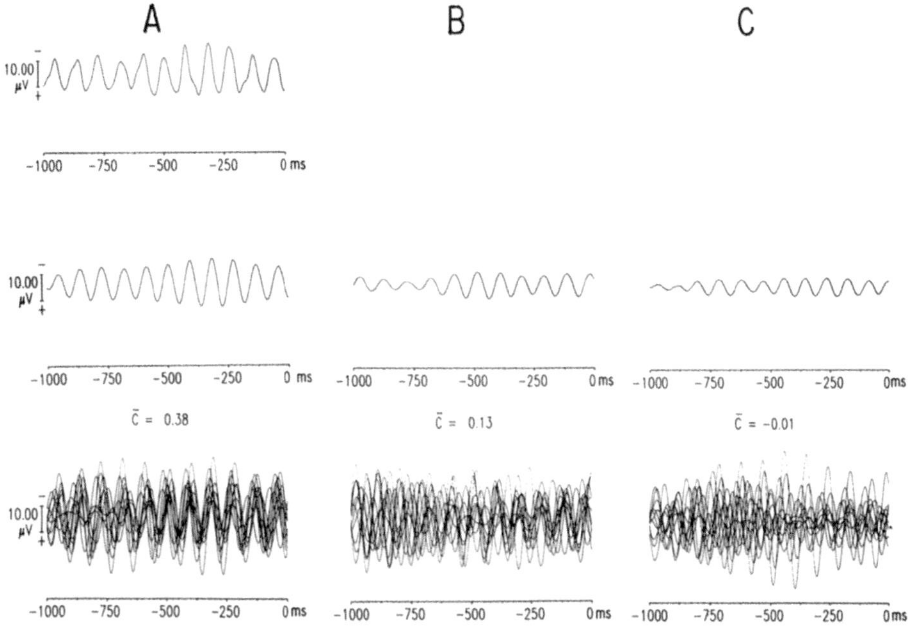

Fig. 12 A–C. Pretarget EEG of subject J. K. (experiment no. 3) during the easiest paradigm, every 4th signal omitted. EEG segments were filtered in the frequency range of 8–13 Hz. The time scale from −1000 ms to "0" indicates 1 s recording time prior to the target (omitted tone). (A) Ten single EEG samples at the beginning of the experimental session (*bottom*). Mean value curves of ten sweeps (*middle*). Broad-band mean value curve from ten sweeps (*top*); filter range, 1–30 Hz. (B) Ten EEG samples in the middle of the experimental session (*bottom*). Mean value curve from ten sweeps (*top*). (C) Ten EEG samples at the end of the experimental session (*bottom*). Mean value curve from ten sweeps (*top*). The correlation coefficients $\bar{C}$ evaluated from three ensembles of ten sweeps are shown at the top of each ensemble. $\bar{C}$ here describes only the period between −500 to 0 ms, i.e., 500 ms prior to target. Subject's report: (A), good performance; (B) and (C), bad performance

the rhythms were less regular and the congruency among sweeps almost disappeared. Also during this stage, 10-Hz EEG with larger amplitudes were observed in comparison with the resting EEG which is shown in Fig. 11. The correlation coefficient decreased drastically against the end of the experiment: At the beginning, where the subjects reported a good performance the correlation coefficient was high $\bar{C}=0.38$, later it was diminished ($\bar{C}=0.13$ and $\bar{C}=-0.01$).

Figure 13 illustrates a similar experiment with the subject J. K. a few months later. Here again he reported that just at the beginning of the experiment he was able, with ease, to mark the target mentally; during the experiment he lost his ability to follow the target, but near the end he again had better control in marking the target. Figure 13 B shows the decrease in congruency and the diminishing of the correlation coefficient. In Fig. 13 C the congruency was better ($\bar{C}$ is $=0.28$). On the following day we started the first experiment with the most difficult paradigm (every third to seventh stimulation omitted) and proceeded next to the easiest paradigm.

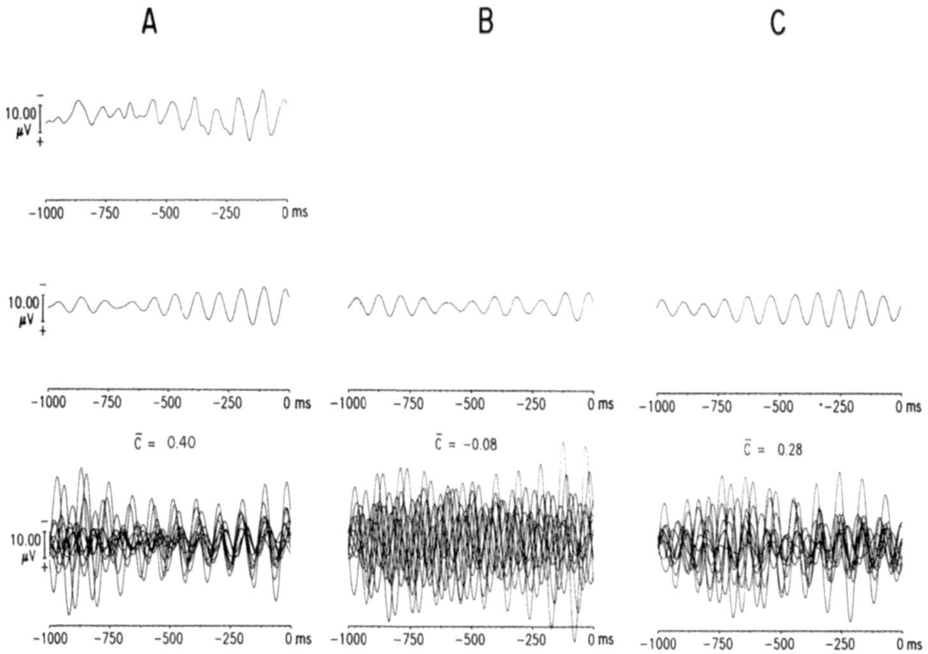

Fig. 13 A–C. Pretarget EEG of subject J. K. (experiment no. 15) during the same (easiest) paradigm. Explanation as for Fig. 12; this shows a repetition after a few months. Subject's report: (A) and (C), good performance; (B), bad performance

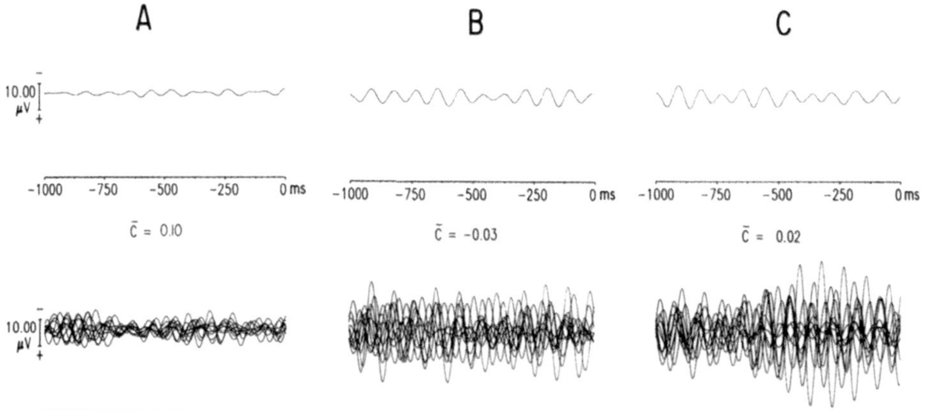

Fig. 14 A–C. Pretarget EEG of subject J. K. (experiment no. 16) during the most difficult paradigm, every 4th to 7th signal omitted. EEG segments were filtered in the frequency range of 8–13 Hz. The time scale from −1000 ms to "0" indicates 1 s recording time prior to the target (omitted light). (A) Ten single EEG samples at the beginning of the experimental session (*bottom*); mean value curve (*top*). (B) Ten EEG samples in the middle of the experimental session (*bottom*); mean value curve from ten sweeps (*top*). (C) Ten EEG samples at the end of the experimental session. The correlation coefficients $\bar{C}$ here describes only the period from −500 to 0 ms. Subject's report: Tried to do well

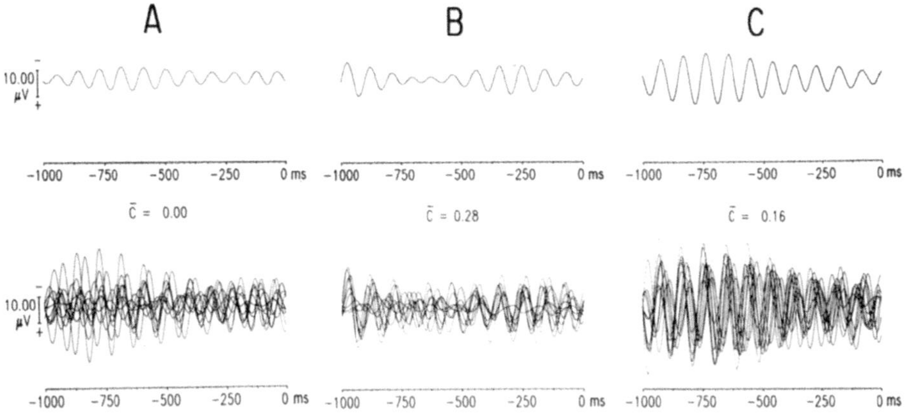

Fig. 15 A–C. Pretarget EEG (8–13 Hz) of J. K. (experiment no. 19) during the easiest paradigm. (A) EEG samples at the beginning of the experimental session (*bottom*). (B) EEG samples in the middle of the experimental session. (C) EEG samples at the end of the experimental session. Mean value curves in the *top* rows. Subject's report: Performance bad at beginning (A), increasingly good toward end of experiment (B, C)

During the most difficult paradigm the subject J. K. reported that at the beginning of the experiment he felt unsure whether he could follow the rhythmicity of the light signals; but in the last two-thirds of the experimental period he was able to mark a larger number of the target signals. Figure 14 A illustrates the beginning and Fig. 14 B the middle stage of the experiments.

The amplitudes of the EEG increased during the experiments but not the correlation coefficient. During another experiment with the easiest paradigm, the subject J. K. reported that this time he had not performed well at the beginning (Fig. 15). However, toward the end of the experiment he definitely had better control in marking the target. Comparison of Figs. 12 and 15 shows that here an opposite effect occurred: In the experiment shown in Fig. 15 the congruency between single curves was better toward the end of the experiment and $\bar{C}$ increased from 0.00 to 0.16.

In five subjects the EEG measurements during the easiest paradigm using light signals were taken after those with the most difficult paradigm. But, during such an experimental session with the most difficult paradigm, congruency of single rhythms like the epochs of Figs. 12 A, 13 A, and 15 C were not observed. Further, the correlation coefficient, calculated during the four stages of the experiment, remained in all cases around 0.05; it never reached values around 0.4.

Remarks on the Comparison of the Easiest and the Most Difficult Paradigms

We want to mention again why comparison of results using the easiest and the most difficult paradigms for the same subject is important to formulate a judgement concerning the existence of event-related pretarget rhythms. For the same subject we have the possibility to increase the probability of the occurrence of the target by up to 100%. The increase in the EEG amplitude and the tendency to regularity and phase-ordering is reflected in correlation coefficients. If the probability of the occurrence of a target is then decreased, one would expect a "less good" or even "bad performance." In the latter case it might be expected that the phase-ordering of the EEG and the tendency to a repeatable pattern would diminish. On the basis of this reasoning, we applied both paradigms to five subjects on the same days, and always obtained comparable results which were similar to the results from the subject J. K.

In the Appendix we describe the interpretation of the correlation coefficient which reflected the comparison of ten EEG signals during the same experimental period. The increase in correlation coefficient means the increase in similarity of single epochs. The fact that subjects who had reported good performance produced a mean correlation coefficient of up to $\bar{C}=0.4$ shows that the EEG can attain good phase-ordered patterns which is contrary to the case of recordings with less probability of occurrence. We also have to emphasize that the recording of almost repeatable EEG patterns during defined experiments with cognitive targets requires a large number of experiments and good cooperation with the subjects.

Different Time Windows

In Figs. 11–15 we have consistently considered the time window between 500 and 0 ms prior to target. Although the EEG of most of the subjects depicted a phase-ordering starting 1000 to 700 ms prior to target signals, the time scale of −500 to 0 ms is the most common one for a rough preliminary evaluation. In order to avoid errors of visual inspection we started an analysis with some moving time windows prior to target. This means that we chose six time windows at various points along the time axis of −1000 to 0 ms. The narrowest window had a duration of 300 ms. Larger correlation coefficients are to be expected for the smaller windows.

Let us consider the ten EEG sweeps illustrated in Fig. 12A. For the 500 ms before stimulation $\bar{C}=0.40$. As Table 2 shows, $\bar{C}$ takes different values depending on the length and position of the time window. For a time window of −300 to 0 ms before stimulation $\bar{C}$ has the highest value, whereas the window of −700 to −300 ms has a much lower value. The same Table 2 shows correlation coefficients of control EEG sweeps of Fig. 11 B. During recording of EEG sweeps where the subject did not report a good performance the correlation coefficients are not much higher even by choosing narrow time windows (the mean value is −0.05). The control EEG of the same subject (sweeps from Fig. 11 A) does not show significant values of $\bar{C}$ even after choosing narrow time windows.

For all experiments performed, the search with different time windows has been carried out; the results are similar to those in Table 2 indicating always highly increased mean values of correlation coefficients during good performance sessions compared with the resting EEG or bad performance sessions. Evaluation of all the subjects under study gave similar results, showing that with analysis of time the correlation coefficient is always highest during the easiest paradigm.

In future experiments the evaluation of correlation with moving windows will be used to search for periods of higher phase-ordering without using up the time in plotting single sweeps for the initial visual inspection.

Table 2. Correlation coefficients $\bar{C}$ of the EEG of left occipital recording for the subject J.K. for several time windows

Time window (ms)	$\bar{C}$ (Experiment no. 3) Good performance (Fig. 12A)	Bad performance (Fig. 12C)	$\bar{C}$ (Experiment no. 15) Good performance (Fig. 13A)	Bad performance (Fig. 13B)	$\bar{C}$ (Experiment no. 19) Good performance (Fig. 15C)	Bad performance (Fig. 15A)	$\bar{C}$ Control EEG (Fig. 11B)
−1000 to 0	0.26	−0.07	0.15	−0,04	0.16	0.01	−0.01
− 700 to 0	0.35	−0.04	0.3	−0.04	0.15	0.03	−0.05
− 500 to 0	0.38	−0.1	0.4	−0.06	0.12	0	−0.04
−1000 to −500	0.1	0.1	−0.07	−0.07	0.18	0.02	−0.01
− 700 to −300	0.35	−0.07	0.14	−0.06	0.14	0.04	−0.01
− 300 to 0	0.31	0.03	0.46	−0.06	0.11	−0.01	−0.02
Mean value of $\bar{C}$ based on 6 time windows	0.29	−0.04	0.23	−0.05	0.14	0.01	0

4.2 Long-Standing Experiments with Subject A. F.

The subject A. F. is a technical assistant in our research group. From the beginning of the experiments he had a great interest in being a subject and gave detailed reports after every measurement session. The analysis of the single sweeps by visual inspection correlated highly with his reports in most cases. As he reported that he had been able to mark the target signal mentally during a given measurement session, the single pretarget EEG curves showed, as a rule, good congruence. The agreement with his report was about 80%. Taking this degree of reliability into account, we performed seven experiments with A. F. in a period of about 3 months. The measurements and the report were made as follows:

1. The resting EEG prior to application of a paradigm was measured.
2. The easiest paradigm with repetitive light stimuli was applied.
3. After application of 30 light stimuli (and ten omitted signals) A. F. wrote a report to describe the session as one with *"good performance"* or with *"bad performance"*.
4. The single sweeps of pretarget EEG were plotted and the reliability of the subject's report was checked by means of analysis with correlation coefficients which give the degree of single sweep congruence.

In six of the seven experiments A. F. reported that he had measurement periods with "good performance" as well as "bad performance". In one experiment he reported only "bad performance."

In Table 3 the correlation coefficients $\bar{C}$ in the frequency band of 8–13 Hz are listed for the resting EEG and for *good* as well as *bad performance* periods during the easiest paradigm. Experiment no. 32 is the one where the subject could achieve a good performance in none of the measurement periods. The correlation coefficient for the resting EEG had values no higher than 0.05 and averaging <0.002. The periods with bad performance had values averaging -0.02. The periods with

Table 3. Correlation coefficients of ten single epochs of EEG in left occipital recordings of the subject A. F. during the easiest paradigm, and of the control EEG

Experiment no.	Easiest paradigm		Control EEG	
	Good performance $\bar{C}$	Bad performance $\bar{C}$	Experiment no.	$\bar{C}$
24	0.16	0.01	22	0
27	0.3	−0.02	25	−0.04
32	–	−0.06	30	0.02
37	0.04	−0.05	35	0.02
42	0.13	0.01	40	−0.05
47	0.1	−0.06	45	0.01
52	0.24	0.06	50	0.03
Mean value	0.16 ($N=6$)	−0.02 ($N=7$)		0.00 (N=7)

Analyzed recording period: 700 ms before the target signal and 700 ms control EEG with arbitrary "0" time. The experiments with "good performance" and "bad performance" were selected only by considering subjects' oral reports. Correlation coefficients are described from all experimental sessions performed during the 7 different days of experimenting.

good performance vary between 0.04 and 0.3, with a mean of 0.16. This mean value shows high variability for one subject. It is mainly noteworthy only as an important indicator of the trend in the experiments with A. F.

Experiments involving varied target probabilities, evaluation of correlation coefficient, and plotting of single EEG curves were performed with six subjects in similar sequences. Each subject was studied on at least two different days. All showed behavior similar to J. K. and partially to A. F.

Statistical evaluation, including correlation coefficients, of all the experiments and results from each subject with various time windows will be reported in another publication.

Important Remark About the Method

In the conventional P300 type of experiments, subjects have the task of counting the target stimuli. Therefore, their task can be correctly checked by comparing the actual number of target signals with the number registered by the subjects. In experiments to measure the readiness potential, prestimulus EEG which precedes motor tasks can be precisely correlated with physical movements. During our experiments the subject has no exact possibility to check his performance in marking the target mentally in the time scale. Most of the subjects (especially subject E. R., one of the authors of this study), reported that they used the first light stimuli after the omitted signal as the reference signal. In other words, if the appearance of the first light stimulation was correctly guessed, the subject concluded by having a good performance in hitting the target (omitted signal). When a subject had a successive number of well-guessed trials during a recording period, he defined it as a good performance.

The following question arises: How does the evoked potential look in such a situation when the subjects guess the time of occurrence of a light signal? This question will be handled in the next section.

4.3 Does the Pretarget Activity Interact with Evoked Potentials?

According to the methodological remark in the previous section, we have systematically evaluated the evoked potential elicited by the first visual stimulation following the target (omitted) signal in order to see what happens when the subjects expect a light stimulation and when it appears exactly at the expected time. Figure 16 illustrates the results from subject E. R. who has, immediately after the experiment, given the following report: "At the beginning of the experiment I got confused and could not mark the omitted signals mentally. But towards the end of the experiment (approx. the last 20 sweeps) I could perform well; could check my performance by guessing exactly the time point of a light stimulation." In other words the subject E. R. has precisely anticipated the light signal. Figure 16 B and C illustrates the filtered activity after the last 20 omissions, just before and after the first flash following the omission. In the last 250 ms prior to stimulation a tendency to synchronization with light stimulation was observed. This tendency is reflected in the mean value curve which shows almost sinusoidal behavior. The RMS value of the filtered average curve 250 ms before stimulation in curve A is 70% smaller than that of curve B. After the stimulation there is no significant response or enhancement in the mean value curve. At the top of Fig. 16 A the filtered evoked potential of subject E. R. is illustrated. The EP (average of 20 sweeps) was recorded upon randomly applied light stimulation. Here a significant enhancement above the baseline is to be observed. The prestimulus EEG does not show the synchrony of the sweeps at the bottom. It can be stated that the EP recorded during the period where the subject focussed his attention on the light stimulation (Fig. 16 B) is highly attenuated and almost disappears in comparison with the normal EP (Fig. 16 A). For 20 sweeps and 250 ms before stimulation $\bar{C}=0.15$ for the experiment in Fig. 16 C and $\bar{C}=0.00$ for sweeps before the EP in Fig. 16 A.

The digital filter used in our evaluation might have ringing effects of maximally 5% in the alpha frequency range. Accordingly, we exclude here the possibility of filtering artifacts which might affect the measurements. However, the observed effect can be demonstrated also by using the wide-passband filter of 1–30 Hz. At the bottom of Fig. 17 once again the average of approximately 20 sweeps from the subject E. R. is shown. The EEG prior to stimulation shows the same dominant 10-Hz activity. In this case there is no significant record of evoked potential following the light stimulation in comparison with the normal EP shown at the top with the usual N100–P200 deflection typical for visual-evoked potentials.

The changes in sensory-evoked potentials during the described paradigm, which also emphasize the interindividual variability, will be described in another

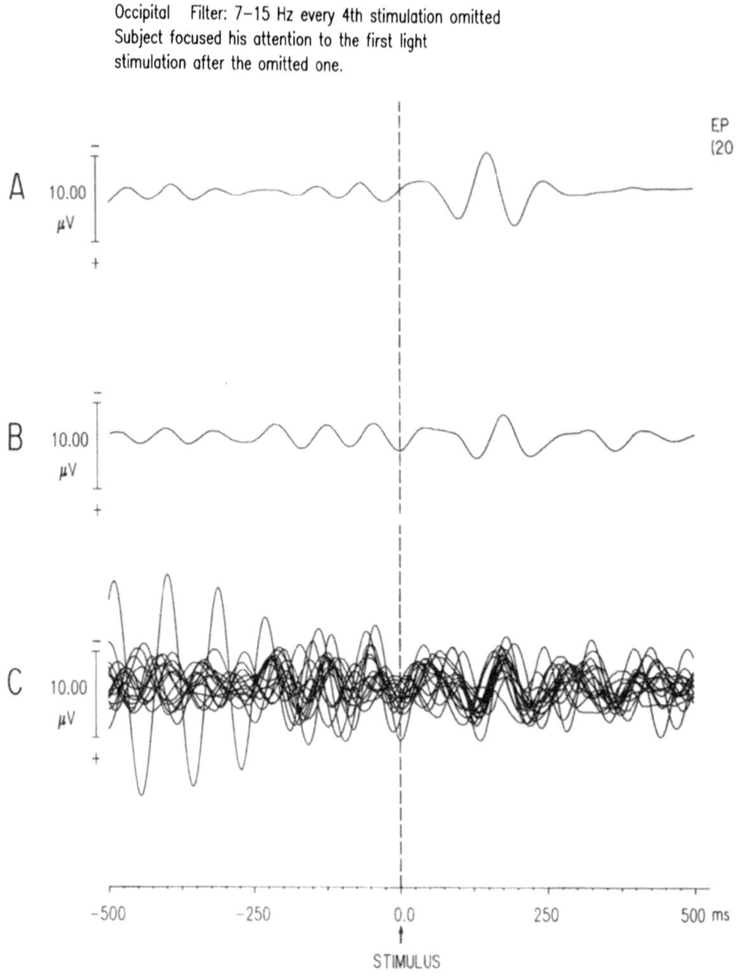

Fig. 16 A–C. The subject E. R. focussed his attention on the first light stimulation after the omitted one. (A) Prestimulus EEG and EP from a successive experiment with the same subject in which the light stimulation was randomly applied. Mean value curve of 20 sweeps. Frequency range of digital filters, 7–15 Hz. (B) Mean value of the ensemble of single sweeps presented in (C). (C) Approximately 20 single EEG sweeps before and after the first light stimulation following omission filtered with band-pass filters of 7–15 Hz. $\bar{C}=0.15$ for the 250 ms prior to stimulation

publication. Here we mention that the EP analysis shown in Figs. 16 and 17 can be given as a model (or prototype) for 80% of the measurements under study.

With these results we can come back to the essential question which was raised in the Introduction of this paper: Can we find cognitive states of the brain using the spontaneous EEG which is so time-locked to an expected event and so synchronized that the evoked potential is reduced in amplitude or almost disappears? The experiments described above show that the occurrence of the light signal ex-

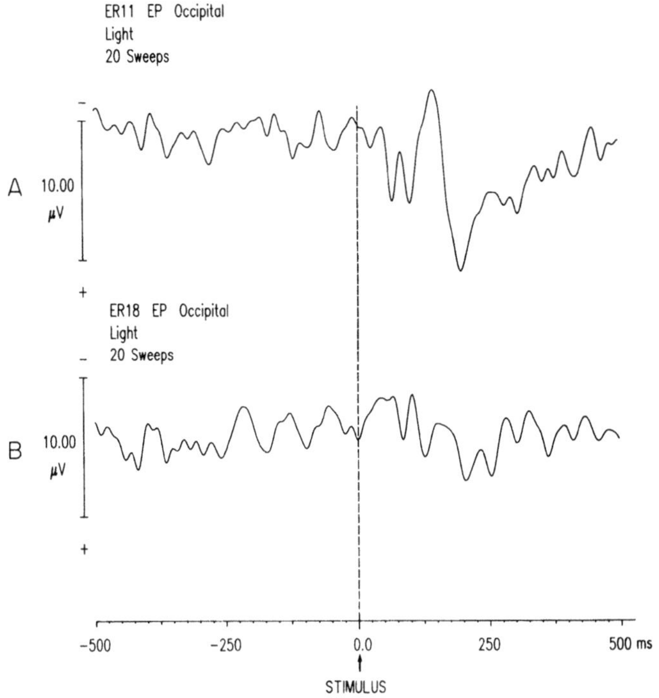

Fig. 17 A, B. A comparative illustration of evoked potentials of subject E. R. The evoked potentials were evaluated from the same ensembles of sweeps presented in curves (A) and (B) of Fig. 16, only the filter range here is different. Passband filters applied, 1–30 Hz

actly at the same time point at which the subject expects it can lead to the fact that the visual EP is not elicited for that flash or is highly attenuated. These results allow the tentative explanation that neural populations which are activated during cognitive tasks of expectation (and/or imagination) for the occurrence of a sensory signal, and the neural populations which are activated upon the actual sensory stimulation must share some common pools.

5 Discussion

5.1 Quasi-deterministic EEG, Cognitive States, Dynamic Memory

The experiments described in the previous sections and our earlier results (Başar 1988) have shown that during cognitive tasks it is possible to measure almost reproducible EEG patterns in subjects expecting defined repetitive sensory stimuli. The use of modern computer techniques makes it possible to speedily search coherent states during mental tasks. During such coherent states single sweeps of EEG are time-locked to a target signal, at least, for periods of 10 min [1] during which the subjects are able to mark a cognitive target mentally in a recurrent manner. We use the expression quasi-deterministic EEG for the recurrently emitted,

[1] A recording session with 10 target signals (10 omitted stimuli and 30 physical stimulations) requires at least 700 s or about 3 min). Some of the subjects can reach the same performances during 30–40 omitted stimuli, in other words around 10 min.

almost reproducible, EEG patterns. The specific brain function is related in this case to a type of short learning process and short-term memory which we tentatively denote as "dynamic memory" (Başar 1988).

In the present paper we have focussed our emphasis on the 10-Hz frequency range although other frequency ranges were briefly demonstrated in Section 4. Tentatively the expression "dynamic memory" could be extended to cover all relevant EEG frequency ranges and most probably combinations of several patterns. The use of the expression "quasideterministic EEG" finds its legitimacy in several results reported by Babloyantz (this volume), Röschke and Başar (Başar et al. 1988), and Freeman and Skarda (1985), who have shown that the EEG reflects the properties of a strange attractor.

5.2 EEG and Behavior, P300, and Selective Attention

Berger (1938) believed that the beta process was the main physiological activity of the living brain and was always present. The alpha process, he thought, could be modified according to the level of consciousness. He explained the alpha blockade during apprehension and orientation as an inhibitory effect of a "work center" that should influence all simultaneous physical activity. His suggestion was, therefore, that alpha activity, rather than describing a resting state, had better be identified as describing a brain in a "passive" state as contrasted with the "active" brain in which alpha disappears and beta or faster activity increases.

Hillyard and Picton (1979) used the term "selective attention" or simply "attention" as a construct which has a rather broad but circumscribed set of meanings, being clearly distinguished from nonselective CNS processes such as arousal or alertness. Attentional processes are those CNS functions which enable perceptual or motor responses to be made selectively to one stimulus category or dimension in preference to others. Irrelevant stimuli that are not being attended are either partially or completely rejected from perceptual experience, entry into long-term memory, and control over behavior. Furthermore, these authors state that attention refers to selective aspects of sensory processing. Accordingly all experimental demonstrations of attention must measure the responsiveness of the organism to more than one category of stimulus. The initial stages of sensory processing are generally thought to proceed rather inflexibly and to consist of an initial afferent registration and feature analysis of incoming sensory data. This information persists on accessible form for about 1 s in a high capacity sensory memory which has been termed the "buffer stores", "iconic" (visual) memory, and "echoic" (auditory) memory.

During our experiments the subjects of the experiments had to pay attention to an omitted stimulus. During this process, if they were able to be attentive enough to mark the target mentally, they anticipated with 10-Hz waves which were time-locked to the target, showing almost reproducible patterns. Depending on the performance of a subject the coherency time of such reproducible wave packets ranged usually from 300 to 1000 ms, most of it prior to the time the omitted stimulus was due.

In earlier publications Başar (1983a, b) has tentatively assumed that the evoked potentials manifest the bifurcation of the strange attractor EEG to a limit cycle attractor of short duration. A strange attractor is manifested by its activity which appears to be random. However, the activity of a strange attractor is deterministic and reproducible if the input and initial conditions can be replicated. The experiments with human subjects which were described in Section 3 and 4 of this study showed that by increased certainty due to expectation of repetitive sensory signals and accordingly by increased attention-stages, the subjects seem to generate internal cognitive inputs to the CNS. These cognitive inputs are probably due to repetitive similar mental efforts. According to our experimental results, if a subject can not mentally predict the occurrence of the expected target signal (omitted stimulation), there is no averaged synchronization of the EEG in the 10-Hz frequency range (see experiments with subject J. K.) or in the 40-Hz frequency range. (In Section 3.2 only global findings were shown.)

Further, our results showed that the 10-Hz, 40-Hz, and 4-Hz EEGs go from disordered states to ordered coherent states during defined cognitive inputs to the CNS. Here lies a similarity to evoked potentials elicited by exogenous sensory stimuli which also show a transition from disordered to ordered coherent states. Moreover, the analysis of the present study shows the following phenomena: when the certainty of the occurrence of a target signal increases, the 10-Hz EEG goes to coherent phase-ordered states without application of physical stimulation. There are replicable phase-ordered (or time-locked) patterns to a target signal (Fig. 15 A).

Here we have to add a word of caution concerning the frequency bands of the filtered EEG. Bullock (1988; see also this volume, Part 1) assumes that the brain is a wide-band generator, and that even in the special cases where one frequency for a time is higher than others, if it has oscillators, it has many at all frequencies. If Bullock's assumption is true, we would have to take the variability of the frequencies in a given frequency range more into consideration. In fact, we see oscillations of about 11–12 Hz during the tasks prior to visual stimulation and of 9 Hz during tasks with auditory stimulation, so that one can not talk about a definite component of alpha activity but about several components in the 10-Hz range. We selected the limit of our filters in the range of 8–13 Hz in order to keep filtering artifacts to a minimum. The errors were kept in this frequency around 5% due to several precautions with recording lengths. Moreover, as is the case in Figs. 12 and 13 we give the broad-band filtered prestimulus EEG to show that the coherent dominant 10-Hz rhythmicity can be detected even without digital filtering.

5.3 What Is New in the "Dynamics of Time-Locked EEG Patterns"?

Observation of changes in the amplitude of the EEG, enhancement or blocking of alpha activity during mental tasks (solving of arithmetic problems or hearing music) is not a new discovery and relevant examples have been described in the literature (Creutzfeldt 1983; Giannitrappani 1985; Petsche et al. 1987). The new observation reported in this study is the fact that alpha waves can be time-locked to a defined cognitive target and that in good defined experiments reproducible alpha patterns can be emitted from the human brain. Freeman (1975) and Freeman and Skarda (1985) have shown that the EEG from olfactory bulb and cortex in awake, motivated rabbits and cats manifests a characteristic temporal pattern consisting of a burst of 40- to 80-Hz oscillations, superimposed on a surface-negative baseline potential shift synchronized to each inspiration. In the present study we have limited our discussion to our findings in the 10-Hz frequency range although the global analysis of the results indicates that similar self-organizing trends in theta, 40-Hz, and beta frequency ranges also exist.

There is an inverse relation between sensory response and successful cognitive efforts. The mental prediction of a sensory stimulation reduces the sensory response drastically (Figs. 16 and 17). Therefore, we tentatively concluded that neural populations which are activated during defined cognitive tasks of visual or auditory expectation display the property of busy channels (or busy populations) and cannot be excited further. Accordingly, the question which we raised at the beginning of the report is partly answered: by cognitive loading of subjects, stages of EEG can be reached so that the (subjects') evoked potentials are highly reduced or partly abolished.

In addition to carrying out conventional analysis as described in the established P300 studies by Sutton et al. (1965) and Galambos and Hillyard (1981), we encountered new developments in the analysis of ERPs in relation to missing auditory stimuli (Nakamura et al. 1986; Gauthier et al. 1986; Takasaka 1985; Friedman 1984; McCallum 1980; Simson et al. 1976). These authors' results were mostly limited to the description of latencies of potentials following the omitted stimuli and to topographic distributions of the family of P300 waves. Gauthier et al. (1986) and McCallum (1980) reported the contingent negative variation (CNV) changes during special experimental conditions. Synchronized pretarget event-related rhythms, however, were not described by these authors, since they did not use small segments of EEG or filter methods, and the subjects were not asked to perform the same mental effort to mark the target as omitted stimuli.

We also repeat here the question raised by D. Lehmann in this volume: Are there, over time, distinguishably different basic units at work, i.e., "building blocks" or "atoms" of thinking or other brain functions, which reflect the characteristics of perception, cognition, and emotion?

Lehmann says that functional states may be recognized for relatively short epochs of EEG depicting systematic relations with types of thought processes. According to his description, the duration of spatially stable microstates covers a range of about 500 ms. Certainly it is attractive to think of the brain microstates as being building blocks of higher information processing, which qualify as areas consciousness only beyond a certain duration. According to our preliminary results the microstates would have various durations depending on the basic rhythmic activity. Sensory-cognitive microstates in the alpha frequency range might have durations of about 300 ms, theta microstates about 500 ms. On the contrary, 40-Hz microstates should have shorter durations of 20 to 50 ms.

The description of spatial distribution is limited in our results. The visual inspection of the coherent states in our curves indicated that for visual-cognitive experiments mostly the occipital area showed the marked behavior. During experiments with acoustical stimuli the congruency was marked in the vertex. Therefore, we presented the experiments with light stimuli for occipital recordings and acoustical stimuli with vertex recording. Before judging the problem of spatial distributions, an extended analysis with different time windows to be analyzed has to be achieved. An investigation with multichannel MEG detectors would provide a more appropriate approach.

Stowell (1987) has confirmed and extended our previous results on preparation rhythms and mentioned the possibility of "evoked potential ghosts". Stowell's elegant psychopsychological paradigms allow a sophisticated view on cognitive cerebral activity.

6 Summary

Experiments on combined analysis of EEG, sensory-evoked potentials, and event-related potentials are presented using an ensemble of paradigms during specific cognitive tasks of human subjects. During repetitively presented acoustical and optical stimuli the subjects were asked to mark mentally an omitted signal which is called "target." The occurrence of the target signal was variable. With increasing probability of the target occurrence the subjects reported more confidence in marking the target signal. Based on subjects' oral reports, periods of "good performance" and "bad performance" were classified, so called from their subjective estimates, not actual measures of performance. About 1 s prior to target signals the subjects' EEG showed coherent phase-ordered rhythmic patterns with a dominant frequency of around 9–10 Hz at the vertex during acoustical stimuli and 11–12 Hz in the right occipital lead during visual stimulation. With the increasing probability of target occurrence, the mean correlation coefficient between single pretarget EEG signals during experiments with some subjects may reach values up to 0.4, showing a tendency to produce almost reproducible patterns which are denoted as "quasideterministic EEG." The results further globally indicate that upon sensory stimuli during such sessions where subjects' EEG goes to coherent stages, the light stimuli do not evoke any ample evoked potentials. The results and their interpretation show that the analysis of an ensemble of parameters like EEG, filtered EEG, EPs, and ERPs are useful to understand cognitive microstates of the brain.

Appendix

The correlation factor ϕ of two time series f(t) and g(t) is defined as the mean value of their product over time $(-T, T)$, provided that f(t) and g(t) do not have DC components.

$$\phi = \overline{f(t)g(t)} = \lim_{T \to 1} \frac{1}{2T} \int_{-T}^{T} f(t)g(t)\,dt$$

or

$$\phi = \frac{1}{T_2 - T_1} \int_{T_1}^{T_2} f(t)g(t)\,dt\,,$$

if f(t) and g(t) are to be considered only between time points T_1 and T_2 and f(t) and $g(t) \neq 0$.

It can be shown that the maximal value of the correlation function is $\phi_{max} = \tilde{f} \cdot \tilde{g}$ where $\tilde{f}$ and $\tilde{g}$ denote the RMS values of f and g.

The value

$$C=\frac{\phi}{\phi_{max}}$$

is denoted as the correlation coefficient. Its value fluctuates only within the limits of -1 and $+1$. If $C=\pm 1$, it is stated that $f(t)$ and $g(t)$ are (positively or negatively) correlated. This is only the case, if $f(t)$ and $g(t)$ are correlated as in the following:

$$g(t)=af(t) \quad (a \text{ being a constant factor}).$$

If $C=+1$ ($a>0$) this means that f and g are completely in phase, and if $C=-1$ ($a<0$) f and g are in counter phase.

If $C=0$, $f(t)$ and $g(t)$ are not at all correlated. This is the case for two sine waves with different frequencies, and also for the noise signals originating from two different sources providing the epoch analyzed is long enough; for the case of noise, C converges gradually toward 0 over many cycles of the lowest frequency in the passband.

In all other cases, this means if $0<C<1$, $f(t)$ and $g(t)$ are piecewise or imperfectly correlated.

The correlation coefficient is a measure to indicate the similarity of two time series. The aim of our analysis by using correlation coefficients between two EEG oscillations is to try to establish the similarities in the EEG activity prior to a task. By simple superpositions of ten filtered signals, we can often recognize a similarity by visual inspection. The phase-ordering of the EEG signals during a stage with a defined cognitive load can be described in a mathematical way by using the correlation function. Since we usually considered ten EEG epochs (focussing on the period just prior to the target) we used a mean value of the correlation coefficient which is the mean value of the correlations of all the ten EEG signals. In other words, for N signals

$$\bar{C}=\frac{2}{N(N-1)}\sum_{\substack{i,j=1\\ i<j}}^{N} C_{ij}$$

C_{ij} being the correlation coefficient of EEG epochs i and j, belonging to the ensemble of analyzed sweeps and N the maximal number of analyzed sweeps.

References

Başar E (1980) EEG-brain dynamics. Relation between EEG and brain evoked potentials. Elsevier/North-Holland, Amsterdam

Başar E (1983a) Synergetics of neuronal populations. A survey on experiments. In: Başar E, Flohr H, Haken H, Mandell AJ (eds) Synergetics of the brain. Springer, Berlin Heidelberg New York, pp 183–200 (Springer series in synergetics, vol 23)

Başar E (1983b) Toward a physical approach to integrative physiology. I. Brain dynamics and physical causality. Am J Physiol 245(4):R510–R533

Başar E (1988) EEG-dynamics and evoked potentials in sensory and cognitive processing by the brain. In: Başar E (ed) Sensory and cognitive processing by the brain. Springer, Berlin Heidelberg New York, pp 30–55

Başar E, Stampfer HG (1985) Important associations among EEG-dynamics, event-related potentials, short-term memory and learning. Int J Neurosci 26:161–180

Başar E, Başar-Eroglu C, Rosen B, Schütt A (1984) A new approach to endogenous event-related potentials in man: relation between EEG and P300-wave. Int J Neurosci 24:1–21
Başar E, Rosen B, Başar-Eroglu C, Greitschus F (1987) The associations between 40 Hz-EEG and the middle latency response of the auditory evoked potential. Int J Neurosci 33:103–117
Başar E, Başar-Eroglu C, Röschke J (1988) Do coherent patterns of the strange attractor EEG reflect determnistic sensory-cognitive states of the brain? In: Markus M, Müller S, Nicolis G (eds) From chemical to biological organization. Springer, Berlin Heidelberg New York
Berger H (1929) Über das Elektroencephalogramm des Menschen. Arch Psychiatr Nervenkr 87:527–570
Berger H (1938) Das Elektroencephalogramm des Menschen. Nova Acta Leopold 6:38
Bullock TH (1988) Compound potentials of the brain, ongoing and evoked: perspectives from comparative neurology. In: Başar E (ed) Dynamics of sensory and cognitive processing by the brain. Springer, Berlin Heidelberg New York, pp 3–18 (Springer series in brain dynames, vol 1)
Bullock TH, Başar E (1988) Comparison of ongoing compound field potentials in the brains of invertebrates and vertebrates. Brain Res Rev 13:57–75
Creutzfeldt OD (1974) The neuronal generation of the EEG. In: Renard A (ed) Handbook of electroencephalography and clinical neurophysiology. Elsevier, Amsterdam
Creutzfeld OD (1983) Cortex cerebri. Springer, Berlin Heidelberg New York
Fessard A (1961) The role of neuronal networks in sensory communications within the brain. In: Rosenblith WA (ed) Sensory communication. MIT Press, Cambridge
Freeman WJ (1975) Mass action in the nervous system. Acedemic, New York
Freeman WJ, Skarda CA (1985) Spatial EEG patterns, non-linear dynamics and perception: the neo-Sherringtonian view. Brain Res Rev 10:147–175
Friedman D (1984) P300 and slow wave: the effects of reaction time quartile. Biol Psychol 18(1):49–71
Galambos R, Hillyard SA (1981) Electrophysiological approaches to human cognitive processing. Neurosci Res Program Bull 20(2)
Gauthier P, Hamon JF, Gottesmann C (1986) Etude de la variation contingente negative (VCN) et des composantes tardives des potentiels evoqués au cours du sommeil chez l'homme. Rev Electroencephalogr Neurophysiol Clin 15(4):323–330
Giannitrapani D (1985) The electrophysiology of intellectual functions. Karger, Basel
Hillyard SA, Picton TW (1979) Event-related brain potentials and selective information processing in man. In: Desmedt JE (ed) Cognitive components in cerebral event-related potentials and selective attention. Prog Clin Neurophysiol 6:1–52
McCallum WC (1980) Brain slow potential changes elicited by missing stimuli and by externally paced voluntary responses. Biol Psychol 11(1):7–19
Nakamura K, Toshima T, Takemura I (1986) The comparative and developmental study of auditory information processing in autistic adults. J Autism Dev Disord 16(2):105–118
Petsche H, Rappelsberger P, Pockberger H (1987) EEG-Veränderungen beim Lesen. In: Weinmann HM (ed) Zugang zum Verständnis höherer Hirnfunktionen durch das EEG. Zuckschwerdt, Munich, pp 59–74
Ramos A, Schwartz E, John ER (1976) Evoked potential-unit relationship in behaving cats. Brain Res Bull 1:69–75
Röschke J, Başar E (1988) The EEG is not a simple noise: strange attractors in intracranial structures. In: Başar E (ed) Dynamcis of sensory and cognitive processing by the brain. Springer, Berlin Heidelberg New York, pp 203–216 (Springer series in brain dynamcis, vol 1)
Simson R, Vaughan HG, Ritter W (1976) The scalp topography of potentials associated with missing visual or auditory stimuli. Electroencephalogr Clin Neurophysiol 40(1):33–42
Stampfer HG, Başar E (1985) Does frequency analysis lead to better understanding of human event-related potentials. Int J Neurosci 26:181–196
Stowell (1987) Cerebral slow waves and time parsing. Int J Neurosci 32:861–874
Sutton S, Braren M, John ER, Zubin J (1965) Evoked potential correlates of stimulus uncertainty. Science 150:1187–1188
Takasaka Y (1985) Expectancy-related cerebral potentials associated with voluntary time estimation and omitted stimulus. Folia Psychiatr Neurol Jpn 39(2):167–172

Microstates of the Brain in EEG and ERP Mapping Studies

D. Lehmann

1 State Dependency of Brain Functions

In the present paper we focus on a particular aspect of brain information processing and the resulting behavior control: its embeddedness into, and thus dependency on, the momentary global functional state of the brain. The state (Ashby 1960) might be defined by anatomical, biochemical, electrophysiological, behavioral, and other measures (Koukkou et al. 1980).

The general dependency of brain work on the global functional state of the brain is evident in such different states as wakefulness in adults vs sleep in adults, or childhood wakefulness vs adult wakefulness. These states are easily distinguished by very many measures which covary: temporal EEG patterns (spectral distribution of EEG power), evoked potential waveforms, perceptual threshold, goal selection, consciousness, speed (reaction time) and accuracy of behavioral responses, and modes of thinking (dreaming, fantasizing), for example, differ markedly between these gross states (Koukkou and Lehmann 1983). State dependency not only concerns brain processing strategies; it is very important that memory storage and access to information (facts and strategies) for recall from memory are also known to be state dependent (Overton 1978; Eich 1980). System-inherent periodic changes of state are controlled by "internal clocks"; information-driven changes of state are caused by the results of information processing, and executed via the "orienting reaction" (see Ohman 1979). A model of brain functioning based on state-dependent processes has been discussed in detail elsewhere (Koukkou et al. 1980; Koukkou and Lehmann 1983).

2 Brain States and Brain Electric Fields

Recording of the brain electromagnetic field permits a noninvasive and continuous assessment of the functional state of the brain. The technique yields a time resolution down to the millisecond range which is uniquely suited for studies of brain information processing. Studies in humans have demonstrated the very high sensitivity of brain electric field recordings (EEG and event-related potentials) to induced or spontaneous conditions of the subject, including parameters and content of input information, relevance of input information, vigilance level, EEG state, attention, task, performance quality, spontaneous thoughts, mood, metabolism, maturational age, and many pathological conditions, including the

Springer Series in Brain Dynamics 2
Edited by Erol Başar and T. H. Bullock

mental disorders of schizophrenia and depression. Numerous papers have reported such results; a few are quoted here, centering on problems investigated by our group (Callaway 1962; Lehmann et al. 1965; Williams et al. 1966; Keesey and Nichols 1967; Berkhout et al. 1969; Koukkou and Lehmann 1976; Başar 1980; Katada et al. 1981; Koukkou 1982; Skrandies and Lehmann 1982; Lehmann et al. 1983; Gath et al. 1983, 1985; Bohdanecky et al. 1984; Brandeis and Lehmann 1986; Skarda and Freeman 1987).

3 Momentary Brain Field Maps and Information Processing

The brain electric field changes over time. If the field is recorded with a sufficient number of simultaneous channels, the changes of the spatial configuration of the field can be viewed as a series of maps of the momentary (instantaneous) spatial distributions of the electric potentials, i.e., as series of momentary "electric landscapes" (Lehmann 1971, 1972, 1981). The landscape of such a map depends on the geometric configuration and strength of the neural generator populations which momentarily are active. Accordingly, a given map landscape manifests a given global, functional state of the brain. The global state consists of a mosaic of local substates. Different map landscapes must have been caused by the activity of at least partially different neural populations. Assuming that given contents or steps or modes of information processing necessitate the activity of at least partially different neural populations, we suggest that examining the mapped landscapes obtained in adequately designed experiments leads to the possibility of establishing durations, postinput times, types, sequencing, and relations between the different constituents of brain information processing.

Traditional waveform analyses of event-related EEG potentials have addressed the question of processing steps of incoming information by identifying "components," i.e., subsecond epochs of the wafeforms, i.e., times of highest voltage between preselected pairs of recording locations. In addition to the problem that the analyzed waveforms typically constitute a small subset of all possible waveforms, changes of waveform components need not be associated with changes of landscapes, and changes of landscape might not be reflected by conventionally defined waveform components. Further discussions of the problems are offered in Lehmann and Skrandies (1984) and Brandeis and Lehmann (1986). The present paper concentrates on the spatial configuration of the brain electric fields, their "landscapes."

4 Are There Basic Building Blocks of Brain Information Processing?

The possible characteristics of different constituents of brain information processing, however, can be established only if there are indeed different map configurations or functional brain states which persist for certain times, recognizable on formal and on functional grounds (they would be supposed to change in a discontinuous fashion, stepwise). In other words, the question is: Are there, over time, distinguishable different basic units of functioning, i.e., "building blocks" or "atoms" of thinking or other higher brain functions, which still reflect the characteristics of perception, cognition, and emotion?

The conceivable alternative is that the different functional states blend into each other in a graded, continuous fashion, establishing a never-steady stream without natural demarcations.

An indication that higher order functional states might change in a noncontinuous way is provided by the common observation that one is either awake with reflecting self-awareness, or asleep and without reflecting self-awareness, conditions in between being at best very very brief: Even lightly "dozing off" appears to involve the step from self-awareness to its disappearance. Likewise, the classical observation of fluctuations of perceptual threshold ("fluctuations of attention," Woodworth and Schlosberg 1954) in the range of seconds points to distinguishable time packaging of higher order states.

5 What Is the Duration of the Putative Building Blocks?

Functional states may thus be recognized for restricted time epochs. For instance, the electric characteristics of relatively short epochs of EEG recording show systematic relations with types of thought processes: Different types or classes of spontaneous thoughts have been found to be reliably associated with different EEG spectral measures obtained during the 16 s before reporting the thoughts (Koukkou and Lehmann 1976; Ehrlichman and Wiener 1980; Lehmann et al. 1983). Even shorter are stationary epochs in the range of seconds which are formally identified by adaptive segmentation of single channel EEG data (Gath et al. 1983; Lopes da Silva 1981; Barlow 1985). But, how short can the analyzed EEG epochs be chosen so that they still reflect higher-order functional properties such as thought types, percepts, emotions, and processing speed and accuracy? In other words, if there are such putative basic functional units or building blocks of identifiable higher nervous functions, what is their duration?

6 The Task: Adaptive Segmentation of Map Series of Brain Electric Fields

In order to class successive maps of brain electric fields as same or different, a global decision might be useful which produces a one-number statement about the degree of dissimilarity between two maps. We have proposed the sum of the standard deviations of the mean voltages at each electrode (using the average reference: Bertrand et al. 1985), divided by the number of electrodes; we called this measure "Global Dissimilarity between Maps" (Lehmann and Skrandies 1980; Lehmann et al. 1987). The advantage of the procedure lies in its unbiased approach, and the ease with which maps with any number of measured points can be treated. The disadvantage of such a global dissimilarity measure is that landscape changes which are small between each pair of successive maps, but which are systematic over subsequent pairs, will let continous changes of the landscape go unnoticed. On the other hand, if high values of dissimilarity are found at certain times and low values at other times, then it must be concluded that there are times of relative stability and of relative instability of the map landscapes. The latter result has been found in spontaneous (Lehmann et al. 1987) as well as in event-related (Lehmann 1987) human brain electric map series, supporting the idea of discontinuous changes of functional state.

7 Recognizing Brain Field Landscapes for Adaptive Segmentation

A more specific approach has to assess the individual landscapes and their differences with some pattern recognition procedure, in order to later test functional differences between different classes of map landscapes. A most simple description of a landscape is the extraction of the locations of its peak and trough (the extreme points) and of the amplitude difference between them (Lehmann and Skrandies 1984; Brandeis and Lehmann 1986). This reduction of the map landscape is related to the popular strategy to represent it by a best-fitted equivalent model dipole generator (e.g., Grandori 1984), but without making assumptions about constraining conditions.

We have used the two-extrema-extraction method successfully in several studies (see Lehmann and Skrandies 1980, 1984; Lehmann 1981; Brandeis and Lehmann 1986). Many other approaches for assessing and comparing the maps are possible (cf. picture processing and recognition procedures), but they are considerably more complicated, also in terms of later analysis requirements.

The extracted extrema are representative for the brain electric map landscape: It can be shown empirically that, as expected theoretically, the spatial distribution map of the occurrence of the extracted extrema locations of a time series of momentary maps (two extrema from each momentary map) is very similar to the spatial distribution map of spectral power (vs the average reference) computed from all available data points in time and space of the same series (Lehmann et al. 1987).

8 Adaptive Segmentation of Event-Related Potential Map Series

Let us first consider the case of event-related potential map series. There is agreement in the literature that the polarity of these potential distributions is functionally important and therefore needs to be considered by a segmentation based on map landscapes. The traditional recognition of components of event-related potential wafeforms (see Sect. 3) has also stressed the importance of polarity. "Spontaneous" brain electric activity, on the other hand, has largely been treated as a phenomenon in which periodic polarity reversal is an inherent characteristic, and absolute polarity has been found to play a minor role, if any at all; the space-based adaptive segmentation will have to consider this (see later). The case of event-related potential maps is thus more basic, and is therefore treated first.

When comparing two event-related potential maps for segmentation (Lehmann and Skrandies 1984; Brandeis and Lehmann 1986), the decision "similar vs nonsimilar" might use a hysteresis window, so that some deviation from perfect congruence is accepted within a steady state, taking into account the usual noise problem. Accordingly, when the two-extrema-extraction method is used for map assessment, a spatial window is erected around the positive and around the negative extremes of the first data time point (map) in the series, e.g., "plus one electrode distance." For each subsequent time point, the new locations of the extrema are checked; as soon as one or both of the extremes leaves its spatial window, the segment is terminated, and the next segment is started by resetting both spatial windows around the new extrema positions.

If the segmentation procedure has to treat the mean of map series from several subjects (or conditions), the extrema can first be extracted from each of the series, and the mean locations of the extrema over subjects (or conditions) can be used for segmentation. This gives the advantage that mean extrema locations can occur at locations where no electrodes were located, even though the analysis has only used actually measured data at the electrode locations; for these mean locations which are not restricted to electrode locations, spatial windows which are smaller than one electrode distance can be used (Brandeis and Lehmann 1986).

The size of the spatial window might be based on an optimization: The number of segments determined within a given analysis epoch (the maximal analysis epoch for averaged event-related potential map series is the interstimulus interval) will generally decrease with increasing size of the spatial window. However, this decrease in the number of determined segments is not linear; the number of segments tends to be constant for several steps of increased window size and to decrease with still larger windows, indicating a "natural" optimum of window size. Another more theoretical window size determinator might be based on the highest spatial frequency which is present in the data.

As alternative strategy for segmentation, the measure of Global Dissimilarity between Maps (Sec. 6) might be utilized. In a pilot study (Fig. 4 in Lehmann 1987), interesting results have been obtained which suggest preferred times of high dissimilarity in the ERP map sequence, regardless of stimulus condition, as if an inherent timing pattern invited changes of map landscape at certain intervals, and stability of the landscapes during other times.

9 Segmentation of ERP Map Series, Investigating Brain Mechanisms of Voluntary Attention and Figure Perception

Based on psychological data, attentional processes have been hypothesized to be decisively involved in visual perception (Treisman and Gelade 1980; Pritchard and Warm 1983). If this is so, figure stimuli and attended nonfigure stimuli should

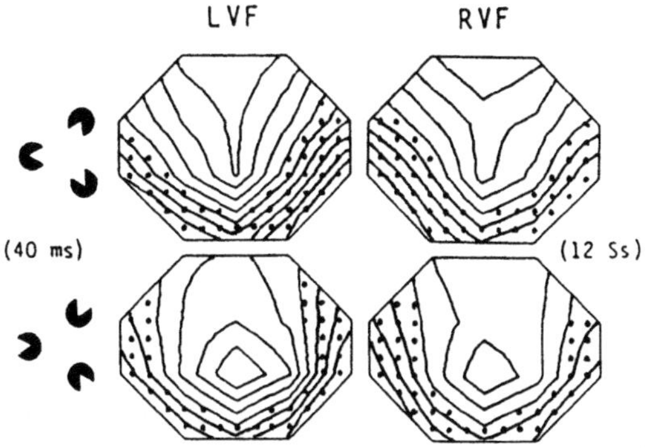

Fig. 1. Maps of mean event-related potential data over 12 subjects during the adaptively determined segment between 200 and 232 ms latency, after the visual 40-ms stimuli shown to the *left* (*upper maps:* stimulus was a KANIZSA-type illusionary figure, a triangle; *lower maps*: stimulus consisted of the same elements, but in a different combination which does not produce an illusionary figure). The event-related maps are clearly (and significantly) different for the two percepts, with maximal positivity more anterior for the triangle figure stimulus (*dotted*: negative, *white*: positive). They are also different for left (*LVF*) and right (*RVF*) visual hemifield stimuli, showing significant lateralizations of the mapped landscapes. (D. Braneis, Thesis Nr ETH 8187, Federal Institute of Technology, Zürich, 1986)

evoke similar electric map landscapes. This hypothesis has been supported by ERP map results (Brandeis and Lehmann 1986): From 168 to 376 ms after the stimulus presentations, and persisting during four segments of the map series, voluntary attention paid to visual stimuli was associated with a significant increase in the anterior-posterior distance between the locations of the maximum and minimum potential in the map landscapes (12 subjects). A similar increase in distance between the map extrema was observed if a visual illusory figure stimulus (of the Kanisza type) was presented instead of a closely related nonfigure stimulus (Fig. 1).

The results also show that attentional brain processes do not simply activate more of the same brain activity which processes the information without attention: Since the ERP map landscapes with and without attention are different, it must be concluded that at least partially different and thus additional neural populations must be activated by voluntary attention.

10 Adaptive Segmentation of Spontaneous EEG Map Series

Continuous recordings of EEG data as waveform time series are impressively dominated by periodic phenomena, known in the trade as "EEG waves." Viewed in the form of series of maps of the field distributions (Lehmann 1971, 1981; Lehmann et al. 1987), the periodicity appears as periodic polarity reversal of stable map landscapes during epochs of varying durations. (Incidentally, the map series do not show wavefronts or continuous propagations of major map features across space; the maps are more suggestive of volcano landscapes.)

The traditional view that a functionally stationary state of brain activity continues during spontaneous activity as long as the temporal, polarity–periodic pattern remains stationary might be extended to the spatial pattern (Lehmann and Skrandies 1980, 1984; Lehmann 1984). As long as a map landscape persists while periodically reversing its polarity, a constant population of neural elements might be assumed to be active in a periodic manner. Accordingly, the task for adaptive segmentation of the spontaneous map series then is to recognize epochs of stationarity of landscapes while disregarding the periodic polarity reversals. We have used the two-extrema-extraction method to characterize the landscapes for segmentation. In this case of spontaneous activity, as a first step, the maps at times of maximal global field power are selected so that only maps with a good signal-to-noise ratio are used. (For the so-called alpha activity of the human brain when recorded during relaxation, there are about 20 peak values of maximal global field power per second.) As a second step, the extrema locations are determined in the first selected map, and the spatial windows are set. The extrema locations of the subsequent maps are checked for possible violation of the window areas, disregarding the polarities. If a violation occurred, a new segment is started.

An alternative strategy for segmentation could use the measure of "Global Dissimilarity between Maps" (Sect. 6). In order to disregard polarity in spontaneous activity, the direct comparison between the successive maps and a comparison with one map of the pair reversed in polarity must be computed, and the smaller dissimilarity value is accepted. A running computation of this measure between successive pairs of maps of the series of maps selected at times of maximal global field power will permit identification of the times when the landscape changes, without considering local values separately.

11 Results of Segmentation of Spontaneous EEG Map Series

The two-extrema strategy discussed above was used to segment spontaneous EEG data recorded simultaneously from 16 electrodes during 2 min of relaxation with eyes closed in six healthy subjects (Lehmann et al. 1987). Segmentation was performed on digitally alpha-band filtered (FIR filter, 8–12 Hz temporal frequency) maps. Each segment was described by the two most frequent locations of the extrema voltages during the segment duration (see Fig. 2). On the average over subjects, the mean segment duration was 210 ms; segments longer than 323 ms covered 50% of the total time, and segments over 600 ms covered 18% of the total time. Occurrence frequency across segment classes varied strongly: on the average only 60 of the 120 theoretically possible classes were observed, and the segment class which covered the largest percentage (20%) of total analysis time therefore represented only about 1.5% of all segment classes.

The sequence of the different microstate segment classes did not appear to follow recognizable simple rules in a first examination, reminiscent of the linear states and nonlinear transitions discussed by Wright et al. (1985). Figure 2 illustrates the sequence problem in an example of successive segments during a relaxation period. A formal mathematical analysis of such EEG data (Fuchs et al.

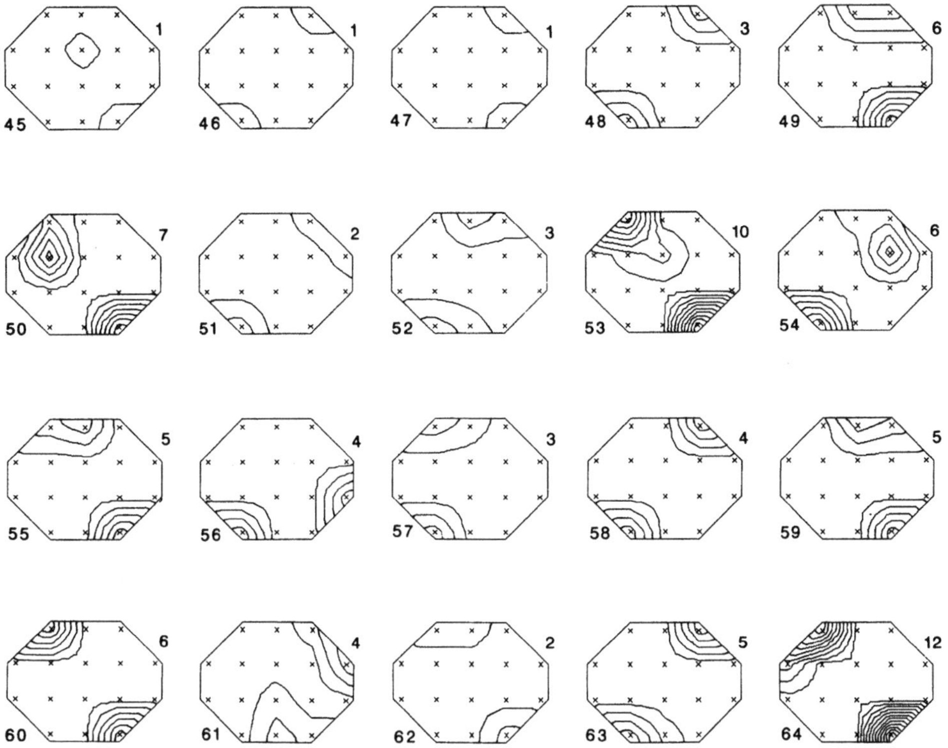

Fig. 2. Adaptively segmented epochs of stable map landscapes during spontaneous EEG of a normal volunteer while resting with closed eyes. Each map represents one spatially stable segment. The sequential segments covered a total time of 4.7 s recording time. Each map shows the accumulated occurrence locations of the voltage extrema during the segment's duration, indicated by isofrequency-of-occurrence lines. Head seen from above, left ear left. Sixteen-electrode recording; *crosses* show electrode sites. The figures in the *right upper corner of each map* multiplied by 50 give the duration of the segment; the *numbers* to the *lower left* are running segment numbers. Data were digitally FIR filter bandpassed to temporal 8–12 Hz. (Lehmann et al. 1987)

1987) in a pilot study along the lines of synergetics (Haken 1983) has shown that five independent variables describe the spatial pattern at each time step.

12 Functional Significance of Adaptively Determined Segments in Spontaneous EEG Data

In Sect. 4 we asked whether there are recognizable basic units of brain activity. This has now been answered positively as far as *formal recognition* of the stationary microstates and their quick, noncontinuous changes are concerned. Whether these formally distinguishable microstates are associated with *functional differences* needed further studies. We decided to investigate the possible dependence of a simple behavioral measure, selective reaction time, from the different microstates defined as EEG segment classes (Lehmann et al. 1987). Eight normal subjects were presented series of brief tones of constant pitch (1000 Hz) at regular intervals (800 ms). At random, a standard tone was replaced by a higher-pitch 1250-Hz "target" tone; overall, about 20% were target tones, and 337 were presented to each subject. The subjects were instructed to press a micro switch in response to each target tone. The segment class which existed immediately before the delivery of each target tone was established, and the reaction time determined. The reaction times differed significantly for different segment classes which accidentally existed at stimulus arrival (Brown-Forsythe ANOVA $P<0.0001$); moreover, the differences in reaction times associated with segment classes were similar over subjects for the 26 segment classes which occurred in all subjects (Friedman ANOVA $\chi^2=40.6$, $df=25$, $P<0.025$).

We conclude that the different brain momentary functional microstates as determined by the EEG landscape at the moment of stimulus arrival have a significant relation with the simple behavioral measure of selective reaction time which results from the stimulus.

A still more ambitious goal is to determine the expected relations between momentary brain electric microstates and types of ongoing thoughts, i.e., types of subjective, private experiences which are reported by the subject. Preliminary results were encouraging (Lehmann and Koukkou 1984), and are presently being followed up.

13 Conclusions

Biologic organisms generally function in a state-dependent manner. More specifically, the global functional state of the brain influences processing, storage, and retrieval of information. The state is continuously readjusted in response to the nature of the momentarily treated informations, but the adjustments are constrained in a recursive fashion by the momentary state.

This paper concentrated on the moment-to-moment variations of functional state, in conjunction with examinations of the brain electric field, which is a sensitive, high-resolution indicator of the state. On formal grounds, the presented results show that stable microstates can be identified in the time range of subseconds, and that these states change in a noncontinuous manner. The microstates are conceived to be superimposed on the well-known gross brain states such as wakefulness and sleep. The reaction time measurements have supported the hypothesis that these global, brief, stationary microstates are directly relevant to behavior. The observed durations of spatially stable microstates cover a range which includes Libet's (1982) minimal time for conscious experience (500 ms) and other time measurements related to mechanisms of conscious experiences (e.g., 400 ms for switching of attention: Reeves and Sperling 1986; see also Michael and Turvey 1979). Spatially stable electric states of even shorter durations were found in analyses of event-related brain potential map series. It is attractive to think of the brain microstates as basic building blocks (steps, modes) of higher information processing, which qualify for access to consciousness only beyond a certain duration.

The state dependency of processing (and recall/storage) of information provides to the brain a multilayered analysis repertoire which allows for a wide array of strategies in dealing with new information, or in dealing with the end result of processing of information. The momentary state is determined not only by the various physiological parameters and momentary motivations, but indirectly also by earlier experiences which had resulted in the construction of long-term goals and by immediately preceding contextual experiences which contribute to the assessment of feasibility of pursuing long- or short-term goals.

14 Summary

Brain functions (processing strategy and mode, speed, access to different portions of memory storage, conscious experience) are viewed as constrained by and defining the momentary global brain functional state. The functional state is manifest as the spatiotemporal pattern of the brain electric field. The building blocks of brain work are hypothesized to be short-lasting epoches (of varying durations) of functional microstates which are characterized by stationary or periodically stationary patterns of the spatial distribution of the brain electric potential field (brain landscape maps). Adaptive segmentation of ongoing human brain electric activity into segments of stationary landscapes yielded segments in the subsecond range. Different classes of segments were associated with different behavioral characteristics. Adaptive segmentation of information-induced sequences of brain electric activity identified even shorter segments where automatic perception of visual figure and voluntary attention were associated with partially similar map landscapes, indicative of related brain mechanisms for the two functions. The conceivable role of brain functional microstates as basic building blocks of brain work and, in particular, of conscious experiences is discussed.

Acknowledgments. Partial financial support for the research work discussed in this paper was received from the Swiss National Research Foundation, the EMDO Foundation, and the Hartmann-Müller Foundation.

References

Ashby WR (1960) Design for a brain. Wiley, London

Barlow JS (1985) Methods of analysis of nonstationary EEGs, with emphasis on segmentation techniques: a comparative review. J Clin Neurophysiol 2:267–304

Başar E (1980) EEG-brain dynamics. Elsevier, Amsterdam

Berkhout J, Walter DO, Adey WR (1969) Alterations of human electroencephalogram induced by stressful verbal activity. Electroencephalogr Clin Neurophysiol 27:457–469

Bertrand O, Perrin F, Pernier J (1985) A theoretical justification of the average reference in topographic evoked potential studies. Electroencephalogr Clin Neurophysiol 62:462–464

Bohdanecky Z, Bozkov V, Radil T (1984) Acoustic stimulus threshold related to EEG alpha and non-alpha epochs. Int J Psychophysiol 2:63–66

Brandeis D, Lehmann D (1986) Event-related potentials of the brain and cognitive processes: approaches and applications. Neuropsychologia 24:151–168

Callaway E (1962) Factors influencing the relationship between alpha activity and visual reaction time. Electroencephalogr Clin Neurophysiol 14:674–682
Ehrlichman H, Wiener MS (1980) EEG asymmetry during covert mental activity. Psychophysiology 17:228–235
Eich JE (1980) The cue-dependent nature of state-dependent retrieval. Memory and Cognition 8:157–173
Fuchs A, Friedrich R, Haken H, Lehmann D (1987) Spatio-temporal analysis of multi-channel alpha EEG map series. In: Haken H (ed) Computational systems – natural and artificial. Springer, Berlin Heidelberg New York, pp 74–83
Gath I, Lehmann D, Bar-On E (1983) Fuzzy clustering of EEG signal and vigilance performance. Int J Neurosci 20:303–312
Gath I, Bar-On E, Lehmann D (1985) Automatic classification of visual evoked responses. Comput Methods Programs Biomed 20:17–22
Grandori F (1984) Dipole localization methods (DLM) and auditory evoked brainstem potentials. Rev Laryngol Otol Rhinol 105 [Suppl]:171–178
Haken H (1983) Advanced synergetics. Springer, Berlin Heidelberg New York
Katada A, Ozaki H, Suzuki H, Suhara K (1981) Developmental characteristics of normal and mentally retarded children's EEG. Electroencephalogr Clin Neurophysiol 52:192–201
Keesey UT, Nichols DJ (1967) Fluctuations in target visibility as related to the occurrence of the alpha component of the encephalogram. Vision Res 7:859–879
Koukkou M (1982) EEG states of the brain, information processing and schizophrenic primary symptoms. Psychiatry Res 6:235–244
Koukkou M, Lehmann D (1976) Human EEG spectra before and during cannabis hallucinations. Biol Psychiatry 11:663–677
Koukkou M, Lehmann D (1983) Dreaming: the functional state-shift hypothesis. A neuropsychophysiological model. Br J Psychiatry 142:221–231
Koukkou M, Lehmann D, Angst J (1980) Functional states of the brain: their determinants. Elsevier, Amsterdam
Lehmann D (1971) Multichannel topography of human alpha EEG fields. Electroencephalogr Clin Neurophysiol 31:439–449
Lehmann D (1972) Human scalp EEG fields: evoked, alpha, sleep and spike wave patterns. In: Petsche HH, Brazier MAB (eds) Synchronization of EEG activity in epilepsies. Springer, Vienna New York, pp 307–325
Lehmann D (1981) Spatial analysis of evoked and spontaneous EEG potentials. In: Yamaguchi N, Fujisawa (eds) Recent advances in EEG and EMG data processing. Elsevier, Amsterdam, pp 117–132
Lehmann D (1984) EEG assessment of brain activity: spatial aspects, segmentation and imaging. Int J Psychophysiol 1:267–276
Lehmann D (1987) Mapping and analysing maps of evoked potentials. In: Barber C, Blum T (eds) Evoked potentials III. Butterworths, London, pp 91–99
Lehmann D, Koukkou M (1984) Physiological and mental processes during sleep: a model of dreaming. In: Bosinelli M, Cicogna P (eds) Psychology of dreaming. CLUEB, Bologna, pp 51–63
Lehmann D, Skrandies W (1980) Reference-free identification of components of checkerboard-evoked multichannel potential fields. Electroencephalogr Clin Neurophysiol 48:609–621
Lehmann D, Skrandies W (1984) Spatial analysis of evoked potentials in man – a review. Prog Neurobiol 23:227–250
Lehmann D, Beeler GW, Fender DH (1965) Changes in patterns of the human electroencephalogram during fluctuations of perception of stabilized retinal images. Electroencephalogr Clin Neurophysiol 19:336–343
Lehmann D, Meier B, Meier CA, Mita T, Skrandies W (1983) Sleep onset mentation related to short epoch EEG spectra. Sleep Res 12:180
Lehmann D, Ozaki H, Pal I (1987) EEG alpha map series: brain microstates by space-oriented adaptive segmentation. Electroencephalogr Clin Neurophysiol 67:271–288
Libet B (1982) Brain stimulation in the study of neuronal functions for conscious experience. Hum Neurobiol 1:235–242

Lopes da Silva FH (1981) Analysis of EEG ongoing activity: rhythms and nonstationarities. In: Yamaguchi N, Fujisawa (eds) Recent advances in EEG and EMG data processing. Elsevier, Amsterdam, pp 95–115
Michaels CF, Turvey MT (1979) Central sources of visual masking: indexing structures supporting seeing at a single, brief glance. Psychol Res 41:1–61
Ohman A (1979) The orienting response, attention, and learning. In: Kimmel HD, von Olst EH, Orlebeke JF (eds) The orienting reflex in humans. Lawrence Erlbaum, Hillsdale, pp 443–472
Overton DA (1978) Major theories of state-dependent learning. In: Ho BT, Richards DW, Chute DL (eds) Drug discrimination and state-dependent learning. Academic, New York
Pritchard WS, Warm JS (1983) Attentional processing and the subjective contour illusion. J Exp Psychol Gen 112:145–175
Reeves A, Sperling G (1986) Attention gaiting in short-term visual memory. Psychol Rev 93:180–206
Skarda CA, Freeman WJ (1987) How brains make chaos in order to make sense of the world. Behav Brain Sci 10:161–173
Skrandies W, Lehmann D (1982) Spatial principal components of multichannel maps evoked by lateral visual half-field stimuli. Electroencephalogr Clin Neurophysiol 54:297–305
Treisman AM, Gelade G (1980) A feature-integrating theory of attention. Cogn Psychol 12:97–136
Williams HL, Morlock HC, Morlock JJ (1966) Instrumental behavior during spleep. Psychophysiology 2:208–215
Woodworth RS, Schlosberg H (1954) Experimental psychology. Holt, New York
Wright JJ, Kydd RR, Lees GL (1985) State-changes in the brain viewed as linear steady states and non-linear transitions between steady states. Biol Cybern 53:11–17

Spatiotemporal Aspects of Synergetic Processes in the Auditory Cortex as Revealed by the Magnetoencephalogram

M. Hoke, K. Lehnertz, C. Pantev, and B. Lütkenhöner

1 Introduction

Synergetics is a discipline concerned with the cooperation of individual parts of a system that produces spatial, temporal, or functional structures (see, e.g., Haken 1977, 1980a, b). The macroscopic behavior of such a system can change dramatically when certain external parameters are varied. The classical physical example of a synergetic system is the laser, which exhibits quite a number of ordering effects. It was Haken (1977) and Başar (1983) who emphasized the similarity between the behavior of an excited neuronal population and that of a laser when it is pumped: The external stimulus pushes the spontaneous electric activity of the brain – which can be regarded as a set of weakly coupled oscillators – into the coherent state of a single harmonic oscillator, ringing with an intrinsic frequency, for a certain coherence time ("coherence length"), after which fluctuations set in again. Various brain structures were found to have different intrinsic frequencies (Başar 1983). The higher the entropy of the system, i.e., the more independently the individual neurons are firing, the higher is the "excitability" or "response susceptibility" of the neuronal population and vice versa: During a regular firing (state of low entropy), either maintained by stimulation or occurring spontaneously, the neuronal population is less or not at all responsive to a (second) stimulus.

Başar (1980) quantified the effect of the exciting stimulus in terms of an "enhancement factor" which is derived from the filtered time-domain waveform of an individual epoch ("EPogram") containing prestimulus EEG activity and evoked potential (EP). This measure, however, does not allow us to decide whether the EP is just an effect of synchronizing the discharges of the neuronal population with the signal energy remaining constant, or whether it contains additive energy, i.e., whether more neurons are recruited into the active population. Sayers et al. (1974, 1979; Sayers and Beagley 1974) concluded from their experiments that EPs are due to a stimulus-induced reordering of the phases in each realization of the background EEG rather than to the evoking of a response which is additive to the background EEG. In their later studies, however, they observed additive energy at high stimulus levels (Beagley et al. 1979), which is in accordance with the results of Jervis et al. (1983), who found that all of the EPs studied contained additive energy in at least one Fourier component.

All the investigations mentioned above have in common that only the EEG recorded from one single position is considered, i.e., spatial aspects have been neglected. Furthermore, temporal aspects of synergetic processes of the brain have

Springer Series in Brain Dynamics 2
Edited by Erol Başar and T. H. Bullock

not been studied so far (except for the comparison of the pre- and perstimulatory period of the EEG epochs). In the present paper, both spatial and temporal aspects of synergetic processes triggered by auditory stimuli are considered. For that purpose magnetoencephalographic (MEG) measurements were performed which permit a considerably higher spatial resolution than do electroencephalographic (EEG) recordings. This is due to the fact that a current dipole located inside a conductive body of axial symmetry produces a magnetic field whose tangential component, measurable at the surface of the body, is almost exclusively generated by "impressed" currents, whereas the electric potential measurable at the surface of the body is brought about by volume currents (e.g., Kaufman et al. 1981). With respect to brain magnetic activity, the impressed current forming a current dipole is represented by the intracellular current flowing within excited dendrites. Consequently, neuromagnetic measurements exhibit a much higher spatial resolution than, and thus are superior to, electric potential measurements, which implies that they are an excellent means for studying selectively, over the respective projection areas, cortical excitation processes occurring either spontaneously or in response to modality-specific stimulation.

2 Stimulation, Data Collection, and Processing

Two normally hearing subjects with no case history in hearing disorders as well as three patients with different pathologies participated in this study. Measurements and full processing of data were repeated in one normal individual to test whether the results are reproducible. In this article, only data of that particular subject will be presented.

Auditory stimuli were tonebursts with a carrier frequency of 1000 Hz, a duration of 500 ms, and rise/decay times of 15 ms each; they were presented, with an intensity of 60 dB HL and an interstimulus interval of 4 s, to the ear contralateral to the hemisphere investigated. Measurements were carried out in an acoustically and electrically but not magnetically shielded room. The magnetic field component normal to the scalp was measured at 60–70 positions ("sampling positions") over the supralateral surface of the scalp, given by a rectangular sampling matrix (7 rows × 10 columns, 15 mm spacing) with its reference in T3 or T4 (Pantev et al. 1986a, b, 1988). The neuromagnetic measurements were done sequentially in random order using a single-channel DC SQUID (BTi) equipped with a second-order gradiometer. At least 128 individual stimulus-related MEG epochs (prestimulus interval 240 ms) were recorded from each sampling position (sampling frequency 250 Hz, bandwidth 0.1–40 Hz) and stored in a Prime computer system for off-line analysis.

The raw data were subjected to a comprehensive statistical analysis which consisted of both a time-domain analysis and a frequency-domain analysis (short-time spectral analysis) of the stimulus-related epochs. Only data of a few individuals could be processed in their entirety because the complete processing of data required an extremely long time (it took almost 1 month to complete all computations and to generate all plots for just one subject since computer time was available during night hours only).

2.1 Time-Domain Analysis: Magnetic Field Distribution

The individual epochs were averaged, and matrices comprising the data from all sampling positions were compiled for each sampling instant (i.e., every 4 ms) between −80 and 400 ms relative to stimulus onset. The matrices were interpolated into 64 × 32 matrices to allow for two-dimensional digital filtering using FFT (Fast Fourier Transform) algorithms. Finally isocontour plots of the magnetic field distribution were generated. A gray scale served to encode magnetic field strength and direction of the magnetic flux.

The parameters of an equivalent current dipole (moment, direction, and location in space) were calculated by approximating the distribution of a single dipole located in a semi-infinite volume conductor to the observed field distribution (Pantev et al. 1986a).

2.2 Short-Time Spectral Analysis

To study time-dependent aspects of evoked cortical activity, a method described by Schafer and Rabiner (1973) for short-time speech analysis and synthesis was adopted [which was also used by Hoke et al. (1984) for a time- and intensity-dependent filtering of brain-stem auditory evoked potentials]. The method is characterized by stepwise shifting a Hanning window over the individual epochs and performing a Fourier analysis of each window. The limitations of this procedure are given by the uncertainty principle: Spectral and temporal resolution are interrelated; a higher temporal resolution (shorter window) decreases the spectral resolution and vice versa. The selected window length of 160 ms turned out to be an acceptable compromise (Fig. 1). Of the total power, 85.4% comes from the interval ±40 ms of the center time of the window while the leading and trailing edges

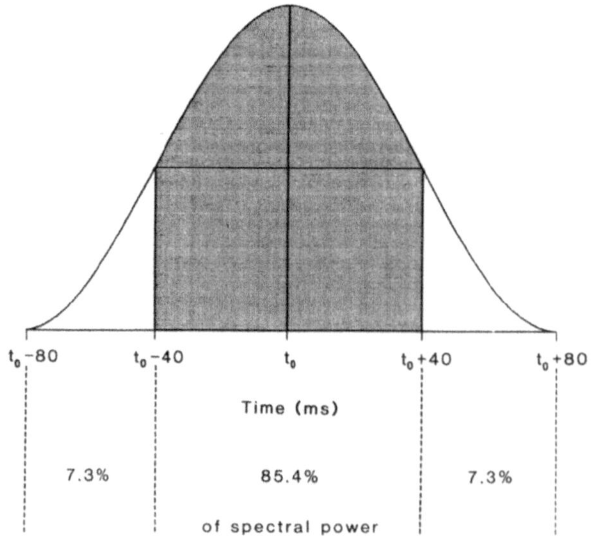

Fig. 1. Hanning window used for the short-time Fourier analysis. The spectral parameters calculated for each window are related to the center time of the window (t_0)

contribute only 7.3% each. That is, if the window is centered, for example, over the maximum of wave M 100, there will be only a negligible contribution of wave M 200 since both peaks are approximately 80 ms apart. (The waves of the time-domain waveform of the evoked magnetic field are referred to as M 50, M 100, M 200, and M 250, according to their approximate latencies.) To allow for an FFT, the window function (40 samples) was padded with zeroes to obtain a 64-sample epoch. The spectral resolution hence amounts to 3.906 Hz. In view of the signal bandwidth, only the lower six Fourier components were processed (3.91–23.44 Hz).

Data processing for one sampling position is schematically shown in Fig. 2. The first step consisted in calculating, after windowing and FFT, the individual amplitudes $a_{i,j,H}$ and phases $\theta_{i,j,H}$ of the lower six Fourier components for each epoch i, where j is the number of sampling points by which the window was shifted, and H the ordinal number of the Fourier component. The window shift introduces a phase shift $\Delta\theta_{j,H}$ which is

$$\Delta\theta_{j,H}=2\pi Hj/k \tag{1}$$

with k being the number of samples of the window (64). By correcting the individual phases for that shift they were related to the original MEG epoch. From the individual amplitudes and phases of each Fourier component, obtained for the whole ensemble of n epochs, the mean amplitude $A_{j,H}$ was calculated according to

$$A_{j,H}=\left[\frac{1}{n}\sum_{i=1}^{n} a_{i,j,H}^{2}\right]^{1/2} . \tag{2}$$

Furthermore, the mean phases $\bar{\theta}_{j,H}$ and the interepochal phase coherences $R_{j,H}$ were calculated according to

$$\bar{\theta}_{j,H}=\operatorname{atan}(\bar{S}_{j,H}/\bar{C}_{j,H}) \tag{3}$$

$$R_{j,H}=(\bar{C}_{j,H}^{2}+\bar{S}_{j,H}^{2})^{1/2} \tag{4}$$

with

$$\bar{C}_{j,H}=\frac{1}{n}\sum_{i=1}^{n}\cos\theta_{i,j,H} \tag{5}$$

$$\bar{S}_{j,H}=\frac{1}{n}\sum_{i=1}^{n}\sin\theta_{i,j,H} \tag{6}$$

(cf. Mardia 1972).

The interpochal phase coherence R is the complement to 1 of the circular variance S, i.e., $R=1$ (resp. $S=0$) if the phases are identical in all epochs,while $R=0$ (resp. $S=1$) if the phases are uniformly distributed, i.e., the epochs do not contain any coherent signal but only noise.

While shifting the window over the individual epochs in steps of one sampling point, these calculations were repeated for each window position. From the resulting values (mean amplitude, mean phase, and interepochal phase coherence for each Fourier component), time functions were composed by associating these

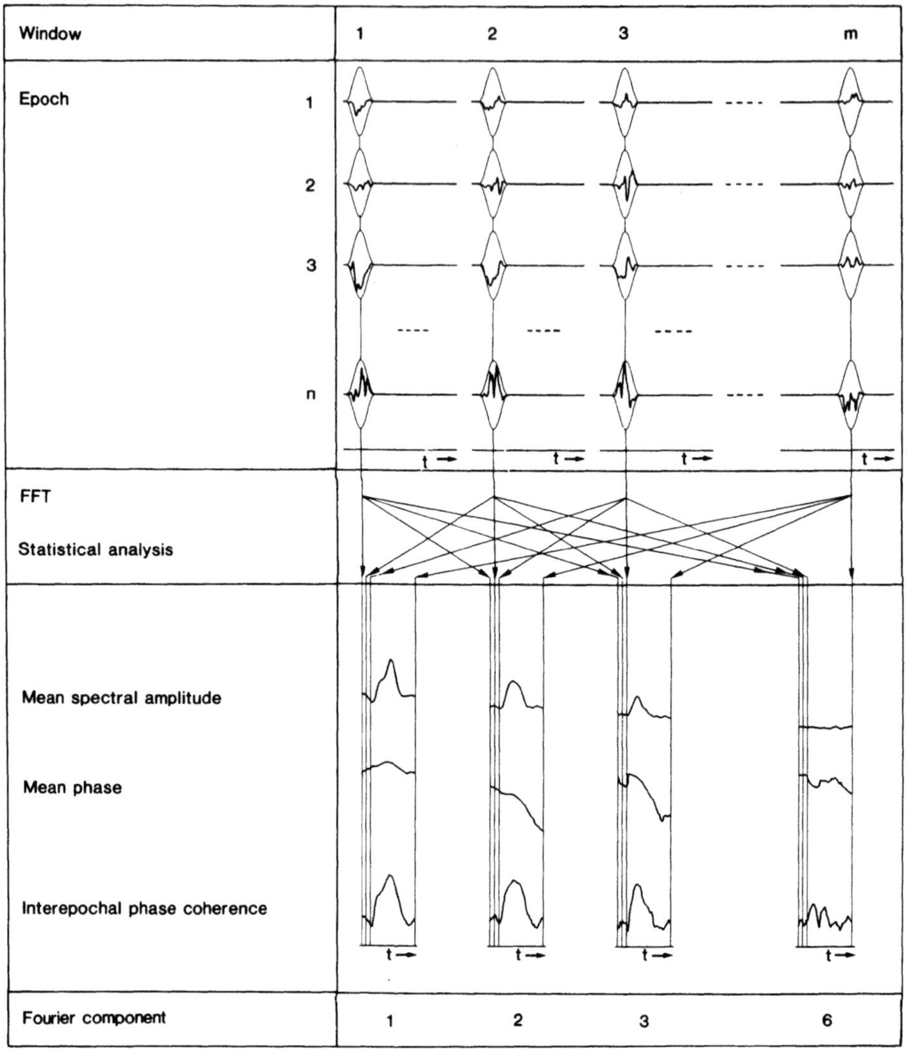

Fig. 2. Principles of data processing for one sampling position. The Hanning window is shifted stepwise (point by point) over all (*n*) individual epochs (*top*). From the values resulting from the statistical analysis (mean amplitude, mean phase, and interepochal phase coherence of the lower six Fourier components), time functions are composed (*bottom*)

values with the center times of the windows. Because six Fourier components were considered, a total of $3 \times 6 = 18$ additional time functions were obtained in this way for each sampling position. Similar to the procedure applied to the averaged waveforms of the magnetic field (time-domain analysis), matrices were compiled from the time functions of all sampling positions, for each window center time between -80 and 400 ms relative to simulus onset. Finally, after interpolation into a 64×32 matrix, isocontour plots were generated. By this means it became possible to compare the instantaneous magnetic field distribution with those of the statistical parameters of each Fourier component. Similar to the generation of isofield maps, a gray scale was used to encode spectral amplitude and interepochal phase coherence, but a different pattern had to be used to encode spectral phase since it represents a circular variable.

2.3 Redefinition of Phases

Before computing the phase maps it was necessary to redefine the mean phases in each of the matrices since otherwise rapid phase changes were faked whenever the mean phase, at two adjacent sampling positions, shifted across the beginning or end of the (arbitrarily) defined phase range spanning from $-\pi$ to π. To avoid those ostensible phase jumps, a straightforward procedure for a redefinition of phases was developed. It is based on the assumption that, given a dipolar source and a significant interepochal phase coherence, there should exist two major areas with opposite phase, located on both sides of the dipole projection onto the scalp. That this assumption is justified is shown in Fig. 3, in which we plotted, for one

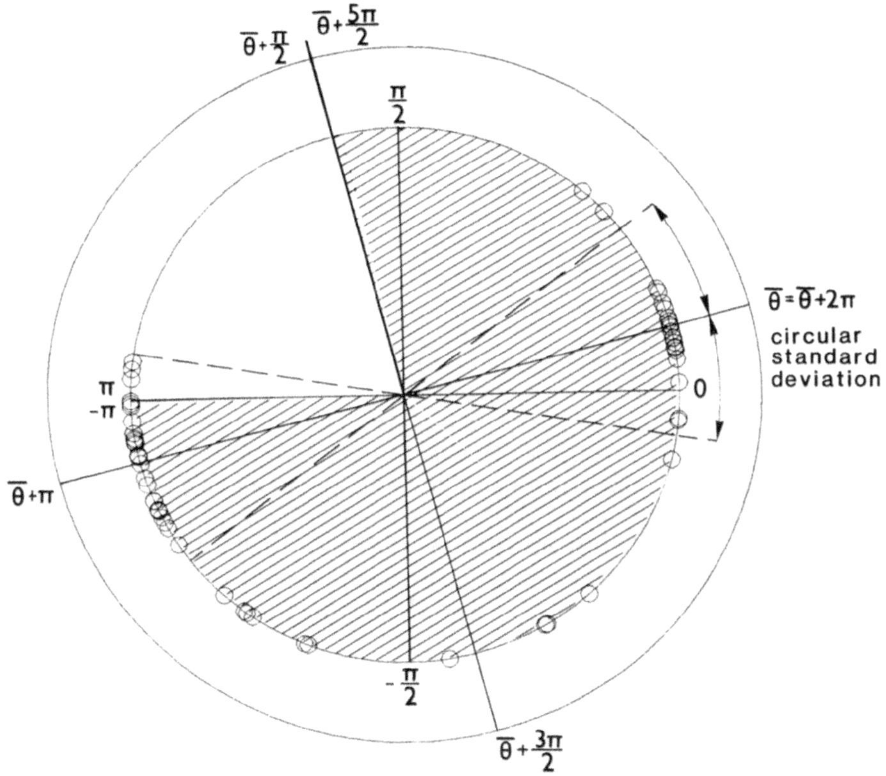

Fig. 3. Bimodal distribution of mean phases of one Fourier component from those sampling positions at which the interepochal phase coherence was significant. $\bar{\theta}$ and $\bar{\theta}+\pi$, "grand mean phases" of both modes; *dotted lines*, range of circular standard deviation. The (mod 2π) discontinuity was placed at $\bar{\theta}+\pi/2$. *Hatched area*, phases to be corrected by adding 2π

specified Fourier component and one given window, the mean phases of all those sampling positions at which the interepochal phase coherence was significant. The mean phases exhibit a clear-cut bimodal distribution, with the two maxima being π radians apart.

The redefined phase matrices were interpolated into a 64×32 matrix, as was done with all the other matrices, and isocontour lines were computed. Areas of equal phases were filled with a hatch pattern rather than with a tone pattern, with the angle of the hatch lines corresponding to half that of the mean phase. This allowed the viewer to recognize at a glance larger areas with opposite polarity (phase difference of π radians) by their hatch lines being perpendicular to each other.

A bimodal distribution with modes π radians apart can easily be transformed into a unimodal distribution (Mardia 1972) by

$$\theta^* = \theta (\mathrm{mod}\ \pi) \tag{7}$$

If $f(\theta)$ is the probability density function for θ, then the probability density function for θ^* is

$$h(\theta^*) = f(\theta^*) + f(\theta^* + \pi), \quad 0 < \theta^* < \pi . \tag{8}$$

Since θ^* is in the range $[0, \pi]$, it is easy to calculate the circular variance and the "grand mean phase" $\bar{\theta}^*$ from which the "grand mean phases" of the two modes of $f(\theta)$ are obtained as $\bar{\theta} = \theta^*$ and $\bar{\theta} = \theta^* + \pi$. The phases can then be redefined by placing the 2π discontinuity exactly in the middle between the "grand mean phases" of both modes, namely in that segment on which fewer phase values were located, i.e., $\bar{\theta} + \pi/2$ or $\bar{\theta} + 3\pi/2$.

3 Results

Figures 4 and 5 give, for all sampling positions and two selected Fourier components, a synoptic view of the time functions of the averaged magnetic field (EF), the mean amplitude (MA), the mean phase (MP), and the interepochal phase coherence (IPC). The EF – which is identical in both figures – reflects the typical field distribution with a polarity reversal to both sides of an imagined projection of an equivalent current dipole (ECD) onto the scalp. The distribution of the MA of the second Fourier component (Fig. 4) is basically similar to that of the EF (with the exception that there cannot exist a polarity reversal): positions of higher field amplitudes also exhibit higher spectral amplitudes. The MA assumes its maximum at approximately 160 ms, i.e., around the peak of wave M200. The IPC functions are basically unimodal functions. The distribution of the IPC over the scalp is much more extended: Even in marginal positions where no increase of spectral energy can be found there is still a significant IPC. The MP is a continuous function as long as the IPC is significant ("coherence time" according to Başar 1983). The coherence length can be as long as some 150 ms.

The findings obtained for the fifth Fourier component (Fig. 5) are different. A significant increase of the mean amplitude can be observed at only a very few

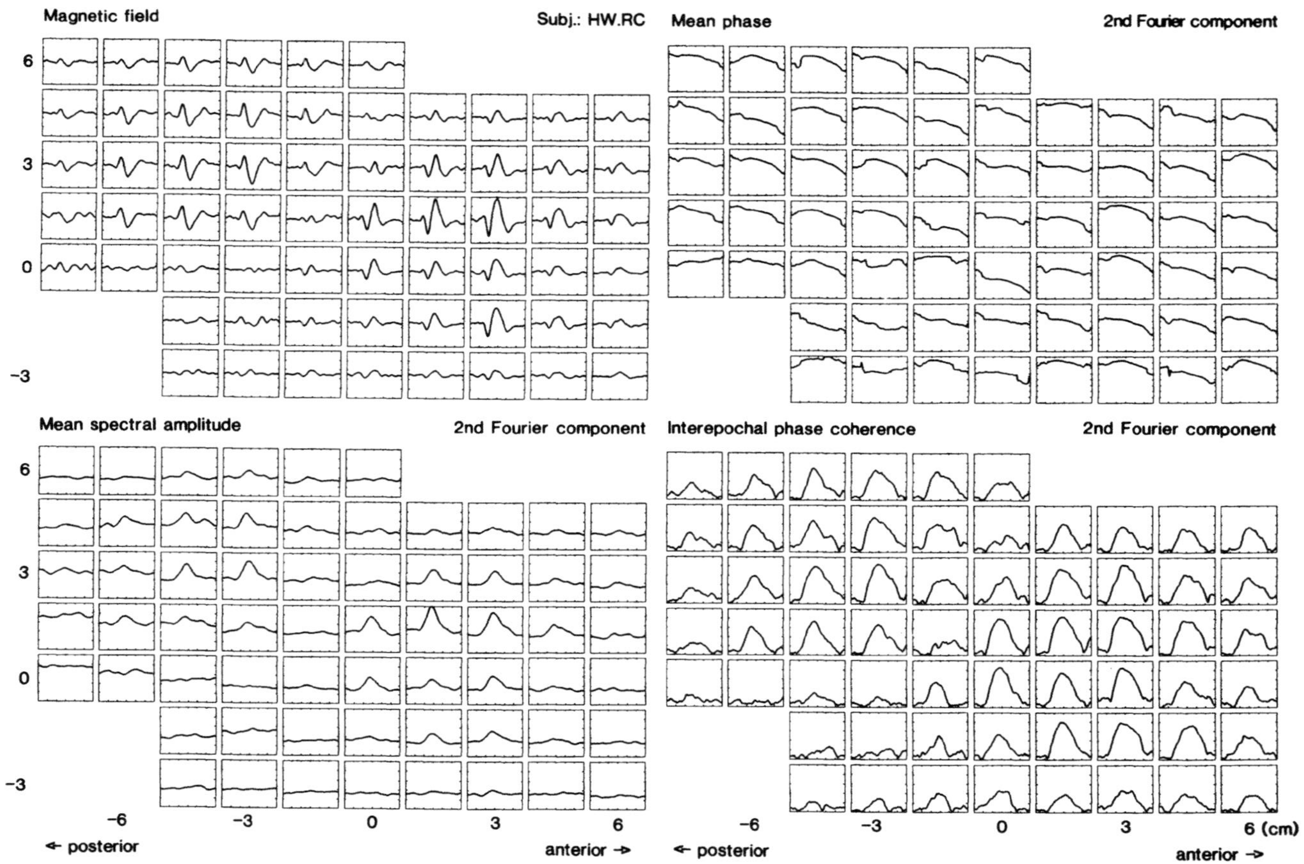

Fig. 4. Synoptic view of all time functions computed for all sampling positions: averaged magnetic field (*top left*) as well as mean amplitude (*bottom left*), mean phase (*top right*), and interepochal phase coherence (*bottom right*) of the second Fourier component. Scaling: time (*horizontal axis*) −80–400 ms relative to stimulus onset; magnetic field: full scale = 1200 fT; mean spectral amplitude: full scale = 550 fT; mean phase: π radians/division; interepochal phase coherence: full scale = 1

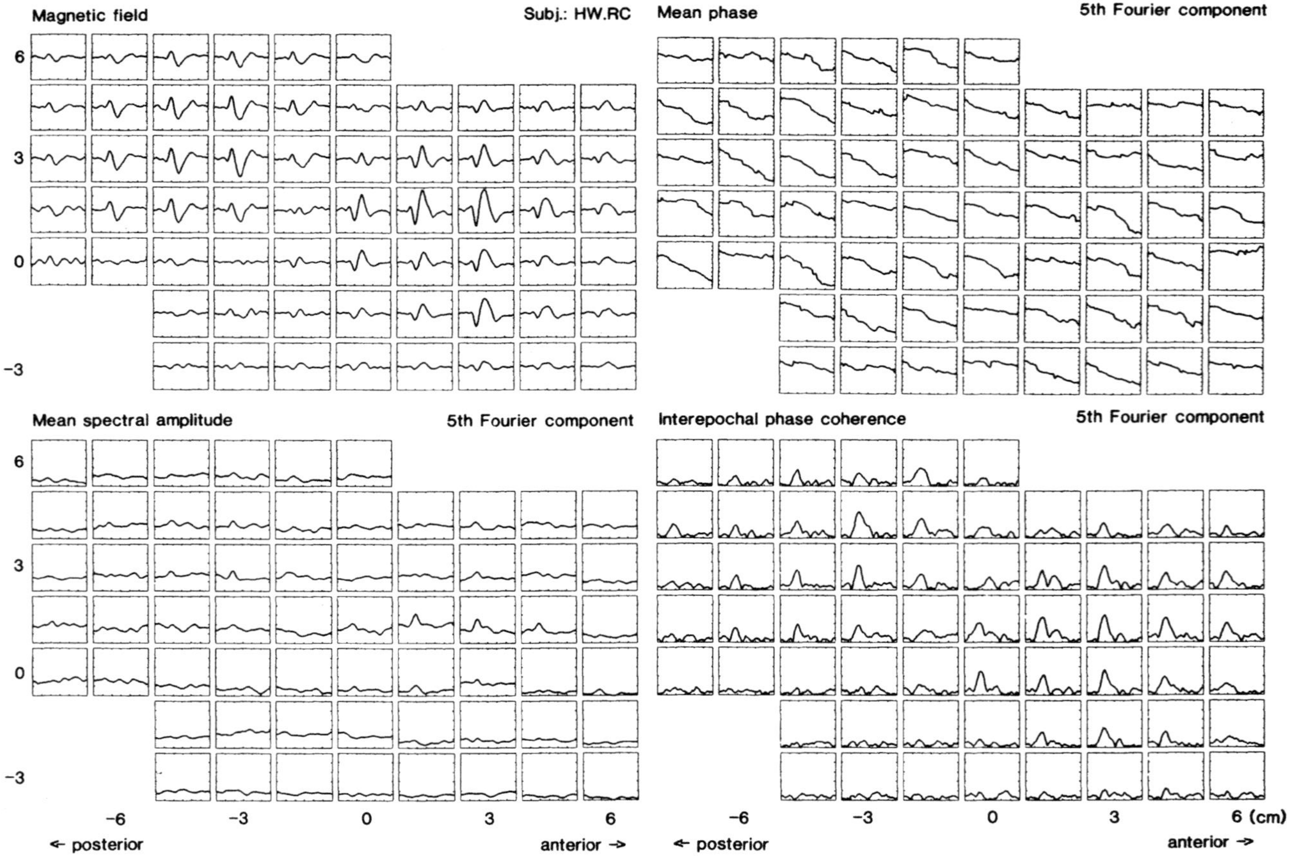

Fig. 5. Synoptic view of all time functions computed for all sampling positions: averaged magnetic field (*top left*) as well as mean amplitude (*bottom left*), mean phase (*top right*), and interepochal phase coherence (*bottom right*) of the fifth Fourier component. Scaling: time (*horizontal axis*) −80–400 ms relative to stimulus onset; magnetic field: full scale = 1200 fT; mean spectral amplitude: full scale = 300 fT; mean phase: π radians/division; interepochal phase coherence: full scale = 1

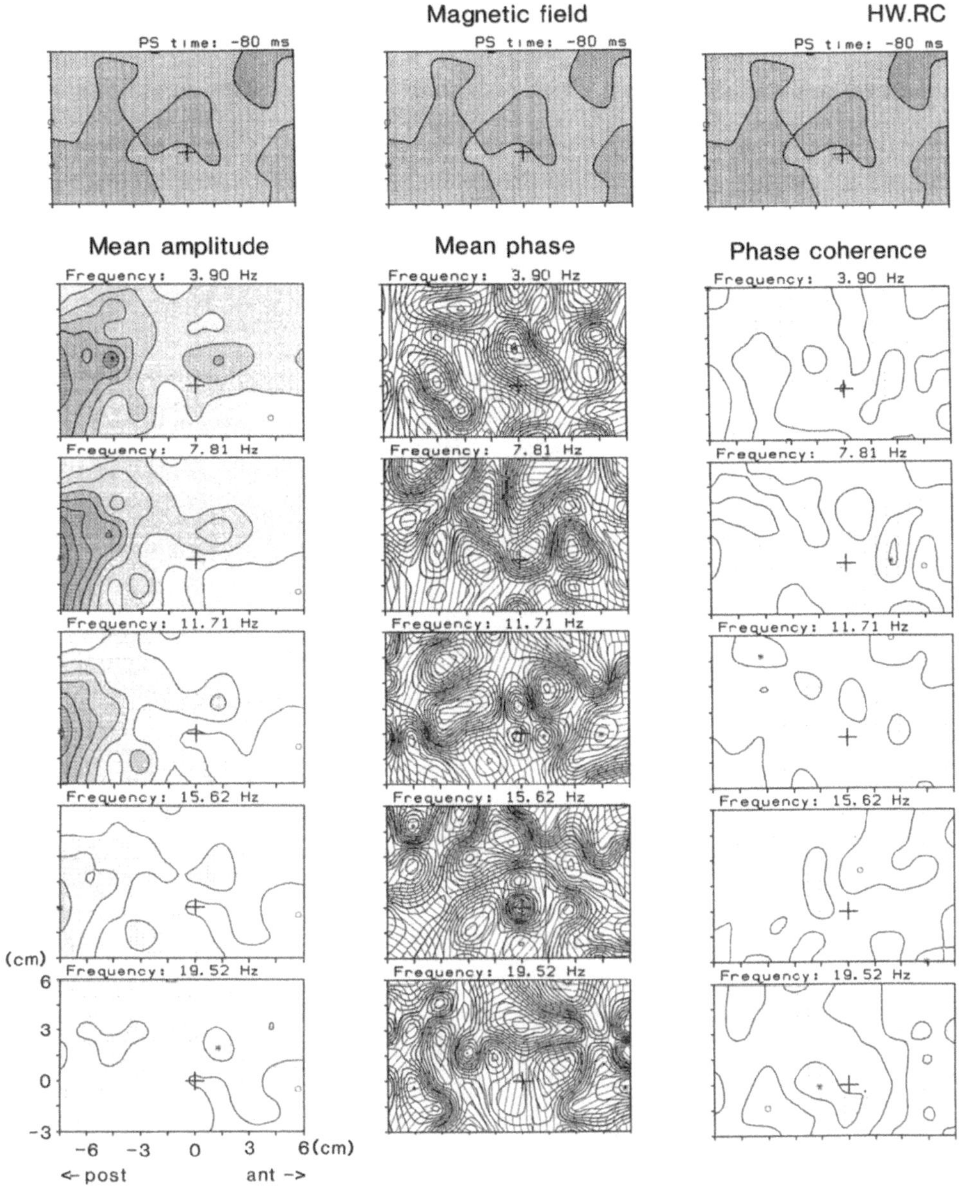

Fig. 6. Instantaneous distributions, 80 ms prior to stimulus onset, of the magnetic field (*top of each column*) and of mean amplitude (*left column*), mean phase (*middle column*) and interepochal phase coherence (*right column*) of the first to the fifth Fourier components. Spacing between iso-contour lines: magnetic field, 40 fT; mean amplitude, 30 fT; mean phase, $\pi/9$ radians (20 degrees); interepochal phase coherence, 0.05

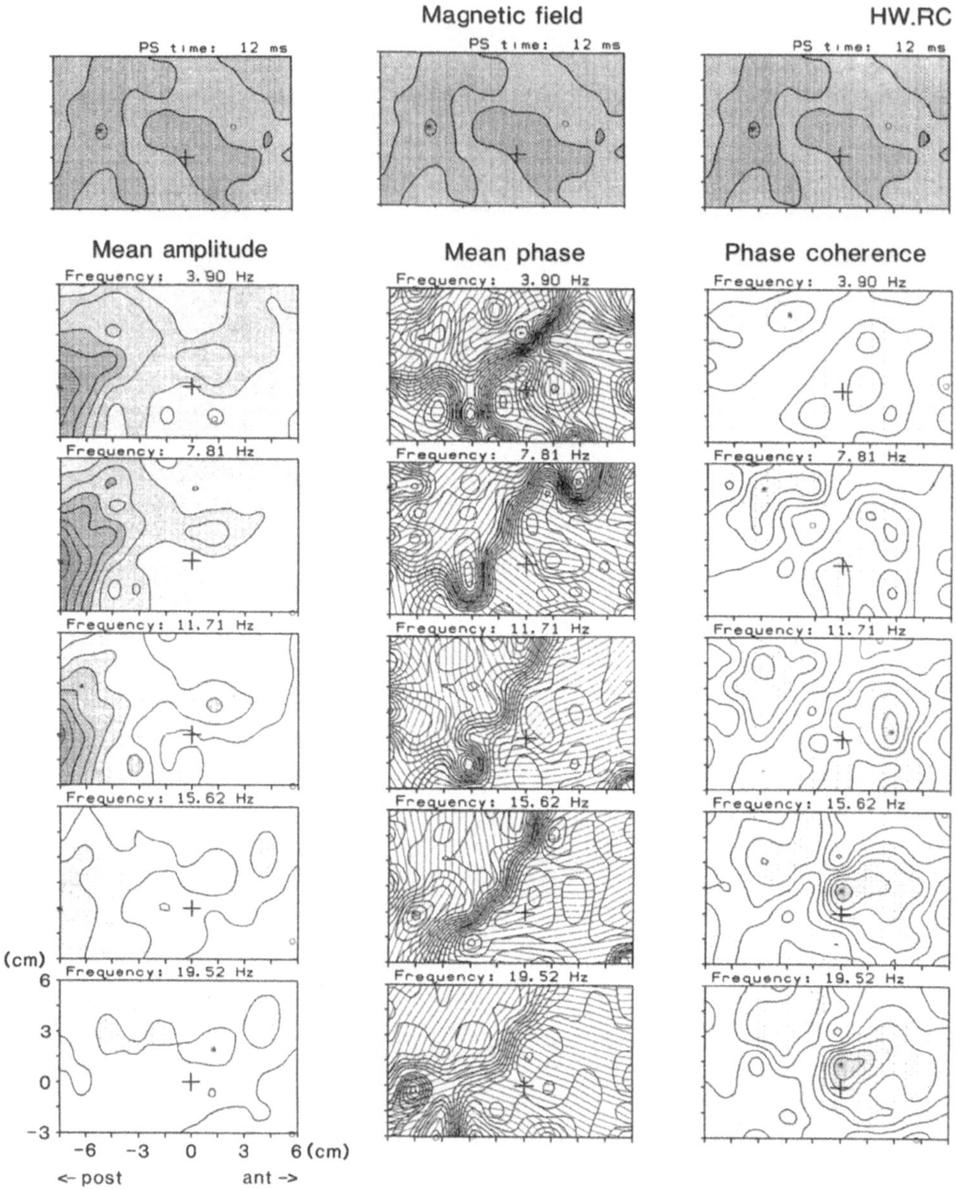

Fig. 7. As Fig. 6, but 12 ms after stimulus onset. See legend to Fig. 6 for explanation

sampling positions. The amplitude maxima now occur at approximately 80 ms (wave M100), while no significant energy seems to be present around the peak of wave M200. The time course of the IPC is also different: There is basically a bimodal waveform whose first peak coincides with the peak in the MA functions (wave M100). As expected, the coherence length in the phase functions is reduced;

there may occur two successive periods of coherence with a phase jump between the two at positions where the IPC function is bimodal.

Figures 6–12 allow an exact comparison of the EF distribution with that of the quantities MA, MP, and IPC of the lower five Fourier components at specified instants. Eighty milliseconds before stimulus onset (Fig. 6), the distributions of the EF, the MP, and the IPC are merely stochastic patterns. Only the MA distributions of the four lowest Fourier components exhibit an accumulation of spectral energy at the posterior edge of the scanned area of the scalp (which remains invariant in the perstimulatory period).

First signs of a stimulus-induced organization are obvious as early as 12 ms after stimulus onset (Fig. 7). It is to be emphasized that, due to the finite length of the Hanning window, activity of wave M40 may also contribute. However, the distinctly different patterns of IPC distribution computed for 12 ms and 40 ms (Fig. 8) after stimulus onset demonstrate that the temporal resolution of the windowing technique is sufficient. Both the EF and the MA distributions are basically similar to that 80 ms prior to stimulus onset. The distribution of MP reveals the formation of a phase gradient running diagonal in a posterior/inferior direction. On both sides of this gradient, fields of increased IPC begin to develop, which is most obvious in the third to fifth Fourier components. For the sake of clarity, the areas on both sides of an imagined projection of the ECD direction onto the scalp will be referred to as posteror and anterior fields. Whereas the IPC in the posterior field is distinctly less significant and the pattern is fairly similar for the second to the fourth Fourier components, the position of the maximum of the third Fourier component in the anterior field differs from that of the fourth and fifth Fourier components by almost 3 cm.

The next instantaneous picture was taken at the maximum of wave M50, approximately 40 ms after stimulus onset (Fig. 8). The EF distribution reflects the small amplitude of the evoked magnetic field. The MA distribution of evoked activity merges with that of spontaneous activity in the posterior field; the MA pattern is basically similar in all Fourier components though there is only a marginal augmentation of spectral energy in the fifth Fourier component. The IPC distribution is quite different from that of the MA and EF; especially the pattern in the anterior field differs, exhibiting a deep indentation in all five Fourier components. Quite different from the distributions of EP, MA, and IPC is that of the mean phase. In the case of all five Fourier components, the whole scalp area scanned is divided into two major fields of opposite polarity (phase difference of π radians) to both sides of the imagined projection of the dipole direction onto the scalp. A reordering of phase is present in all Fourier components even at those marginal positions where the IPC is no more significant.

Figure 9 shows the situation 80 ms after stimulus onset, when wave M100 assumes its maximum. The MA distribution exhibits the same asymmetry as the EF: The anterior field shows a higher magnetic field strength and a higher spectral amplitude as compared with the posterior field. The shape of the MA distribution is, at least in the anterior field, very similar to that of the EF, while the distribution of evoked activity in the posterior field merges with that of spontaneous activity. The second Fourier component is the strongest one, and there is no evidence of additive spectral energy for the fifth Fourier component. The distribu-

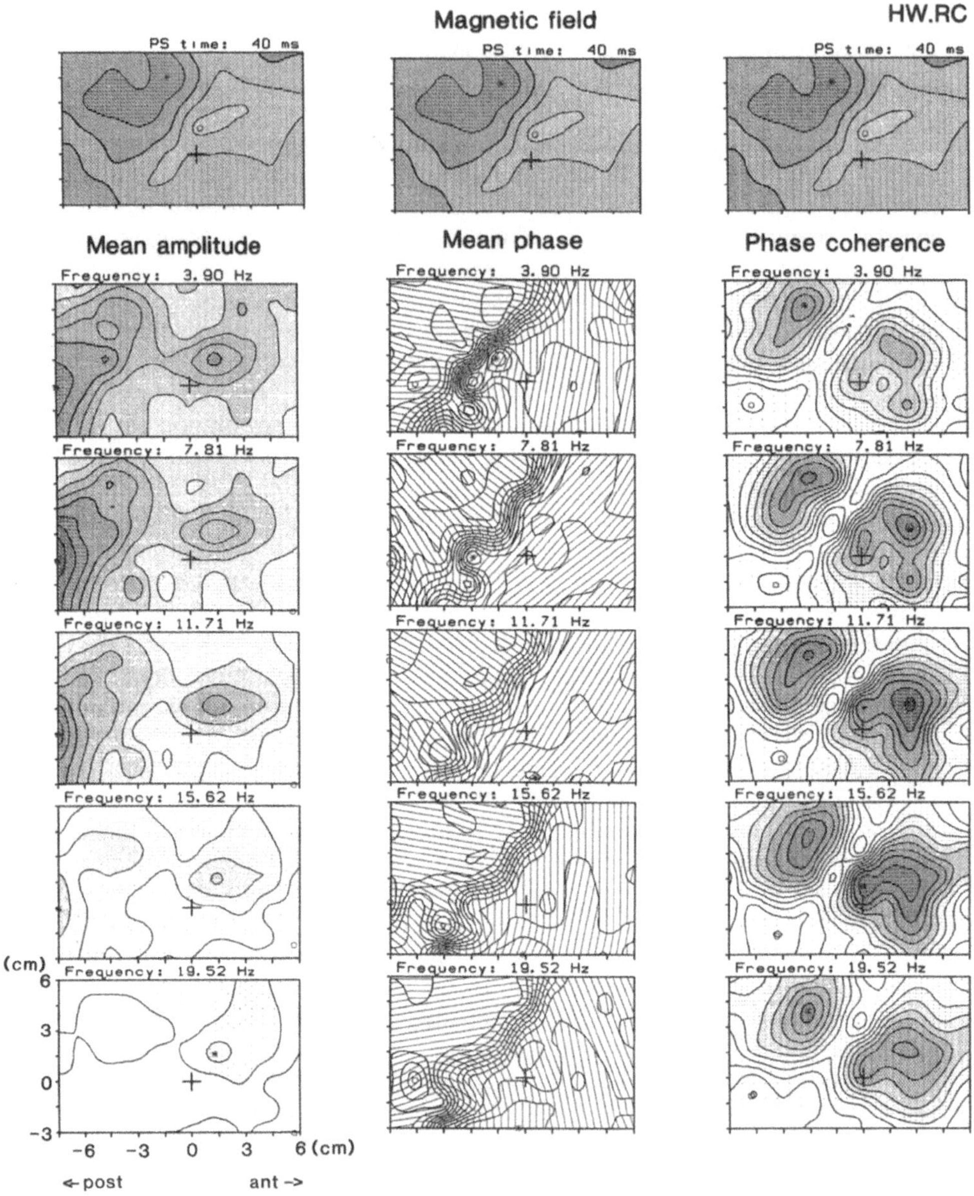

Fig. 8. As Fig. 6, but 40 ms after stimulus onset. See legend to Fig. 6 for explanation

tion of the MP as well as that of the IPC is basically similar to that obtained 40 ms after stimulus onset (Fig. 8). As in Fig. 8, the fifth Fourier component still exhibits a significant IPC though the increase of spectral energy is marginal.

At 108 ms, the time of field reversal in the EF waveform (Fig. 10), the distribution of the MA is similar to that 80 ms after stimulus onset. The lower Fourier

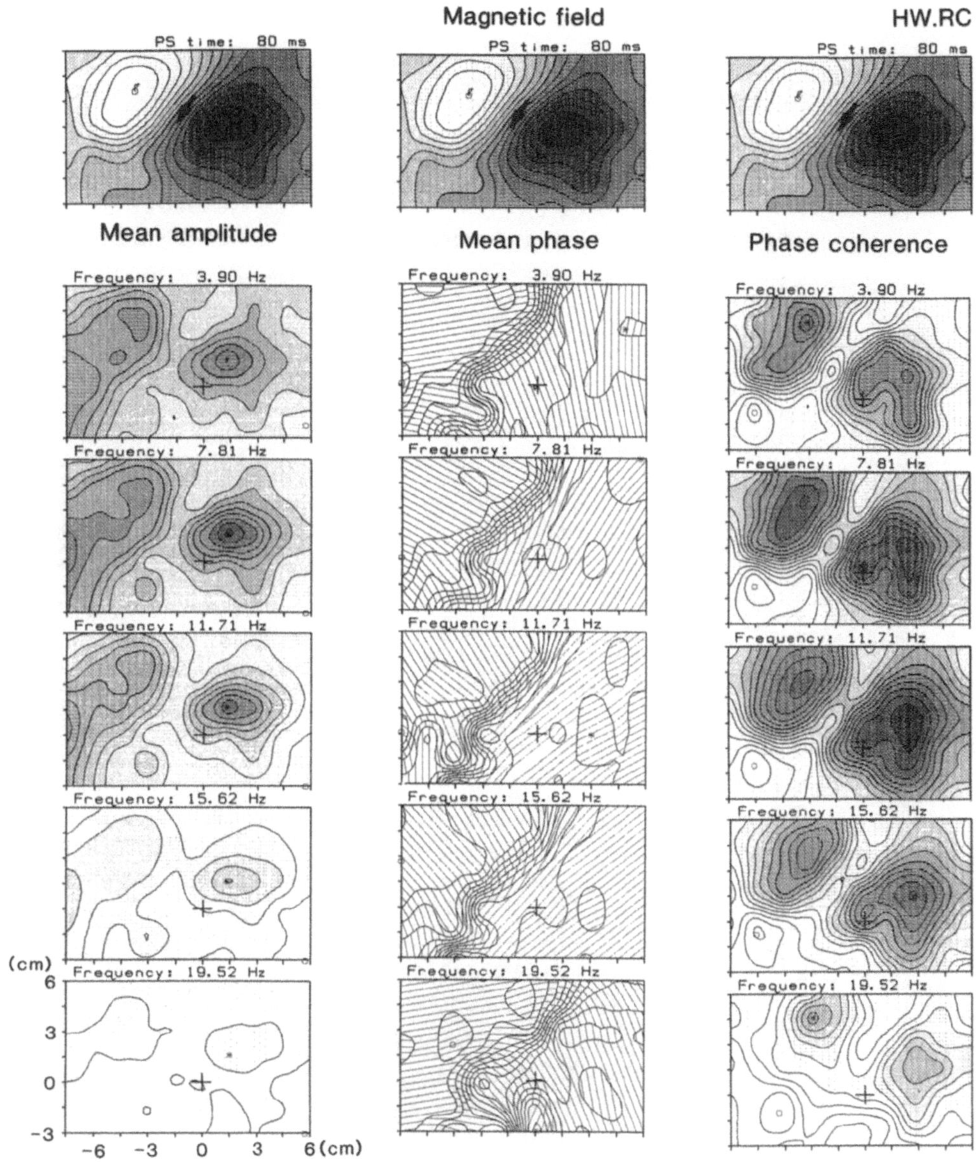

Fig. 9. As Fig. 6, but 80 ms after stimulus onset. *Arrow*, projection of the equivalent current dipole onto the tangential plane with length proportional to dipole moment (*arrowhead* = 20 nAm). See legend to Fig. 6 for further explanation

components are augmented while the higher ones are diminished. The phase pattern of the fifth Fourier component might be regarded as stochastic again. The phase gradient across the boundary between both fields of opposite polarity in the lower three Fourier components is steeper than was observed 80 ms after stimulus onset. The distribution of the IPC resembles that observed 80 ms after

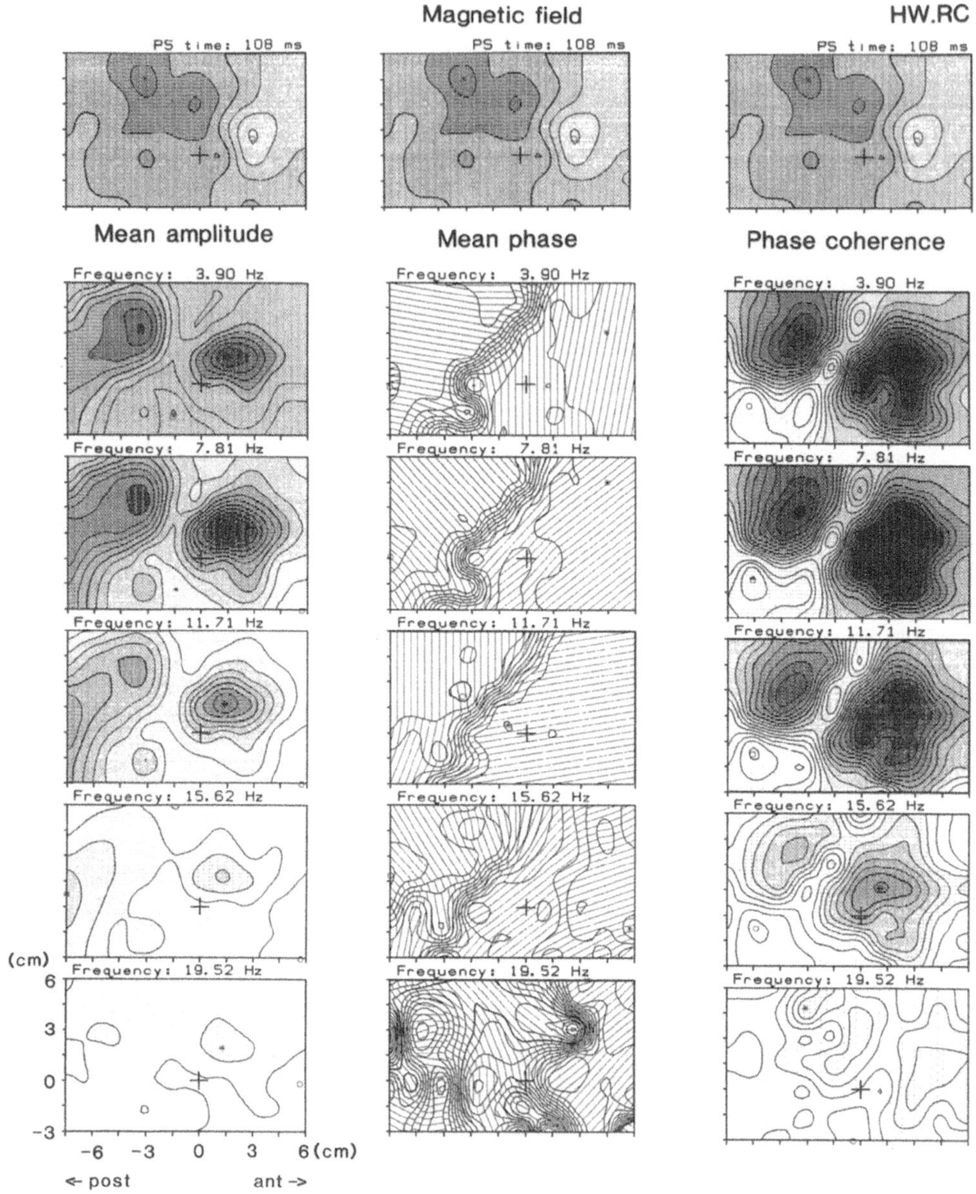

Fig. 10. As Fig. 6, but 108 ms after stimulus onset. See legend to Fig. 6 for explanation

stimulus onset (indentation in the anterior area), but there is no significant IPC in the case of the fifth Fourier component, as can also be deduced from the total decomposition of phase organization.

Figure 11 is the instantaneous situation at the time when wave M200 is peaking. The shape of the MA distribution is basically like that of the EF, with the

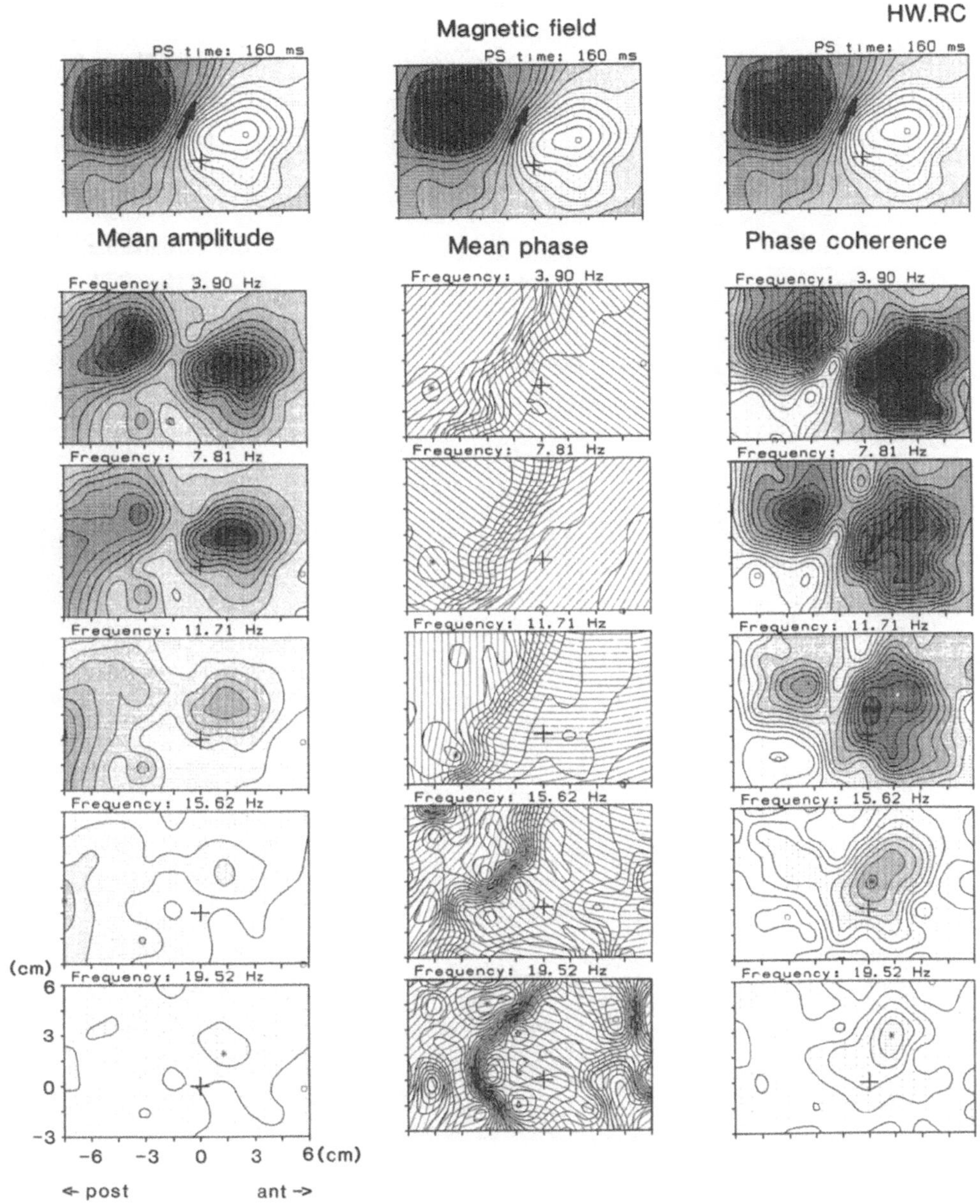

Fig. 11. As Fig. 6, but 160 ms after stimulus onset. See legends to Figs. 6 and 9 for explanation

first Fourier component now being the strongest one. The phase gradient across the boundary between both fields of opposite polarity becomes less steep again, and even the MP distribution of the fifth Fourier component now exhibits a formation of two larger fields of opposite polarity. Whereas the IPC distribution for the lower three Fourier components is not substantially different from the previ-

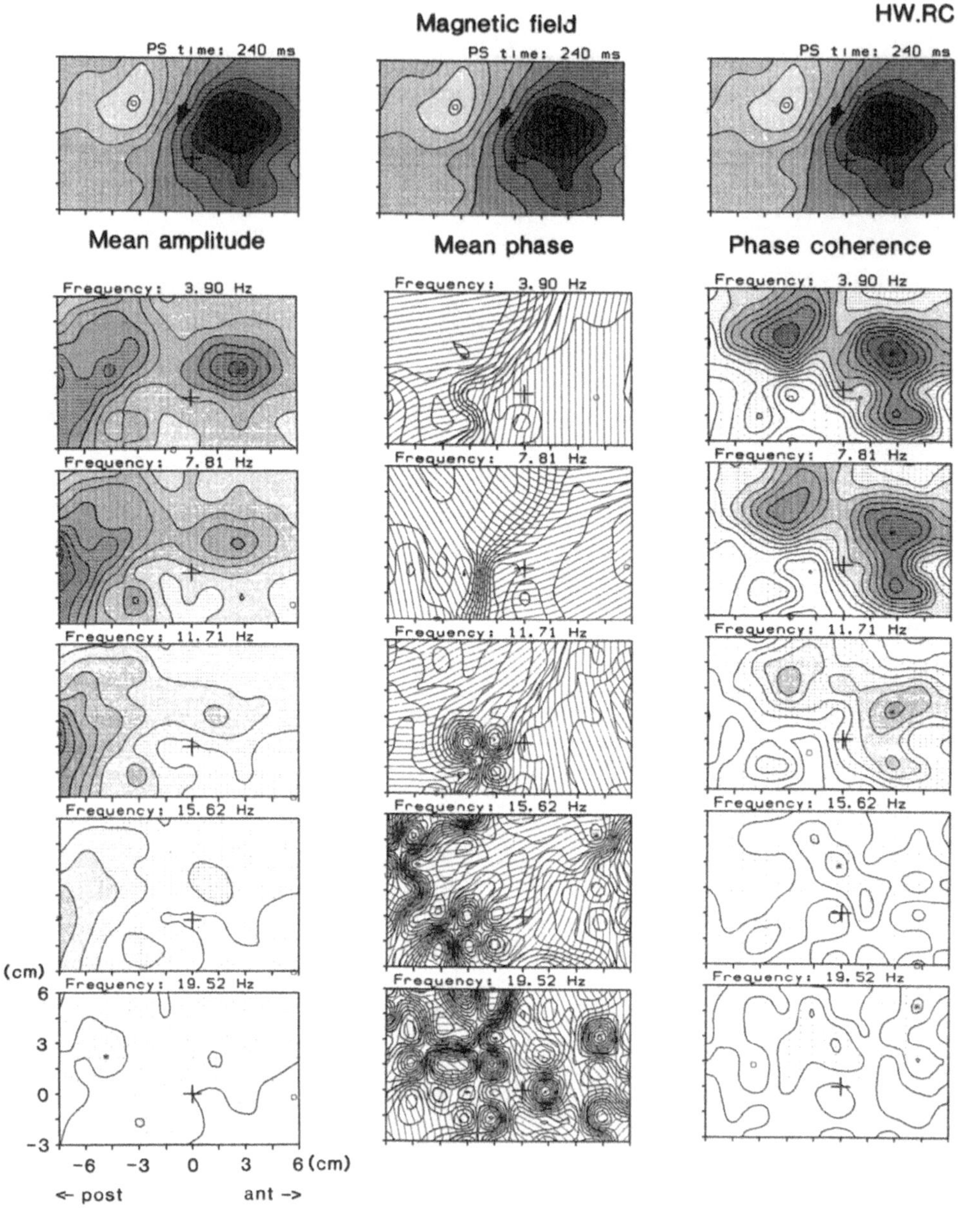

Fig. 12. As Fig. 6, but 240 ms after stimulus onset. See legends to Figs. 6 and 9 for explanation

ous distributions, the IPC distribution for the fourth and fifth Fourier components no longer looks dipolar.

In the last instantaneous picture taken 240 ms after stimulus onset (Fig. 12), only an increase in spectral energy of the lower two Fourier components can be observed. The MP distribution shows a stochastic pattern for the fifth Fourier component, and a decomposition of phase organization is imminent for the third and fourth Fourier components. The IPC distribution of the lower two Fourier components again is different from the previous distributions; the fourth and fifth Fourier components no longer exhibit a significant IPC.

4 Discussion

4.1 Mean Amplitude

There obviously exist three different sources of a circumscribed aggregation of spectral energy. The first is due to the spontaneous activity of generators of alpha rhythm located in parietal areas of the cortex. The alpha activity extends over the posterior portion of the scanned area, where it becomes evident in the lower three Fourier components with its spectral maximum at 7.81 Hz. The alpha activity remains invariant in place and amplitude during stimulus presentation. However, as the MP and IPC distributions reveal, the alpha rhythm, which solely represents incoherent activity before stimulus onset, is distinctly synchronized by the auditory stimulus. This suggests that sensory stimulation is capable of synchronizing natural brain rhythms even if they do not originate from modality-specific projection areas.

Another, though marginal, circumscribed aggregation of spectral energy is detectable in the first three Fourier components prior to stimulus onset, located approximately 2 cm superior–anterior from T4. This activity is probably due to a modality-specific contribution to the expectancy wave since auditory stimulation was done periodically. Like the alpha activity, this second category also reflects completely incoherent activity, which becomes especially evident if we compare the instantaneous distributions of MA and IPC of the third Fourier component 80 ms prior to and 12 ms or 240 ms after stimulus onset. While the MA patterns are almost congruent at all three instants, solely the IPC patterns after stimulus onset reflect a distinct organization which is missing in the prestimulus period.

The third category of spectral energy is due to the exciting stimulus. The patterns are basically similar at all instants in all Fourier components where additive energy is present. The distributions show the same asymmetry as the EF: the anterior field exhibits steeper gradients than the posterior field. While the MA distribution in the anterior field is circumscribed and fairly oval shaped, the distribution of evoked activity in the posterior field merges with that of spontaneous alpha activity. Comparison of the distribution of spectral amplitudes, including the magnetic field distribution, reveals that the patterns are fairly congruent at different instants, except for the first and second Fourier components at 160 ms and 240 ms after stimulus onset, where the maxima are slightly displaced in an anterior direction. (The dipole location in the tangential plane does not closely reflect the displacement of field pattern because of the greater depth of the ECD of the later waves.) The amplitude maximum of the first Fourier component occurs at 160 ms (maximum of wave M200), while that of the second Fourier component occurs around 108 ms (in between the peaks of waves M100 and M200); the time functions of both Fourier components are unimodal. Higher Fourier components exhibit bimodal or multimodal time functions in which the two first prominent peaks coincide with those of waves M100 and M200.

Spatiotemporal short-time spectral analysis has revealed that the spectral amplitude of evoked activity is not that insignificant, as was stated by Sayers et al. (1974, 1979; Sayers and Beagley 1974). The difference between their data and the results presented here is with high probability due to the fact that Sayers et al. (1974, 1979; Sayers and Beagley 1974) and Jervis et al. (1983) analyzed EEG signals rather than MEG signals, and their analysis window was much longer than that employed in our technique. Consequently, a circumscribed, short-term increase in spectral energy could hardly be detected because it was spread over the longer epoch and superimposed upon by EEG contributions from remote cortical areas; MEG measurements and shorter analysis windows have increased both the spatial and the temporal resolution.

4.2 Mean Phase

The phase distribution is completely different to the distributions of EF and MA: The whole area scanned is divided into two fields by a boundary running in an inferior/superior direction along an imagined projection of the ECD onto the scalp. Across the boundary a phase gradient of approximately π radians exists which is steepest in the early stage of organization (12 and 40 ms) and gradually deteriorates as time elapses. It is most surprising that an organization is detectable in the phase maps even before a significant interepochal phase coherence is present (cf. 12 ms after stimulus onset, Fig. 7). In the presence of high phase coherence (cf. second Fourier component at 108 ms, Fig. 10), two large fields of practically uniform phase exist on both sides of the boundary. This implies that, as already mentioned in the context of alpha activity, a phase reordering occurs even in those remote areas where no interepochal phase coherence is detectable. The strong spontaneous alpha activity, for example, is completely synchronized 108 ms after stimulus onset (cf. Fig. 10 and 11, phase maps of the lower three Fourier components). Hence, stimulus-induced synergetic processes in the brain seem not to be restricted to the modality-specific cortical projection areas but also affect surrounding brain areas. Consequently we suggest the introduction of the concept of a "coherence area" to describe the spatial aspect of the synergetic process, in analogy to the concept of a "coherence length" which describes the temporal aspect of a synergetic process. The comparison of a phase pattern composed of just two large coherence areas with the stochastic phase pattern found before stimulation most clearly demonstrates the "laser effect" due to the exciting stimulus, i.e., the transient synchrony of neural generators in larger brain areas.

The coherence length is greatest in the second Fourier component (the first Fourier component is disregarded because the windowing introduced strong phase distortions in the lowest and highest Fourier components); the phase in both fields remains practically invariant between poststimulus times of 40 and 240 ms. The coherence length becomes smaller the higher the Fourier component. Higher Fourier components also sometimes show two or more periods of a shorter coherence length with a distinct phase jump in between, which may point to the existence of different generator populations being subsequently activated. So, for instance, the third Fourier component shows two successive coherence periods of shorter coherence length (40–80 ms, 108–160 ms), after which a phase jump of about 90 degrees occurs. The phases in both fields remain fairly constant during the coherence length, whereas the transition between two successive periods of coherence is relatively abrupt, demonstrating that the technique of short-time spectral analysis is superior to using sharply tuned filters which cannot follow, due to their longer decay times, abrupt phase changes occurring within one-third of the period of the Fourier component in question.

4.3 Interepochal Phase Coherence

While the phase maps are that representation which first reveals the beginning of organization, the IPC maps seem to reveal information which is not evident in the EF and MA maps. First, whereas MA maps are fairly congruent with the EF maps, the IPC distribution is not at all congruent with that of the MA. This suggests that fields exist in which both additive energy and phase reordering are present while there exist other fields in which a distinct phase reordering is present without a substantial change of spectral energy. Secondly, there is possibly not just one homogeneous area of phase reordering in the anterior field but three separate areas which become visible as early as 40 ms after stimulus onset (cf. Fig. 8, IPC plot for 7.81 Hz). It should be emphasized that the described incongruence of MA and IPC patterns was found to be reproducible in the one individual tested twice, and the IPC patterns again revealed three subfields of highest phase coherence located superior, superior–anterior, and inferior–anterior to T4. Although in the absence of further investigations the conclusion is highly speculative, it seems that the analysis of IPC maps of auditory evoked magnetic fields allows, for the first time, the separation of cortical evoked activity arising simultaneously from functionally separable structures.

5 Summary and Conclusions

A method of spatiotemporal short-time spectral analysis for studying synergetic processes of the brain has been developed and applied to ensembles of MEG epochs sampled, coherent with the auditory stimulus, at 60–70 positions over the temporal area of the scalp. The first step of this method consists in processing the ensemble of individual epochs taken at all sampling positions. The time-dependence of spectral features is retrieved by stepwise shifting a Hanning window over the individual epochs of the respective ensembles in steps of one sampling point. A Fourier analysis of each window is performed, and the amplitudes and phases of the Fourier components of interest are computed. In a second step, for each window position, the mean amplitude, the mean phase, and the interepochal phase coherence are computed from all individual amplitudes and phases of the respective ensembles. The third step consists in associating these values with the center times of the windows so that, for all Fourier components of interest, time functions of mean amplitude, mean phase, and interepochal phase coherence can be composed for the respective sampling positions. Finally, as done for the time-domain averages of the MEG epochs, isocontour plots can be generated, for each window center time, from the respective time functions of all sampling positions. By this means it becomes possible to compare the instantaneous magnetic field distribution with those of the statistical parameters of each Fourier component, i.e., mean amplitude, mean phase, and interepochal phase coherence.

This technique, whose high spatial resolution is due to the fact that the magnetic field component measured normal to the scalp is solely produced by intracellular currents, obviously allows one to distinguish between three different categories of accumulated activity. The first comprises the spontaneous activity of alpha generators, which remains invariant in place and amplitude during stimulus presentation. However, as the distributions of the mean phase reveal, the activity of the alpha rhythm, which is incoherent before stimulus onset, is distinctly synchronized by the auditory stimulus. This suggests that sensory stimulation is capable of synchronizing natural brain rhythms even if they do not originate from modality-specific projection areas. This synchronization is so high that even remote areas are affected where no significant interepochal phase coherence exists. In analogy to the concept of a "coherence length" describing temporal aspects of the synergetic process, the introduction of the concept of a "coherence area" to describe spatial aspects of the synergetic process is suggested.

Another, though marginal, circumscribed accumulation of energy is likewise detectable prior to stimulus onset. This activity, completely incoherent like alpha activity, probably reflects endogenous activity, namely the modality-specific contribution to the expectancy wave, since auditory stimulation was done periodically.

The third category of accumulated energy is directly related to the exciting stimulus. The distribution of interepochal phase coherence, which is, during stimulation, not at all congruent with that of the evoked field, suggests that fields exist in which both additive energy and phase reordering are present while other fields exist in which a distinct phase reordering is not associated with a substantial increase in spectral energy. Secondly, there possibly exist multiple areas of phase organization, especially in the anterior field. Though, without further investigations, it is highly speculative, it seems that the analysis of IPC maps of auditory evoked magnetic fields allows, for the first time, the separation of cortical evoked activity arising simultaneously from functionally separable structures. If this is the case, the short-time spectral analysis technique will be extremely helpful in improving source models of evoked magnetic fields, which are still the weakest part of present dipole localization methods.

Acknowledgments. This work has been supported by grants from the *Deutsche Forschungsgemeinschaft* (Ho 847/2-2).

References

Başar E (1980) EEG-brain dynamics: relation between EEG and brain evoked potentials. Elsevier/North-Holland, Amsterdam

Başar E (1983) Synergetics of neuronal populations. A survey on experiments. In: Başar E, Flohr H, Haken H, Mandell AJ (eds) Synergetics of the brain. Springer, Berlin Heidelberg New York, pp 183–200

Beagley HA, Sayers BMcA, Ross JA (1979) Fully objective ERA by phase spectral analysis. Acta Otolaryngol (Stockh) 87:270–278

Haken H (1977) Synergetics: an introduction. Springer, Berlin Heidelberg New York

Haken H (ed) (1980a) Dynamcis of synergetic systems. Springer, Berlin Heidelberg New York

Haken H (1980b) Lines of developments of synergetics. In: Haken H (ed) Dynamics of synergetic systems. Springer, Berlin Heidelberg New York, pp 2–21

Hoke M, Wickesberg B, Lütkenhöner B (1984) Time- and intensity-dependent low-pass filtering of auditory brain stem responses. Audiology 23:195–205

Jervis BW, Nichols MJ, Johnson TE, Allen E, Hudson NR (1983) A fundamental investigation of the composition of auditory evoked potentials. IEEE Trans Biomed Eng 30:43–49

Kaufman L, Okada Y, Brenner D, Williamson S (1981) On the relation between somatic evoked potentials and fields. Int J Neurosci 15:273–282

Mardia KV (1972) Statistics of directional data. Academic, New York London

Pantev C, Hoke M, Lütkenhöner B, Lehnertz K, Spittka J (1986a) Causes of differences in the input-output characteristics of simultaneously recorded auditory evoked magnetic fields and potentials. Audiology 25:263–276

Pantev C, Hoke M, Lehnertz K (1986b) Randomized data acquisition paradigm for the measurements of auditory magnetic fields. Acta Otolaryngol [Suppl] 413:21–25

Pantev C, Hoke M, Lehnertz K, Lütkenhöner B, Anogianakis G, Wittkowski W (1988) Tonotopic organization of the human auditory cortex revealed by transient auditory evoked magnetic fields. Electroencephalogr Clin Neurophysiol 69:160–170

Sayers BMcA, Beagley HA (1974) Objective evaluation of auditory evoked EEG responses. Nature 251:608–609

Sayers BMcA, Beagley HA, Henshall WR (1974) The mechanism of auditory evoked EEG responses. Nature 247:481–483

Sayers BMcA, Beagley HA, Riha J (1979) Pattern analysis of auditory-evoked EEG potentials. Audiology 18:1–16

Schafer RW, Rabiner LR (1973) Design and simulation of a speech analysis-synthesis system based on short-time Fourier analysis. IEEE Trans Audio Electroacoust 21:165–174

II. Theoretical Approaches to Brain Function (Linear and Nonlinear)

Introductory Remarks

E. Başar

The sessions of the second brain dynamics conference in which most of the authors of Part II of the book gave their talks were entitled "Strange Attractors in the Brain?," "Models: Linearity versus Nonlinearity," and "Hypergeometries of Strange Attractors Revealed by Computer Analysis of EEG." The papers covered various topics on nonlinearities in brain function and some linear mathematical approaches.

An early nonlinear approach to understanding the CNS function was undertaken by Erich von Holst (1939) who introduced the concept of relative coordination to neurophysiology. Von Holst developed a hydrodynamic model to simulate fish motion. What von Holst had done was an initial step to analyze the nonlinear interaction of coupled oscillators by observing dynamic patterns of fish motions such as the separate beating of the dorsal and pectoral fins. The analyzed oscillatory patterns of coordinated movements observed by von Holst are what we today call dynamic nonlinear patterns. Unfortunately, his experiments and interpretations could be understood only in a limited manner. [1]

Some of the pioneering papers which describe the dynamic patterns in neurophysiology have been collected by Katchalsky et al. (1974). This was a conference volume initiated to discuss the concept of Katchalsky, who took the Belousov-Zhabotinskii reaction as a model to understand the electrical oscillations of the brain. Later, in 1982 a conference entitled "Nonlinearities in Brain Function" became one of the most important landmarks in this direction. I particularly want to mention that during this conference the participants seriously expressed the concept of *strange attractors* related to brain function. Nonlinear cooperative phenomena in the brain were discussed in a conference entitled "Synergetics of the Brain." Again here a number of proposals were made to consider the strange attractors in brain function although the final word about the EEG strange attractor could not yet be pronounced (Başar et al. 1983).

During the first Brain Dynamics Conference in Berlin papers by Babloyantz, Freeman, and Röschke and Başar discussed the concept of the "strange attractor" and integrated it as an important initial step during the conference. One year later, due to the efforts of A. J. Mandell, nonlinearities in brain research were dis-

[1] In order to further the ideas of Erich von Holst an extended and more general electric model of nonlinear relaxation oscillators was developed with the aim of understanding some of the EEG properties by the use of dynamic patterns (Başar 1976, 1980). It should be mentioned that the idea of relaxation oscillators and dynamic patterns in the EEG research has also been expressed by Barlow (1962) who published his results in MIT Progress Reports which were distributed only to a limited number of scientists.

Springer Series in Brain Dynamics 2
Edited by Erol Başar and T. H. Bullock

cussed extensively in Bethesda and during this conference the nonlinear characteristics of EEG were extensively viewed (Koslow et al. 1987).

During the second Brain Dynamics Conference (this volume) we had three sessions in which the concepts of nonlinear dynamics were discussed in an extended manner. As the Epilogue of the present volume written by T. H. Bullock also emphasizes, the search for strange attractors in the brain has been one of the important trends of the conference. It has to be stated that this trend had its own intrinsic dynamics because, as one sees in the discussions throughout the book, not only the nonlinear trend but also the linear approaches to brain function have been discussed extensively in the conference. It ought to be mentioned here that these introductory remarks to Part II are not written with the aim of giving a complete picture of this development, covering all major contributions reported in conference proceedings. This short narrative review is intended to point out that the field dealing now with the general expression of "nonlinear dynamcis" emerged first from the necessity to recognize cooperative phenomena and nonlinear dynamic patterns, and only later, by using the correlation dimension as a powerful descriptor.

Following the pioneering work by Babloyantz on dimensionality of EEG during sleep (Babloyantz et al. 1985), several investigators started to evaluate the correlation dimensionality of the EEG in various experimental conditions. The results of this part of the book already include some of the earliest results on magnetoencephalography of healthy subjects and epileptic patients (Saermark), measurements during epileptic discharges and Creutzfeld-Jakob disease (Babloyantz), and EEG of the awake human (Babloyantz, Graf and Elbert, Röschke and Başar). By using a frequency window Röschke and Başar searched for high frequency attractors in the cerebellum and brain stem of the cat during the waking state (100–1000 Hz range). Skinner et al. used the experimental methodology of W. J. Freeman and evaluated the dimensions of the olfactory bulb electrical activity in rabbits. The following chapters include most of the current work done in this new avenue of brain research. In reviewing the articles in this part of the book it is noteworthy to mention that all the results should as yet be interpreted with some caution. They should be considered as pioneering works to be evaluated later with greater precision.

The paper by Rössler and Hudson gives important thoughts on the detection of strange attractors. These authors point out that quasiperiodic signals contaminated by noise look a little like a chaotic attractors, accordingly, the title "The self-similarity in hyperchaotic data." Of great importance is the question whether or not the brain can indeed be expected to yield data that reveal a "hidden" self-similar structure in phase space. Rössler and Hudson further opine that a "maintenance activity" of periodic or quasiperiodic type would not be surprising in the brain.

In another approach, one of the most important advances would be to define the microstates of the brain (see also Lehmann, Part II) by using several other indicators, for example, the power spectrum of the EEG, as Röschke and Başar indicate. Small changes in the balance of frequency components which are in the spontaneous activity can give rise to drastic changes in the correlation dimension. This may explain the huge differences in correlation dimensions of human EEG during the waking state as measured by different laboratories. Furthermore, it must be important to differentiate states with abundant alpha and those without alpha activity. But this state definition remains unsatisfactory since often alpha activity is accompanied by bursts of 40-Hz activity which are not registered in conventional EEG devices lacking special 40-Hz monitoring. In other words, for an adequate and complete description of brain waves, linear descriptors are still useful and complementary. Linear approaches as presented by the chapters by Wright et al. and Blinowska and Franaszczuk and the consideration of the superposition principle are of considerable importance for the description of brain waves.

Additionally to papers in Part II of this book it is also important to mention the analysis of strange attractors by Freeman (Part V of this book). Freeman comments that we can expect within a short time to witness an explosive development with many as yet unforeseen applications in "applied hypergeometry." The essay of Mpitsos about chaos and the problem of nonstationarity (also in Part V) complements the picture.

Finally, it is noteworthy that the interaction between theoretical scientists and experimenters who measure real biological data can be most useful as the active discussion following the paper by de Callataÿ on "Artificial intelligence" (p 214) shows.

Since the field of nonlinearities in biological function is a newly emerging discipline, the following references are offered with some suggested readings.

References

Babloyantz A, Nicolis C, Salazar M (1985) Evidence of chaotic dynamics of brain activity during the sleep cycle. Phys Lett [A] 111:152–156

Barlow JS (1962) The relationship among "photic driving" responses to single flashes, and the resting EEG. Quarterly progress report no 64, Res Lab Electronics. MIT Press, Cambridge

Başar E (1976) Biophysical and physiological systems analysis. Addison-Wesley, Reading

Başar E (1980) EEG Brain Dynamics. Elsevier, Amsterdam

Başar E (1988) Dynamics of sensory and cognitive processing by the brain. Springer, Berlin Heidelberg New York (Springer series in brain dynamics, vol 1)

Başar E, Flohr H, Haken H, Mandell AJ (eds) (1983) Synergetics of the brain. Springer, Berlin Heidelberg New York (Springer series in synergetics, vol 23)

Katchalsky K, Rowland V, Blumenthal R (1974) Dynamic patterns of brain cell assemblies. Neuroscience Research Program Bulletin. MIT Press, Cambridge

Koslow SH, Mandell AJ, Shlesinger MF (1987) Perspectives in biological dynamics and theoretical medicine. Ann NY Acad Sci 504

Mandell AJ, Whybrow PC, Yates FE (eds) (1983) Nonlinearities in brain function. Am J Physiol 245:R 453–R 533

von Holst E (1939) Die relative Koordination als Phänomen und als Methode zentralnervöser Funktionsanalyse. Ergeb Physiol 42:228–306

Suggested Reading

Abraham RH, Shaw CD (1981) Dynamics – the geometry of behavior. Part 3. Global behavior. Aerial, Santa Cruz

Abraham RH, Shaw CD (1987) Dynamics – the geometry of behavior. Part 1. Periodic behavior. Aerial, Santa Cruz

Babloyantz A (1986) Molecules, dynamics, and life: a introduction to self-organization of matter. Wiley, New York

Freeman WJ, Skarda CA (1985) Spatial EEG patterns, non-linear dynamics and perception: the neo-Sharringtonian view. Brain Res Rev 10:147–175

Markus M, Müller S, Nicolis G (1988) From chemical to biological organization. Springer, Berlin Heidelberg New York

Mayer-Kress G (1985) Dimensions and entropies in chaotic systems. Springer, Berlin Heidelberg New York

Schuster HG (1984) Deterministic chaos. An introduction. Physik, Weinheim

Thom R (1975) Structural stability and morphogenesis. An outline of a general theory of models. Benjamin, Massachusetts

Winfree AT (1980) The geometry of biological time. Springer, New York

Self-Similarity in Hyperchaotic Data

O. E. RÖSSLER and J. L. HUDSON

1 Introduction

The brain is a giant dynamic system. Its complexity should exceed e.g. that of a dripping faucet (Rössler 1977), which is one of the work-horses of nonlinear science (Martien and Shaw 1985). There are numerical methods for estimating the embedding dimension of a signal generated by a dynamic system (Froehling et al. 1981) which have been successfully applied to many kinds of data including brain-generated ones (see Mayer-Kress 1986 for many references).

From partial differential equations – which are nothing other than infinitely many coupled ordinary dynmic systems, all equal – it is known that frequently the dynamics involves but a few (finitely many) variables according to the center manifold theorem (Ruelle and Takens 1971). (Compare in this context a quiet lake on a wind-still evening: it is virtually at rest, that is, it has a point attractor.) Therefore it is conceivable in principle that the brain, with all of its 10^{11} neurons, 10^{14} synapses, and thousands of miles of cable (excitable PDEs; Winfree 1980), might likewise generate a cooperative signal that involves but a few variables (that is, dimensions) during part of its waking activity.

On the other hand, there is an important objection possible. A huge TV set – or a computer, for that matter – might also generate fairly low-dimensional (for example, multiple-periodic) signals when its internal circuitry is tapped either at a point chosen at random or in a global fashion analogous to scalp leads. These signals would clearly have to be interpreted as a "maintenance activity" that reveals almost nothing about the essential, meaningful activity of the machine in question. The latter activity would be determined by the machine's programs and/or its visual inputs rather than by its internal clocks and scanning generators. The signals of the last-mentioned type might nevertheless be so preponderant, quantitatively, that the main activity only imposes itself as a sort of superimposed noise.

Similarly with the brain, low dimensional quasiperiodic signals – as discovered most clearly by Albano et al. (1987; see their beautifully resolved Fig. 5) – would have to be differentiated from a chaotic attractor of a similarly low dimensionality. Unfortunately, quasiperiodic signals contaminated by noise look a little like a chaotic attractor. To be more specific, the number of dimensions may be the same in both cases. Therefore, the question arises of how one might distinguish between these two extreme possibilities. As it turns out, the presently available methods of determining the dimension of a time series are not optimally suited to making this crucial distinction. There may be ways to improve their per-

Springer Series in Brain Dynamics 2
Edited by Erol Başar and T. H. Bullock

formance such that individual subdimensions governed by a periodic signal – possessing a close-to-zero underlying Lyapunov exponent – can be differentiated from other, genuinely chaotic subdimensions characterized by a positive Lyapunov exponent. This task, identification of the zero-LCE subdimensions, has, however, yet to be implemented.

An alternative possibility is to positively test for the presence of more or less large, positive Lyapunov characteristic exponents. Such positive exponents, while not proving chaos (Farmer 1981), nevertheless are a strong indicator that the dimension in question involves an unstable dynamics and thus may be part of a low-dimensional chaotic attractor. Again, the existing dimension-determining algorithms will have to be modified to permit picking up such subdimensions individually. While methods for identifying one or several positive exponents are available (Wolf et al. 1985; Stroop and Meier 1988), the just-proposed combination of breaking up the dimensions found into three subsets: trivial (periodic), nontrivial (chaotic), and a residue stemming from modulating (high-dimensional) noise, is apparently unavailable for the time being.

Interestingly, one of the three tasks of identification just described, the picking up of deterministic (hyper-)chaos contained in the data, can also be solved pictorially rather than algebraically. This new method is to be described in the following with an example.

2 The Self-Similar Fingerprints of Chaos

Chaos as is well known is closely related to the formation of a Cantor set. The latter constitutes the simplest self-similar fractal (Mandelbrot 1982). A chaotic attractor like that governed by the Hénon map (Hénon 1976),

$$\begin{aligned} x_{n+1} &= ax_n(1-x_n)-by_n \\ y_{n+1} &= x_n , \end{aligned} \tag{1}$$

possesses a Cantor-like vertical cross-section (in the y direction) as is well known. Under linear magnification, as shown by Hénon in his original paper (1976), layers upon layers appear. However, this sandwich structure is not suitable for diagnostic purposes when the amount of noise is not negligible. The reason is that unless the attractor is very weakly attracting (that is, b, the constant contraction factor, is close to unity), the sheet-like structure is locally very thin ("1-D like"). Very little noise then suffices to blur the distinction between layers.

There is, however, a very robust "second" Cantor set hidden in the attractor, in the x direction. This fact was exploited by Smale (1967) when he first described his "basic set" (that is, the uncountable set of periodic solutions of differing periodicities hidden in such horseshoe-shaped maps). As shown in Fig. 1, the basin to the left of the saddle S "invades" the region to its right in a set of stripes that indeed "eat out" a Cantor set of vertical excisions. This happens in Eq. (1) whenever the parameter a exceeds a value close to 4, as is well known (Mira 1987). The attractor then of course also disappears (to let only the basic set survive). However, the very same structure even "lives" within the unexploded attractor.

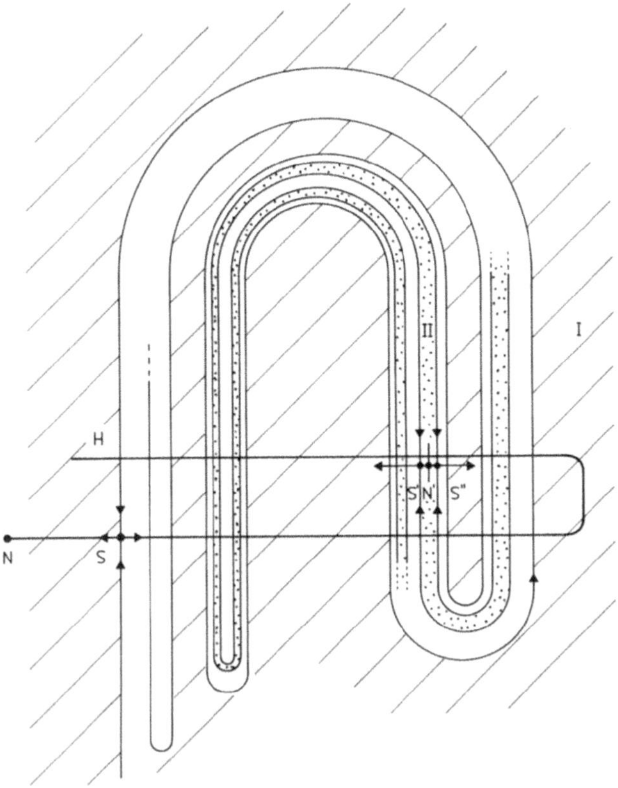

Fig. 1. Vertical "biting out of stripes" from an exploded former chaotic attractor. The attractor is exploded because (as happens in Eq. 1 with $a > 4$) the right-hand unstable manifold of the main saddle S has crossed the separatrix to the left-hand basin, at point *H*. The black (*dashed*) basin of the atractor *N* then "invades" the region of the former chaotic attractor, leaving only the "basic" set of periodic points (not shown) intact. Also shown is a second basin, of the attractor *N'*, which is embedded in the former chaotic attractor. Both basins then intermingle in a very complicated fashion. (See Mira 1987, pp 296–297)

It can easily be recovered pictorially. This fact is shown in Fig. 2, (b), where we see the original Hénon attractor (a), rendered in an unusual way. Every point that next arrives in bin I has been plotted in black and every point that instead arrives in bin II first is shown in white (blank). Evidently, the "same" Cantor stripes as seen previously in Fig. 1 appear. The reason for this is that the dynamics of Eq. (1), with $a < 4$, still contains the Cantor structure of Fig. 1 "symbolically." The technique used in Fig. 2b is an example of symbolic dynamics. Symbolic dynamics looks at sequences of digits appearing in chaotic maps (see Crutchfield and Packard 1982). In the present case, it was instead used as a "coloring scheme."

Why is this possibility of staining data worth mentioning? The reason has to do with the fundamental stretching and folding process that underlies chaos (Rössler 1976; see Rössler 1983 and 1987a for pictures of a taffy puller). This repetitive stretching is responsible for the fact that, going back in time, every point

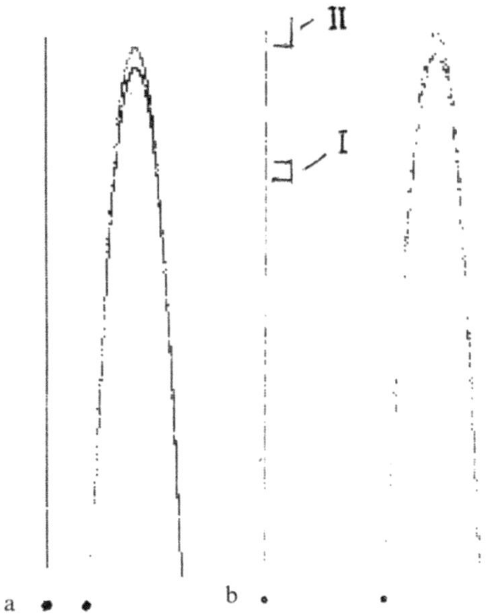

Fig. 2. (a) The Hénon attractor, Eq. (1) with $a=3.94$ and b positive ($b=0.05$). The x axis is shown here pointing up, whereas in Fig. 1 it pointed to the right. In this way, both a complete 2-D view to be compared with Fig. 1 can be presented (*on the right*) and a simplified but more informative 1-D top-down view (*on the left*), jointly as a single stereo pair. (The same convention will apply in Fig. 3.) The origin, point 0,0, is shown *bold.* Initial point 0.1, 0; 1200 iterates are shown at 11-digit accuracy. (b) The same data, replotted using a two-color convention. All points that the next time first land in bin I rather than bin II are shown in *black*, while those that do the opposite are erased (shown in white as *blanks*). Compare with Fig. 1. Bin I, $x>0.95$; bin II, $0.72<x<0.75$; 2300 iterates

in the "dough" can be classified according to how it will, in the future, behave relative to its neighbors. Hence any pair of "escape hatches" (regions or bins in the dough from which there is no return) necessarily have antecedents, characterizable by the "color" of the bin of their destiny, at every level of resolution, within the taffy. This fact is a direct consequence of the exponential mixing process (Rössler 1987a) that takes place within the taffy (attractor). Unexpectedly, this result is robust when more than one dimension of stretching is involved. In this "hyperchaos"-generating situation (Rössler 1979), the pattern becomes even more distinctive. The simplest equation for an invertible map generating hyperchaos is a straightforward generalization of Hénon's 2-D map to three dimensions. It reads (Rössler et al. 1988):

$$\begin{aligned} x_{n+1} &= ax_n(1-x_n)-by_n \\ y_{n+1} &= ay_n(1-y_n)-bz_n \\ z_{n+1} &= x_n \,. \end{aligned} \tag{2}$$

This equation, too, is a globally smooth map (a diffeomorphism), whose constant contraction factor per step (Jacobian determinant) this time is equal to b^2 (rather than b). Its points could again have been generated by a smooth dynamic system, like the even simpler continuous hyperchaos generator of Rössler (1979) or its chemical analogue (Killory et al. 1987) for example.

Figure 3b shows a picture analogous to that of Fig. 2b. It was obtained using the two bins labelled I and II. Instead of a Cantor *stripes* structure, we have a Cantor *crosses* structure ("iterated Swiss flag"). If in place of the two bins chosen in Fig. 3b, two other bins are chosen arbitrarily, the point patterns of Fig. 4 arise e.g. Thus, the fingerprint of hyperchaos is not a *particular* self-similar "unmixing" structure, but rather the fact that a self-similar structure arises at all whenever each point is forced to eventually go into one of two (or more) final patches.

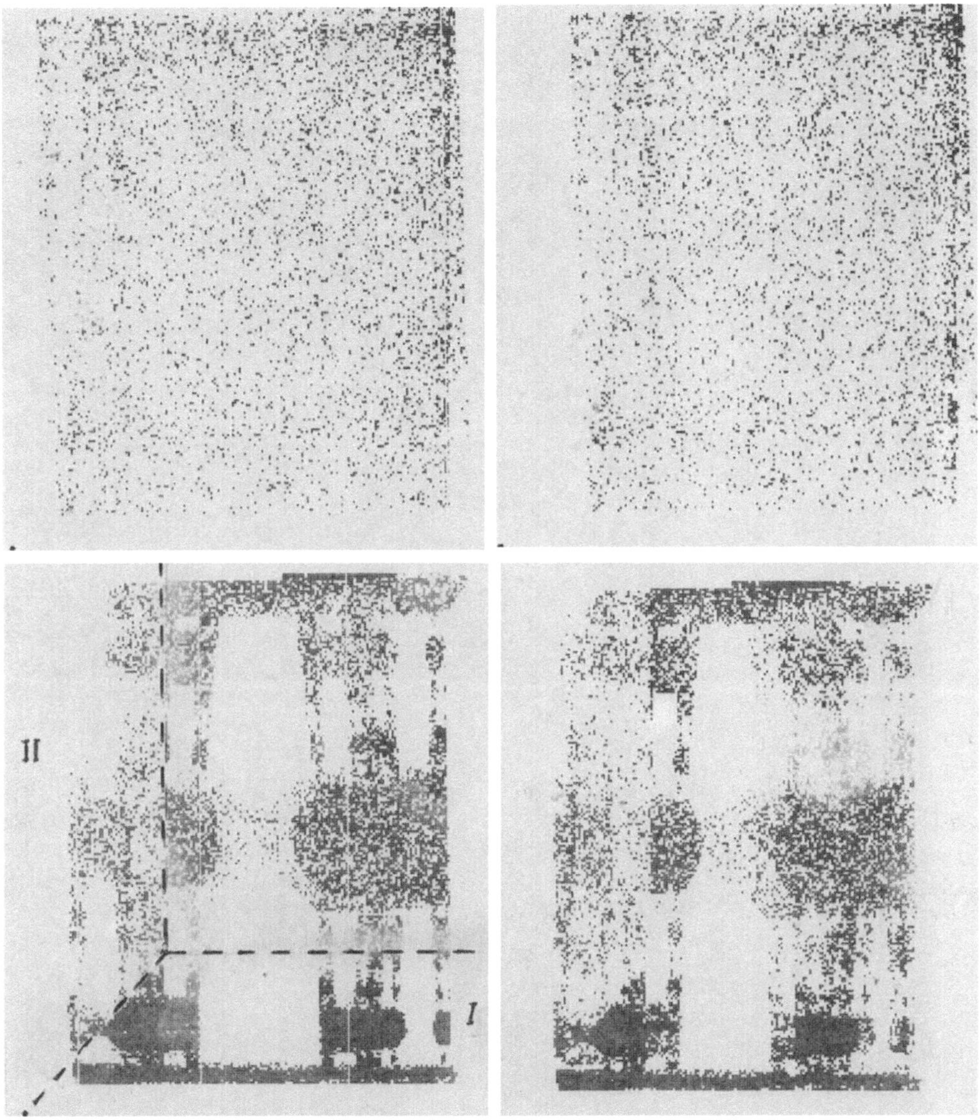

a

b

Fig. 3. (a) The hyperchaotic attractor, Eq. (2) with parameters *a* and *b* as in Fig. 2. Stereo pair as in Fig. 2. The *left-hand picture* is the top-down view (x, y), the *right-hand picture* has $-0.1\ z$ added to the horizontal component (x). Initial condition 0.1, 0, 0.1. The origin 0, 0, 0 is bold; 10000 iterates. Eventually everything becomes black. (b) The same data, replotted using a two-color convention. As in Fig. 2b dependent on whether the present string of points has bin I or bin II coming ahead next, it is made visible or suppressed, respectively. The prototype ("iterated Swiss flag"; see Rössler et al. 1986, 1988), obtaining for $b \to 0$, is still recognizable. Resolution can be improved by taking only a thin slice rather than the full z thickness. Bin I, $x < 0.3$ with $y < x$; bin II, $y < 0.3$ with $x < y$; 600000 iterates. Note: Choosing the two parameters *a* differently in Eq. (2) will leave the result qualitatively unchanged

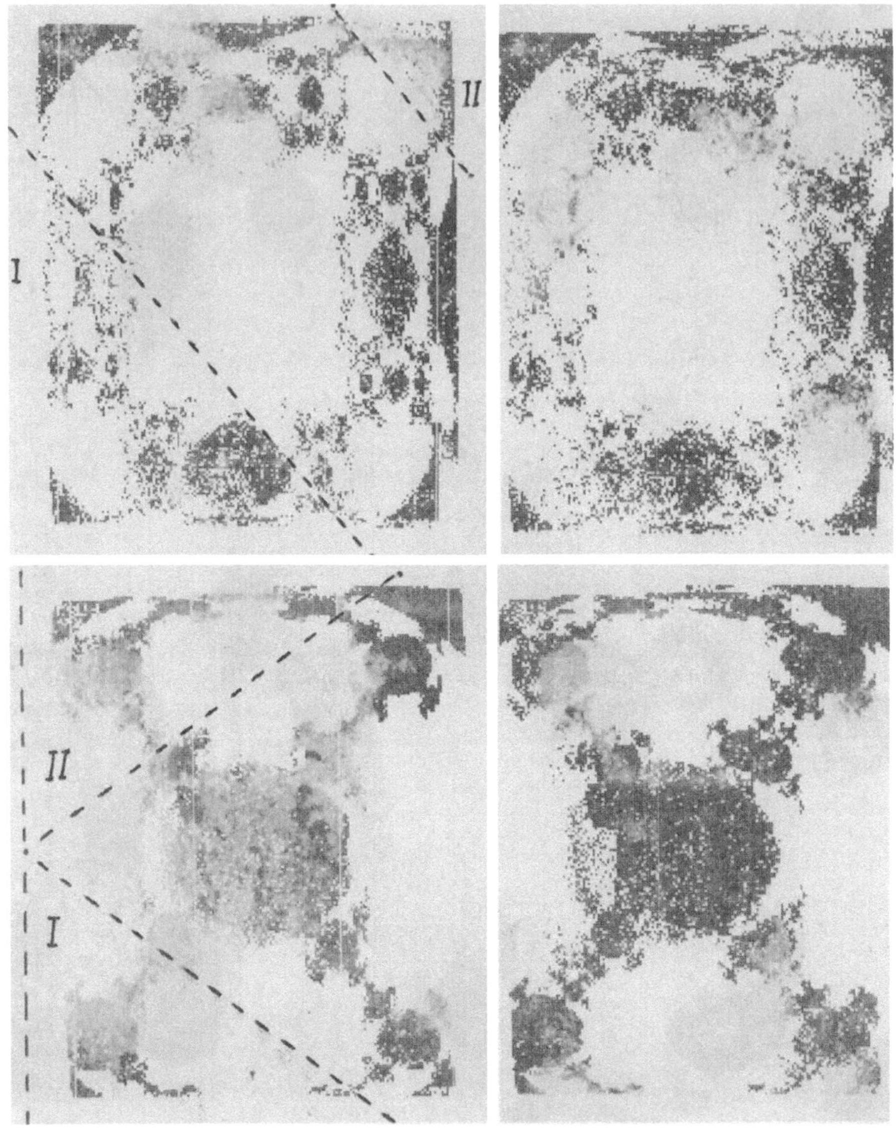

a

b

Fig. 4. (a) The data of Fig. 3a replotted in a different way. Bin I, $x+y<0.8$; bin II, $x+y>1.7$; 110000 iterates. (b) The data of Fig. 3a replotted in a third way. Bin I, $y<0.5-0.6x$; bin II, $y>0.5+0.6x$; 320000 iterates

3 Future Developments

The above examples are not yet real-life. Instead of maps, continuous flows will have to be used next to check how readily the self-similar structure survives in a Poincaré cross section when (a) the two Lyapunov characteristic exponents are no longer more or less equal (as was the case in Fig. 3), and (b) masking noise is added. Such data could be generated, for example by using the 4-variable system of Rössler (1979) calculated at low accuracy.

Nevertheless we expect the method to be fairly robust. Test runs showed that when the second sheet-like dimension contained a zero exponent, for example, the self-similarity indeed once more becomes locally stripe-like. Moreover, by varying the relative orientations and widths of the two bins, it should be possible even with strongly unequal positive Lyapunov characteristic exponents (stretching factors) to recover an easily recognizable self-similar pattern. If this prediction bears out, the main asset of the method could become apparent, namely its high resistance to contamination by noise. This is because any self-similar pattern can already be discerned with fairly few levels of repetition. Actually, two would virtually suffice for a "living proof" that a deterministic mixing process is at work in the data. Moreover by plotting point densities over a certain neighborhood rather than individual points, arbitrarily high noise levels become acceptable if only the data sequence is long enough.

4 Discussion

The above "staining" techniques has yet to be applied to real-life data, that is, to a cross-section obtained from the reconstructed attractor as the same time signal is used may times over with a delay (work in preparation). However, once such data are stored in a computer, by definition it will be as easy to subject them to the staining decision as the data generated by Eq. (2), for example. Also "cross sections through the cross section," in case the attractor is higher dimensional, will be just as easily obtained as was the case in Fig. 3 (with only a slice being plotted). The technique therefore appears to form a viable option to the experimentalist; this at least is what we hope.

This technical side of the staining proposal is, however, not the main point in the present context. More important is the question of whether or not the *brain* can indeed be expected to yield data that reveal a "hidden" self-similar structure in phase space (as opposed to the time domain; cf. Başar, this volume). Here only the future can yield the answer. However, to venture a guess, we would not be surprised if no clearcut positive evidence showed up. The reason for this scepticism was already given in the Introduction. While it would not be surprising to find a "maintenance activity" of periodic or quasiperiodic type in the brain, the possibility of a low-dimensional deterministic chaotic process being implied in the data has yet to be demonstrated. Of course, a maintenance activity of a low-dimensional chaotic type is also a possibility that cannot be ruled out a priori; compare this idea with the data on the smell brain (Freeman, this volume). In addition, we know that chaos is among the behavioral capabilities of whole animals (Todt 1969) – with Eq. 3 of Rössler (1980) forming an appropriate model system, cf. Rössler (1981) – but also of a larger system formed, e.g. by two coupled formal brains (Rössler 1987b). Moreover, it has been proposed that the brain as an organ may be subject to chaotic variations in its bulk activity (Braitenberg 1984), whereby, one might add, the local or not so local blood flow may form an essential participatory variable. Such "metabolic chaos" too could be considered as a type of chaotic maintenance activity of the organ in question. In a similar vein, it is conceivable that a chaotic pacemaker of hemodynamic and/or localized neuronal type (cf. Glass and Mackey 1979) may, in deep sleep or even in moments of idling, acquire the role of a "nonsignal" internally. Again, one could speak of a chaotic maintenance activity in that case. However, in spite of this possibility (cf. Rössler 1983), it would still come as a surprise if the fingerprints of low-dimensional hyperchaos of the type seen in Fig. 3 above could be shown to be present in scalp-EEG data, for example. In other words, the open question is whether or not the beautiful data available today (see Babloyantz, this volume) will be robust enough to allow for the detection beyond doubt of more than one positive Lyapunov characteristic exponent, either by using the present or by other techniques (e.g., Wolf et al. 1985; Stroop and Meier 1988).

To conclude, a new method has been proposed which may prove useful in the future analysis of real-life data generated by reaction-kinetic or biological systems that produce low-dimensional attractors. If everything goes well, the method may be of help in discriminating between deterministic hyperchaos generators, on the one hand, and other low-dimensional, but closer to quasiperiodic, time sequences of a similar phenomenological dimensionality on the other. The brain is but one out of many signal generators (cf. Bullock, this volume) for which one would like to have an answer to this type of question.

5 Summary

A new "staining technique" based on symbolic dynamics is proposed for investigating data generated by low-dimensional systems. The method is expected to help answer the question of whether or not the working brain is a generator of low-dimensional deterministic chaos.

Acknowledgments. We thank Walter Freeman, Erol Başar, Ted Bullock, Agnes Babloyantz, Paul Rapp, Gottfried Mayer-Kress, Jim Crutchfield, Helen Killory and Claus Kahlert for discussions. The work was supported in part by the N. S. F.

References

Albano AM, Mees AI, de Guzman GC, Rapp P (1987) Data requirements for reliable estimation of correlation dimensions. In: Degn H, Holden AV, Olsen LF (eds) Chaos in biological systems. Plenum, New York, pp 207–220

Braitenberg V (1984) Vehicles. MIT Press, Cambridge, p 65

Crutchfield JP, Packard NH (1982) Symbolic dynamics of one-dimensional maps: entropies, finite precision and noise. Int J Theor Phys 21:433–466

Farmer D (1981) Unpublished preprint

Froehling H, Crutchfield JP, Farmer D, Packard NH, Shaw R (1981) On determining the dimension of chaotic flows. Physica 3D:605–617

Glass L, Mackey MC (1979) Pathological conditions resulting from instabilities in physiological control systems. Ann NY Acad Sci 316:214–235
Hénon M (1976) A two-dimensional mapping with a strange attractor. Commun Math Phys 50:69–77
Killory H, Rossler OE, Hudson JL (1987) Higher chaos in a four-variable chemical reaction model. Phys Lett 122A:341–345
Mandelbrot B (1982) The fractal geometry of nature. Freeman, San Francisco
Martien P, Shaw R (1985) The dripping faucet. Phys Lett 110A:399–402
Mayer-Kress G (ed) (1986) Dimensions and entropies in chaotic systems. Quantification of complex behavior. Springer, Berlin Heidelberg New York (Springer series in synergetics, vol 32)
Mira C (1987) Chaotic dynamics, from the one-dimensional endomorphism to the two-dimensional diffeomorphism. World Scientific, Singapore
Rössler OE (1976) Chaotic behavior in simple reaction systems. Z Naturforsch 31 a:259–264
Rössler OE (1977) Chemical turbulence – a synopsis. In: Haken H (ed) Synergetics. Springer, Berlin Heidelberg New York (Springer series in synergetics, vol 1)
Rössler OE (1979) An equation for hyperchaos. Phys Lett 71A:155–157
Rössler OE (1980) Chaos and turbulence. In: Haken H (ed) Dynamics of synergetic systems. Springer, Berlin Heidelberg New York, pp 147–153 (Springer series in synergetics)
Rössler OE (1981) An artificial cognitive-plus-motivational system. Prog Theor Biol 6:147–160
Rössler OE (1983) The chaotic hierarchy. Z Naturforsch 38a:788–801
Rössler OE (1987a) Anaxagoras' idea of the infinitely exact chaos. In: Marx G (ed) Teaching nonlinear phenomena, vol 2. Chaos in education. Hungarian Publications in Physics Education, National Center for Educational Technology, Veszprém, Hungary, pp 99–113
Rössler OE (1987b) Chaos in coupled optimizers. Ann NY Acad Sci 504:229–240
Rössler OE, Kahlert C, Parisi J, Peinke J, Röhricht B (1986) Hyperchaos and Julia sets. Z Naturforsch 41a:819–822
Rössler OE, Klein M, Hudson JL, Wais R (1988) Self-similar basin boundary in an invertible system (folded-towel map). In: Kelso JAS, Mandell AJ, Shlesinger MF (eds) Dynamic patterns in complex systems. World Scientific, Singapore, pp 209–218
Ruelle D, Takens F (1971) On the nature of turbulence. Commun Math Phys 20:167–195
Smale S (1967) On differentiable dynamical systems. Bull Am Math Soc 73:747–829
Stroop R, Meier PF (1988) Evaluation of Lyapunov exponents and scaling function from time series. J Opt Soc Am[B]5:1037–1045
Todt DJ (1969) On the control of irregular behavior sequences: results of an analysis of the song of the blackbird, *Turdus merula* (in German) In: Marco H, Farber G (eds) Kybernetik 1968. Oldenbourg, Munich, pp 465–485
Winfree AT (1980) The geometry of biological time. Springer, Berlin Heidelberg New York
Wolf A, Swift JB, Swinney HL, Vastano JA (1985) Determining Lyapunov exponents from a time series. Physica 16D:285–317

Estimation of Correlation Dimensions from Single and Multichannel Recordings – A Critical View

A. Babloyantz

1 Introduction

One of the most intriguing features of the nonlinear dynamics of the past decades is the realization that even deterministic dynamics involving only a few degrees of freedom may generate random behavior. The work of Packard et al. (1980) and Takens (1981) showed that from a seemingly random behavior, valuable information could be extracted for the characterization of the dynamics underlying the system which produces the signal. Algorithms were developed whereby from the measurement in time of a single parameter of a system, phase spaces could be constructed, and the correlation dimensions of the attractors, the Lyapunov exponents and the Kolmogorov entropies evaluated (Grassberger and Procaccia 1983a; Farmer et al. 1983, Wolf et al. 1985). Thus, one can distinguish between determnistic dynamics and random processes and classify systems according to their degree of coherence.

Such an approach opened a wide new area of investigation and at last one could tackle problems that were difficult to handle before. The investigations extended to such diverse fields as hydrodynamics (Brandstater et al. 1983), physics (Jeffries 1985), chemistry (Roux et al. 1983), climatic variability (Nicolis and Nicolis 1984), and physiology (Babloyantz et al. 1985).

Electroencephalograms (EEGs) in general show rather random behavior, although a given pattern may characterize a well-defined stage of cerebral activity. Therefore, it was natural to use these new formalisms of nonlinear dynamics for the elucidation of activities underlying the EEG.

During the first meeting of "Dynamics of Sensory and Cognitive Processing of the Brain," Babloyantz et al. (1985), Babloyantz (1988), and Başar et al. (1988) using the formalism of nonlinear dynamics showed that some stages of human and animal EEG may be described by deterministic chaotic dynamics involving only a few degrees of freedom and exhibiting a low value of correlation dimension.

Two years later, at a meeting of the same group, not only the number of papers dealing with deterministic chaos had increased substantially, but a workshop was devoted to the subject of nonlinear dynamics and its relevance to the study of the brain.

The aim of the present paper is to discuss the modality and validity of these methods of nonlinear dynamics for the study of the EEG. Among these methods spectral analysis, time autocorrelation function, and phase-space construction are qualitative methods, whereas Lyapunov exponents, Kolmogorov entropies,

Springer Series in Brain Dynamics 2
Edited by Erol Başar and T. H. Bullock

and the attractor dimension give quantitative information. Here we restrict ourselves to the evaluation of the correlation dimension using the by now standard algorithm proposed by Grassberger and Proccacia (1983b). The concepts of fractal dimension, strange attractors, and Grassberger and Proccacia algorithms, could be found in Schuster (1984) and also in the paper by Başar et al. (this volume). We also show how one can extend the dimensional analysis to the multichannel recordings.

2 Phase-Spaces

The first step in any dynamic analysis is the construction of the phase portrait. Once the phase portrait is available, its topological dimension may be readily evaluated. In the absence of any knowledge about the details of the system's dynamics, a phase-space topologically equivalent to the actual system may be constructed from time series provided from the variation of a single parameter of the system.

The standard phase space construction necessitates only a single time series. All the relevant variables of the system are constructed from this time series. The construction relies on the fact that if $X(t)$ is the original time series, then the space spanned by the variables $X(t)$, $X(t+\tau)$, $X(t+2\tau)$ ... $X(t+(n-1)\tau)$ is at least topologically equivalent to the original phase space and therefore represents many of its dynamic properties. Moreover, n is restricted by the relation $n>2d+1$ where d is the lowest integer greater than the dimension of the attractor (Grassberger and Porcaccia 1983b). We refer to such a procedure as the "lagging method."

The lagging method is simple and straightforward. However, although in principle, for infinite time series every value of the lag τ is acceptable, in reality great care must be taken in the choice of τ. We shall come back to this question in the next section where we show the importance of the value of τ in the determination of the correlation dimension.

Recently Eckman and Ruelle (1985) have conjectured that the phase space spanned by several simultaneous measurements of an observable variable in various sites of the system $X_1(r_1,t)$, $X_2(r_2,t)$ ... $X_1(r_n,t)$ may also yield a phase portrait which is topologically equivalent to the portrait obtained from the variables X_1, ...X_m of the system. Such a phase space may be constructed very easily from multichannel EEG recordings. In this construction the lag τ disappears and is replaced by another subjective quantity, namely the interelectrode distance which is bounded by a minimum length imposed by the physical reality.

A more recent technique for phase-space construction using a single time series was introduced by Broomhead and King (1986). It is based on singular value decomposition which is a noise-reducing procedure. In essence, in this technique one diagonalizes the covariance matrix constructed from the phase-space vectors of the previous constructions and one obtains orthogonal eigen vectors which may be used to reconstruct a phase portrait and evaluate its correlation dimension. Singular values decomposition may be applied also to multichannel recordings.

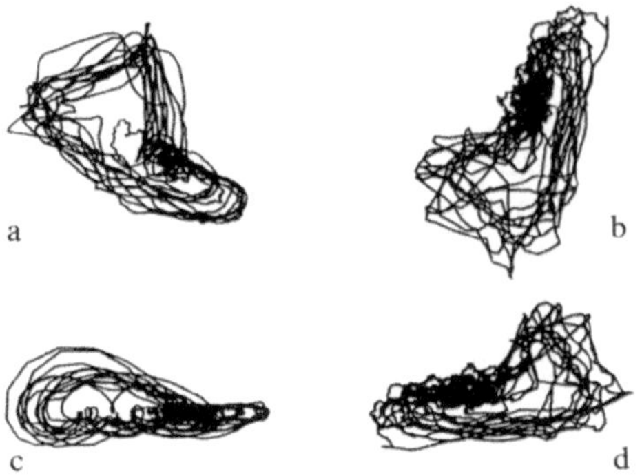

Fig. 1 a–d. Phase portraits of the Creutzfeld-Jakob attractor. A similar structure is seen from four different three-dimensional constructions: (a) Lagging method from one lead, (b) multichannel phase portrait, (c) and (d) are respectively the (a) and (b) portraits constructed using the singular eigen vectors

Figure 1 shows the phase portrait obtained from the EEG of a patient suffering from Creutzfeld-Jakob disease (Babloyantz and Destexhe 1988). All four procedures outlined above have been used. We note that the lagging method, multichannel portrait, multichannel data, after singular value decomposition, yield very similar phase portraits. On the contrary, the portrait obtained with the help of the singular eigen vectors and from a single time series (Fig. 1 c) looks different and is flat and smoother as noise and also probably some of the finer twists of chaotic trajectories have been eliminated.

3 Attractor Dimension

In order to discriminate between a deterministic and random activity, one evaluates the dimension D of the attractors which have been constructed in the phase space. Several definitions for the dimension exist. In this paper we shall concentrate on correlation dimension D_2 which can be evaluated for all attractors shown in Fig. 1.

The procedure for the evaluation of the correlation dimension from a single time series has been reported in several papers and will not be reproduced here. However, substantial error in the value of D_2 may appear if the recorded data have not been handled carefully. Several sources of error are possible.

1. Sampling frequency. The choice of the sampling frequency may be a source of error. Since the computation time approximately scales as $N \times (N-1)$, oversampling has to be avoided. In the case of spectral analysis, the usual rule adopted to avoid any loss of frequencies due to digitization is derived from the Shannon rule which says that the lowest sampling frequency is $2 \times f_{max}$, where f_{max} is the higher peak in the power spectrum. Can similar rules hold for dimensional analysis? We studied the same alpha waves sampled at three different frequencies (200, 400, and 1200 Hz) (Babloyantz and Destexhe 1986). In these cases, the power spectrum does not show significant peaks higher than 100 Hz. We observe that oversampling leads to substantial deformation of the spatial correlation integrals and to an underestimation of the correlation dimension. Recent algorithms have been proposed which prevent the onset of such deformations (Theiler 1986). We found that a sampling frequency of the order of 200 Hz and 15 s duration seems to be the most appropriate sampling rate for the evaluation of the dimension of the alpha waves.

2. *Time delay* τ. Although in principle all values of the time delay τ are acceptable for an infinite number of data points, in practice only a narrow range of τ will give correlation integrals with sufficiently large stretches of linear regions.

In spite of lack of rigorous theoretical evidence, procedures have been proposed in order to find an acceptable value for the delay τ (Fraser and Swinney 1986). These simply postulate that the best value is the one for which the variables are sufficiently uncorrelated such that minimum common information is present in the dynamics. Unfortunately this procedure does not always furnish the best τ especially for the EEG data. As the best value of τ, some authors prefer to choose the first zero of the autocorrelation function. In any case a check for stationarity of this parameter is strongly recommended. A wide range of values of τ must be considered before D_2 as a function of τ remains constant.

3. *Data length*. In order to get a good evaluation of D_2, ideally one must consider several hundred thousand points. However, this is not always possible, especially, in the case of the EEG where the dynamics do not remain stationary for very long. If the phenomenon is of short duration and if the dynamics are characterized by a rather low dimension, acceptable results may be found with several thousand points. In other cases, short time series may seriously underestimate the value of D_2. For example when analyzing the alpha waves according to the length of the time series, values as low as $D_2 = 2.6$ and as high as $D_2 = 6.6$ may be found. Here also one must use a data length such that a very substantial increase in the time series does not change the value of D_2. Of course the time must not exceed the duration of the stationarity of the phenomena.

4. *Upper limit of* D_2. When analyzing REM sleep and β waves which look like highly random signals, surprisingly, one finds respectively values of $D_2 = 9.7$ and $D_2 = 8.9$. We are not sure that such findings indicate the presence of a very highly deterministic chaos in these two stages of brain activity. Our suspicion is founded in the fact that even random white noise saturates at high embedding dimensions and gives a correlation dimension of the order of nine and ten. To be sure that the Grassberger and Proccacia algorithm does not saturate for random signals, calculations must be performed with time series completely free of any type of correlation. Until then we recommend that any value of $D_2 > 8$ be considered as suspicious.

If the abovecited precautions are taken into account, the correlation dimension may be evaluated reliably by digitizing the EEG. Such an analysis has been applied to the alpha waves, several stages of the sleep cycle, Creutzfeld-Jakob disease and petit mal type epilepsy. All these stages show the presence of deterministic chaos (Babloyantz 1988; Babloyantz et al. 1985; Babloyantz and Destexhe 1986; 1987 a, b, 1988 a).

The EEG probably manifests synchrony stages of neural masses. Low amplitude EEG reflects relatively desynchronized states, whereas high amplitude waves are an indication of synchrony between neural masses. In Fig. 2 the correlation dimension of these stages of the brain activity is plotted against a measure of the amplitude of the EEG. To obtain this measure, the signal is computed at 1-s in-

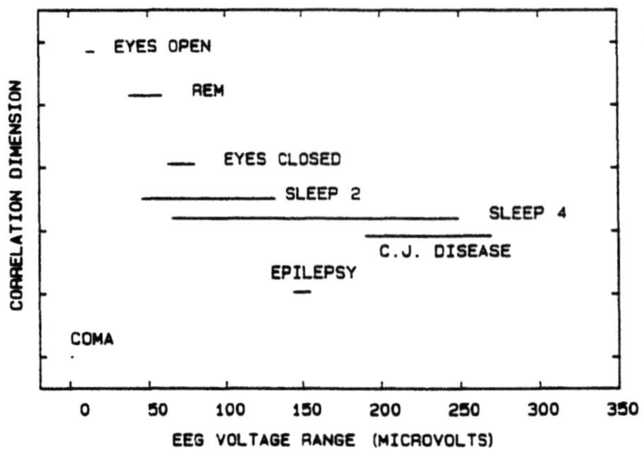

Fig. 2. Dimension-amplitude plot. The variations of EEG voltage observed at the optimal sampling frequencies (see text) and in successive stretches of 1 s are represented for each type of EEG behavior. The synchronization between neurons which occur in pathologies is reflected by high amplitudes and low dimensions

tervals. The highest and lowest amplitudes are included in the horizontal bars. During the normal states of brain activity, the dimension of the chaotic attractors decreases as the amplitude of the waves increases. The reverse situation is seen in pathological conditions. Creutzfeld-Jakob disease is characterized by a higher amplitude and a higher dimension than the epileptic seizure (Babloyantz and Destexhe 1987a, b, 1988a).

Multichannel recording is a common practice in most EEG laboratories. Therefore it would seem that multichannel analysis is much more economical and more suitable for study of the brain. However, here also one faces a new dilemma. How small or large should the interelectrode distance be?

If deterministic chaos is detected from a single-channel recording, it indicates the presence of chaotic activity in the recorded site. Nothing guarantees that the time series from an adjacent lead will show a chaotic activity or if it does, whether we are dealing with the previous chaotic attractor. Therefore, prior to the use of multichannel analysis, each channel must by studied by the lagging method to make sure that all of them belong to the same dynamics.

Presently we are engaged in such multichannel study of the EEG data. The analysis could be applied to multielectrode recording either in a horizontal setup or in a vertical insertion. In the case of the human EEG recorded from the scalp, many channels are contaminated with noise originating from muscular activity. This fact makes it difficult to find several channels that yield the same D_2 when the lagging technique is used. For example, in the case of Creutzfeld-Jakob disease, the values of D_2 in different electrodes range from $D_2=3.7$ to $D=5.4$. This discrepancy may indicate that the disease does not involve a single attractor, or the unique attractor is highly contaminated with artifacts. If these leads are used in a multichannel analysis we find a value of $D_2=3.8$.

Better results could be obtained if one uses the EEG recorded in animal studies where electrodes are either inserted into the cortex or the activity is recorded from the surface of the cortex. However, in order to illustrate and test the procedure we consider multichannel electrical signals which are obtained by measuring the electrocardiogram (ECG). In this case there is no doubt that all leads must reflect the same dynamics, namely the activity of the cardiac muscle. The multichannel ECG of a normal individual, comprising 32 leads, is first analyzed chan-

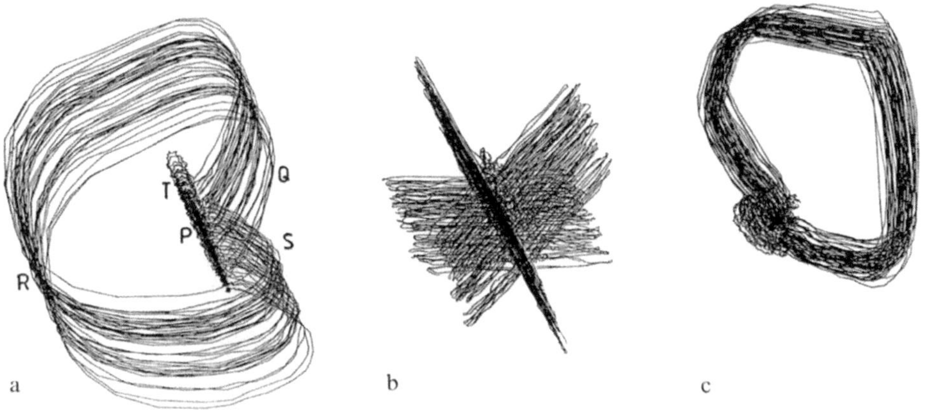

Fig. 3 a–c. Phase portraits of human ECG constructed in a three-dimensional space. A two-dimensional projection is displayed for two values of the delay, (a) 12 ms and (b) 120 ms; (c) represents the phase portrait constructed from the three simultaneous leads

nel by channel using the usual lagging method. Six leads of 240 s duration (60000 points) gave $D_2 = 3.6 \pm 0.01$. Multichannel analysis was performed by using a combination of five leads chosen among those channels analysed by the lagging method. We found a value of $D_2 = 2.9 \pm 0.1$, which is in good agreement with previously reported results (Babloyantz an Destexhe 1988).

In this example and in all the other examples analyzed by us, we see a systematic discrepancy between the lagging technique and multichannel analysis. The latter always furnishes a value of D_2 smaller than the one obtained by the lagging method. The reason for this may be found in the following observation. The phase portrait obtained from a single channel using small τ has a striking resemblance to the one obtained from multichannel time series (see Fig. 3 a and c) whereas for the optimal τ the phase portrait is quite different (Fig. 3 b). The discrepancies between the two methods may be understood in the following manner. In the lagging procedure, as τ is increased, the attractor is gradually unfolded (the correlation intergrals show several simultaneous linear regions) showing its full complexity for the "best" τ after stationarity is reached. This gives a rather high value of D_2. On the contrary, when using space vectors, some of the variabilities of the dynamics are smoothed out and we observe a lower value of D_2.

The use of space vectors has many advantages over the usual lagging method: (a) one avoids costly trials before finding the appropriate time lag τ; (b) one sees a single, long and smooth linear region. Thus, the evaluation of the slope, D_2, is less ambiguous and more accurate.

4 A Visual Representation of Chaos

In the analysis of the physiological signals, sometimes, one may encounter almost periodic signals which, however, show obvious deviations from a purely limit cycle type of behavior. This is the case, for example, in the ECG. It is important

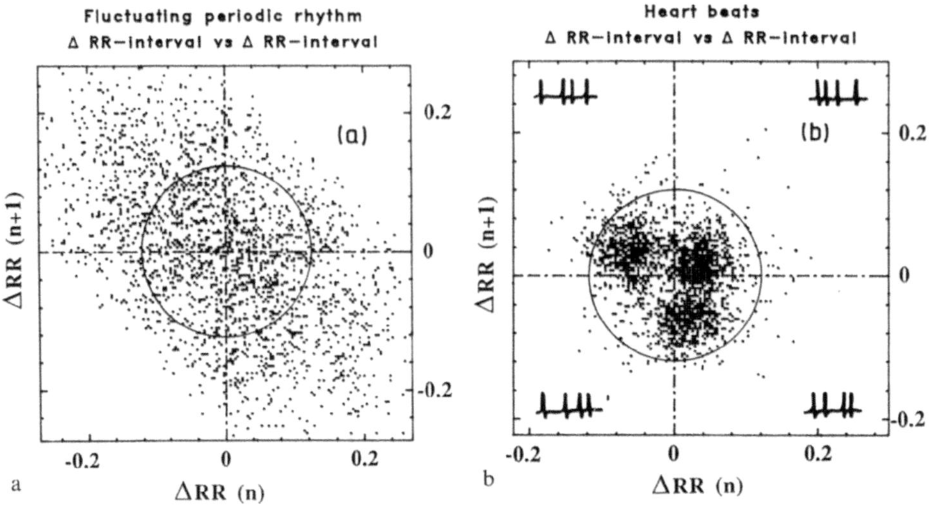

Fig. 4a, b. Representation of the interbeat interval correlation. The difference between two successive beat-to-beat intervals is plotted against the same difference one beat later. (a) Pseudorandom fluctuations around a periode oscillator. (b) Behavior of a normal heart. We see an obvious correlation between the variabilities of three consecutive RR intervals. (A digital filtering of the ECG below 0.5 Hz before the extraction of the RR intervals does not alter this structure)

to distinguish whether one is in the presence of a deterministic chaos or the purely periodic signal is contaminated by random noise. Such a distinction is possible with the help of the following analysis:

Let RR(n) represent the time interval between two successive peaks of the almost periodic signal. One constructs a map of interval variation Δ RR($n+1$) versus Δ RR(n). This construct reveals the correlations between the three successive peaks. For a perfect oscillator, the origin is the only possible value of Δ RR, whereas a noise-prone oscillator will exhibit a uniform distribution (Fig. 4) reflecting the statistical independence of the successive values of Δ RR. Figure 3 b maps the variability of the normal heart. An obvious "structure" is seen in the fluctuations of the RR intervals which is an unambiguous sign of coherence in the interbeat variability (Babloyantz and Destexhe 1988).

5 Discussion

Brain activity is a time-evolving process and the only way to extract information from the EEG is to use methods which look at the dynamic aspect of the cerebral cortex. The nonlinear dynamics have furnished several methods of analysis suitable for such a study. In this paper we have only concentrated on the correlation dimensional analysis. However, today, from the same time series, we may compute other dynamic quantities such as autocorrelation functions, Kolmogorov entropies, Lyapunov exponents, and also construct Poincaré sections which are usable only for low-dimensional attractors. All these methods must be used together with dimensional analyses outlined in this paper. As we saw, the latter may give divergent results if various parameters of the time series are not chosen with care.

With this type of analysis, from the time measurement of a single variable, it is possible to assess the degree of coherence of the system, to discriminate between noise and deterministic dynamics, and to establish a hierarchy between several states of the same cortex. The finding that several key stages of brain activity follow deterministic chaotic dynamics sheds some light on the mechanism of information-processing capabilities of the cortex. This capability stems from the fact that chaotic attractors are sensitive to the initial conditions.

Moreover, these powerful tools of data analysis may help to bridge the gap between experimental and theoretical approaches to complex networks. It imposes constraints on model construction. A given model must finally reproduce dynamic properties of the real system as computed from the notions mentioned above.

The principal aims of this paper have been: (a) to show the pitfalls of the usual lagging method and how to avoid them; (b) to test for the first time, using physiological data, a conjecture put forward whereby one avoids some of the difficulties of the lagging method by using multichannel recordings; (c) to apply to physiological data the singular decomposition technique; and (d) to introduce a new way of characterizing more regular signals. However, research in this promising field of nonlinear dynamics is progressing rapidly. Algorithms exist which enable the evaluation of the number of local dimensions of attractors. A comparative study of all the existing methods and their use in the study of physiological signals may be found in a recent paper by Destexhe, Sepulcre, and Babloyantz (1988).

6 Summary

Recent developments in the theory of nonlinear dynamics provide several means for the study of complex systems. From the measurements of a single variable in the form of a time series, several important properties of the system can be assessed. Phase space construction, time autocorrelation function, and spectral analysis are qualitative methods, whereas correlation dimension, Lyapunov exponents and Kolmogorov entropy may be evaluated quantitatively. We discuss the relevance of such an approach to the study of EEG, evaluating the advantages and shortcomings of existing algorithms. We show the sensitivity of the correlation dimensions to the choice of the length of the data sets and also to the value of the lag. We applied these methods to the study of various stages of EEG in normal as well as in pathological instances. In normal individuals we show that as the wake-sleep cycle sets in, some stages of the brain activity oscillate between several deterministic dynamics of chaotic nature characterized by various degrees of coherence. The β waves need more than ten degrees of freedom for their description, whereas α waves show a chaotic attractor of rather high correlation dimension $D_2 = 6.1$ which may be described by deterministic dynamics with few degrees of freedom. Sleep stages two and four are respectively characterized by chaotic attractors with $D_2 = 5.03$ and $D_2 = 4.05$–4.40. As REM sleep sets in, again the coherence of the brain is lost and chaotic attractors vanish. In the case of severe pathologies such as coma and "petit mal" type epilepsy, the dimension of the chaotic attractor drops to a low value of $D_2 = 2.05$.

References

Babloyantz A (1988) Chaotic dynamics in the brain activity. In: Başar E (ed) Dynamics of sensory and cognitive processing of the brain. Springer, Berlin Heidelberg New York (Springer series in brain dynamics, vol 1)

Babloyantz A, Destexhe A (1986) Low dimensional chaos in an instance of epilepsy. Proc Natl Acad Sci USA 83:3513

Babloyantz A, Destexhe A (1987a) Strange attractors in the human brain. In: Rensing L, an der Heiden U, Mackey MC (eds) Temporal disorder in human oscillatory systems. Springer, Berlin Heidelberg New York, p 48 (Springer series in synergetics, vol 36)

Babloyantz A, Destexhe A (1987b) Chaos in neural networks. In: Proceedings of the first IEEE international conference on neural networks, San Diego, June 1987

Babloyantz A, Destexhe A (1988a) The Creutzfeld-Jakob disease in the hierarchy of chaotic attractors. In: Markus M, Müller S, Nicolis G (eds) From chemical to biological organization. Springer, Berlin Heidelberg New York p 307 vol 39

Babloyantz A, Destexhe A (1988b) Is the normal heart a periodic oscillator? Biol Cybern 58:203

Babloyantz A, Nicolis C, Salazar M (1985) Evidence of chaotic dynamics during the sleep cycle. Phys Lett 111A:152

Brandstater A, Swift J, Swinney HL, Wolf A, Farmer JD, Jen E, Crutchfield JP (1983) Low dimensional chaos in a hydrodynamic system. Phys Rev Lett 51:1442

Broomhead DS, King G (1986) Extracting qualitative dynamies from experimental data. Physica 20D:217

Destexhe A, Sepulchre JH, Babloyantz A (1988) A comparative study of the experimental quantification of deterministic chaos. Phys Lett [A]132:101

Eckmann JP, Ruelle D (1985) Ergodic theory of chaos and strange attractors. Rev Mod Phys 57:617

Farmer JD, Ott E, Yorke JA (1983) The dimension of chaotic attractors. Physica 7D:153

Fraser AM, Swinney HL (1986) Using mutual information to find independent coordinates of strange attractors. Phys Rev 38A:1134

Freeman WJ (1986) Petit mal seizure spikes in a factory bulb and cortex caused by runaway inhibition after exhaustion by excitation. Brain Res Rev II:259

Grassberger P, Procaccia I (1983a) Characterization of strange attractors. Phys Rev Lett 50:346

Grassberger P, Porcaccia I (1983b) Measuring the strangeness of strange attractors. Physica 9D:189

Nicolis C, Nicolis G (1984) Is there a climatic attractor? Nature 311:529

Packard NH, Crutchfield JP, Farmer JD, Shaw RS (1980) Geometry from a time series. Phys Rev Lett 45:712

Roshke J, Başar E (1988) The EEG is not a simple noise: strange attractors in intracranial structures. In: Başar E (ed) Dynamics of sensory and cognitive processing by the brain. Springer, Berlin Heidelberg New York (Springer series in brain dynamics, vol 1)

Roux JC, Simoyi RM, Swinney HL (1983) Observation of a strange attraction. Physica 8D:257

Shuster H (1984) Deterministic chaos. Physik, Weinheim

Takens F (1981) Detecting strange attractors in turbulence. In: Rand DA, Young LS (eds) Dynamical systems and turbulence, Warwick 1980. Springer, Berlin Heidelberg New York, p 366 (Lectures notes in mathematics, vol 898)

Theiler J (1986) Spurious dimensions from correlation algorithms applied to limited time series data. Phys Rev 34A:2427

Wolf A, Swift JB, Swinney HL, Vastano JA (1985) Determining Lyapunov exponents from a time series. Physica 16D:285

Correlation Dimensions in Various Parts of Cat and Human Brain in Different States

J. RÖSCHKE and E. BAŞAR

1 Introduction

The analysis of deterministic chaos is currently an active field in many branches of research. Mathematically, all nonlinear dynamical systems with more than two degrees of freedom can generate chaos and, therefore, become unpredictable over a longer time scale. In order to describe periodic, aperiodic, or even chaotic behavior of nonlinear systems, several approaches have been used. In 1963 Lorenz applied concepts of nonlinear dynamics to the convection phenomenon in hydrodynamics in order to describe atmospheric turbulence (Navier-Stokes equation). He demonstrated the possibility that the unpredictable or chaotic behavior observed in an infinite-dimensional system might be caused by a three-dimensional dynamical system. Our research group's first nonlinear approach consisted in comparing the relation between EEG and evoked potentials by considering the Duffing equation as an adequate nonlinear model (Başar 1980). Later, assuming the EEG to be a chaotic attractor and mentioning the possibilities of applying the Navier-Stokes equation for comparison, we described that the EEG might reflect properties of a strange attractor (Başar 1983; Başar and Röschke 1983).

Since the pioneering analysis by Babloyantz et al. (1985), several investigators have undertaken the nonlinear analysis of the EEG using Grassberger and Procaccia's algorithm in order to evaluate the correlation dimension D_2 (see for example Babloyantz et al., Saermark et al., Graf and Elbert, Skinner et al., this volume). Our first analysis involved the evaluation of D_2 for EEG activities of intracranial structures of the cat brain (Röschke and Başar 1985a, b, 1988; Başar et al. 1988). Although the present study also includes our earlier results in slow-wave sleep (SWS) for comparative analysis, our major aim was to look at a new frequency window – the range between 100 and 1000 Hz. In this frequency range the spontaneous activity of the cerebellum and the brain stem depicts ample electrical activity and resonances evoked by sensory stimulation (Adrian 1935; Başar 1980). Further, a first evaluation of D_2 during the transition stages was undertaken. An important trend in our analysis is also description of the dimensionality of EEG including the large fluctuations of D_2 during transitions. However, we submit that the use of other descriptors (for example, description by means of power spectra) is necessary to complement the interpretation of dimensionality.

Springer Series in Brain Dynamics 2
Edited by Erol Başar and T. H. Bullock

2 Definitions, Concepts and Explanations

A set of first-order differential equations

$$\dot{X}_i = F_i(x_i, t) \qquad (i = 1, \ldots, n)$$

is called a dynamical system. If time does not appear explicitly in the function F_i, the system is called *autonomous*. It is a well-known fact that the nonlinearity of the F_i is a necessary (but not sufficient) condition for the generation of deterministic chaos. Deterministic chaos means that the behavior of the system is not predictable over longer periods. Nevertheless, there exists a prescription, for example, in terms of differential equations, for calculating future behavior from given initial conditions. Often investigations of dynamical systems have been performed in so called phase-space.

In general, phase-space is identified with a topological manifold. An n-dimensional phase-space is spanned by a set of n-independent linear vectors. This requirement is usually sufficient. Following a proposal made by Takens (1981) we span a 10-dimensional phase-space by $x(t), \ldots, x(t+9\tau)$ where τ is a fixed time increment. Every instantaneous state of a system is therefore represented by a set $(x_1, \ldots, x_n)$, which defines a point in the phase-space. The sequence of such states over the time scale defines a curve in the phase-space, called a trajectory. As time increases, the trajectories either penetrate the entire phase-space or they converge to a lower dimensional subset (Fig. 1). In this latter case, the set to which the trajectories converge is called an attractor.

In relation to the topological dimension of the remaining attractor, one can deduce various properties of the investigated system. If the Hausdorff dimension of an attractor is a noninteger, the attractor is called a strange attractor. All strange attractors which have been encountered up to now have a fractal dimension. In order to give a geometric explanation of the computation of the dimension, we will illustrate schematically a proposal made by Grassberger and Procaccia (1983) to compute the so-called correlation dimension D_2. Assuming that the

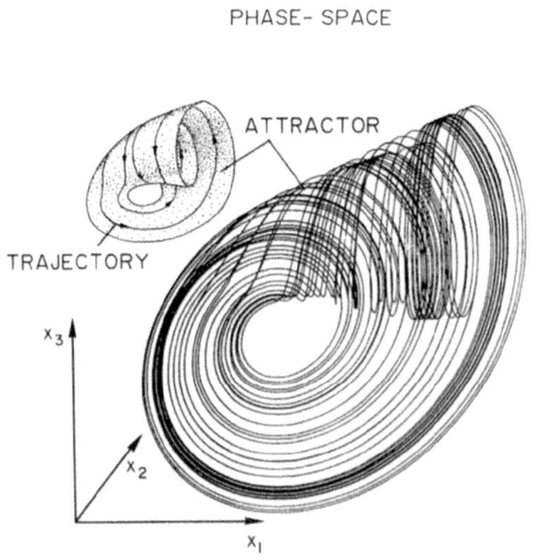

Fig. 1. Trajectories converge to a lower dimensional subset called an attractor (Modified after Abraham and Shaw 1983)

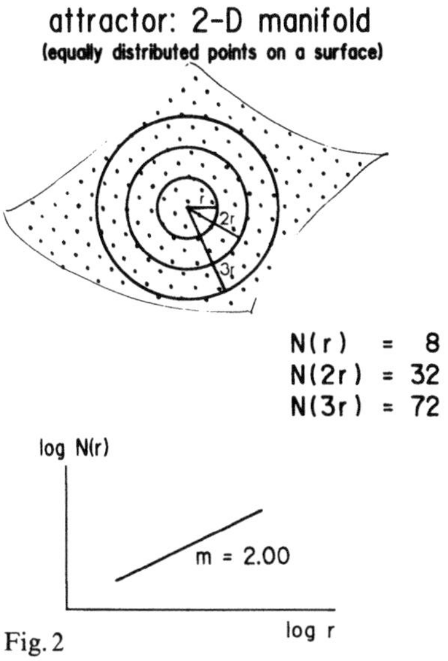

Fig. 2

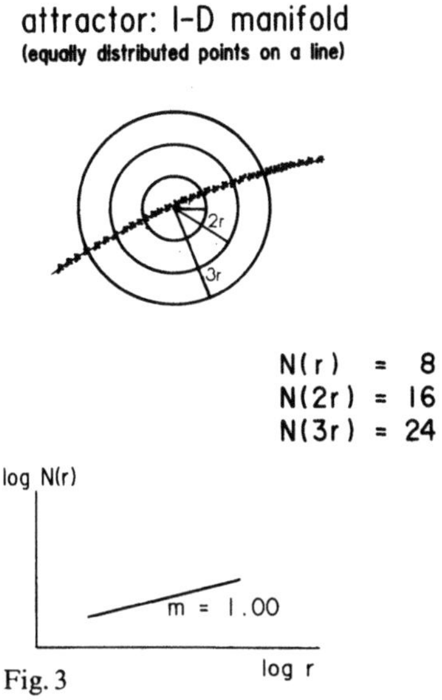

Fig. 3

Fig. 2. For every point of the 2-D attractor the number N(r) of points lying inside a circle of radius r are counted. By plotting log N(r) versus log r a straight line with slope $m=2.00$ is obtained

Fig. 3. By counting the number of points N(r) lying inside a circle of radius r and plotting log N(r) versus log r a straight line with slope $m=1.00$ is obtained

attractor represents a two-dimensional manifold in the phase-space, it is possible to evaluate the dimension in the following manner. Figure 2 illustrates the basic features of the computational algorithm. For every point of the attractor, the number of points lying inside a circle are counted (or in the case of a three-dimensional phase-space inside a ball) which have a radius of r_0, $2r_0$, $3r_0$, etc. For the two-dimensional manifold $N(r_0)=8$, $N(2r_0)=32$, $N(3r_0)=72$, ... Now, if a plot of log (r) versus log N(r) is performed, a straight line is registered. The slope of this line is $m=2.00$. This is exactly the dimension of the attractor. This result never changes, even if a higher dimensional phase-space is considered.

It can be shown that if the attractor is not a two-dimensional manifold but a simple curve in the phase-space, the evaluation of the dimension leads to a value of $D_2=1.0$ (Fig. 3). The number of points lying inside the circle with radius r is $N(r_0)=8$, $N(2r_0)=16$, $N(3r_0)=24$, and so on. By plotting log N(r) versus log r, one finds a straight line with slope $m=1.00$. This is exactly the dimension of the simple curve, which we assumed to be the attractor. In fact, Grassberger and Procaccia's algorithm counts the number of points lying inside the circle for every point of the attractor and averages the results. This method is therefore simple, but time-consuming. Grassberger and Procaccia have shown that it gives reliable results, even though the attractor is a strange attractor. Therefore, it is possible to calculate fractal dimensions as we describe in the following.

One of the most important properties of a strange attractor is its sensitive dependence on initial conditions. This means that points which are arbitrarily close initially, become macroscopically separated after sufficiently long times. In other

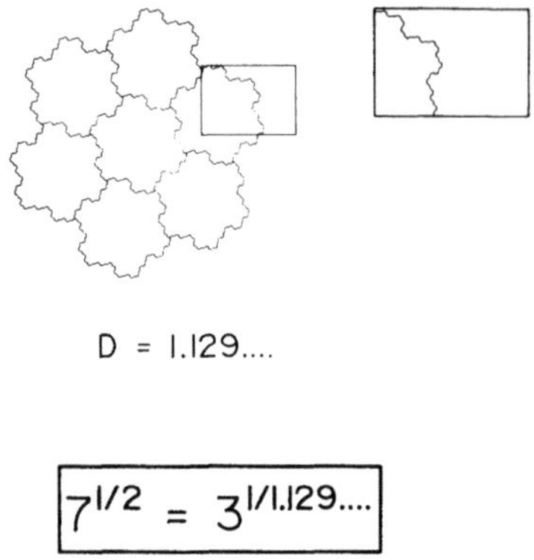

Fig. 4. A self-similar object with a fractal dimension of $D = 1.129$ ... (After Schroeder 1986)

words, similar causes do not produce similar effects. This is an extensive statement, which seems to damage the causality principle of natural philosophy. However, by examining the properties of a strange attractor more precisely, one finds that a strange attractor may have a strong conformity, called self-similarity, which is an invariance with respect to scaling.

Self-similar objects possess a fractal dimension. And "fractals," so called because their fractal or Hausdorff dimension exceeds their topological dimension and show strange properties, which are not known in linear systems (see Schuster 1984).

Let us now consider Fig. 4. The main pattern is obviously similar to its seven parts. The area ratio is 7:1 and the perimeter is three fold the perimeter of one of the seven similar pieces (Schroeder 1986).

Euclid's geometry tells us that areas of similar figures are proportional to their linear dimensions squared. However, 3 squared equals 9 and not 7. Nowadays, we have to point out that geometric objects like the one shown in Fig. 4 do not obey Euclid's rules of geometry. The boundaries of the illustration in Fig. 4, while everywhere continous, are nowhere differentiable. This object is a fractal with a Hausdorff or fractal dimension of $D = 1.129$ ...

In this case we can reformulate Euclid's theorem: For similar figures, the ratios of corresponding measures are equal when reduced to the same dimension. In our case:

$$7^{1/2} = 3^{1/1.129}$$

Another example, which shows the strange properties of self-similar objects is presented in the following digital set:

10010110011010010...

This infinite binary sequence has the property that it is self-similar. In fact, striking out every second term the sequence reproduces itself

1~~0~~0~~1~~0~~1~~1~~0~~0~~1~~...

Surprisingly, although the binary sequence shown is aperiodic, its Fourier transform is not at all noise-like (Fig. 5). The self-similarity of the sequence induces

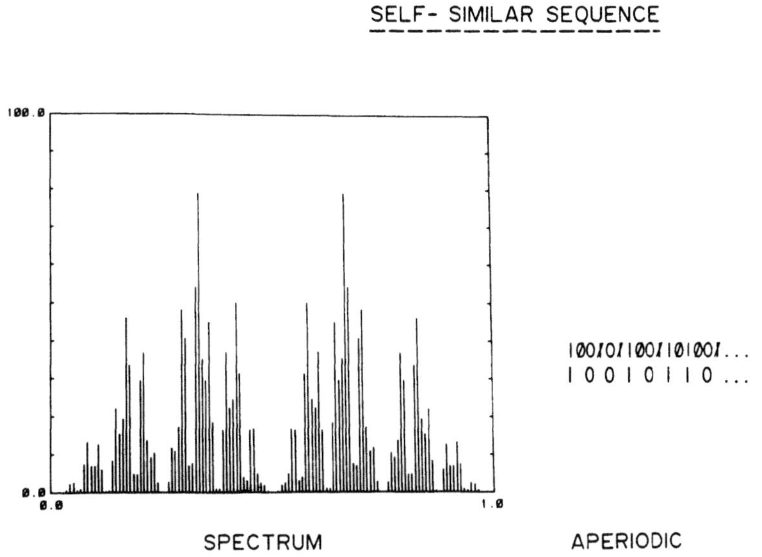

Fig. 5. Spectrum of the self-similar sequence (after Schroeder 1986), striking out every second term, the sequence reproduces itself

long-range correlations and causes its spectrum to be highly structured with many peaks as if it were periodic. In other words, self-similar sequences can be both aperiodic and have spectra resembling those of a periodic sequence.

3 Methods

3.1 Mathematical Methods

One of the oldest concepts of dimension is that of a topological dimension D_T. For a point, $D_T=0$; for a line, $D_T=1$; and for a plane $D_T=2$. A first generalization is the Hausdorff dimension or fractal dimension D_F. For simple sets, for example, a limit cycle or a torus, the fractal dimension D_F is an integer and is equal to the topological dimension D_T. For an n-dimensional phase-space, let $N(\varepsilon)$ be the number of n-dimensional balls (or cubes) of radius ε required to cover an attractor. Then the fractal dimension D_F is defined as

$$D_F=\lim_{\varepsilon\to 0}\frac{\log N(\varepsilon)}{|\log\varepsilon|}$$

For the self-similar geometric object (see Fig. 4), one can choose: $\varepsilon=(\frac{1}{3})^n$, then $N(\varepsilon)=(1/\sqrt{7})^n$ and it follows that

$$D_T=\lim_{\varepsilon\to 0}\frac{\log N(\varepsilon)}{|\log\varepsilon|}=\lim_{n\to\infty}\frac{\log 3^n}{\log\sqrt{7^n}}=\frac{\log 3}{\log\sqrt{7}}=1.12915\ldots$$

A generalization of the fractal dimension is introduced in information theory. The Renyi information of order q is defined as

$$I_q = \frac{1}{1-q} \log \sum_{i=1}^{N(\varepsilon)} p_i^q \qquad (q \neq 1)$$

$$I_q = - \sum_{i=1}^{N(\varepsilon)} p_i \log p_i \qquad (q = 1)$$

Let p_i be the probability that an arbitrary point (of an attractor) falls into ball i with radius ε and let $N(\varepsilon)$ be the number of nonempty balls. The generalized dimension D_q of order q is given by

$$D_q = \lim_{\varepsilon \to 0} \frac{I_q(\varepsilon)}{\log 1/\varepsilon}$$

For $q=0$ we find $D_0 = D_F$. D_1 is called the "information dimension" and D_2 is called the "correlation dimension." In practice, the correlation dimension D_2 is the generalized dimension easiest to be estimated from attractors generated by experimental data, because

$$I_2 = -\log \sum_{i=1}^{N(\varepsilon)} p_i^2 = -\log C(R)$$

where $C(R)$ is a measure of the probability that two arbitrary points x_i, x_j will be separated by distance R. $C(r)$ is called the "correlation integral" and can be easily computed:

$$C(R) = \lim_{N \to \infty} \frac{1}{N^2} \sum_{\substack{j, i=1 \\ \neq j}}^{N} \theta(R - |x_i - x_j|)$$

where θ is the Heavyside function.

It then follows that

$$D_2 = lim_{R \to 0} \frac{\log C(R)}{\log R}$$

or $C(R) \propto R^{D_2}$.

The main point is that $C(R)$ behaves as a power of R for small R. By plotting $\log C(R)$ versus $\log R$ one can calculate D_2 from the slope of the curve. For an attractor, whose dimension is not known, it is necessary to calculate $C(R)$ for several embedding dimensions. In fact one should choose the maximal dimension n of the embedding phase-space twice the dimension of the attractor. The evaluation of the calculated correlation dimension D_2 should converge towards a saturation value. That means $\Delta D_2/\Delta n = 0$ for some $n > n_0$.

3.2 Technical Considerations

The first step is to find the phase-space description of the digitized data. We have followed Taken's proposal (1981) and used the time-shift method. This

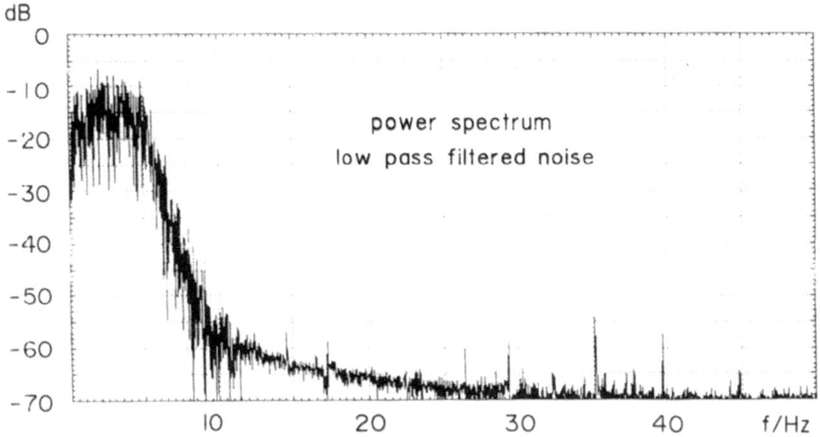

Fig. 6. Spectrum of a low pass filtered noise (cut-off frequency 5 Hz, 48 dB/octave)

means that a point x^i of the k-dimensional phase space is defined as

$$x^i = (x_i, x_i + \tau, \ldots, x_i + (k-1)\tau)$$

where $x_s = x(s \cdot \Delta t)$ and Δt is the sampling interval.

Here, one of the most important questions is the choice of τ. In principle, the calculation of the dimension of an attractor should be independent of the choice of τ. In practice, this independence is given in an interval $\tau_1 < \tau < \tau_0$, where τ_0 is the first zero in the autocorrelation function.

In order to choose the adequate τ, we analyzed a low pass filtered noise, the spectrum of which is shown in Fig. 6. The calculation of the correlation integral was performed for $\tau = (10, 20, 30, 40, 50, 60, 70, 80, 90, 110, 130, 150$ ms). Figure 7 illustrates the dependence of D_2 on different values of τ for three embedding dimensions ($n=2$, $n=5$, $n=7$). One can detect that for different embedding dimensions n, the higher the value of τ is, the higher the calculated correlation dimension D_2. It is obvious that the highest value of D_2 is determined by the embedding dimension n. But for a certain value of $\tau = \tau_{\min}$ no increase of D_2 is observed. In Fig. 8 a plot of $\tau_{\min}$ versus the embedding dimension n is illustrated. Obviously, for every n the maximal value of D_2 is calculated if $\tau > 90$ ms. The autocorrelation function analysis (ACF) of the low pass filtered noise shows that the first zero of the ACF is $\tau_0 = 90$ ms, too. This means that there should be no influence on the calculation of D_2 in a range $\tau_1 < \tau < \tau_0$.

In all of our computations, data sets of $N = 4096$ points have been used. In principle, it should be sufficient to consider only one point x^* of the attractor and evaluate the correlation integral $C(r)$ for all distances $|x^* - x^j|$. In our computations *every* distance $|x^i - x^j|$ $(i \neq j)$ has been taken into account and the results have been averaged. The norm $\| \;\; \|$ we used is the Euclidean norm

$$\| \;\; \| = (\Sigma(x_i - y_j)^2)^{1/2} .$$

Identical results will be obtained if the maximum norm is chosen

$$\| \;\; \| = \max |x_i - y_j|$$

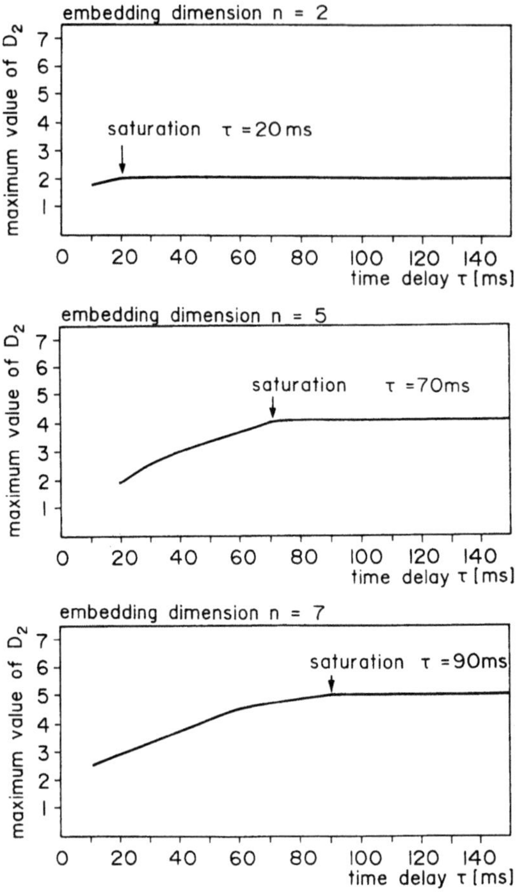

Fig. 7. For three embedding dimensions ($n=2$, $n=5$, $n=7$), the higher the value of τ, the higher the calculated correlation dimension D_2. It is evident that D_2 is lower than or equal to n. For a certain saturation value τ_{min}, no increase in the calculation of D_2 is observed

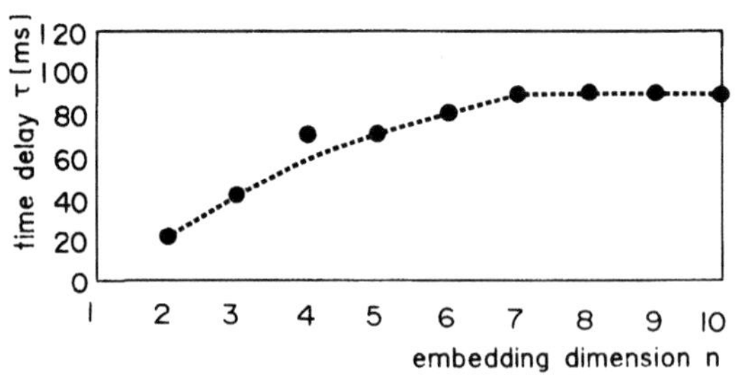

Fig. 8. The saturation value of τ_{min}, from which no further increase in the calculation of D_2 is observed, depending on the embedding dimension n

The slopes of the curves in the plot of "$C(r)$ versus log r" have been evaluated using the linear regression method. The logarithmic scale of the radii r was usually about a range of two decades, which was separated into 64 equal parts, though we evaluated the slope of the curves from 12–24 points. If the correlation coefficient of the linear regression was smaller than 0.98, the logarithmic scale of the radii was changed. Then the dimension D_2 was evaluated for a second time by separating a range of one decade or less into 64 equal parts.

4 Results

4.1 Early Results During Slow-Wave Sleep

In order to analyze the dimensionality of field potentials during slow-wave sleep (SWS), five cats with chronically implanted electrodes were studied. The chronic electrodes were needles of small exposed tips, implanted cortically (subdural) in the auditory cortex (GEA), and subcortically in the hippocampus (HI), and reticular formation (RF). The 15 experimental trials were recorded during SWS. The intracranial EEG signals were low-passed to 50 Hz and then digitized using a 12-bit AD converter and stored in the memory of an HP 1000-F computer. the sampling frequency was $f_s = 100$ Hz for all trials. Dimensions of the EEG signals were evaluated over a time period of about 20 s ($N = 2048$) and 40 s ($N = 4096$). Details of the software have been described elsewhere (Röschke 1986; Röschke and Başar 1988). The phase-space was constructed using the time-delayed coordinate proposed by Takens.

The evaluation of the correlation dimension for the auditory cortex is shown in Fig. 9, which represents an evaluation of the activity from the cat named "Toni" in SWS. One can detect that the slopes of the curves by plotting $N(r)$ versus log r converge against a saturation value of about $D_2 = 5.00$. The time lag τ was $\tau = 50$ ms. Identical results are obtained, if a time lag between 20 ms and 80 ms is chosen.

In order to underline the advantage of this procedure, the power spectrum of the EEG is illustrated (Fig. 10). This spectrum resembles that of colored noise. On the contrary, the case of the phase-space description and the evaluation of the correlation dimension lead to a convergence. Such a saturation is not reached in the

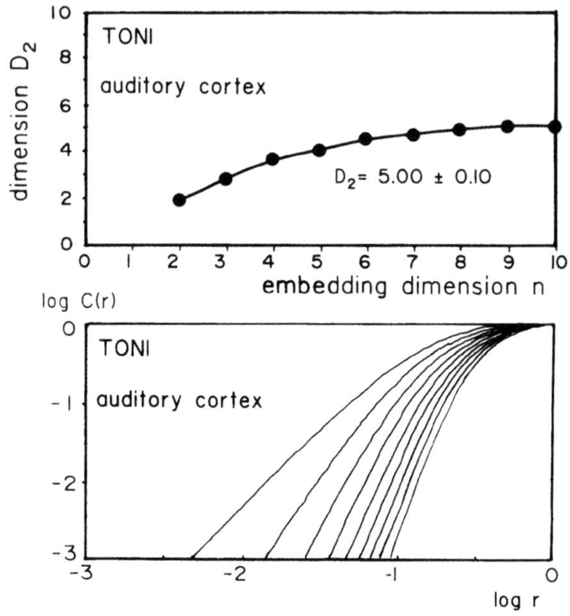

Fig. 9. The plot of log C(r) versus log r (*at the bottom*) leads to a convergence of the slopes of the curves with a saturation value of $D_2 =$ 5.00 (*at the top*); 4096 data points, $\tau = 50$ ms, cat "Toni," SWS, auditory cortex

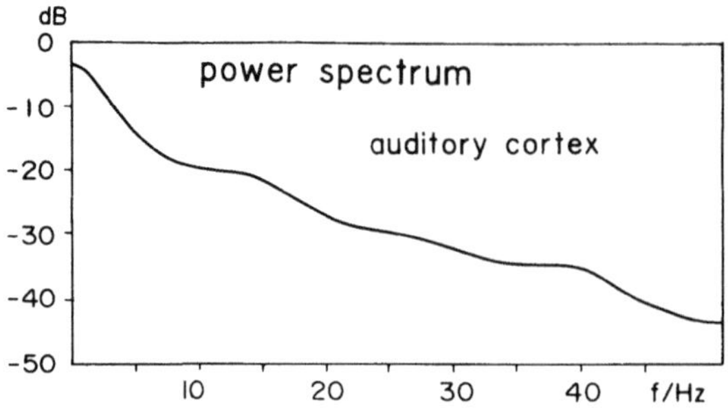

Fig. 10. Power spectrum of EEG epochs analyzed over a period of 40 s (analyzed in Fig. 9). The spectrum was smoothed by means of a previously published linear prediction method (Röschke 1986)

Table 1. Mean values of 15 trials (five cats) from the auditory cortex, reticular formation, and hippocampus

D_{GEA}	$=5.06 \pm 0.31$
D_{RF}	$=4.58 \pm 0.38$
D_{HI}	$=4.37 \pm 0.36$

case of a noisy signal. This is the difference between noise and strange attractors.

The mean values from 15 trials, each of 40 s measured in the acoustic cortex, the reticular formation, and the hippocampus of the cat brain of "Toni" during SWS, are shown in Table 1.

The standard deviations lie within the same range and are in all cases smaller than 10% of the mean. Significant differences between cortical and subcortical structures of the cat brain could be demonstrated. Detailed information on the results from five cats have been presented by Röschke and Başar (1988).

It is to be emphasized that the dimension D_2 of the investigated system, in our case the brain, varies from 4.37 in the hippocampus to up to nearly 5.00 in the cortex and that the dimensions of the different attractors are also relatively stable in experiments with individual cats. Moreover, in nearly 90% of the trials the maximal dimension is detected in the auditory cortex.

4.2 High Frequency Behavior of the Cat Cerebellar Cortex and Brain Stem

Although the spontaneous electrical activity of the brain which we call EEG has its most ample components in a frequency region between 1 and 50 Hz, it is well known that a number of higher frequency components can be observed in the field potentials of the brain.

Usually, the electroencephalographer turns the EEG low-pass filter down to 30 Hz routinely and only frequency regions of theta, alpha, and beta waves are analyzed or interpreted. It was a long-standing interest of our research group to try to describe the brain's spontaneous field potentials in the brain stem (reticular formation and inferior colliculus) and the cerebellum. In these structures we were able to demonstrate resonance phenomena and spontaneous field potential power

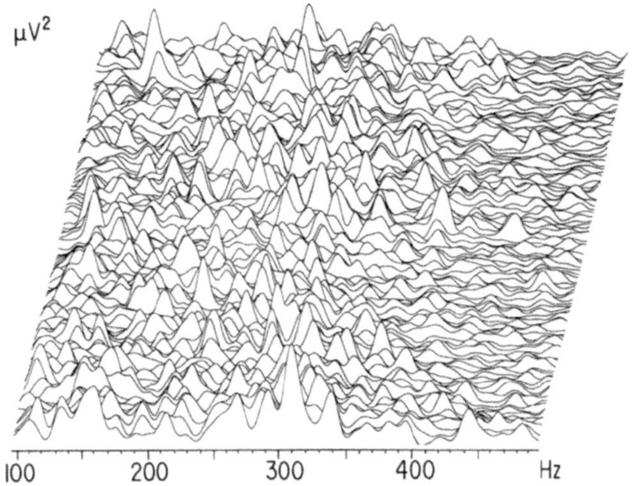

Fig. 11. Power spectral analysis of the electrical activity (cerebellar cortex) during waking stage from cat "Kati"

peaks in a frequency region between 100 and 1000 Hz (Başar 1980). Since the days of Adrian (1935) it has been well known that the cerebellum depicts a high frequency component between 180- and 300-Hz activity. Figure 11 shows a power spectral plot of the activity of the cerebellar cortex measured from a chronically implanted cat during waking stage. Spectral peaks in the frequency range over 100 Hz are observed in all the studied cats ($N=4$). Mostly, activity is dominant in the 200- to 300-Hz frequency range.

One of our questions is how to consider the principle of self-similarity along the frequency scale of the brain's spontaneous activity. Is it possible to observe a self-similarity of brain waves in the higher frequency regions between 100 and 1000 Hz by comparing them with the lower frequency region of 1–100 Hz? According to this question we have undertaken the following evaluation: We translate the time scale, depending on the higher frequency, by a factor of 100. Time periods of 400 ms were, therefore, investigated, choosing a sampling frequency of 10 kHz and a low pass of 5 kHz. In order to eliminate the influence of the lower frequency components, the data have been high-pass filtered with a cut-off-frequency of 100 Hz. Field potentials of four freely moving cats with chronically implanted electrodes in the cerebellar cortex and the inferior colliculus were investigated during the waking stage. For the cerebellum a mean value of the correlation dimension of about $D_{CE}=7.05 \pm 0.15$ and for the inferior colliculus a mean value of $D_{IC}=6.70 \pm 0.20$ was obtained. Table 2 presents the correlaton dimension D_2 computed for all the experimental data from four cats and all experiments. In both cases an ensemble of 65 trials was investigated. But a convergence of the slopes, which is required to determine the correlation dimension, was observed only in approximately 25% of the studied data. In other words, in 75% of the investigated time periods the EEG signal cannot be distinguished from a noise. Only in 25% of the recordings can it be concluded that the EEG signal was a deterministic signal.

Table 2. Correlation dimension D_2 of high frequency components of ongoing activity in the cerebellum and inferior colliculus of several cats (100–5000 Hz) recorded from implanted semi-microelectrodes during the waking stage

Cerebellum

Cat	Successive samples of 400 ms						
Flic	7.14	7.01	7.14	7.08	7.14	7.14	
Maus	7.14	6.94	7.14	7.19	6.94	7.10	7.05
Kati	6.80	6.95					
Lucy	7.20	7.35	7.15	6.84	6.75		

Mean value: $D_2 = 7.05 \pm 0.15$

Inferior colliculus

Cat	Successive samples of 400 ms					
Flic	6.75	6.75	6.82	6.77	6.75	6.75
Maus	6.99	6.41	6.92			
Kati	6.57	6.57				
Lucy	6.60	6.66	6.50			

Mean value: $D_2 = 6.70 \pm 0.16$

The important difference from the results of SWS-EEG is this: For the SWS-EEG in nearly 75% of the results a convergence was observed, whereas for the waking state and high frequencies a real plateau is rarely found. In the waking state every movement artifact influenced the high frequency EEG recordings considerably; inevitable complications cannot be avoided by artifact rejection. It is important to note that the observations in the high frequency range depict a significant difference between the dimension of the cerebellum and the inferior colliculus. Even in this preliminary evaluation a difference between the two structures can be demonstrated.

If the cats fell into the SWS stage during the experiment, a reduction of the dimension of the cerebellar attractor to values near $D_{CE} = 6.80$ was observed. A systematic analysis for the inferior colliculus has not yet been undertaken.

We have to make a technical restriction on the interpretation of the results. Could the rejection of frequencies lower than 100 Hz by high-pass filtering influence our results concerning the strange attractor properties of the higher frequency EEG? In order to evaluate this point we have used a noise-generator (Wavetek-Rockland) and filtered the signals with a high-pass filter of 100 Hz (48 dB/octave). Figure 12 presents slopes of the curves by plotting $C(r)$ versus log r for the high-pass filtered noise in comparison with the cerebellar signal. For the noise no saturation of the slopes is observed. Embedding in successively higher dimensional phase-spaces, it was impossible to detect any subspace (with finite dimension) somehow describing the dynamics of the noise signal. Contrary to the cerebellar signal the high-pass filtered noise does not show convergence and accordingly no dimensionality. Therefore, we tentatively conclude that the filtering procedure does not drastically influence our results on attractor properties of the

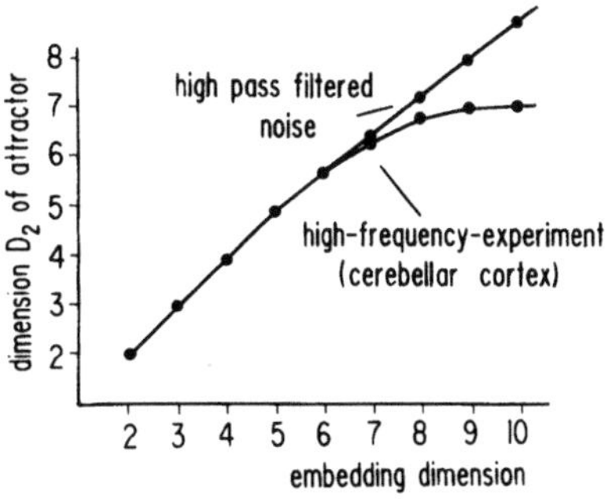

Fig. 12. The calculation of D_2 for a high pass filtered noise (high pass filter, 100 Hz) and the EEG signal from the cerebellar cortex. The EEG signal converges to a saturation value, whereas the noise which is filtered in the same frequency range does not show any convergence

cerebellum and inferior colliculus. Certainly the results indicate that the frequency region between 100–1000 Hz does not have properties of noise. More exact evaluation of D_2 requires more experiments and better mathematical and technical tools.

4.3 Are Transition Stages Also Reflected in the Correlation Dimension?

A survey of the correlation dimension D_2, which is described in contributions presented in this volume, shows a variability between 2 and 11. This variation depends on the state of the experimental subjects, on pathology, and on the evolutionary stage of the brains under study – and no doubt other factors as well. An important goal would be to interpret this range of values. As Freeman comments in the final discussion section "How Brains May Work" (Part V of this volume), it would be most useful to pay attention to the state changes, in other words, to bifurcations of brain activity rather than to merely classify the dimensionality. According to this view we have performed some experiments with freely moving cats and with human subjects who displayed abundant alpha activity with eyes closed.

Figure 13 shows the power spectra of the cat hippocampus with a dominant theta activity during the waking stage. The results indicate that if the cat brain state has a transition from waking stage to SWS stage, the maxima of power jumps down from about 5 Hz to between 1–3 Hz with several minor peaks. When the cat woke up, again a marked theta activity was observed and in this stage, D_2 was relatively high ($D_2 = 5.24$). During the SWS stage the dimension was reduced and had a value of $D_2 = 3.96$. Following the SWS stage D_2 again reached higher values of around 5.

Figure 14 illustrates the power spectra of the EEG activity of the subject J. K. who showed regular alpha activity with eyes closed. Here again it can be seen that the EEG activity reached slightly different states during the same recording session. In the present case it is impossible to describe exactly in which cognitive state the brain is. Elsewhere in this volume Başar et al. have described the same subject J. K. reaching long-standing alpha activity periods depicting marked time coherency depending on cognitive tasks; because of the short duration of the cog-

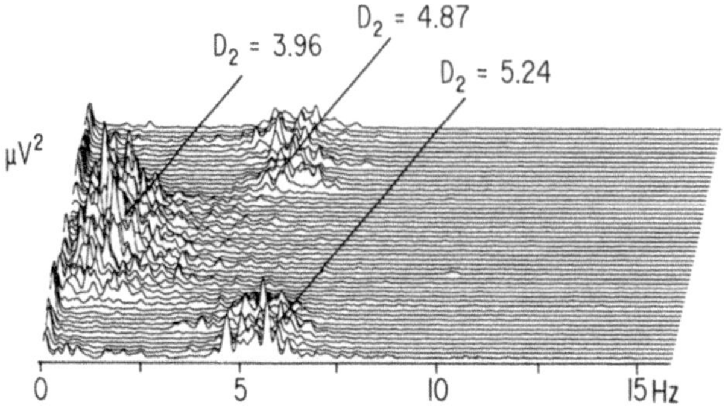

Fig. 13. Power spectra (5-s epochs) of the hippocampus (cat "Jenny") and correlation dimension D_2 (for 40-s epochs) during theta activity (waking stage) and slow-wave sleep

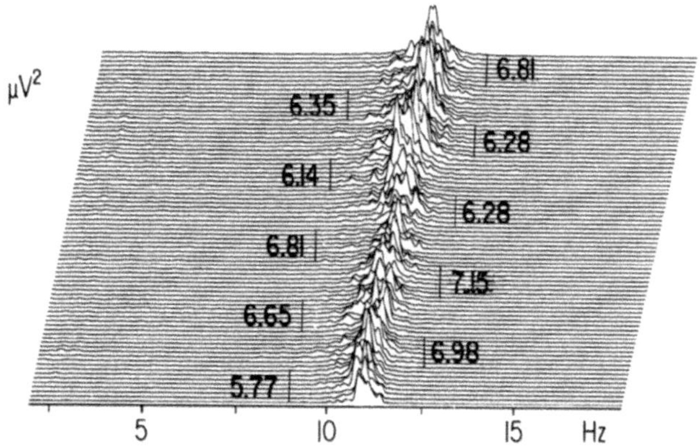

Fig. 14. Power spectra of EEG activity for 5-s epochs in subject J. K. (occipital recording) and correlation dimension D_2 for 40-s epochs

nitive tasks the evaluation of D_2 was not made. However, preliminary results indicate a trend towards a lower dimension during these brief tasks.

What conclusion is to be drawn from such a measurement where we observed a transient EEG activity? In this paper, we propose that together with a description of correlation dimensions one should use other parameters in parallel to D_2, for example, the power spectra which describe not only the overall power but also the quality of EEG signals in terms of its several frequency components. Another descriptor consists in estimating the RMS values in the conventional EEG frequency bands. Other important points to be analyzed are differences between closed eyes and opened eyes experiments, the range of filters and of EEG amplifiers, and also the classification of subjects as "alpha type" or "beta type". In the EEG studies another important but neglected point is the amount of 40-Hz activity which can reach maximal amplitudes of 10 to 14 μV depending on experimental conditions (Başar et al. 1987). This 40-Hz frequency is mostly masked in the conventional EEG analysis by visual inspection and also in the power spectral analysis. Digital filters can be used to describe the trend of this activity approximately. We do not yet know what the contribution of 40-Hz activity is to the EEG dimensionality. Only after clearly defining the experimental conditions and the distribution of the EEG in various stages, can use of the correlation dimension be valuable as a descriptor which combines and extends some other important properties of brain waves.

5 Discussion

Our research group's interest in the nonlinear analysis of the EEG dates back to 1972. Başar (1972) described a simple model of nonlinear interactions in the spontaneous activity of the brain. This model was a generalization of Erich von Holst's analysis of relative coordination in the CNS. Later, this tentative model concept was extended, by mentioning the Duffing equation, to describe evoked potentials as deterministic forced oscillators and EEG as randomly forced oscillations (Başar et al. 1979; Başar 1980, 1983). Başar and Röschke (1983) assumed that EEG components in several frequencies could be manifested as part solutions of Navier-Stokes equations. Shortly after the pioneering study of Babloyantz et al. (1985), Röschke and Başar (1985 b) reported the dimensionality of substructures of the cat brain during slow-wave sleep.

The implications of the expression "strange attractor EEG" have been described in an increasing number of studies (Röschke and Başar 1988; Başar et al. 1988; Babloyantz 1988; Freeman 1988; Başar 1983, 1988, Epilogue). We also want to refer to the Epilogue written by Bullock and other contributions in the present volume (see Rössler, Freeman, Babloyantz, Saermark et al., Graf and Elbert, this volume).

In this discussion we will present some remarks concerning the new data and make some comments regarding future research. The ensemble of data presented here might help to elucidate in part a few points that still remain obscure in this important new trend to describe the dimensionality of EEG.

5.1 The Variability of D_2 in the Human EEG

The description of dimensionality of the human EEG continues to show enormous fluctuations in studies by several authors (see Babloyantz, Graf and Elbert, Saermark, this volume). The correlation dimension of human EEGs varies between 5 and 11 in analyses so far known to us.

In the present study we reported that during a measuring period of 25 min the correlation dimension of the EEG of the same subject varies between 6 and 7, more than 20% during a short period. It is perhaps understandable that D_2 could show larger differences between subjects, during undefined states, with various electrode locations, digital filters, and types of EEG recording (50-Hz noise, surrounding, sound-proof room). We highly recommend the use of standard experimental conditions taking into account all the above criteria. At least investigators should describe experimental conditions exactly. This is important, since the serious use of D_2 in future studies dealing with the observation of fluctuations (transition changes) can be enormously useful.

5.2 Bifurcations

In Section 4 of the present study it is illustrated that if a cat goes from the slow-wave sleep to the waking state the power spectral peaks shift from delta activity to theta activity and the correlation dimension increases from about 4 to 5. This is a consistent observation of the hippocampal activity where changes of D_2 are always well correlated with the delta→theta transition. If the dimension D_2 is an important measure to describe states of the brain, the determination of transitions (or bifurcations) by using D_2 as an indicator would gain importance for studies of behavior of the CNS. We mention a word of caution for this type of description. The correlation dimension D_2 of the hippocampus during the waking state and D_2 of the auditory cortex during SWS are approximately 5. Accordingly, there is no difference between these two different brain states. Therefore, it would be extremely inadequate to consider D_2 as a universal indicator of the brain state. It provides experimenters, for the time being, with an abstract quantification which can be helpful in connection with several other biological descriptors of the brain state.

5.3 Power Spectra

The use of power spectra and digital filtering of the EEG seems to be a useful mathematical tool to complement the interpretation of the correlation dimension D_2. As indicated above (Sect. 5.2) the dimension of the cortex during sleep as well as wakefulness and of hippocampus during the waking stage give the same values of $D_2 = 5$. In that case, spectral analysis enables the experimenter to classify the brain waves. If the dimension of a human EEG during the waking stage is $D_2 = 6$, this might be the result of a mixture of theta, delta, and alpha waves and 40-Hz activity. We do not have models which describe, step by step, the distributions of spectral peaks, their weights, and the resulting correlation number. The use of power spectra might also help to reconcile discrepancies between results of various research groups, especially from records during the waking stage.

5.4 High Frequency Activity of the Cerebellum and Brain Stem

The high frequency spectral activity of the cerebellum and inferior colliculus have been mostly neglected by investigators. The experimental data described in this report demonstrate that in a frequency range between 100 Hz and 1000 Hz the brain has a high frequency attractor which is not as stable as the lower frequency attractor during SWS. A comparison with a noise signal which has been filtered with the same high-pass digital filters showed a completely different behavior. Therefore, it can be concluded that the high frequency cerebellar and collicular activity are not due to noisy signals but most probably stem from deterministic activities.

We want to mention that the brain's electrical activity might show self-similarity comparing the attractors in frequency regions of 1–100 Hz and 100–1000 Hz, although the evidence is still no more than suggestive. We would echo Adey's comments on "Bioinstrumentation: cutting edge and limiting factor in the future of brain research (this volume) concerning the need for looking to higher frequency windows.

From the mathematical viewpoint we do not assume or demonstrate the existence of two strange attractors in the frequency regions of 1–100 Hz and of 100–1000 Hz. For the time being the analysis undertaken by considering two frequency windows is only an approximation of the most appropriate analysis. The better procedure would be to analyze only one window between 1–1000 Hz (or higher). Unfortunately, this is still technically impossible due to computer limitations. Accordingly, the analysis in the higher frequency window has to be considered for the present as an indicator that the high frequency activities also point to deterministic chaos and that the activity between 100 and 1000 Hz is not simply noise.

The use of the brain state matrix concept which is proposed by Başar (Part V of this book) including several linear and nonlinear indicators seems necessary to us in studies related to the evaluation of D_2 according to the description given in Section 5.3.

6 Summary

We describe and compare the correlation dimension (D_2) of EEG activity during waking and sleep stages in various intracranial structures of the cat brain and in scalp activity of the human brain. D_2 is evaluated using Grassberger and Procaccia's algorithm. The following results were obtained: (1) During slow-wave sleep, D_2 has values between 4 and 5 depending on the brain structure under study; highest values were recorded in the auditory cortex (D_2 ca. 5) and lowest in the hippocampus (D_2 ca 4). (2) When the hippocampal EEG of a sleeping cat shows a transition from slow-wave sleep to the waking stage with dominant hippocampal theta waves, D_2 jumps from a value of approximately 4 to 5. (3) The cerebellar cortex and inferior colliculus of the cat brain depict a less stable strange attractor in the high frequency range between 100 and 1000 Hz (D_2 varies between 6 and 8). (4) During a recording session with a human subject showing abundant alpha activity, the D_2 is likely to show large fluctuations within a short period; in the example presented D_2 fluctuated between 6 and 7. (5) EEG activity gives a highly different result from that of white noise. (6) Power spectra of the EEG is a necessary and complementary brain descriptor to D_2. The correlation dimension can be more valuably used as one of several descriptors of the brain state; using only the correlation dimension to describe the brain state can lead to misinterpretations.

References

Abraham RH, Shaw CD (1983) Dynamics – the geometry of behavior. Part 2: chaotic behavior. Aerial Press, Santa Cruz

Adrian ED (1935) Discharge frequencies in cerebral and cerebellar cortex. J Physiol 83:32–33

Babloyantz A (1988) Chaotic dynamics in brain activity. In: Başar E (ed) Dynamics of sensory and cognitive processing by the brain. Springer, Berlin Heidelberg New York, pp 196–202 (Springer series in brain dynamics, vol 1)

Babloyantz A, Nicolis C, Salazar M (1985) Evidence of chaotic dynamics. Phys Lett[A]:152–156
Başar E (1972) Remarks on mathematical signal processing by the brain during rhythmic neurophysiological stimulation. Int J Neurosci 4:71–76
Başar E (1980) EEG-brain dynamics. Relation between EEG and brain evoked potentials. Elsevier/North-Holland, Amsterdam
Başar E (1983) Toward a physical approach to integrative physiology. I. Brain dynamics and physical causality. Am J Physiol 245(4):R510–R533
Başar E (1988) EEG-dynamics and evoked potentials in sensory and cognitive processing by the brain. In: Başar E (ed) Dynamics of sensory and cognitive processing by the brain. Springer, Berlin Heidelberg New York, pp 30–55 (Springer series in brain dynamics, vol 1)
Başar E, Röschke J (1983) Synergetics of neuronal populations. A survey on experiments. In: Başar E, Flohr H, Haken H, Mandell AJ (eds) Synergetics of the brain. Springer, Berlin Heidelberg New York, pp 199–200 (Springer series in synergetics, vol 23)
Başar E, Demir N, Gönder A, Ungan P (1979) Combined dynamics of EEG and evoked potentials. I. Studies of simultaneously recorded EEG-EPograms in the auditory pathway, reticular formation and hippocampus of the cat brain during the waking stage. Biol Cybern 34:1–19
Başar E, Rosen B, Başar-Eroglu C, Greitschus F (1987) The associations between 40 Hz-EEG and the middle latency response of the auditory evoked potentials. Int J Neurosci 33:103–117
Başar E, Başar-Eroglu C, Röschke J (1988) Do coherent patterns of the strange attractor EEG reflect sensory-cognitive states of the brain? In: Markus M, Müller S, Nicolis G (eds) From chemical to biological organization. Springer, Berlin Heidelberg New York
Freeman WJ (1988) Nonlinear neural dynamics in olfaction as a model for cognition. In: Başar E (ed) Dynamics of sensory and cognitive processing by the brain. Springer, Berlin Heidelberg New York, pp 19–29 (Springer series in brain dynamics, vol 1)
Grassberger P, Procaccia I (1983) Measuring the strangeness of strange attractors. Physica [D]9:183–208
Lorenz EN (1963) Deterministic nonperiodic flow. Atmos Sci 20:130
Röschke J (1986) Eine Analyse der nichtlinearen EEG-Dynamik. Doctoral dissertation, University of Göttingen
Röschke J, Başar E (1985a) A phase portrait analysis of the EEG and evoked potentials. Electroencephalogr Clin Neurophysiol 61:S114
Röschke J, Başar E (1985b) Is EEG a simple noise or a "strange attractor"? Pflugers Arch 405(2):R45
Röschke J, Başar E (1988) The EEG is not a simple noise. Strange attractors in intracranial structures. In: Başar E (ed) Dynamics of sensory and cognitive processing by the brain. Springer, Berlin Heidelberg New York, pp 203–216 (Springer series in brain dynamics, vol 1)
Schroeder MR (1986) Number theory in science and communication. Springer, Berlin Heidelberg New York
Schuster HG (1984) Determnistic chaos. An introduction. Physik, Weinheim
Takens F (1981) Detecting strange attractants in turbulence. In: Rand DA, Young LS (eds) Dynamical systems and turbulence, Warwick 1980. Springer, Berlin Heidelberg New York, pp 366–381 (Lecture notes in mathematics, vol 898)

Magnetoencephalography and Attractor Dimension: Normal Subjects and Epileptic Patients

K. SAERMARK, J. LEBECH, C. K. BAK, and A. SABERS

1 Introduction

In recent years methods from nonlinear dynamics analysis have been applied to examine the human electroencephalogram (EEG) with respect to attractor dimensions, Lyapunov exponents, etc.; using the same methods, Röschke and Başar (1988) have performed an analysis of EEG activity in cortical and subcortical structures of chronically implanted cats with respect to the fractal dimension of an EEG attractor during slow wave sleep. For the human EEG, Babloyantz et al. (1985) were able to determine an attractor dimension for two sleep stages (stages 2 and 4) whereas for REM sleep and for the awake state no attractor dimension was quoted. The dimensions found for the two sleep stages were close to 5 (stage 2) and 4 (stage 4). The analysis was based on data for two and three subjects, respectively, the experimental data consisting of time series of 4000 points with a sampling time of 10 ms. Similarly, for the chronically implanted cats (electrode positions: acoustic cortex, hippocampus, and reticular formation), Röschke and Başar quote a value of 4–5 for the EEG attractor during slow wave sleep. Further works by Babloyantz and Destexhe (1986) and Layne et al. (1985) quote still other values for attractor dimensions obtained under a number of conditions: anesthetized subjects, epileptic patients, alpha traces, etc.

In analyzing the EEG data for the attractor – correlation dimension the above references make use of the Grassberger-Procaccia algorithm (Grassberger and Procaccia 1983). Here one calculates the correlation function, $C(r)$, as defined in the above references and examines its behavior for small values of r. $C(r)$ behaves like $\sim r^d$ and if a plot of d versus the embedding dimension n shows a saturation value for d, then this value is taken as the correlation dimension for the attractor. It can be argued, however, that the Grassberger-Procaccia algorithm is not ideally suited for the determination of the dimension of high-dimensional attractors. Thus Layne et al. (1985) discussed problems associated with dimensional analysis of EEG data (T3-C3 and P3-O1 electrodes) and found – even for long data sets containing 15000 points – that the dimension could be ascertained only with a very high degree of uncertainty (of the order of 50%–100%). Based on calculations for shorter subsections (4000 data points) they also showed that the dimension depends on the number of data points used in the analysis and on the position of the subsection of data points within the larger set of data points. Hence their conclusion "The EEG is both non-stationary and high-dimensional. Therefore, we cannot speak of dimension in the strict sense. Our calculations of 'dimension' have a meaning only in a comparative sense," is well taken and the word

Springer Series in Brain Dynamics 2
Edited by Erol Başar and T. H. Bullock

"dimension" could appropriately be replaced by their "apparent dimension" or "dimensional complexity." However, even with this caution in mind it still appears very tempting to use the Grassberger-Procaccia algorithm in the determination of correlation dimensions of EEG and magnetoencephalographic (MEG) data. This is especially so since there are very few other methods for performing "global" analyses of EEG and MEG recordings. It should be remarked, however, that the criticism with respect to nonstationarity is especially significant for EEG and MEG recordings for epileptic patients. Such recordings are normally interictal recordings and the interictal state is in general not a stationary state. A dimensional analysis of MEG recordings was briefly presented previously (Saermark 1986). In this paper we present a number of results for normal subjects and for epileptic patients. In particular we point out that the present experimental data indicate that the correlation function $C(r)$ – as calculated in the Grassberger-Procaccia algorithm – for epileptic patients shows a behavior which is different from that shown for normal subjects.

2 Results

The MEG recordings discussed in this paper were obtained for two groups, one population consisting of five normal subjects (two female, three male; age 21–25 years) and another population consisting of four epileptic patients (all female; age 20–29 years).

The magnetic field was measured by means of a first order SQUID gradiometer (BTI) located in magnetically quiet surroundings. The base line of the gradiometer is 6.3 cm, and the lower sensing coil (diameter 2.4 cm) is situated 1.2 cm above the bottom of the cryostat tail. The latter was separated from the subject's skull by a distance of 1–2 mm. Measurements of the magnetic field component normal to the skull were performed for both right and left hemisphere (RH, LH) positions. Each MEG recording consisted of 8192 points with a sampling time of 10 ms, yielding a total recording epoch of nearly 82 s. For each normal subject the MEG was recorded in 10–15 positions, all of which lay above the temporal lobe or the immediately adjacent areas of the brain. For each of the epileptic patients 50–60 MEG recordings were taken at positions evenly distributed across the hemisphere suspected for epileptic foci, with an additional 10–15 recordings being taken at the opposite hemisphere.

In analyzing the MEG data for "attractor dimension" we used the Grassberger-Procaccia procedure and thus calculated the correlation function $C(r)$ as a function of r for a number of embedding dimensions: $n = 2, 4, 6, \ldots 20$. Generally 4000 data points were used in the analysis; however, in a number of cases this was increased to 8000 data points. For each subject only two or three measuring positions were used in order to keep the computer expenses within reasonable limits. For normal subjects positions close to the auditory cortex were chosen while for epileptic patients positions where the recordings showed a clear prominence of low-frequency waves (3–6 Hz) were chosen.

In Fig. 1 we show a plot of $\log_2 C(r)$ versus $\log_2(r/r_0)$, r_0 being a reference value for r, for a normal subject and for an MEG recording. We note that this

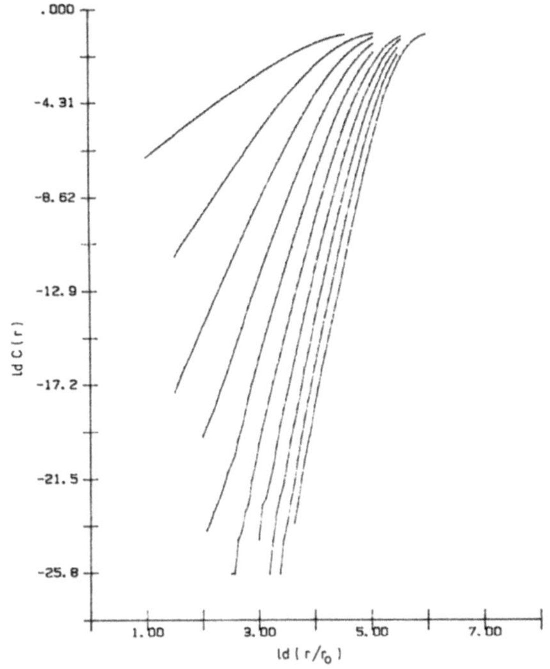

Fig. 1. The correlation function $C(r)$ for a normal subject based on MEG data. *Abscissa:* $\log_2 (r/r_0) \equiv$ *ld* (r/r_0), where r_0 is a reference r value. *Ordinate: ld* $C(r)$. Embedding dimensions 2, 4, 6, ... 20

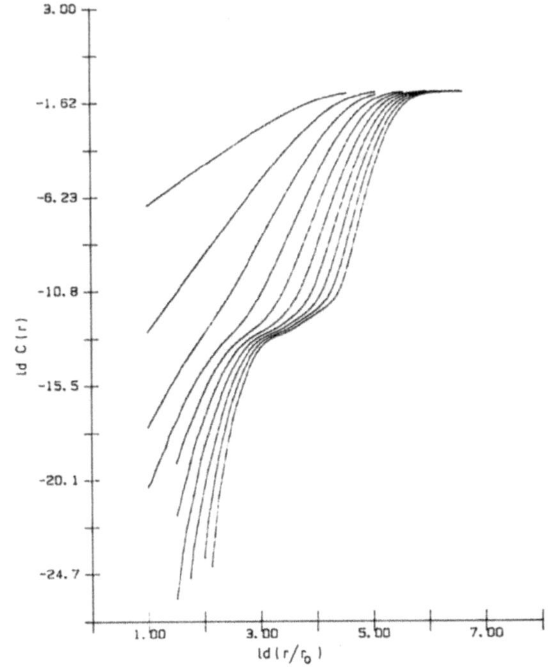

Fig. 2. The correlation function $C(r)$ for interictal MEG data from an epileptic patient. Axes and embedding dimensions as in Fig. 1

subject showed particularly well developed alpha waves with maximal amplitudes of the order of 2500–3000 fT. For all of the embedding dimensions used, the curves for $C(r)$ show a linear portion with a slope which converges to a limiting value. For epileptic patients, on the other hand, $C(r)$ behaves quite differently, as is clearly seen in Fig. 2. For small values of the embedding dimension, say $n<6$, the curves do show a linear behavior; however, for increasing values of the embedding dimension an inflection point, with a decreasing value of the slope of the tangent, gradually develops for abscissa values around 2–3. It follows that in this region of n values the curves show no overall linear behavior, but can be characterized by the presence of two inflection points: the marked one just mentioned and in addition one occurring at a larger r value. Further, at an r value smaller than the one for the marked inflection point, the tangent to the C(r) curve shows a maximum slope slightly greater than the slope of the tangent at the larger r value. The behavior of $C(r)$ shown in Fig. 1 (normal subject) was consistently seen for all normal subjects (including a third group of five normal subjects for whom the analysis was not carried through in a systematic way), while the behavior of $C(r)$ shown in Fig. 2 (epileptic patient) was consistently seen for all epileptic patients. In comparing Figs. 1 and 2 it should be emphasized that the measuring instrument and the measurement procedure were identical for the two groups of subjects and that the raw data were used in the analysis, i.e., no filtering or other data handling was carried out before the dimensional analysis was performed. We therefore believe, until evidence is available to the contrary, that the difference in the behavior of $C(r)$ for the two groups of subjects is a genuine difference. To bring out this difference clearly, Fig. 3 shows $C(r)$ for an epileptic patient ($n=18$) and for a normal subject ($n=20$).

For the group of normal subjects we may plot the slope d of the linear region of $C(r)$ versus the embedding dimension, whereas for the group of epileptic patients this cannot be done. We have here chosen to plot the slope of the tangent at the inflection point occurring at the larger r value and also the slope of the tangent at the marked inflection point. In Fig. 4 the crosses indicate the population average for the group of normal subjects while the rhombohedral symbols show the results for a single epileptic patient. The vertically hatched symbols refer to an analysis based on MEG recordings (4000 data points) while the open symbols refer to an analysis based on EEG recordings taken 3 months after the MEG recordings (again based on 4000 data points). The cross-hatched symbol (for $n=20$) refers to an analysis based on MEG recordings and 8000 data points. As the analysis of the EEG data also gave rise to the presence of the marked inflection point, in this figure we have chosen to show the results from the analysis of the EEG data, i.e., the horizontally hatched rhombohedral symbols which virtually coincide with the results from the analysis of the MEG data. From Fig. 4 one notes, first of all, that there is good agreement between the analysis of the MEG data and the EEG data, with respect to both the slope of the tangent at the upper inflection point and the slope of the tangent as the marked inflection point; further, for the epileptic patient there may be an apparent dimension of around 7, while the data for the population of normal subjects may indicate an apparent dimension of around 11. Finally, the tangent slope at the marked inflection point approaches, rather rapidly, a value of 1 for increasing embedding dimensions; thus

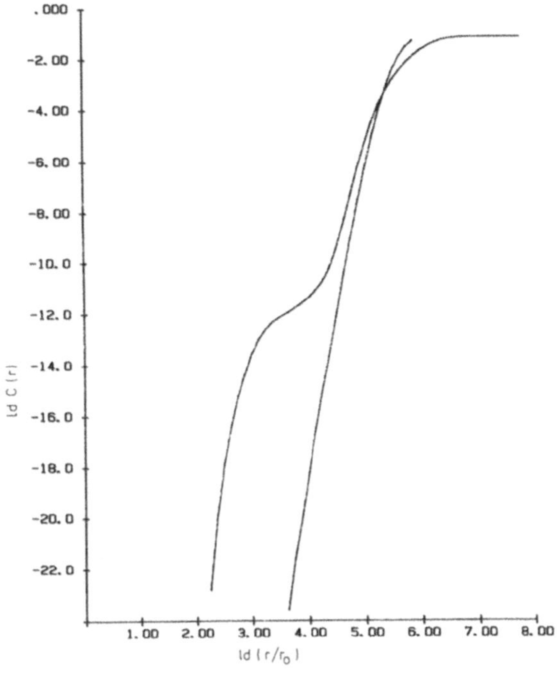

Fig. 3. Comparison of correlation function $C(r)$ for a normal subject ($n=20$) and an epileptic patient ($n=18$). The marked inflection point is visible for the epileptic patient, but is not present for the normal subject

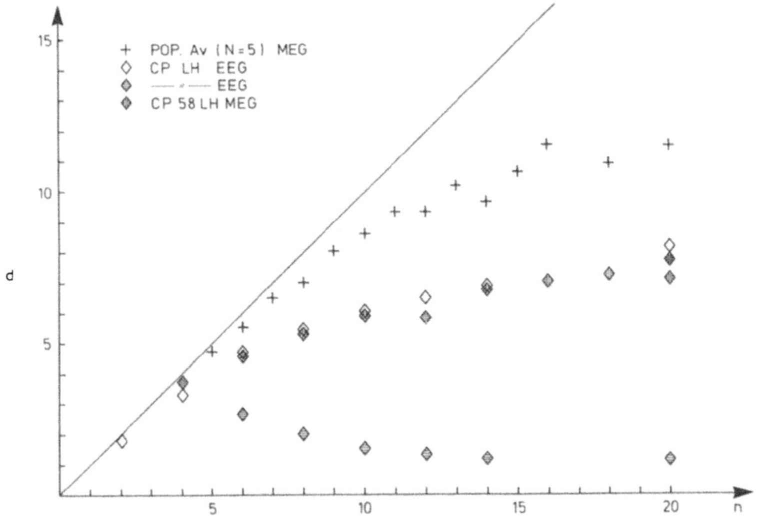

Fig. 4. Plot of the exponent d (see text) versus the embedding dimension for a population of normal subjects and for an epileptic patient CP. For the latter both MEG and EEG data have been used. See text for explanation of symbols

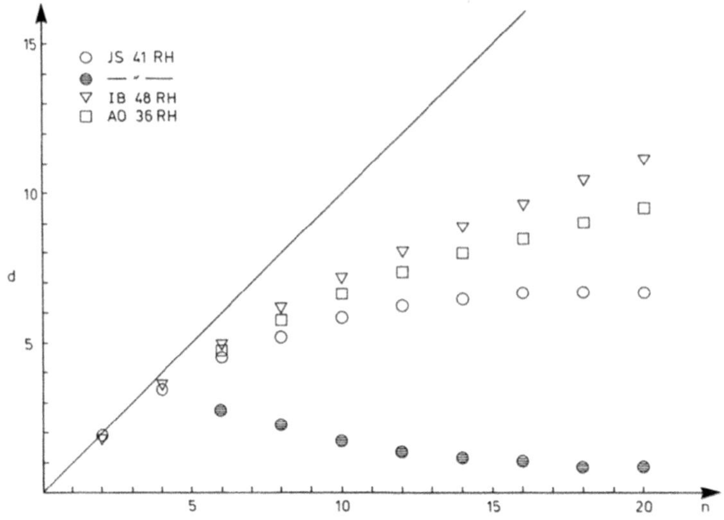

Fig. 5. Plot of the exponent *d* (see text) versus the embedding dimension for three epileptic patients (RH measuring position)

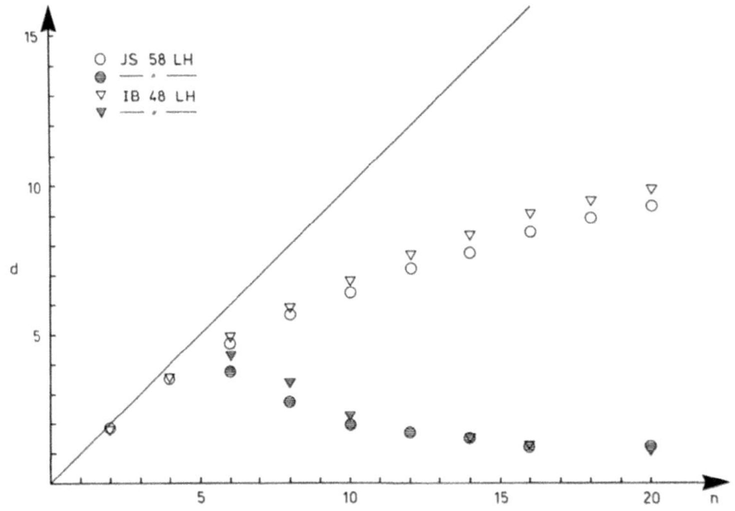

Fig. 6. Plot of the exponent *d* (see text) versus the embedding dimension for two epileptic patients (LH measuring position)

a bifurcation in *d* versus *n* appears to occur for an embedding dimensions of around 5–6 in the case of epileptic patients but not in the case of normal subjects.

In Figs. 5 and 6 we show results obtained for the three other epileptic patients considered in this paper. For patient JS one notes an apparent dimension between 6 and 7 for the RH measuring position (Fig. 5) whereas the LH measuring posi-

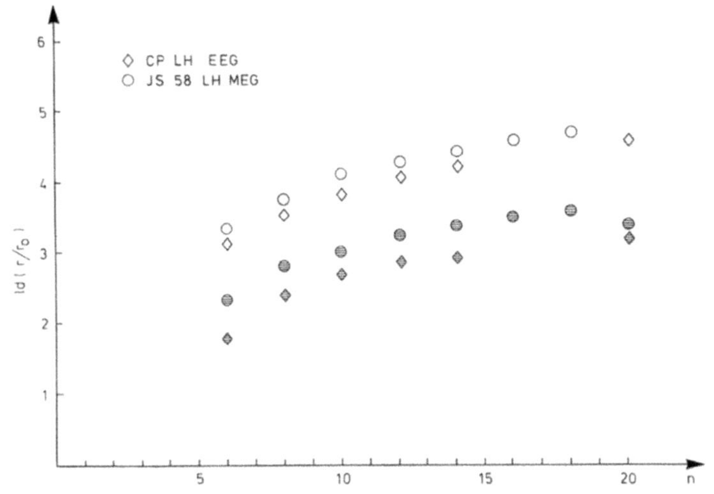

Fig. 7. Plot of the *r* value for the position of the marked inflection point (*open symbols*) and the upper inflection point (*hatched symbols*) versus embedding dimension for two epileptic patients. Note the nearly constant difference between the two positions

tion (Fig. 6) does not give rise to an apparent dimension. The latter is true also for the two remaining patients IB and AO (Fig. 5, 6). In all cases, however, there is a bifurcation point – the position of which may vary slightly from patient to patient – and in all cases the slope of the tangent at the marked inflection point approaches 1 for increasing values of the embedding dimension.

Finally, in Fig. 7 we show the *r* values for the position of the two inflection points. For the sake of clarity we show only data for two patients; however, the remaining two patients showed identical behavior. For the patient JS we have used MEG data and for the patient CP, EEG data. One notes a clear tendency toward a constant ordinate difference between corresponding open and hatched symbols. As the ordinate is a logarithmic scale this implies a constant ratio (slightly larger than 2) between the *r* values for the position of the upper inflection point and the marked inflection point. This was also found for the two other epileptic patients.

3 Discussion

We have here presented a number of results from a dimensional analysis of MEG data and some EEG data. In view of the cautioning remarks concerning the applicability of the Grassberger-Procaccia algorithm (see Sect. 1) with respect to the determination of the "dimension" of high-dimensional attractors, however, all conclusions must be very tentative. It is especially disturbing that Layne et al. [4] estimate the uncertainty in the dimension *d* to be as large as 50%–100%. On the other hand, in the experiments by Röschke and Başar (1988), where five chronically implanted cats were examined, the intersubject variation in d was considerably lower, i.e., of the order of 10%. Similarly, for the group of normal subjects considered here the intersubject variation in d for a given value of the embedding dimension n is also considerably smaller: less than 20%. From this point of view one could with some justification argue that the data for normal subjects may indicate an apparent dimension of the order of 11.

For the group of epileptic patients a surprising feature emerged in the correlation function $C(r)$, in the form of what we here denoted as "a marked inflection point" (see Figs. 2, 3). As a consequence one finds a bifurcation point in a plot of d versus n, as shown in Figs. 4–6; however, this bifurcation is only observed for the group of epileptic patients, not for the group of normal subjects. The slope of the tangent at the marked inflection point rapidly approaches the value 1 for increasing embedding dimension for all epileptic patients, whereas the slope of the tangent at the upper inflection point (see Figs. 4–6), as a function of the embedding dimension, behaves differently from patient to patient. In two cases (Fig. 4, CP 58 LH MEG; Fig. 5, JS 41 RH MEG) one does find an apparent dimension. In these two cases, visual inspection of the MEG recordings shows the presence of prominent low-frequency waves (mainly 3- to 5-Hz waves) for the whole time epoch; these records may therefore be considered nearly stationary and the finding of an apparent dimension be ascribed some significance. It is also noteworthy that for the patient CP (see Fig. 4) the dimensional analysis of EEG data – which, as mentioned earlier, were taken 3 months later than the MEG data – was in excellent agreement with the analysis of the MEG data. For the three remaining patients no EEG data were available to us; thus we cannot claim that the two types of data will always yield an identical apparent dimension. In this connection it is to be remarked that EEG data depend on the existence of volume currents whereas the MEG data are predominantly determined by intracellular currents. It is therefore not obvious that the two types of data should yield the same apparent dimension. For the three remaining epileptic patients one cannot claim the existence of even an apparent dimension (see Figs. 5, 6). It is a characteristic feature, however, that for the epileptic patients all points fall considerably below the points representative for the group of normal subjects (see Fig. 4), which could indicate that an apparent dimension for epileptic patients will be lower than for a normal subject, but, at the same time, that the present interictally recorded MEG data represent a nonstationary state. In this paper we have chosen to perform the dimensional analysis on the basis of the raw experimental data; it is conceivable, however, that by first applying bandpass filtering (e.g., 2–8 Hz) one might find an apparent dimension for all the epileptic patients.

The present data lead, as pointed out earlier, to the existence of a bifurcation point in a plot of d versus n. This could possibly be interpreted in the following way. For sufficiently large values of the embedding dimension, say $n > 6$, the phase-space points show a tendency to fall into two separate groups of a certain extent and separated by a certain distance. If this is the case, there will be a range of r values where $C(r)$ will show a tendency to flattern out, then (for increasing values of r) start to increase rapidly again before ultimately bending over and becoming constant for r going to infinity (in the logarithmic scale used in Figs. 1–3 the constant value is -1). The fact that (see Fig. 7) the r values for the positions of the two inflection points both appear to approach a limiting value for increasing values of n and the fact that there appears to be a constant ratio between these r values, i.e., a ratio nearly independent of the embedding dimension for $n > 6$, may support this hypothesis. We emphasize that this ratio appears to be nearly the same for all the epileptic patients examined in this study. It is, of course, quite a different matter to find out what are the neurological implications of such a hypothesis.

4 Summary

In this paper we present results obtained by applying the Grassberger-Procaccia algorithm to a dimensional analysis of MEG and in part also EEG data for a group a five normal subjects and for a group of four epileptic patients. On the basis of the data presented, it is concluded that the correlation function $C(r)$, as defined by Grassberger and Procaccia, behaves differently for the two groups. For one epileptic patient a comparison between an analysis based on MEG data and an analysis based on EEG data is performed, the comparison showing an excellent agreement. In a number of cases the analyses show a trend to an "apparent dimension," whereas in other cases they indicate a nonstationarity of the recorded experimental time series.

References

Babloyantz A (1988) Chaotic dynamics in brain activity. In: Başar E (ed) Dynamics of sensory and cognitive processing of the brain. Springer, Berlin Heidelberg New York

Babloyantz A, Destexhe A (1986) Low-dimensional chaos in an instance of epilepsy. Proc Natl Acad Sci USA 83:3513–3517

Babloyantz A, Nicolis C, Salazar M (1985) Evidence of chaotic dynamics of brain activity during the sleep cycle. Phys Lett [A]111:152–156

Grassberger P, Procaccia I (1983) Measuring the strangeness of strange attractors. Physica [D]9:183–208

Layne SP, Mayer-Kress G, Holzfuss J (1985) Problems associated with dimensional analysis of electroencephalogram data. In: Mayer-Kress G (ed) Dimensions and entropies in chaotic systems. Springer, Berlin Heidelberg New York

Röschke J, Başar E (1988) The EEG is not a simple noise: strange attractors in intracranial structures. In: Başar E (ed) Dynamics of sensory and cognitive processing by the brain. Springer, Berlin Heidelberg New York

Saermark K (1986) On physical models of elementary neural sources. International Workshop an Functional Localization: A Challenge for Biomagnetism. Sept 1–6, Torino

Chaotic Attractors in a Model of Neocortex: Dimensionalities of Olfactory Bulb Surface Potentials Are Spatially Uniform and Event Related

J. E. Skinner, J. L. Martin, C. E. Landisman, M. M. Mommer, K. Fulton, M. Mitra, W. D. Burton, and B. Saltzberg

1 Introduction

The olfactory bulb has both the cell types and the neurochemicals intrinsic to neocortex, but it has a much simpler and better understood neurophysiological structure (Shepherd 1970). The amplitude of the field potential that occurs on the top of each columnar unit is linearly related to the firing probability of the immediately underlying output cell (Freeman and Schneider 1982; Gray et al. 1984, 1986). Thus the recording of all such surface potentials, at the spatial frequency of the functional units, makes possible knowledge of the total output of the bulb without having to make massive microelectrode penetrations.

The amplitudes of these simultaneously recorded surface potentials form a pattern on the bulbar surface that is altered by a novel or classically conditioned odor (Freeman and Schneider 1982). This alteration is prevented by interstitial beta-receptor antagonists (Gray et al. 1984, 1986). Beta-receptor antagonism also blocks synaptic potentiation in hippocampal cortex (Bliss et al. 1983; Hopkins and Johnston 1984), an alteration which is thought to be a model for learning and memory (Anderson and Wigstrom 1980; Swanson et al. 1982). Thus the study of the spatial patterns on the bulbar surface (the output) during controlled stimulus events (the input) may lead to an understanding of how information is acquired and stored in a neural system composed of large numbers of neurons.

The learning-dependent changes in the surface patterns, although statistically significantly different during pre- and postodor intervals, are quite noisy in appearance (Gray et al. 1986). Such complex signals suggest an underlying chaotic dynamical process (Farmer 1982). Other investigators have reported fractional dimensionalities of surface potentials from neocortex (Babloyantz and Destexhe 1986, 1987; Röschke and Başar, this volume; Graf and Elbert, this volume). Our present data indicate that: (a) a single, global chaotic attractor may represent the total bulbar activity spatially, (b) an odor causes this widespread process to shift to a new attractor, and (c) the shift is to an attractor with a larger Hausdorff number if the odor is novel and a lower one if the odor is familiar.

2 Methods

Rabbits were surgically prepared, instrumented with an 8×8 electrode array on the lateral surface of the olfactory bulb, and behaviorally trained according to a

Springer Series in Brain Dynamics 2
Edited by Erol Başar and T. H. Bullock

previously described protocol (Gray et al. 1986). The spacing between each electrode was at the spatial frequency of the surface activity (500 μm). The bandpass (3 dB) was set to 20–200 Hz and the digitizing rate (12 bits) set to 500 kHz with 640 samples/second per channel. The bandpass eliminates the large event-related slow potential (ERSP), but leaves the predominant 40- to 80-Hz frequencies unaltered. The ERSP was recorded with DC amplifiers (Beckman type R) from platinum-black electrodes, each of which was spaced at 0.5-mm intervals along the dorsal longitudinal surface and referenced to an electrode in sagittal bone.

In all cases the animal was *thoroughly* familiarized with the laboratory, personnel, and procedures before data were collected. Data were obtained and edited into 500-ms epochs for the 3 s before and after the beginning of the first inspiration of a faint novel odor (butanol); such an inspiration is detectable because it invariably has a shorter inspiratory interval compared to those observed during the previous 3-s control period. Respiration was monitored with a thoracic bellows and strain gauge.

The animal was accustomed to the delivery of several other faint odors (wintergreen and clove oil) for 2 weeks preceding the collection. An habituated odor (wintergreen) was presented 9 times during each of 10 days preceding the experiment. This habituated odor was presented for six trials during the same day preceding the novel stimulus (butanol). The intertrial intervals ranged from 3 to 10 min. The same experiment was repeated for 3 additional days to make possible observation of habituation to the butanol. A second novel odor (benzaldehyde) was used in a similar manner to that of butanol for observation of within-subject consistency.

The algorithm of Grassberger and Procaccia (1983) was used to analyze the data. The estimate of the the saturated correlation dimension (Hausdorff-Besicovitch number, HN) was calculated for each of the chosen data epochs. In most cases, 15 embedding dimensions were used. In one rabbit the simultaneously recorded epochs from 32 electrodes were each analyzed using this maximum number of embedding dimensions. We calculated the known HN of the Koch curve (log 4/log 3; Mandelbrot 1983) to test our use of the algorithm; the data were the apex and base points of all triangles generated, using three to eight iterations. Analysis of *white noise* epochs filtered to the bandpass (20–200 Hz), power spectrum (50–120 Hz), or principal components (50–80 Hz) of the bulbar data all had HNs that were clearly distinguished from those of the actual data passed through the same filter.

Each 500-ms epoch recorded from the bulb consisted of 320 digitized voltages produced by an A/D converter running at 500 kHz, but sampling at 640 Hz. Stationarity was presumed for each epoch. Although brief compared to epochs used by others, the 500-ms interval was chosen to increase the likelihood that stationarity would be maintained. Additionally, this choice seemed a reasonable interval because it takes approximately 500 ms for the cortex to process information, as demonstrated by (a) the duration of the short-latency event-related potentials (Haider et al. 1981) and (b) the minimum interval for neocortex to process low intensity input (Libet 1985).

Each of the voltage vectors in the epoch was subtracted from the value of all the rest to provide a set of absolute differences for dimensional analysis. The number of absolute differences, N accumulated within a range, R (of a given embedding dimension), is related to the correlation dimension, Dc, by the expression, $N = R^{Dc}$. If a linear relationship is detected in a plot of $\log N$ vs $\log R$, then Dc is quantifiable as the slope. The effects on slope produced by large and small R values (i.e., finite sets and near-neighbor correlations, Theiler 1986) were eliminated by visually determined cut-offs. When the embedding dimension is increased, the slope representing Dc will eventually saturate; that is, if an attractor exists. This saturated value of the slope is the estimate of the HN.

The tau interval is used to define the coordinates for the vectors for each embedding dimension. We selected tau by two methods: (a) the interval of the first zero-crossing of the autocorrelation function of the surface potential for each specific 500-ms epoch, or (b) one-half the mean zero-crossing interval for each epoch. For the data analyzed in this study, the two methods produced almost identical tau values, ranging from to 3.6 to 4.2 ms. No attempt was made to study the empirical dependency between tau and the saturated value of the HN. Convergence was clearly demonstrated in our data; the tau intervals used were small compared to the epoch lengths.

3 Results

Figure 1 A shows a 100-ms sample epoch (time) recorded simultaneously from 32 electrodes (space) during a preodor control condition. Four rows of eight electrodes each are shown. Note that during the control condition smooth voltage gradients are present, in both space and time (voltage is the upward coordinate). Figure 1 B shows the activities of the same electrodes 500 ms after inspiration of a novel odor. Note that the amplitude variations in time and space are no longer smooth. For example, the fusiform shape of the temporal signal seen in the leftmost trace in Fig. 1 A is not apparent in 1 B, and the peaks of the activities are often spatially isoelectric in 1 B, but decline smoothly, from row to row, in 1 A.

Figures 2–4 show the extensive analyses of data from a single animal. The traces in Fig. 2 show (from top to bottom): (a) the respiratory waveform, (b) the sniffing behavior (instantaneous change in respiratory rate), (c) the analog signal from a single sample electrode, and (d) the HNs for all of the 500-ms epochs for this single electrode. These data span the period just before and after the novel odor. Figure 3 shows the spatial distribution of the HNs calculated from the signals from 32 simultaneously recorded electrodes. These spatial data cover the same odor event as the temporal data shown in the previous figure. Note the stability of the HNs in the temporal domain in Fig. 2 and the homogeneity of calculated values in the spatial domain in Fig. 3. Only the electrodes in the upper left corner of Fig. 3 show any spatial heterogeneity. Postmortem examination revealed that these electrodes extended beyond the bulb and were not in contact with its surface, as were all of the rest in the array.

Figure 4 is similar to Fig. 2, but shows the HNs when the same animal has sniffed the butanol for the third time. In both cases the odor was injected into the

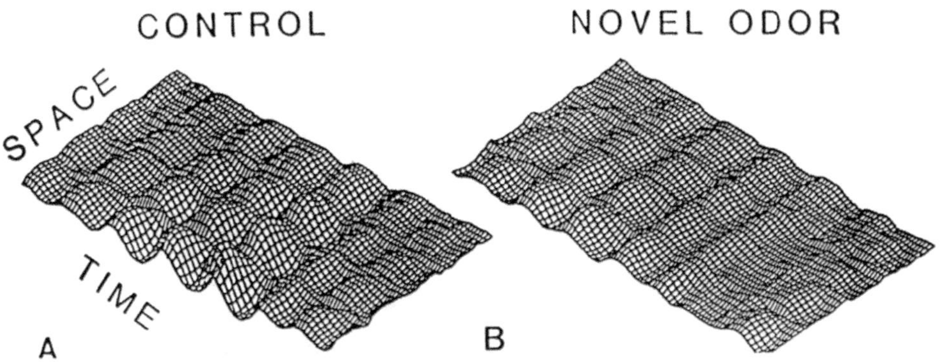

Fig. 1 A, B. Simultaneously recorded data from 32 electrodes in an 8 × 8 array on the lateral surface of the right olfactory bulb. Four adjacent rows of eight electrodes each are observed in both space and time. The upward deflection is negative potential and the maximum peak-to-peak voltage shown is 325 μV. (A) A representative 100-ms epoch approximately 500 ms before the beginning of the first inspiration of a faint novel odor (butanol). (B) For the same electrodes, a 100-ms epoch approximately 1 s later

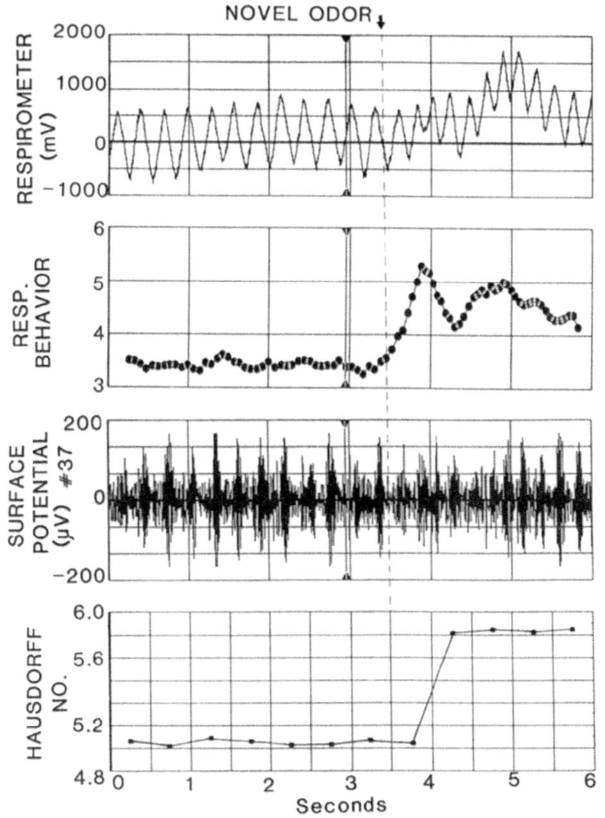

Fig. 2. Event-related alterations in respiration (*Respirometer*), instantaneous respiratory rate (*Resp. Behavior*), bulbar electric activity (*Surface Potential, position 37*), and surface potential dimensionality (*Hausdorff No.*). The perturbations were evoked together, in a conscious rabbit, by the first presentation of a novel odor (butanol). The algorithm used to analyze the respiratory signal is sensitive to instantaneous changes in slope and the units are respirations per second. The HNs are estimated for each of the 500-ms epochs indicated on the time scale, and each calculation was based on 15 embedding dimensions

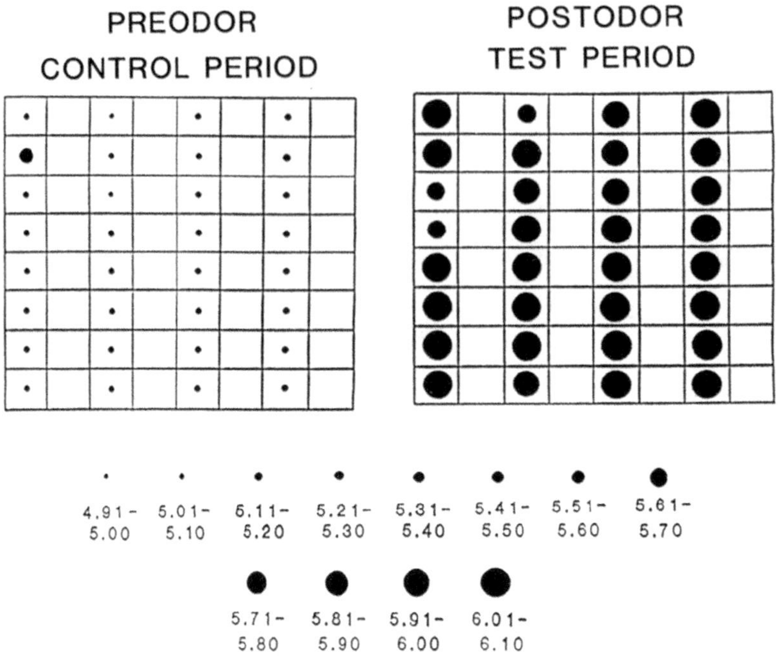

Fig. 3. Spatial distribution of HNs calculated for the surface potentials from 32 simultaneously recorded electrodes on the lateral surface of the olfactory bulb of the conscious rabbit, before and after the first presentation of a faint novel odor (butanol). The scale at the bottom identifies the interval (0.1 units of correlation dimension) in which each HN was embedded. Each preodor epoch was 500 ms in duration and began 1.0 s before the beginning of the first inspiration of the novel odor; each postodor epoch was 500 ms in duration and began 1.0 s after the beginning of the first inspiration of the novel odor

airstream of the nose cone at 3.0 s. Note in Fig. 4 that the behavioral reaction (Resp. Behavior) does not occur until the epoch beginning at 4.0 s. When the odor was novel both the behavior and the HN shifted together during the epoch beginning at 3.50 s, as seen in Fig. 2. In Fig. 4, however, there is a "droop" in the HN trace that begins at 3.50 s and anticipates the behavioral reaction.

Figure 5 A shows a sample plot of log *R* vs log *N* for all of the embedding dimensions used for one of the 500-ms epochs. Figure 5 B shows the plot of the slopes (i.e., the correlation dimension) vs the embedding dimensions. The linear portion of the slopes was determined visually, as is generally done by other investigators (Babloyantz and Destexhe 1986; Röschke and Başar, this volume; Graf and Elbert, this volume). Note that the linear portion of each slope and the saturation level of the group are both easily distinguished. Precision in determining the saturated value has been presumed, as the embedding dimensions number more than twice the value of the estimated HN.

Our version of the Grassberger and Procaccia algorithm was tested on several continuous functions, the dimensionalities of which are known. When used on Gaussian noise, no saturated value of the correlation dimension was observed within 15 embedding dimensions. The theoretical trajectory, using random

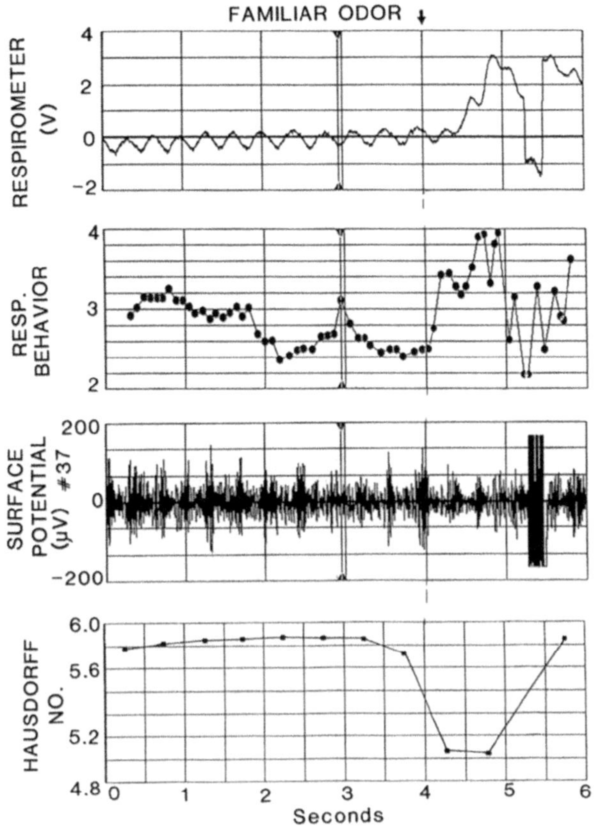

Fig. 4. Respiratory signal (*Respirometer*) and bulbar surface potential from a single electrode (position 37) and their analyzed perturbations (*Resp. Behavior* and *Hausdorff No.*) evoked in a conscious rabbit by the third presentation of an *habituated* odor (butanol). The data are from the same animal shown in Fig. 2. Note the difference in scale of the upper two traces. The HNs are estimated for each of the 500-ms epochs indicated on the time scale, and each calculation was based on 15 embedding dimensions; the epoch between 5.0 and 5.5 s was not analyzed because of the recording artifact seen in the surface potential record, caused by the animal shaking its head

numbers, is indicated in Fig. 5B (NOISE). When used to analyze a sine wave, the saturated level occurred at HN = 1.000. When used to estimate the HN of the Koch curve (Fig. 6C), the calculated value was very close to the theoretical value (i.e., 0.1265E+01 was observed and 0.1262E+01 is the theoretical value). It was determined empirically that at least 250 data points were needed for a close match between calculated and theoretical values, and that, for the Koch curve, increasing the number above 257 (four iterations) produced little additional improvement.

Table 1 presents the statistical evaluation for the changes in dimensionality evoked by a novel odor (butanol). A similar effect is observed, in the same animal, for the alternate novel odor (benzaldehyde), as shown in Table 2. Also seen in Table 2 is the effect of experience (habituation) on the direction of the dimensionality shift evoked by butanol. Note the small amount of variance detected, both temporally and spatially, for the control and odor conditions. Table 3 shows that between-subject variance is similar to that within-subjects.

A novel odor also produces the ERSP, a response which accompanies the spatial pattern changes on the surface of the olfactory bulb. Figure 7 shows an example of this response, which was evoked by butanol as the novel odor. Note that in all previous figures the bandpass filtered out this giant potential.

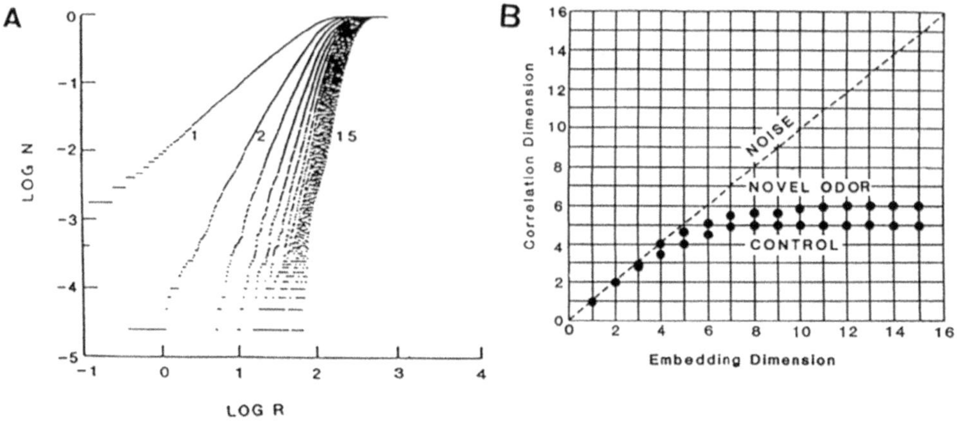

Fig. 5. (A) Sample data from a postodor epoch (first presentation of butanol) showing the accumulation of data points (*N*) as the range of the embedding dimension (*R*) is incremented within the set of difference scores. The latter were differences between all pairs of digitized voltages that occurred within the 500-ms epoch. The embedding dimension is indicated for the 1st, 2nd, and 15th traces. Units of *R* are the largest digitized value/1000 (approximately in units of 100 mV) and units of *N* are the accumulated number divided by the total number of data points. (B) Plot of the correlation dimension (i.e., slopes of the linear portion of each trace seen to the left) vs the embedding dimension. The preodor control data are also shown. The line labeled *NOISE* indicates the curve that would occur if the analysis had been made of a signal containing Gaussian noise. The asymptotic or saturated value indicates the HN as a point on the correlation dimension, which is a continuous scale. The mean and standard deviation of the last five points is our calculated estimate of the HN

Table 1. Hausdorff numbers[a] computed for 32, simultaneously recorded, 500-ms epochs during the control (C) and novel (N) stimulus conditions (EP, electrode position in the 8 × 8 array)

EP	C[b]	N[c]	EP	C	N	EP	C	N	EP	C	N
01[d]	5.03	6.02	17[d]	4.99	5.65	33	5.01	5.95	49	5.03	6.05
02[d]	5.63	6.04	18	5.03	6.02	34	5.01	5.98	50	5.11	6.05
03[d]	5.02	5.70	19	5.00	6.00	35	5.02	6.00	51	5.08	6.03
04[d]	5.03	5.07	20	5.03	6.03	36	5.02	6.06	52	5.03	6.03
05	5.05	6.01	21	5.04	6.01	37	5.03	6.03	53	5.07	6.01
06	5.05	6.04	22	5.04	6.05	38	5.05	6.04	54	5.03	6.06
07	5.05	6.06	23	5.05	6.04	39	5.05	6.04	55	5.07	6.06
08	5.05	6.06	24	5.04	6.00	40	5.06	6.06	56	5.05	6.04

[a] Numbers were calculated by the Grassberger and Procaccia (1983) algorithm; this algorithm was also used to calculate the dimensionality of the Koch curve. The latter is generated by a fractal function with a known HN, which is log(4)/log(3) or 1.265. The calculated value was found to be, within 5 digits of precision, the same as the known value.
[b] The 27 C values, with the five electrodes indicated by d excluded, have a mean and standard deviation of 5.04 ± 0.02, which is statistically significantly different from 5.00 ± 0.02 ($P < 0.01$).
[c] The 27 N values, with the five electrodes indicated by d excluded, have a mean and standard deviation of 6.03 ± 0.02, which is statistically significantly different from 5.04 ± 0.02 ($P < 0.01$, with or without the electrodes indicated by d included).
[d] These electrodes were found not to be in contact with the surface of the olfactory bulb.

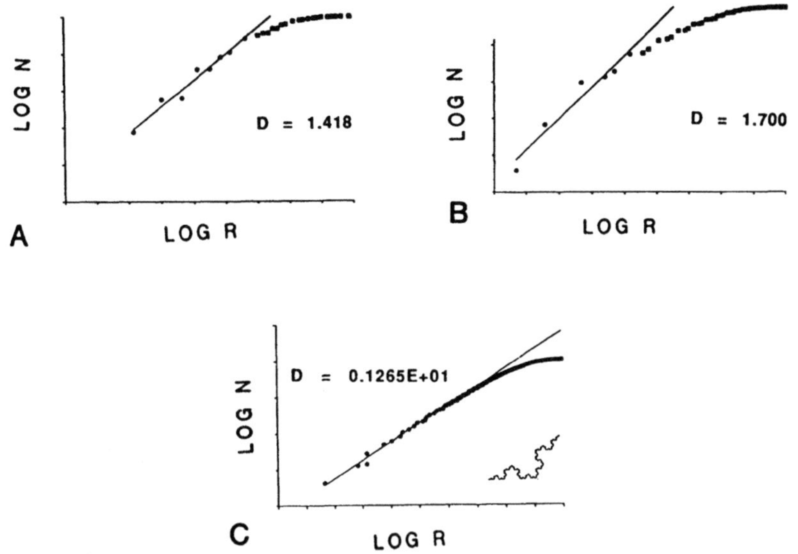

Fig. 6. Slope (D) of the plot of $\log N$ vs $\log R$ for the zero-crossing intervals in the 500-ms epochs just before (A) and 500 ms after (B) a novel odor (butanol). The slopes shown are for the first embedding dimension after saturation (embedding dimension of 2). The magnitude of the event-related increase in dimensionality for this animal is larger than that for the first rabbit. The same algorithm was used to calculate the slope for the data from the Koch curve function (C, 257 data points). For the latter, note that D is very close to the theoretical value, $\log 4/\log 3$, which is 0.1262E+01

Table 2. Effects of experience, within the same rabbit, on the direction of the event-related shift in the HN in response to the same and to different odor stimuli

Odor	Electrode position in array	HNs for control epochs[a]		HNs for test epochs[b]	
		C1	C2	T1	T2
The odors are novel (1st trial)					
Butanol	35	5.07	5.04	6.01	6.04
	37	5.03	5.03	6.02	6.04*
Benzaldehyde	35	6.06	6.04	7.08	7.08
	37	6.06	6.05	7.06	6.99*
The odor is familiar (3rd trial)					
Butanol	35	6.04	6.04	5.05	5.04
	37	6.07	6.05	5.06	5.04*

[a] C1 = 500-ms epoch 1.5 s before inspiration of odor. C2 = 500-ms epoch 1.0 s before inspiration of odor.
[b] T1 = 500-ms epoch 0.5 s after inspiration of odor. T2 = 500-ms epoch 1.0 s after inspiration of odor.
* $P < 0.01$, for all C and T comparisons during each odor.

Table 3. Within- and between-subject effects of a novel odor on the HNs associated with the pre- and postodor data (to reduce processing time, the dimensionality was calculated for the zero-crossing intervals of the analog signal, not its digitized amplitudes)

	Novel #1[a]		Novel #2[b]	
	HN, control	HN, test	HN, control	HN, test
Animal 1	0.90±0.004	0.95±0.002*	1.03±0.01	1.06±0.01*
Animal 2	1.41±0.004	1.70±0.004[c]*		

[a] Novel #1 = first presentation of butanol; HN, control and HN, test = brief epochs just before and after the inspiration of the novel odor.
[b] Novel #2 = first presentation of benzaldehyde.
[c] The estimate of the standard deviation was presumed to be the same as that of animal 1, as no suitable variance estimate was available for this rabbit.
* $P<0.01$ (between control and test data from the same rabbit).

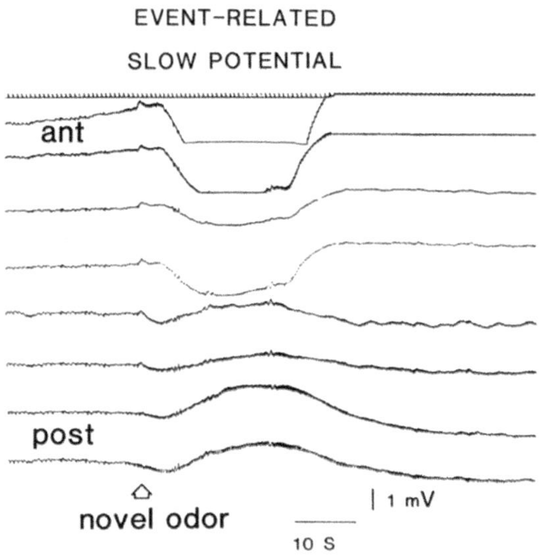

Fig. 7. Effects of a faint novel odor (butanol) on the direct-coupled surface potentials of the olfactory bulb of the conscious rabbit. Note the large amplitude of the ERSPs. The recordings were made by DC-amplifiers from monopolar platinum-black electrodes (referenced to sagittal bone) distributed in an anterior–posterior (*ant*, *post*) line across the dorsolateral surface of the bulb; interelectrode spacing was at 0.5-mm intervals. Negative polarity is upwards

4 Discussion

Our current data, obtained from a simple-system model of cortical tissue, confirm previous observations (Babloyantz and Destexhe 1986, 1987; Röschke and Başar, this volume; Graf and Elbert, this volume) that a low-dimensional chaotic attractor can be constructed from a time-varying surface potential. These previous studies also showed that reduced cortical functioning, produced by (a) closing the eyes, (b) going to sleep, or (c) having an epileptic seizure, will result in the attractor having a reduced dimensionality. We showed that presenting a familiar odor to an awake alert rabbit also leads to a reduced dimensionality.

The unique feature of our data is that they show that the same odor can also evoke an *increase* in the dimensionality of the attractor, if it is novel rather than familiar. The direction of the evoked shift by the specific odor depends upon the context in which it is presented. That is, the dimensionality shifts of the attractor are *event related* and change in a direction related to the *learned* significance of the stimulus and not to its inherent attributes.

Our data show another feature not previously demonstrated for biological tissue: the same HN is observed *globally*, at all simultaneously recorded electrode sites. Even though the analog signals are all different at each site (Fig. 1), the HNs are all the same (Fig. 3). This spatial homogeneity appears to be the case generally, during all of the behavioral states, whether control or event related. It would therefore seem that a single, global, chaotic attractor exists everywhere throughout the olfactory bulb and that the dynamical process shifts to a new attractor (i.e., the HN changes) in relation to the odor event (novel vs habituated) and not to the specific odor itself.

The stability of the HNs in both the spatial and the temporal domains cannot be explained by the fact that they are all recorded monopolarily from a single reference electrode. Previous experience has shown that sagittal bone is a very good ungrounded referent; it manifests no potential above the amplifier noise level (Skinner and Yingling 1977).

5 Theoretical Considerations

5.1 Higher Cognitive Information Processing: Event-Related Shifting Among Chaotic Attractors

It is generally the event-related dimensionality increase or decrease that seems to be the focus for current neurobiological applications of mathematical dynamics to "higher cognitive functioning." The fact that the HNs generally appear to be fractions (i.e., have fractal dimensionalities) must also be of great importance. Why should HNs be both fractal and event related?

A reductionistic explanation for the fractal activity in the olfactory bulb dictates that the underlying molecular processes are themselves fractal and chaotic. This view is confirmed by the findings that physicochemical reactions (Bak 1985, 1986), enzymatic processes (Markus et al. 1985; Hess and Markus 1987), single ion-channel kinetics (Liebovitch et al. 1987; Liebovitch and Sullivan 1987), and membrane potentials (Chay and Rinzel 1985) all manifest behaviors characteristic of chaotic attractors. Furthermore, Liebovitch et al. (1987) state that their fractal model of ion-channel data is superior to previous stochastic models because it has fewer adjustable parameters, is more consistent with the dynamics of protein conformations, and fits better with their single-channel recordings from corneal endothelium and hippocampal neurons.

These nuts-and-bolts explanations, however, do not tell us *why* the fractal processes occur in cognitive mechanisms – was this selection by nature merely adventitious or was there a strategy involved? We propose that the reason for the existence of fractal dimensionalities of processes in cognitive systems is that only this type of dynamical process enables *data compression* before information storage. This hypothesis is supported by the realization that two coordinates given to the fractal function generating, say, the Mandelbrot set, can be deterministically projected into ten megabytes of exact pixel information. The problem is, how to do the *inverse;* that is, to compress the enormous amount of pixel information into the two coordinates and its underlying chaotic dynamical system.

Barnsley et al. (1987) have shown that geometric affine mapping applied to graphics data from "natural" objects (e.g., data depicting fern leaves or steam clouds) results in the determination of several different templates for the single data set. These abstracted templates of inherently fractal data can then be stored or transmitted at reduced bit rates and later used to reconstitute, deterministically, the original data set. The data-compression ratio can be as high as $10^6:1$. We propose that this property (i.e., the "inverse" process mentioned above) may explain why nature selected fractal dynamical systems in biological tissues to process cognitive information.

A stimulus (e.g., an odorant) has fixed physical attributes, but its accompanying context (e.g., the trial number) is variable each time the stimulus is presented to the subject. When the stimulus-plus-context is novel, more information must be stored than when the stimulus exists within a familiar context. This is because the thalamic gating mechanism *lets more information through to the cortex* when the stimulus is novel than when it is familiar (Skinner and Yingling 1977). Once the stimulus is either completely conditioned or habituated, evidence shows, inhibitory thalamic gating suppresses the ascent of irrelevant sensory information (Skinner and Yingling 1977). We now propose that when the stimulus is novel a higher-dimensional chaotic attractor is required to store the greater amount of information. This hypothesis is based on our present observations.

The affine mapping and attractor reconstitution mechanics described and demonstrated by Barnsley and Sloan (1988) require a relatively massive parallel processor to handle the fractal data. Cortical tissues, including the olfactory bulb, are well known massive parallel processors (Shepherd 1970). We propose that parallel processing in cortical tissues is not present for "backup" in case some circuits are damaged, as is often suggested, but rather is present because of the need for rapid on-line data compression.

Massive parallel processing by highly interconnected elements (i.e., "connectionist models") have emergent properties that resemble real behavioral learning (Hopfield and Tank 1986; Sejnowski and Rosenberg 1987). Skarda and Freeman (1987) suggested that fractal dynamics occur in the olfactory bulb, the structure of which conforms to the connectionist models. They claim, however, that features not attributable to the connectionist models are present in their data. They propose that making the connectionist models operate as chaotic dynamical systems may yield even more realistic emergent properties.

What is the next step, now that we know the olfactory bulb is a connectionist and chaotic dynamical system? We propose that biological empiricism may now afford the most rapid breakthrough in our understanding of the properties of this simple system, as it learns and remembers. Now that we have the idea that we are dealing with a single chaotic attractor whose dimensionality is event related, we may be able to generate a number of new hypotheses that will give us new insights. For example:

1. Does beta-receptor antagonism prevent the event-related attractor shifting?
2. Does a spatially fixed HN now allow us to "map" other regions of the attractor network, lying perhaps outside the bulb, each of which is also participating in the process?
3. If data compression is the raison d'être for a chaotic dynamical process, and if information storage is by a "compressed representation" and not a specific "engram," will we finally be able to recognize biologically stored information?

We seem to have come full circle to the neuroscience of the 1960s, but things are different; now we are armed with a theoretical perspective that incorporates a recently recognized strategy of nature.

5.2 The Brain–Heart Connection: A Common Pharmacology and Physiology?

Studies in comparative animal physiology suggest that the higher centers of the brain became a focus for natural selection and evolved to enable an organism to attend to its environment and simultaneously to prepare advance autonomic support for life-saving behaviors (Cannon 1931). Our laboratory has shown that the frontal lobes may be such an orchestrator of the sensory and autonomic channels (Skinner 1971; Skinner and Yingling 1977; Yingling and Skinner 1977). Bilateral output pathways from the frontal lobes to the thalamus selectively gate sensory input during behavioral attention. Separate descending projections to the brain stem initiate autonomic reactions in anticipation of the behavior they support. Of medical importance are our observations that the neural activities in the frontocortical–brain stem pathways are both necessary and sufficient to trigger lethal cardiac arrhythmias (Skinner and Reed 1981) and to maintain blood pressure elevations in experimental hypertension (Szilagyi et al. 1987).

Many electrochemical activities in the frontal lobes are event related (Skinner 1971; Skinner and Yingling 1976, 1977; Yingling and Skinner 1977). These include: extracellular slow potential shifts, norepinephrine release, cyclic AMP accumulation, potassium inactivation, and membrane depolarization (Skinner 1984; Skinner et al. 1983). Beta-receptor antagonists, which are cardiovascular drugs, reduce the magnitudes of all of these cortical responses; these same drugs also normalize blood pressure elevations and/or prevent lethal cardiac arrhythmias in human patients and animal models (Skinner 1971; Skinner and Yingling 1977; Yingling and Skinner 1977).

Long-term potentiation (LTP) is currently thought to be a useful model for synaptic learning and memory (Anderson and Wigstrom 1980; Swanson et al. 1982). Different cardiac drugs with separate modes of action (beta-receptor antagonism, catecholamine depletion) *disable* the development of LTP (Hopkins and Johnston 1984; Bliss et al. 1983). Our laboratory (Krontiris-Litowitz et al. 1985) has shown that yet another cardiac drug, with yet another mode of action (muscarinic agonism), also has an anti-LTP effect. Why both brain (LTP) and heart (arrhythmogenesis) processes are blocked in parallel by a variety of cardiac drugs constitutes a fascinating puzzle.

We have just demonstrated that novel odors evoke in bulbar surface potentials a dimensionality shift to a chaotic attractor that exists globally throughout the olfactory bulb. This result means that the spatial pattern of activities will be different between the control and postevent conditions. Our laboratory has shown that intrabulbar infusion of the cardiac antiarrhythmic drug, propranolol, will prevent the spatial pattern change that normally occurs between the control and postodor (novel odor) epochs (Gray et al. 1986). This latter observation is evidence that propranolol also blocks event-related dimensionality shifts. Could an arrhythmia be the result of a sudden shift in a global cardiac attractor?

The cellular and subcellular oscillators that produce the cardiac action potential are known to be characterized by a chaotic attractor (Guevara et al. 1981). The nonoscillatory cardiac tissues also manifest nonlinear dynamics (Chialvo and Jalife 1987). Goldberger et al. (1984) suggested that cardiac arrhythmias could arise because of a change in a *parameter* of the chaotic attractor in the heart. We propose that the attractor itself changes, i.e., its dimensionality shifts.

Based on the nonretraction theorem in topology, Winfree (1983, 1987) suggested that the orthogonal overlap of voltage and phase gradients in myocardial tissue will result in at least one point, called a phase singularity, which is chaotic; i.e., chaotic in the sense that application of a suitably graded stimulus in this neighborhood puts the tissue in a state from which it cannot spontaneously recover to normal beating at a predictable time. He showed with his model that this singularity is surrounded by rotating waves of isochrons (loci of equal activation time); these waves in fact have been observed in cardiac tissue during arrhythmogenesis. He further demonstrated that the relative size of the phase singularity can be increased. This increase, he suggested, could lead to an increase in the vulnerability of the heart of arrhythmogenesis.

Three observations from our laboratory demonstrate ways to increase cardiac vulnerability to arrhythmogenesis. These observations may suggest ways to increase the relative size of Winfree's phase singularity.

1. Descending activity in cardiac nerves is necessary for the occurrence of the lethal arrhythmia, ventricular fibrillation (Skinner and Reed 1981). Winfree's model suggests that a voltage gradient, probably produced by nerves and orthogonal to the isochron gradient, is necessary to produce the phase singularity; without current being injected into the heart by the nerves, the singularity could not exist.
2. Stimulation of sympathetic and parasympathetic activity combined produces the most malignant type of descending neural activity (Skinner et al. 1983); such dual activation of the heart certainly would change the shape of the phase singularity and it could be expected to increase its relative size.
3. The event-related increase in the dimensionality of an attractor in the brain (represented by the present data) may be projected to the heart (e.g., through the frontocortical–brain stem pathway), in which case the relative size of the phase singularity may be increased – i.e., dimensionality increase may be related to phase singularity increase.

6 General Conclusions

All of the olfactory bulb appears to be used all of the time for the processing of odor signals. This result is an electrophysiological example of Lashley's well-known principle of mass action. The main problem in studying the neurophysiology of mass action is that we do not yet know how to deal with the enormous number and complexity of interactions that occur among neurons during a "higher cognitive process." Nonlinear mathematical dynamics may be the tool we need to study such intense interactive processes. When we gain a clearer understanding of the biological significance of fractal processes and dimensionality shifts, comparable to our understanding of some physical systems (Gleick 1987), then we may come to a completely different (perhaps even revolutionary) view of the basic cognitive process. The same process may operate in the heart as well as in the brain, and, if understood, could lead to new insights and interventions for dealing with the problems of cardiovascular malfunction.

7 Summary

After familiarization with the recording chamber, each of two rabbits was presented with a novel odor while recording from an 8×8 electrode array on the surface of the olfactory bulb (bandpass: 20–200 Hz, 3 dB). The Hausdorff number (HN) is a mathematical descriptor of the number of independent dimensions that appear to generate an observed dynamical process; it can be fractional. Preodor epochs (500 ms each) had HNs that were the *same,* both spatially (mean $\text{HN} = 5.04 \pm 0.02$ #SD) and temporally (mean $\text{HN} = 5.04 \pm 0.02$). After the odor, the HNs briefly increased ($P < 0.01$) to new values (6.03 ± 0.02), lasting for some seconds, before returning to control. Following only two trials of habituation, there was a *decrease* in the odor-evoked HNs. All HNs observed were fractional, with statistical significance. All electrodes showed the same HNs at all times. All HNs had clearly saturated correlation dimensions (using 15 embedding dimensions); this convergence occurred even though only 320 digitized data points were in the epoch. These results suggest that the electric activities throughout the bulb may be characterized spatially by a single global chaotic attractor, the dimensionality of which is event related.

References

Anderson P, Wigstrom H (1980) Possible mechanisms for long-lasting potentiation of hippocampal synaptic transmission. In: Tsuko Y, Aghanoff BW (eds) Neurobiological basis of learning and memory. Wiley, New York, pp 17–47

Babloyantz A, Destexhe A (1986) Low-dimensional chaos in an instance of epilepsy. Proc Natl Acad Sci USA 83:3513–3517

Babloyantz A, Destexhe A (1987) Strange attractors in the human cortex. In: Rensing L, an der Heiden U, Mackey MC (eds) Temporal disorder in human oscillatory systems. Springer, Berlin Heidelberg New York, pp 48–56 (Springer series in synergetics, vol 36)
Bak P (1985) Mode-locking and the transition to chaos in dissipative systems. Phys Scr 9:50–58
Bak P (1986) The devil's staircase. Phys Today 86:38–45
Barnsley MF, Sloan AD (1988) A better way to compress images. Byte 13:215–223
Barnsley MF, Massopust P, Strickland H, Sloan AD (1987) Fractal modeling of biological structures. NY Acad Sci 504:179–194
Bliss TVP, Goddard GV, Rives M (1983) Reduction of long-term potentiation in the dentate gyrus of the rat following selective depletion of monoamines. J Physiol (Lond) 334:475–491
Cannon WB (1931) Again the James-Lange and the thalamic theories of emotion. Psychol Rev 38:281–295
Chay TR, Rinzel J (1985) Bursting, beating, and chaos in an excitable membrane model. Biophys J 47:357–366
Chialvo DR, Jalife J (1987) Non-linear dynamics of cardiac excitation and impulse propagation. Nature 330:749–752
Farmer JD (1982) Dimension, fractal, measures, and chaotic dynamics. In: Haken H (ed) Evolution of order and chaos. Springer, Berlin Heidelberg New York
Freeman WJ, Schneider WS (1982) Changes in spatial patterns of rabbit olfactory EEG with conditioning to odors. Psychophysiology 19:44–56
Gleick J (1987) Chaos: making a new science. Penguin, New York
Goldberger AL, Shabetai R, Bhargava V, West BJ, Mandel AJ (1984) Nonlinear dynamics, electrical alternans, and pericardial tamponade. Am Heart J 107:1297–1299
Grassberger P, Procaccia I (1983) Measuring the strangeness of strange attractors. Physica [D]9:183–208
Gray CM, Freeman WJ, Skinner JE (1984) Associative changes in the spatial amplitude patterns of rabbit olfactory EEG are norepinephrine-dependent. Soc Neurosci Abstr 10:121
Gray CM, Freeman WJ, Skinner JE (1986) Chemical dependencies of learning in the rabbit olfactory bulb: acquisition of the transient spatial pattern change depends on norepinephrine. Behav Neurosci 100:585–596
Guevara MR, Glass L, Shrier A (1981) Phase locking, period-doubling bifurcations and irregular dynamics in periodically stimulated cardiac cells. Science 214:1350–1353
Haider M, Groll-Knapp E, Ganglberger JA (1981) Event related slow (DC) potentials in the human brain. Rev Physiol Biochem Pharmacol 88:125–197
Hess B, Markus M (1987) Order and chaos in biochemistry. Trends Biochem Sci 12:45–48
Hopfield JJ, Tank DW (1986) Computing with neural circuits: a model. Science 233:626–633
Hopkins WF, Johnston D (1984) Frequency-dependent noradrenergic modulation of long-term potentiation in the hippocampus. Science 226:350–352
Krontiris-Litowitz J, Skinner JE, Birnbaumer L (1985) A muscarinic agonist (Ethmozine) prevents long-term potentiation in the hippocampal slice. Soc Neurosci Abstr 11:781
Libet B (1985) Unconscious cerebral initiative and the role of conscious will in voluntary action. Behav Brain Sci 8:529–566
Liebovitch LS, Sullivan MJ (1987) Fractal analysis of a voltage-dependent potassium channel from cultured mouse hippocampal neurons. Biophys J 52:979–988
Liebovitch LS, Fischbarg J, Koniarek JP, Todorova I, Wang M (1987) Fractal model of ion-channel kinetics. Biochim Biophys Acta 896:173–180
Mandelbrot BB (1983) The fractal geometry of nature. Freeman, New York
Markus M, Kuschmitz E, Hess B (1985) Properties of strange attractors in yeast glycolysis. Biophys Chem 22:95–105
Sejnowski TJ, Rosenberg CR (1987) Parallel networks that learn to pronounce English text. Complex Syst 1:145–168
Shepherd GM (1970) The olfactory bulb as a simple cortical system: experimental analysis and functional implications. In: Schmitt FO (ed) The neurosciences: second study program. Rockefeller University Press, New York, pp 539–552
Skarda CA, Freeman WJ (1987) How brains make chaos in order to make sense of the world. Behav Brain Sci 10(2):161–173

Skinner JE (1971) Abolition of a conditioned, surface-negative, cortical potential during cryogenic blockade of the nonspecific thalamo-cortical system. Electroencephalogr Clin Neurophysiol 31:197–209

Skinner JE (1984) Central gating mechanisms that regulate event-related potentials and behavior. In: Elbert T, Rochstroh B, Lutzenberger W, Birnbaumer N (eds) Self-regulation of the brain and behavior. Springer, Berlin Heidelberg New York, pp 42–58

Skinner JE, Reed JC (1981) Blockade of a frontocortical-brainstem pathway prevents ventricular fibrillation of the ischemic heart in pigs. Am J Physiol 240:H1156–H1163

Skinner JE, Yingling CD (1976) Regulation of slow potential shifts in nucleus reticularis thalami by the mesencephalic reticular formation and the frontal cortex. Electroencephalogr Clin Neurophysiol 40:288–296

Skinner JE, Yingling CD (1977) Central gating mechanisms that regulate event-related potentials and behavior: a neural model for attention. In: Desmedt JE (ed) Progress in clinical neurophysiology, vol 1. Kárger, Basel, pp 30–69

Skinner JE, Beder S, Entman ML (1983) Psychological stress activates phosphorylase in the heart of the conscious pig without increasing heart rate and blood pressure. Proc Natl Acad Sci USA 80:4513–4517

Swanson LW, Teyler TJ, Thompson RF (eds) (1982) Hippocampus long-term potentiation: mechanisms and implications for memory. Neurosci Res Program Bull 20:613–769

Szilagyi JE, Taylor AA, Skinner JE (1987) Cryoblockade of the ventromedial frontal cortex reverses hypertension in the rat. Hypertension 9:576–581

Theiler J (1986) Spurious dimension from correlation algorithms applied to limited time-series data. Phys Rev [A]34:2427–2432

Winfree AT (1983) Sudden cardiac death: a problem in topology. Sci 248:144–161

Winfree AT (1987) When time breaks down: the three-dimensional dynamics of electrochemical waves and cardiac arrhythmias. Princeton University Press, Princeton

Yingling CD, Skinner JE (1977) Gating of thalamic input to cerebral cortex by nucleus reticularis thalami. In: Desmedt JE (ed) Progress in clinical neurophysiology, vol 1. Karger, Basel, pp 70–96

Dimensional Analysis of the Waking EEG

K. E. GRAF and T. ELBERT

1 Introduction

The theory of nonlinear systems has become increasingly useful and relevant to the study of empirical dynamics. When a system produces irregularity in one or more of its variables, it is of interest whether this behavior results from *randomness* (meaning that the number of degrees of freedom is infinite) or whether a finite, and possibly small, number of degrees of freedom has produced the chaos (meaning that the system is deterministic). Our understanding of deterministic systems was greatly enhanced when Lorenz (1963) discovered that a simple system with as few as three differential equations can generate totally irregular fluctuations of the system's variables – a phenomenon nowadays generally referred to as *deterministic chaos*. The prominent features of chaos are *unpredictability* over extended time periods, and *sensitive dependence on initial conditions*. Once started with specific initial values, the system's future might be totally different from what it would have been if it had been started under slightly different initial conditions. Chaos may not be the ultimate description for a system's irregular dynamic. As outlined by Rössler (1983), more complex structures "beyond chaos" may await discovery.

The scientific areas in which analyses in terms of chaotic but deterministic behavior have been useful range from physics (e.g., the dynamics of fluids, Swinney and Gollub 1986), lasers (Albano et al. 1986), and chemistry (e.g., nonequilibrium reactions, Roux et al. 1983), to biology (the dynamics of population growth, Schaffer and Kot 1986) and physiology (e.g., certain kinds of altered breathing and cardiac rhythms, Glass et al. 1986; cellular metabolism, Rapp 1986). The suggestion that the activity of single neurons as well as cell assemblies may be described by means of nonlinear system analysis (Rapp et al. 1985) leads us to the dimensional analysis of the EEG.

The EEG originates primarily in cerebral cortex. It reflects summed electrical activity of billion of neurons. Their interconnections form the most complex system of which we are aware. It has been argued that the brain has so many degrees of freedom that we never will be able to construct appropriate models which might elucidate the workings of our mind. But consider a gas in a closed chamber in which innumerable molecular collisions are taking place all the time. Each particle travels along a different path. Nevertheless, there are a number of macroscopic relationships which are quite able to describe properties of such a system like pressure, temperature, and their relationship in the *general law for gases*. This law was discovered and described on a purely macroscopic level. On the otherhand, if we look at a small particle being pushed around in Brownian motion by molecules within the gas, we cannot detect the rules which govern such a process

Springer Series in Brain Dynamics 2
Edited by Erol Başar and T. H. Bullock

on a macroscopic basis alone, although the motion is visible on the macroscopic level.

Extracting parameters from the EEG corresponds to a macroscopic measurement. The attempts to set up rules among EEG patterns and brain wave components or between such measures and behavioral elements have not been particularly successful in the past. This has led pessimists to view the EEG like the noise of the brain motor, very much like the seemingly random Brownian motion. But even for optimists who untiringly try to uncover relations which might predict EEG behavior, the (so far primary) dimensional analysis of the EEG has exceeded their prior expectations. Babloyantz (Babloyantz et al. 1985; Babloyantz and Destexhe 1986) was the first to report low dimension values in the human EEG. She detected dimensions of 4.05 and 4.27 for slow wave sleep stage 4 and 4.99–5.03 for non-REM stage 2. Others (Mayer-Kress and Layne 1986; Albano et al. 1986; Dvorak and Siska 1986; Rapp et al. 1986) reported dimensions lower than 10 during the relaxed waking state with eyes either closed or open. If such analyses are confirmed, one could conclude that the EEG results from a "rather simple" deterministic system with only a couple of variables to be determined. Based on neuroscientific evidence, Elbert and Rockstroh (Elbert and Rockstroh 1987; Rockstroh and Elbert 1988; Elbert 1987) have outlined ways by which a chaotic time series of electrical activity might be generated by the brain's regulatory processes. They conclude that nonlinear system analysis of the EEG is worthy of further study.

The present paper represents such a study. After a brief methodological introduction we will compare two ways of evaluating the time lag needed to construct the "equivalent" state space of the EEG generating system, and then go on to estimations of the EEG dimensionality.

2 Methods I: Optimal Time Lag and Data Preparation

The *state space* is a useful concept for visualizing the system's dynamic behavior. It is an abstract space whose coordinates are the degrees of freedom of the considered system, i.e., the number of independent variables. We cannot, however, construct directly such a state or phase space for the part of the brain which generates the EEG, since we do not know its specific coordinates or its degrees of freedom. Fortunately, it is apparently possible to employ a technique to reconstruct an equivalent state space, since a single time series of one variable [like the EEG(t)] can be interpreted as containing a condensed image of the whole system which has generated the time series. [1]

[1] The EEG refers to a voltage which is measured between any two electrodes on the head as a function of time. All this electrical activity could arise from one single system (in the sense of system theory). Then the information, contained in EEG time series recorded from different locations would be largely redundant, but the EEG record with the highest amplitudes would provide most information. For this reason we chose vertex–ear recordings for the present analyses. On the other hand, multiple sources in different and rather independent systems of the brain may contribute to the EEG. Then differing electrode locations may pick up activity predominantly generated in different systems, although volume conduction will always cause some overlap in recorded activity, simulating again a single complex system, rather than an ensemble of different subsystems.

Given a single time series like EEG(t) from just one location, how is it possible to resurrect the complete dynamics of the generating system which possibly includes quite a number of independent variables? Takens (1981) and Packard et al. (1980) proposed reconstructing the state space by means of *time delays* τ: The values measured at fixed time delays EEG(t), EEG($t+\tau$), EEG($t+2\tau$) ... are treated as though they characterize new variables. The delayed values then define a single point in a multidimensional state space and thus a vector. Let EEG(t_1), EEG(t_2), EEG(t_3), ..., EEG(t_n) be the empirical scalar time series of the digitized EEG. The reconstructed vectorial time series will then be:

$$\left\{ \begin{pmatrix} \mathrm{EEG}(t_1) \\ \mathrm{EEG}(t_1+\tau) \\ \mathrm{EEG}(t_1+2\tau) \\ \mathrm{EEG}(t_1+3\tau) \\ \vdots \\ \mathrm{EEG}(t_1+(d_e-1)\tau) \end{pmatrix} \begin{pmatrix} \mathrm{EEG}(t_2) \\ \mathrm{EEG}(t_2+\tau) \\ \mathrm{EEG}(t_2+2\tau) \\ \mathrm{EEG}(t_2+3\tau) \\ \vdots \\ \mathrm{EEG}(t_2+(d_e-1)\tau) \end{pmatrix} \begin{pmatrix} \mathrm{EEG}(t_3) \\ \mathrm{EEG}(t_3+\tau) \\ \mathrm{EEG}(t_3+2\tau) \\ \mathrm{EEG}(t_3+3\tau) \\ \vdots \\ \mathrm{EEG}(t_3+(d_e-1)\tau) \end{pmatrix} \ldots \right\}.$$

with d_e as the *embedding dimension* to reconstruct the dynamics. Connecting the points subsequent in time creates a *trajectory* in the state space. The set of such trajectories for all of the possible initial conditions forms a phase portrait. If trajectories starting from different initial values are attracted to a discrete region within the state space with a lower dimension (i.e., a subspace), this region is called an *"attractor."* (An attractor may have a fractal dimension, and is then called "chaotic" or "strange"; see Crutchfield et al. 1986; Graf 1986, for an introduction to chaotic attractors.) The attractor characterizes the dynamically stable solution which the system will adopt in the long run. Hence the attractor is a manifestation of the system's self-organization. As shown by Whitney (1936), a complete embedding of the system's dynamics is ascertained for values of d_e greater than 1 plus 2 times the dimension of the dynamics, i.e., the attractor. In many cases, however, a smaller value of the embedding dimension d_e is sufficient. The lowest embedding dimension which is possible for a given attractor determines the number of degrees of freedom of the system which has generated the corresponding time series.

It is clear that the time delay τ plays a crucial role when reconstructing the attractor. For practical reasons, τ is chosen as a multiple of the sampling time Δt, which is the interval between two successive measurements. Takens (1981) has based his proof for the adequacy of the time delay technique on two assumptions: the time series is infinite, and the signal is not contaminated with noise. Thus the value of the time delay should have no effect on the outcome. These suppositions, however, can be hardly satisfied by empirical data. Therefore, only a certain range of time delays will adequately represent the dynamics of the system considered. More precisely, the assumption that successive measurements contain new information, irrespective of the length of the time intervals between them, is not valid, since precision of measurement is limited. If the time lag τ is too short, successive points in phase space will appear along a straight line, stretching the attractor artificially along the diagonal. Choosing too large a time lag τ, on the other hand, will make recognition of the attractor difficult, since it is a property of chaotic attractors that small errors in measurement blow up exponentially over time.

There are two possibilities for obtaining an estimate for a valid τ value from a measured time series. The first one is to set τ equal to the smallest time lag for which the *autocorrelation function* is zero or to the first local minimum if that is earlier than the zero point. In the latter case the local minimum will be positive. The second possibility is to choose the first local minimum of the *mutual information function*.

The concept of mutual information as developed by Fraser and Swinney (1986) evolved from information theory in an attempt to answer the question: Given the measured value of a function $f(t)$ at time t, how many bits of $f(t+\tau)$ can be predicted on the average? $f(t)$ and $f(t+\tau)$, the coordinates of the state space, will be most independent where the mutual predictability reaches a minimum. The calculation of mutual information basically works in terms of *simple* and *conditional entropy:*

Assume a signaling system S sends a series of messages. All the messages sent are drawn from a certain set of possible messages $s_1, s_2, s_3, \ldots, s_n$. By the frequency of each of the possible messages s_i, the associated probabilities $P_s(s_1), P_s(s_2), \ldots, P_s(s_n)$ can be determined. The average amount of information gained from a detection that specifies a message of S is called the *entropy* H of the signaling system S:

$$H(S) = -\sum_{i=1}^{n} P_s(s_i) \cdot \log P_s(s_i) \, .$$

If the base of the logarithm is 2, the units of $H(S)$ are bits. In other words, $H(S)$ specifies how many yes/no questions are needed in order to identify one given message, provided the questioning strategy is optimal. As a next step let us consider two coupled signaling systems $[S, Q]$. If the message of S is known, a prediction of the message of Q will be possible. The goal here is to quantify the precision of the prediction, so we must examine the conditional entropy $H(Q/S)$ first. Based on the detection of S, we will make a prognosis of Q, and $H(Q/S)$ will tell us the average increment for the information in bits, which we will obtain from a detection that specifies Q. According to Fraser and Swinney (1986) we can write

$$H(Q/S) = H(S, Q) - H(S) \quad \text{whereby}$$

$$H(S,Q) = -\sum_{i,j} P_{sq}(s_i, q_j) \cdot \log P_{sq}(s_i, q_j)$$

P_{sq} is the joint probability distribution of the coupled systems. Computing the difference between the simple entropy $H(Q)$ and the conditional entropy $H(Q/S)$, i.e., subtracting the residual information content of Q – given the knowledge of S – from the total information content of Q, we obtain the mutual information $MI(Q,S)$.

$$\begin{aligned} MI(Q,S) &= H(Q) - H(Q/S) \\ &= H(Q) + H(S) - H(S,Q) = MI(S,Q) \, . \end{aligned}$$

Table 1. Subject characteristics

Sex	Age	Medication (per day)	Local minimum (τ in ms)	Results illustrated in
Epileptic patients				
m	17	4 Tegretal 4 Zentropil 2 Luminal	60	
m	43	2 Orphiril 3 Tegretal	70	
m	35	3 Tegretal 2 Zentropil	70	Figs. 1 (bottom) and 5
f	15	1.5 Phenhydan 3.5 Timonil Retard 5 Soxinutin	60	
m	16	2 Tegretal 3 Ergenyl	90	Fig. 6
m	25	1 Maliasin 1 Petinutin 1 Orphiril 1 Phenhydan	50	
f	15	None	30	Figs. 1 (bottom), 2, and 3
Healthy controls				
m	28	–	60	
m	44	–	50	Fig. 4

This value provides the answer to our question; namely, it quantifies how precisely on the average, the message of Q can be predicted from knowing the message of S. If the two systems are closely coupled, a precise prediction of the messages of one system will be possible from the messages of the other system. If the dependency among systems is low, their mutual predictability will be poor, and *MI* will be small. A detailed procedure for the calculation of *MI* has been provided by Fraser and Swinney (1986). We have applied this procedure, making the assignment $[S, Q] = [\mathrm{EEG}(t), \mathrm{EEG}(t+\tau)]$.[2]

The advantage of the mutual information over the autocorrelation function lies in the fact that the former considers all kinds of relations, not only the linear ones, as the latter does.[3]

[2] The procedure requires the length of the time series to be a power of 2. We have chosen $2^{13} =$ 8192 data points.

[3] For embedding dimensions greater than 2 it is required to consider all the possible dependencies, now among a much larger number of coordinates, in order to find the optimal time delay. Therefore mutual information could be generalized the redundancy, which deals with any number of variables (Fraser 1986). But then a very high number of data points would be required, which is generally not available, i.e., such calculations are generally not possible. The solution most often accepted is to assume one and the same time lag τ for all of the embedding dimensions. Then τ can be calculated by using an embedding dimension of 2.

Both methods of estimating the optimal time lag τ were applied for 81.92-s segments of EEG selected to be free of movement or eye movement artifacts, but not blink artifacts. Our data were recorded from Cz vs. linked earlobes during the waking states of two healthy human subjects and six epileptic patients. Vertical eye movements were recorded from electrodes below and above the left eye. EEG and EOG were amplified by means of a Beckmann type R polygraph set for a band pass from 0.005-30 Hz (30 s time constant). All patients but one were under long-term treatment with antiepileptic medication (see Table 1). All were drug resistant, having from one to several seizures per week. After electrode hook up, subjects were told to relax for 5 min. The session started with another 5-min relaxation period, followed by 3 min of hyperventilation and a final 3-min rest period. Subjects were instructed to keep their eyes open and to fixate a 3×3 cm cross positioned 2 m in front of them. The EEG segments were selected from the initial 5-min relaxation period and for patients additionally during times when seizure-like activity appeared. EEG records were also obtained while the subjects were told to relax with their eyes closed. The EEG was sampled at a rate of 100/s. (The results for the conventional analysis of these data are described in more detail by von Bülow et al. 1988.)

3 Results for the Time Lag τ

The mutual information function was calculated for three 81.92-s long EEG segments, chosen from the beginning (I), the middle part (II), and the end (III) of the initial 5-min relaxation period. How stable is the first minimum during such a time period? Figure 1 (top) depicts the mutual information of an unmedicated patient's EEG, plotted against the time lag τ. For all the three periods, the minimum lies around 30–40 ms. This value is much shorter than the one of a medicated epileptic patient, whose mutual information function is also shown in Fig. 1 (bottom). His minimum is also stable; it lies, however, between 60 and 70 ms. For the six patients under medication, the values range from 50 to 90 ms (see Table 1). Despite the considerable variance between subjects, the minimum turned out to be completely stable within subjects, at least within the 10 ms resolution used. As indicated in Fig. 2, this stability also holds over a period of 5 weeks, at least for a given experimental condition.

For 22 EEG segments recorded during the waking state, we compared the two possible ways the time lag τ was selected: from the mutual information function or from the autocorrelation function. The results were consistently the same: mutual information will reach its first local minimum aproximately at that point where the autocorrelation intersects with the abscissa. If there is a first (positive) minimum of the autocorrelation function, as in Fig. 3, mutual information will have its minimum at this value of τ. For some cases, however, the mutual information function does not exhibit a pronounced local minimum during its slow decay with increasing time lag.

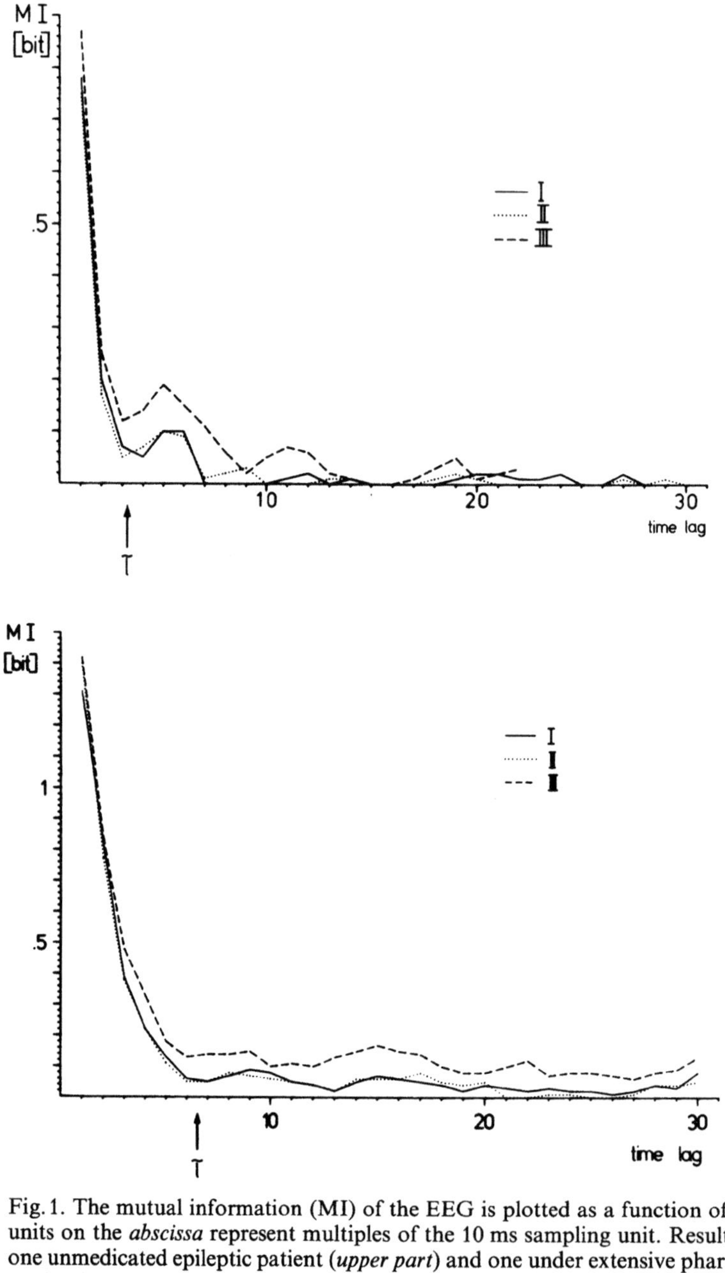

Fig. 1. The mutual information (MI) of the EEG is plotted as a function of the time lag τ. The units on the *abscissa* represent multiples of the 10 ms sampling unit. Results are illustrated for one unmedicated epileptic patient (*upper part*) and one under extensive pharmacotherapy (*lower part*). The three different lines in each plot represent EEG segments of one subject chosen from the beginning (I), middle part (II), and end (III) of a 5-min rest period. The length of the EEG segments is 81.92 s, corresponding to 8192 points. *Arrows* mark the optimal time lag τ, i.e., the first local minimum

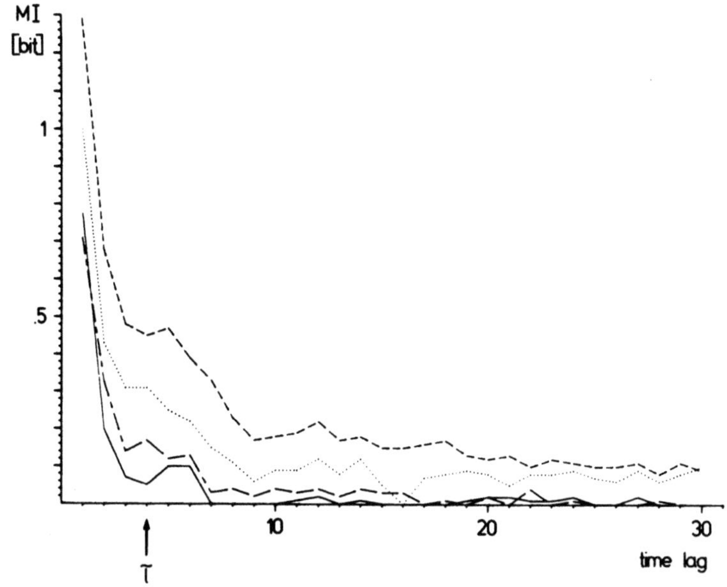

Fig. 2. Mutual information as a function of the time lag τ, calculated from four different EEG segments which were recorded from an unmedicated patient on four different occasions over a period of 5 weeks

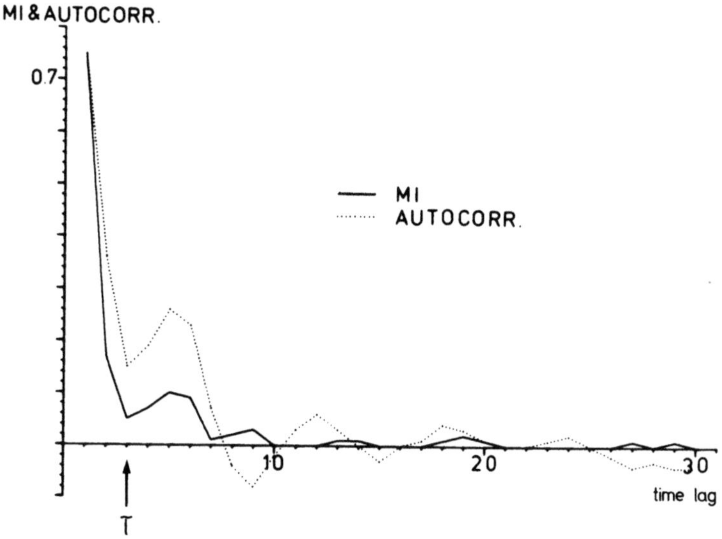

Fig. 3. Example of the mutual information function (units in bits) and the autocorrelation function (units correspond to correlation coefficients) computed for the same EEG segment. Both methods yield the same result for the optimal time lag τ

3.1 Discussion

The two criteria for choosing the optimal time lag τ yield equal results, at least during the waking state. Obviously, nonlinearity in the autocorrelations of the time series does not play a major role in the EEG analyses. This is not necessarily the case for other irregular time series as shown, for example, for the Roux attractor (by Fraser and Swinney 1986).

It seems sufficient to extract optimal EEG time delays from the EEG autocorrelation function, which also circumvents the problem of identifying τ when there is no distinct local minimum in the MI function.

As indicated in Table 1, the optimal time lag τ appears to be longer in patients under anticonvulsant medication. Although the present data base is too small for a definite conclusion, this result fits with observations of delayed event-related potentials, a generally slowed EEG, and a slower reaction time under anticonvulsants (Rockstroh et al. 1987, von Bülow et al. 1988). It is quite possible that a kind of cycle time, inherent in the regulation of brain activity (Elbert 1987; Elbert and Rockstroh 1987), is slowed down in subjects receiving antiepileptic medication. It is interesting that τ is shortest for the unmedicated epileptic patient, being below the values for the normal controls, whereas medication brings the values back towards the range for healthy controls. According to this, the patient with $\tau =$ 90 ms must have been overmedicated. As a matter of fact, after obtaining the EEG result, analyses of the plasma drug levels of this patient showed concentrations in the toxic range. Consequently the dose was reduced.

4 Methods II: Dimensional Analysis

After selecting the time lag τ, we can construct the equivalence of the state space and search for the attractor of the system.

A chaotic motion in state space generally has a *fractal dimension*, a feature which allows its recognition. Periodic or quasiperiodic motions possess integer values (Fig. 7). The procedure widely used is the one proposed by Grassberger and Procaccia (1983), and Grassberger (1983): Within the reconstructed state space of dimension d_e, a reference point on the trajectory is chosen at random. Then the number of points which lie in a hypercube with radius r around this chosen point is counted. This counting is performed for subsequently larger radii r, until ultimately all points of the time series lie within this hypercube. In order to obtain a representative sample, the same procedure is repeated for other reference points. The counts are averaged for each radius r, and plotted against r using a double logarithmic scale. The resulting function has a ∫-like shape. A linear fit is performed on the straight segment, the slope of which determines the desired "correlation dimension" d_c, provided the embedding dimension d_e is high enough. In order to test this supposition, the whole procedure must be repeated for successively higher embedding dimensions d_e until an independence of the calculated dimension d_c from the embedding dimension d_e is reached.

Let $N_{\vec{x}(i)}(r)$ be the number of adjacent points inside a hypercube of radius r which is centered around a reference point $\vec{x}(i)$ in the reconstructed phase space of a certain dimension d_e. The correlation dimension d_c will then be

$$d_c = \lim_{n_{data} \to \infty} \lim_{r \to 0} \frac{\log \left[\frac{1}{n_{ref}} \frac{1}{n_{data}} \sum_{i=1}^{n_{ref}} N_{\vec{x}(i)}(r) \right]}{\log r}$$

where n_{data} is the number of points in the reconstructed vectorial time series and n_{ref} is the number of reference points.[4]

Farmer et al. (1983) proposed an alternative method, called (averaged) *"pointwise dimension"* (d_{pw}) or *"information dimension."* A calculation of the dimension is done separately for every reference point. Then these estimations of the dimensionality are averaged:

$$d_{pw} = \lim_{n_{data} \to \infty} \lim_{r \to 0} \frac{1}{n_{ref}} \sum_{i=1}^{n_{ref}} \frac{\log \left[\frac{1}{n_{data}} N_{\vec{x}(i)}(r) \right]}{\log r}$$

Two points will be stressed here which suggest a preference for the latter method: Not all the reference points are equally well qualified for a good estimation since a chaotic attractor is generally nonuniform. If the straight region in the ʃ-like curve becomes relatively small, the confidence of the estimation becomes low. Using the method of Farmer et al. (1983), such points can be easily excluded from the averaging process, thus enhancing the quality of estimation. The standard deviation of the dimensionality, for instance, will become smaller, as demonstrated by Holzfuß and Mayer-Kress (1986), only if a selection of the best reference points is used for the computations. The second point is also related to the nonuniformity of the attractor. If a reference point lies in a region with a low density of points, a relatively large radius r is needed to encompass most other points. On the other hand, a relatively short radius r will be sufficient if the reference point lies in a region with a high density of points. The effect is a vertical displacement of the ʃ-like curves. In this case the averaging in the formula of d_c will result in a false, too flat slope, as demonstrated by Holzfuß and Mayer-Kress (1986). Furthermore, the averaged curve has a smaller scaling region and consequently the uncertainty in determining the slope increases. However, estimating each slope separately and then averaging across these estimates gives adequate accuracy. Only the averaged

[4] In order to be comparable with the subsequent dimension formula this formula may be written in a different form. Usually the original equation of Grassberger and Procaccia is cited as:

$$C(r) = \lim_{n_{data} \to \infty} \left\{ \frac{1}{n_{data}^2} \sum_{i,j=1}^{n_{data}} [\vartheta(r - |\vec{x}(i) - \vec{x}(j)|)] \right\}$$

with $\vartheta(x)$ as the Heavyside function [$\vartheta(x)=0$, if $x \leqq 0$, and $\vartheta(x)=1$, if $x>0$]. $|..|$ is the Euclidian norm or – as a simplification – the maximum norm.

Due to the so-called scaling law $C(r) \approx r^{d_c}$ which holds for small r, we end up with

$$d_c = \lim_{r \to 0} \frac{\log C(r)}{\log r}.$$

pointwise dimension takes into account the nonuniformity of the attractor with its various density of points in phase space.

Both methods give the same results when applied to analytic time series (Holzfuß and Mayer-Kress 1986). When applied to the EEG, however, dimensional values will become markedly different, as shown by Mayer-Kress and Holzfuß (1987).

We have used the method of averaged pointwise dimension for the present analyses, including only the best 20% of the reference points. First, 200 reference points were chosen at random, but only the best 40 points were kept for the determination of the pointwise dimension d_{pw}. By the best 40 points we mean those giving the relatively largest scaling regions, i.e., regions to which straight lines can be fit. This selection is made for every one of the embedding dimensions.

The embedding dimension d_e runs from 1 to 30. The dimensionality of the EEG, i.e., the dimension of an attractor, is determined by the asymptotic value reached by d_{pw} for larger embedding dimensions. (No saturation is obtained for purely random time series, i.e., noise, see Fig. 7).

5 Results for the Dimensional Analyses

An example of a selected EEG segment from a healthy, awake subject, with eyes open and closed, is illustrated in Fig. 4. It shows the pointwise dimension d_{pw} as a function of the embedding dimension d_e. The higher proportion of α-waves in the condition with eyes closed causes a decay in the dimensionality from about 12 to about 10. Figure 5 illustrates the results for a single patient. Whereas there seems to be no saturation in the eyes-closed condition, an attractor with a (fractal) dimension around 9 seems to dominate the EEG recorded while the subject held his eyes open. In contrast to the healthy control subject (Fig. 4), the patient (Fig. 5) failed to demonstrate clear α-waves in the eyes closed condition. Rather, the EEG became even more desynchronized.

We could not detect any systematic differences between patients and controls in the dimensionality of seizure-free EEG records. Generally, the dimensionality lies around 10. An exact calculation of such high dimensionalities requires much longer time series; therefore, these values should be interpreted only in relation to each other rather than in an absolute sense.

The EEG was also examined from a patient undergoing an absence seizure. The EEG segment, depicted in Fig. 6, shows slow waves, large amplitudes, and occasional spike patterns. The dimensional analysis yields lower values for the ictal than for the interictal period (Fig. 6).

In order to validate our computational strategies, we applied the analysis procedure to a random time series with the same length as the EEG segments. As shown in Fig. 7, there is no indication of saturation. d_{pw} climbs above 16 as d_e approaches 30. Comparisons of Figs. 4–6 with Fig. 7 demonstrate that there is a clear difference between a random, undetermined time series and the EEG, with its finite and probably fractal dimensions.

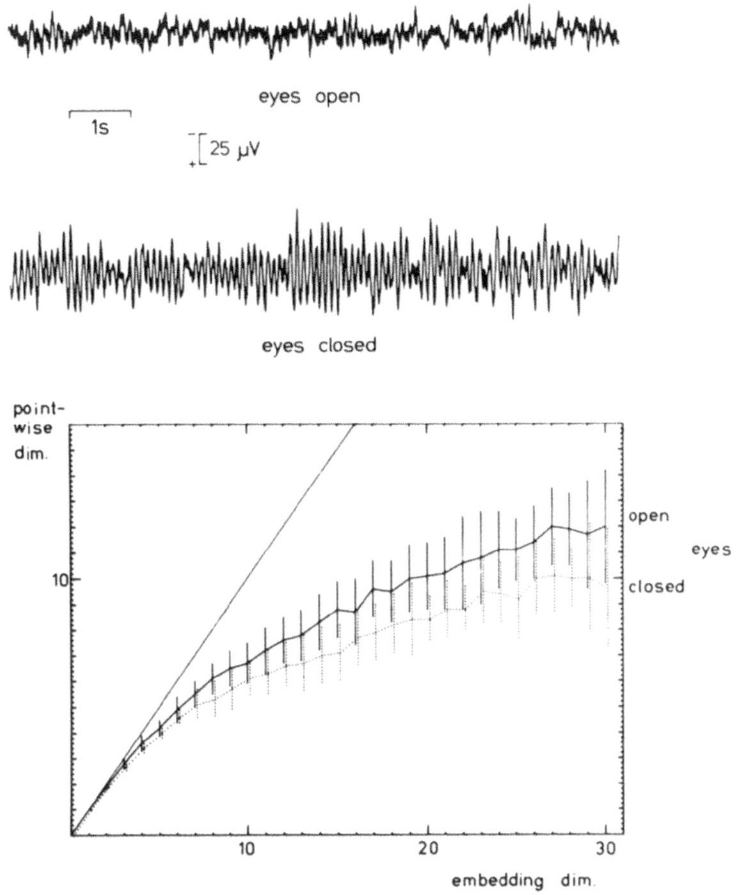

Fig. 4. EEG segments obtained from a healthy subject, as illustrated on the *top panel*, were analyzed to give the dimensional values illustrated in the *bottom panel*

5.1 Discussion

Dimensional analyses of the EEG have been presented before. Mayer-Kress and Layne (1986) reported a fractal dimension of 8.7 for the waking EEG, which even decreased to 5.1 after the subject kept his eyes closed and relaxed for a while. The difference between eyes open and eyes closed conditions is very much in line with our results for the healthy subjects. Our overall values, however, are higher than those previously reported. The EEG segments chosen for the present analyses are relatively long, which means that the time series could be less stationary. Or, the EEG generating system might switch between different modes across extended periods, exhibiting more of its complexity. It is also possible that the brain's regulating systems which generate the EEG might work in simpler modes at times, but switch to more complex modes when necessary. The relaxed waking state generat-

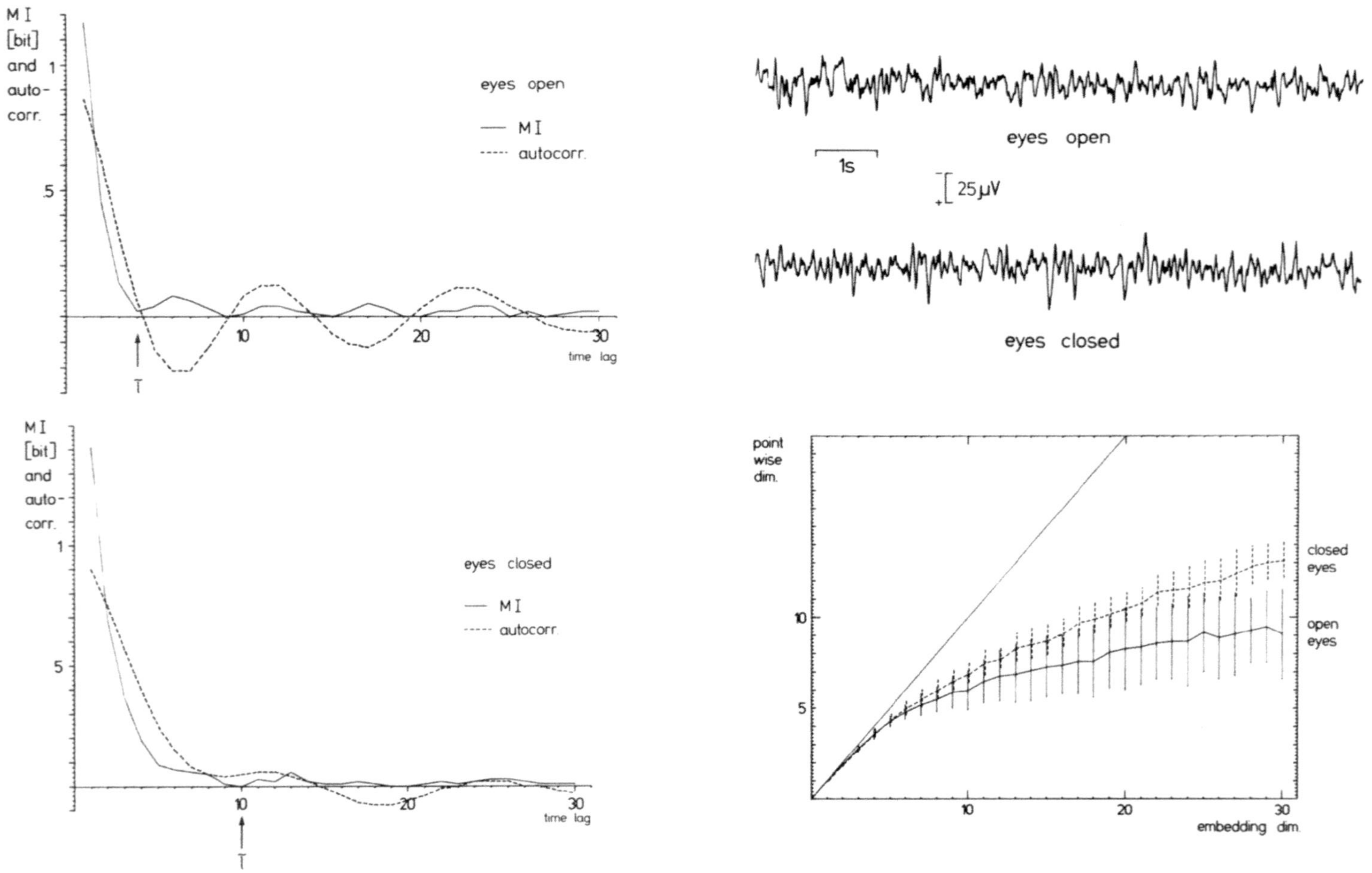

Fig. 5. Same analyses as in Fig. 4 for the EEG of an epileptic patient during a seizure-free period (*right side*). The panels on the *left side* show the mutual information function (*solid lines*) and the autocorrelation (*dotted lines*) for the illustrated EEG segments

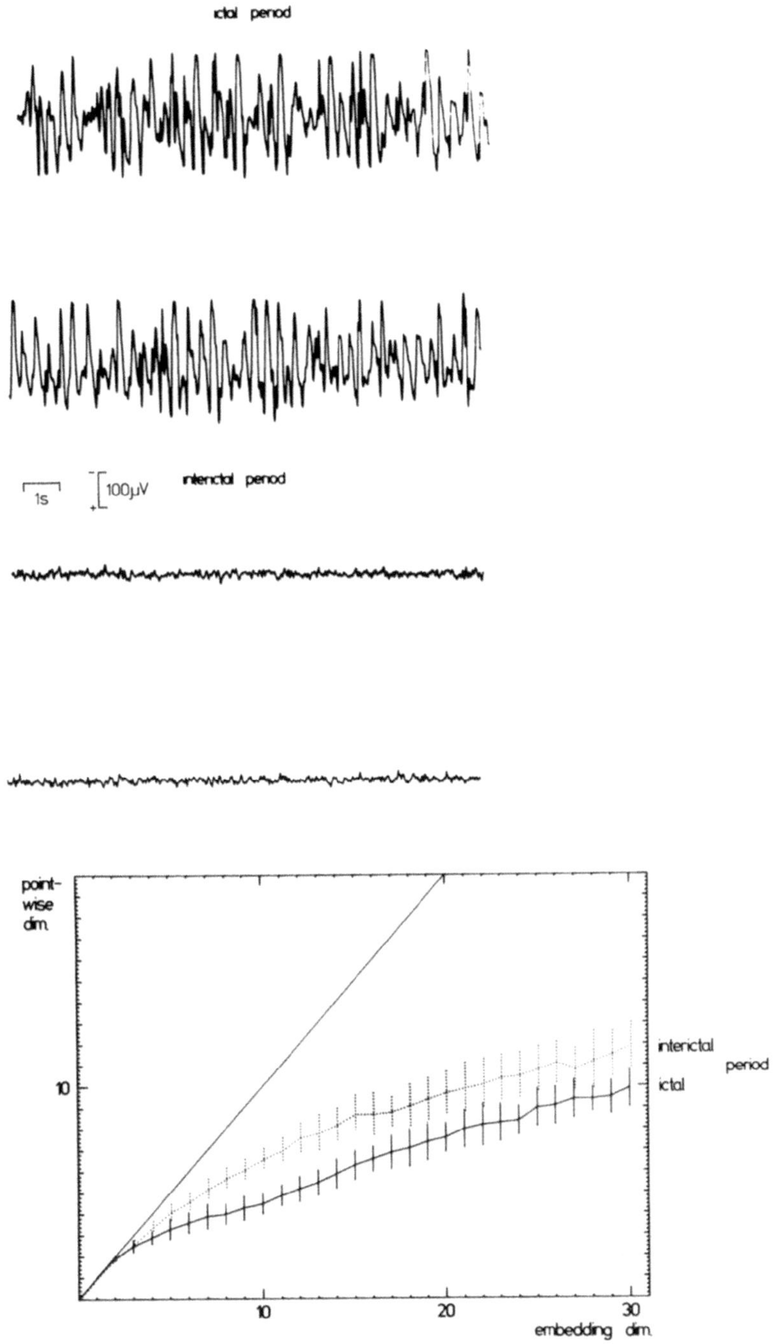

Fig. 6. Dimensional analyses of the EEG recorded during a seizure-free and during an ictal period of an epileptic patient. (Time series 4096 points.) The *upper panel* shows the segments of the raw EEG; the *lower panel* demonstrates the lower dimensionality of the EEG during the ictal period

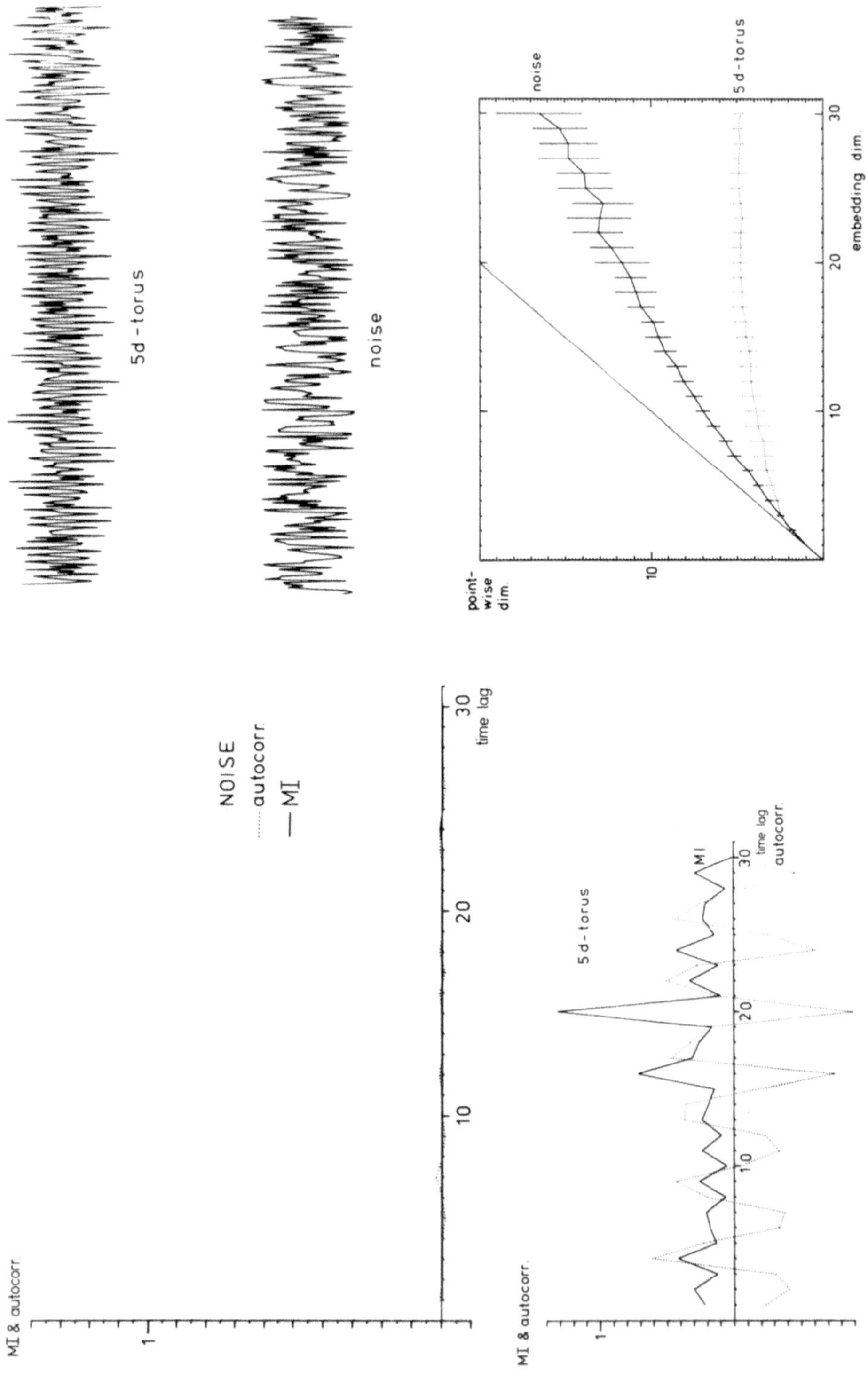
5d-torus
noise
point-wise dim.
10
embedding dim.
10
20
30
NOISE
autocorr.
MI
MI & autocorr.
1
time lag
10
20
30
5d-torus
MI
time lag
autocorr.

ing pronounced α-waves, seems to be such a "simpler mode" in terms of dimensionality. Of course, a condition with eyes closed could give a variety of results depending on what is going on in the subject's mind. This is illustrated by the example in Fig. 5: The EEG itself shows that this patient did not achieve a relaxed state with predominant α-waves. But, whereas visual inspection of the EEG makes it difficult to detect any difference between eyes closed and eyes open conditions, the dimensional analysis reveals the more complex mode for the eyes closed condition.

Epileptogenic EEG seems to display reduced dimensionality. An attractor with a fractal dimension between only 2 and 3 has been found by Babloyantz and Destexhe (1986) for an instance of a petit mal attack. We have found somewhat lower dimensional values during ictal (as compared to interictal) periods as well. A saturation of the pointwise dimension, however, with increasing embedding dimension, is not obvious (Fig. 6). One problem which has to be solved for further analyses is to obtain stationary, extended EEG time series.

6 Summary

Research on nonlinear systems theory has proven to be essential for the understanding of biological systems. The present contribution examines the usefulness of nonlinear system theory for the description of electrical brain activity. The issue addressed is how to characterize the system's (fractal) dimension. Grassberger and Procaccia (1983) have suggested a "correlation dimension", and Farmer et al. (1983) a "pointwise dimension." The latter seems to be preferable and therefore has been applied in the present paper (comparison of these two quantities in Holzfuß and Mayer-Kress 1986).

In every case, an optimal time lag must first be chosen in order to reconstruct the phase space. The autocorrelation function and the "mutual information" function (Fraser and Swinney 1986) were compared for their effectiveness in evaluating the optimal time lag. The latter has been used extensively in previous research since it accounts for nonlinearities. Nevertheless, for the 22 EEG-segments studied, both methods yielded equal results. We conclude that, at least for the waking state, nonlinearities do not play a major role in EEG sequences, and hence, the autocorrelation function with its shorter calculation time may be used.

EEG data from seven epileptic patients and two healthy controls indicated that anticonvulsant medication expands the optimal time lag. This is in con-

◀

Fig. 7. Same analysis as in Fig. 4 for noise, and for a five-dimensional quasiperiodic motion (5 d-torus). The time series for the torus is generated by $10\sin(\sqrt{2}x)+5\sin(\sqrt{3}x)+3\sin(\sqrt{5}x)+4\sin(\sqrt{7}x)+6\sin(\sqrt{11}x)$. The overlapping periods are chosen not to be integral multiples of each other. Therefore, the dimensionality of the motion equals five, the number of component periods. (The results, the same as reported by Holzfuß and Mayer-Kress, confirm the adequacy of the procedures.)

formity with delayed event-related potentials and reaction times in medicated patients.

The averaged pointwise dimension of the waking EEG was typically found to be nine or higher. If α-waves emerged while the eyes were closed, the dimension decreased, otherwise it increased. Regarding the dimensionality, we could not detect any systematic differences between the seizure-free EEG recorded from epileptic patients and the EEG of controls. As expected, the EEG during an absence seizure is characterized by lower values than the one recorded during interictal periods.

Acknowledgments. We appreciate the helpful comments of Drs Brigitte Rockstroh, Walton R. Roth, and Otto E. Rössler. This work was supported by grants from the Max Kade Foundation and the Deutsche Forschungsgemeinschaft (SFB 307).

References

Albano AM, Abraham NB, de Guzman GC, Tarroja MFH, Bandy DK, Gioggia RS, Rapp PE, Zimmerman ID, Greenbaun NN, Bashore TR (1986) Lasers and brains: complex systems with low-dimensional attractors. In: Mayer-Kress G (ed) Dimension and entropies in chaotic systems. Springer, Berlin Heidelberg New York, pp 231–252

Babloyantz A (1985) Strange attractors in the dynamics of brain activity. In: Haken H (ed) Complex systems – operational approaches in neurobiology, physics, and computers. Springer, Berlin Heidelberg New York, pp 116–122

Babloyantz A, Destexhe A (1986) Low-dimensional chaos in an instance of epilepsy. Proc Natl Acad Sci USA

Babloyantz A, Salazar JM, Nicolis C (1985) Evidence of chaotic dynamics of brain activity during the sleep cycle. Phys Lett 111(3):152–156

Bülow I v., Elbert T, Lutzenberger W, Rockstroh B, Birbaumer N, Canavan A (1988) Effects of hyperventilation on EEG-frequency and potentials in epilepsy. J Psychophysiol (in press)

Crutchfield JP, Farmer JD, Packard NH, Shaw RS (1986) Chaos. Sci Am 255(6):46–57

Dvorak J, Siska J (1986) On some problems encountered in the estimation of the correlational dimension of the EEG. Phys Lett 118 A(2):63–66

Elbert T (1987) Regulation corticaler Erregbarkeit – Im EEG ein deterministisches Chaos? In: Weinmann HM (ed) Zugang zum Verständnis höherer Hirnfunktionen durch das EEG. Zuckschwerdt, München, pp 93–107

Elbert T, Rockstroh B (1987) Threshold regulation – a key to the understanding of the combined dynamics of EEG and event-related potentials. J Psychophysiol 1:317–333

Farmer JD, Ott E, Yorke JA (1983) Dimension of chaotic attractors. Physica 7D:153–180

Fraser AM (1986) Using mutual information to estimate metric entropy. In: Mayer-Kress G (ed) Dimension and entropies in chaotic systems. Springer, Berlin Heidelberg New York, pp 82–91

Fraser AM, Swinney HL (1986) Independent coordinates for strange attractors from mutual information. Phys Rev A 33(2):1134–1140

Glass L, Shrier A, Bélair J (1986) Chaotic cardiac rhythms. In: Holden AV (ed) Chaos. – Nonlinear science: theory and application. Manchester University Press, Manchester, pp 237–256

Graf KE (1986) Grundlagen der experimentellen Anwendung nichtlinearer dynamischer Systemtheorie. Universität Tübingen

Grassberger P (1983) Generalized dimensions of strange attractors. Phys Lett 97 A(6):227–230

Grassberger P, Procaccia I (1983) Characterization of strange attractors. Phys Rev Lett 50(5):346–349

Holzfuß J, Mayer-Kress G (1986) An approach to error estimation in the application of dimension algorithms. In: Mayer-Kress G (ed) Dimension and entropies in chaotic systems. Springer, Berlin Heidelberg New York, pp 114–122

Lorenz EN (1963) Deterministic nonperiodic flow. J Atmos Sci 20:130–141

Mayer-Kress G, Holzfuß J (1987) Analysis of the human electroencephalogram with methods from nonlinear dynamics. In: Rensing L, an der Heiden U, Mackey MC (eds) Temporal disorder in human oscillatory systems. Springer, Berlin Heidelberg New York

Mayer-Kress G, Layne SP (1986) Dimensionality of the human electroencephalogram. Submitted to Proceedings of New York Academy of Sciences conference on perspectives in biological dynamics and theoretical medicine, Bethesda

Packard NH, Crutchfield JP, Farmer JD, Shaw RS (1980) Geometry from a time series. Phys Rev Lett 45(9):712–716

Rapp PE (1986) Oscillations and chaos in cellular metabolism and physiological systems. In: Holden AV (ed) Chaos – nonlinear science: theory and application. Manchester University Press, Manchester, pp 179–208

Rapp PE, Zimmerman ID, Albano AM, de Guzman GC, Greenbaun NN (1985) Dynamics of spontaneous neuronal activity in the simian motor cortex: the dimension of chaotic neurons. Phys Lett 110 A(6):335

Rapp PE, Zimmerman ID, Albano AM, de Guzman GC, Greenbaun NN, Bashore TS (1986) Experimental studies of chaotic neuronal behaviour: cellular activity and electroencephalographic signals. In: Othmer HG (ed) Nonlinear oscillation in biology and chemistry. Springer, Berlin Heidelberg New York, pp 175–205

Rockstroh B, Elbert T (1988) On the regulation of excitability in the cerebral cortex – a bridge between EEG and attention? In: Pickenhain L (ed) Cortical DC-potential shifts and human performance (in press)

Rockstroh B, Elbert T, Lutzenberger W, Altenmüller E, Diener HC, Birbaumer N, Dichgans J (1987) Effects of the anticonvulsant carbamazepine on event-related brain potentials in humans. In: Barber C, Nodar H (eds) Evoked potentials III. Butterworth, London, pp 361–369

Rössler OE (1983) The chaotic hierarchy. Z Naturforsch 38 a:788–801

Roux JC, Simoyi RM, Swinney HL (1983) Observation of a strange attractor, Physica D 8:257–266

Schaffer WM, Kot M (1986) Differential systems in ecology and epidemioloy. In: Holden AV (ed) Chaos –nonlinear science: theory and application. Manchester University Press, Manchester, pp 158–178

Swinney HL, Gollub JP (1986) Characterization of hydrodynamic strange attractors. Physica 18 D:448–454

Takens F (1981) Detecting strange attractors in turbulence. In: Rand DA, Young LS (eds) Lecture notes in mathematics 898. Springer, Berlin Heidelberg New York, pp 366–381

Whitney H (1936) Differentiable manifolds. Ann Math 37:645

A Model of the Generation of Electrocortical Rhythms

K. J. Blinowska and P. J. Franaszczuk

1 Introduction

The investigation of brain electrical activity is usually approached in one of two ways. In the first approach the methods of analysis of stochastic signals are used to extract the characteristic features of the EEG, without any attempt to elucidate the physiological basis of its generation (Gersh and Yonemoto 1977; Gersh et al. 1977; Bodenstein and Praetorius 1977). In the second approach, models based on the neurophysiological data are created, but very often the results of the modeling are difficult to compare directly with the experimental data and only general features of the EEG are described (Aninos and Zenone 1980; Wilson and Cowan 1973; Zetterberg 1973).

The aim of our recent work has been to make both approaches contiguous by connecting the coefficients of the stochastic model obtained from the recorded EEG with the parameters of a simple linear model of brain activity generation. Our line of reasoning was similar to that of Freeman (1975), who developed a formalism based on the estimation of transfer functions of the neural sets. The transfer function obtained by us via an autoregressive model directly from the EEG time series led us to the equations of the damped oscillator type. The description of the EEG time series by oscillator equations was postulated by Başar (1983) and Wright and Kydd (1984). We have connected parameters describing the solutions of the oscillatory equations in the form of the EEG waves with the parameters of a simple physiological model similar to the model developed by Lopes da Silva (1976).

We have applied our method to the analysis of the electrocortical activity recorded in the rabbit brain, but the formalism developed by us can be applied to a wide class of neurophysiological investigations. Its advantage lies in the fact that parameters extracted directly from given EEG series can be confronted with the parameters of physiological models. This makes possible testing of models and on the other hand parameters describing EEG signals can be interpreted in terms of the mechanisms of generation of the electrical activity.

2 Outline of the Method

We assume that the EEG time series can be described by means of the autoregressive (AR) model:

$$\sum_{j=0}^{p} a_j x_{n-j} = \varepsilon_n \qquad (1)$$

Springer Series in Brain Dynamics 2
Edited by Erol Başar and T. H. Bullock

where a_j = coefficients of the model ($a_0 = 1$), x_n = the value of the sampled signal at moment n, and ε_n = zero mean uncorrelated white noise process.

In the framework of an AR model we can consider our system as a linear filter with a white noise input. We have shown (Franaszczuk and Blinowska 1985) that the transfer function of this system can be expressed in the form:

$$H(s) = \sum_{j=1}^{p} \frac{D_j}{s - \alpha_j}, \tag{2}$$

where s = Laplace complex frequency; D_j, α_j = parameters which are real or occur as a complex conjugate pairs, and can be determined from AR model coefficients; and p – AR model order.

Introducing new parameters:

$$\begin{aligned} \beta_j &= Re(\alpha_j), \quad & \omega_j &= Im(\alpha_j), \\ \varphi_j &= Arg(D_j), \quad & B_j &= 2|D_j|, \end{aligned} \tag{3}$$

we can write the impulse response function in the form:

$$h(t) = \sum_{j=1}^{m} B_j \exp(-\beta_j t) \cos(\omega_j t + \varphi_j). \tag{4}$$

We have assumed that the transfer function has only single poles, which is true when the model order has not been overestimated. If multiple poles are present, the exponential terms in the formula should be multiplied by polynomials.

3 Application of the Derived Method of the Θ Rhythm in Rabbit

The electrocortical activity was measured from electrodes implanted in four brain structures of the rabbit: motor sensory cortex (MSC), ventroposterolateral thalamic nuclei (NUPL), periaqueductal gray (PAG), and midbrain reticular formation (MRF). The ECoG signal was registered for spontaneous activity and after the application of a pain stimulus (with a delay of 2.5 s after the end of the stimulus). A more detailed description of the experiment has been published elsewhere (Blinowska et al. 1988). The AR model was applied to the EEG signals and coefficients a_j, transfer function, and impulse response function were found. The model order was estimated as 7 by means of Akaike criterion, which corresponds to the differential equation of order 7 [discussion of the model order determination is given in Franaszczuk and Blinowska (1985)]. The interpretation of the system described by this kind of equation seems difficult. Since $H(s)$ can be expressed in different ways corresponding to the system of filters relevant to the considered process (Rabiner and Gold 1975), it seems appropriate in our case to represent the system as a set of parallel filters (Fig. 1) corresponding to the rhythm generators:

$$H(s) = H_1(s) + H_2(s) + H_3(s) + H_4(s). \tag{5}$$

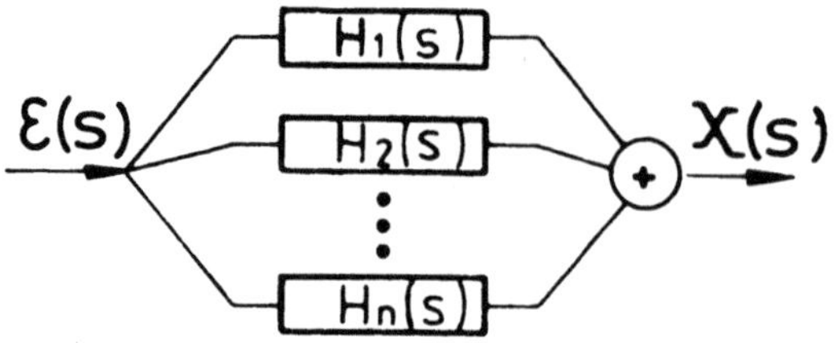

Fig. 1. The transfer function in the form of parallel filters

Now $H_1(s)$, $H_2(s)$, and $H_3(s)$ are the transfer functions of the second order systems, $H_4(s)$ of the first order system. This kind of representation is possible when function $H(s)$ has three complex conjugate poles and one real pole. This was the case for the most of the data records of the analyzed EEG.

The impulse response function takes the form:

$$h(t) = \sum_{j=1}^{3} B_j \exp(-\beta_j t) \sin(\omega_j t + \varphi_j) + \tfrac{1}{2} B_4 \exp(-\beta_4 t) \tag{6}$$

and the corresponding transfer function can be expressed as:

$$H(s) = \sum_{j=1}^{3} B_j \frac{(s+\beta_j)\cos\varphi_j - \omega_j \sin\varphi_j}{s^2 + 2\beta_j s + \beta_j^2 + \omega_j^2} + \tfrac{1}{2} B_4 \frac{1}{s+\beta_4}. \tag{7}$$

The differential equations of the system described by this kind of $H(s)$ are the damped oscillatory equations:

$$\frac{d^2}{dt^2} x_j(t) + 2\beta_j \frac{d}{dt} x_j(t) + (\beta_j^2 + \omega_j^2) x(t) = y_j, \quad j = 1, 2, 3,$$

$$\frac{d}{dt} x_4(t) + \beta_4 x_4(t) = y_4. \tag{8}$$

Parameters of Eqs. 6 and 7 can be interpreted as: B_j = amplitudes, β_j = damping factors, ω_j = frequencies, and φ_j = phases of the brain waves connected with the hypothetical rhythm generators. The first three of these parameters, called by us FAD (frequency, amplitude, damping), bear a resemblance to Zetterberg's SPA parameters (Zetterberg 1973). Zetterberg connected SPA parameters only with the power spectrum properties such as spectral peak frequency, its bandwidth, and power. SPA parameters were calculated from autocorrelation function and their disadvantage was the fact that for a model order higher than 5, the parameter describing height of spectral peak sometimes became negative (Isaksson et al. 1981). In our approach parameter B_j, connected with the wave amplitude and the spectral peak height, never becomes negative (by definition – Eq. 3).

We have calculated FAD parameters for ECoG of several animals with the aim of investigating their statistical properties. Figure 2 shows parameters β and ω in the root loci. For each data record three points falling into different frequency bands were plotted. One can see that the fluctuations of frequencies were much smaller than the fluctuations of β values. In order to describe the average spectral properties of the EEGs and their spectral characteristics, we have pro-

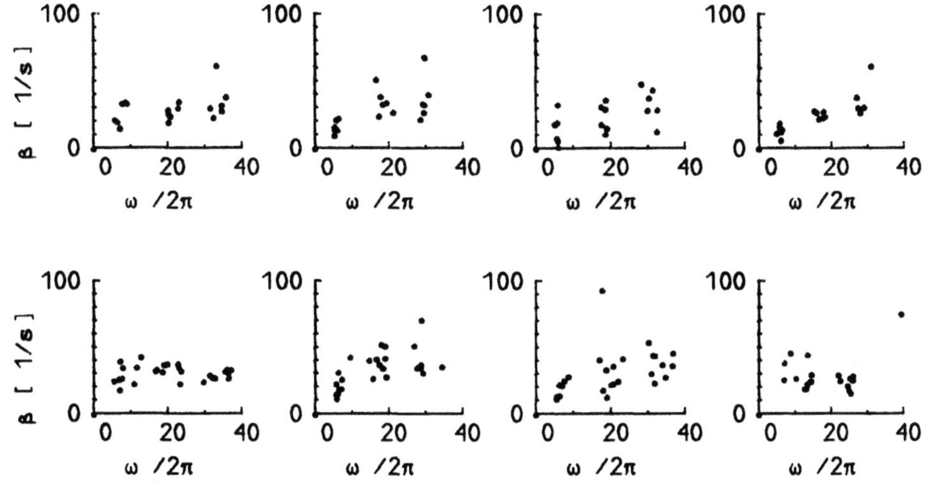

Fig. 2. Parameters ω and β of impulse response function in the complex plane – for spontaneous activity (*bottom*) and after stimulus (*top*)

posed a method of rhythm diagrams (Mitraszewski et al. 1987; Blinowska et al. 1988). They are constructed in the following way: FAD parameters are found for the consecutive EEG data records. The frequency range of the investigated signal is divided into bins. For each characteristic ω we plot the rectangle of the width corresponding to the bin width and height proportional to $B_j\exp(-\beta_j)$. The examples of the rhythm diagrams and average power spectra corresponding to the same data are shown in Fig. 3. Although both plots are calculated from the same AR coefficients and represent the same information, it is easy to see that the EEG rhythms are much better distinguished in the rhythm diagrams. This is true especially for the weaker components, which in the power spectra are usually masked by the slopes of the lines of greater intensity. We have also applied the method of rhythm diagrams to the human EEG (Fig. 4), obtaining very clear division into Θ, α, β_1, and β_2 components (Blinowska et al. 1988).

4 FAD Parameters and Model of Brain Rhythm Generation

In Sect. 2 we evaluated the transfer function in the form of the polynomials ratio. The transfer function of the model assuming populations of excitatory and inhibitory neurons connected into feedback loop is expressed by the formula (e.g. Lopes da Silva et al. 1976).

$$H(s) = \frac{H_e(s)}{1 + C_e C_i q_e q_i H_i(s) H_e(s)}, \tag{9}$$

where C_e and C_i are coupling constants corresponding to the number of active connections between excitatory and inhibitory neural populations, and q_e and q_i

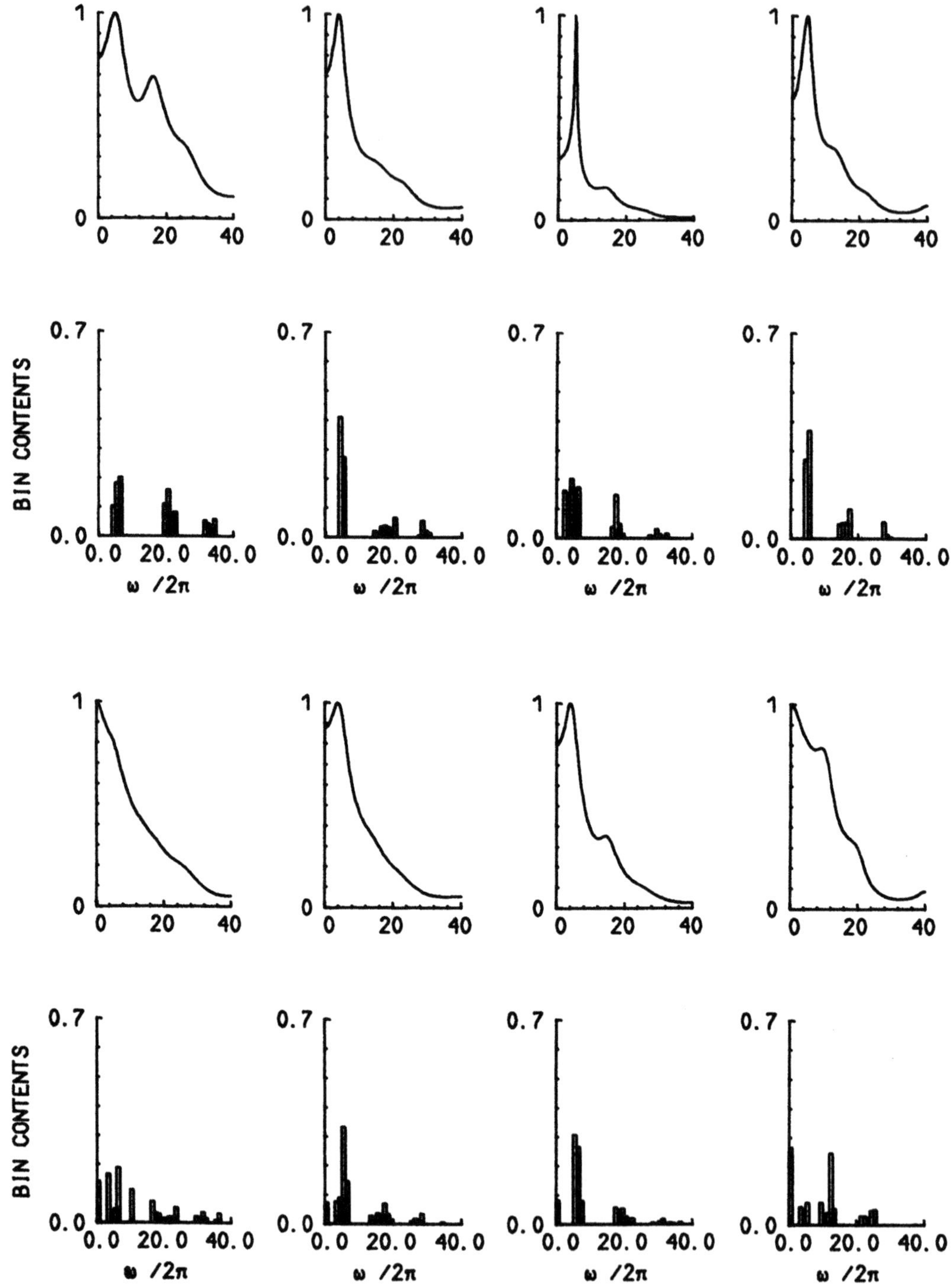

Fig. 3. Rhythm diagrams and corresponding averaged power spectra – for spontaneous activity (*bottom*) and after stimulus (*top*)

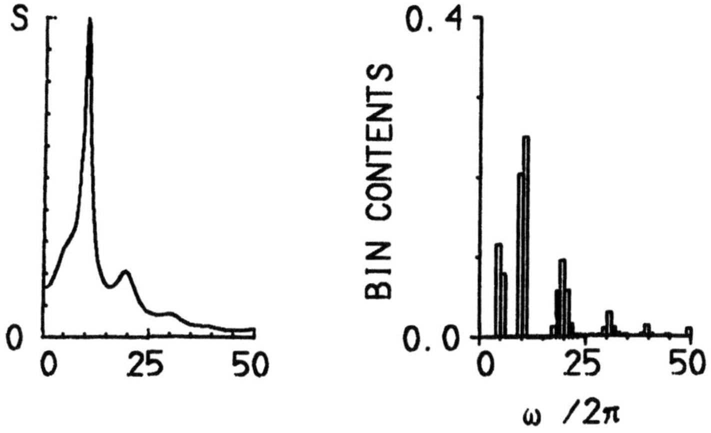

Fig. 4. Rhythm diagram and power spectrum for human EEG

are connected with the function of wave to pulse conversion in the neural sets. This relationship is described by nonlinear function $f(U)$. q_e and q_i are the inverse of the coefficients of expansion (only first order terms are taken). $H_e(s)$ and $H_i(s)$ are connected with the delays in the propagation of the electrical activity in the neural masses. Freeman (1975) gives the impulse response function of the neural set as:

$$h(t) = A\exp(-at) + B\exp(-bt) + C\exp(-ct), \tag{10}$$

where $a \approx 1000\,\mathrm{s}^{-1}$, $b \approx 100\,\mathrm{s}^{-1}$, and $c \approx 20\,\mathrm{s}^{-1}$. The first two components describe the rising and falling phases of PSP and the third component is responsible for the lumped delay of the input.

The EEG signal analyzed by us was measured with the gross electrode ($\varphi = 0.3$ mm), and sampled at a rate of 100 Hz (which means a 10 ms distance between samples), so we were not able to observe the fast components. Therefore we can approximate function $h(t)$ by the expression:

$$h(t) = C\exp(-ct). \tag{11}$$

We assume that the time course of this function is the same for excitatory and inhibitory populations. Applying the Laplace transform we obtain:

$$H_e(s) = H_i(s) = \frac{A}{s+c}. \tag{12}$$

Inserting the above expression to Eq. 9 we get:

$$H(s) = \frac{A(s+c)}{(s+c)^2 + C_i C_e q_i q_e A^2}. \tag{13}$$

Substituting $C_i C_e q_e q_i A^2 = \mathrm{K}$ in the above equation, we obtain the formula:

$$H(s) = \frac{A(s+c)}{s^2 + 2as + a^2 + \mathrm{K}}. \tag{14}$$

Comparing Eq. 7 and Eq. 14 it is easy to see that each of the components of the second order in the sum in Eq. 7 has the denominator of the same form as the above equation; therefore we can write: $c = \beta_j$, $K = \omega_j^2$.

Values of parameters β_j obtained for three observed rhythms are of the order of magnitude of parameter c estimated by Freeman (1975). The value of β obtained for the strongest component agrees with value c within the limits of experimental results.

In all our studies concerning EEG activity of rabbit brain structures we have obtained three frequency components (ω_1, ω_2, and ω_3) with energies of approximately 6, 18, and 30 Hz. Their identity was confirmed by cluster analysis (Mitraszewski et al. 1987). The presence of higher frequency components may explain the fact that the brain Θ rhythm is not sinusoidal, but has a characteristic saw-teeth shape. Because of their low intensity ω_2 and ω_3 are not well distinguished in the power spectra, but our method of rhythm diagrams and FAD parameters reveals them without doubt. The three observed components can be explained either as independent rhythms generated by three different neural assemblies or as a result of the existence of one generator in the presence of nonlinearities. In the second case, the results obtained could be explained as a basic rhythm and its odd harmonics. If we take into account that function f(U) for the population of excitatory neurons is symmetric with respect to stability point – the working point of the population – the model will have nonlinearity in the form of odd function, which explains the existence of odd harmonics.

In order to check whether higher order components are really harmonics we calculated the bispectrum. Within the limits of experimental errors we were not able to confirm that ω_2 and ω_3 are harmonics of ω_1. We investigated the dependence of ω_2 and ω_3 on ω_1 by means of linear regression. Regression coefficients indicated dependence of higher components on basic rhythm, especially for EEG registered after stimulus and component ω_2. The inclination of the regression lines was similar for ω_2 and ω_3, and close to 1 for NUPL and MRF and 0.7 for MSC and PAG. These last results are in disagreement with the hypothesis of higher components being odd harmonics of basic rhythm. Nevertheless, it seems that there is some dependence between the three observed rhythms. The problem of their origin is still open.

5 Discussion of Modeling Results

We have investigated the ECoG of rabbits under changing experimental conditions, i.e., for spontaneous activity, after pain stimulus, and under the influence of drugs changing the synaptic transmission in cholinergic and GABA-ergic systems. Physiological aspects are discussed elsewhere (Blinowska et al. 1988); here we would like only to comment on the general trends observed in that study. In hardly any of the cases studied did the three observed frequencies change significantly. Rather it was the pattern of multiple and partial coherences which changed upon stimulus and drug application (Franaszczuk et al. 1985). The stimulus changed the bandwidth of components but not their frequencies. The analysis of variances showed that the parameters which changed most after the drug application were parameters β and B, describing damping and amplitude of EEG waves. These findings support the conjectures of Wright et al. (1985) concerning the constancy of frequencies and variation of damping of brain rhythm generators.

In the above simple model the frequency of rhythm depends on the constant K; β is connected with the delays in the neural tracts. It seems plausible that the time dispersion in the propagation of neural activity is subject to the considerable changes due to statistical fluctuations or the influence of external factors, which would explain our observations concerning larger statistical fluctuations of β values and variations of parameter β under medication. On the other hand, stability of observed rhythms can be connected with the rather fixed connections in the neural population and small changes of the "working point" on the nonlinear f(U) function.

6 Conclusions

Presented in this paper is a method of analysis and modeling of electrical activity of brain structures that makes it possible to relate the results of statistical analysis of EEG signals with the models of generation of the rhythms. In the framework of the applied AR model even for short data strings FAD parameters describing EEG rhythms can be found; therefore the evolution of the signals over time (e.g. EEG after stimulus application) can be followed. Introduced parameters characterizing EEG waves (frequencies, amplitudes, and damping factors) make possible independent estimation of the frequency components of the signal and can serve as input values for multivariate statistical analysis.

Analysis of the vast body of experimental data confirms the existence of well distinguished EEG rhythms whose identity was difficult to establish by means of conventional spectral analysis. The proposed method bypasses the difficulties connected with the arbitrary division of the spectrum into the frequency bands. Weak components usually masked in the spectra by the stronger ones are clearly distinguished in the proposed method. The averaged spectral properties of the signal are presented in the form of the rhythm diagrams, which are characterized by better frequency resolution than power spectra and are very useful for the comparison of the signals under changing experimental conditions. The described method made possible the distinction of three persistent frequency components in the ECoG of rabbits. Its application to the human EEG seems very promising – all rhythms defined by electroencephalographers are clearly separated in the rhythm diagrams.

The introduced formalism makes possible the construction of linear dynamics models of electrical activity of neural populations in connection with autoregressive models describing the EEG signals. The results of modeling can be directly compared with the definite time series. The proposed simple model of Θ rhythm generation connects the parameters describing EEG rhythms with the delays and coupling constants in the neural tracts. The observed stability of rhythm frequencies supports the model.

Within the framework of the proposed approach, models of different complexity can be created by choosing the appropriate configuration of the linear filters defining the transfer function. Since function $H(s)$ can be decomposed into factors or components of arbitrary degree, it is possible to analyze systems described by the differential equations of an order higher than 2; for example, the models of the neural assemblies based on the equations of the third degree proposed by Freeman (1975) can be considered.

In the formalism of our model, spontaneous as well as evoked activity can be treated as outputs of the same system. The superposition of the frequency components of the impulse response function taken with the appropriate time delays should give the time course of the evoked potential. It seems that the possibilities of the proposed model have only begun to be explored.

7 Summary

A model connecting the parametric description of the EEG time series with rhythm generation mechanisms was developed. From the autoregressive coefficients a transfer function and an impulse response function of the system were calculated. The transfer function, in the form of parallel filters, was proposed for the description of the EEG rhythms. Parameters describing the signal in terms of frequencies, amplitudes, and damping factors (FAD) of the EEG waves were estimated and used for the analysis of electrocortical activity measured in the brain structures of rabbit. The existence of three well established rhythm components characterized by small fluctuations of frequencies supported the conjecture of the existence of brain rhythm generators. The simple linear model of EEG generation based on the assumption of two neural populations connected in a feedback loop was proposed and its parameters were connected with FAD parameters. The method of rhythm diagrams was proposed for presenting the average properties of the EEG time series, and its advantages in comparison with power spectra were demonstrated.

Acknowledgments. We are grateful to Dr. M. Kowalczyk and Dr. W. Szelenberger for making available their EEG data. The work was supported by the grant CPBR 11.9 from the Polish Academy of Sciences.

References

Aninos PA, Zenone S (1980) A neural net model for the α-rhythm. Biol Cybern 36:187–191

Başar E (1983) Toward a physical approach to integrative physiology. I: Brain dynamics and physiological causality. Am J Physiol 245:R510–R533

Blinowska KJ, Kowalczyk M, Franaszczuk PJ, Mitraszewski P (1988) The application of a new method of parametrization of EEG time series in the study of nociception. Acta Neurobiol Exp (in press)

Bodenstein G, Praetorius HM (1977) Feature extraction from the electroencephalogram by adaptive segmentation. Proc IEEE 65:642–657
Franaszczuk PJ, Blinowska KJ (1985) Linear model of brain electrical activity – EEG as a superposition of damped oscillatory modes. Biol Cybern 53:19–25
Franaszczuk PJ, Blinowska KJ, Kowalczyk M (1985) The application of parametric multichannel spectral estimates in the study of electrical brain activity. Biol Cybern 51:239–247
Freeman WJ (1975) Mass action in the nervous system. Academic Press, New York
Gersh W, Yonemoto J (1977) Parametric time series models for multivariate EEG analysis. Comput Biomed Res 10:113–125
Gersh W, Yonemoto J, Naitoh P (1977) Automatic classification measure and the eigenvalues of parametric time series model feature. Comput Biomed Res 10:297–318
Isaksson A, Wennberg A, Zetterberg LH (1981) Computer analysis of EEG signals with parametric models. Proc IEEE 69:451–461
Lopes da Silva FH, Van Rotterdam A, Barts P, Van Heusden E, Burr W (1976) Models of neuronal populations: the basic mechanisms of rhythmicity. In: Corner MA, Scrab OF (eds) Perspectives of brain research, pp 281–308 (Progress in brain research, vol 45)
Mitraszewski P, Blinowska KJ, Franaszczuk PJ, Kowalczyk M (1987) A study of stability of electrocortical rhythm generators. Biol Cybern 56:255–260
Rabiner LR, Gold B (1975) Theory and application of digital signal processing. Prentice Hall, Englewood Cliffs
Wilson HR, Cowan JD (1973) A mathematical theory of the functional dynamics of cortical and thalamic nervous tissue. Kybernetik 13:15–80
Wright JJ, Kydd RR (1984) A linear theory for global electrocortical activity and its control by the lateral hypothalamus. Biol Cybern 50:75–82
Wright JJ, Kydd RR, Lees GJ (1985) State changes in the brain viewed as linear steady states and non-linear transitions between steady states. Biol Cybern 53:11–17
Zetterberg LH (1973) Experience with analysis and simulation of EEG signals with parametric description of spectra. In: Kellaway P, Petersen I (eds) Automation of clinical electroencephalography. Raven, New York, pp 227–234
Zetterberg LH, Kristiansson L, Mossberg K (1978) Performance of a model for a local neuron population. Biol Cybern 31:15–28

Linearity and Nonlinearity in Electrocortical Waves, and Their Elementary Statistical Dynamics

J. J. WRIGHT

1 Introduction

A major theme of this conference is the modeling of electrocortical activity by powerful new tools applicable to nonlinear dynamic systems. Neuronal elements are highly nonlinear, and successful identification of low dimensional attractors in EEG has been attained in some circumstances (Babloyantz et al. 1985, Babloyantz and Destexhe 1986, 1987; Freeman 1987; Rosche and Başar 1987). A question arises as to whether linear systems analysis still has a place in the identification of brain dynamics. The two approaches are complementary, as was pointed out by Freeman during the final session of the first ICSCB. The present chapter attempts to define some aspects of this complementarity.

Linear systems theory depends on the principle of linear superposition of waves, which is in turn dependent on linearity of the system's differential equations. Techniques are well developed to identify and analyze linear systems in conditions involving driving by noisy input signals. Conversely, nonlinear attractor theory applies best to conditions in which recursive nonlinear transformations operate within comparatively autonomous systems. No low-dimensional attractor could be demonstrated to be present if there was strong ongoing noise-like input to the system.

In an endeavor to resolve some of the problems which arise in linear and nonlinear techniques, this paper discusses the following related assertions:

1. Linear and nonlinear approaches to gross electrocortical activity may best apply to limits on a continuous spectrum, from high to low conditions of cortical arousal.
2. In conditions of strong brain stem driving of the telencephalon, electrocortical waves appear well accounted for as linear resonant phenomena, with associated damping factors characterizing activity for the ensemble.
3. Simple considerations from thermodynamics can explain those whole-brain dynamics which are manifested by electrocortical activity during cortical "activation."
4. The assumptions upon which a linear treatment depends fail outside certain limits, including increasing electrocortical coherent states. Then chaotic and other phenomena may readily ensue, e.g., in the olfactory bulb or during sleep or low vigilance conditions.

Springer Series in Brain Dynamics 2
Edited by Erol Başar and T. H. Bullock

2 A Model for Electrocortical Activity

Theoretical aspects of this work have been dealt with in detail elsewhere (Wright and Kydd 1984a; Wright et al. 1985c). These references are not further specifically cited, for the sake of brevity. The model described is intended to apply to conditions in which the telencephalon is being driven by input from the reticular activating system, and also from diffuse inhibitory pathways (e.g., the mesotelencephalic dopaminergic and noradrenergic pathways). The broadest notions of the brain organization envisaged are diagramed in Fig. 1.

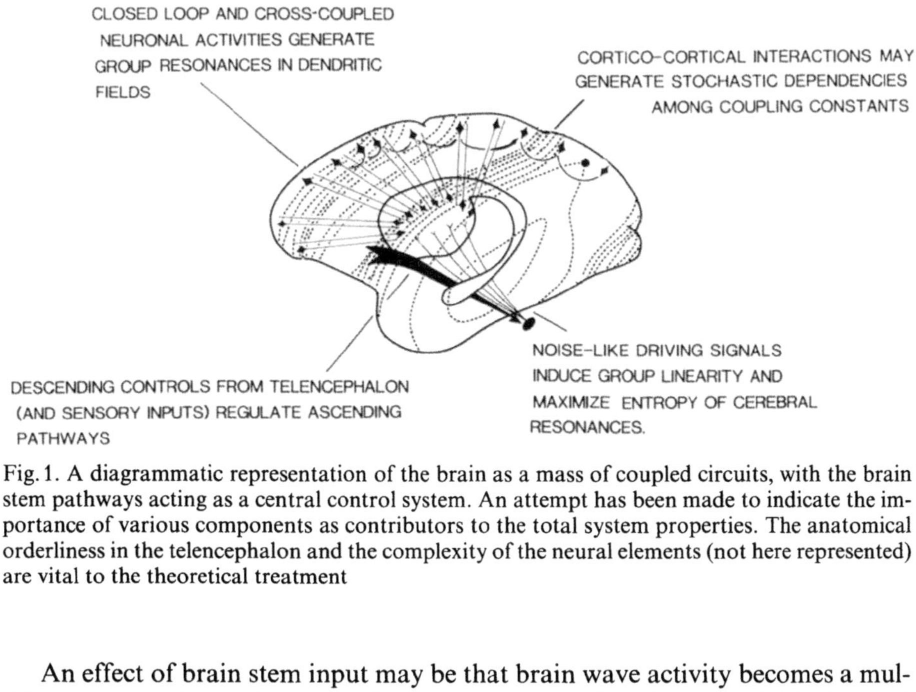

Fig. 1. A diagrammatic representation of the brain as a mass of coupled circuits, with the brain stem pathways acting as a central control system. An attempt has been made to indicate the importance of various components as contributors to the total system properties. The anatomical orderliness in the telencephalon and the complexity of the neural elements (not here represented) are vital to the theoretical treatment

An effect of brain stem input may be that brain wave activity becomes a multilinear process in one limit, despite the extreme nonlinearity of individual neural units, as outlined below.

2.1 Application of Quasilinear Differential Equations to a Nonlinear System

This model is developed from a very general treatment of coupled nonlinear oscillators, to which restrictions are imposed which render the description analogous to conditions in the brain. The mathematical treatment is based on the *form* of the linear differential equation for forced oscillation of a damped harmonic oscillator, i.e.:

$$\ddot{x}+D\dot{x}+N^2x=x_0 ,$$

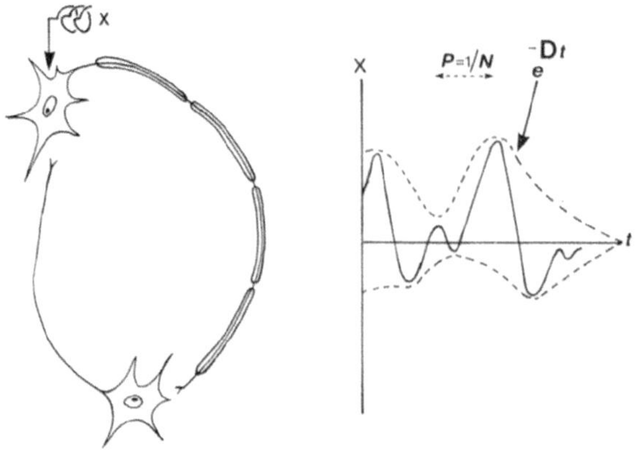

Fig. 2. One possible elementary component from the telencephalon, from Fig. 1; a closed neural loop containing a series of nonlinear components. Any such rhythmic element contributing to the surface EEG may have its time-varying signal described as solutions of a quasilinear second-order differential equation with time-varying parameters. The physical meaning that can be attached to the parameters is also indicated, for the instance of free system behavior after an initial impulse

where x, $\dot{x}$, and $\ddot{x}$ are displacement from rest and its first and second differentials with respect to time, D is the damping coefficient, N is the natural frequency, and x_0 is the driving force at time t.

The simple linear equation can be applied in *quasilinear* form to any pattern of electrical activity arising from a Lagrange-stable system, as in Fig. 2. The stochastic differential equation

$$\ddot{x}+D(t)\dot{x}+N^2(t)x=x_0$$

represents the nonlinearities as equivalent to time variation in the parameters D and N.

A large ensemble of coupled elements is then described by the family of equations:

$$\ddot{x}_i+D_i\dot{x}_i+N_i^2(t)x_i=\sum^{n} K_{i,j}(t)x_j \qquad (i=1\cdots n,\ i\neq j)$$

and

$$D_i(t)=\sum^{n} F_{i,j}(t)x_j \qquad (i=1\cdots n).$$

Notice that multiplicative as well as additive couplings are permitted, again quasilinear in representation. The time-varying coefficients need to be recognized as themselves being complicated functions of the state variables, i.e.

$$D_i(t)=f_i(x_1,\dot{x}_1\cdots x_n,\dot{x}_n),$$

$$K_{i,j}(t)=g_{ij}(x_1,\dot{x}_1\cdots x_n,\dot{x}_n),$$

$$F_{i,j}(t)=h_{ij}(x_1,\quad \dot{x}_1\cdots x_n,\dot{x}_n),$$

$$N_i(t)=l_i(x_1,\dot{x}_1\cdots x_n,\dot{x}_n).$$

The above system of general, coupled, nonlinear oscillators can be represented by the state-variable method, in a dynamic equation of the form

$$\dot{\mathbf{z}}=[\mathscr{A}_0+\Gamma\mathscr{B}_0\mathbf{z}']\mathbf{z},$$

where $\mathbf{z}$ is the EEG state vector

$$\mathbf{z}=[x_1,\quad \dot{x}_1,\cdots x_n, \dot{x}_n]^T, \qquad \dot{\mathbf{z}}=\frac{d\mathbf{z}}{dt},$$

$$\mathbf{z}'=[\dot{x}_1,\quad x_1,\cdots \dot{x}_n, x_n]^T.$$

To simplify the algebraic representation, a convention is here adopted in which Γ is an operator indicating that the column vector $\mathscr{B}_0\mathbf{z}'$ is redefined as the main diagonal of a $2n\times 2n$ diagonal matrix $\Gamma\mathscr{B}_0\mathbf{z}'$, and

$$\mathscr{A}_0=\begin{bmatrix} 0 & 1 & 0 & 0 & & \\ -N_1^2 & 0 & K_{1,2} & 0 & K_{1,3} & \\ 0 & 0 & 0 & 1 & & \\ K_{2,1} & 0 & -N_2^2 & 0 & & \\ & & & & \cdot & \\ & & & & & \cdot \\ & & & & & & \cdot \end{bmatrix},$$

$$\mathscr{B}_0=\begin{bmatrix} 0 & F_{1,1} & 0 & F_{1,2} & 0 & \\ 0 & 0 & 0 & 0 & 0 & \\ 0 & F_{2,1} & 0 & F_{2,2} & 0 & \\ & & & \cdot & & \\ & & & & \cdot & \\ & & & & & \cdot \end{bmatrix},$$

where $\mathscr{A}_0$ and $\mathscr{B}_0$ are matrices also of order $2n\times 2n$.

By assessing the determinant

$$|\lambda I-[\mathscr{A}_0+\Gamma\mathscr{B}_0\mathbf{z}']|=0,$$

where λ are eigenvalues, and I is an identity matrix, we obtain the system characteristic equation, a polynomial of form

$$a_m(t)\lambda^m+a_{m-1}(t)\lambda^{m-1}+\cdots+a_0(t)=0,$$

where $m=2n$.

For all systems in which a_i $(i=0\cdots m)$ are constants, the system represented is a time-invariant linear system. This will not be true for systems of the quasilinear form described above, except in special circumstances. It is now necessary to apply assumptions which constrain the general description in ways applicable to a realistic brain. This can be done in the way shown in Table 1, in which anatomical and physiological constraints are transformed to appropriate mathematical assumptions by appeal to physical meanings of parameters in the case of linked oscillators. This gives a broad indication of physical significance in the quasilinear treatment.

Table 1. Transformation of anatomical and physiological properties to mathematical descriptions

Anatomical and physiological properties	Mathematical equivalent descriptions
Nerve network is anatomically highly structured into roughly identical repeating units with repeated coupling systems (cortical columns, cortico-cortical connections, etc.)	Coupling coefficients and cycle times (N_i, $K_{i,j}$, $F_{i,j}$) each have finite mean values describing interlinkage over sufficient epochs, and these mean values are clustered, reflecting repetition of similar units
Microscopic structures exhibit local variation in structure, are functionally complicated, and are nonlinear	Instantaneous values of couplings and cycle times are stochastically independent, as a consequence of nonlinear noise-driven perturbations about their mean values

2.2 Emergence of Group Linearity

Under the above assumptions, for systems in which n tends to infinity (as for billions of linked neurons), within any stable domain in $\mathbf{z}$, $a_i(t)$ will be relatively constant over time, compared to $a_i(t)$ for systems and conditions in which strong transient stochastic dependencies develop between the coupling parameters, *Kij* and *Fij*. This effect depends on absence of covariance terms in sums of products of terms formed in expansion of the determinant given above, from which $a_i(t)$ derive. The linear equivalence is most marked for eigenvalues and eigenvectors associated with low natural frequencies.

Although highly nonlinear group properties can still occur in some systems fitting the above restrictions, it can be shown that the tendency to group linearity will be marked when (a) most *Kij, Fij(t)* are zero, (b) all nonzero *Kij, Fij* have values less than 1 at all times, and (c) $N_i(t)$ have high mean values with comparatively small variances. These three additional conditions appear applicable to neurons in the mammalian brain, where (a) all neurons have connections to only a small fraction of the total neuronal pool, (b) these couplings are usually individually weak, and (c) cycle-times in neuronal circuits are associated with high frequencies. That is, during stable states *the EEG will exhibit linear properties of wave superposition, and invariance of system natural frequencies. However, a nonlinear residual,* $B_0\mathbf{z}'$, *in the system dynamic equation permits the appearance of time-varying damping of the resonant modes.*

Physically interpreted, this means that the telencephalic surface may be considered as a closed linear wave medium, on which linear wave patterns could form, including standing waves. Many metastable states can occur for such a system, each associated with a set of damping parameters for each of the possible resonant modes.

It is further notable that the brain stem pathways are so placed as to exert strong control on both the strength of the driving signals and the temporal damping of resonances, while the natural frequencies and eigenfunctions of the resonant modes are determined by the constancy of circuit times and the boundary conditions set by the cortical surface.

2.3 Evidence for Validity of Theory, by Critical Tests

The outcome of critical tests for this theoretical treatment have been reported earlier, including at the first meeting of this conference (Wright and Kydd 1984b, c; Wright et al. 1984, 1985a, b, 1987). Unilateral manipulation of brain stem pathways, within a controlled comparison, repeated measures design, permitted certain independent spectral properties to be checked for the interrelationships expected if a linear wave description with variable temporal damping was physically valid. The hypotheses were repeatedly confirmed, but the range and precision of a linear description remained unclear.

2.4 Application of Autoregression Analysis

Blinowska and her group (Franaszczuk and Blinowska 1985; Mitraszewski et al. 1988) have reported a treatment of electrocortical activity which represents the recorded signal in terms similar to the physical description of wave formation described above. By fitting an autoregressive model (optimised by Akaike criterion) of the form

$$\sum_{j=0}^{p} a_j x_{n-j} = \varepsilon_0 ,$$

where a_j = model coefficients, x_n = sampled signal at n, and ε_0 = noise, to 2-s epochs of the raw EEG signal, then taking the z-transform and evaluating the system roots in the z-plane, they obtain the equivalent natural frequencies and damping coefficients for a linear system. These exhibit comparatively invariant natural frequencies clustered about the frequencies of the great cerebral rhythms, and exhibit time variation in damping parameters from epoch to epoch. Detailed interpretation and application of this method is the subject of further work in progress. For present purposes, this technique conveniently represents certain gross properties of system dynamics, outlined in the following section.

3 "Thermodynamics" of a Neural Network with Group Multilinear Properties. Statistical Dynamics of Cortical Arousal

Within such a multilinear, multistable ensemble, elementary statistical dynamic properties appear to account for gross aspects of electrocortical activation and suggest something of the brain's global organization as a finite-state machine.

Two limiting cases can be considered: firstly, conditions of strong, diffuse input by brain stem projections to telencephalic systems; secondly, evolution of system properties when input from the brain stem diminishes. These two cases prove sufficient to account for much of the sequence of EEG states in the awake animal, and also suggest a limit in which the above, group multilinear properties would prove insufficient to explain brain dynamics.

3.1 Conditions of Strong Input from Brain Stem to Telencephalon (Cortical Activation)

The assumed system properties defined in Table 1 require us to consider that during cortical activation the telencephalon is being strongly excited by a spatially diffuse wide-bandwidth noise input to some or all of its constituent subsystems. It is simultaneously subject to inhibitory influences (partly generated by inhibitory ascending brain stem pathways). Thus the telencephalic system is analogous to a thermal system in contact with a reservoir with which energy is being exchanged.

It can be shown using elementary concepts of equilibrium thermodynamics that each resonant mode can be associated with a property analogous to internal, or stored, energy in a mechanical system, given by

$$U_i = \frac{|x_0|^2}{4\mathscr{D}_i} \tan^{-1}\left[\frac{\mathscr{D}W}{\mathscr{M}_i^2 - W^2}\right],$$

where U_i = internal energy of ith mode, W = bandwidth of driving noise, $|x_0|^2$ = square amplitude of noise component at any frequency interval, $\mathscr{M}_i$ = natural frequency of ith mode, and $\mathscr{D}_i$ = damping coefficient of ith mode.

Entropy per mode may be defined as σ, where

$$\sigma = \left(1 + \frac{U_i}{\varepsilon}\right) \log\left(1 + \frac{U_i}{\varepsilon}\right) - \frac{U_i}{\varepsilon} \log\left(\frac{U_i}{\varepsilon}\right)$$

where ε is a small value and the ensemble "temperature", τ is equal to the average energy per mode. These energy and entropy expressions applied to Gibbs' equation lead to the definition of free energy, F, as

$$F = \sum^{n} U_i \quad - \tau \sum^{n} \sigma_i .$$

"Thermal" equilibrium occurs with equipartition of energy between resonant modes, with maximum entropy, and with minimum free energy. For conditions in which W is large compared to $\mathscr{M}$, equilibrium will occur with *equal temporal damping at each resonant mode*, since this meets the above requirements for thermal equilibrium.

In practice, we would expect near equilibrium conditions to prevail, with energy per mode (and damping) to be distributed about a mean value in a Maxwellian fashion.

Figure 3 shows a plot of natural frequencies and damping coefficients for the electrocorticogram of a rat, in conditions of cortical desynchronization. In con-

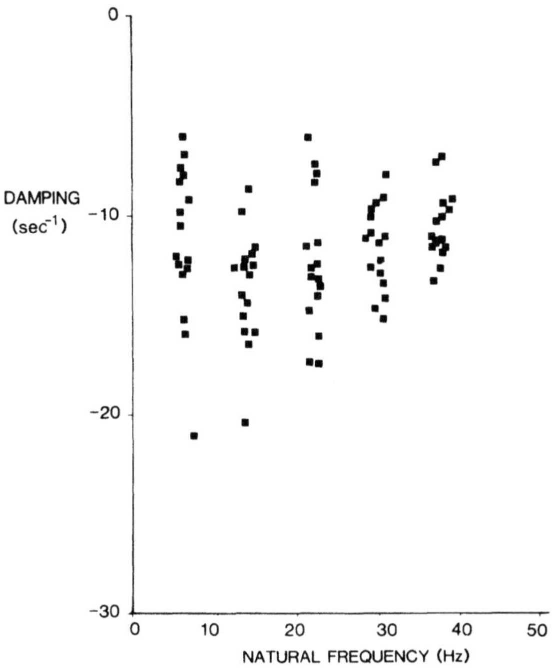

Fig. 3. Damping and natural frequencies of resonant modes in conditions of "cortical desynchronization." The plot shows the superposition of estimates obtained from 15 sequential 2-s segments of posterior cortical electrocorticogram from an adult rat. An autoregressive model has been fitted (in accord with method of Franaszczuk and Blinowska 1985) and converted to equivalent linear resonance parameters

trast to Fig. 4, the damping coefficients for each natural frequency are distributed about roughly equal group mean values.

3.2 Evolution of System Energies when Brain Stem Input Subsequently Diminishes

In the equilibrium condition, the system is maximally "open" to determination of future system behavior by external input signals, since no resonant modes are preferentially "tuned." Conversely, the withdrawal of input from the brain stem permits the emergence of determinate processes of intermode interaction which can define whole system behavior through a Markov process, in which resonant modes "compete" by exerting reciprocal influences on each other's damping, thus governing the "next-state function" of the whole brain. The most common path of evolution from near-equilibrium Maxwellian excitation of resonant modes will be toward predominant unimodal activity, usually at a low frequency, since it can be shown that the most stable sequence of mode interactions (starting from "thermal" equilibrium) tends to occur when the higher frequency modes progressively become highly damped. If total energy is roughly conserved, this will mean a shift to lower relative dampings at the lowest frequency modes. Since more stable states are (by definition) more likely to persist and therefore inherently more probable with the passage of time, we can expect a drift of the system towards increasing power at the lowest frequencies in most sequences of transitions. New external input will tend to return the system toward maximum entropy – i.e., "thermal" equilibrium.

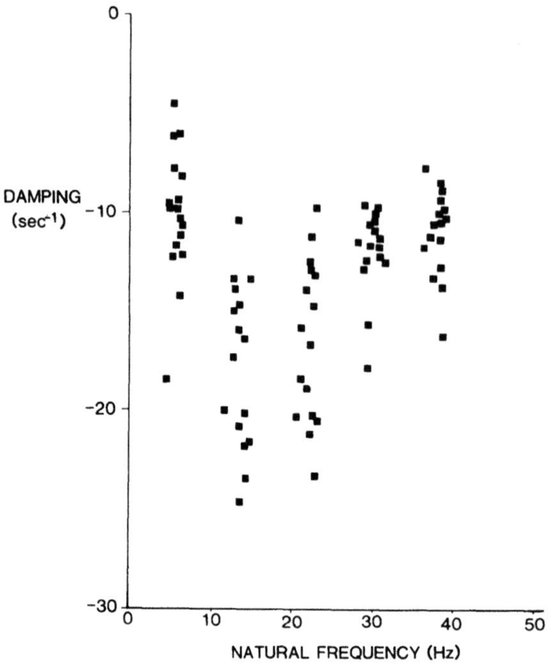

Fig. 4. Damping and natural frequencies in conditions of lowered arousal. Calculations were made as for Fig. 3

This emergence of relatively low damping at low frequency modes during lowered activation is shown in a representative example in Fig. 4. A detailed analysis of an extended data set will be reported elsewhere.

Continuation of this trend toward predominant low frequency activity will bring brain dynamic properties toward a limit of the applicability of such a group linear description.

3.3 Limitations of the Multilinear Domain – The Emergence of Nonlinear Waves

All of the above treatment rests heavily on the assumption of stochastic independence between the instantaneous values of the quasilinear (time varying) coupling parameters applied within the model. This assumption is expected to be roughly true only in the presence of sufficient diffuse driving noise, arising from the brain stem, guaranteeing an effective "uncoupling" of long-range dependencies between neural elements. As any resonant mode increases in internal energy, so will time and space dependencies in neural coherence increase. Thus mode evolution will in itself lead to the release of long-range coupled behavior in which wave activities might remain linear but exhibit new eigenfunctions [as in the treatment of Nunez (1981) for cortico-cortical interactions] or become highly nonlinear, entering chaotic or limit cycle domains. A most important exception to the linear treatment is seen in the Freeman model of olfactory bulb dynamics (Freeman 1985, 1987). In this case EEG is at all times outside the group linear domain proposed for the excited telencephalon, but, interestingly, exhibits periodic behaviour under maximum sensory input.

4 Conclusion

The present treatment can embrace a number of phenomena, including the stationary EEG landscapes reported by Lehmann et al. (1987) (interpreted as interacting standing waves), some properties of the late evoked potential (presumed to be the group system impulse function), and the EEG changes typifying alerting and lowered vigilance.

While electrocortical activity may thus be composed of linear waves, these waves are linear in a stochastic sense, i.e., they may show group linearity. The canonical equation for the system is an equation containing a nonlinear residual expressed as time-varying damping. Yet again, the coupling parameters in this model may be considered chaotic or noisy, and this may be reflected in neuronal firing rates. Thus the class of dynamics applicable depends on the level of analysis, as well as the condition of brain stem input.

The above aspects of linear versus nonlinear dynamics are quite separate from modeling techniques which utilize linearization of nonlinear phenomena over a particular operating range of a discrete system variable. Notably, the Nunez (1981) model involves linearization of nonlinear describing equations for coupled neurons. Nunez gives a rationale for wave linearity over a considerable operating range in terms of the saturation curve of depolarization of a large cell mass.

There is therefore a need to consider (a) the level (gross or microscopic) at which we are discussing cellular interactions, and (b) the domain of applicability – particularly when we are considering differing behavioral states or periods of EEG stationarity versus transition. Also, the anatomical locus being modeled needs to be distinguished carefully. There is no reason why brain dynamics need conform to any particular domain of theory over all conditions, yet determination of the appropriate paradigms for differing conditions is important for the development and testing of theoretical models.

If the brain exhibits behaviors falling into a multilinear, near equilibrium range at one end of the arousal continuum and nonlinear behaviors (including chaotic states) at the other end, then both linear and nonlinear methods of system identification are respectively applicable, and could converge to a common target of whole brain neural modeling.

5 Summary

Electrocortical macropotentials can exhibit properties of linear resonance. This group linearity for gross waves in the telencephalic dendritic fields contrasts with the known nonlinearity of the underlying elements. Linear conditions may occur only during high to moderate brain stem input to the telencephalon. Time variation of damping coefficients for the linear resonances can account for a multitude of near-stationary states of the EEG signal. Tests of hypotheses have been supportive, and autoregression analysis of EEG yields results consistent with expectation.

Elementary principles of thermodynamics can also be applied to the behavior of electrocortical waves. It is shown that in conditions of strong brain stem excitation of the telencephalon, linear macropotentials approach maximum entropy with equal temporal damping of each resonant mode. With lowered brain stem driving, lower frequency activity progressively predominates. Experimental demonstration of these effects, in terms of changes in damping factors, is shown from autoregression analysis. Deterministic properties governing whole brain information processing are implied by the EEG interpretation.

An attempt to unify linear and nonlinear approaches to the EEG is made. It is accepted that microscopic neural interactions are nonlinear and may be chaotic, and also that nonlinear dynamic analysis may better apply to EEG in states of low arousal. The detailed cellular models proposed by Freeman (for olfactory processing) and Nunez (for cortico-cortical interaction) are considered in relation to the present findings.

References

Babloyantz A, Destexhe A (1986) Low dimensional chaos in an instance of epilepsy. Proc Nat Acad Sci USA 83:3513–3517

Babloyantz A, Destexhe A (1987) Strange attractors in the human cortex. In: Rensing L, an der Heiden J, Mackey MC (eds) Temporal disorder in human oscillatory systems. Springer, Berlin Heidelberg New York, pp 48–56 (Springer series in synergetics, vol 36)

Babloyantz A, Nicolis C, Salazar M (1985) Evidence of chaotic dynamics during the sleep cycle. Phys Lett AllI:152–156

Başar E (1980) EEG brain dynamics, relation between EEG and brain evoked potentials. Elsevier, Amsterdam

Franaszczuk PJ, Blinowska KJ (1985) Linear model of brain electrical activity – EEG as a superposition of damped oscillatory modes. Biol Cybern 53:19–25

Freeman WJ (1975) Mass action in the nervous system. Academic, New York

Freeman WJ (1987) Simulation of chaotic EEG patterns with a dynamic model of the olfactory system. Biol Cybern 56:139–150

Lehmann D, Ozaki H, Pal I (1987) EEG alpha map series: brain micro-states by space-oriented adaptive segmentation. Electroencephalogr Clin Neurophysiol 67:271–288

Mitraszewski P, Blinowska, KJ, Franaszczuk PJ, Kowalczyk M (1987) A study of stability of electrocortical rhythm generators. Biol Cybern 56:225–260

Nunez PL (1981) Electric fields in the brain. The neurophysics of the E.E.G. Oxford University Press, New York

Röschke J, Başar E (1987) The EEG is not a simple noise. Strange attractors in intracranial structures. In: Başar E (ed) Dynamics of sensory and cognitive processing by the brain. Springer, Berlin Heidelberg New York

Wright JJ, Kydd RR (1984a) A linear theory for global electrocortical activity, and its control by the lateral hypothalamus. Biol Cybern 50:75–82

Wright JJ, Kydd RR (1984b) A test for constant natural frequencies in electrocortical activity under lateral hypothalamic control. Biol Cybern 50:83–88

Wright JJ, Kydd RR (1984c) Inference of a stable dispersion relation for electrocortical activity controlled by the lateral hypothalamus. Biol Cybern 50:89–94

Wright JJ, Kydd RR, Lees GJ (1984) Amplitude and phase relations of electrocortical waves regulated by transhypothalamic dopaminergic neurones: a test for a linear theory. Biol Cybern 50:273–283

Wright JJ, Kydd RR, Lees GJ (1985a) Quantitation of a mass action of dopaminergic neurones regulating temporal damping of linear electrocortical waves. Biol Cybern 52:281–290

Wright JJ, Kydd RR, Lees GJ (1985b) Contributions of noradrenergic neurones of the locus coeruleus to the temporal damping of linear electrocortical waves. Biol Cybern 52:351–356
Wright JJ, Kydd RR, Lees GJ (1985c) State-changes in the brain viewed as linear steady-states and non-linear transitions between steady-states. Biol Cybern 53:11–17
Wright JJ, Kydd RR, Lees GJ (1987) Gross electrocortical activity as a linear wave phenomenon with variable temporal damping, regulated by ascending catecholamine neurones. Int J Neurosci 33:1–13

Can Artificial Intelligence Help in Finding How Brains May Work?

A. de Callataÿ

1 Introduction

Here I will set out the research methods of artificial intelligence (AI) workers, and why they believe they can contribute to discovering how brains may work. I explain three main AI principles with simple examples: symbolic computation, qualitative physics to control artificial animals, and mechanisms of automatic deduction. AI builds neural-like computers whose components can be speculatively mapped to the neurons, suggesting their functions. I illustrate AI researches with examples from my work.

2 AI Methods

Figure 1 places AI in relation with other disciplines studying brains.

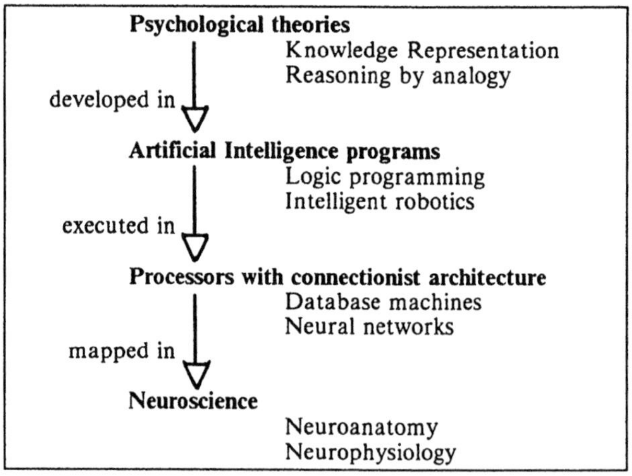

Fig. 1. The place of AI in the multidisciplinary research on brains. AI has implemented on computers the laws of thought, i.e., the laws of the mental world. Special computers are now designed for AI; they have many processing elements connected by wires, allowing a direct analogy with brains. When computer scientists want to study whether two kinds of computers are similar, they call "mappings" a set of correspondences ensuring that both machines have an equivalent behavior. Speculative mappings between a type of connectionist computer and a brain predict which processing functions are done by each biological element. These functional properties are theories which could be falsified by physiological experiments. The relations between the above disciplines show how AI may fill a gap between neuroscience and psychology

Springer Series in Brain Dynamics 2
Edited by Erol Başar and T. H. Bullock

2.1 Reviving the Old Theories of Knowledge

AI implements on computers the classical theories of knowledge developed in philosophy and psychology. The laws of thought, also called the laws of the mental world, were taught for almost 2000 years. Can they still be useful? I think AI scientists rediscovered these theories when they implemented problem solving and learning on computers.

2.2 Neural-Like Computers

AI programs are executed on current computers first designed by von Neumann for numerical computation, but AI works with symbolic computation, a different method which I will explain later.

Computers are made of logic circuits, tiny electrical wirings and transistor-like devices performing the logical functions of AND and OR gates, memories, etc. The arithmetic and logic units (ALU) are complex assemblies of logic circuits. ALUs have no equivalent structures in the brain.

Many new computer designs are studied today. Some computers are better adapted for brain modeling because they have more processing elements and connection wires, allowing more operations to be done in parallel.

Neural-like networks are computers made of devices more similar to neural systems. They have no ALUs, no addressable memory, and no disk memory. Each ALU is replaced by an analog processor, a simple example being cells summing the inputs modulated by the synapses. The neural-like systems are also called brain models.

2.3 Computer Science

Mathematics is the study of what can be done with numbers and additions, and computer science is the study of what can be done by combining logic circuits. Both sciences use logic theory. An example of a computer science theorem is the demonstration that two methods of computing give the same results, a demonstration saving the need of an experimental verification that the claim is true for every input. In principle, one could build different machines having an identical behavior, or create an artificial controller with a behavior similar to a conceptual brain.

Formulas can predict the speed of programs executed on a new type of computer in project. I have shown an example of particular applications (retaliation problem, in Callataÿ 1986) where my brain model, with operations taking 15 ms, can go faster than sequential computers with a cycle of one μs.

2.4 Models and Reality

To explain the analogy between computers and brains, let us take a related example. A chemical reaction can be described by an analogy with recipients communicating through a small aperture. The analogy means that both systems can be described by the same abstract formulas. Both have the same mathematical model permitting a simulation. The chemical concentration is replaced by water level, or both are mapped into a numerical variable. Can we draw similar analogies between AI and behavior? Yes, but instead of using formulas for numerical computations, we must use rules for all-or-none operations, like inferences. For example, we may show that an AI program, tested on a conventional computer, is a model for a specialized neural-like computer.

We assume frictionless bodies to study the movement laws. Similarly we assume situations corresponding to their abstract mental representation to make logical inferences. "Abstractions are real parts, phases or elements of things or their relations" (Cohen 1944).

2.5 Mapping

A mapping is an operation putting in correspondence each element of two systems. A mapping preserves the functions of the overall system when all mapped components have the same abstract properties.

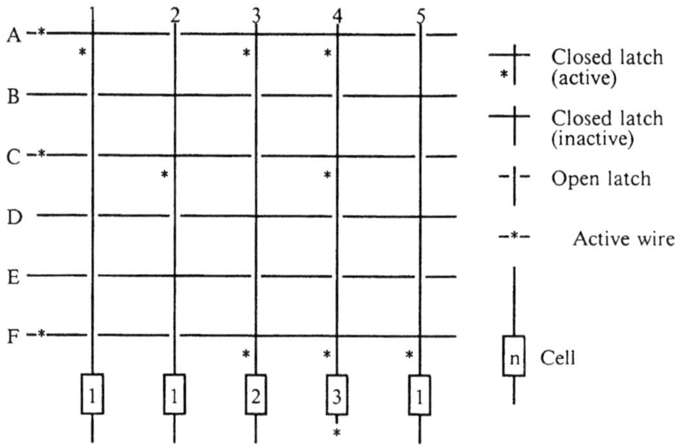

Fig. 2. Content addressable memory. The system defined here includes a group of cells, each one with its output wire (axon), and many input wires, each connecting each cell of the group through a latch (synapse) with an all-or-none conduction. The latches can be irreversibly closed, but less than 1% are normally closed. As an example of processing, let us explain how this neural-like machine can retrieve documents. Each input wire is identified by a keyword such as those describing a document. Each cell corresponds to a document described by keywords. These keywords are here the content of the document. The query activates its keyword wires. All cells concurrently add the impulses received from the active wires through their closed latches. The most activated cell corresponds to the best document. A classifier is a device ensuring that the winner cell takes all. The classifier is mapped to basket and chandelier cells (Callataÿ 1969). A query sends the content of the memorized information to find the cell storing this content. The device is therefore a content addressable memory (CAM). In artificial animals, the CAMs mainly find the concepts, images, words, events, or rules corresponding to given properties or features (reproduced with modifications from Callataÿ 1986)

In the previous example of mathematical models, water level and chemical concentration have the same properties, being a value at the same places in the formulas. Mapping is more precise than an analogy when a complex system is mapped into a simple device or conversely. For example, a Fourier transform will map sinusoids to values, so that an irregular wave will become a frequency spectrum, but the operations in both systems may not be equivalent when the sinusoidal waves do not interact in the same way as the frequency values. In Fig. 2, I have mapped the document retrieval algorithm to a content addressable memory (CAM) and this one to a group of neurons.

2.6 AI as an Exploratory Discipline

Computers are wonderful toys. It is fascinating to write programs exploring their limits. Program authors are frequently surprised by some unpredictable behavior. Programs can learn, create, and behave irrationally.

In most domains, models and prototypes are designed and experimented with before the discovery of the explanatory theories. The ellipsoidal model of planet trajectories was found by Kepler, preparing the discovery of Newton's law of gravity. Similarly I hope that AI exploration will help to discover interesting models of thought.

2.7 AI as an Empirical Science

Can exploration be called science? This can only happen when the reproducible results can be formalized. AI scientists claim to be experimenters (Newell and Simon 1976). Their experimental apparatus is the computer. The empricial results are the observed behavior of the programs.

2.8 Cognitive Science

Programmers have designed many computer applications and have acquired an expertise in understanding the internals of most kinds of reasoning, i.e., how the problem must be represented, which knowledge should be involved, and how the reasoning process can be simulated. This problem concerns cognitive science, a science now studied by developers of expert systems.

Cognitive science has two complementary descriptions;

- A psychological discipline, opposed to behaviorism, where mental intentions are accepted, and where introspection is an allowed method
- An AI discipline, studying knowledge representation and processing

2.9 AI Goals

The goals of AI are not to understand brains, but intelligence. Most AI scientists believe they will find a better way to produce intelligence than by copying the brain. Engineers did not copy the bird's flight but found propellers and jets which are easier to build. Similarly they found that solid-state technology produces faster and more reliable computers than biological technology (at least with the present state of the art).

Most AI researchers are studying computer systems for industrial purposes; the few studying brains or related topics have diverse opinions. The neural network community studies neural models with progressively modifiable synapses (McClelland and Rumelhart 1986). I think that all-or-none connections are required for symbolic processing (Callataÿ 1986), and I have pursued this idea with obstinacy since my first paper in 1969. Many of my conclusions are similar to those of Minsky (1986). Any developed model of the brain must integrate many models based on single principles.

3 Symbolic Computation

I have selected a few AI principles, which are frequently misunderstood, to explain AI, but the complete theory is more complex than the following simplistic examples may show. More complete descriptions relevant to brain modeling can be found in Anderson (1983), Callataÿ (1986), Johnson-Laird (1983), Minsky (1986), and Simon (1976). To understand AI, one must clearly see the difference between analog and symbolic computation (Fig. 3).

3.1 Physical Symbol

There are two ways to define symbols: by their uses or by their properties.

A typical line in a computer program is: price = quantity × unit price. The names of entities (price, quantity, and unit price) are symbols. A "symbol" is

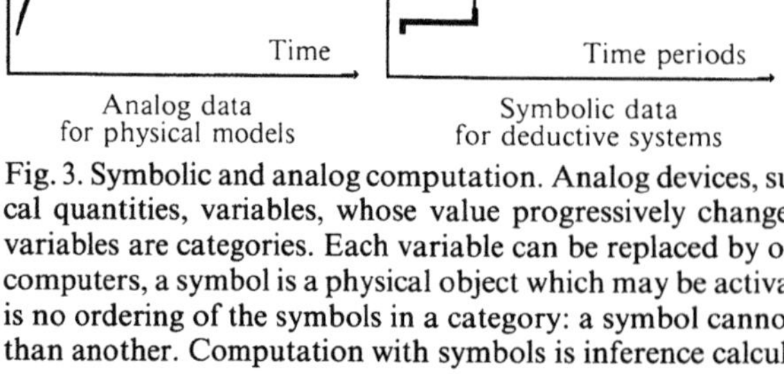

Fig. 3. Symbolic and analog computation. Analog devices, such as a servo-mechanism, use physical quantities, variables, whose value progressively changes. In contrast, in symbolic devices, variables are categories. Each variable can be replaced by one of a finite number of symbols. In computers, a symbol is a physical object which may be activated. It is ON or OFF. Usually, there is no ordering of the symbols in a category: a symbol cannot be said to be nearer to one symbol than another. Computation with symbols is inference calculus, and not numerical calculus

"something concrete standing for something else which cannot be as easily represented."

A "physical symbol" is defined as a simple component of a physical device (Newell 1980):

- It is unique (we can give it a unique identifying name)
- It has a state, i.e. an activation value; a state is a transient memory which is set by external processes
- The state is readable by other processes; for example, the state of a "logic gate" produces a voltage through an amplifier, so that the current drawn to read the state does not modify it.

An intention is a physical symbol structure which "must exist in the thinker, but not necessarily in the real world" (Brentano, in Chisholm 1966).

AI programmers manipulate "physical symbols," for which they know the meaning, but a "physical symbol" may also exist independently of a programmer.

Despite Eccles's hypothesis of another world, the AI researchers implicitly agree that their programs represent and process the mental world. Quine (1985) wrote that he has no objection to saying that the objects processed by logic belong to the mental world. In a sense, computer tests are empirical studies of the mental.

The physical symbols of my brain model represent events, phenomena, transformations, relations, rules, or properties rather than individual things (Callataÿ 1988).

I have mapped physical symbols to neurons with long axons. One could study whether the mapping of each symbol to an attractor in a chaotic nervous system (Prigogine and Stengers 1984) is compatible with the other requirements of brain models.

3.2 Production Rule

A rule – the technical term is production rule – is a kind of equation with the following format: If the physical symbols A, C, and F are active, then the physical symbol L may be activated. A content addressable memory processes rules (Fig. 2): If the content (A, C, and F) is searched, then it is in the cell L. In behavior control, the rule is: If the pattern (A, C, and F) is matched, then do the action L. A rule permits inferences in logic: If the conditions A, C and F are true, then conclude L.

Let us compare mathematics and logic. Addition is the main operation of mathematics, just as inference is the main operation of problem solving. Analog computations are arithmetic operations described with formulas containing numbers structured with few primary digits in columns. Deductive computations are logic operations described with rules structured with many primary symbols.

There are few numerical rules which are well known. The pattern matching rules are more complex, and there are several ways to define approximate matching. There are few physical laws (equations). In contrast, there are many knowledge rules, and some may be fuzzy, false, or unknown. The meaning of numbers is well understood. The relation between a symbol and its correspondent in the environment is poorly defined.

A simple rule activates a physical symbol and is stored in a cell of a connectionist network. Newell and Simon (1972) have already done this neural mapping: they said that the rule fires when it is executed. They have concluded, after a study of high level thinking, that what brains learn are production rules.

3.3 Robot Control with Methods of Qualitative Physics

Most present robots are dumb machines repeating some programmed behavior. More intelligent machines are constructed with sophisticated controller systems. I call these last robots artificial animals.

Figure 4 illustrates a traditional AI experiment with a simple robot. After the discretization of the sensory inputs, the only data which are processed are physical symbols. The numerical values are lost. For any symbolic computation, the complex inputs should be put into few categories, a major problem known as categorical perception. The knowledge of the skill is represented by rules. A rule may be:

When the angular position is far left
and the angular speed is slow right
and the cart position is near left
then the cart speed should be slow left.

Periodically the rule corresponding to the instantiated symbols is executed (Callataÿ 1969). A more sophisticated method of robot control uses degrees of membership: a measurement is coded with two symbols; e.g., 40% far left, and

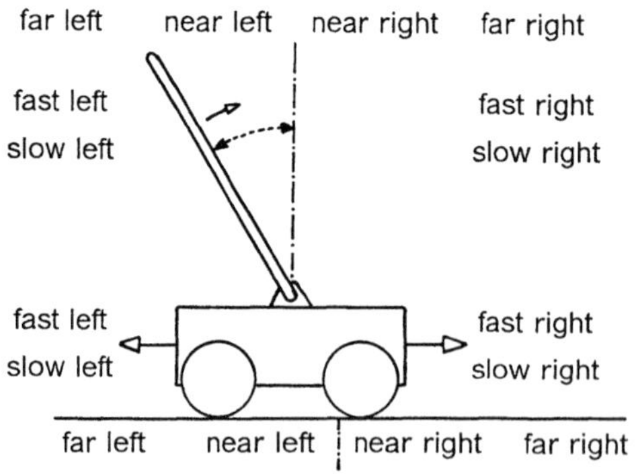

Fig. 4. Broom balancing. The robot must balance an inverted broom on a cart and hold it motionless in a central position. The physical measurement of the broom angular position is categorized into qualitative symbols: e.g., far left, near left, near right, far right. The broom angular speed can be: fast left, slow left, slow right, fast right. Relative to the center, the cart position is: far left, near left, near right, far right. An engine applies one of the following speeds to the moving cart: fast left, slow left, slow right, fast right. Symbolic reasoning is studied in qualitative physics

60% near left. This code can be represented by two neurons with different firing frequencies. Then several rules are selected with various intensity coefficients. Each selected rule generates a command. The theories of interpolation or of fuzzy sets give the formulas for rule combination, from which a precise numerical value is computed for the resulting commands. Acceptable performance can be achieved with few categories and long periods. Robots controlled by numerical formulas can also keep brooms in equilibrium. Which is the best method? Dynamic formulas give smoother movements, but I believe that rules are required for a general purpose learning method of complex decisions in unexpected situations.

An artificial animal carries out hybrid computations, analog and symbolic. Some of its processor types are listed in the following. The sensory analog data are periodically sampled and converted into mental symbols by feature extractors. These input symbols are processed by the symbolic processor. Subsequently, servo-mechanisms convert into analog forces the symbols which the central symbolic processor has stored in their setpoints. The above operations provide the interface between the physical and mental worlds. The sensory data are preprocessed in feature extractors, but I conjecture that these AI devices do not learn.

3.4 Behavior Rule

What is learned should be always valid. What kind of invariant information could be learned from experience? An artificial learns the following production rule after each trial delimited by a standard rhythm: If the body was in position B and the executed muscular command was M, then the body reaches a new position A. This observed transition constraint is stored as: To go from B to A, do M. Therefore, when the subsequent position A is given as an intention, the corresponding muscular command can be spontaneously retrieved and executed.

Compared with conventional rules, the behavior rules require an intention and associate the states before and after an event. New rules are added in the long term memory and do not modify existing information.

The behavior rule describes a short story. This story is an event produced by physical laws, an event that will recur in the same context because the physical laws are repetitive.

An expert system has two components: an inference engine and a knowledge base of rules. When connectionist hardware is designed as an inference engine, the addition of rules is sufficient to build a comprehensive machine which solves problems in the domain of these rules. In artificial animals, no false rules can be loaded (as programmers frequently do), because the behavior rules can only be learned if they are executed, and thus verified in real life.

A behavior rule may still be erroneous because the perception may wrongly categorize the sensory data and ignore significant details. I think that we can design complex artificial animals which are learning behavior rules and building knowledge most of whose rules are and will remain relevant.

When a theory is falsified by an experience, there is no logical way to discover which part of the theory is wrong: more knowledge is required. In the same way, only the entire learned event is true, as the machine cannot know whether the rule is still true when a feature of the sensed event is changed. It was found impossible to have generalized axioms explaining common sense (McCarthy 1987). The only source of spontaneous knowledge is the experiments of life, recorded as short stories. Because the generalization by induction from these examples gives rules which can always be falsified by a new event, knowledge is kept in the original rules.

A behavior rule is decomposed by the CNS partitions into a hierarchical structure of simple, partial rules stored in cells. There are several cooperative hierarchies in my brain model (Callataÿ 1986). The study of the organization and validity of these structures has only begun (Callataÿ 1988).

3.5 The Reasoning Process

For AI cognitivists, a representation is physically present when a particular neural activation pattern normally produced by a sensation is generated in another way. This representation is an intention when it can generate the muscular action to reproduce the corresponding situation.

The cognitivists change the rule of the strict behaviorists, stimulus → response, to: stimulus → intention → response. Introspection shows that we must represent intentions and will them before acting. This intention/resolution procedure is analogous to problem solving in logic programming.

With the above principles, the reasoning of an artificial animal is done in a repetitive sequence of two phases, each one having its own method of reasoning and learning. In the first phase, an intention is set with the conventional stimulus-response method. In an artificial animal, the stimulus-response system is applied in the top of the event structure: stimulus being replaced by global sensation (perception) and response by intention. The processor model for classical conditioning is usually a perceptron driven by the situation and the autonomic drives. Then this intention becomes the problem to be solved as shown above. Only this second phase needs the methods of deductive logic (Callataÿ 1986). For complex or chained behaviors, low-level intentions are set by high-level processing.

Although all sensory data are watched in a parallel machine, most sensory data are ignored in further processing. Only those in the attention focus are included in the behavior rule. In my brain model, the focus is the set of enabled processors. An intelligent deductive processing is assumed to select a relevant focus according to the intentional context.

There are two steps for storing a behavior rule after an event has occurred:

- Finding the neural network nodes which can associate the inputs from the focused sensory-motor parts (the memory allocation is assumed to be decided in the hippocampus which transmits its commands by the cingulum, and by the amygdala and mammillary circuits).
- Activating or reactivating the cells in the enabled areas in which the relevant relations describing the event will be stored (done in the selected cerebral minicolumns).

3.6 Is Human Intelligence Qualitatively Similar to Animal Intelligence?

Perhaps we should start by studying animal intelligence and later learn how human intelligence can emerge in a larger machine. What I did is, in a sense, the contrary. I have studied whether the complex methods of the theory of knowledge can be used for animal behavior. I have built mental controllers for simple operations of robotics, such as the management of a humanoid robot arm learning to reach a target. A prototype has shown that this symbolic method permits fast and efficient learning (Callataÿ 1986).

So I assume that infrahumans reason with mental methods, and that most of our automatic behavior uses symbolic processing. It does not follow that consciousness has the same meaning in infrahumans, children below 4 years old, and adult humans. Most symbolic processes do not leave traces in memory and are therefore unconscious (Callataÿ 1986).

4 Natural Reasoning and Learning

In the following, I discuss the reasoning methods in mammalian brains, whatever their size.

Human reasoning normally produces well-defined and efficient actions. Therefore a principle should force brains to achieve true results. That we notice frequent exceptions to this principle just confirms that it exists (Klages, reported in Thibon 1936). Can logic provide this principle? Logic includes a deductive method to make inferences. For example, if an artificial animal knows the rules: To go from X to Y, take the path A, and to go from Y to Z, take the path B, and wants to go to Z, being in X, it can at once discover the path A followed by B, although it has never taken both in sequence. Although this logic inference is an elementary operation in logic programming, it requires a complex computational device.

While the deductive process solves problems, it reaches bifurcation points where several rules can fire, but only one can be chosen. These alternatives are called (mental) search paths. Some paths change the attention focus on the feature extractors or on the short-term memory. Complex problems, such as puzzles, have many paths, and require a long search.

4.1 Is There a Universal Algorithm for Reasoning?

Several AI researchers have succeeded in finding general algorithms to solve problems. The best method to explain this class of algorithms to my students is to teach the programming language Prolog, and to explain its inference method. Some systems have algorithms to evaluate states and choose paths (breadth-first search). The parallel processing of these algorithms is studied. Other developments allow processing of fuzzy and incomplete knowledge (described in Callataÿ 1986).

4.2 Spontaneous Programming

In contrast to computers, brains are not programmed, apart from instincts and reflexes: they learn from experience. The experimental test of each behavior rule before memorization and the logical consistency of deduction in the neural computer might explain the quality of brain programming. People can drive, surf, and climb mountains. These tasks require more robust programs than those that computer-controlled robots can now provide when unexpected events occur.

Brains can be indirectly programmed by a teacher who asks the student to do an exercise. What is learned by this experiment is a useful rule which will automatically be reused in appropriate contexts.

A type of hierarchy divides the brain into independent agents (Minsky 1986). This hierarchy can be compared to an industrial society (Callataÿ 1986). Some agents are workers receiving sensory data and acting. Most others are managers reading the summarized state of the workers or other managers, and giving them intentions. The managers can program the workers like a teacher: they give commands to do some tasks.

Beside commands and intentions, managers also send commands to focus on some feature extractors, some segments of the short-term memory, and some rule domains. In complex expert systems, this organization is the blackboard management delimiting the context of each subproblem and coordinating the resolution steps. How this is controlled without programmers is not solved (Callataÿ 1988).

4.3 Meaning

The inputs of connectionist networks fire rules propagating further activations and creating a complete brain state. When linguistic tasks are processed in these networks, the cells represent words and expressions. The machine becomes a semantic network, in which the evoked state is the meaning of the few initial instantiations. I have written a neural-like program to show how a perceptual state with multiple options is built from few input data and how further processing focuses the attention on the few additional feature extractors which may help in discriminating this ambiguous perception (Callataÿ 1987).

4.4 Is Logical Abstraction a Good Model for Human Reasoning?

Demonstrations of machine intelligence are not very convincing models of animal intelligence: observers comment that the method applies to artificial cases like chess-playing, but not to real-life situations. Expert systems are only good for artificial problems concerning regulations, mechanisms, or taxonomies which were conceived by men.

A symbol belongs to a category defined by its potential types of properties. To reason with abstractions, one must transform symbols by ignoring some of their properties. In my model, this is done by deducing that other symbols can be instantiated within the agents of more general categories. But what is learned is first put in concrete categories, and the spontaneous reasoning is done in this concrete environment (Johnson-Laird 1983). I guess that human brains have more hierarchical levels than infrahumans, allowing the adults to do abstract reasoning when their top levels have matured. This can correspond to the Piaget's stages of development (discussed in Callataÿ 1986).

4.5 The Reasoning Method of Expert Systems

Because of difficulties with reasoning from generalized axioms (McCarthy 1987), many AI researchers have conjectured that people use heuristic knowledge. A heuristic rule is an isolated knowledge rule which is not associated with explanations on its validity. An example of an old heuristic rule is "Do not eat pork." The rule was applied when its justification was not known and was at that time a superstition. Superstitious behavior is a normal animal behavior (Lorentz). "Re-do what was successful," "Avoid what was painful" are generally efficient methods. The habituation mechanism can prevent too many repetitions (Callataÿ 1986).

The present expert systems are software programs simulating the surface reasoning of a human expert. Interviews with the expert lead to collecting his/her rules. When the rules are consistent, the machine blindly applies this knowledge. The expert systems have a superstitious behavior taught by humans.

"An expert does not reason, he knows" (Wright, cited in Minsky 1986). Is applying knowledge a method where one understands? I cannot answer, but the above discussion shows that intelligent behavior is not defined: a behavior is judged intelligent until we know how it is done (Minsky 1986). It is likely that the routine behavior of humans is based on the same principle as expert systems: apply knowledge. The above methods require the memorization of most events. This is the mnesic causation defined as the main feature of brains by Bertrand Russell (1921). Most parts of my parallel processor are uselessly scanned at each period, but this huge processing may sometimes discover a threatening heuristic rule applicable to the present situation.

4.6 Similarity-Based Reasoning

Problem solving with strict logic has the following defect: when no rule exactly applies, a program is halted. Animals must take a decision, even a bad one. So my hypothesis is that the brain, when faced with an unknown situation, searches the most similar previous event, and takes it as a valid rule. Events probably never occur twice, so the use of strict logic is almost impossible, but similar events are found at once by the document-retrieval algorithm.

Two events are analogous when they have many features in common. When a rat is trained to run into a path and finds it flooded, it may still recognize this path which is its most similar stored perception. It has made a kind of analogy, a frequently useful method. My neural-like computer spontaneously reasons by analogy, a method which is not allowed by strict logic, but which humans frequently use (Cohen 1944).

An analogy is valid when it concerns a behavior prediction in a domain where the different features of both events are irrelevant for the intended action. In neural network, knowledge is partitioned into categories restricted to some properties. A CAM finds the best analogy with the locally available information.

The scientific reasoning done by educated people is not a spontaneous, natural behavior and is not addressed here. The spontaneous reasoning of a similarity-based inference system may be false and is therefore better described by the word intuition.

Until better principles are discovered, I assume that the document-retrieval algorithm is the simplest and most efficient algorithm for pattern matching in natural reasoning because: (a) it is logical when there are rules strictly applicable; (b) it recognizes similarities; (c) it learns new rules without destroying previous ones; (d) it efficiently categorizes inputs (Callataÿ 1986).

The mechanisms for logical resolution are the same as for analogies in CAM; therefore the two methods are taken as equivalent in the following developments.

4.7 Can a Processing Principle Be Preserved when Complexity Increases?

A similar question concerns mathematics: Does a very long and complex computation produce a valid result? The explanation is that mathematics is based on a theory demonstrating that the results are true when the generalized formulas are valid and when all operations are done according to the laws of mathematics. Similarly in logic, when all inferences are done according to the laws of logic, and when the axioms correspond to reality, the results are true whatever the complexity of the deduction.

If the rules for adding devices to a logical computer are such that a processing principle for logical resolution is preserved, any computation thus performed in such an arbitrarily large computer will provide true results. A speculative rule of neurogenesis may create these logical neural networks (Callataÿ 1986).

Since artificial animals are hybrid devices, a good logical brain is not sufficient: efficient innate reflexes and perceptual systems are also needed for fast learning, and relevant choice of intentions favoring survival.

4.8 Reliability and Redundancy

When a neuron is killed, one rule is lost, but the other nerve cells normally have redundant or similar rules which can still provide actions with acceptable performances. When the almost redundant rules are processed in parallel, the selection of the best rule is done without performance loss. With ten redundancies, one could destroy most of the brain before noticing behavior anomalies. This may overcome Lashley's objection and make possible a behavior control with redundant grandmother neurons (Callataÿ 1986).

4.9 What Happens When the Computer Size Increases?

Traditional mechanical machines have thousands of components. Computers have millions of components which are the logical circuits on the chips. The behavior of the mechanical automata of the eighteenth century was stereotyped. Modern robots are less stereotyped. What would happen when the robot controllers will be millions of times more complex than current computers?

Humans feel insulted when they are compared to machines because our current machines are quite primitive, but Minsky's play on words invites us to consider the dignity of being a machine with a trillion interacting parts (in Letovsky 1987).

5 An Example of Neurophysiological Mapping

I have studied the consequences of assuming symbolic processing for a brain model. These AI studies have suggested the following principles.

Because the activated state of a physical symbol is held during some time until another cell is more activated, we can define this period as a time interval. Assuming some regularity and synchronization of these intervals, the first consequence is that the brain functions rhythmically (Callataÿ 1969). A second consequence comes from observations on AI programs: only a few symbols can be simultaneously activated, otherwise there is data confusion: less than some ten thousands at any moment in a machine sized like the human brain. Therefore cell firing cannot be cell activation. A third consequence results from the requirements that the rules be permanent and be learned from a single event. What the brain learns, production rules, must be stored with all-or-none irreversible latches. In a continuously changing living organism, the spine shape is supposed to be maintained by a homeostasis which is destroyed when memorization occurs. Learning is by selection (Changeux 1974).

AI researchers can only modify their machine (hardware or software) when the transformation keeps or improves its intelligent behavior. Therefore the discrepancies of mapping between the nervous system and the machines cannot be rectified until better processors or programs are discovered. AI scientists can map machines to brains, but not the contrary.

As an amateur neuroscientist, I have guessed at a mapping which leads to the following speculations. A priori, I suppose they are each unlikely (I explain in Callataÿ 1986, how I have deduced the mappings from the references):

- In a computer, a physical symbol activation provokes a voltage shift which can be read by connected elements. In the brain, this is equivalent to a logical impulse emitted by the active neuron to other neurons. In the mapping, a logical impulse is a high frequency burst in an isolated cell, as shown in Fig. 5 (epilepsy as a model of memory was suggested by Goddard 1967; Barondes 1969). Bursts could be more efficient than the simple addition of impulses (Gamble and Koch 1987). With localized memory traces, learning must be prevented in more than 99.999% of the cells. This control could be mapped to regulatory systems preventing epilepsy.

- The latch closure is done by the resorption of the dendritic spine in an all-or-none manner (Connor and Diamond 1982; Koch and Poggio 1983; Lynch 1986).
- A cell bursts only after a strong protracted dendritic shaft activation (Llinas 1979).
- Each group of cells works as the CAM described in Fig. 2. The present CAM with partial matching is more useful than the perfect CAM described in Callataÿ (1969).
- The rules are learned in a CAM by activating simultaneously the input wires and the cell which has produced the rule action. Then latches at the conjunction of input/output activation are closed. This principle is called conjunction by Eccles (1973). It was discovered by Hebb (1949) and applied by Brindley (1967), Callataÿ (1969), and Marr (1969).

6 An Example of Neuroanatomical Mapping

The network of Fig. 5 has three cell groups. Human-like networks would have millions of groups, of which thousands are enabled. A complex network is equivalent to a program. The idea is to create specialized networks to execute a type of AI program. Some networks are particularly interesting: those which can learn from experience and resolve problems. Those are innate programs which can spontaneously learn additional programs (Callataÿ 1986, 1988).

So far the most complex simulated system is a two-dimensional prototype of a humanoid robot arm spontaneously learning how to reach a target. For this task, the robot's controller needs less than a million latches, equivalent to den-

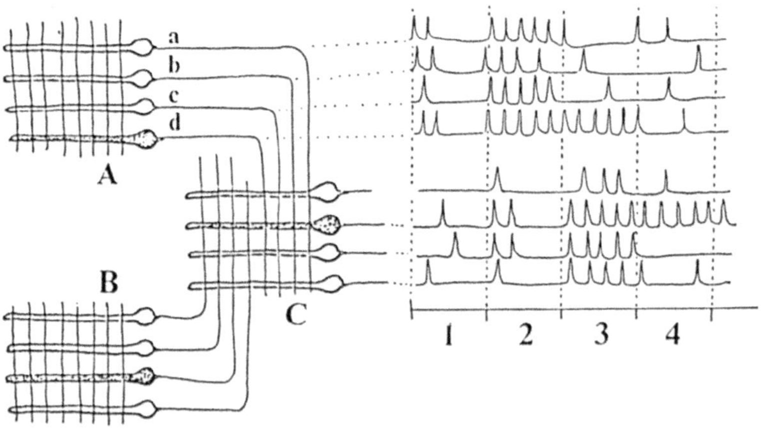

Fig. 5. Processing of a rule in a neural-like network. The drawing shows how the inference with the production rule: "if *Ad* and *Bc* then *Cb*" is implemented. *Ad* and *Bc* each transmit a logical impulse to the cell group *C*. The cell *Cb* has closed latches with *Ad* and *Bc*: *Cb* is the most excited cell of the group and bursts at the subsequent cycle (reproduced with modifications from Callataÿ 1986)

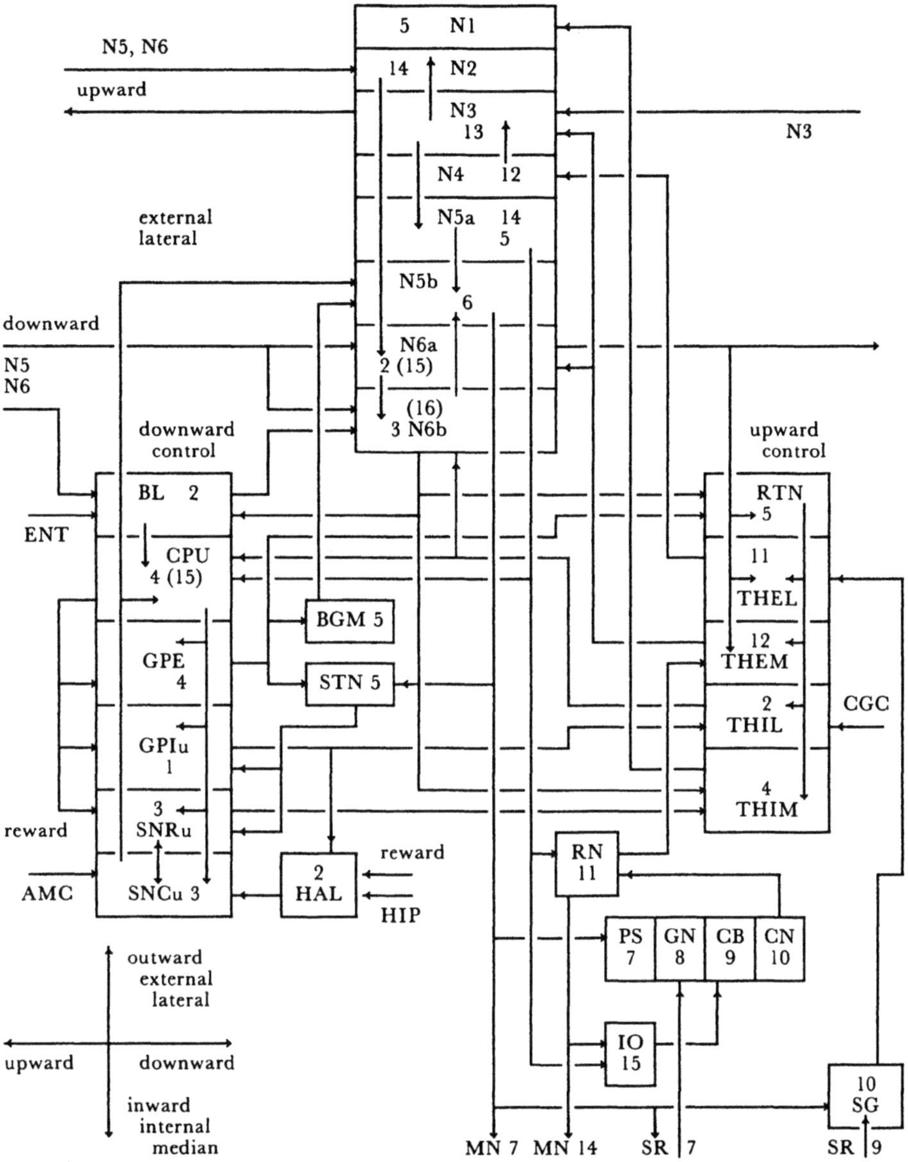

Fig. 6. A network program corresponding to a cerebral column. As an example of AI neural-like computer, I show a drawing reproduced from Callataÿ (1986), where the machine is explained in detail. Each *box* is a content-addressable memory (CAM). Each *arrow* represents connections between CAMs. The drawing defines a computer module. An interconnected group of these modules can be mapped to a computer program which solves problems. Its similarity with the mammalian cerebral minicolumn is intended to suggest a possible analogy of processing

dritic spines. A controller with 10^{14} spines could learn millions of behavior rules per day without resorbing more than 3% of its spines after 100 years. I have computed that recognition becomes blurred above 1% to 6% use, depending on device characteristics. About 1000 spines per second can be resorbed on average.

The above AI methods permit designing engineered machines, like the generalized, deductive, learning program shown in Fig. 6. Its drawing is a sketch of a network/program which is written but not yet tested. For this project, we must also have an artificial animal which can train itself.

The above model suggests a biological sequence of activation which can be falsified. Because there are several ways to write programs giving the same results, this program, even if it is successful, would not necessarily be that of the brain.

7 Conclusion

The computing requirements to produce an animal behavior were studied assuming a symbolic processing model. Using AI methods quite different from those of biological experimentation, I have guessed a sketchy model suggesting how brains may work, and I conjecture that there is no other conceptual way to implement mental processes, although other mappings can be imagined.

AI projects related to brains can be classified into: (a) creating or improving intelligent programs and neural-like computers (brain models); (b) supporting models with experimental simulations showing a relevant behavior of artificial animals; (c) explaining results of negative simulations with theoretical proofs that a model cannot do what was intended (e.g., the study of perceptron in Minsky and Papert 1969).

The mapping of brain models to nervous systems may be done in several ways, and AI alone cannot discover these speculative mappings. Each mapping can be falsified by biological experimentation.

I hope that by progressively improving the AI machine models, researchers will arrive at a precise mapping which may predict how some brain circuits work. AI studies obviously have a limited scope because they do not use a real brain. But, even if physiologists knew everything about the functioning of each neuron, AI studies could still help in interpreting these results.

8 Summary

Using artificial intelligence (AI) methods quite different from those of biological experimentation, I have proposed a sketchy model suggesting how brains may work. Most AI studies are empirical: just as one studies natural phenomena, one observes the behavior of computer programs or artificial animals. Cybernetics uses analog computation while AI is done with qualitative inferences. My method is to investigate how simple animal behavior could be managed by a hybrid machine using the AI mechanisms of high level thinking coupled with cybernetics devices providing the interface between the robot and the physical world. Reasoning has been decomposed into four phases, each one having its own mechanisms: for choice of an intention representing a desired sensation; for resolution of the logical problem, Which actions permit achieving the given intention?; for learning by storing in long term memory the life stories, i.e., the association of the action and the sensations before and after it; for selection of the attention focus and of a subset of the short term memory to ensure processing continuity. Symbolic and adaptive neural-like networks are machines quite different from present industrial computers and controllers. Mapping these machines to brains leads to paradoxical predictions on how some neurophysiological circuits work. The neural net is similar to the mammalian brain connections discovered in the 1970s. The paper shows the AI methods, some AI principles, and the theories integrated in my brain model, giving some examples of brain mappings.

References

Anderson JR (1983) The architecture of cognition. Harvard University Press, Cambridge

Barondes SH (1969) The mirror focus and the long term memory. In Jasper HJ, Ward AA, Pope A (eds) Basic mechanisms of the epilepsies. Little Brown, Boston, pp 371–374

Brindley GS (1967) The classification of modifiable synapses and their use in models for conditioning. Proc R Soc (Lond) [Biol]168:361–376

Callataÿ A de (1969) Brain model with periodic processing. Curr Mod Biol 2:307–319

Callataÿ A de (1986) Natural and artificial intelligence: processor systems compared to the human brain. North-Holland, Amsterdam

Callataÿ A de (1987) Un noyau de système expert basé sur la recherche documentaire. Journal des Ingénieurs, FABI 19:5–9

Callataÿ A de (1988) Logic programs directly processed in a network of content addressable memories. Future Generations Computer Systems 4:117–131

Changeux JP (1974) Some biological observations relevant to a theory of learning. In: Bresson F, Mehler J (eds) Current problems in psycholinguistics. CNRS, Paris

Chisholm RM (1966) Brentano, F: The true and the evident. Routledge, London

Cohen MR (1944) A Preface to Logics. Holt (reprinted Dover, New York, 1977)

Connor JR, Diamond MC (1982) A comparison of dendritic spine number and type on pyramidal neurons of the visual cortex of old adult rats from social and isolated environments. J Comp Neurol 210:99–106

Eccles JC (1973) The cerebellum as a computer: patterns in space and time. J Physiol 229:1–32

Gamble E, Koch C (1987) The dynamics of free calcium in dendritic spines in response to repetitive synaptic input. Science 236:1311–1315

Goddard GV (1967) Development of epileptic seizures through brain stimulation at low intensity. Nature 214:1020–1021

Hebb DO (1949) The organization of behaviour. Wiley, New York

Johnson-Laird PN (1983) Mental models. Cambridge Univesity Press, New York

Koch C, Poggio T (1983) Electric properties of dendritic spines. Trends Neurosci 7:80–83

Letovsky S (1987) AI and the human mind: conference report. AI Fall:63–69

Llinas R (1979) The role of calcium in neuronal function. In: Schmitt FO, Worden FG (eds) The neurosciences: fourth study program. MIT press, Cambridge, pp 555–572

Lynch G (1986) Synapses, circuits, and the beginnings of memory. MIT Press, Cambridge

McCarthy J (1987) Generality in artificial intelligence (postscript to his Turing Award Lecture in 1971) Comm ACM (December) 30:1030–1035

McClelland JL, Rumelhart DE (eds) (1986) Parallel distributed processing: explorations in the microstructure of cognition. MIT Press, Cambridge

Marr D (1969) A theory of cerebellar cortex. J Physiol 202:437–470

Minsky M (1986) The society of mind. Simon and Schuster, New York

Minsky M, Papert S (1969) Perceptrons. MIT Press, Cambridge

Newell A (1980) Physical symbols systems. Cogn Sci 4:135–183

Newell A, Simon HA (1972) Human problem solving. Prentice Hall, Englewood Cliffs

Newell A, Simon HA (1976) Computer science as empirical inquiry: symbols and search (1975 ACM Turing Award Lecture). CACM 19(3):113–126 [reprinted in Haugeland J (ed) (1981) Mind design: philosophy – psychology – artificial intelligence. MIT Press, Cambridge]
Prigogine I, Stengers I (1984) Order out of chaos: man's new dialogue with nature. Bantam Books, Toronto
Quine WW (1985) States of mind. J Philos 82:5–8
Russell B (1921) The analysis of mind. Allen and Unwin, London
Simon HA (1976) The information storage called human memory. In: Rosenzweig MR, Bennett EL (eds) Neural mechanisms of learning and memory. MIT Press, Cambridge
Thibon G (1936) La Science du Caractère (L'Oeuvre de Ludwig Klages) Desclée de Brouwer, Paris

Discussion Following Remarks by A. de Callataÿ

Freeman: Not everyone in connectionism deals with symbols and the kinds of schematic diagram you've demonstrated. The Hopfield network broke away from rule-based symbol manipulation into a computable energy function, which is a different theory of information processing. This introduced nonlinear dynamics – a set of attractors as stored memories, basins as the categories for classes of input, and continuous-time operation – for solutions of formerly intractable travelling salesman problems (TSPs). Here is an alternative to rule-based symbol manipulation, for character recognition, pattern classification, etc., which offers a model closer to neuroscience than the traditional PDP perceptrons.

de Callataÿ: Yes, you are right. That kind of neural network is close to analog processing, but the Hopfield network has not yet produced a complex AI program; I hope they will. This is another approach which I have not described.

Bullock: At this point I remember the old horsebreeder story, which I claim for my hometown racetrack of Del Mar: A millionaire wanted to improve his breed and consulted scientists. The veterinarian told him it was just a matter of getting the right relations between muscles and cardiac output. The biochemist said the horse was a bundle of enzymes and the proper sequencing of enzymes was the thing. The psychologist said the horse was a bundle of nerves, motivations, and sensations, so one should study what was going on in its little mind. But the physicist said, "Let us consider the horse as a sphere."

Babloyantz: In studying the brain we are in big trouble, because in all previous scientific endeavor we had an object in front of us and we studied it using our brains, whether that object was called "nature," "phenomena," or "ideas" – coming from our brains. But in studying the brain we are in a circular process; for the first time in scientific adventure the brain has to measure itself. So I wonder if we can ever "understand" the brain. I suppose for most of us this means putting that brain out in front of us, like an object, and applying all the physicochemical methods we know. But we should not forget that both physicochemical and mathematical concepts evolve, and we may yet one day have enough of the right tools to say how the brain works as a physical-chemical-biological entity.

To forestall pessimism, let me recall how things have actually evolved in this century. At the beginning matter was "atoms" – the Bohr nucleus with its orbiting electrons. In less than a half century we have photos of atoms as a commonplace, though such pictures still amaze me. Maybe in another half century we'll see a photo of the dynamic brain in action – who knows? Now, on the chemical side,

Springer Series in Brain Dynamics 2
Edited by Erol Başar and T. H. Bullock

similar mathematical tools have opened up the concept of self-organization. Mathematics evolved to help us do a job: the first extant computational document is a laundry bill; calculus was invented to assist navigation; twentieth century mathematics was invented to help us discover elementary particles, and this mathematics predicted particles which were only afterwards observed in experiments. Coming closer to the brain and biology, remember that in 1921 Bray's oscillating chemical reaction was called crazy by other chemists, or an artifact, and he was forgotten until 1966 when self-organizing chemical systems became respectable.

In Brussels our group has spent 20 years on a simple, two-variable model to show how matter might become alive, with just two simple equations which later became several hundred papers! That's how self-organization started, and it now looks as if there was something in it. But 20 years ago you would have laughed at us when we talked about cooperativity. Now things are evolving faster and faster, mathematics also. At the beginning of the century we were using the mathematics developed a century or two before, and with the concepts of limit-cycles and attractors 20 years ago we were using Poincaré's nineteenth century mathematics. But strange attractors and chaos are new, and they depend on the theory of nonlinear mathematics which is new and evolving daily; so now we are beginning to run in parallel with the mathematicians – we peddlers between mathematicians and neuroscientists. We have to keep our eyes on both parties. We must be interdisciplinary: physicists, biologists, chemists, and so on; we must all work together, to understand the brain.

AI has now caught up with self-organization and is joining us, to make better machines. From your side we need more sophisticated experiments and ideas. Maybe then we shall have some idea, within 50 or 100 years, of how the brain works. Maybe that will be a scary day, like science fiction. That's my defense for theoretical aspects in the study of the brain, not just because I happen to be doing it, but because I think these theoretical aspects are going to become more and more important and so will be more interesting to everybody.

Freeman: A bit more historical perspective: McCulloch and Pitts' (1943) work developed the logical calculus of neural operations, based on an abstraction of the neuron to a digital switch. They got their ideas from Lorente de Nó who had postulated reverberating circuits from his neuroanatomical diagrams. He later disavowed the term "reverberating" but his Golgi diagrams clearly showed the possibility for circulation of impulses in networks; but even in the late 1940s neurophysiologists ignored McCulloch's ideas and concluded he was crazy. It was John von Neumann who understood the logical calculus and used the neural metaphor in many of his papers. The idea of the brain as a logical switching device emerged from neuroanatomical and neurophysiological studies. Von Neumann's posthumous book *The Computer and the Brain* concluded that the brain is not a digital switching device but an analog-hybrid and that the arithmetical depth of the digital computer is matched by the logical depth of the brain; and that there is a language of the brain that we don't know, not explicit mathematics.

Now, in your description, Dr. Babloyantz, we have the emergence of nonlinear dynamics largely done by simulation on digital computers, and hard-core mathematicians claim that this is not mathematics! There's a delicious irony in the fact that one of the key mathematicians of his age said that the language of the brain was not mathematics, and in an oblique way mathematicians are now agreeing with him, because numerical simulation is now called a branch of experimentation rather than mathematics. But this is our tool to describe brain operations.

Mpitsos: A brief history of my experience with nonlinear dynamics: I've studied neurophysiology and behavior for 22 years, and mathematics for the past several, before going to mathematicians. As you know from my own talk in this conference, I believe that chaos is occurring in the dynamics of neural and other biological systems but I am disturbed to find that the available mathematics does not adequately apply to the behavior of biological systems. Yet I see this strong tendency to have biology comply with mathematical concepts, and that is not the direction to go. Present knowledge of the nonlinear dynamics of chaos is highly inadequate to answer the questions I want to answer. There's no way I am going to be able to fulfill the same criteria as are fulfilled for a Rössler equation. I think mathematicians need to observe and participate in physiological experiments, in order to see what is really needed.

Babloyantz: I did not say that what we have today is going to solve everybody's problems. I meant to say that we are on a road that might take us somewhere if we all go together. Chaos and nonlinear dynamics entered the study of the brain only recently. The concept of self-organization led to experiments that are documented. We only ask you neuroscientists to ask some new questions, but you are the ones to do the experiments. Take the example from traditional chemistry, where chemists thought that everything could be explained by equilibrium thermodynamics, whereas today we know that that was nonsense. It took half a century to look at open systems far from equilibrium, and now self-organizing reactions turn up every month. All I'm saying is, "Look and you may find something." If you have any kind of network or set of connections (forget about neurons), just put ten identical things together in contact with each other, and you can have so many behaviors; then imagine 10^7 similar units interconnected. Think in this direction that's all.

Mpitsos: Neurobiologists and behaviorists have been saying this for some time; it's not new, nor is it a mathematical concept. If mathematicians require your level of verification, we're wasting our time, because there is no way I am going to make my animal act like a Rössler equation.

Babloyantz: I'm surprised, because mathematics is simply a way of avoiding the endless repetition of an unnecessary experiment for every single particular case. Newton's laws saved us the fatigue of observing and measuring the fall of every single apple on earth, and allowed us to make verifiable predictions. Let me give you another example from our own experience: A few years ago all we knew about was attractors – the limit-cycle and the torus. Now we have a new toy we are playing with, called chaos. Strange attractors have incredible variety, and someone may come along next week to tell us there are strange-strange attractors. Maybe such things will be more helpful. So far, wherever we've progressed in science was when we could cast things in mathematical terms. Mathematics is a crystal ball, because words cannot help from one point to the next.

Sheer: Isn't mathematics just a form of logical notation, and if you have a logical notation in a set system where the rules are set up, you follow the rules of that logical notation? It's a question of setting the system that applies to what you're doing and then following the rules of that logical notation, isn't it?

Babloyantz: Yes, but also it has internal mechanisms. For example, differential equations have their own internal mechanism – initial conditions, boundary conditions, etc. – but when you set up your problem, you can set whatever you want in a differential equation.

Sheer: The trick comes in fitting one's observational data into the mathematical system that best describes them.

Babloyantz: Yes!

Grüsser: This remark is a bit heretical, because I am an old fashioned positivist: Mathematical tools are just tools to enable an experimenter to predict reality in physiology or physics. I was very impressed some years ago by Prigogine's encouragement of apply his self-organization principles to neurophysiological experiments. Do you know of an experiment where the data were modeled according to Prigogine's ideas and where the outcome of a new, *not done,* experiment was predicted, and this prediction was either verified or falsified? So far I think all the applications have been post hoc.

Babloyantz: Our group has done this for chemical systems.

Grüsser: This I know, and Prigogine gave very nice examples. But his idea was that neuronal networks might follow similar mathematical rules. My question is: Has someone applied Prigogine's rules to neural systems?

Babloyantz: We are just starting. We have had someone from AI who was looking for self-organization in neural nets ...

Anon. from the floor: Some years ago a researcher from McGill took Purkinje cells from the chick and plated them in cell culture. When you plate them they start beating, with spike discharges in a fairly regular way. He then perturbed the system with a stimulating electrode and did a phase-resetting curve. From this he applied the mathematics of chaos, which says that if you stimulate this spontaneously beating dish in certain phases with certain frequencies you can destabilize the natural beating. This was predicted by the math. They then stimulated at those frequencies, latencies, and phases, and that spontaneously beating cell culture became very irregular (Glass et al. 1984; see also Winfree 1987).

Babloyantz: That reminds me that some strange-attractor phenomena, of the kind we are familiar with from chemistry, have been seen in slime mold aggregations.

Freeman: I must say that my book *Mass Actions in the Nervous System* followed a year after my first visit to Prigogine's laboratory in 1974. What I learned from Prigogine was how to recast my linearized differential equations into a form which he had pioneered along with his group, of which Dr. Babloyantz is a member. I had to make certain crucial changes; for example, to throw out the diffusion term and its distance-dependent delay, and change the nonlinearity from the multiplicative type which he had to the static sigmoid nonlinearity which is characteristic of the nervous system. Nevertheless, the basic concepts which I've been using for the past 12 years have come straight out of his laboratory, and have proved very useful for predictive purposes.

Babloyantz: For your nonlinearity you replace diffusion by connectivity. Connectivity is a discrete form of diffusion, but basically the ideas are the same.

Freeman: No, they are not. The newer axon assures that they are not.

References

Freeman WJ (1975) Mass action in the nervous system. Academic, New York

Glass L, Guevara MR, Belair J, Shrier A (1984) Global bifurcations of a periodically forced biological oscillator. Phys Rev [A] 29(3):1344–1357

McCulloch WS, Pitts WH (1943) A logical calculus of the ideas immanent in nervous activity. Bull Math Biophys 5:115–133

Neumann J von (1958) The Computer and the Brain. New Haven, Yale University

Winfree A (ed) (1987) Cardiac arrhythmias. Princeton University Press

III. Animal Models in Sensory and Cognitive Processes

Introductory Remarks

T. H. BULLOCK

Collected together under this heading are a set of diverse contributions representing the whole domain of studies upon nonhuman species. Some of these are original data papers, others more or less broad overviews of an area.

Buchwald, for example, takes up the long-worrisome question of distinguishing the electrical signs of the sensory from the cognitive events in the brain. Bullock surveys the animal kingdom for evidence of evolution of brain waves and evoked potentials. Eckhorn, Bauer, and Reitboeck report new results with a technique doubly unusual in using up to 19 individually advanced electrodes and in analyzing both spike and slow wave activity; these uncover cell clusters as discontinuities in the representation of visual information in cortical areas 17 and 18. Csepe, Karmos, and Molnar report cognitive-event related potentials (ERPs) – "mismatch negativity" to deviant tones, from cortical and subcortical regions in the waking and even the sleeping cat. Sasaki and Gemba also analyze ERPs in nonprimate subjects. Pockberger, Speckmann, and Walden review the use of the "penicillin model" in the rat to show that epileptic phenomena are a special form of neocortical activity with inhibitory as well as excitatory components.

To these chapters may be added Mpitsos's contribution (in Part V) and much of Adey's (I and V), Başar and coworkers (I and V), Roeschke and Başar's (II), Skinner's (II), and Freeman's (V) papers.

It may rightly be noted that few studies on nonhuman species undertake the paradigms associated with higher cognitive ERPs in humans. More than that, the animal models in use so far are heavily weighted by a few laboratory species so that differences according to phylogenetic position have hardly been noticed and the opportunity to study evolution in dynamic operations of the most evolved organ is mainly still before us. Perspectives for future research are suggested, therefore, not only by the discoveries reported in these chapters, but also by the gaps.

Springer Series in Brain Dynamics 2
Edited by Erol Başar and T. H. Bullock

Comparisons of Sensory and Cognitive Brain Potentials in the Human and in an Animal Model

J. S. Buchwald

1 Introduction

An increasing variety of chemical substances have been shown to be produced by the brain which are capable of influencing its activity. Originally termed transmitters, over the past decades hormones have also been shown to be produced by and to act upon the brain, as have other "modulator" substances such as the endorphins and enkephalins. Thus, the distinction between transmitters, hormones, and modulators has become increasing blurred as a function of contemporary neuroscience.

I should like to suggest that there may be an analogy in such terminology as "sensory" and "cognitive" brain processes and that the territories of these terms have progressively merged and overlapped as a result of both experimental and clinical observations. Let us consider, for a moment, what is involved in a normal conversation. The listener must concurrently engage a number of brain processes in order to: (1) spatially localize the speaker, (2) determine emotional content, and (3) translate the linguistic message. If the conversation is in a room full of people, even more processing is necessary, as different speakers come and go and the background noise changes in intensity and/or significance. Where do cognitive processes begin in the brain's analysis of this multifaceted barrage of incoming information; where do sensory processes stop?

As the acoustic environment becomes richer and more complex, there is clearly increased stimulation of the cochlear hair cells and increased transmission of action potentials by the acoustic nerve. These are indisputably sensory functions. Thereafter, however, like a series of shotgun explosions, outputs project from the cochlear nucleus to numerous target loci in the brain stem; these in turn project to other brain stem and forebrain targets, which in turn project elsewhere. Thus, from the level of the first central relay, diverging, converging, and parallel pathways with a variety of synaptic interruptions increase in profusion as the brain stem and forebrain are ascended. Some of these pathways function to transport and distribute different components of the acoustic signal to specific processing systems while others modulate and gate such distribution as a consequence of feedback loops. How do we determine the "cognitive" versus "sensory" functional components of such projections?

Springer Series in Brain Dynamics 2
Edited by Erol Başar and T. H. Bullock

2 The Scalp-Recorded Evoked Potential Sequence

As one means of assessing sensory and/or cognitive processing of acoustic information, we have for some years been interested in the possibilities offered by surface-recorded auditory evoked potentials. In the human, the sequence of scalp-recorded potentials elicited by clicks or tones shows a remarkable similarity across large numbers of individual subjects or across repeated recording sessions for any one subject. The consistency with which these positive and negative voltages appear within characteristic latency ranges reflects a predictable pattern of brain activation by the acoustic stimulus. Furthermore, under appropriate parametric conditions, particular response components can be related to sensory or cognitive modes of brain function. If the evoked potential generators can be identified, such brain functions can be linked to relevant generator substrates, and the evoked potential sequence can be used as a means of temporally and spatially tracking the brain's processing of acoustic information.

Within this context, we have been particularly interested in the identification of evoked potential generators. For such studies, animal models become highly desirable insofar as they permit systematic exploration of the brain system postulated as relevant to a particular evoked potential component. The utility of animal models became particularly clear with the advent of the auditory brain stem responses (ABRs). Through a number of different animal experiments, the ABRs were proved to be neurological phenomena rather than electronic artifacts and to originate from the brain stem auditory pathway. As a result, theoretical dogma yielded to empirical evidence, and the ABRs have become accepted as far-field responses originating from distant brain stem generators and volume conducted to the scalp (reviewed by Buchwald 1983).

The period of intense investigation of the ABR origins provided a watershed for current study of evoked potential generators. The ABR data clearly showed (1) that scalp-recorded evoked potentials are not necessarily generated by the cortex, and (2) that animal models can be crucially important in determining the sources of these potentials. Since that time, a number of animal models of other evoked potential components have been and are being developed. In the subsequent discussion of sensory and cognitive evoked potentials, however, I shall draw primarily upon our own work using the cat as an animal model.

Clearly there are differences between the brain of the cat and human, particularly in the proportionate representation of neocortical areas such as the frontal and parietal lobes. The cognitive capacities of the two species are obviously different and the scalp recorded evoked potential components are not identical in latency and waveform. In our use of the cat as a model of certain human evoked potential components, however, we have attempted to define functional characteristics of specific components in each species, e.g., recovery cycle, effects of sleep–arousal states, etc. In this way we have been able to compare specific components with similar functional characteristics across species and to proceed, in the experimental animal, with systematic controlled investigations of the generator substrates of these components.

3 Sensory Brain Potentials

"Sensory: connected with the reception and transmission of sense impression."

3.1 The Auditory Brain Stem Responses

In the human and in a number of animal species, the ABRs have been recorded under many different parametric, clinical, and experimental conditions (see Moore 1983). There is a consensus that these short-latency, scalp-recorded responses to auditory stimuli are generated by the brain stem relays of the primary auditory projection pathway from the level of the acoustic nerve through the level of the inferior colliculus. The ABRs seem to qualify as "sensory" because they change markedly in amplitude and latency as a function of stimulus parameters, such as intensity, but are unaffected by changes in the internal state of the subject. Moreover, the ABRs are generated within the system anatomically defined as the primary auditory pathway. Behaviorally, the presence of the ABRs is generally correlated with the ability to "hear" the acoustic stimulus; the subjective hearing threshold usually approximates the ABR threshold and the louder the stimulus becomes subjectively, the larger the amplitude of the ABRs. The use of these responses as an objective measure of "hearing" and brain stem function has been extensive for more than a decade in neonatal nurseries, on neurological wards, in mental institutions, and in other situations in which the subject is unable to respond to conventional audiometric examinations.

3.2 Middle Latency Responses: Component Pa

Within the 10–100 ms latency range, components between the ABRs and the longer latency event-related potentials have been termed middle latency responses (MLRs). The human Pa component, a moderately large positivity with a latency of approximately 30 ms, has heretofore received most of the attention directed toward the MLRs (see Kileny et al. 1987 for overview). This component resembles the ABRs in a number of parametric dimensions: it persists at relatively fast click rates of 10/s, i.e., it shows a rapid recovery cycle (Erwin and Buchwald 1986a), and it does not change significantly as a function of arousal levels, i.e., it is constant in amplitude and latency across states of wakefulness, slow wave sleep (SWS), and rapid eye movement (REM) sleep (Erwin and Buchwald 1986b).

In the cat, a middle latency evoked potential component, wave 7, has similarly been shown to resemble the ABRs and the human Pa, insofar as it also is resistant to click rates up to 10/s and does not change significantly across states of wakefulness, SWS, and REM sleep (Chen and Buchwald 1986). This vertex-recorded potential is parametrically correlated with the primary auditory evoked response generated in auditory cortex in terms of latency, waveform, and recovery cycle, and appears to be causally dependent upon primary auditory cortex insofar as it disappears with ablation of this cortical area (Buchwald et al. 1981).

These animal data support a number of clinical studies which report a selective absence of Pa in subjects with lesions of the superior temporal gyrus (reviewed in Kileny et al. 1987). The Pa component has been considered a sensory potential because of its physiological parameters, its postulated generation in auditory cortex of the superior temporal gyrus, and since, in the presence of intact ABRs hearing deficits may occur which are associated with depression or disappearance of Pa.

4 Cognitive Brain Potentials

"Cognitive: knowing or apprehending."

4.1 Middle Latency Responses: Component P1

Within the 10–100 ms latency range considered part of the territory of the MLRs lies the P1 component, which has been studied relatively little in comparison to the previously discussed Pa potential. In part this may be because the peak latency of P1 is in the 50–65 ms period, too long for the 50 ms time base generally used to study Pa. Additionally, the usual 10/s rate of stimulation used in studies of Pa causes P1 to disappear. Based on our animal studies, however, we have hypothesized that a "cognitive" human MLR component exists, i.e., P1, and I shall briefly describe these animal data and the consequent human experiments.

In the cat we found that a surface-recorded positive evoked potential recorded in the 20–25 ms latency range, "wave A," did not change during complete neuromuscular paralysis but disappeared during barbiturate anesthesia. When rates of click stimulation exceeded 1/s, wave A diminished and disappeared without change in the earlier latency ABRs and wave 7. In a series of lesion studies, we found that wave A persisted after bilateral removal of the cerebral hemispheres (including basal ganglia, hippocampus, and all neocortex) (Buchwald et al. 1981). Thus, its essential generator substrate was within the remaining brain stem–diencephalic tissue, which included the thalamus and midbrain. In a mapping study, we subsequently found field potentials and single units with the same general latency and recovery cycle characteristics as wave A, which could be recorded from the brain stem reticular formation, ventral to the inferior colliculus, and rostrally through the midbrain tegmentum to the intralaminar, center median, and anterior nuclei of the thalamus (Hinman and Buchwald 1983). These data suggested that components of the ascending reticular activating system might be involved in the generation of wave A. If so, we hypothesized that wave A should reflect changes in arousal levels; this was tested in another series of experiments. We found that wave A diminished and disappeared during increasing depths of SWS but suddenly reappeared during episodes of REM sleep (Chen and Buchwald 1986). This directly linked mechanisms causing sleep–wakefulness to wave A generation.

Other studies have shown that reticular activating system (RAS) units in the area of the midbrain pedunculopontine (PPT) nucleus slow during SWS and in-

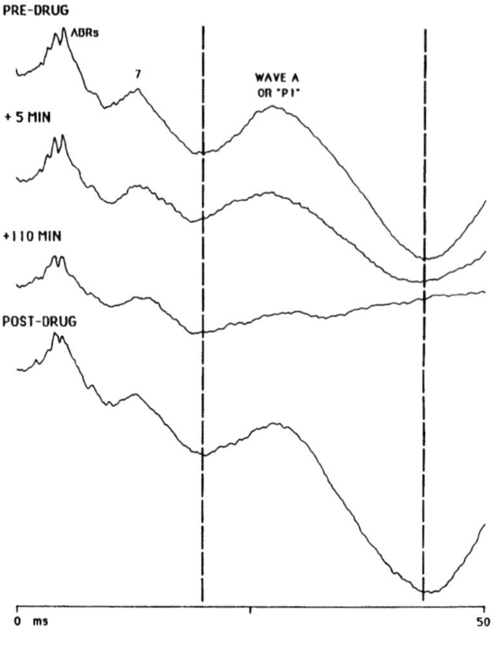

Fig. 1. Traces showing the effects of scopolamine upon wave A in one cat. We have now replicated scopoloamine dose–response series (0.1, 0.2, 0.3, and 0.4 mg/kg s.c.) in six cats with essentially identical results. In all cases the scopolamine produces a disappearance of the wave A component with little effect on the preceding ABRs. This effect is concurrent with the appearance of EEG slow waves; it shows a more rapid onset and is more intense and long lasting at the higher drug doses. The traces illustrated are subsequent to a 0.4 mg s.c. dose of scopolamine. (Dickerson et al. 1986)

crease their firing frequencies during REM (Steriade et al. 1982). Based in part upon these data as well as upon choline-acetyltransferase immunohistochemistry and the effects of carbachol iontophoresis upon single unit discharge, increased activity of PPT cholinergic cells has been correlated with periods of REM sleep, and decreased activity in these cells has been correlated with periods of SWS (Gillin et al. 1978; Vivaldi et al. 1980).

We hypothesized that, since wave A increased during REM and decreased during SWS, it might likewise be related to a cholinergic component of the RAS. In a subsequent series of experiments we recorded wave A during cholinergic (muscarinic) receptor blockade with scopolamine and found that wave A disappeared even at minimal doses (0.15 mg/kg), an effect not shared by any of the other surface-recorded potentials (Fig. 1) (Dickerson et al. 1986). This result suggests that a cholinergic component of the RAS is essential for the generation of wave A. We postulate: (1) that this substrate is represented by cholinergic cells in the midbrain PPT nucleus which project to intralaminar, center median, and anterior thalamic nuclei, and (2) that the postsynaptic activation of cholinoceptive neurons in these nuclei is volume conducted to the scalp and recorded as wave A.

As these data were unfolding for the cat wave A, we began to look for a similar potential in the human. In the existing literature, however, none of the human MLR potentials resembled wave A. Thus, we began to replicate the stimulation and recording parameters used in the animal experiments in our human recordings. Under these conditions, a potential did emerge which resembled wave A in all respects, the P1 component. These data will be briefly summarized.

The P1 positivity shows a peak latency range of 50–65 ms in normal adults and significantly diminishes or disappears as click rates exceed 1/s, whereas the

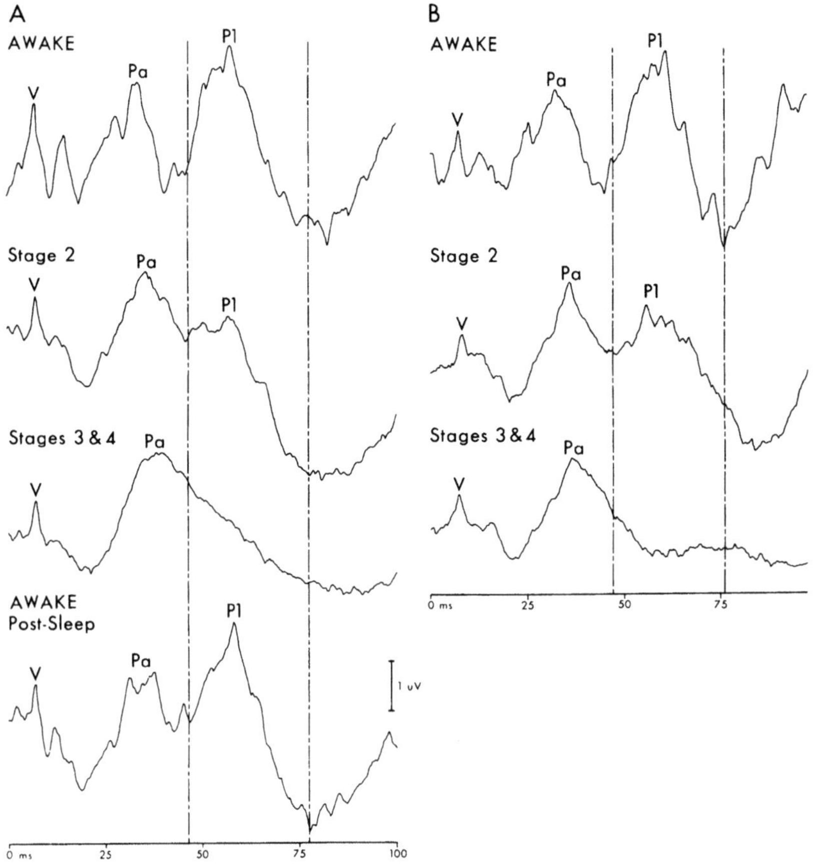

Fig. 2 A, B. The averaged wave forms of one subject recorded during wakefulness and sleep stages in two separate sessions conducted 3 months apart. The first session (a) consisted of an afternoon nap (2 h). The P1 component was substantially reduced in amplitude during sleep stage 2 and disappeared completely during sleep stages 3 and 4. The component reappeared at its original amplitude during postsleep recordings. The second session (b) consisted of an overnight sleep (8 h). The P1 amplitude changes across sleep stages were similar to the previous afternoon nap recording. A final wake-up recording was not obtained in this instance. The time base is 100 ms and the calibration is 1 μV positive up. (Erwin and Buchwald 1986 b)

ABRs and Pa do not change (Erwin and Buchwald 1986 a). As subjects shift from wakefulness to deep SWS, the P1 gradually disappears (Fig. 2); during the occurrence of REM episodes, the P1 abruptly reappears. Concurrently, there are no changes in the ABRs or in Pa, nor do longer-latency potentials show the significant changes in amplitude manifested by P1 (Erwin and Buchwald 1986 b). In very preliminary studies, one subject has shown a marked reduction in P1 amplitude following a low dose of scopolamine delivered transdermally.

Thus, in the surface-recorded human evoked potential sequence P1 is unique in a number of functional characteristics, characteristics which are shared by the cat wave A. While the absolute latencies of these components differ in the cat and human, there is no reason to require them to be the same in two different species with different brain size and complexity of organization. Thus, based on data relevant to the generator substrate of wave A, we postulate that P1 may be generated by a cholinergic component of the RAS, which in turn postsynaptically activates thalamic cells and produces a field potential which is volume conducted to the scalp as P1.

4.2 Event Related Potentials: Component P300

A second cognitive potential in which we have been particularly interested is the "P300," a long duration positivity which usually occurs with a peak latency in the 300–500 ms range. This response is produced by a rare stimulus, to which the subject must attend and respond, randomly presented in combination with a frequent stimulus of the same modality. While the amplitude and latency of the P300 are not closely related to physical parameters of the stimulus, as are the ABRs, P300 amplitude is dependent upon stimulus probability and task requirements and, if discrimination between the rare and frequent stimulus is made more difficult, P300 latency increases. Because of these characteristics, the P300 is considered to be an index of discrimination, sequential information processing, and short-term memory (Fig. 3) (Squires et al. 1976; Donchin 1981). In normal aging, without associated disease, the P300 progressively increases in latency and diminishes in amplitude, but there is significantly more change, e.g., the response may disappear, in age-matched Alzheimer's disease subjects (Pfefferbaum et al. 1984; Picton et al. 1984; Syndulko et al. 1982).

The fact that the P300 reflects rather specific cognitive processing functions and that it is differentially affected in certain diseases has made the question of its origin particularly important. However, little consensus presently exists. Electromagnetic surface mapping studies, as well as depth recordings in the human hippocampus, entorhinal cortex, and amygdala, have implicated components of the limbic system in P300 generation (Halgren et al. 1980; McCarthy et al. 1982; Okada et al. 1983). On the other hand, topographical mapping of voltages and electromagnetic recording of current sources have suggested temporal–parietal cortex as a primary generator source of the P300 (Simson et al. 1977; Richer et al. 1983). Although human depth recordings provide a direct opportunity to test hypotheses of generator origins, interpretation of the resultant data is problematic since this procedure must necessarily be carried out only in very abnormal brains, must be restricted to target loci most relevant for a particular diagnosis, and may be accompanied by scalp P300 responses which are unstable or missing altogether.

In an attempt to pursue the generator question more extensively, a number of animal models of the P300 have been developed (cat: Wilder et al. 1981; Buchwald and Squires 1982; Harrison and Buchwald 1985; monkey: Arthur and Starr 1983; Paller et al. 1984; rat: O'Brien 1982). In the cat and monkey, endogenous responses have been elicited which resemble the human P300 along a number of critical dimensions: (1) dependence upon task relevance, (2) dependence upon stimulus probability, (3) independence of physical properties of the stimulus, and (4) changes in latency and amplitude as a function of aging.

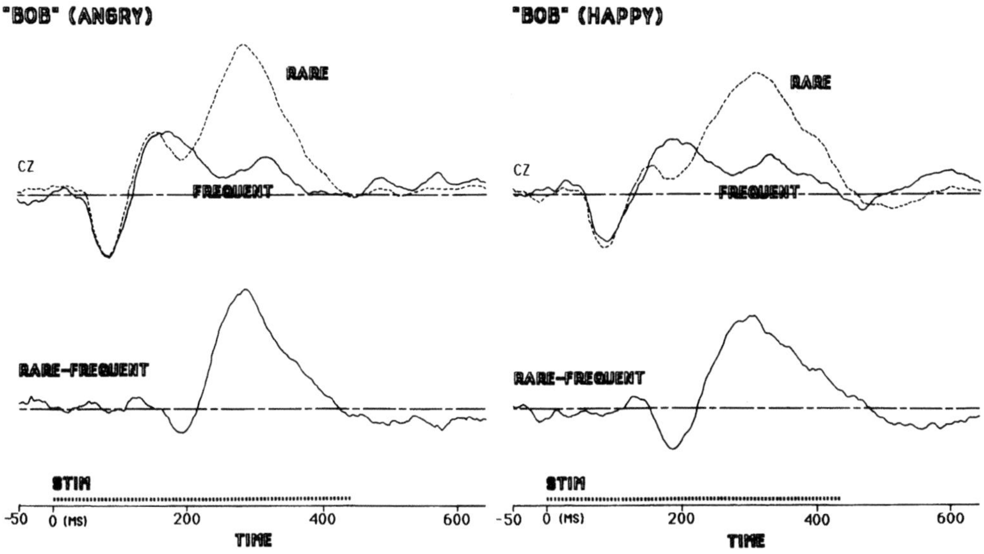

Fig. 3. An ongoing study in our laboratory showing P300 grand averages from 14 adult subjects (seven male, seven female, 20–30 years) presented with affective prosodic stimuli. The word "Bob" has been digitized with either an angry or happy intonation and normalized with respect to duration and peak intensity. Rare stimuli are presented on 20% of the trials, frequent stimuli on 80%; a conterbalanced design reverses stimulus probability on the subsequent block of trials. The subject is asked to press a button to each rare stimulus presentation. Difference potentials resulting from subtraction of frequent from rare responses are shown in the *lower traces*. The residual large positivity in the 200–400 ms latency range is the P300 component

We are presently using the cat P300 model in two series of experiments to test the hypothesis of cortical generation of the P300 and the hypothesis of limbic system generation of the P300. Data from our investigations thus far will be briefly summarized.

4.2.1 Cortical Generation

An initial experimental series was directed toward the role of primary auditory cortex in the generation of the cat P300 produced by rare and frequent auditory stimuli. Recordings carried out over 12 sessions before and after complete bilateral aspiration of primary auditory cortex (areas AI, AII, and Ep) showed no postoperative changes in the cat P300 (Fig. 4) (Harrison et al. 1986). A second study was directed toward the role of association cortex in P300 generation. Complete bilateral aspiration of the major polysensory association cortices (Irvine and Phillips 1982) (pericruciate, middle suprasylvian, and anterior lateral gyri) was carried out and recordings from 12 preoperative and 12 postoperative sessions were compared. Again, no significant postoperative changes in the cat P300 occurred (Harrison and Buchwald 1986). Thus, we presently have no indication that the processing of auditory stimuli by either primary auditory or polysensory association cortex is necessary for this cognitive potential.

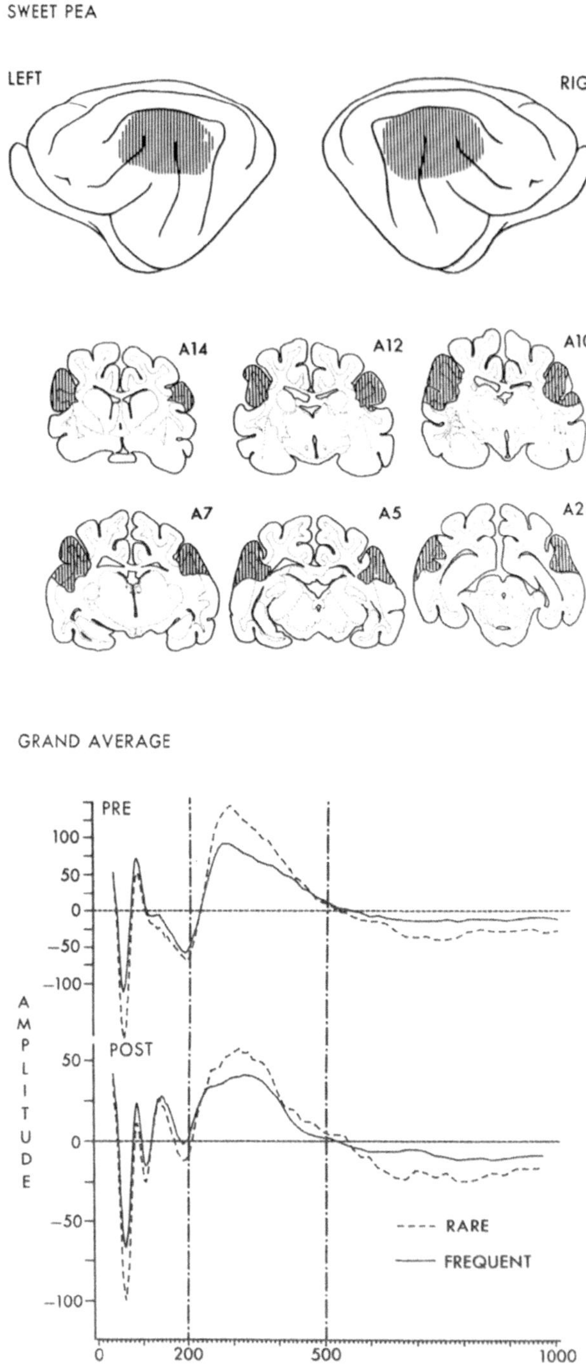

Fig. 4. Extent of cortical ablation and evoked potential grand averages for cat "Sweet Pea." As indicated by the surface diagrams and coronal sections, AI, AII, and Ep were removed to the depth of the underlying corona radiata. Grand averages reflect 12 pre- and 12 postoperative recording sessions of responses to the "rare" loud click (*dashed lines*) and "frequent" loud click (*solid lines*). In both cases, paired *t*-tests showed significant differences in the 200–500 ms period between the rare and frequent click responses. (Harrison et al. 1986)

4.2.2 Limbic System Generation

We are also carrying out a series of studies to investigate the role of the limbic system in generation of the cat P300. Because markedly abnormal P300s and a characteristic loss of cholinergic cells in the nucleus basalis occur in individuals with Alzheimer's disease we initially attempted to replicate a portion of this clinical finding by lesioning the cholinergic cells of the medial septal nucleus which provide a major projection to the hippocampus. When cat P300 recordings over 12 preoperative sessions were compared with those over 12 sessions after bilateral lesioning of the medial septal area, a significant decrease or disappearance of the postoperative response was found (Fig. 5) (Harrison et al. 1988). In these animals there was marked acetylcholinesterase (AChE) depletion in the hippocampus, a reflection of the destruction of the cholinergic projection cells. Control lesions rostral to the lesioned septal area resulted in no change in the P300 or in hippocampal AChE staining density. These data suggested an important role for the septohippocampal system in modulating and/or generating the P300.

In other experiments, depth mapping of field potentials was carried out concurrently with the usual surface recording of the cat P300. Systematic recordings from the rostral to caudal extent of the hippocampus indicated the presence of a potential which was larger to the rare than to the frequent stimulus and which corresponded in latency to the surface P300 (reviewed in Buchwald 1989). Inversion of the potential (from positive to negative re O DC) occurred as the electrode passed through the hippocampus and appeared to be correlated with recordings above or below (positive) compared to within (negative) the pyramidal cell layer.

In a second mapping study, similar correlations were sought between the surface-recorded P300 and depth field potentials in the medial septal nucleus. Systematic mapping of the septum again revealed responses which were larger to the rare than to the frequent stimulus and which were similar in latency to the surface P300. Depth potential inversions from positive to negative (re O DC) occurred within the dorsal half of the septum in a number of cases (Fig. 6) (Harrison and Buchwald 1987).

Brain areas in which field potentials are consistently shown to invert are generally considered to be probable loci of potential generation or modulation. Thus, these mapping studies are consistent with the septal lesioning study in suggesting an important role for septohippocampal components of the limbic system in the modulation and/or generation of the P300.

5 Sensory Versus Cognitive Brain Potentials

In the preceding discussion, there has been an emphasis upon the postulated generator substrates of "sensory" and "cognitive" evoked potentials, particularly as suggested by data obtained from the cat model of component waveforms in the human. But at what point do "sensory processes" disappear and "cognitive processes" emerge? Does the answer relate to anatomy: responses within the primary

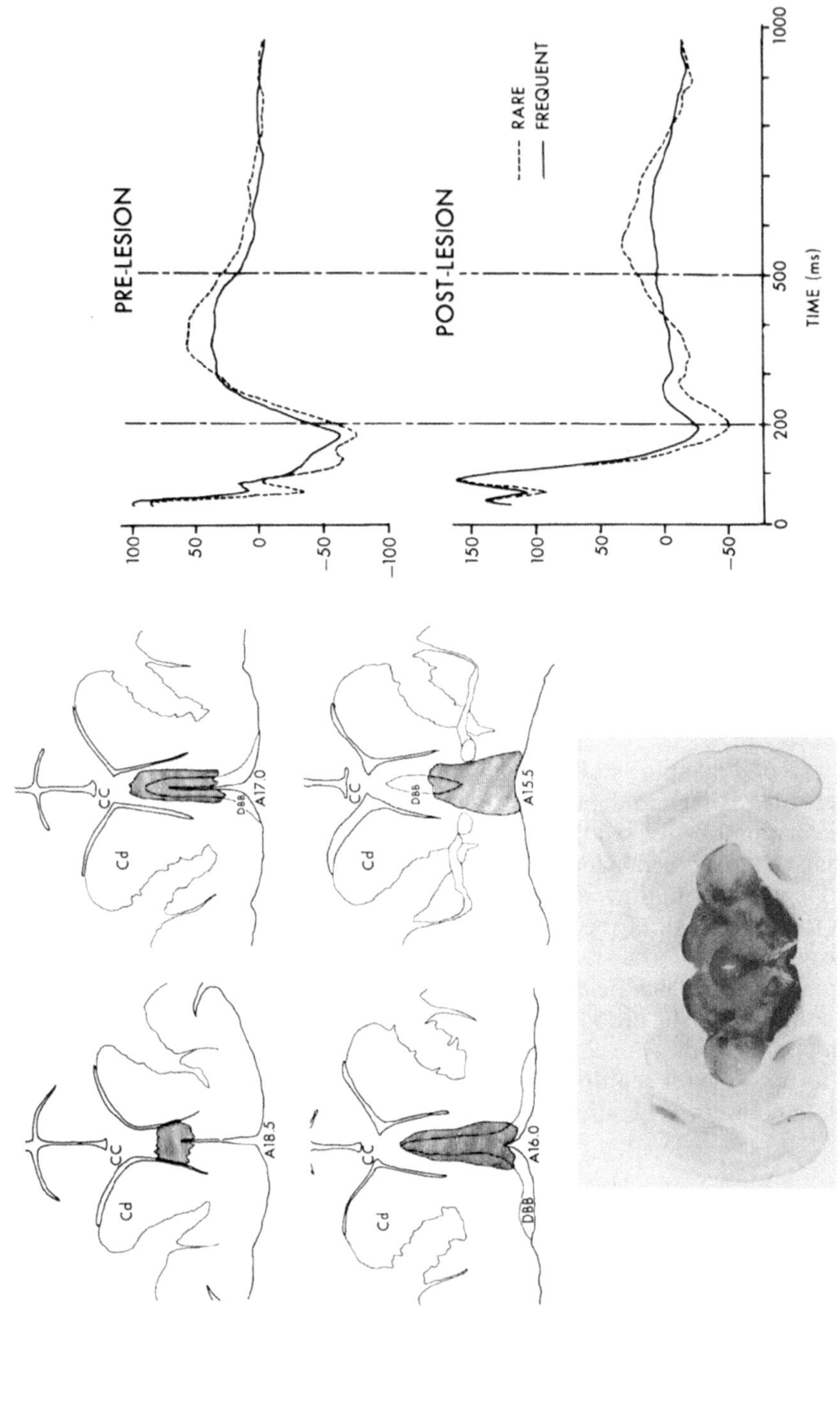
PRE-LESION
POST-LESION
RARE
FREQUENT
TIME (ms)
0
200
500
1000
100
50
0
-50
-100
150
100
50
0
-50
Cd
CC
DBB
A18.5
A16.0
A17.0
A15.5

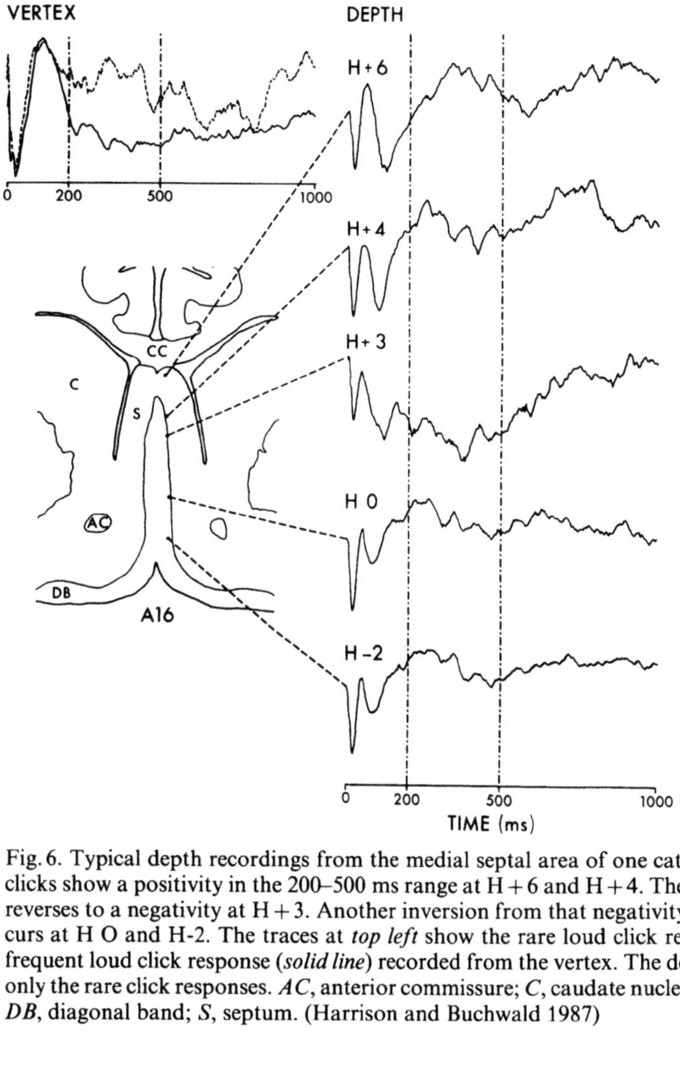

Fig. 6. Typical depth recordings from the medial septal area of one cat. Responses to rare loud clicks show a positivity in the 200–500 ms range at H + 6 and H + 4. The polarity of the response reverses to a negativity at H + 3. Another inversion from that negativity to a small positivity occurs at H O and H-2. The traces at *top left* show the rare loud click response (*dashed line*) and frequent loud click response (*solid line*) recorded from the vertex. The depth recordings illustrate only the rare click responses. *AC*, anterior commissure; *C*, caudate nucleus; *CC*, corpus callosum: *DB*, diagonal band; *S*, septum. (Harrison and Buchwald 1987)

Fig. 5. Typical effects of bilateral septal lesions upon the cat P300 response and upon hippocampal AChE. Data from one cat show the extent of bilateral lesions of the medial septum and diagonal band and the resultant bilateral hippocampal AChE depletion. Evoked potential data represent grand averages of the 12 pre- and 12 postlesion recording sessions for this cat. On the *ordinate,* O refers to zero DC voltage between the active and reference electrodes at the analog to digital converter; 25 units are equal to 1.9 μV. Paired *t*-test comparisons of data points for "rare" and "frequent" click response blocks indicate significant rare vs frequent differences ($P < 0.05$) in the prelesion data over the 353–438 ms latency range but no significant postlesion differences. *CC*, corpus callosum; *Cd*, caudate nudeus; *DBB*, diagonal band of Broca. (Harrison et al. 1988) auditory pathway are sensory, and responses elsewhere are not? Does the answer relate to physiology: auditory responses with constant latency and morphologies are sensory, whereas unpredictable responses are not? Or does the answer relate to behavior: orienting and startle reflexes require a sensory but not necessarily a cognitive component?

In this paper, the term "sensory" has been applied to those short and middle latency components which are constant in latency and amplitude regardless of the subject's internal state, which change in direct relation to changes in the physical properties of the stimulus, and which appear to be generated within the primary auditory pathway from the level of the acoustic nerve to the auditory cortex. In the cat, these potentials are represented by the ABRs and wave 7 and in the human by the ABRs and, we believe, Pa. Thus, both anatomical and physiological data provide support for the "sensory" appellation.

Behaviorally the case is not so clear. The ABRs have been used to screen for hearing deficits, as previously indicated. In both the pediatric and the adult clinical literature, however, exceptions have been reported in which "hearing" may be markedly abnormal in the presence of normal ABRs or, conversely, the ABRs may be missing with little or no concurrent effect on hearing threshold. Even when both ABRs and hearing thresholds are normal, however, what of the sleeping individual who shows no behavioral responses, e.g., no orienting and no "hearing," but whose ABR and Pa amplitudes are the same as during wakefulness? Clearly we must make ourselves stay awake to "hear"; so, is "hearing" a sensory or a cognitive process? In these situations cognitive mechanisms seem to be intimately interrelated with sensory processes. While the auditory system may be uniquely complex in terms of the variety and complexity of its signal processing capacity, this blending of sensory and cognitive functions occurs in all "sensory" systems.

In this paper two human cognitive potentials have been discussed, which we believe can be modeled in the cat and for which we believe there is now a considerable body of data regarding their generators. These cognitive potentials show more variance in their latencies and amplitudes than do the sensory potentials, relatively little dependence upon the physical properties of the stimulus and great dependence upon stimulus significance and the subject's state, i.e., levels of arousal and attention.

The evoked potential which uniquely reflects changes in the arousal levels of sleep–wakefulness is a middle latency component, wave A in the cat and P1 in the human. The animal data indicate that this potential does not require the primary auditory pathway above the level of the lateral lemniscus and that it is generated by medial projections from the ascending RAS to targets in the intralaminar, centrê median, and anterior thalamic nuclei. Thus, neither the laterally projecting auditory forebrain system nor cortical circuitry is necessary for the generation of this cognitive potential.

In both the human and the cat model of the P300 event-related potential, response amplitude changes as a function of stimulus probability, and latency increases as a function of discrimination difficulty. These characteristics suggest that sequential information processing and short-term memory systems of the brain are engaged by the P300 protocols. In the cat, these systems do not appear to require the participation of primary auditory cortex or of polysensory association cortex insofar as bilateral removal of these cortical areas does not significantly change the cat P300 response. On the other hand, subcortical limbic structures, particularly the septohippocampal system, do seem to be important in the modulation and/or generation of the cat P300 since septal lesions which deplete AChE staining in the hippocampus are accompanied by diminution and disappearance of the cat P300. While we can still only speculate about the details of this generator system in the cat and human, the animal data are consistent with the limited depth recording and electromagnetic data obtained from the human limbic system during P300 production. Thus, we may again suggest that neither the laterally projecting auditory forebrain system nor the activation of cortical circuits is importantly involved in the generation of this potential.

Such cognitive potential data require us to consider the importance of alternative, i.e., nonprimary, routings for sensory information. While there has long been anatomical and physiological evidence for a variety of medial and parabrachial pathways which project from auditory brain stem nuclei through midbrain and forebrain levels, little functional significance has been attached to such systems. We believe that exploration of the surface-recorded evoked potentials both in animal models and in the human may begin to unfold new and fascinating details regarding the anatomical substrates and functional levels at which cognitive and sensory processes interact.

6 Summary

The meanings of "sensory" and "cognitive" brain processes have been questioned. Are distinctions made on the basis of anatomical, physiological, or behavioral criteria? In our laboratory the surface-recorded evoked potential sequence has been used as one means of assessing sensory and/or cognitive processing of acoustic information. Experimental procedures in animal models have been combined with parametric recordings in human subjects for each of a number of surface-recorded evoked potential components which have different functional characteristics.

"Sensory" has been applied to human short latency ABR and middle latency Pa evoked potentials, and to the cat ABR and wave 7 models of these components, all of which are constant in latency and amplitude regardless of the subject's internal state, which change with the physical properties of the stimulus, and which appear to be generated within the primary auditory pathway from the level of the acoustic nerve to the auditory cortex.

"Cognitive" has been applied to the human middle latency P1 and long latency P300 evoked potentials, and to the cat wave A and P300 models of these components, all of which are variable in their latencies and amplitudes, show little dependence upon the physical properties of the stimulus, but are highly dependent upon stimulus significance and upon the subject's state, i.e., levels of arousal and attention. Data from the cat model suggest: (1) that these potentials do not require neocortex of the forebrain primary auditory pathway, (2) that P1 is dependent upon a thalamic generator driven by terminal projections from the ascending RAS, and (3) that P300 is dependent upon septohippocampal components of the limbic system for generation and/or modulation.

Overall, these data suggest that cognitive processes may begin at subcortical levels far in advance of those levels at which sensory processes terminate.

References

Arthur DL, Starr A (1983) Task-relevant late positive component of the auditory event-related potential in monkeys resembles P300 in humans. Science 223:186–188

Buchwald JS (1983) Generators. In: Moore EJ (ed) Bases of auditory brainstem evoked responses. Grune and Stratton, New York, pp 157–195

Buchwald JS (1989) Animal models of cognitive event-related potentials. In: Rohrbough JW, Parasuraman R, Johnson R (eds) Event related potentials of the brain. Oxford University Press, New York

Buchwald JS, Squires NS (1982) Endogenous auditory potentials in the cat: a P300 model. In: Woody C (ed) Conditioning: representation of involved neural function. Raven, New York, pp 503–515

Buchwald JS, Hinman C, Norman RJ, Huang EM, Brown KA (1981) Middle and long-latency auditory evoked responses recorded from the vertex of normal and chronically lesioned cats. Brain Res 205:91–109

Chen BM, Buchwald JS (1986) Midlatency auditory evoked responses: differential effects of sleep in the cat. Electroencephalogr Clin Neurophysiol 65:373–382

Dickerson L, Buchwald J, Chen BM (1986) Midlatency auditory evoked responses in the cat: effects of scopolamine, a postsynaptic antagonist of muscarinic cholinergic neurotransmission. Soc Neurosci Abstr 12:1130

Donchin E (1981) Surprise! ... Surprise? Psychophysiology 18:493–513

Erwin RJ, Buchwald JS (1986a) Midlatency auditory evoked responses: differential recovery cycle characteristics. Electroencephalogr Clin Neurophysiol 64:417–423

Erwin RJ, Buchwald JS (1986b) Midlatency auditory evoked responses: differential effects of sleep in the human. Electroencephalogr Clin Neurophysiol 65:383–292

Gillin JC, Mendelson WB, Sitaram N, Wyatt RJ (1978) The neuropharmacology of sleep and wakefulness. Annu Rev Pharmacol Toxicol 18:563–579

Halgren E, Squires NK, Wilson CL, Rohrbaugh JR, Babb TL, Crandall PH (1980) Endogenous potentials generated in the human hippocampal formation and amygdala by infrequent events. Science 210:803–805

Harrison J, Buchwald J (1985) Aging changes in the cat P300 mimic the human. Electroencephalogr Clin Neurophysiol 62:227–234

Harrison J, Buchwald J (1986) Cat P300: effects of association cortex ablation. Neurosci Abstr 12:1131

Harrison J, Buchwald J (1987) A cat model of the P300: searching for generator substrates in the auditory cortex and medial septal area. Electroencephalogr Clin Neurophysiol [Suppl] 40:473–480

Harrison J, Buchwald J, Kaga K (1986) Cat P300 present after primary auditory cortex ablation. Electroencephalogr Clin Neurophysiol 63:180–187

Harrison J, Buchwald J, Kaga K, Woolf NJ, Butcher LL (1988) Cat P300 disappears after septal lesions. Electroencephalogr Clin Neurophysiol 69:55–64

Hinman CL, Buchwald JS (1983) Depth evoked potential and single unit correlates of vertex midlatency auditory evoked responses. Brain Res 264:57–67

Irvine DRF, Phillips DP (1982) Polysensory "association" areas of the cerebral cortex: organization of acoustic input in the cat. In: Woolsey CN (ed) Cortical sensory organization, vol 3. Multiple auditory areas. University of Wisconsin Press, Madison, pp 111–156

Kileny P, Paccioretti D, Wilson AF (1987) Effects of cortical lesions on middle-latency auditory evoked responses (MLR). Electroencephalogr Clin Neurophysiol 66:108–120

McCarthy G, Wood CC (1982) Intracranially recorded event-related potentials during sentence processing. Soc Neurosci Abstr 8:976

McCarthy G, Wood CC, Allison T, Goff WR, Williamson PD, Spencer DD (1982) Intracranial recordings of event-related potentials in humans engaged in cognitive tasks. Soc Neurosci Abstr 8:976

Moore EJ (1983) History. In: Moore EJ (ed) Bases of auditory brain-stem evoked responses. Grune and Stratton, New York, pp 3–14

O'Brien JH (1982) P300 wave elicited by a stimulus-change paradigm in acutely prepared rats. Physiol Behav 28:711

Okada YW, Kaufman L, Williamson SJ (1983) The hippocampal formation as a source of the slow endogenous potentials. Electroencephalogr Clin Neurophysiol 55:417–426

Paller KA, Zola-Morgan S, Squire LR, Hillyard SA (1984) Monkeys with lesions of hippocampus and amygdala exhibit event-related brain potentials that resemble the human P300 wave. Soc Neurosci Abstr 10:849

Pfefferbaum A, Ford JM, Wenegrat BG, Roth WT, Kopell BS (1984) Clinical application of the P3 component of event-related potentials. 1. Normal aging. Electroencephalogr Clin Neurophysiol 59:85–103

Picton TW, Stuss DT, Champagne SC, Nelson RF (1984) The effects of age on human event-related potentials. Psychophysiology 21:312–325

Richer R, Johnson RA, Beatty J (1983) Sources of late components of the brain magnetic response. Soc Neurosci Abstr 9:656

Simson R, Vaughan HG Jr, Ritter W (1977) The scalp topography of potentials in auditory and visual go/no go tasks. Electroencephalogr Clin Neurophysiol 43:864–875

Squires KC, Wickens C, Squires NK, Donchin E (1976) The effect of stimulus sequence on the waveform of the cortical event-related potential. Science 193:1142–1146

Steriade M, Oakson G, Ropert N (1982) Firing rates and patterns of midbrain reticular neurons during steady and transitional states of the sleep-waking cycle. Exp Brain Res 47:37–51

Syndulko K, Hansch EC, Cohen SN, Pearce JW, Goldberg Z, Montan B, Tourtelotte WW, Potvin AR (1982) Long-latency event related potentials in normal aging and dementia. In: Courjon J, Mauguiere R, Revol M (eds) Clinical applications of evoked potentials in neurology. Raven, New York, pp 459–462

Vivaldi E, McCarley W, Hobson JA (1980) Evocation of desynchonized sleep signs by chemical microstimulation of the pontine brain stem. In: Hobson JA, Brazier MAB (eds) The reticular formation revisited. Raven, New York, pp 513–529

Wilder MB, Farley GR, Starr A (1981) Endogenous late positive component of the evoked potential in cats corresponding to P300 in humans. Science 211:605–607

Evolution of Compound Field Potentials in the Brain

T. H. BULLOCK

1 Introduction

Brain waves must have evolved! This seems a reasonable assertion given the prevailing view that the field potentials recorded in the brain are the compound vector sums of many cellular including neuronal generators, influenced to some degree by the differentiation of cell types and their histological organization. Relevant anatomical features that have changed markedly in evolution include ramification of dendrites, palisade arrangement of polarized cells, and lamination; physiological features that are possibly correlated, though not known to be so, include synchronization, field effects, and glial influence. Although it is not clear even in the best studied species whether the spatiotemporal pattern of the compound field potentials is merely an epiphenomenon, a sign of underlying transactions, or has some degree of causal significance, it merits comparison among representatives of the highly diverse taxa of animals, since the nervous system has changed so drastically, at least anatomically and perhaps in the behavior it mediates.

This chapter briefly summarizes the state of knowledge for ongoing background activity and evoked potentials of higher invertebrates and lower as well as higher vertebrates. To maximize comparability attention is confined to the so-called micro-EEG, recorded with semimicroelectrodes within the brain extracellularly or in contact with its pial surface, and to the activity of the highest levels of the CNS in the waking, alert but inactive animal, not intentionally stimulated except for the study of evoked potentials. Considering the well known diversity of EEG patterns in different parts of the forebrain of mammals and their sensitivity to state and ontogenetic stage, one might despair of comparing species. These considerations certainly limit our ability to see small differences, but the tentative conclusions recapitulated here, as far as they go, seem to be justified. A more extensive review has recently appeared (Bullock and Başar 1988).

2 Ongoing Electrical Activity in Invertebrate Cerebral Ganglia

In spite of the basic similarities among all higher invertebrates and vertebrates in the cellular potentials, these two great groups of species exhibit quite different spatiotemporal patterns of ongoing, compound field potentials. Instead of the

Springer Series in Brain Dynamics 2
Edited by Erol Başar and T. H. Bullock

smooth, slow waves, mostly below 30 Hz, typical of higher centers of vertebrates, semimicroelectrodes in the neuropile of the brain of annelids, insects, crustaceans, and gastropods see either low-voltage, spiky, very broad-band activity or activity dominated by one or two spiking units of high voltage; a small movement of the tip may change from one pattern to the other. It is much easier to record large spikes than in any vertebrate brain and the further comparisons made here are based on positioning the electrode so that it is not dominated by one or a few units. Broad band in this context means that the power spectrum has not substantially fallen, at or below frequencies of 100 Hz or more, in striking contrast to typical vertebrate telencephalic, diencephalic, or tectal recordings, which fall steeply above ca. 20 Hz.

Invertebrate cerebral activity varies widely in the amount of power at low frequencies (ca. 1–15 Hz); under quite special conditions, such as those sampled by Adrian (1931, 1937) and Adrian and Matthews (1934), bouts of such low frequency can resemble mammalian alpha waves – however, the conditions necessary are difficult to duplicate. A different set of conditions, which has been replicated a number of times (Jahn and Crescitelli 1938; Crescitelli and Jahn 1942), causes 20- to 40-Hz rhythms to appear for a few seconds in the optic ganglia of some insects. These unusual situations aside, slow rhythms are notably lacking in arthropods and gastropods. One further feature that may be characteristic of invertebrates, although so far measured only in the cerebral ganglia of the gastropod, *Aplysia,* is the lack of synchrony in a population or volume of tissue; coherence between two recording sites 1 mm apart is generally indistinguishable from that of two independent noise sources.

Cephalopods may be different from other invertebrates. Octopus has been reported (Bullock 1984) to resemble vertebrates more closely than it does other molluscs or arthropods in the character of its ongoing, background activity, when it is in an active state. The peak of its power spectrum is perhaps typically 20–25 Hz and spike activity is extremely weak. Most remarkable is its ability to switch back and forth, without behavioral sign, between this "active" state and an "inactive" one in which the neuropile of the vertical lobes is virtually silent electrically.

3 Ongoing, Background Activity in Higher Levels of Vertebrates

Ongoing, background activity is a prime example of an emergent property of the population, where even an extensive knowledge of cellular function could not predict or even explain post hoc the signs of organized assembly function. Seen as a biological phenomenon, two features are particularly notable. One is the relatively smooth, slow character and widespread coherence – which would not be expected on a simple cellular model. The other is the similarity of this spatiotemporal pattern from fish to mammal; no systematic differences have been noted even in widely separated classes, including elasmobranchs, teleosts, amphibians, reptiles, birds, and mammals, at least in gross wave form and spectral density distribution. Although within the same individual differences are conspicuous in various states as well as places, between taxa with major changes in histology the differences in electrical activity, if any, are smaller.

Apparent contradictions to this are the species differences among well studied mammals and even between individual humans. For example, some humans have a prominent alpha wave in a certain state, with eyes closed; some dogs also show this, but it is not characteristic of rats, cats, monkeys, other laboratory mammals, or vertebrates. Theta waves are prominent in some laboratory mammals under certain conditions, but they have not been found in nonmammalian species as yet. Slow wave sleep may be peculiar to mammals. Arousal in fish does not cause low-voltage-fast but high-voltage-fast waves. Nevertheless, in spite of such differences, so far it is not possible to recognize a sample record from an alert vertebrate as coming from a human or a carp.

It is likely that one general difference is the amplitude of the ongoing electrogram. The variability of measurements is great, for a long list of known as well as unknown reasons; therefore a general statement cannot be made from the limited number of measurements available. It appears probable, however, that the maximum root-mean-square (RMS) voltage of the micro-EEG averaged over 30 s from the place in the forebrain with the largest amplitude in a shark, ray, or carp is typically less than a quarter of that in a rat, rabbit, or cat.

The lowest values may serve as estimates of the expectation from the null hypothesis, namely that a large number of generators are quite independent and add their extracellular currents algebraically, as noncooperative, random, vector sums. This permits us to say that any amplitude above this minimum represents either some degree of cooperativity, synchrony, nonrandom orientation, or altered power spectrum of the cellular elements. The limitation on this otherwise convenient comparison is that we cannot assume the average cell is equally active in two states or species.

Even less certain is the generalization that coherence anywhere in the 1- to 40-Hz band, which normally declines with distance between recording electrodes, typically does so more slowly in mammals than in fish and is intermediate in reptiles. This can be seen as a measure of synchrony since it reflects the volume of tissue at or above some level of coherence. The limited evidence says that lower vertebrates exhibit less synchrony than higher vertebrates in the ongoing, background mode, but the problems of variance and comparability are severe. Even in mammals the evidence suggests that synchrony in the ongoing EEG is low, in respect to volume of tissue above, say, 0.5 coherence, compared with the possible synchrony during electrical seizures.

Looking for additional descriptors that might reveal differences in the micro-EEG, which as a complex time series could obscure important systematic differences, Bullock and Başar (1988) introduced the envelope of narrow-band-filtered components of the normal, wide-band record. The finding was that narrow bands of the EEG show waxing and waning such that rarely has any band a maintained amplitude for as long as a second. The amplitude modulation, computed as an envelope, gives a time series that can be examined for its power spectrum, for agreement with or departure from the expectation based on white noise instead of EEG, and for cross-correlations between bands. So far, no reliable differences between taxa have been found except that *Aplysia* has little or no cross-correlation of waxing and waning between bands, whereas the vertebrates examined have a modest degree of cross-correlation, with high variance.

4 Evoked Potentials in Invertebrates and Vertebrates

Invertebrates have hardly been studied with respect to evoked compound field potentials (EPs), whereas many authors have recorded single unit spike responses. In preliminary studies on *Aplysia,* recording from loci in the cerebral neuropile, where single unit spikes do not dominate the activity, Başar and I have observed that synchronized activation by brief shock to a connective can elicit a large evoked wave with an onset latency of 15–30 ms, a peak latency of 35–50 ms or more, and a duration of 50 ms or more. This means the power spectral components are essentially the same as those of, say, a flash-evoked primary sensory EP in the cat cortex. I have also recorded even longer lasting EPs in the stellate ganglion of *Octopus* to stellar nerve shock and in the optic ganglion to a light flash. Brief, compound spike-like EPs, however, are also seen in *Aplysia,* perhaps depending on the combination of recording locus and stimulated nerve; these may possibly be more typical for invertebrate cerebral ganglia. Event-related potentials (ERPs) with long latency (several hundred milliseconds), in situations which would be likely to involve a cognitive event in humans, have not yet been reported from invertebrates, and probably have not been looked for.

EPs have been reported in a number of studies of lower vertebrates. Confining attention to those from loci in the telencephalic pallium that give the largest responses, we have data, mainly from this laboratory, from elasmobranchs and teleosts after afferent nerve shock and physiological stimulation of receptors by acoustic tone bursts, flash, water wave, and electric fields. The general result is much like that in mammals – a succession of fairly slow waves with peak latencies in the tens of milliseconds. There are specific differences among modalities, loci, and species but no general differences in EP characteristics have yet emerged between lower vertebrates and mammals. ERPs elicited by stimuli designed to involve more endogenous or cognitive processes have not yet been reported from any nonmammal, and probably have not been looked for.

5 Discussion

The situation is not so simple as a dichotomy in which vertebrate higher centers have ongoing activity whose power spectrum reaches a peak below 20 Hz and falls at least 20 dB by 50 Hz, and the invertebrates a spectrum that extends strongly into the hundreds of Hertz, most energy being above 50 Hz. Lower levels of the vertebrate brain and cord show intermediate spectra; insects under special conditions can show strong low-frequency rhythms. Since the part of the brain, the physiological state, the ontogenetic stage, and unknown factors all contribute to variability, it is astonishing that even the tentative approximation of the stated dichotomy can be made. Such a gross difference so far exceeds those due to region of the cerebrum and state, that the dichotomy holds almost without regard to state not only for the forebrain but also for the tectum.

A biological look at the phenomenon of ongoing background activity brings out the descriptive fact that, with only special exceptions in unusual cases, this activity cannot be called rhythmic. It is usually characterized by a broad spectrum without consistent, sharp peaks; there is substantial energy at every frequency over many octaves and when the energy in a given frequency band is isolated by filtering, it is found to be amplitude modulated more deeply than the same band in white noise right up to the frequency of the filter. "Brain rhythms" is a common expression that may have utility for special cases such as the alpha and theta waves found in some mammals, but it is not a generally applicable synonym for the chaotic mixture of waxing and waning activity, probably largely the vector sum of discrete events in cellular units rather than maintained rhythms or resonances at each frequency.

The results of comparison of representative species of gastropods and arthropods and of fishes and mammals raise two general questions straight away: (1) Why are the records from telencephalon and tectum of lower vertebrates so similar to those of mammals, in spite of differences in lamination and organization – more similar than are different states of the same individual, or different places such as olfactory bulb and cerebellum? (2) Why are the records from most higher invertebrates so different from those of vertebrates, in spite of the similarities in cellular potentials? In particular, why do the invertebrates, apart from octopus, have such spiky activity and weak slow waves? Are there new or unfamiliar descriptors for ongoing activity that would contribute something toward clarifying these questions? We cannot answer these questions but can exclude some factors and permit others.

Excluding many possible explanations, the stated features of the species studied are evidently not due to electrode characteristics; they are observed with a wide range of electrode types (including pipettes and bare metal) and sizes at least up to 0.2 mm diameter. Brain size is not a good correlate: the typical vertebrate electrogram is seen in mammalian brains from mouse to dolphin, in small reptiles, amphibians, and fish, and even in an isolated fragment of the frog olfactory lobe; the typical invertebrate form of ongoing activity is seen from the tiny cerebral ganglion of the earthworm to the huge compound ganglion of large *Limulus*, lobster, and 5-kg *Aplysia*. Cell size is not a good correlate: the recordings are selected against sites near large cells in favor of sites in the finer textured neuropile and in any event, most of the cerebral ganglion cells in invertebrates are not huge like the much studied giant cells. A wide range of densities of neuronal elements occurs in each of the groups concerned, particularly in the neuropile of invertebrates, but also in the pallium of anamniotes.

Lamination is not a good correlate. The pallium of elasmobranchs, many teleosts, and amphibians is poorly laminated but grades through reptiles into the isocortex of mammals, without any obvious electrogram correlate in the form and frequency structure. There is a possible positive correlation of lamination in total power or electrogram voltage. This requires further evidence because of the wide variance among records and the number of variables that can influence observed voltage. What can be said is that in the virtual absence of lamination the typical vertebrate electrogram wave form and spectral density distribution is found. Essentially the same form and spectrum is seen in the optic tectum, which is usually much more laminated than the pallium but which also exhibits a range from almost no lamination in elasmobranchs to 14 sharply defined layers in some teleosts, reptiles, and birds. Invertebrates have conspicuous lamination in some optic ganglia but otherwise it is rare outside of cephalopods. Cells with a common orientation are quite general, however; the rind of unipolar cells at the surface of most ganglia has the neurite pole generally facing the neuropile. It remains to be learned whether those parts of the optic ganglia that have especially laminated neuropile in insects, crustaceans, and cephalopods show a different form of ongoing activity or more voltage than the less differentiated or oriented regions from which most of our recordings come.

The power spectrum of activity of single cells seems unlikely to be the cause of the vertebrate–invertebrate difference. Variance from sample to sample between cells and between states is so large that a definitive statement about the typical cell spectrum in each taxon cannot be made, let alone a prediction about the expected compound field potential which is the summed resultant of a natural population of cells. We can say that the spectrum is generally very broad, having substantial energy at all frequencies from DC up to at least several hundred Hertz. Spikes and synaptic potentials and the several other forms of cellular events are not narrow band phenomena. Basically the same variety of cellular events is known in each group of animals.

I conclude that the differences in form and spectrum of ongoing activity are primarily due to cooperative properties of the assemblage – or their absence, in the case of independence of the cellular generators.

Synchrony is one hypothesis for the relative weakness of slow waves in most invertebrates: they might have little synchrony among neurons. Synchrony in a population of neurons is not a simple parameter. The term synchronization has no generally agreed definition in neurophysiology. Bullock et al. (1984) have proposed to use synchrony, not primarily to mean coincidence of spikes or of peaks of spike rate, but to refer to a degree of covariance among a population of cells, including slow events. I emphasize the covariance within a constellation or volume of cells as distinct from agreement between two points which might be far apart but under some common influence.

Synchrony in this sense can be estimated by the rate of decline in coherence with distance between loci, measuring extracellular field potentials in given frequency bands. Semimicroelectrodes can be considered as sampling volumes of neural tissue with many generators (synapses, dendrites, somata, axons, terminals, glia). Although the volume is not accurately knowable or everywhere the same, it is empirically demonstrable that loci 100 μm apart *can* have very low coherence (Elul 1962). We underline that synchrony is not, in our view, adequately measured by coherence or correlation alone, but that the diameter of a population of generators above some level of coherence is a better estimator. Synchrony will depend on several factors. These include the degree of dependence or independence of the component frequencies, e.g., the prevalence of repeated compound wave forms, the fraction of the population in a given volume that maintains a fixed phase relation for stated frequencies with a certain steadiness of the phase lock among units, and the fraction of the variable total energy that occurs in each of the relevant frequencies (at least 1–20 Hz in our systems). We do not know whether each of the cellular generators produces the whole spectrum of frequencies. We do not know where the graded parameters that influence synchrony are situated. These are among the questions for future study.

The absence of any significant coherence in *Aplysia* between electrodes only 0.7–1.5 mm apart, over the whole frequency range 3–100 Hz (excluding loci with one or a few large spiking units), may explain in large measure the relative weakness of slow waves in ongoing activity. In the vertebrates, coherence at 1 mm is generally high, although widely varying among recordings; even at 2 mm it has usually not fallen below 0.6 in our experience. Elul (1962) reported "no similarity" between closely adjacent microelectrodes but I question how representative this is. This character of *Aplysia* activity together with its drastic change upon moving the electrode tip 0.1 mm creates the impression, difficult to prove, that it is the summation of a population of independent generators with more or less similar, wide-band spontaneous activity.

The vertebrate forebrain electrogram, in contrast, has a significant degree of synchrony over volumes several millimeters in diameter, in the low-frequency range (up to ca. 20 Hz). This does not mean that the synchrony is high relative to the possible synchrony; it is probably quite low compared, for example, to that exhibited during seizures.

6 Summary

Comparison of ongoing and evoked compound field potential fluctuations of higher brain centers (the micro-EEG of some authors) in higher invertebrates and lower and higher vertebrates, although still preliminary, leads to some surprising puzzles. Recordings were made with semimicroelectrodes in the neuropile of the cerebral ganglion of the molluscs *Aplysia, Helix,* and *Octopus* and the arthropods *Limulus, Melanoplus,* and *Cambarus,* and in or on the cerebral cortex or pallium and optic tectum of rays, sharks, bony fish, turtles, lizards, cats, and rabbits, in unstimulated resting or generalized arousal states. Manipulations of state did not alter the main findings.

Power spectra, looking only above 1 Hz, in the cerebral ganglia of the annelids, arthropods, and gastropods are similar; typically there is very little slow (<30 Hz) rhythmic wave activity; the record is fast and spiky. Integrated energy above 50 Hz exceeds that from 2–50 Hz and falls slowly with frequency; in *Aplysia* the power spectrum falls <10 dB between 10 and 300 Hz. Vertebrates from fish to mammals show remarkably similar activity, mainly slow (<30 Hz); it commonly falls >20 dB between 10 and 50 Hz. Cephalopods are exceptional among invertebrates; their ongoing activity is quite vertebrate-like, though the power maximum is about an octave higher and *Octopus* can switch to virtual electrical silence without visible behavioral sign.

Amplitude is low in invertebrates and lower vertebrates. RMS voltage over the whole spectrum is typically <10–$25\ \mu V$; in the reptiles sampled it seems to be typically higher, up to $30\ \mu V$; in the cat and rabbit cortex it is generally $>85\ \mu V$. In the vertebrates amplitude does not change greatly with small shifts in electrode position, as it does in invertebrates.

The slope of the decline of coherence with distance between surface electrodes is used as an estimator of synchrony. Averaged coherence between loci 1 mm apart is negligible in *Aplysia* in any band from 3 to 100 Hz; in the ray tectum it is low, 0.25–0.5 between 3 and 16 Hz. In the turtle dorsal pallium it is higher, 0.6–0.75 at 2 mm in this band. In the rabbit cortex coherence is even higher, typically >0.7 at 1 mm, and >0.3 at 4 mm in this band.

Narrow-band-pass filtered electrograms typically show constant waxing and waning in each band; amplitude is not maintained even for a second. Amplitude modulation envelopes have been used as another descriptor of the EEG; they have continuous spectra extending down to <1 Hz. Envelopes of the several bands often cross-covary positively, though weakly, in the vertebrates but not in *Aplysia*.

We interpret the ongoing compound field potential in *Aplysia* and most other invertebrates, and perhaps vertebrate spinal cord and lower brain stem gray matter, to approximate that of a population of independent, spontaneously active neurons. Vertebrate forebrain and tectum, in contrast, have a significant degree of synchronization of low-frequency components. Normally the degree is low for frequency components above 25 Hz and separations between loci above 3 mm; it falls with distance to insignificant levels at ca. 8 mm (apart from special cases such as alpha or theta waves or EPs).

Evoked potentials to simple sensory stimuli are quite similar in all vertebrates; no general differences between fish and mammals, for example, have been noted. They can also look similar in invertebrates in respect to latency and duration or spectral components, but it is possible that more commonly the invertebrates exhibit brief, short-latency EPs, like those common in the medulla. ERPs in situations that would be likely in humans to involve a cognitive event have not been noted, or seriously looked for, in invertebrates or nonmammalian vertebrates; my guess is that they will be found and that the difficult task will be to discern the qualitative or quantitative differences from ERPs of mammals – differences that must be there.

Detailed inspection of the evidence suggests that the differences described between vertebrates and most invertebrates are not importantly due to electrode characteristics, brain size, cell size, number, or density, lamination, or single cell power spectrum but are primarily due to assembly properties, that is, to cooperativity (see further, Chap. 2). I believe the ongoing compound field potentials are information-rich if we examine them, not only with many simultaneous electrodes, but in the appropriate scales, which are millimeters or less in the spatial dimension and fractions of a second in the temporal dimension. I believe that new descriptors will turn up additional significant differences among the vertebrates, especially between brains of different grade of organization, such as those of fish and mammals, giving scope for discovery of correlations with histological organization, cell differentiation, and above all, degrees and kinds of cooperativity.

References

Adrian ED (1931) Potential changes in the isolated nervous system of *Dytiscus marginalis*. J Physiol (Lond) 72:132–151

Adrian ED (1937) Synchronized reactions in the optic ganglion of *Dytiscus*. J Physiol (Lond) 91:66–89

Adrian ED, Matthews BHC (1934) The interpretation of potential waves in the cortex. J Physiol (Lond) 81:440–471

Bullock TH (1984) Ongoing compound field potentials from octopus brain are labile and vertebrate-like. Electroencephalogr Clin Neurophysiol 57:473–483

Bullock TH, Başar E (1988) Comparison of ongoing compound field potentials in the brains of invertebrates and vertebrates. Brain Res Rev 13:57–75

Bullock TH, Lange GD, McClune MC (1984) A measure of synchrony in the cortical EEG: the slow wave drowsy state is slightly more synchronized horizontally than the low voltage fast state. Soc Neurosci Abstr 10:1143

Crescitelli F, Jahn TL (1942) Oscillatory electrical activity from the insect compound eye. J Cell Comp Physiol 19:47–66

Elul R (1962) Dipoles of spontaneous activity in the cerebral cortex. Exp Neurol 6:285–299

Jahn TL, Crescitelli F (1938) The electrical response of the grasshopper eye under conditions of light and dark adaptation. J Cell Comp Physiol 12:39–552

Discontinuities in Visual Cortex and Possible Functional Implications: Relating Cortical Structure and Function with Multielectrode/Correlation Techniques

R. ECKHORN, R. BAUER, and H. J. REITBOECK

1 Introduction

In spite of its obvious partitioning in areas and layers, for a long time the cortex was assumed to be a rather homogeneous structure. Modern anatomical and physiological methods revealed, however, that the neocortex is a highly organized, structurally discontinuous system which cannot yet be described by a generally accepted model. In the following we want to contribute to this "model" discussion some of our recent results.

Discontinuities are found in the visual cortex on different levels of structural and temporal organization: *Structural discontinuities* exist on the lowest level as discrete patches within the layers, containing afferents and efferents, and intra- as well as extra-areal projection neurons of similar specificities. On the medium level there are functional columns, the hypercolumns, with bands and patches defined by the distinct sensory coding properties of the neurons. On the third level, the global structure shows an organization of distinct, specialized maps, the visual areas.

Temporal discontinuities of cortical information processing can be seen more readily at the level of group activities than at the single cell level: Signal correlations, as they were found in various combinations of the signals of single cells, multiple units, and local slow waves, showed abrupt transitions from stochastic uncoupled states (maintained activity) to stimulus-driven states with highly ordered activities, including states of coherent oscillations.

The question is why the CNS "renounces" the continuous spatial and temporal representations of sensory information at several levels of functional organization in favor of discontinuous ones.

Evolution may provide a clue: The cortex has to perform its tasks optimally at minimal costs. This means, with a limited number of neurons, connections, and levels of organization, a percept of the "important" ranges of the stimulus variables and temporal periods has to be constructed (Sakitt and Barlow 1982). Theoretical considerations suggest that these limitations and constraints can be overcome by discrete representations. Support for the discontinuous representation of continuous variables with minimal redundancy and a minimum number of neurons is provided by systems and information theories. Continuous band limited signals can be represented completely by a limited number of discrete amplitude samples (sampling theorem). This rule might apply to the spatial stimulus aspects represented in the visual cortex (e.g., position and orientation of a line segment). In addition, different sensory dimensions could be mapped in an interdigi-

Springer Series in Brain Dynamics 2
Edited by Erol Başar and T. H. Bullock

tated (structurally discontinuous) way and, thus, the required number of neurons and connections would be greatly reduced. Such a discrete sampling representation, from which intermediate values can be reconstructed by neural interpolation (Ballard 1987), also seems to play a role in the temporal domain: Stimulus-related temporal structuring, like rhythmic phase-locked activations of groups of neurons with similar stimulus specificities, was frequently found in our recordings. We suggest that transient signal correlations provide the basis of an effective code that can define global relations by linking related figural aspects.

Information reduction from millions of bits per second in the optic tract (Eckhorn and Pöpel 1974, 1975; Eckhorn et al. 1976) to the low bit rate at the level of conscious perception seems to be one of the main tasks of the visual system. Highly structured discrete solutions for pattern representation, therefore, are likely to be favored by the visual cortex. Transient spatial and temporal structuring within the representation of a visual scene, combined with integrations by a global code, might achieve this compression of redundant information. We started to analyze some of the structural and functional discontinuities of the visual cortex that might be involved in coding aspects of this information reduction.

2 Experimental Results of Our Group

2.1 Methods

Experiments were performed in areas 17 and 18 (A17 and A18) of the visual cortex of the semichronically prepared, lightly anesthetized cat (Lohmann et al. 1986; Leferink et al. 1987; Eckhorn and Reitboeck 1987; Eckhorn et al. 1987; Bauer et al. 1987). Multiple electrode techniques (Reitboeck 1983) were combined with newly developed signal correlation methods (Schneider et al. 1983; Eckhorn et al. 1986a). With an array of 7–19 individually movable fiber microelectrodes [PtWo-core quartz insulated; 70 μm shaft; 3–7 MΩ; Reitboeck (1983)], single- and multiple-unit (MUA) activities as well as local slow wave potentials were recorded, often simultaneously by the same electrode. The electrode spacings in our manipulators were either 250 or 330 μm in the plane tangential to the electrode tracks, although of course the electrode tips were not always at the same depth. This electrode separation is well sited for sampling the discrete cortical patch structures that generally extend over 250–1000 μm.

2.2 Experimental Results: Structural Aspects

2.2.1 Local Clustering

We consistently found that most sensory coding properties of visual cortical neurons (A17 and A18) remain almost unchanged over several 100 μm of electrode advancement and then change abruptly. This functional clustering was observed both in perpendicular and in tangential electrode tracks; similar observations have been mentioned by other investigators (e.g., Payne et al. 1981; Weyand et al. 1987). Correspondence between the clustered appearance of single-unit coding properties and the anatomically demonstrated patches/blobs, however, has only been shown in a few examples up to now.

2.2.2 Different Termination Sites of ON and OFF Afferents in Area 18

It is well known that the ON and OFF pathways in the retina are functionally and spatially separate. We asked whether these signal streams terminate separately in the visual cortex or whether they are already combined in the geniculate nucleus. The local spread of signals induced by the ON and OFF streams in A18 (Eckhorn et al. 1986a) and their influence on cortical neurons was studied by measuring the laminar distribution of the current source densities (CSDs; e.g., Mitzdorf 1987) under "fast-bright" and "fast-dark" stimulation (Fig. 1). We found separate functional stratifications of the Y-ON and Y-OFF axon terminals in the upper and lower parts of layer IV, respectively (Eckhorn et al. 1986a).

2.2.3 The Vertical Cortex Column: Iso- and Cross-orientation Columns

The constancy of the orientation coding within a vertical column is a central element in the columnar model of visual cortex (Hubel and Wiesel 1962). In contrast to this concept our own measurements have revealed frequent orientation shifts

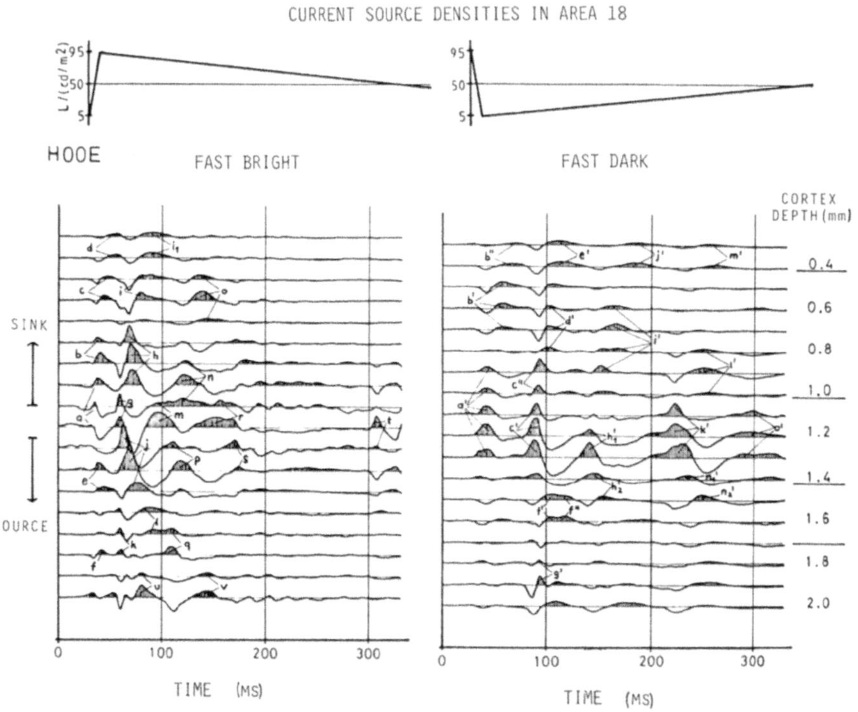

Fig. 1. CSD depth profiles in cat visual cortex (A18), obtained during specific ON and OFF stimulation. (from Eckhorn, Lohmann and Reitboeck 1986a)

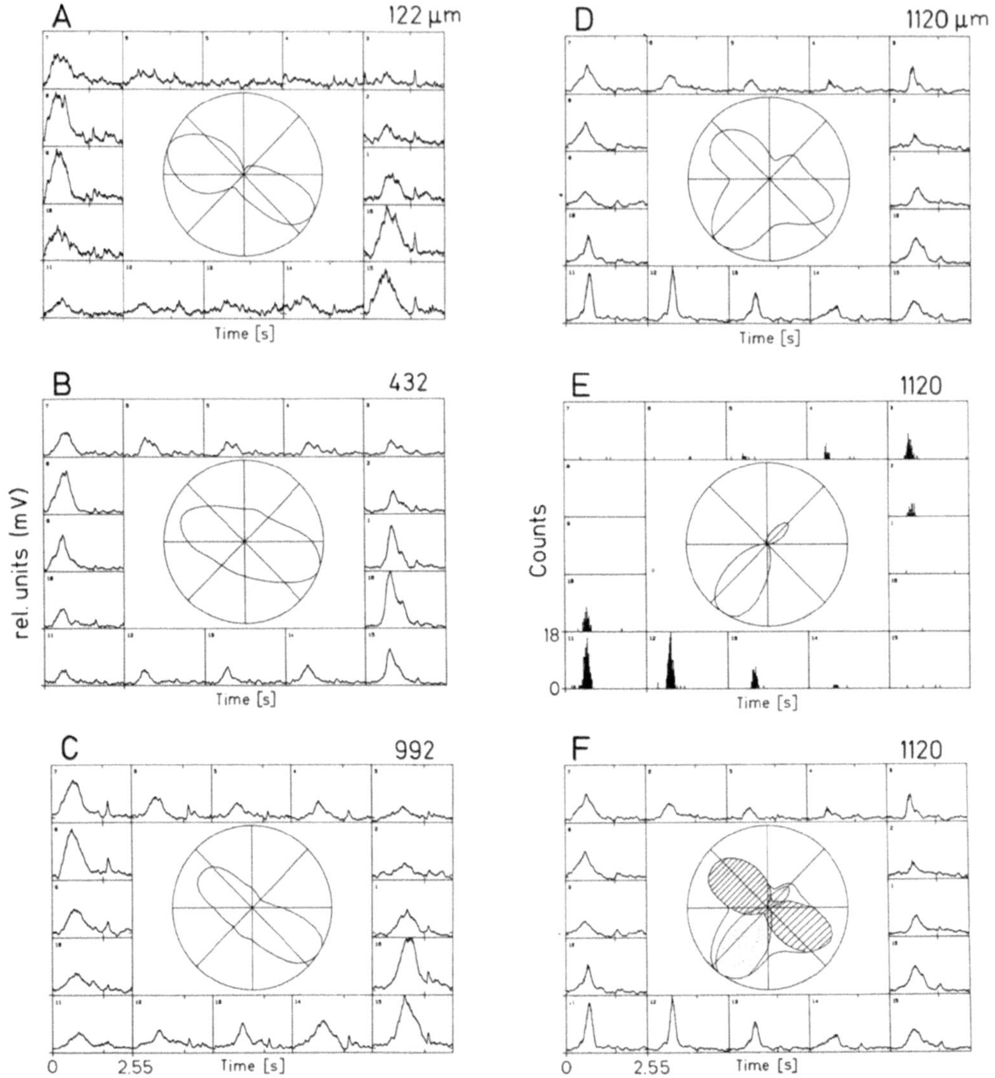

Fig. 2A–F. Cross orientation column in cat primary visual cortex: Orientation/direction tuning curves measured in different cortical depths with multiple-unit (A, B, C, D, F) or single-unit activities (E). *Frames:* PSTHs; *centers:* tuning curves

of 70°–90° between supra- and infragranular layers, both in monkey (Bauer et al. 1983) and in cat (Bauer 1982; Bauer and Fischer 1987). These results led to scientific controversies (e.g., Murphy and Sillito 1986) which may be settled now: With sequences of recordings of MUA and single-unit spike trains at 40- to 60-µm spacings we did not miss any larger neuronal sublayer. In about two-thirds of these perpendicular penetrations, however, a shift approximately to the orthogonal stimulus orientation was found (Fig. 2). It occurred suddenly at the border between layers IV/V or in lower layers V and VI within 50 µm of electrode advancement (Bauer et al. 1987). It is concluded that in lower layers discrete clusters of cells exist which have orthogonal orientation tuning to the correspondent cells in upper layers and that, therefore, two orientation systems coexist: iso- and cross-orientation columns.

The frequency and the distributions of the observed iso- and cross-orientation columns suggests that they are arranged in a systematic way. This arrangement may be related to the different radial and concentric global maps that have been discovered in the visual system. Several such maps, whose trajectories of increased orientation and movement sensitivity run orthogonal to each other, seem to be established in primary visual cortex (Bauer et al. 1987).

2.2.4 Global Discontinuities: Multiple Visual Cortex Areas

The visual cortical areas are linked by a system of reciprocal connections which are arranged in a "patchy" way (e.g., Symonds and Rosenquist 1984; Bullier et al. 1984). We were interested in basic mechanisms of cooperativity between different cortical areas mediated via these discrete structures using A17/18 couplings as an example. We recorded with a seven-channel array from neurons in A17 of the cat and with a separate electrode from A18. The latter semi-microelectrode was also used for intracortical electrical stimulation (Leferink et al. 1987).

In recording positions where the RFs from both areas overlapped, we found about 0.4 mm wide regions in the supragranular layers of A17 where neurons could electrically be excited with short latencies (Fig. 3). Neurons in intervening and surrounding positions often were inhibited for 50–150 ms after the stimulus. After HRP injection at the stimulation site in A18, patches of labeled cells occurred exactly at those positions in the A17 supragranular layers where cells could

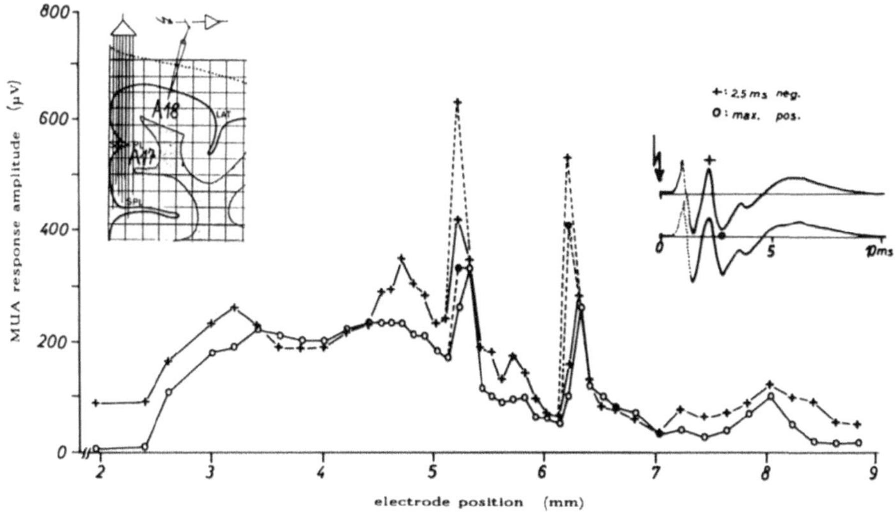

Fig. 3. Visual area 18-to-17 excitation. Electrical stimulation of the upper layers in A18 with a semi-microelectrode revealed a spatial MUA response profile in A17 with two peaks. At the peak positions single cells could be driven from A18 with minimal current. *Inserts: left,* stimulation and recording arrangement; *right,* single sweep MUA responses. (from Leferink et al. 1987)

be excited transsynaptically from A18. If these preliminary data are confirmed, they would indicate that the reciprocally projecting systems between A17 and A18 lie in register down to the level of discrete patch organization.

Cross-correlation analysis of maintained activities gave an additional clue regarding structure/function discontinuities of corticocortical connections: several "patches of enhanced correlations" were found within that range of A17 where the RFs of A17 overlapped with the RF of a single A18 recording. In our small set of data, we have so far found mainly synergistic correlation components of the medium and broad type, i.e., correlated waves of medium (20–60 Hz) and low (2–10 Hz) frequencies were in phase in both areas. Both components had maximal correlation amplitudes at those recording sites where their RFs overlapped (Leferink et al. 1987).

2.3 Experimental Results: Temporal Aspects

2.3.1 Correlated Activities as a Potential Code

Generally, near-synchrony of firing can be detected by single neurons. We tested whether the neural response index can be generalized from the magnitude of a neural signal to the correlation strength between two neural signals: Would such a correlation measure result in appropriate tuning curves? Our own recent observations (Fig. 4) and those of others (Bloom and Gerstein 1984; Gerstein et al.

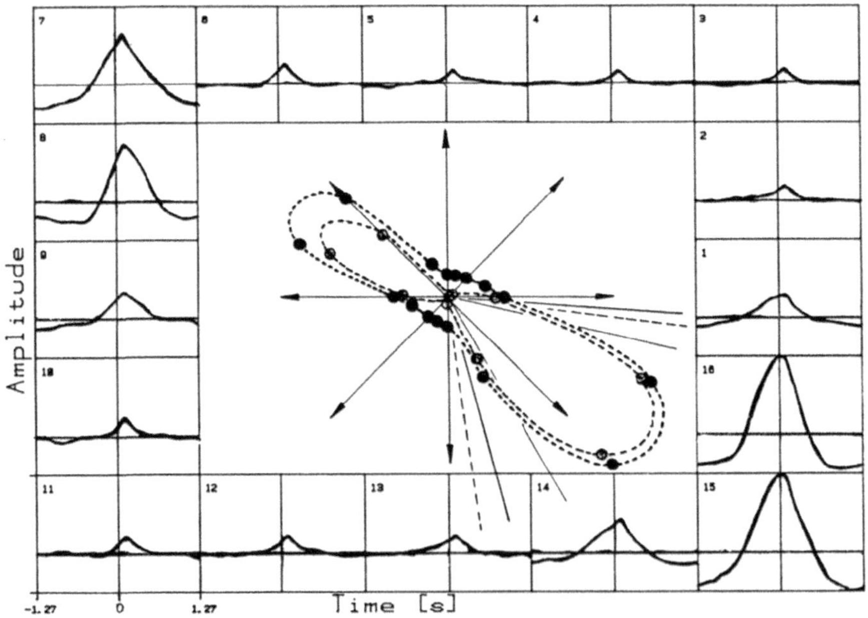

Fig. 4. Correlation as a tuned response property. *Frame:* Cross-correlograms obtained with MUA recordings from the upper layers of the same column in cat primary visual cortex stimulated with a moving light bar in 16 different directions. *Center:* Directional tuning derived from the correlation amplitudes. Tuning width is about half those obtained in the conventional way

1986) show that correlation of firing between two neurons or between two separate groups of neurons can itself be a tuned quantity in circumstances where such tuning is much weaker in the single cell signals.

2.3.2 Correlations Within a Vertical Cortex Column

The cross-correlations between signals of supra- and infragranular layers analyzed by us up to now, have revealed synergistic medium and broad range activations in both iso- and cross-orientation columns. The functional significance of this common type of synergistic activation may be as follows:

Medium range correlations have correlation peak widths in the range of about 10–50 ms (see Fig. 6) and they might be due to stimulus-specific resonance phenomena in local neuron assemblies. For this assumption we had some evidence from evoked local slow wave field potentials (LFPs) which were recorded by the same electrode, together with the action potentials. The LFPs showed oscillations

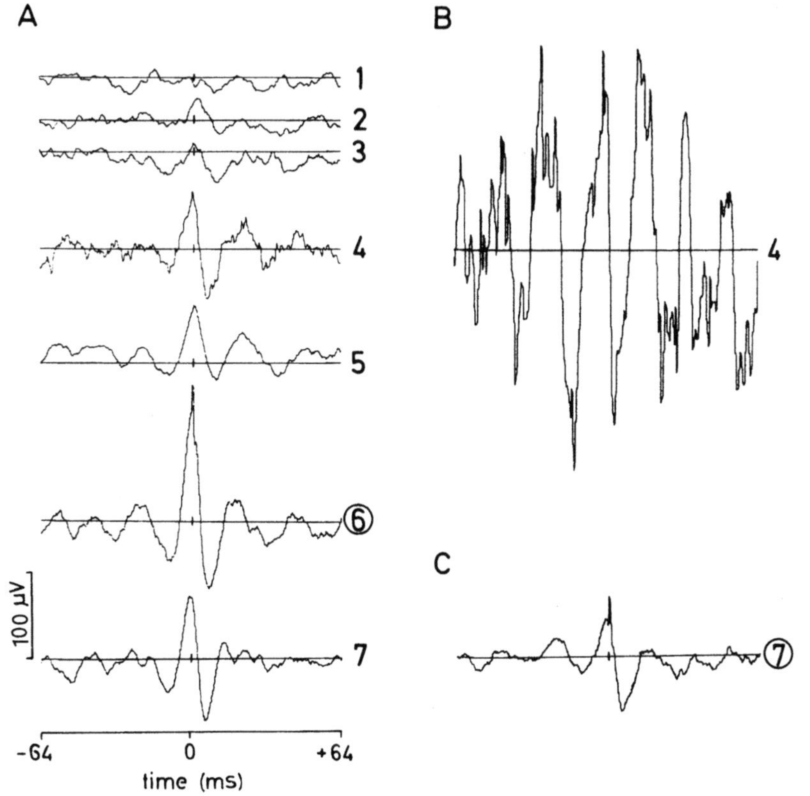

Fig. 5 A–C. Stimulus specific intracortical evoked slow wave potentials (LFPs) in cat area 17 show *resonances* at 40–60 Hz (here during stimulus movement). (A) Spike-triggered LFP averages (STA); recordings from layers II/III with seven-electrode linear array; electrode spacing 250 µm, trigger spike on electrode 6. (B) Single sweep LFP from electrode 4. (C) STA from the same recording position as A; trigger spikes and LFPs from electrode 7. All STAs resemble the average of 100 LFP epochs

at about 40–60 Hz in response to those specific visual stimuli required to evoke spikes at this same position (Fig. 5). We recorded them in all layers simultaneously: there was no phase shift between the LFPs of different layers in this "high-frequency" range. The spikes of some cells, recorded simultaneously with the same electrode, were tightly phase-locked to the oscillating slow wave potentials (Fig. 5; see also Gray and Singer 1987). These oscillations might serve to establish transiently the connections within and between the local columnar subsystems, i.e., the oscillatory modulations might be required for the generation of more general coding properties beyond the classic receptive field by linking related aspects of a visual scene and signaling these relations to the coactivated columns of the same and other cortical and subcortical structures.

Broad range correlations have correlation peak widths in the range of about 50–500 ms (see Fig. 6) and might code even more global aspects of sensory representation compared to the medium-range couplings. We found broad correlations – at least during visual stimulation – to be related to the amplitude envelopes of the medium range oscillating responses. The time course of the oscillatory potentials' envelope was often found to be similar to the PST histograms of spike trains recorded simultaneously at the same location.

2.3.3 Stimulus-Specific Variations of the Effective Neural Connections

We were interested in the temporal continuity of cortical connections with and without stimulation and in their possible stimulus-induced or intrinsic variations. For this we tested whether correlations between cells or cell groups can rapidly change in type and relative strength. Such neural cooperative behavior can be studied when correlations due to a common stimulus are "artificially" separated from correlations due to internal synaptic interactions. The separation can be achieved by subtracting the appropriate expectation values which contain only the stimulus-induced correlations ("shift predictors"; Gerstein and Perkel 1972; Schneider et al. 1983).

We found that visual activation often changed the internally generated correlations specifically within milliseconds (Eckhorn et al. 1987). For example, during the preferred stimulus direction the cross-correlogram indicated symmetric coactivation, while in the opposite direction antisymmetric activation/inhibition was present. Such transient context-dependent changes in effective coupling appear to be a general property of neural assemblies (see also Gerstein et al. 1986; Epping and Eggermont 1987).

2.3.4 Distinct Temporal Correlation Ranges

We obtained some additional clues supporting the existence of temporal discontinuities. In cat primary visual cortex, we were able to separate distinct correlation ranges:

Precise couplings (nearly synchronous spikes) were found in single-unit signals recorded from neighboring cells. Correlations of *medium precision* (correlation peak width: 10–20 ms) were often apparent at short and medium recording distances (1) between supra- and infragranular layers of the same cortex column (2)

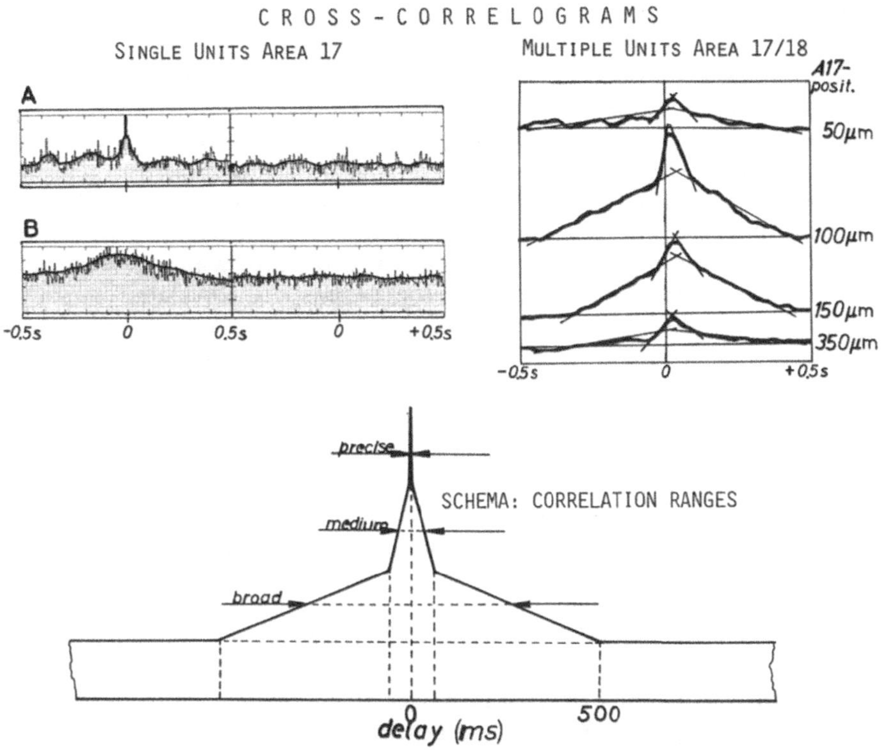

Fig. 6. Schema of superimposed temporal correlation ranges found in cat visual cortex: "precise", "medium," and "broad." Examples of cross-correlograms of single-unit (*A, B; upper left*) and multiple-unit activities (*upper right*; from Eckorn et al. 1987)

between neighboring columns, and (3) between visuotopically equivalent positions of different areas. Broad correlations (50–500 ms) were found over a wide range of electrode distances, including short, medium, and far away recording positions.

These results from correlation measurements can also be interpreted to be the result of coherent activations of functionally related assemblies occurring in different frequency bands. Our recent measurements of local intracortical slow wave field potentials and those of other groups which found stimulus-related oscillating slow waves point in this direction (Başar 1983; Freeman and Skarda 1985; Gray and Singer 1987; Başar et al., this volume).

3 Hypotheses for Cortical Function: Spatial and Temporal Aspects – A Hierarchical Model of Neural Correlators with Feedback Loops

Starting from the observation of coherent activations in visual cortex which were seen in three distinct temporal ranges of correlations (Fig. 6), we suggest that these *interaction ranges* might be related to distinct levels of structural and coding organization within a hierarchical cortical network (Eckhorn et al. 1987; Reitboeck et al. 1987). Structurally, our model is organized in several layers of discrete *correlation units* of increasing complexity. The elements of the lowest layer are single neurons; those of the second layer are local neuron assemblies, and the third layer consists of a system of connected distributed assemblies (further stages might be added). Within this hierarchical organization, each structural level has its specific spatial and temporal interaction ranges. The correlators with high spatial and temporal precision feed their signals to the medium range correlation units and these in turn signal to the next layer having the broadest interaction ranges. The higher level structures have excitatory feedback loops to the lower levels so that more global correlations (of higher levels) can support the lower level correlation systems, e.g., by sensitizing them rhythmically. Such reciprocal connections and their stimulus-specific rhythmical activations are also a feature of the concept of "reentrant signaling" (Edelman and Finkel 1984) and the "model of coherent computation" of Grossberg (1983). Horizontal in-layer connections might be added at each stage: sensitizing connections can support the positive feedback from higher layers for "linking together" network elements that are activated by similar stimulus attributes. The following functional implications might be attributed to the different levels of the correlator network model: narrow-range units are activated by precisely correlated signals generated by local details of a visual pattern; medium-range units establish relations between local groups in order to combine or separate related figure elements; broad-range units can exert global region definitions and possibly sensory and attentive cognitive integrations. Such a cooperative mechanism of a neural network, which can transiently link together neuron groups via correlated activities, is supposed to be a powerful mechanism for information compression (Haken 1987), and it supports region definition and figure/ground separation (see also Reitboeck et al. 1987; von der Malsburg 1983).

4 Summary

The purpose of this investigation was to search for coding and data compression strategies that enable the visual system to reduce the enormous amount of incoming information into a structured representation that can be perceived at the limited capacity of conscious visual perception. We started to investigate cortical coding at several levels of structural and temporal organization in the primary visual cortex (areas 17 and 18) of the lightly anesthetized cat. A multielectrode arrangement of 17–19 individually advancable fiber microelectrodes (250 or 330 μm apart) was used to record simultaneously both single and multiple spikes as well as slow waves. A variety of geometric recording positions could thus be chosen. Neural interactions were analyzed with signal correlation techniques.

The discontinuous neural representation of visual information, both in the structural and temporal domains, was a prominent finding of this investigation. *Local cell clusters* were often found to have nearly constant coding properties over regions extending up to 200 μm, changing abruptly at the border to neighboring clusters. These neural assemblies are postulated to represent discrete "sampling nodes" of the visual scene, with their members participating in a common computational task. The assemblies are transiently activated by appropriate visual stimulation, and they change from a stochastic state of maintained uncoupled activity to a more coherent state of *correlated activities* (e.g., stimulus-evoked coherent oscillations).

We assume that the coherence of the neural activities transiently links together assemblies that represent common features of the actual visual scene (e.g., attributes belonging to a moving object). Visually related activities are, thus, transiently labelled by a temporal code that signalizes their momentary association.

Acknowledgments. Experiments and data evaluations were supported by our colleagues M. Arndt, Dr. W. Epping, Dr. C. Habbel, W. Jordan, Dr. F. Krause, J. Leferink, Dr. H. Lohmann, and Prof. H.-J. Wagner. This project had financial support from the Deutsche Forschungsgemeinschaft Ba 636/4-1, 2 and Re 547/2-1 and from Stiftung Volkswagenwerk VW I/35695.

References

Ballard DH (1987) Interpolation coding: a representation for numbers in neural models. Biol Cybern 57:389–402

Başar E (1983) Synergetics of neuronal populations. A survey on experiments. In: Başar E, Flohr H, Haken H, Mandell A (eds) Synergetics of the brain. Springer, Berlin, Heidelberg New York, pp 183–200

Bauer R (1982) A high probability of an orientation shift between layers 4 and 5 in central parts of the cat striate cortex. Exp Brain Res 48:245–255

Bauer R, Fischer WH (1987) Continuity or incontinuity of orientation columns in visual cortex: a critical evaluation of published and unpublished data. Neuroscience 22:841–847

Bauer R, Dow BM, Snyder AZ, Vautin RG (1983) Orientation shift between upper and lower layers in monkey visual cortex. Exp Brain Res 50:133–145

Bauer R, Eckhorn R, Jordan W, Arndt M (1987) Iso- and cross-orientation columns in cat striate cortex: a re-examination with simultaneous single- and multi-unit recordings. Proc European Brain and Behavior Society Conf, Novi Sad, T10

Bloom G, Gerstein G (1984) Soc Neurosci Abstr 10:245

Bullier J, Kennedy H, Salinger W (1984) Branching and laminar origin of projections between visual cortical areas in the cat. J Comp Neurol 228:329–341

Eckhorn R, Pöpel B (1974) Rigorous and extended application of information theory to the afferent visual system of the cat. I. Basic concepts. Biol Cybern 16:191–200

Eckhorn R, Pöpel B (1975) Rigorous and extended application of information theory to the afferent visual system of the cat. II. Experimental results. Biol Cybern 17:7–17

Eckhorn R, Reitboeck HJ (1987) Assessment of cooperative firing in groups of neurons. In: Başar E (ed) Springer series in brain dynamics, vol 1. Springer, Berlin, Heidelberg New York, pp 219–227

Eckhorn R, Grüsser O-J, Kröller J, Pellnitz K, Pöpel B (1976) Efficiency of different neuronal codes: information transfer calculations for three different neuronal systems. Biol Cybern 22:49–60

Eckhorn R, Lohmann H, Reitboeck HJ (1986a) Laminar spread of ON- and OFF-signal streams in cat area 18 investigated with a combined single unit and field potential analysis with Ganzfeld stimuli. Proc European Neurosci Assoc Satellite Vision Conf, Lyon, p 26

Eckhorn R, Schneider J, Keidel R (1986b) Real-time covariance computer for cell assemblies is based on neuronal principles. J Neurosci Methods 18:371–383

Eckhorn R, Bauer R, Jordan W, Epping W, Arndt M (1987) Correlated activity and coding properties of neuron groups within vertical columns of the visual cortex: a multi-electrode study. 2nd IBRO-Congress. Neuroscience 22 [Suppl]:1306P
Edelman GM, Finkel LH (1984) Neuronal group selection in the cerebral cortex. In: Edelman GM, Gall WE, Cowan WM (eds) Dynamic aspects of neocortical function. Wiley, New York, pp 653–695
Epping WJM, Eggermont JJ (1987) Coherent neural activity in the auditory midbrain of the grassfrog. J Neurophysiol 57:1464–1483
Freeman W, Skarda CA (1985) Spatial EEG patterns, non-linear dynamics and perception: the neo-Sherringtonian view. Brain Res Rev 10:147–175
Gerstein GL, Perkel DH (1972) Mutual temporal relationships among neuronal spike trains. Statistical techniques for display and analysis. Biophys J 12:453–473
Gerstein GL, Aertsen A, Bloom M, Espinosa I, Evanczuk S, Turner M (1986) Multi-neuron experiments: observation of state in neural nets. Proceedings symposium synergetics, Springer, Berlin Heidelberg New York
Gray CM, Singer W (1987) Stimulus-dependent neuronal oscillations in the cat visual cortex area 17. 2nd IBRO-Congress Neuroscience [Suppl]:1301P
Grossberg S (1983) The quantized geometry of visual space: the coherent computation of depth, form, and lightness. Behav Brain Sci 6:625–692
Haken H (1987) Information compression in biological systems. Biol Cybern 56:11–17
Hubel DH, Wiesel TN (1962) Receptive fields, binocular interaction, and functional architecture in the cat's visual cortex. J Physiol 160:106–154
Leferink J, Eckhorn R, Nelson JI, Wagner H-J (1987) Characterization of patchy projection systems between cortical areas: a combined multi-electrode and HRP study in cat's visual cortex. 2nd IBRO-Congress, Neuroscience 22 [Suppl]:1305P
Lohmann H, Eckhorn R, Reitboeck HJ (1986) Receptive fields evaluated from visual evoked intracortical potentials. Proc ENA Satellite Vision Conf, Lyon, p 40
Mitzdorf U (1987) Properties of the evoked potential generators: current source-density analysis of visually evoked potentials in the cat cortex. Intern J Neurosci 33:33–59
Murphy PC, Sillito AM (1986) Continuity of orientation columns between superficial and deep laminae of the cat primary visual cortex. J Physiol (Lond) 381:95–110
Payne BR, Berman N, Murphy EH (1981) Organization of direction preferences in cat visual cortex. Brain Res 211:445–450
Reitboeck HJ (1983) A 19-channel matrix drive with individually controllable fiber microelectrodes for neurophysiological applications. IEEE SMC 13:676–682
Reitboeck HJ, Eckhorn R, Pabst M (1987) A model of figure/ground separation based on correlated neural activity in the visual system. In: Haken H (ed) Springer, Berlin Heidelberg New York
Sakitt B, Barlow HB (1982) A model for the economical encoding of the visual image in cerebral cortex. Biol Cybern 43:97–108
Schneider J, Eckhorn R, Reitboeck HJ (1983) Evaluation of neuronal coupling dynamics. Biol Cybern 46:129–134
Symonds LL, Rosenquist AC (1984) Laminar origins of visual cortical connections in the cat. J Comp Neurol 229:39–47
Von der Malsburg C (1983) In: Başar E, Flohr H, Haken H, Mandell AJ (eds) Synergetics of the brain. Springer, Berlin Heidelberg New York
Weyand TG, Malpeli JG, Lee C, Schwark HD (1987) Cat area 17: IV. Two types of corticotectal cells defined by controlling geniculate inputs. J Neurophysiol 56:1102–1108

Subcortical Evoked Potential Correlates of Early Information Processing: Mismatch Negativity in Cats

V. Csépe, G. Karmos, and M. Molnár

1 Introduction

A negative component of event-related brain potentials which is specific to stimulus change (Näätänen and Gaillard 1983; Näätänen 1986), the "mismatch negativity" (MMN), seems to reflect an automatic comparing process. This comparison is thought to be between the neural representation of standard stimuli – the "neuronal model of the stimulus" (Sokolov 1960) – and the neuronal pattern elicited by deviant stimuli. Although electrical scalp recordings (Simson et al. 1977) and magnetoencephalographic data (Elberling et al. 1980, 1982; Hari et al. 1982, 1984) suggest that the generators underlying the MMN involve several mechanisms in the auditory cortex, they give no information about other possible mechanisms on which the MMN may depend.

Our previous experiments (Csépe et al. 1987) show that MMN exists in cats both in wakefulness and in slow wave sleep (SWS). The MMN in cats appeared in the middle latency negative waves, the N50 and N60 components of the auditory cortical evoked potentials (EPs), and in the N30 wave of the vertex responses as a second negative peak. Our data suggested that a genuine comparing process is active during the whole sleep–waking cycle and that MMN can be regarded as independent not only of conscious perception but to some extent also of vigilance.

On the basis of vertex and association cortical EP changes it could not be excluded that far-field components originating from some subcortical structures appeared in cortical recordings. Therefore the objective of the present study was to follow some intracerebral EP changes to determine whether consistent EP changes appeared that could be correlated with MMN changes recorded from the auditory and association cortices and from the vertex. In particular, we wanted to follow the EP changes recorded in the medial geniculate body (MGB) and in the dorsal hippocampus, regarded as a key structure in the "working memory" (Marr 1971; Teyler and DiScenna 1986). Our hypothesis to be tested using these subcortical recordings was that a multistage comparison process with participation of neuronal models in different structures of the brain may exist.

2 Methods

Five adult cats weighing 3.0–3.9 kg were used as subjects. The recording electrodes were implanted under Nembutal anesthesia (35 mg/kg body weight), while

Springer Series in Brain Dynamics 2
Edited by Erol Başar and T. H. Bullock

the animals were placed in a stereotaxic head-holder. The cortical electrodes, made from 0.23 mm diameter stainless steel enamel-coated wire, were implanted epidurally above the AI (middle ectosylvian gyrus) and AII (ventral region of the ectosylvian gyrus) areas of the auditory cortex, and the association cortex (middle suprasylvian gyrus). The electrodes used for vertex recording were placed 5 mm posterior to the bregma in the midline.

The subcortical recording electrodes, made from 0.15 mm diameter stainless steel enamel-coated wires, were implanted into the MGB (stereotaxic coordinates: A5, L10, V0.5) and into the dorsal hippocampus (A4.5, L4.5, V8) according to the atlas of Snider and Niemer (1961). Stainless steel electrodes were implanted bilaterally into the peribulbar tissue and into the neck muscles to record the electro-oculogram (EOG) and muscle activity (EMG) respectively. A stainless steel screw of 1.5 mm diameter, placed 10 mm anterior to the bregma in the midline above the bone covering the frontal sinus, served as reference electrode. All the electrodes were connected to a small Winchester connector mounted on the skull. The electrodes and the connector were fixed to the skull by dental acrylic in which the place for the miniature bone conductor was formed as well. This device, used for constant acoustic stimulation, was fixed on the animal's forehead during the experiments (Karmos et al. 1970).

The experiments were carried out in a sound-attenuated chamber. The auditory stimuli were short tones generated by sine wave bursts of 1 ms duration (rise–fall time, 0.5 ms) with an equivalent intensity of 80 dB SPL. Standard tones (STs) of 4 kHz and deviant tones (DTs) of 3.6 kHz pitch frequency were presented at 2 Hz repetition rate in blocks of 256 stimuli. The probability of DTs in a block was 33%, 20%, 10%, and 5%. The different blocks were given in random order during the experiments.

The brain electrical activity was amplified with bandpass setting 0.16–1000 Hz. The EEG activity recorded from the auditory, motor, visual, and association cortical areas, the vertex, the hippocampus, and the MGB, the EMG, and the EOG were monitored on paper and simultaneously on FM tape for off-line analysis. The off-line processing of the EPs was performed by a TPA 1140 computer (sampling rate 1 kHz). No averaging across animals was carried out; all of the figures represent EPs recorded in individual experiments. Evoked responses elicited by STs (SEPs) and those elicited by DTs (DEPs) were separately averaged. The averaging was made from 32 potentials. For the statistical comparison between the amplitudes of the various components of DEPs and SEPs a two-sampled t-test was used. The equivalent latency points of the SEPs and DEPs were compared by the serial t-test performed by the computer.

The cortical and subcortical MMN changes were studied in the quiet waking state and in SWS. The level of vigilance was judged on the basis of EEG, EMG, and EOG activities. Quiet wakefulness was characterized by a low-voltage, high-frequency cortical EEG, sustained EMG activity, and frequent eye movements in the EOG. In the quiet state the animal was sitting in the experimental cage without any apparent signs of paying attention to the environment. SWS was characterized by a high-voltage, low-frequency EEG, sustained tonic EMG activity, and little or no EOG activity. During SWS the animals were lying in a curled up position in the experimental cage. There were no signs of a wakening effect of the low probability DTs on the animal's behavior or on the EEG spectra.

The characteristic components of the EPs analyzed in this study corresponded to those described earlier by Karmos (1982) and Molnár et al. (1986). Therefore in the present paper, the EP components appearing during wakefulness and sleep are not described in detail.

3 Results

3.1 Surface Recordings

The EPs recorded in the quiet waking state from the AII area of the auditory cortex (A.c.), from the vertex, and from the association cortex (As.c.) are shown in Fig. 1 (For comparison the MGB and hippocampal EPs can also be seen on this figure.) The SEPs and DEPs elicited by 5% probability DTs were separately averaged.

The amplitude change of the characteristic middle latency negative wave, the N45, could be seen on the A.c. DEPs. A second negative peak, the N60, appeared

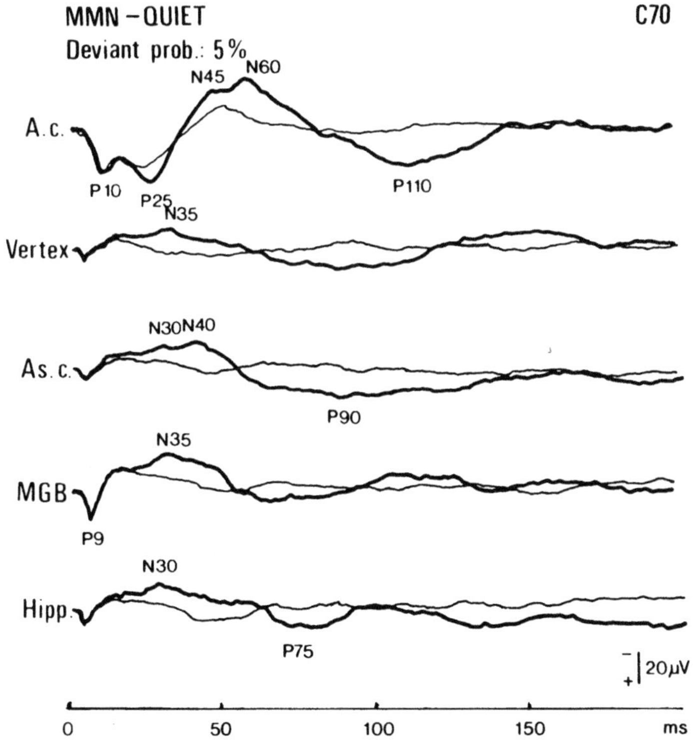

Fig. 1. EPs recorded in the quiet waking state from the AII area of the auditory cortex (*A.c.*), the vertex, the association cortex (*As.c.*), the medial geniculate body (*MGB*), and the hippocampus (*Hipp.*) at 5% deviant probability. EPs evoked by standard tones (SEPs) are drawn with a *thin line* and those elicited by deviant tones (DEPs) with a *thick line* in all figures. (In this and in the all the figures the superimposed EPs are averages of 32 responses.)

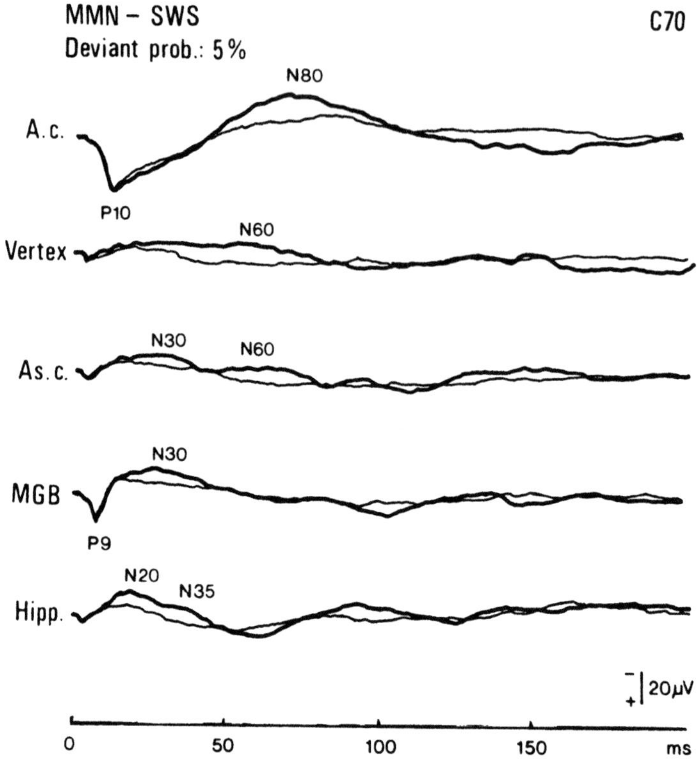

Fig. 2. EPs recorded in SWS from cortical areas (*A.c.*, *As.c.*) and intracerebral structures (*MGB*, *Hipp.*) at 5% deviant probability. (All abbreviations as in Fig. 1.)

in addition to the increased amplitude N45 component. This double-peaked negativity was followed by a positive wave, the P110.

On the vertex DEPs, elicited by the 5% probability DTs, the amplitude and the duration of the N35 wave was enhanced. This broadened negativity was followed by an increased positive deflection as in the auditory cortical DEPs. The As.c. DEPs showed similar changes to those seen in the vertex responses. The amplitude of the N30 wave increased and was overlapped by a second negativity of 40–50 ms latency. This increased negativity was followed by a characteristic positive wave (P90), as in the other surface DEPs.

The A.c., vertex, and As.c. SEPs and DEPs recorded in SWS are shown in Fig. 2. In the A.c. DEPs a longer latency negativity appeared, the amplitude of which also increased. In SWS only one peak of the middle latency negativity could be identified. In the vertex and association cortical DEPs the amplitude of the N60 wave increased.

3.2 Subcortical Recordings

Medial geniculate body responses recorded during the quiet waking state in the experimental blocks of different probability DTs are summarized in Fig. 3. The

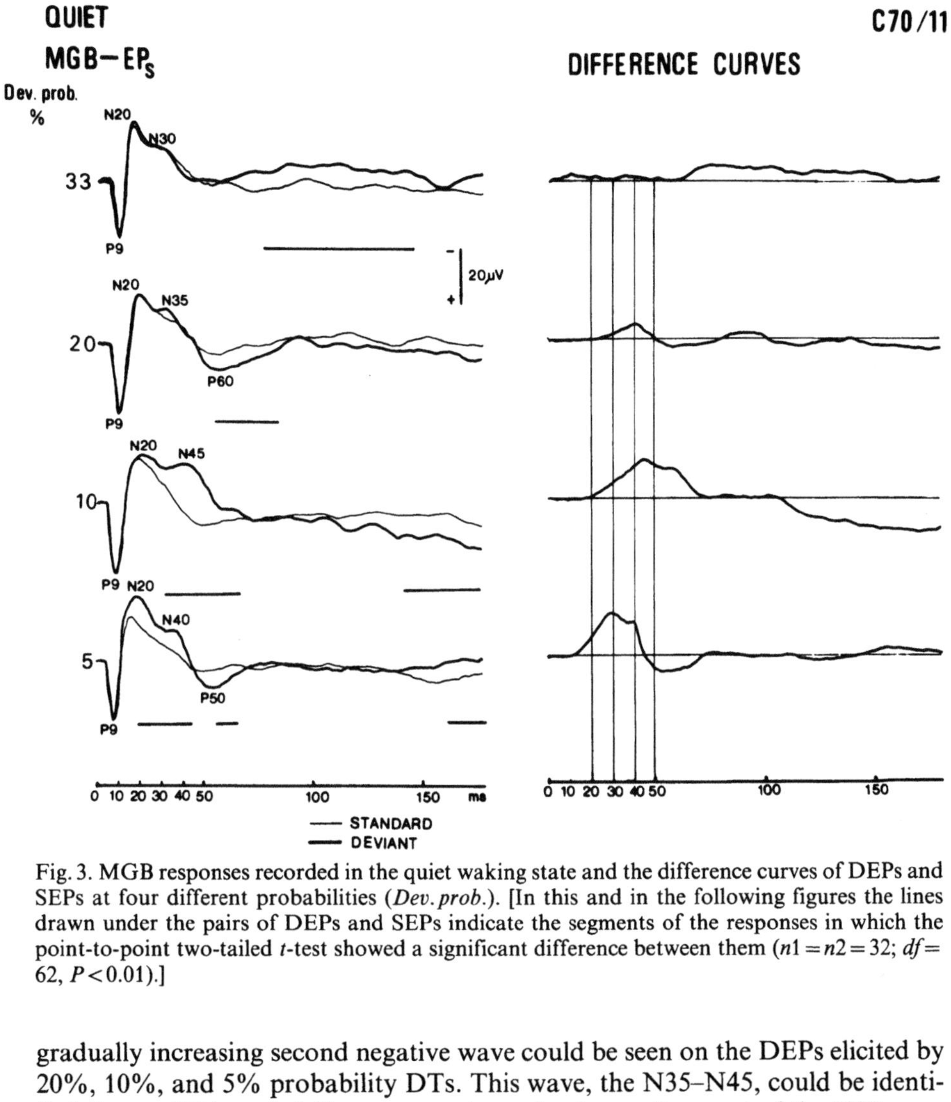

Fig. 3. MGB responses recorded in the quiet waking state and the difference curves of DEPs and SEPs at four different probabilities (*Dev. prob.*). [In this and in the following figures the lines drawn under the pairs of DEPs and SEPs indicate the segments of the responses in which the point-to-point two-tailed t-test showed a significant difference between them ($n1 = n2 = 32$; $df = 62$, $P < 0.01$).]

gradually increasing second negative wave could be seen on the DEPs elicited by 20%, 10%, and 5% probability DTs. This wave, the N35–N45, could be identified first at 20% probability as a notch on the descending part of the N20 component. The MMN was followed by a gradually increasing positivity of 50–60 ms latency. The lines drawn under the SEP–DEP pairs indicate those latency segments where the difference between the samples exceeded the level of $P < 0.01$. Significant differences could be demonstrated in the negative component between 20 and 50 ms at 10% and 5%. The P60 and P50 components of MGB DEPs showed significant differences at 20% and 5%, while at the 10% probability this positivity was melted into the descending part of the N45 wave.

The increase of the N35 wave could be followed in the difference curves drawn on the right side of Fig. 3. At 10% probability a large negativity of 45 ms peak latency could be observed in the difference curves of SEPs and DEPs and at 5% a second peak of 40 ms became identifiable.

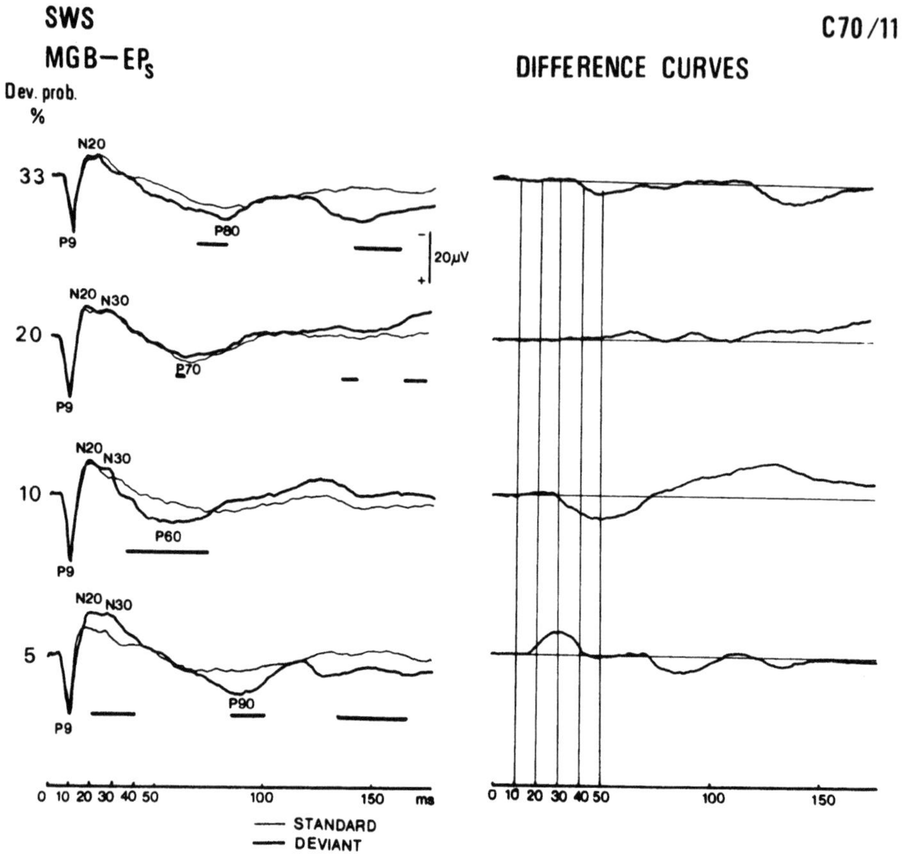

Fig. 4. SEPs and DEPs recorded from MGB in SWS at the different deviant probabilities and their difference curves

Data concerning the MGB recorded during SWS are shown in Fig. 4. The increased amplitude negativity could be observed only at the lowest probability and it was followed by the large amplitude P90 component. At 10% a significantly increased amplitude positivity (P60) appeared on the DEPs. These characteristic differences could be seen on the difference curves as well at the 10% and 5% probability.

The hippocampal EPs were more variable than those recorded from the sensory specific structures. The hippocampal EPs recorded in the quiet waking state at different probabilities are shown in Fig. 5. A gradually increasing amplitude negativity could be observed between 20 and 30 ms when the probability of the DTs was lower than 20%. On the DEPs elicited by 10% and 5% probability DTs an increasing positivity followed the negative wave. At the 5% probability a broad positive wave beginning at 50 ms could be observed, and this segment of the DEPs and the SEPs showed significant difference. The difference curves demonstrated an increasing negativity between 20 and 30 ms at 20%, 10%, and 5%, and a second one between 60 and 90 ms at 20% and 10%. At 5% a positive shift beginning at 50 ms could be seen in the difference curve.

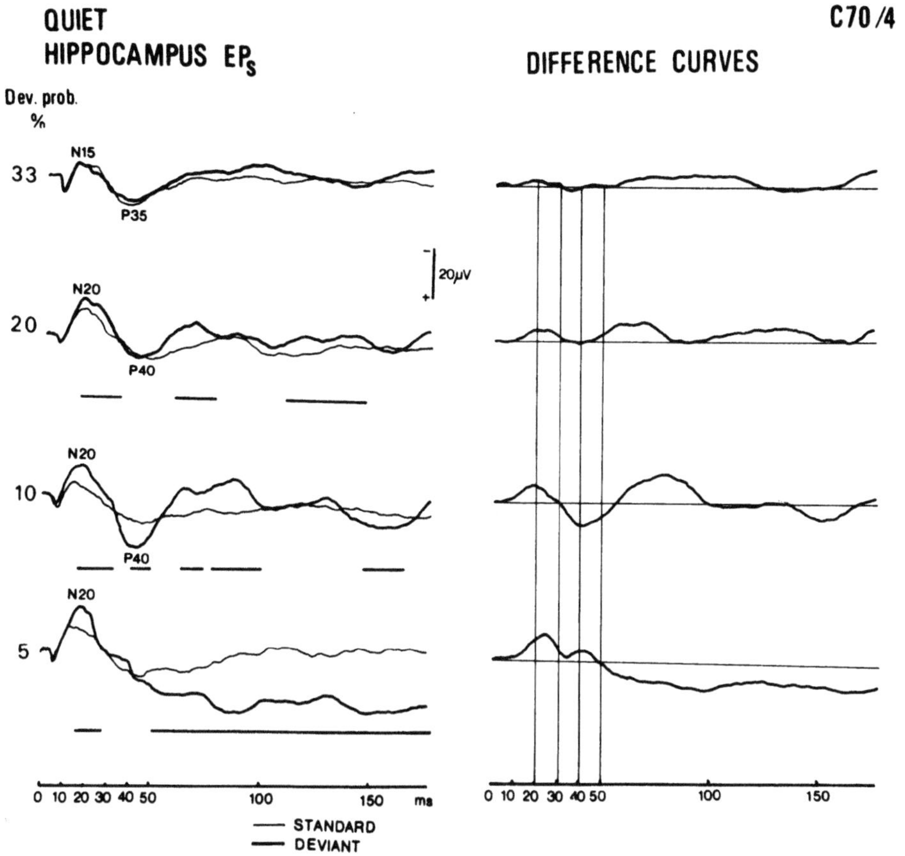

Fig. 5. Hippocampal SEPs and DEPs in the quiet waking state and their difference curves at the different deviant probabilities

The hippocampal SEPs and DEPs elicited by different probability DTs recorded in SWS are shown in Fig. 6. The characteristic EP changes could be observed only at 10% and 5%. A significant difference could be observed between 20 and 45 ms in the negative wave and between 50 and 80 ms in the following positive wave. These characteristic changes could clearly be seen on the difference curves as well.

4 Discussion

The present study indicates that changes of the middle latency negative wave of both the cortical and subcortical EPs closely related to the changes of the probability of deviant stimuli in the series of standard stimuli. Although the changes of vigilance affected the waveform of the cortical and subcortical EPs, MMN-like changes could be observed both in the quiet waking state and in SWS.

In the quiet waking state a double-peaked, increased amplitude negativity was elicited by the low probability DTs on the auditory cortical EPs recorded from

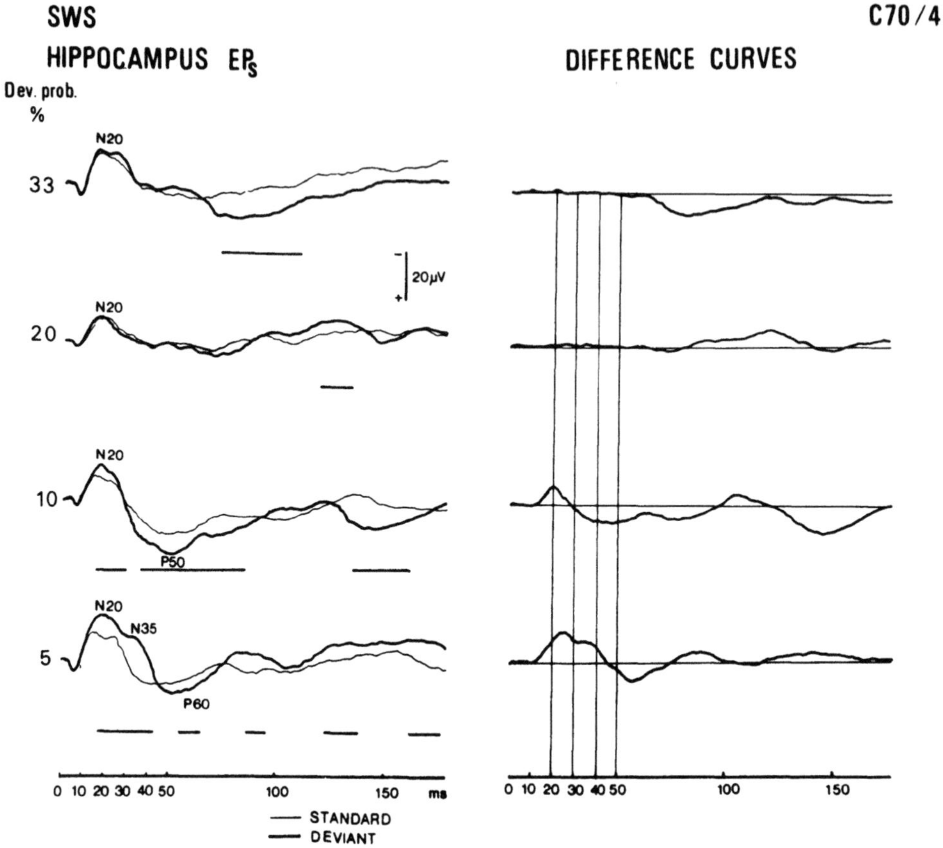

Fig. 6. Hippocampal SEPs and DEPs recorded in SWS at different probabilities and their difference curves

the AII area. On the present representative figures (Figs. 1 and 2) only the EPs recorded in the experimental blocks of lowest probability DTs were demonstrated. These cortical EP changes corresponded to our earlier data (Csépe et al. 1987) which showed that the lower the probability of DTs, the larger the amplitude of middle latency negative wave complex became. The present vertex and association cortical EP data corresponded to our earlier findings as well. These results showed that the increased amplitude middle latency negativity of 30–40 ms latency appeared on the DEPs elicited by the low probability DTs.

In SWS in the auditory and association cortical EPs and in the vertex EPs the amplitude increase of the middle latency negative wave appeared only at the lowest probability. Although the waveform of EPs recorded during SWS was different from that recorded in the quiet waking state, the deviant stimuli elicited a broad negativity with an 80 ms peak latency in auditory cortical EPs and a 20 ms shorter latency negative wave on the vertex and association cortex. These observations are in agreement with our earlier data (Csépe et al. 1987), on the basis of which this negativity was regarded as the equivalent in the cat of the human MMN described by Näätänen et al. (1978, 1980).

The observed latency difference between the vertex, the association cortical MMN, and the auditory cortical MMN suggested that more than a single comparison process might be present. It seems possible that the generation of the 20–30 ms earlier vertex and association cortical MMN is different from that of the auditory cortical components. These data also suggest that a multistage comparison might be activated during the whole sleep–waking cycle. The subcortical EPs recorded for testing this hypothesis showed characteristic changes in the MMN paradigm during the sleep–waking cycle. In the MGB responses a second negative peak with increased amplitude of 35–40 ms latency appeared when the probability of DTs was lower than 20%. This negativity was followed by an enlarged positivity between 50 and 100 ms. The probability which first elicited an MMN in MGB responses was the same as in the case of cortical EPs and the latency of the MGB MMN corresponded to that of the vertex and associational cortical EPs. In SWS the MGB MMN appeared only on the DEPs elicited by the lowest probability. The latency of this negative wave only slightly increased. This significant shift in the probability of DTs eliciting MMN suggested that the comparing process is sensitive only to the very rare deviant stimuli in the MGB as well as in the cortical areas.

On the basis of MGB recording it seems to be proved that not only the auditory cortex generates MMN and that the different acoustic stimuli are already compared at least on the level of the last relay nucleus of the auditory pathway. On the basis of these data it cannot be said that the MMN has a solely auditory cortical origin, as was suggested by human magnetoencephalographic data (Hari et al. 1984; Sams et al. 1985). Our data are in agreement with those early results according to which the frequency discrimination could be relearned after bilateral ablation of auditory cortex (Thompson 1960; Goldberg and Neff 1961; Butler et al. 1957). The implication of these data was that in the frequency discrimination task the MGB has a role partly independent of the functioning of the auditory cortex.

The hippocampal EP data showed that MMN-like changes were generated not only in the specific sensory system. In the quiet waking state an increased amplitude negativity of 20–30 ms latency appeared when the probability of DTs was lower than 20%. This negative wave was followed by a large positivity, especially at the lowest probability. In SWS this increased amplitude negativity could be recorded only at the lowest probability.

Our hippocampal data suggest that MMN might also be generated outside of the auditory system, which is consistent with our hypothesis that the incoming stimuli are processed in a multistage comparison way. The hippocampal MMN may indicate that the hippocampus is a system which may take part in the comparison of different signals, as suggested by Sokolov (1975). The very early hippocampal MMN underlines the possible role of the hippocampus functioning as a "sensory gate" (Lopes da Silva et al. 1985). The sensory information reaching the hippocampus through entorhinal–hippocampal projections (Witter and Groenewegen 1984; Van Groen and Witter 1985) might be processed quite early and affects the transmission of information. Experimental data of Vinogradova (1975) has suggested that the response of hippocampal neurons is particularly dependent on the history of sensory stimuli, especially on their "novelty." Our earlier MMN data (Csépe et al. 1987) suggested that this type of novelty is not particularly important in MMN generation. The long-term habituation of the applied stimuli, used in our experiment, indicated that MMN is not directly related to the triggering of the orientation reaction. However, it seems possible that the hippocampus is active in the comparison process, independent of whether the deviant stimulus in the series of repetitive stimuli has or has not a "novelty" feature.

In conclusion it should be stressed that MMN-like changes could be recorded from both the subcortical sensory specific (MGB) and archicortical (Hipp.) structures. These MMN-like changes could be observed in SWS only at the lowest probability, similar to those observed in the cortical recordings. The investigated subcortical structures, being active in a multistage comparison process, may underly the generation of the MMN, reflecting an early automatic information processing. Further experiments are needed to establish the possible relations between the different stages of the comparison, which could be a crucial process prior to the higher level information processing.

5 Summary

Auditory evoked potentials (EPs) elicited by standard (STs) and deviant tones (DTs) of different probability were studied in freely moving cats in quiet waking state and in slow wave sleep (SWS). In the EPs recorded from the auditory and association cortices, the vertex, the medial geniculate body (MGB), and the hippocampus, the characteristic mismatch negativity (MMN) could be observed. In the auditory cortex an increased amplitude middle-latency negative wave-complex appeared, and in the other cortical areas an earlier MMN could be recorded. Early (25–30 ms) MMN-like changes could be recorded from both the MGB and hippocampus. Subcortical MMN-like changes could be observed in SWS only at the lowest probability, similar to those observed in the cortical recordings.

References

Butler RA, Diamond IT, Neff WD (1957) Role of auditory cortex in discrimination of changes in frequency. J Neurophysiol 20:108–120

Csépe V, Karmos G, Molnár M (1987) Evoked potential correlates of stimulus deviance during wakefulness and sleep in cat – animal model of mismatch negativity. Electroencephalogr Clin Neurophysiol 66:571–578

Elberling C, Bak C, Kofoed B, Lebech J, Saermark K (1980) Magnetic auditory responses from the human brain. A preliminary report. Scand Audiol 9:185–190

Elberling C, Bak C, Kofoed B, Lebech J, Saermark K (1982) Auditory magnetic fields. Source location and "tonotopical organization" in the right hemisphere of the human brain. Scand Audiol 11:59–63

Goldberg JM, Neff WD (1961) Frequency discrimination after bilateral ablation of cortical auditory areas. J Neurophysiol 24:119–128

Hari R, Kaila K, Katila T, Tuomisto T, Varpula T (1982) Interstimulus interval dependence of the auditory vertex response and its magnetic counterpart: implications for their neural generation. Electroencephalogr Clin Neurophysiol 54:561–569
Hari R, Hämäläinen M, Ilmoniemi R, Kaukoranta E, Reinikainen K, Salminen J, Alho K, Näätänen R, Sams M (1984) Responses of the primary auditory cortex to pitch changes: neuromagnetic recordings in man. Neurosci Lett 50:127–132
Karmos G (1982) Auditory cortical correlates of motivation. In: Lissák K, Molnár P (eds) Motivation and the neural and neurohumoral factors in regulation of behaviour. Akadémiai Kiadó, Budapest, pp 141–158
Karmos G, Martin J, Kellényi L, Bauer M (1970) Constant intensity sound stimulation with a bone conductor in the freely moving cat. Electroencephalogr Clin Neurophysiol 28:637–638
Lopes da Silva FH, Groenewegen HJ, Holsheimer J, Room P, Witter MP, van Groen Th, Wadman WJ (1985) The hippocampus as a set of partially overlapping segments with a topographically organized system of inputs and outputs: the entorhinal cortex as a sensory gate, the medial septum as a gain-setting system and the ventral striatum as a motor interface. In: Buzsáki G, Vanderwolf CH (eds) Electrical activity of the archicortex. Akadémiai Kiadó, Budapest, pp 83–106
Marr D (1971) Simple memory: a theory for archicortex. Philos Trans R Soc Lond 262:23–81
Molnár M, Karmos G, Csépe V (1986) Laminar analysis of intracortical auditory evoked potentials during the wakefulness–sleep cycle in the cat. Int J Psychophysiol 3:171–182
Näätänen R (1986) Neurophysiological basis of the echoic memory as suggested by event-related potentials and magnetoencephalogram. In: Klix F, Hagendorf H (eds) Human memory and cognitive capabilities, mechanisms and performances. Elsevier, Amsterdam, pp 615–628
Näätänen R, Gaillard AWK (1983) The orienting reflex and the N2 deflection of the ERP. In: Gaillard AWK, Ritter W (eds) Tutorials in event-related potential research: endogenous components. Elsevier, Amsterdam, pp 119–141
Näätänen R, Gaillard AWK, Mäntysalo S (1978) Early selective attention effect on evoked potential reinterpreted. Acta Psychol (Amst) 42:313–329
Näätänen R, Gaillard AWK, Mäntysalo S (1980) Brain potential correlates of voluntary and involuntary attention. In: Kornhuber HH, Deecke L (eds) Motivation, motor and sensory processes of the brain: electrical potentials, behaviour and clinical use. Elsevier, Amsterdam, pp 343–348, (Progress in brain research, vol 54)
Sams M, Paavilainen P, Alho K, Näätänen R (1985) Auditory frequency discrimination and event-related potentials. Electroencephalogr Clin Neurophysiol 62:437–448
Simson R, Vaughan HG, Ritter W (1977) The scalp topography of potentials in auditory and visual discrimination tasks. Electroencephalogr Clin Neurophysiol 2:528–535
Snider RS, Niemer WT (1961) A stereotaxic atlas of the cat brain. University of Chicago Press, Chicago
Sokolov EN (1960) Neuronal models and the orienting reflex. In: Brazier MA (ed) The central nervous system and behaviour. Macy, New York
Sokolov EN (1975) The neuronal mechanisms of the orienting reflex. In: Sokolov EN, Vinogradova OS (eds) The neuronal mechanisms of the orienting reflex. Erlbaum, Hillsdale, pp 217–235
Teyler TJ, DiScenna P (1986) The hippocampal memory indexing theory. Behav Neurosci 2:147–154
Thompson RF (1960) Function of auditory cortex of cat in frequency discrimination. J Neurophysiol 23:321–334
Van Groen Th, Witter MP (1985) Electrophysiological and tracing study of the septotemporal distribution of entorhinal projections to the hippocampus in the cat. In: Buzsáki G, Vanderwolf CH (eds) Electrical activity of the archicortex. Akadémiai Kiadó, Budapest, pp 107–123
Vinogradova OS (1975) The hippocampus and the orienting reflex. In: Sokolov EN, Vinogradova OS (eds) Neuronal mechanisms of the orienting reflex. Erlbaum, Hillsdale, pp 128–154
Witter MP, Groenewegen HJ (1984) Laminar origin and septotemporal distribution of entorhinal and perirhinal projections to the hippocampus in the cat. J Comp Neurol 224:371–385

"No-Go Potential" in the Prefrontal Cortex of Monkeys

K. Sasaki and H. Gemba

1 Introduction

Various parts of the cerebral cortex reveal different kinds of electrical activities upon execution of simple reaction-time hand movements in response to visual stimuli in the monkey (Fig. 1) (Gemba et al. 1981; Sasaki and Gemba 1982). In the prefrontal cortex, predominant field potentials at a relatively short latency

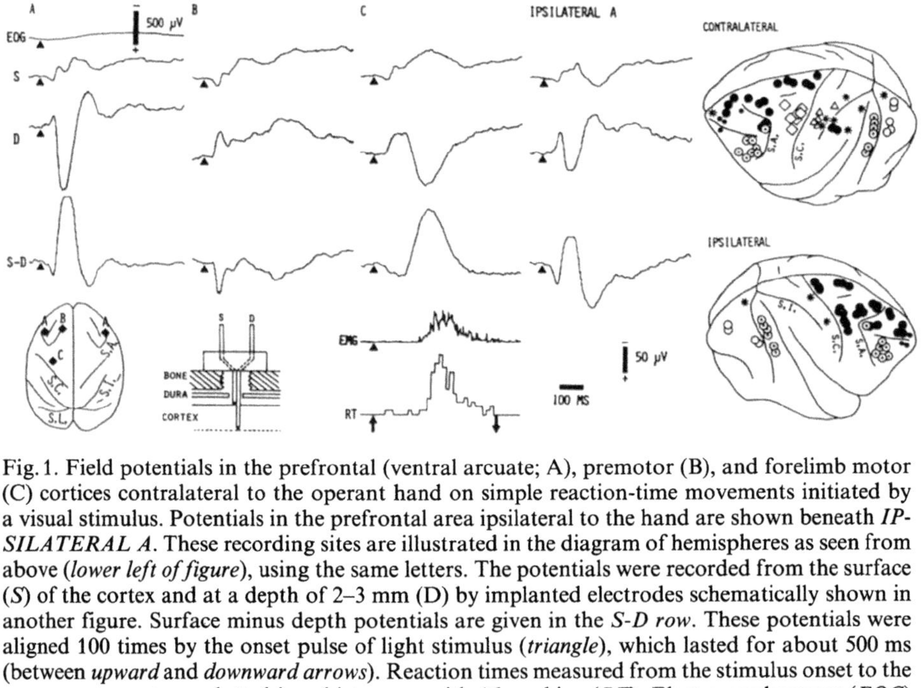

Fig. 1. Field potentials in the prefrontal (ventral arcuate; A), premotor (B), and forelimb motor (C) cortices contralateral to the operant hand on simple reaction-time movements initiated by a visual stimulus. Potentials in the prefrontal area ipsilateral to the hand are shown beneath *IPSILATERAL A*. These recording sites are illustrated in the diagram of hemispheres as seen from above (*lower left of figure*), using the same letters. The potentials were recorded from the surface (*S*) of the cortex and at a depth of 2–3 mm (D) by implanted electrodes schematically shown in another figure. Surface minus depth potentials are given in the *S-D row*. These potentials were aligned 100 times by the onset pulse of light stimulus (*triangle*), which lasted for about 500 ms (between *upward* and *downward arrows*). Reaction times measured from the stimulus onset to the movement onset are plotted in a histogram with 16-ms bins (*RT*). Electro-oculograms (*EOG*) were recorded from the rostrolateral edge of frontal bone above the orbit and averaged. The time scale is 100 ms for all traces. Calibration is 500 μV for EOG and 50 μV for cortical potentials. Electromyogram (*EMG*) of wrist extensor muscles was led from the skin over the muscles, rectified, and averaged. The two diagrams on the *right* present electrode sites on hemispheres contralateral and ipsilateral to the moving hand, summarizing 14 cases. Different symbols indicate different kinds of field potential. *Asterisks* denote sites of no significant potential. *S.A.*, sulcus arcuatus; *S.C.*, sulcus centralis; *S.I.*, sulcus intraparietalis. (Modified from Sasaki and Gemba 1982)

Springer Series in Brain Dynamics 2
Edited by Erol Başar and T. H. Bullock

after the onset of visual stimulus were found in the ventral part of the arcuate area and its continuation ventral to the caudal part of the principal sulcus ("prearcuate area") of both hemispheres (Fig. 1 A, IPSILATERAL A, diagrams to the right) (see Sasaki and Gemba 1982). A gradual increase in the potential in the prearcuate area during learning processes of the simple reaction-time movement was assumed to be correlated with the recognition learning of the movement, i.e., association of the visual stimulus with the hand movement (Sasaki and Gemba 1982). In the well trained monkey, the prearcuate area was found, with a local cooling method, to be one of the important cortical areas which send the motor command of the visually initiated movement to the motor cortex through the neocerebellum and thalamus (Sasaki and Gemba 1987).

Electrical activities in the prefrontal areas more rostral and dorsal than the prearcuate, however, seemed less marked on such a simple reaction-time movement. We supposed that some parts of the prefrontal cortex other than the prearcuate area might be involved in a task requiring somewhat higher integration and information processing (see Rosenkilde 1979). To test this possibility, we adopted go/no-go reaction-time hand movements with color discrimination in this study and found a specific potential on no-go responses in the dorsal bank of the principal sulcus and also in the rostroventral part of the inferior convexity of the prefrontal cortex. In the present paper, development of the no-go potential in the process of acquiring the discrimination and change of the potential in the process of learning the reversed discrimination task will be presented. A short preliminary report has been published (Sasaki and Gemba 1986 b).

2 Methods

Six adult monkeys (*Macaca fuscata*) were used in the present study, and they all showed entirely consistent results. They were first trained to perform a simple reaction-time hand movement (lever lifting with wrist extension) in response to a light stimulus (about 500 ms duration, delivered at various time intervals of 2.5–6.0 s), which was followed with a reward about 600 ms after every appropriate movement that occurred within the 500 ms duration. When a monkey learned the simple reaction-time hand movement, recording electrodes were implanted in the cerebral cortex under Nembutal (pentobarbital sodium) anesthesia.

Silver needle electrodes (250 µm in diameter, insulated except for a pointed tip) were placed on the surface of, and 2–3 mm deep in, cortical areas in both hemispheres; several loci were in the dorsolateral part of the prefrontal cortex, a few in the premotor cortex, a few in the supplementary motor area, a few in the motor cortex, etc. To explore activities within a sulcus, two electrodes arranged to be parallel with each other were inserted in the sulcus so that one was placed on the surface and the other at a depth of 2.0–3.0 mm in the cortical bank buried in the sulcus (see Fig. 2, lower left diagram). Two or three pairs of such parallel electrodes were implanted in both the dorsal and the ventral bank of the principal sulcus (in total, four or six pairs on one side), and two pairs were implanted, in both the rostral and the caudal bank of the arcuate sulcus near to the caudal end (four pairs on one side). The positions of these electrodes were later confirmed by histological examination.

Two or three weeks after the implantation of electrodes, training was started again with the simple reaction-time task. After a monkey began to perform the movement with a success rate of more than 90% in response to the go signal with reward, a no-go stimulus of another color was introduced at the same spot (light emitting diode of two colors), without reward, and with random alternation between the two stimuli. Initially the monkeys also lifted the lever upon the no-go stimulus, but they gradually ceased from responding to it. We gave neither punishment for failure of no-go response with movement nor a reward for a correct no-go reaction without movement. The time required for acquisition of discrimination was different in the six monkeys, ranging from several sessions in 1 day to a few weeks (training of 3–5 days per week). After a monkey had learned to discriminate almost completely between go and no-go signals, the stimuli were reversed, i.e., the former no-go stimulus became the rewarded go stimulus and the former go stimulus became the unrewarded no-go stimulus. Within a period of several days to a few weeks, the monkey gradually learned the new situation and again reached a steady state in which more than 90% of trials were correctly discriminated. Cortical field potentials were recorded during all these processes.

Cortical field potentials were led against indifferent electrodes (silver electrodes implanted in the bone rear to the ear on both sides), amplified (2.0 s time constant), and recorded on magnetic tapes through data recorders. In one session, usually 100 go and 100 no-go signals were delivered at random alternation and random intervals (2.5–6.0 s) and the potentials were averaged separately for go and no-go trials by the onset pulse of stimuli. Shift of the lever by hand movement was recorded through a transducer attached to the lever and averaged 100 times by the stimulus onset pulse. Reaction times measured from the stimulus onset to the movement onset were plotted in a histogram of 16-ms bins (RT) (Figs. 1, 2).

3 Results

3.1 No-Go Potential in the Dorsal Bank of the Principal Sulcus

When a monkey learned the discrimination task, a specific potential to no-go trial was recorded in the caudal part of the dorsal bank of the principal sulcus as shown in Fig. 2 (see Sasaki and Gemba 1986 b). In Fig. 2, diagrams I and II present records on go trials (Go) while diagrams IV and V give those on no-go trials (No-go). Diagram III shows subtracted results of S-D potentials on go trials (II) from those on no-go ones (IV). As illustrated in the inset diagram to the lower left, an electrode in the principal sulcus (PS, S) picked up potentials on the cortical surface, and electrodes implanted in cortices of the dorsal (dPS, D) and ventral (vPS, D) banks yielded potentials as shown in I (Go) and V (No-go). Surface minus depth potentials (S-D) of the banks are given in diagrams II and IV respectively, those of the vPS being in the uppermost row and those of the dPS in the second row.

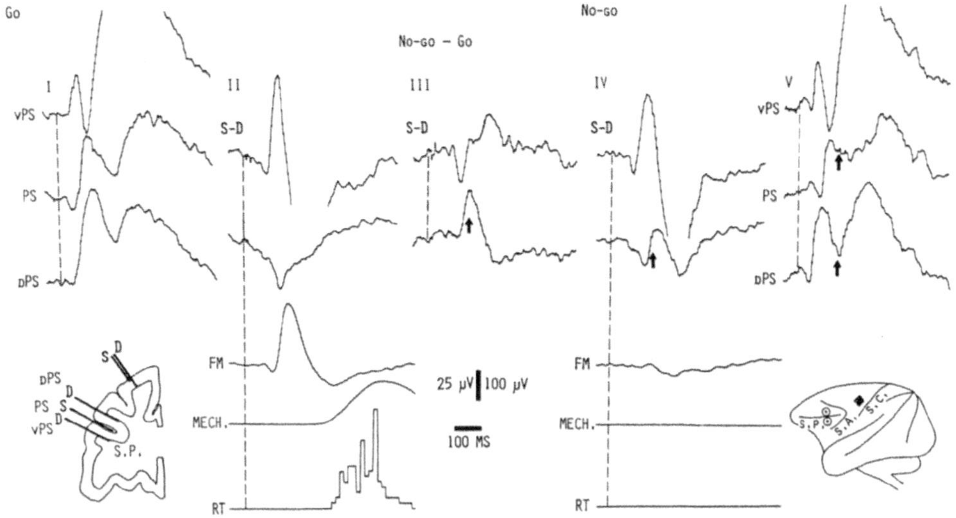

Fig. 2. Field potentials recorded from the ventral and dorsal banks of the principal sulcus and the forelimb motor cortex contralateral to the operant hand. *I* and *II* show go responses and *IV* and *V* no-go responses, III giving the difference between go and no-go responses. *I:* Monopolar records against the indifferent electrode from the depth of the ventral bank (*vPS*), the principal sulcus (*PS*), and the depth of the dorsal bank (*dPS*). They are, respectively, potentials from the cortical depth (*D, vPS*), surface (*S, PS*), and depth (*D, dPS*) as illustrated in the *diagram at the lower left. II:* Differences between records PS and vPS and between PS and dPS in I are given in the *first and second rows,* i.e., surface minus depth (*S-D*) potentials of the ventral and dorsal banks of the principal sulcus. S-D potential of the forelimb motor cortex is shown in the third row (*FM*). Below the FM potential are the mechanogram (*MECH.*) of the lever shift in an arbitrary unit and the reaction time histogram (*RT*) with 16-ms bins. *IV* and *V*: Field potentials recorded in the same situations as II and I but for no-go reactions. *III:* Electrical differences between upper two rows of II and IV (IV minus II). These cortical potentials and mechanograms were aligned 100 times by the onset pulse of either the go or the no-go stimulus. *RT* plots the time from the stimulus onset to the movement onset for the same 100 movements. Electrode sites are illustrated on the left hemisphere of the *diagram at the lower right.* The time scale is 100 ms for all traces. Calibration is 25 μV for the potentials in the principal area and 100 μV for those in the motor cortex. Negativity upwards in I and V. The surface-negative, depth-positive dipole is taken upwards for S-D records

In the forelimb motor cortex (FM) row of diagrams II and IV, surface minus depth potentials recorded from the forelimb motor cortex contralateral to the operant hand are presented for go and no-go trials respectively. Below the FM row, mechanograms of the lever shift were transduced to electrical changes (MECH), and reaction times measured from the onset pulse of light stimulus to the onset time of movement are shown in a histogram of 16-ms bins (RT). These cortical potentials and mechanograms were aligned by the onset pulse of the stimulus for the 100 successful movements that occurred within the stimulus duration. This well trained monkey did not move the hand on no-go stimuli at all, as seen from the flat lines of MECH and RT (No-go). The FM revealed premovement potentials on go trials which consisted of early surface-positive, depth-negative (s-P, d-N) and late surface-negative, depth-positive (s-N, d-P) potentials, the late ones being elicited by the impulse through the neocerebellum and the thalamus as demonstrated previously (Sasaki et al. 1982). On no-go trials, no distinguishable potentials were seen with the FM electrode (FM row of diagram IV).

A remarkable difference between the potentials on go and no-go is noted in the dorsal bank of the principal sulcus: an s-N, d-P potential at a latency of about 120 ms can be seen in the dorsal bank on the no-go trial (second row of diagram IV as marked by an upward arrow). Subtraction of the potential on the go trial from that on the no-go trial gives the potential shown in the second row of diagram III, marked by an arrow. As seen in diagram V, this potential can be attributed to an active dipole due to s-N, d-P potentials in the dorsal cortical bank (marked by arrows on PS and dPS records). In the ventral bank, both go and no-go signals induced early s-P, d-N followed by s-N, d-P potentials, some difference in size being noted merely in the uppermost row of diagrams II–IV.

The no-go potential as shown in Fig. 2 appeared in the caudal part of the dorsal bank of the principal sulcus of all six monkeys in the present study (see Fig. 6). The potential was seen in the dorsal bank of both hemispheres, contralateral and ipsilateral to the operant hand as shown and discussed below. Latency of the potential from the onset of no-go stimulus was 110–150 ms and its summit was at 130–190 ms.

The no-go potential was usually recorded more predominantly in the dorsal bank of the principal sulcus of the hemisphere contralateral to the operant hand in all six monkeys tested, although strict comparison of the potentials at different sites would be difficult with the recording method used in this study. The contralateral predominance was noted by recording the potential of both hemispheres with the operant hand used in the initial trial (five right hands and one left hand). However, when the operant hand was alternated from one to another, the predominance was more or less changed and became complicated afterwards. Change of the operant hand usually increased the activity of the dorsal bank on the side that had been less predominantly activated previously, and the dorsal banks on both sides became more or less active. The laterality might be not so strict in an association cortex like the dorsal bank as in the motor and primary sensory cortices, especially after the cortices on both sides are once recruited by the alternation of the operant hand from one side to another.

3.2 Development, Disappearance, and Redevelopment of the No-Go Potential in the Learning Process and upon Reversal of the Discrimination Task

The no-go potential in the dorsal bank of the principal sulcus gradually increased in parallel with learning of the discrimination task and reached a steady level. When go and no-go signals were reversed, the potential quickly disappeared, then reappeared, and gradually increased in size again during acquisition of the reversed discrimination. These processes are presented in Figs. 3–5 for one monkey. Potentials of surface minus depth in the dorsal bank and forelimb motor cortex on the side contralateral to the moving hand are presented in these figures.

The initial learning of the discrimination task required considerably different times for different monkeys, from several sessions in 1 day to a few weeks.

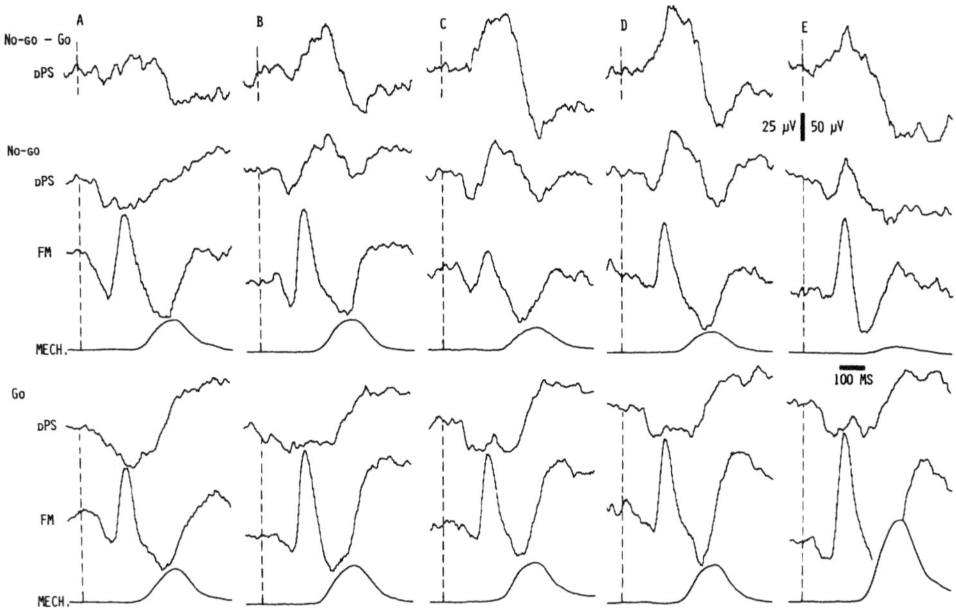

Fig. 3. Changes of field potentials in the dorsal bank of the principal sulcus (*dPS*) and the forelimb motor cortex (*FM*) contralateral to the moving hand with mechanogram of the lever shift (*MECH.*) in learning processes of go/no-go discrimination. The *uppermost row* (*No-go − Go*) gives the difference between no-go and go responses (no-go minus go) in the dPS. No-go and go responses in dPS and FM (S-D) with MECH. are presented respectively in three rows marked by No-go and Go. A–D were taken from serial sessions of the same-day experiment, one session including 100 go and 100 no-go trials. E was recorded 1 week later. The time scale is 100 ms for all traces. Calibration is 25 µV for dPS potentials and 50 µV for FM potentials. Cortical potentials are all S-D and the surface-negative, depth-positive dipole is taken upwards. MECH. is shown in an arbitrary unit

Figure 3 shows the process in the monkey which acquired the discrimination in the shortest time. The monkey had been performing simple reaction-time hand movements on a green light stimulus with 500 ms duration, i.e., the subject lifted a lever within the duration and obtained a reward. When a red light stimulus of 500 ms duration was delivered without reward at random alternation with the green light stimulus, the monkey moved its hand irrespective of the two color stimuli in the initial two sessions (each involving 100 successful movements in response to the go stimulus), the first session being presented in Fig. 3 A. Field potentials in the dorsal bank and forelimb motor cortex, and mechanograms transduced from the lever shift, looked much the same on No-go and Go trials. The difference between the potentials of the dorsal bank on no-go and go trials (electrical subtraction) is presented in the uppermost row.

In the third session of that day (B), the monkey started to discriminate the go from the no-go signal as seen in a slight difference between the mechanograms on no-go and go trials (100 trials respectively). An upward deflection is noted in the potential of the dorsal bank on no-go trials when compared with that on go trials and represents s-N, d-P potentials in the bank. In the following sessions (C and D) on the same day, the s-N, d-P potential grew and its rising phase became

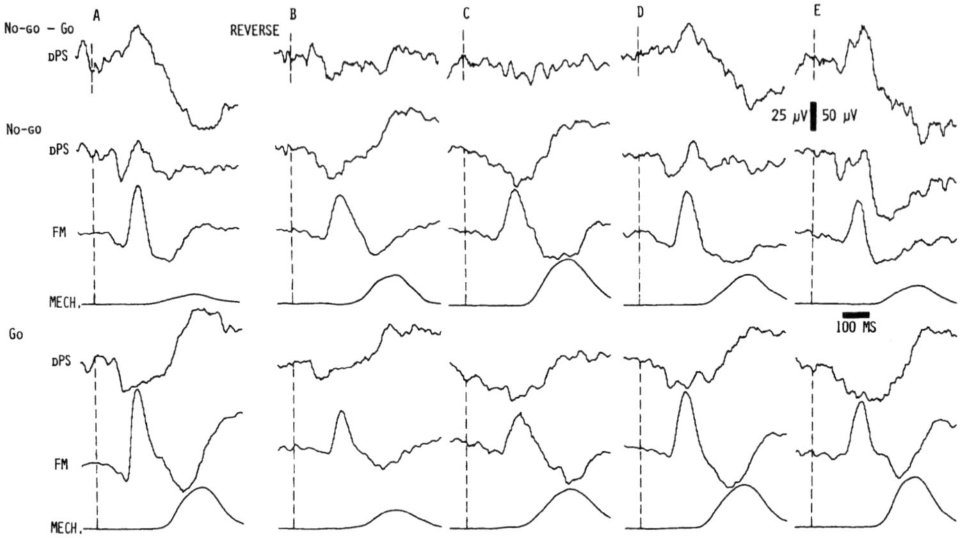

Fig. 4. Records from the same monkey as in Fig. 3. Go and no-go signals were reversed between the sessions A and B (A and B). C–E were taken 2, 5, and 7 days after the reversal. Abbreviations and calibrations are as in Fig. 3, but go and no-go signals are opposite in B–E to A, i.e., green and red lights were, respectively, the stimuli for go and no-go in A, whereas they were the stimuli for no-go and go in B–E

steep, as the difference between mechanograms on no-go and go trials gradually increased. One week later, the records in E were obtained, and the s-N, d-P potential was a little smaller than the previous day, although a much more marked difference between the no-go and go mechanogram was noted. These initial learning processes involving the discrimination task are plotted in the earlier phase of Fig. 5.

After the monkey reached a steady state in the discrimination task, light stimuli of go and no-go were reversed, i.e., the former go signal became no-go without reward and the former no-go changed to go with reward. As seen in Fig. 4, the monkey was completely confused by the reversed signals and responded more frequently to the new no-go stimulus than the go stimulus on the day of reversal (Fig. 4 B). A very small no-go potential could be seen on go trials so that the No-go – Go trace reveals the opposite direction (downward). Two days after the reversal, the monkey lifted the lever almost equally on both no-go and go stimuli and no obvious difference was observed between the dPS potentials on no-go and go trials (Fig. 4 C). Five days after the reversal, the monkey started to discriminate the new signals (compare mechanograms in Fig. 4 D) and the no-go potential reappeared (uppermost row in D). Seven days after the reversal, the discrimination recovered to almost the level it had been before the reversal (Fig. 4 E). These processes are plotted in the middle part of Fig. 5 (days 16–52 on the abscissa).

Such reversals were repeated three times more in the same monkey as plotted in Fig. 5 (days 56–67 on the abscissa). Relearning was repeated on the reversals and the no-go potential disappeared and reappeared as outlined above. The only

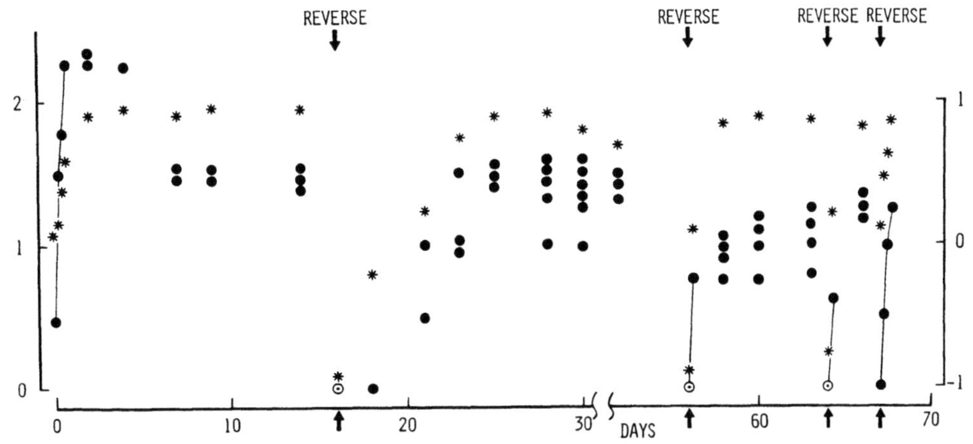

Fig. 5. Height of the no-go potential (difference between no-go and go potentials in dPS) is plotted on the *ordinate* in an arbitrary unit (*left*) by *filled circles* against days of the experimental series on the *abscissa* as exemplified in Figs. 3 and 4. *Open circle with dot in center* indicates that the no-go potential (no-go minus go potential) was below zero on the ordinate and opposite in polarity to ordinary cases (e.g., Fig. 4 B), i.e., the no-go potentials was recorded on go trials immediately after reversal of go and no-go signals. *Circles connected by a line* represent successive sessions in the same-day experiment in which the no-go potential gradually developed. The ratio of the mechanogram on go trials to that on no-go trials is also plotted on the ordinate (*right*) by *asterisks* against days on the abscissa. The height of the mechanogram on go trials (average of 100 times) minus that on no-go trials (average of 100 times) was divided by the former to calculate the ratio. Therefore, zero on the ordinate indicates an equal size of mechanograms on go and no-go trials. Note a value above zero of the asterisk immediately after the fourth reversal, in contrast to minus values of asterisks after the first, second, and third reversals. Every reversal is marked by an *arrow*

difference was a shorter time for the relearning and the reappearance as the reversal was repeatedly experienced (Fig. 5). The last time (fourth reversal), the monkey responded equally on both newly reversed signals immediately after the reversal, in contrast to the more predominant response to the former go than the new go signal which had been seen in the previous reversals (in Fig. 5 note filled circle at 67 days as compared with open circle with dot in center immediately after the previous three reversals). This might indicate that the monkey was quickly able to doubt the reversed signals after having repeatedly experienced such reversal. Three of the other five monkeys in the present study were tested with such reversal experiments, and all three revealed results entirely consistent with those in the monkey discussed above.

3.3 Distribution of the No-Go Potential in the Cerebral Hemisphere

The no-go potential could be recorded in the caudal part of the dorsal bank of the principal sulcus in both hemispheres of the six monkeys. In two of the six monkeys, a similar potential was also observed in a small area of the ventral edge of the inferior convexity in the prefrontal cortex of both hemispheres. Electrode

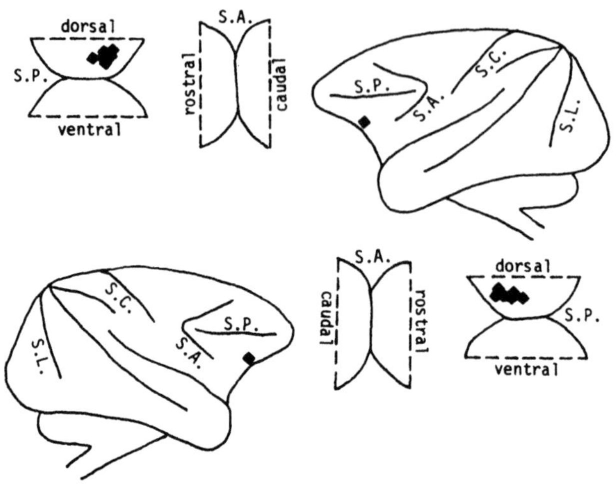

Fig. 6. Electrode sites in which the no-go potential could be recorded, summarized from six monkeys. The lateral view of the left (*upper half*) or right (*lower half*) hemisphere is presented with unfolded banks of the principal (*S.P.*) and arcuate (*S.A.*) sulcus. The rostral and rostrodorsal ends of the unfolded schemata of principal and arcuate sulci are indicated by the abbreviations *S.P.* and *S.A.* respectively. The arcuate sulcus is not only unfolded but also straightened schematically. Many electrodes with which the no-go potential was unrecorded are not marked in the diagrams. *S.C.*, sulcus centralis; *S.L.*, sulcus lunatus

sites where the potential was obtained are plotted in Fig. 6. Further exploration of the potential in the inferior convexity is being undertaken at present and results will be published in detail elsewhere.

4 Discussion

The potential specific to the no-go reaction was found in the two loci in the prefrontal cortex of monkeys performing the go/no-go reaction-time hand movement with color discrimination. The loci, the caudal part of the dorsal bank of the principal sulcus and the rostroventral portion of the inferior convexity, were found in both hemispheres with some predominance of the side contralateral to the operant hand. The "no-go potential" was composed of s-N, d-P deflections at a latency of 110–150 ms after the onset of the no-go light stimulus. The monkeys initially moved their hand on both go and no-go signals but gradually ceased to respond to the no-go stimulus. The monkeys obviously tried to avoid futile movements and learned a more efficient way of obtaining rewards.

Experiments involving reversal of the go and no-go signals demonstrated that the discrimination between go and no-go stimuli was irrespective of colors used for the signals. The no-go potential was also found not to be specific to the color used for the no-go signal. Relearning of the reversed stimuli was usually more difficult than the initial learning of the discrimination task, but it appeared to become easier as the reversal was repeated. These learning and relearning processes were found to be closely correlated with the development of the no-go potential in the prefrontal loci. This suggests that the no-go potential is related to the judgment not to move and suppresses the movement execution.

Studies involving ablation of the prefrontal cortex indicated that lesion of the principal sulcus had no grave effects upon go/no-go performance (Gross 1963; Gross and Weiskrantz 1962; Pribram and Mishkin 1955). On the other hand, ablation of the inferior convexity of the prefrontal lobe did influence it (Butters et al. 1973; Iversen and Mishkin 1970). The latter reports might fit the finding in the present study of the no-go potential in the rostroventral locus of the inferior convexity. The former reports did not seem relevant to the constant and marked no-go potential in the dorsal bank of the principal sulcus demonstrated in this study. However, such a discrepancy between ablation studies and a chronic recording experiment does not necessarily mean that the recorded potential is not of functional importance, since compensatory function of the association cortex can be so great that the functional role of a localized area of the association cortex possibly may be assumed easily by some other locus in the cortex when the area in question is ablated (see Sasaki and Gemba 1984, 1986a).

The s-N, d-P polarity of the no-go potential suggests that the main generators of the potential are EPSP currents elicited in superficial parts of apical dendrites of pyramidal neurons in the dorsal bank of the principal sulcus (Sasaki et al. 1970, 1976). However, the origins of afferent fibers producing the superficial EPSP currents are not yet known, two possibilities being some thalamic nucleus and/or cortical area. In general, superficial cortical layers are more likely to receive afferent inputs from thalamic neurons than from cortical neurons (Sasaki et al. 1981; see Sasaki 1979).

It is also unknown how the assumed EPSP currents at a latency of 110–150 ms from the onset of the light stimulus are related to the no-go reaction. One possibility is that the EPSP activates the cortex to send an inhibitory influence to the motor system. In two monkeys among the six tested in the present study, the premovement superficial thalamocortical (T-C) responses in the motor cortex mediated by the neocerebellum and thalamus were almost entirely suppressed on no-go responses (Fig. 2) (see Sasaki et al. 1982). In the other four monkeys, on the other hand, the superficial T-C responses in the motor cortex were reduced in size on no-go responses, but nevertheless remained to a considerable extent (Figs. 3, 4). Such variability in the size of superficial T-C responses on no-go trials might suggest that inhibitory influences act on both the motor cortex and a neuronal circuit prior to the motor cortex with different proportions in individual subjects. If the inhibition, e.g., IPSP, works directly on pyramidal neurons in the motor cortex, the premovement superficial T-C responses in the motor cortex are little changed, since the T-C responses are mainly composed of EPSP current in the neurons. Alternatively, if the inhibitory effect acts on the neuronal level prior to the motor cortex, the superficial T-C responses must be reduced in size. Another possibility is that the inhibitory effect goes to spinal motoneurons independently of the motor cortex–spinal cord system without touching the superficial T-C responses in the motor cortex. This possibility is unlikely to be mainly involved, because the T-C responses are suppressed at least to some extent in all the subjects.

One could argue that the no-go potential is a kind of efference copy. At present we have no direct evidence to discard or favor this argument. Recently we stimulated the prefrontal loci of no-go potential in monkeys performing the movement on the go stimulus and were able to suppress the movement (Sasaki et al. 1989). This appears to support the assumption that the prefrontal loci have an inhibitory action upon the motor performance.

5 Summary

Monkeys were trained to perform reaction-time hand movements in response to go and no-go light signals of different colors, and field potentials in various cortical areas were recorded and analyzed with chronically implanted cortical electrodes. In the case of simple hand movements with the go light stimulus only, marked potentials were recorded in the prefrontal, premotor, motor, and prestriate cortices, etc., as previously reported. With the discrimination task, the caudal part of the dorsal bank of the principal sulcus and the rostroventral part of the inferior convexity of the prefrontal cortex were found to reveal an electrical activity specific to no-go trials. The activity was composed of relatively sharp surface-negative, depth-positive potentials at a latency of 110–150 ms after the onset of the no-go stimulus. The no-go potential appeared in both hemispheres, though it was often slightly more predominant on the side contralateral to the operant hand. The no-go potential grew in parallel with the gradual acquisition of the no-go reaction during the learning of discrimination tasks. With reversal of the go and no-go signals, the no-go potential at once disappeared, then reappeared, and gradually grew as the discrimination was reestablished on the reversed stimuli, thus verifying that it is seen only on no-go reactions, irrespective of the color used for the stimulus. It is suggested that the activity of these prefrontal areas is closely related to the judgment not to execute the reaction-time movement and presumably suppresses the motor execution.

Acknowledgments. The authors thank Prof. N. Mizuno for histological examination. This research was supported by a Grant-in-Aid for Scientific Research no. 62440024 from the Ministry of Education, Science, and Culture of Japan.

References

Butters N, Butter C, Rosen J, Stein D (1973) Behavioral effects of sequential and one-stage ablations of orbital prefrontal cortex in the monkey. Exp Neurol 39:204–214

Gemba H, Hashimoto S, Sasaki K (1981) Cortical field potentials preceding visually initiated hand movements in the monkey. Exp Brain Res 42:435–441

Gross CG (1963) A comparison of the effects of parietal and total lateral frontal lesions on test performance by monkey. J Comp Physiol Psychol 1:41–47

Gross CG, Weiskrantz L (1962) Evidence for dissociation of impairment on auditory discrimination and delayed response following lateral frontal lesions in monkeys. Exp Neurol 5:453–476

Iversen SD, Mishkin M (1970) Perseverative interference in monkeys following selective lesions of the inferior prefrontal convexity. Exp Brain Res 11:376–386

Pribram KH, Mishkin M (1955) Simultaneous and successive visual discrimination by monkeys with inferotemporal lesions. J Comp Physiol Psychol 48:198–202

Rosenkilde CH (1979) Functional heterogeneity of the prefrontal cortex in the monkey: a review. Behav Neurol Biol 25:301–345

Sasaki K (1979) Cerebro-cerebellar interconnections in cats and monkeys. In: Massion J, Sasaki K (eds) Cerebro-cerebellar interactions. Elsevier Amsterdam, pp 105–124

Sasaki K, Gemba H (1982) Development and change of cortical field potentials during learning processes of visually initiated hand movements in the monkey. Exp Brain Res 48:429–437

Sasaki K, Gemba H (1984) Compensatory motor function of the somatosensory cortex for the motor cortex temporarily impaired by cooling in the monkey. Exp Brain Res 55:60–69

Sasaki K, Gemba H (1986a) Effects of premotor cortex cooling upon visually initiated hand movements in the monkey. Brain Res 374:278–286

Sasaki K, Gemba H (1986b) Electrical activity in the prefrontal cortex specific to no-go reaction of conditioned hand movement with colour discrimination in the monkey. Exp Brain Res 64:603–606

Sasaki K, Gemba H (1987) Effects of cooling the prefrontal and prestriate cortex upon visually initiated hand movements in the monkey. Brain Res 415:362–366

Sasaki K, Staunton HP, Dieckmann G (1970) Characteristic features of augmenting and recruiting responses in the cerebral cortex. Exp Neurol 26:369–392

Sasaki K, Kawaguchi S, Oka H, Sakai M, Mizuno N (1976) Electrophysiological studies on the cerebellocerebral projections in monkeys. Exp Brain Res 24:495–507

Sasaki K, Gemba H, Hashimoto S (1981) Premovement slow cortical potentials on self-paced hand movements and thalamo-cortical and cortico-cortical responses in the monkey. Exp Neurol 72:41–50

Sasaki K, Gemba H, Mizuno N (1982) Cortical field potentials preceding visually initiated hand movements and cerebellar actions in the monkey. Exp Brain Res 46:29–36

Sasaki K, Gemba H, Tsujimoto T (1989) Suppression of visually initiated hand movement by stimulation of the prefrontal cortex in the monkey. Brain Res (in press)

Epileptic Phenomena in the Neocortex: From Activity of Single Neurons to Field Potentials of Neuronal Pools

H. POCKBERGER, E.-J. SPECKMANN, and J. WALDEN

1 Introduction

In the past epileptic phenomena have been identified by such terms as "hyperexcitation" or "hypersynchronization." These terms, which are merely descriptive, come from studies of EEG phenomena such as spikes, sharp waves, and spike–wave complexes. As these electrographic features (graphoelements) imply overwhelming excitatory processes, one was inclined to think of an epileptic focus as an aggregate of neurons totally in a state of excitation and virtually devoid of inhibition.

The study of epileptic activity in the neocortex reveals characteristic phenomena at the cellular level as well as at the level of neuronal pools. In the past many observations have been made at both levels separately; however, the gap between the two levels is still wide. Therefore, it is quite clear that the epileptic phenomena in the mammalian brain await further clarification. This article summarizes results obtained from studies of focal epileptic activity at the single neuron level and the level of neuronal pools.

2 Focal Epileptic Activity – The Penicillin Model

In current epilepsy research many different so-called epileptogenic drugs are used for the induction of epileptic activity. Each of these methods of causing epileptic activity has its own characteristic features with respect to its time course, but none seems to be superior to the other in resembling epileptic activity in humans. At the cellular level a characteristic pathological discharge pattern – the paroxysmal depolarization shift (PDS) – is common to all these models. One of the best studied is the penicillin epilepsy model.

Local application of penicillin to the neocortical surface or at any cortical depth leads to the immediate generation of interictal EEG spikes (ISs), the hallmark of focal epileptic activity. Figure 1 outlines an experimental setup in which penicillin is applied to the surface of the motor cortex of the rat. As the penicillin solution diffuses into the different cortical layers, ISs develop and increase considerably in amplitude. If all cortical layers are in contact with this epileptogenic substance, large negative ISs can be recorded on the cortical surface. Recordings of the surface EEG within the zone of penicillin application and its surround show epileptic ISs of constant shape and repetition rate restricted to the penicillin

Springer Series in Brain Dynamics 2
Edited by Erol Başar and T. H. Bullock

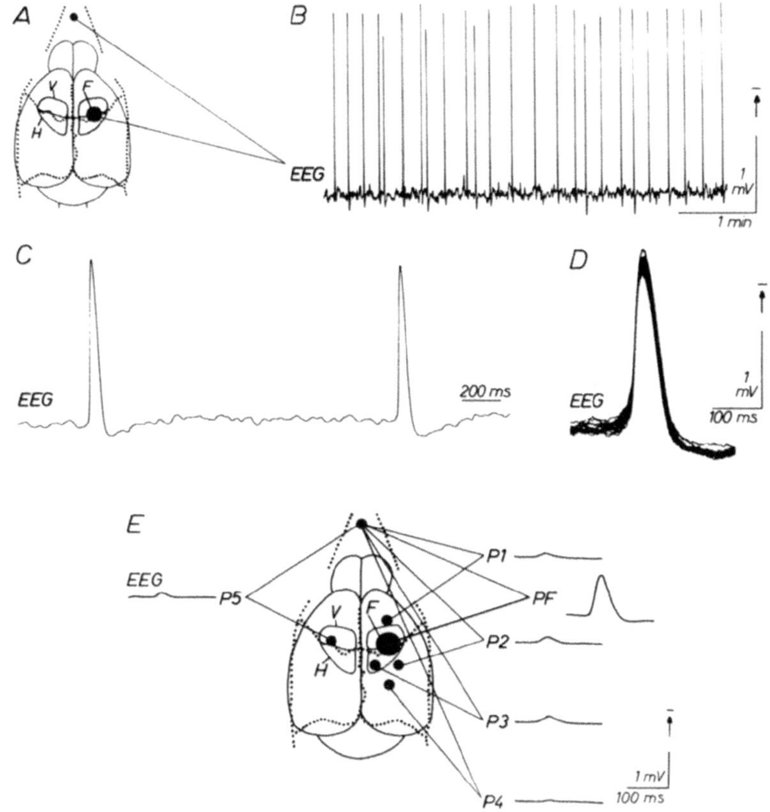

Fig. 1 A–E. Experimental setup of focal seizure activity induced by local application of penicillin. *F*, focus; *V* and *H*, motor representation of anterior and posterior extremities. (A) Recording schema. (B, C) ISs recorded with different time base. (D) Superimposition of 20 ISs. (E) Recordings within and in the surround of an epileptic focus. Note steep potential gradients between recordings from electrode PF and P1–P5. (Speckmann 1986)

application site. Such focal discharges serve as a model for partial seizure activity observed in humans. Because of relative stability of epileptiform discharges in time and space, such a model is particularly suited for the analysis of single cell activity and the electrical activity of neuronal pools.

3 Bioelectrical Activity of Single Neurons

Epileptic responses at the cellular level can be subdivided into excitatory, the PDS, and inhibitory ones, *epileptic inhibition*. These two types of membrane potential change are recorded in different neurons which are spatially separated. To study both types the following experimental setup was chosen: After a single neuron was punctured with a microelectrode, penicillin was applied to the cortical surface. This procedure allows one to analyze the transition of spontaneous cell

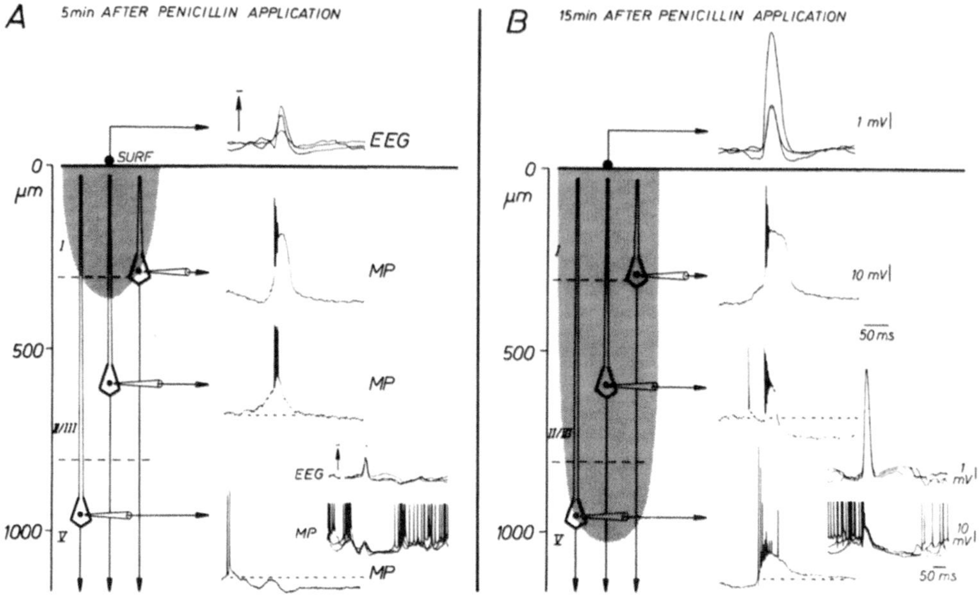

Fig. 2 A, B. Development of epileptic discharges after epicortical penicillin application. (A) Five minutes after penicillin is applied to the cortical surface ISs of 2 mV are recorded at the surface. Intracellular recordings of different neurons at different cortical depths reveal that PDSs occur only in superficial neurons. Neurons in layers V and VI show a massive inhibition simultaneous to each IS. (B) Outline of the situation 10 min later. In this stage epicortical ISs have increased considerably in amplitude; neurons in all cortical layers show PDSs simultaneously with each IS on the surface. (Speckmann 1986). *MP*, membrane potential

activity (EPSPs and IPSPs as well as different discharge patterns of action potentials) into epileptic activity. During the first few minutes after application of penicillin, neurons located in superficial cortical layers generate simultaneous to each IS a PDS (see Fig. 2). This epileptic cell activity, first described by Matsumoto and Ajmone-Marsan (1964), consists of a large depolarization with superimposed repetitive action potentials. At the same time neurons in the middle cortical layers exhibit smaller depolarizations with superimposed bursts of action potentials. Neurons in the deep cortical layers, i.e., layers V and VI, however, are massively inhibited as neurons in the upper layers are in a state of epileptic excitation. Thus, during development of an epileptic focus one observes a vertical arrangement of pronounced excitation and inhibition within the application site of the epileptogenic agent. This phenomenon has been named "vertical inhibition" (Elger and Speckmann 1983).

Subsequently the epileptic activity of single neurons changes as the focus enlarges due to the diffusion of the epileptogenic substance. Then, PDSs are observed in neurons of all cortical layers. Nevertheless, inhibitory processes are still active in the immediate surround of penicillin application. This inhibition is of similar type to the one described above and has been named "surround inhibition" (Prince and Wilder 1967).

The mechanisms responsible for the generation of PDSs have to be sought mainly in the appearance of slow postsynaptic inward currents carried mainly by Ca^{2+} (Speckmann et al. 1986). The ionic nature of vertical and surrounding inhibition is still unknown. As is clearly seen in Fig. 2A, the inhibition recorded in a layer V neuron consists of two phases with an intermediate depolarization. The first, short inhibitory potential is inverted if KCl-filled microelectrodes are used for intracellular recording, whereas the second, longer lasting one, is not. Therefore, one can assume that the ionic mechanisms underlying these inhibitory processes are only in part Cl^- dependent.

The existence of both the massive excitatory and inhibitory processes, the latter surrounding the former, has an important consequence for the understanding of epileptic phenomena – namely the restriction of focal epileptic activity in the so-called interictal phase.

4 Bioelectrical Activity of Neuronal Pools

The study of electrical activity of single neurons within and in the immediate surround of an epileptic focus established the coexistence of inhibition and excitation. Analysis of field potentials generated by neuronal pools provides insight, into the possible dynamic interplay in space and time of the two processes.

4.1 Field Potentials Along the Plane of the Neocortex

It was shown above that epileptic activity of single neurons exhibit some kind of temporal development, i.e., the transition from spontaneous activity to PDS and to *epileptic inhibition*, and furthermore, the transition from *epileptic inhibition* to PDS. With this sequence of events field potentials recorded simultaneously in the different laminae of the neocortex also undergo a transition. In the first few minutes after application of penicillin to the cortical surface, ISs are negative within the superficial cortical layers but positive in the deep layers. With the diffusion of penicillin into the neocortex the negative ISs grow in amplitude while positive potentials in the deep layers diminish increasingly and finally are converted into negative ones. An exact description of this temporal development can be found elsewhere (Pockberger et al. 1984a).

In this context only the spatiotemporal pattern of current source–sink densities underlying the field potentials of an IS recorded 10 min after epicortical penicillin application is shown here (cf. Fig. 3). The sequence of events starts with a sink in layer IV/V, which because of its prominent location has been named the "trigger sink." After this initial event another much more intense sink density develops in layers II/III. This latter sink often exhibits a double peak, the first of the two being of short duration, the second having a rather hump-like shape. These sink densities within layers II/III reflect inward currents at postsynaptic elements. Comparative studies in different neocortical areas of the rabbit proved

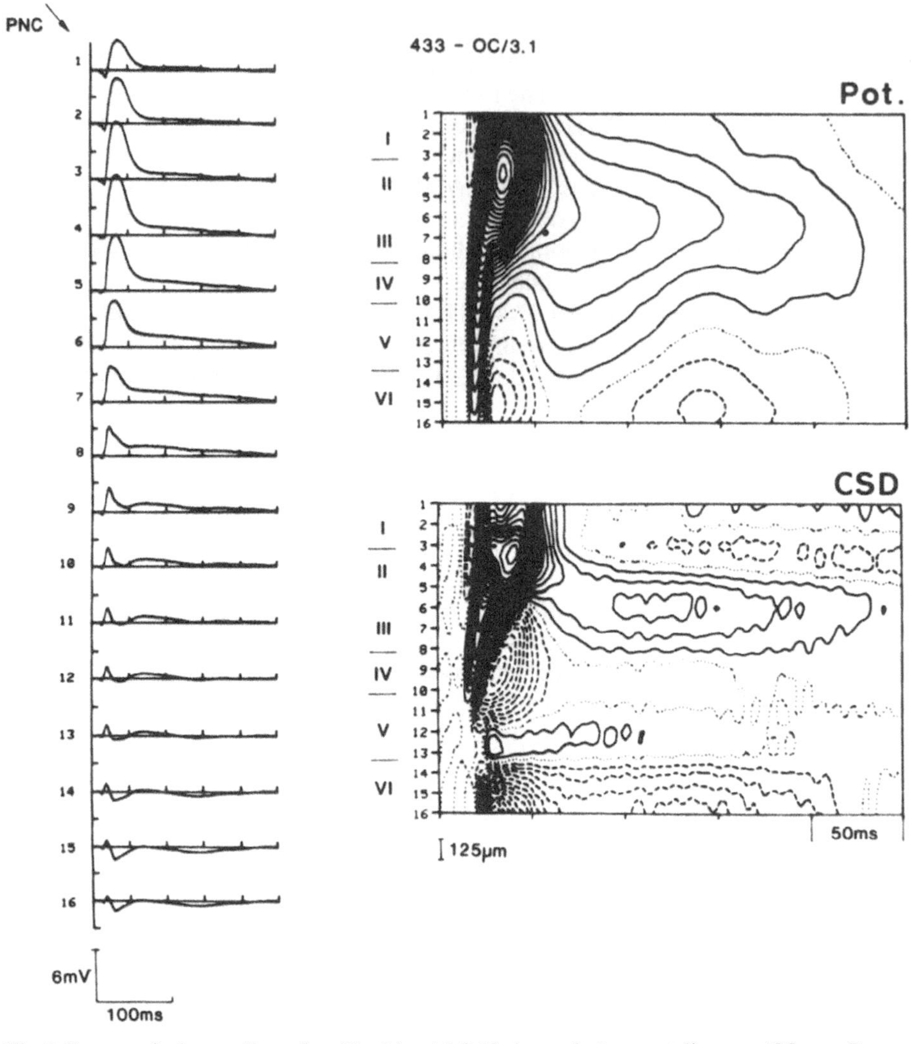

Fig. 3. Intracortical recording of an IS with a 16-fold electrode (contact distance 125 µm; *Roman numerals* in the chronotopograms indicate location of contacts within neocortical layers). *Pot.*, potential chronotopogram – contour line spacing 250 µV; *solid lines*, negative potential; *dashed lines*, positive potential of equal amplitude. *CSD*, current source density plot, contour line spacing 2.5 mV/mm^2; *solid lines*, current sinks; *dashed lines*, current sources of equal strength. *PNC*, penicillin application. This IS starts with a sink density in layers IV/V ("trigger sink") which is followed by more intense sink densities in layers II/III after a brief delay. As these latter sinks develop, sources mainly occur in the deep cortical layers (V, VI). (After Petsche et al. 1987)

that the location of main sink densities in layer II/III always corresponded with the site of the formation of dendritic bundles formed by apical dendrites of layer V and layer II/III pyramidal neurons (Pockberger et al. 1984b). Thus, it has been concluded that the major event underlying ISs is a massive inward current at apical dendrites (cf. Elger and Speckmann 1983).

Concomitantly with the massive sink densities in layers II/III, sources occur in the deep cortical layers. These sources reflect outward currents – on the one hand capacitive ones conteracting the inward currents in the upper cortical layers, on the other hand active currents due to inhibitory potentials. The above interpretation of current sources is supported by findings at the intracellular level which clearly show inhibitory potentials (see also above).

4.2 Field Potentials Recorded at the Cortical Surface

Field potential analysis along the axis of the cortical diameter gives only limited information on how epileptic potentials might spread. Therefore recordings from the cortical surface with an appropriate electrode grid were chosen to study the dynamics of field potentials within and in the immediate surround of an epileptic focus.

The example given in Fig. 4 again demonstrates the steep potential gradients within and in the surround of the focus. In this particular experiment penicillin was applied within the neocortex (electrode 17 – application capillary at 400 μm depth). At the penicillin application site negative monophasic ISs were recorded. Simultaneously biphasic positive–negative potentials occurred at those electrodes nearest the application site (electrodes 6, 7, 10, and 11). Potentials recorded farther from the focus were almost exclusively monophasic and negative. Contour line plotting of the field potentials at equidistant time points revealed that the initial surface positive phase starts out at a distinct site within the focus (near electrode 11), gains in amplitude, and spreads to another site (near electrode 6). While this latter positivity vanishes the negative phase develops at still another cortical volume element (near electrode 7). The subsequent time sequence shows a rise in amplitude and spread of the surface negativity.

From such an analysis of field potentials one can conclude that the different events of epileptic excitation and inhibition show a dynamic interaction in time and space. This is more obvious when studying self-sustained tonic–clonic seizures. Particularly during the tonic phase, which is characterized by discharges of high rhythmicity, positive and negative potential fields exhibit a circling behavior at distinct frequencies (cf. Petsche et al. 1984).

5 Conclusions

In the preceding sections evidence is provided that epileptic phenomena display not only massive excitatory processes but inhibitory ones as well. The main electrical events within and in the surround of an epileptic focus are summarized in Fig. 5. Electrical activity of single neurons within the focus are characterized by the PDS. Within the immediate surround massive inhibitory processes are found – vertical and surrounding inhibition. Electrical activity of neuronal pools, i.e., field potentials, comprises high amplitude negative potentials within the focus and positive ones in the surrounding zone. Neither inhibitory nor excitatory processes, however, are stationary events; rather they display a time- and space-de-

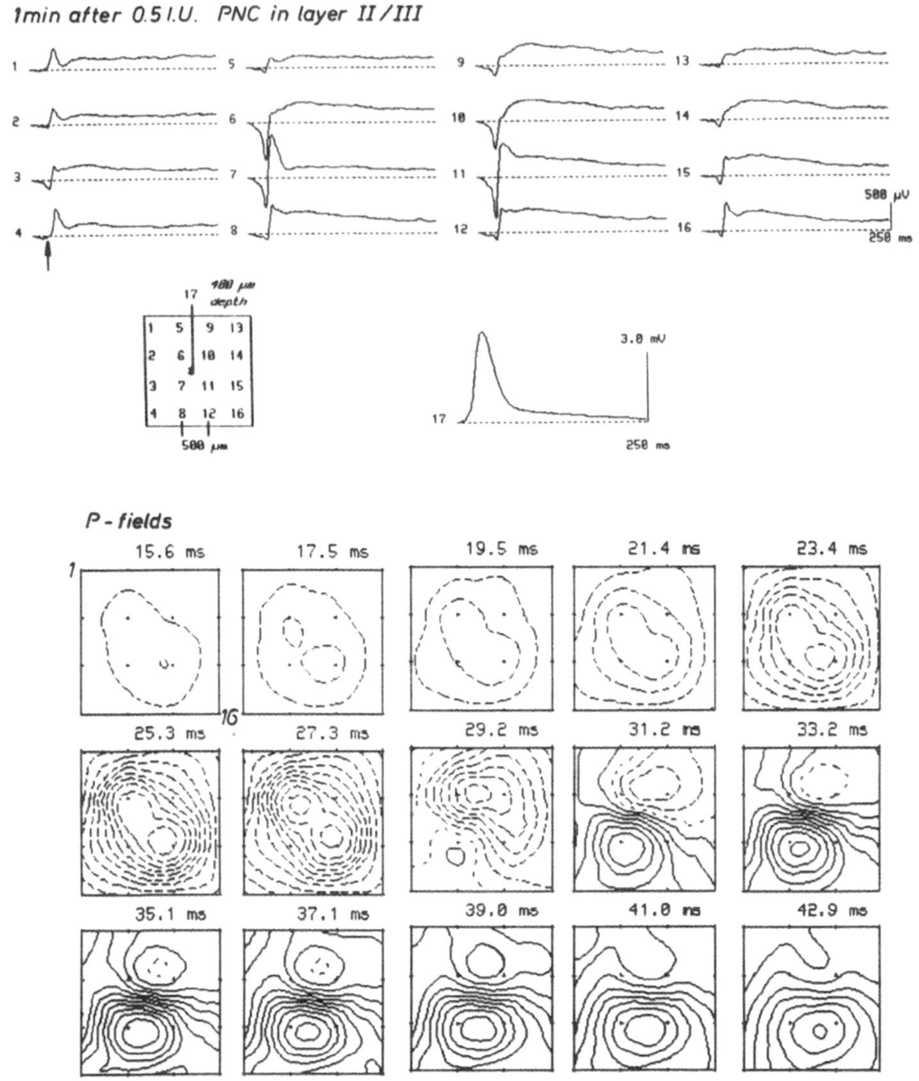

Fig. 4. Field potentials of an IS recorded with 16 electrodes from the cortical surface (*insert on the left* outlines electrode arrangement). Electrode no. 17 was inserted in the middle of the recording grid to a depth of 400 μm and served for penicillin (*PNC*) application. While at the site of PNC application negative ISs occur, biphasic positive–negative potentials are seen at the surface. *P-fields:* field potentials plotted every 1.9 ms from the beginning of IS. IS starts with a positivity (maximum at electrode 11) which spreads to electrode 6. As the positivity diminishes a negative field develops near electrode 7 and spreads from there to adjacent sites

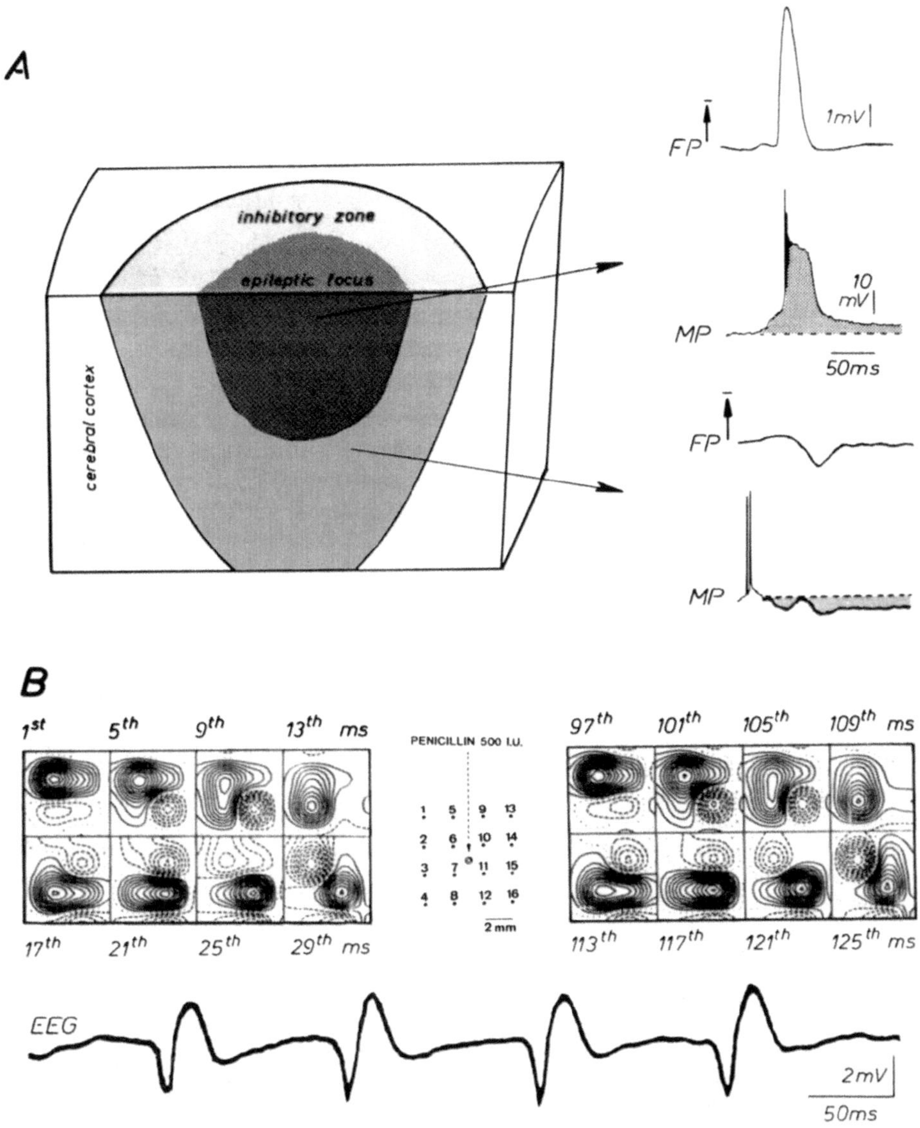

Fig. 5 A, B. Schematic drawing summarizing the main electrical events at the cellular level and at the level of neuronal pools within and in the surround of an epileptic focus. (A) The epileptic focus (*dark shaded area*) is surrounded by an inhibitory zone (*light shaded area*). Within the focus steep negative field potentials (*FP*), i.e., ISs, are accompanied by paroxysmal depolarization shifts of single neurons (*MP*). Within the inhibitory zone mostly positive field potentials are associated with inhibitory potentials at the cellular level. (B) Dynamic interaction of positive (*solid lines*) and negative (*dashed lines*) field potentials recorded on the neocortical surface during tonic seizure activity (*insert* shows electrode arrangement). Note that during the rhythmic 9/s discharges positive and negative field potentials exhibit a circling behavior

pendent pattern even during a single IS (cf. Fig. 4). The most regular spatiotemporal pattern is observed during tonic seizures activity (cf. Fig. 5 B). In this situation a regular circling of positive and negative potential field is observed.

It is concluded that epileptic phenomena represent a special entity of neocortical activity with their own dynamic regularity. The cause of this has to be sought in membrane changes of different types of neuron (inhibitory as well as excitatory and the resulting intercellular communication, which is different from that under "normal" conditions.

6 Summary

Epileptic activity is characterized by typical phenomena at the cellular level as well as at the level of neuronal pools. This chapter summarizes results obtained from studies on both levels of investigation and tries to combine them. To this purpose the classical penicillin model for focal epileptic activity was used.

References

Elger CE, Speckmann E-J (1983) Penicillin-induced epileptic phenomena in the motor cortex. Vertical inhibition. Electroencephalogr Clin Neurophysiol 56:604–622

Matsumoto H, Ajmone-Marsan C (1964) Cortical cellular phenomena in experimental epilepsy: interictal manifestations. Exp Neurol 9:286–304

Petsche H, Pockberger H, Rappelsberger P (1984) On the search of the sources of the EEG. Neuroscience 11:1–27

Petsche H, Pockberger H, Rappelsberger P (1987) Mechanisms leading to the propagation of self-sustained seizure activities. In: Wieser HG, Speckmann E-J, Engel J (eds) The epileptic focus. John Libbey Eurotext, London, pp 59–81

Pockberger H, Rappelsberger P, Petsche H (1984a) Penicillin-induced epileptic phenomena in the rabbit's neocortex. I. The development of interictal spikes after epicortical application of penicillin. Brain Res 309:247–260

Pockberger H, Rappelsberger P, Petsche H (1984b) Penicillin-induced epileptic phenomena in the rabbit's neocortex. II. Laminar specific generation of interictal spikes after applications of penicillin to different cortical depths. Brain Res 309:261–269

Prince DA, Wilder BJ (1967) Control mechanisms in cortical epileptogenic foci. Arch Neurol Psychiatry 16:194–202

Speckmann E-J (1986) Experimentelle Epilepsieforschung. Wissenschaftliche Buchgesellschaft, Darmstadt

Speckmann E-J, Witte OW, Walden J (1986) Involvement of calcium ions in focal epileptic activity of the neocortex. In: Heinemann U, Klee M, Neher E, Singer W (eds) Calcium electrogenesis and neuronal functioning. Springer, Berlin Heidelberg New York, pp 386–395 (Experimental Brain Research Series, vol 14)

IV. Brain Potentials During Sensory and Cognitive Processes

Introductory Remarks

E. BAŞAR

A philosophical description of "memory" stems from Bergson (1939) who stated that: "It is memory which makes the past and future real and therefore creates true duration and true time."

Learning and memory always involve the learning and remembering of something, or some things. Further, although there is as yet no general agreement about just what is learned and remembered (e.g., stimulus-response connections, relationships among stimuli), the things learned and remembered require adequate transduction of such stimulation, and some degree of processing of the result and electrophysiological events within sensory systems (Weinberger 1986). Accordingly, the use of event-related potentials has opened important new windows in the neurophysiology of learning and memory of human subjects.

The collection of papers of this part of the book deal with cognitive potentials and present investigations in a broad range which starts in search of memory localization and extends to semantic properties of brain-compound potentials.

The results of Roy John and coworkers who investigated electrophysiological maps of mental activity suggest the activation of widespread brain regions during cognitive activity rather than localized changes. Furthermore, they state that individuals show different patterns of reorganizatin of brain electrical activity. According to the interpretation of John and coworkers the mental activity is not localized. (The reader should also be referred to earlier works of Roy John and his concept of the hyperneuron; see John 1988.)

Polich gives an overview of the P300 potential by stating that this potential has been associated with memory performance and individual differences in cognitive capability. He introduces a useful comparison of waveforms from identical twins.

It is generally agreed that learning and remembering depend on "attention," the use of an effective strategy, and the cognitive abilities needed to carry out that strategy. The review by Sheer emphasizes "attention." He reports on 40-Hz potentials during sensory and cognitive tasks and provides a useful review of the fast EEG activity which has been largely neglected in publications on cognitive tasks. Of clinical interest are the results of middle-latency response changes in Alzheimer's patients since, as Dr. Sheer commented in the conference, middle-latency responses are mostly related to 40-Hz activity (cf. Changes in middle-latency responses, Buchwald, Part III).

Makeig and Galambos introduce a new approach by recording steady-state evoked potentials, for example, to 40-Hz tone pips maintained for many minutes,

Springer Series in Brain Dynamics 2
Edited by Erol Başar and T. H. Bullock

and than superimposing stimuli that induce cognitive events. This proves to be a sensitive and reproducible measure of the state of the system.

Deecke et al. describe functional investigations into human cortical motor physiology, motor learning, and imaging (thinking of a motor sequence without commanding it) by using electric and neuromagnetic fields or cerebral blood flow. The performance-related DC potential shifts over the primary and secondary sensory areas vary with stimulus modality. It is worth noting that motor-imaging activates primary and secondary visual areas but not the supplementary motor area.

Gevins presents a model that takes into account not only event-related potentials but also considers integrative-motor and somesthetic-motor components. He also describes a model of preparatory and feedback "networks."

Maurer describes P300 cartography during physiological aging and in diseases such as dementia of Alzheimer's type, Wilson's disease, and schizophrenia. The cartography of the test enables a differentiation between dementia and psychosis. The work by Ihl et al. emphasizes the role of the ascending cholinergic system in cognitive processing and memory formation of the brain and shows the increase in P300 activity in frontal to temporal regions under the influence of physostigmine.

Following Tulving's (1972) distinction, semantic memory refers to a person's store of general knowledge, whereas "episodic memory" refers to contextually bound events whose time and place of occurrence form part of the remembered information.

Münte et al. indicate that the N400 wave of the ERPs, first measured by Kutas and Hillyard, is sensitive to semantic processing. They use a lexical-decision paradigm to record differences in ERP as a function of semantic association. Bötzel et al. describe ERP data in search of "face responsive" EP components and conclude that such components do exist, especially most prominently at the midline electrodes.

Rockstroh describes experimental paradigms for examination of area-specific organization of slow potentials and concludes that surface-negative slow potential shifts facilitate performance, representing cerebral resources and that these resources are allocated in an area-specific manner.

To these chapters may be added the contributions by Lehmann (Part I) and Başar and coworkers (Part I), which deal with microstates in EEG and cognitive performance in a broad sense. P300 in the experimental animal brain is described by Buchwald (Part III), extending the picture of cognitive potentials to nonhuman species.

It might be noted that the collection of papers which describe cognitive ERPs represents a broad spectrum and the presented data are of general physiological interest and not solely for psychophysiological research. The papers should appeal to scientists of neurophysiology and behavior in general. The picture can be complemented by reading the papers in Parts I and II.

References

Bergson H (1939) Matière et mémoire. Essai sur la relation du corps à l'esprit. Presses Universitaires de France, Paris

John ER (1988) Resonating fields in the brain and the hyperneuron. In: Başar E (ed) Dynamics of sensory and cognitive processing by the brain. Springer, Berlin Heidelberg New York, pp 368–377 (Springer series in brain dynamics, vol 1)

Tulving E (1972) Episodic and semantic memory. In: Tulving E, Donaldson W (eds) Organization of memory. Academic, New York

Weinberger NM (1986) The neurophysiology of learning: a view from the sensory side. In: Squire LR, Butters N (eds) Neuropsychology of memory. Guilford, New York, pp 489–503

Quantitative Electrophysiological Maps of Mental Activity

E. R. John, L. S. Prichep, and R. J. Chabot

1 Introduction

At the first of the Berlin conferences I presented the results of our attempt to produce metabolic maps of a memory, which have since been published in detail (John et al. 1986). In brief, green and red contact lenses on the eyes of split-brain cats were used to achieve different visual input to the two hemispheres of the brain. During an initial uptake period following the injection of 2-fluorodeoxyglucose (2-FDG) labeled with radioactive carbon [14C], one hemisphere received input about a novel pair of triangles on the two doors of a Y-maze, while the other hemisphere received input about previously learned discriminanda (a circle and a star) on the two doors. The animal used the randomized positions of the discriminanda to select the door behind which it obtained food. During a second uptake period following the injection of 2-FDG labeled with radioactive fluorine [18F], both hemispheres received input about the novel pair of triangles, which had no cue value since food could be obtained behind either door. The amount of time spent in the maze, the distance run, and the food ingested were equated in the two uptake periods.

In the first uptake period, the glucose metabolism of local brain regions in one hemisphere reflected only nonspecific influence on neuronal activation, while the glucose metabolism of regions in the other hemisphere reflected the same nonspecific influence *plus* the activation of neurons engaged in the retrieval of memories about the previously learned specific significance of the discriminanda. In the second uptake period, the glucose metabolism of both hemispheres reflected only nonspecific influences. Serial sections of the brain were obtained immediately after the second uptake period and were each used to obtain two autoradiographs (ARGs), one representing the anatomical distribution of the [14C]-labeled 2-FDG reflecting neuronal activity during the first uptake period and another representing the distribution of the [18F]-labeled 2-FDG reflecting neuronal activity during the second uptake period.

Microdensitometric scanning was used to quantify the metabolic activity of cells contained in 250 μm × 250 μm "pixels" in each ARG. Digital subtraction of the quantified images yielded "difference images" reflecting nonspecific metabolic variability or "noise" on one side of the split brain but the retrieval of previously learned information plus noise on the other side. Using the mean and standard deviation of all pixels in the difference images of the whole brain, these images were Z-transformed to rescale the metabolic difference between the two uptake periods for each pixel in the metric of *probability*. Using this probability

Springer Series in Brain Dynamics 2
Edited by Erol Başar and T. H. Bullock

scaling, it was found that about 100 000 000 cells showed a significant increase in metabolism in one side of the split brain when the memory retrieval process occurred.

Maps of the anatomical distribution of these cells which mediated the process of retrieval of this learned information showed them to be distributed among widespread regions of the brain. These findings were a metabolic confirmation of data previously obtained in studies of EEG rhythms, evoked potential waveshapes, and multiple and single unit discharges recorded with chronically implanted electrodes placed in numerous brain regions, reviewed at our previous meeting. Thus, a self-consistent picture has emerged from a variety of electrophysiological studies using methods with a wide range of anatomical resolution, from large neuronal populations down to single cells, and with an autoradiographic method permitting exhaustive quantification of local metabolic activity. This picture indicates that the system which mediates the storage and retrieval of information in the brain is anatomically diffusely distributed and engages vast numbers of neurons in many different brain regions. This picture cannot be reconciled with notions of narrowly localized neuronal circuits established by a particular experience and dedicated to its unique representation.

In the time since the last meeting, our laboratory has been exploring ways to obtain topographic maps of brain electrical activity associated with the performance of different mental tasks in human subjects, in order to see whether we could obtain evidence of similar brain mechanisms in man by using some new functional imaging techniques which we have developed. The remainder of this chapter should be considered as a discussion of work in progress, which is far from a finished study.

2 Mental Tasks

After a series of pilot studies, we selected a number of different mental tasks and different methods of quantitative analysis for more careful explorations. We have used six different tasks and three different kinds of electrophysiological data to construct several different kinds of map. All of these variations have been studied in six normal subjects. A subset has been further studied in large groups of normal, depressed, and schizophrenic subjects.

Two types of task have been studied: *categorical matching,* or *"semantic"* tasks, and *memory* tasks.

In the semantic tasks, a pair of stimuli is presented every 2 s from the beginning to the end of the recording session. The task requires the subject to count the number of pairs in which both stimuli are in the same category. Four kinds of matching stimuli have been used:

1. Pairs of abstract forms in different orientations; no names can be attached to these forms. This task is referred to as "abstract form," or *AF*.
2. Pairs of pictures of real objects, which have names. This task is referred to as "picture semantic," or *PS*.

3. Pairs of words which are the names of those objects previously seen, printed in various fonts. This task is referred to as "word semantic," or *WS*.
4. Pairs of words like those previously seen, but played through earphones from a tape recording. This task is referred to as "auditory semantic," or *AS*.

These are complex tasks in which decision requires each member of a pair to be categorized using logical or language criteria. *These criteria are also dependent to some extent upon memory.*

Two kinds of memory task have been studied:

1. A *short-term memory* task, which we abbreviate as STM, requires the subject to learn a spoken list of up to ten two-digit numbers. These numbers are repeatedly presented during a brief training session. The subject must rehearse the learned subset during the recording period and must recite them at the end of that time.

2. A *long-term memory* task, which we call LTM, requires the subject to recall sequentially the faces of friends with names whose initial letter proceeds through the alphabet, beginning with the letter A. As each face is visualized, the subject tries to recall as many details as possible about his last interaction with that person. When those details are exhausted, a friend is recalled whose name begins with the next letter of the alphabet. This is a very poorly defined task, in which old memories with unspecified personal affective content are retrieved in a manner that cannot be confirmed. We rely upon the assurances given by our subjects that they actually perform the task. Most of them report it as extremely engrossing and pleasant.

All recordings were made from 19 monopolar derivations referred to linked earlobes, using the 10/20 placement system, plus an EOG electrode pair to detect eye movements. An on-line algorithm was used to reject artifacts. Two minutes of artifact-free EEG recording, or 100 evoked responses, were collected, depending upon the type of mapping used. EEG was collected as 48 segments each 2.5 s long. EP analysis epochs were 500 ms long.

3 Norms, Z-Transforms

A variety of quantitative analyses were performed. Many of the maps I will present are Z-transformed. Whenever this is the case, the data have first been normalized to achieve Gaussianity. Using normative data for the eyes-closed EEG from 120 healthy normal subjects, Z-transformations were then carried out relative to the mean and standard deviation of the normative distribution. Topographic maps were color-coded to indicate the statistical significance of observed task effects relative to baselines (BL), based upon these Z-scores (some figures use only a gray scale). Our previous studies indicate reasonable test–retest replicability within a subject under baseline conditions (John et al. 1983, 1987).

In Fig. 1, the absolute power of the EEG (in microvolts squared) has been divided into four frequency bands: Delta, Theta, Alpha, and Beta. The figure presents data from the eyes-open (EO) resting EEG, the WS, the eyes-closed (EC)

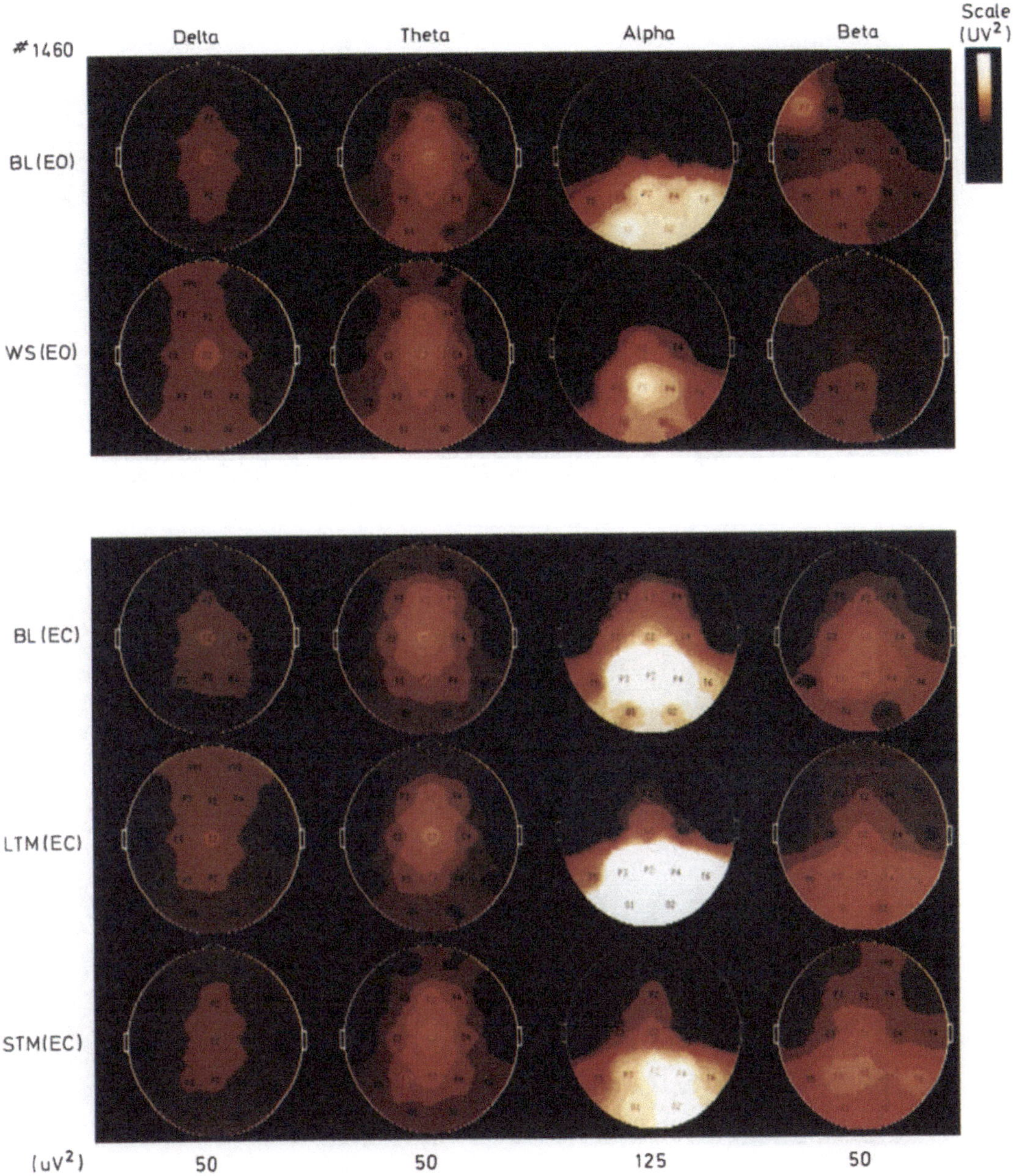

Fig. 1. Absolute power (μV^2) in the delta, theta, alpha, and beta bands recorded during EO EEG BL, WS, EC EEG BL, LTM, and STM

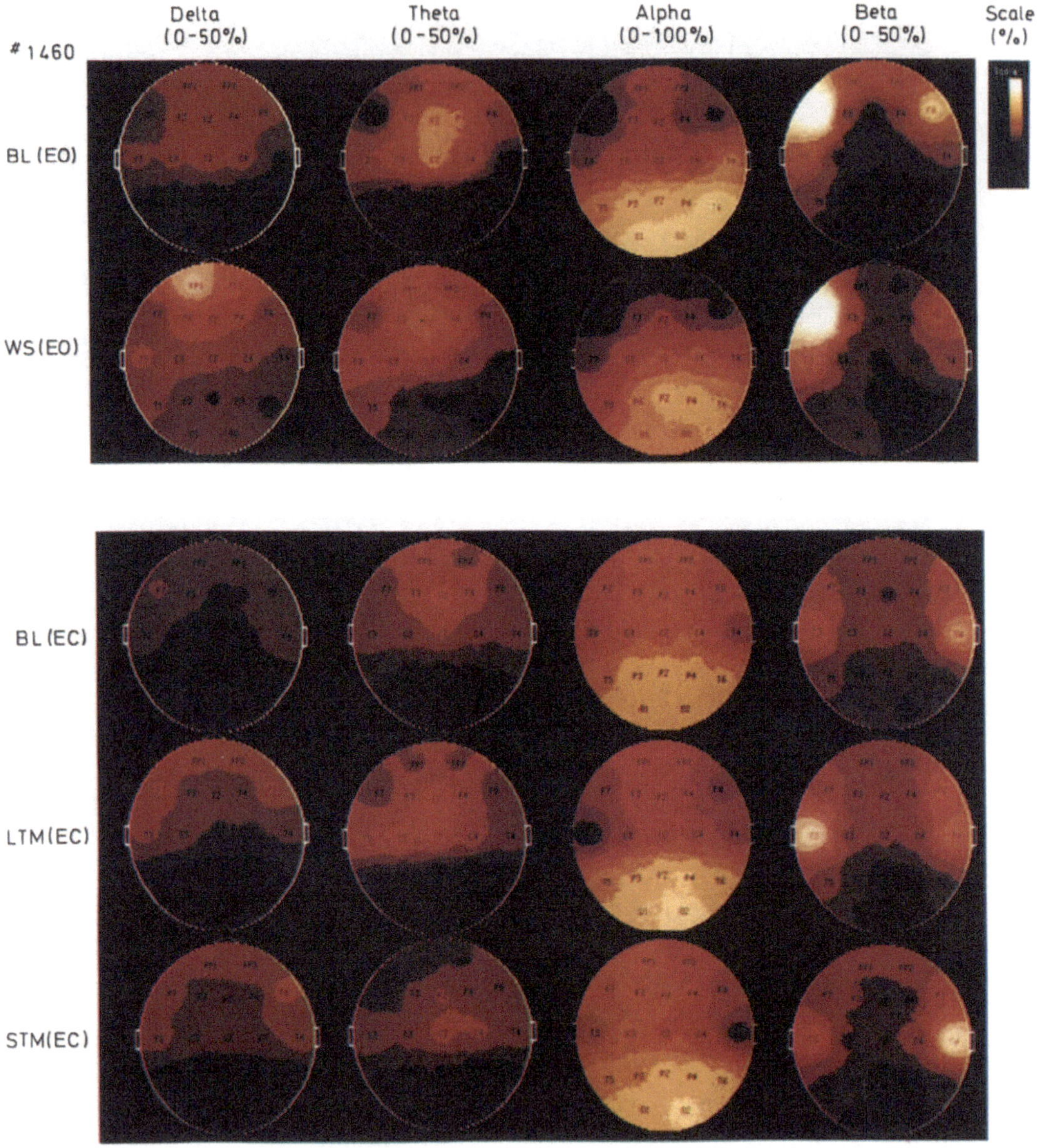

Fig. 2. Same data as in Fig. 1, now shown as relative (%) power in the delta, theta, alpha, and beta bands

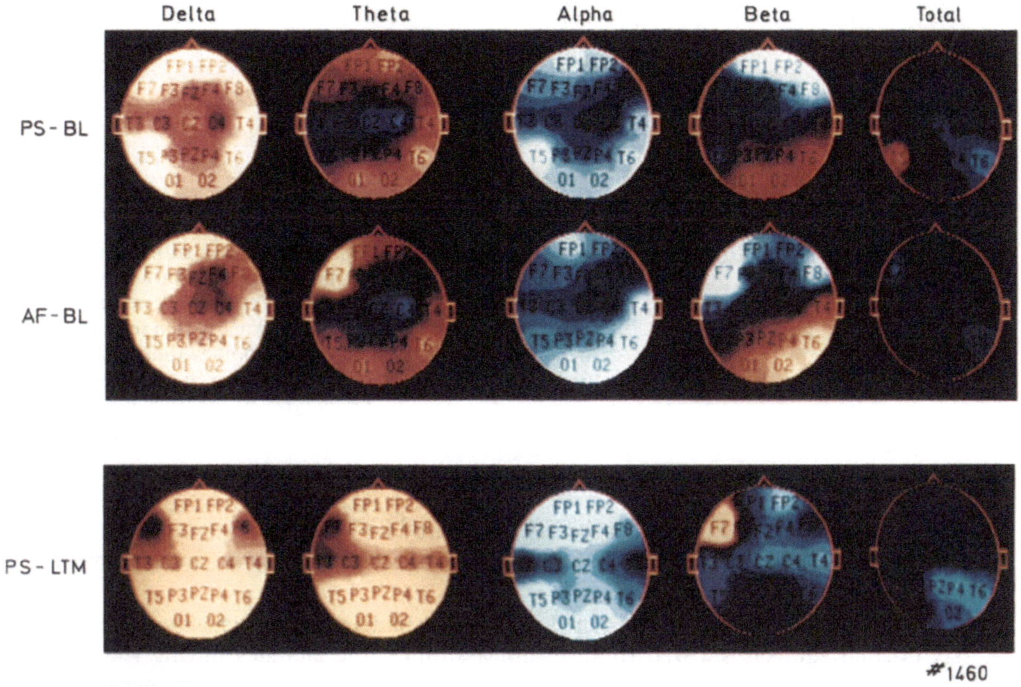

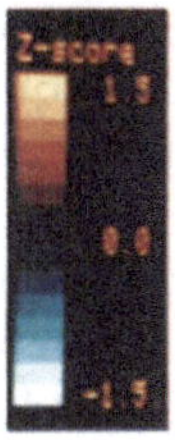

Fig. 3. Same data as in Figs. 1 and 2, now shown as the difference in the Z-score of each value when the measures are Z-transformed against age-regressed normative data and the appropriate baseline values are subtracted from the values obtained during each mental task

resting EEG, the LTM, and the STM tasks. Some changes from baseline (BL) can be seen in each band but are hard to evaluate in this form.

Figure 2 presents the same data shown in Fig. 1, now seen as *relative or percent power in each band.* The changes are now somewhat easier to see but are still hard to evaluate. WS seems different from STM and LTM, which seem somewhat similar.

Two steps can be taken to examine these changes more precisely. First, all results are Z-transformed against age-regressed normative data. Second, the appropriate baseline is subtracted from each mental task, yielding the maps of the *change in Z-score* of electrophysiological features observed during each task.

Figure 3 illustrates results for the same subject whose data were presented in Fig. 1 and 2. The changes during the two categorical matching tasks (PS and AF) are strikingly similar and involve the whole spectrum and most of the cortical regions very significantly. The scale is now coded in standard deviations.

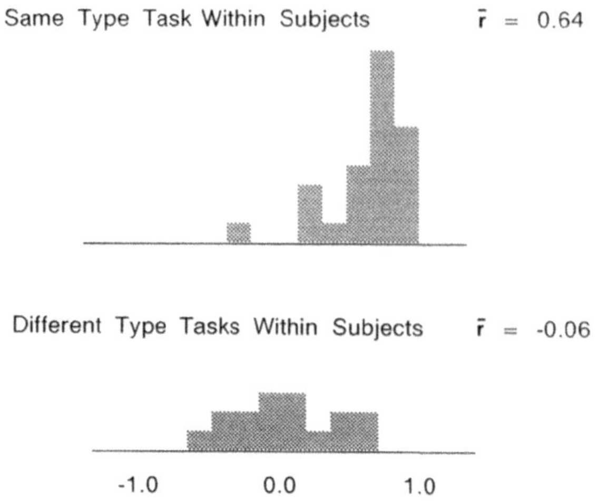

Fig. 4. Histograms showing correlations of Z-transformed relative power (against age-regressed normative data) within individual subjects both within a class of cognitive tasks and across different types of cognitive task

In the third row, the changes during the PS and LTM tasks are compared. The differences between brain activity in the two types of mental tasks, are extremely significant.

Examination of the changes in Z-scores for relative power across a group of six normal subjects showed that, within a subject, the changes during the two memory tasks were very similar. The changes in the four matching tasks were also very similar. Each of these two types of mental task seemed to engage widespread cortical regions very significantly, with changes which involved all four frequency bands. However, the memory tasks engaged the cortex very differently than the matching tasks.

In order to evaluate these changes in relative power during mental tasks in a quantitative manner, we computed correlation coefficients between the changes in Z-scores across the 19 electrode positions in the topographic map when the Z-scores of each task were compared to those of every other task within each subject, after subraction of the appropriate baselines (difference Z-scores).

Figure 4 shows the distributions of correlation coefficients between the changes in Z-scores for relative power in the delta, theta, alpha, and beta bands within the same subjects, when *two tasks in the same class* were compared (upper histogram) and when *two tasks in different classes* were compared (lower histogram). That is, the correlations between STM vs LTM, AF vs PS, WS vs PS, AS vs PS, etc. were pooled in the upper histogram, while the correlations between STM vs AF, LTM vs AF, STM vs PS, LTM vs PS, etc. were pooled in the lower histogram, across all subjects.

The mean correlation $\bar{r}$ among the maps of changes in Z-score for relative power was 0.64, within tasks of the same class. For tasks in different classes, the mean correlation $\bar{r}$ was -0.06. The value of χ^2 for the difference between these two distributions was 8.3, significant at $P<0.005$.

Using the same tasks, data have been collected from two very different additional measures of brain electrical activity. The results, documented in the following two sections, support quite the same conclusions.

4 Cross-spectral Coherence Matrix

In the first method, the average cross-spectral coherence was computed for each lead versus every other lead, separately for the four frequency bands and across the whole EEG spectrum, at 0.05-Hz intervals. Each of 48 2.5-s EEG samples was analyzed and the results were averaged to obtain mean values and standard deviations. The resulting data were normalized to achieve Gaussianity (log COH^2/ 100-COH^2) and Z-transformed against normative data from 120 healthy persons.

In contrast to the relative power features discussed earlier, which quantify the electrical activity *within* local brain regions, these cross-spectral coherence features quantify the relationships of electrical activity *between* all pairs of brain regions.

Figure 5 illustrates the interpretation of the topographie maps of cross-spectral coherence. The analysis of cross-spectral coherence in an individual EEG yields five such maps, representing the average value of each coherence relationship in the delta, theta, alpha, and beta bands and across the whole frequency spectrum. Each map depicts the 19 electrodes of the 10/20 System as 19 small heads, with the front of each head facing upward. Each small head contains one brightened yellow spot representing the position of the fiducial electrode and 18 rectangles representing the coherence between that position and the other 18 positions of the 10/20 System. The color of each rectangle encodes the significance of the deviation from the value of the population norm. All data discussed here were computed as the difference when the Z-score of the appropriate baseline was subtracted from the Z-score obtained from the EEG samples recorded during each mental task (difference Z-scores).

Figure 6 shows such difference Z-score maps (task minus baseline) for a typical subject, observed in the beta band during the AF, WS, STM, and LTM tasks. The correlations between the matrix of difference Z-scores for the two tasks in the same class are shown between the vertical arrows next to the topographic maps. Note that the changes in relationships between brain regions which occur during performance of these mental tasks involve widespread regions of the cortex. Different tasks in the same class cause similar patterns of new relationships to be established ($r=0.80$, AF vs WS; $r=0.90$, STM vs LTM), but different classes of mental activity appear to be mediated by different organization of brain relationships, reflected by the correlations shown between the horizontal arrows ($r=0.41$, AF vs STm; $r=0.60$, WS vs LTM).

These changes in coherence relationships during mental tasks were evaluated quantitatively by computing the correlation coefficients between the difference Z-scores at all positions of the cross-spectral coherence matrices for all frequency bands, obtained during performance of any two tasks by the same subject. Figure 7 presents histograms of the correlation coefficients thus obtained across the set of six normal subjects. The *bottom* histogram shows that each subject displayed consistent patterns of organization of brain relationships within each class of tasks. When *two tasks in the same class* were compared within the same subject, the mean correlation coefficient, $\bar{r}$, was equal to 0.66. Each different class of task within each subject was subserved by a different pattern of brain organization.

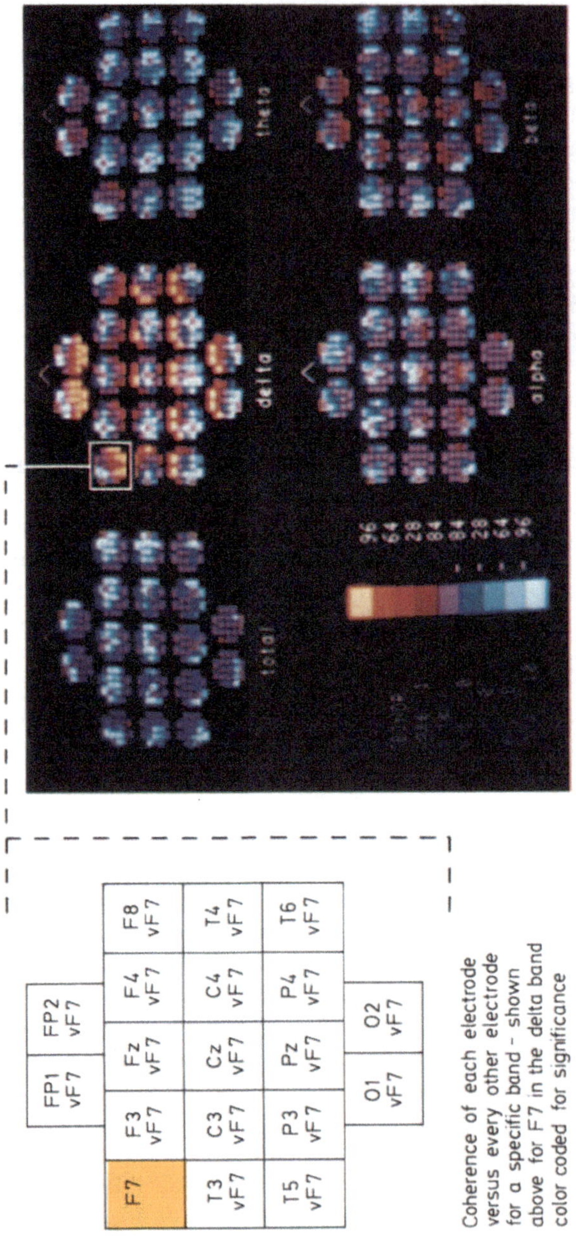

Fig. 5. Schematic explanation of cross-spectral coherence measure using F_7 lead for the delta frequency range. Note each electrode position (*blown-up insert on left*) shows the coherence of the brain region with all other regions

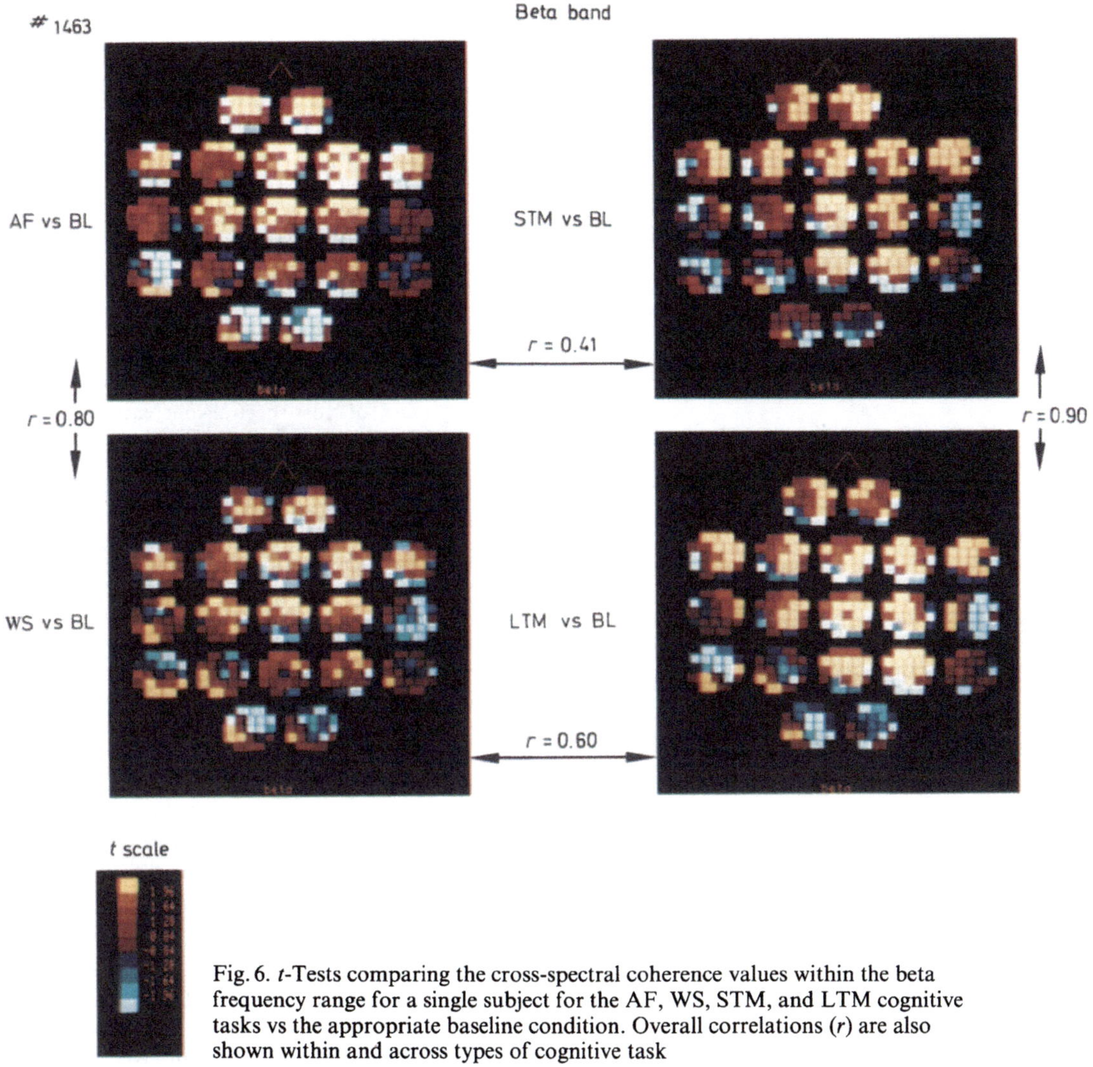

Fig. 6. *t*-Tests comparing the cross-spectral coherence values within the beta frequency range for a single subject for the AF, WS, STM, and LTM cognitive tasks vs the appropriate baseline condition. Overall correlations (*r*) are also shown within and across types of cognitive task

The *middle* histogram shows that when *two tasks in different classes* were compared within the same subject, the mean correlation coefficient, $\bar{r}$, was equal to 0.02. Different subjects appeared to use distinctive cognitive styles. The mean correlation, $\bar{r}$, between different subjects performing the same task was equal to 0.12. By chi-squared, the differences between these distributions were highly significant ($P<0.0005$ for same tasks within subjects vs between subjects).

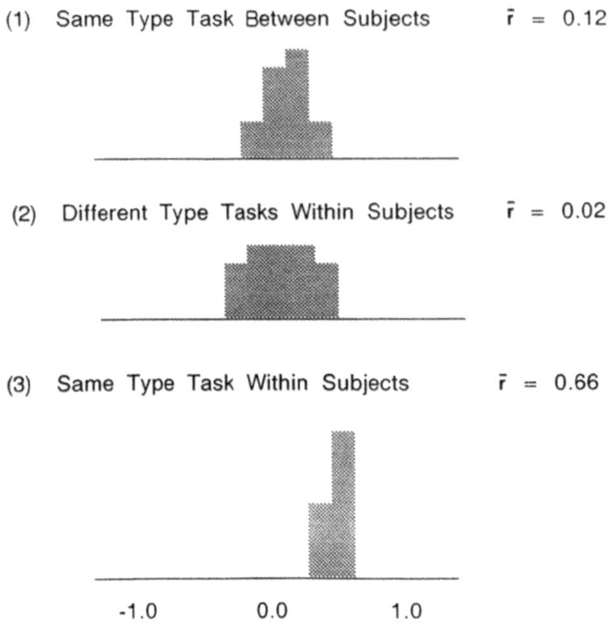

Fig. 7. Histograms showing overall mean correlations ($\bar{r}$) of cross-spectral coherence changes from baseline for total power for the same type of cognitive task between (*1*) and within (*3*) subjects and for the different types of cognitive task within subjects (*2*)

5 Noncontingent Probe Factor Scores

We have analyzed brain electrical activity during these mental tasks with the *noncontingent probe* (or NCP) technique. In NCP, evoked potentials (EPs) are collected to an irrelevant "background" or "noncontingent" stimulus at rest. "Task EPs" collected to the same noncontingent stimulus during performance of a task are then subtracted from these "rest EPs." At the latencies at which the afferent volley from the NCP traverses anatomical regions in which neuronal ensembles are engaged by the task, differences will appear between the rest EPs and the task EPs. Such differences may be in either direction, since local ensembles may be either engaged or disinhibited due to engagement of other ensembles elsewhere.

In order to analyze the differences between the rest EPs and the task EPs quantitatively, we utilized the methods of factor analysis. We factor analyzed the EPs obtained from the 19 electrodes of the 10/20 System for 120 normal subjects, across a variety of visual and auditory stimulus conditions. Averages were based upon 100 stimuli, and the analysis epoch was 500 ms. We used principle component analysis, followed by a Varimax rotation. We obtained seven general factors. The details of these computations have been described elsewhere (John 1988).

The EPs obtained from any electrode can be described precisely as a weighted linear combination of seven factor waveshapes, plus a small reconstruction error. The weightings are *factor scores*. Thus:

$$EP_i(t) = \sum_{j=1}^{7} A_{ij} F_j(t) + R\,, \qquad (1)$$

where $EP_i(t) = EP$ waveshape from electrode i; A_{ij} = factor score giving amount of factor j present in EP_i; $F_j(t)$ = waveshape of factor j; R = RMS reconstruction error.

For each type of stimulus, the mean value and standard deviation of the distribution of factor scores required to reconstruct the *EP* at each electrode in the normative sample has been ascertained. Knowledge of these statistics permits Z-transformation of the factor scores and the residual such that:

$$EP_i(t) = \sum_{j=1}^{7} Z_{ij} F_j(t) + Z_R, \qquad (2)$$

where $EP_i(t) = EP$ waveshape from electrode i; $Z_{ij} = Z$-transform of A_{ij}; $F_j(t)$ = waveshape of factor j; $Z_R = Z$-transform of RMS residual error.
Any set of EPs, recorded from the 19 electrodes of the 10/20 System under specified stimulus conditions in an individual subject, can now be described as a set of seven topographic maps. One map represents each factor, with the factor score indicating the contribution of that factor to the EP from each electrode, *Z-transformed* relative to the normative data and *color coded* proportional to the statistical significance of the observed deviation from the normative mean value. An eighth map represents the root mean square factor Z-score, a ninth map the residual error, and a tenth map the total power in the EP.

Using this method, we calculated the difference in Z-transformed factor score, ΔZ_{ij}, for each factor score needed to reconstruct the rest EP and task EP from the 19 electrodes of the 10/20 System in three category matching tasks in several normal subjects. For the WS and PS tasks, the NCP was a click (CL). For the AS task, the NCP was a flash (FL). The topographic maps of ΔZ_{ij} for one subject are shown in Fig. 8. Note that all difference maps in Fig. 8 were constructed by subtracting the task map from the baseline map.

The similarities among these maps across the different matching tasks is striking. Cross-correlations were computed between the sets of factor Z-score differences obtained in these three cateogory matching tasks. Factors 4, 6, and 7 showed particularity high similarities across these tasks, as seen in Table 1.

Table 1. Overall $\bar{x}$ correlations between Z-transformed factor score changes (shown in Fig. 8) calculated across all subjects comparing the three cognitive tasks (PS, WS, AS) used with the noncontingent probe method

	Factor 1	Factor 2	Factor 3	Factor 4	Factor 5
PS vs. WS	0.48	0.49	0.36	0.90	0.45
PS vs. AS	0.31	0.03	0.26	0.62	−0.39
WS vs. AS	0.1	0.39	0.84	0.71	−0.20

	Factor 6	Factor 7	RMS Z	Reconst.	Residuals	Power
PS vs. WS	0.82	0.79	0.36	0.53	0.01	0.41
PS vs. AS	0.35	0.54	0.23	0.63	0.35	0.18
WS vs. AS	0.56	0.69	0.09	0.30	0.19	0.31

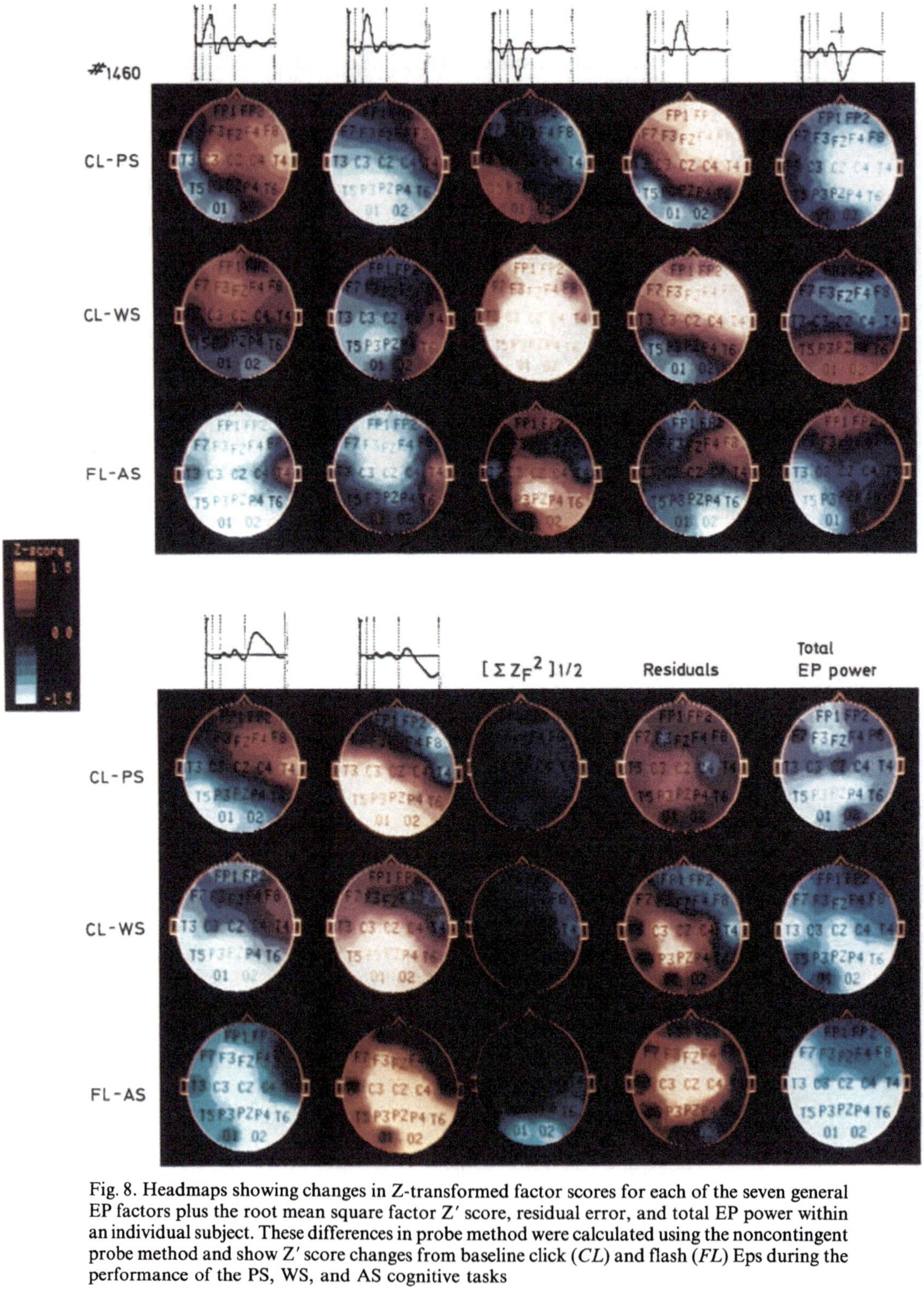

Fig. 8. Headmaps showing changes in Z-transformed factor scores for each of the seven general EP factors plus the root mean square factor Z′ score, residual error, and total EP power within an individual subject. These differences in probe method were calculated using the noncontingent probe method and show Z′ score changes from baseline click (*CL*) and flash (*FL*) Eps during the performance of the PS, WS, and AS cognitive tasks

6 Conclusions

In contrast to the relative power and cross-spectral coherence EEG data, which reflect the activity of superficial cortical regions closely subjacent to the scalp electrodes, the NCP data reflect a wide variety of subcortical processes as well as cortical–subcortical transactions. The NCP data indicate engagement of a substantial portion of the subcortex by these tasks.

Basically similar conclusions are tentatively suggested by the data thus far collected by these three different methods:

1. Within a subject, different mental tasks within a particular class appear to be performed by brain activity organized in a characteristic way.
2. Within a subject, mental tasks in different classes are performed by brain activity organized in different ways.
3. Individuals have different physiological cognitive styles.
4. Mental activity is not localized. Both cortical and subcortical structures participate in each of these mental tasks, and widespread brain regions appear to be engaged by each mental task.

These findings in man are quite compatible with the findings from EEG, EP, and unit recordings, as well as from double-labeled 2-deoxyglucose metabolic maps in experimental animals, revealing the engagement of widespread brain regions in cognitive behavior.

7 Summary

The relative utility of quantitative EEG and EP measures for documenting changes in the organization of brain electrical activity during the performance of different cognitive tasks is explored.

To this end, EEG and noncontingent probe EPs were obtained while subjects performed cognitive tasks which could be classified as either categorical matching or memory tasks. Categorical matching tasks required "same" vs "different" decisions about pairs of visually presented words, line drawings, or abstract forms, and auditory stimulation consisting of word pairs. Memory tasks required the continual short-term memory activaton of sequences of two-digit numbers or the continual search and retrieval of autobiographical information from short-term memory.

Quantitative EEG changes from baseline control conditions were examined separately for the delta, theta, alpha, and beta frequency bands for absolute power, relative power, Z-transformed relative power, and for the cross-spectral coherence across 19 monopolar leads. Noncontingent probe EP changes were assessed using Z-transformed factor scores. No consistent replicable patterns of EEG changes with cognitive activity were found for absolute or relative power. Consistent replicable EEG changes occurred in Z-transformed relative power and cross-spectral coherence within a class of cognitive tasks within individual subjects, and this pattern was replicated for three of the factors derived from the noncontingent probe EP data. All consistent changes found suggest the activation of widespread brain regions during cognitive activity rather than localized changes, and that individuals show different patterns of brain electrical activity reorganization.

References

John ER (1989) Topographic brain mapping. In: Maurer K (ed) Topographic brain mapping of EEG and evoked potentials. Springer, Berlin Heidelberg New York

John ER, Prichep L, Ahn H, Easton P, Fridman J, Kaye H (1983) Neurometric evaluation of cognitive dysfunctions and neurological disorders in children. Prog Neurobiol 21:239–290

John ER, Tang Y, Brill AB, Young R, Ono K (1986) Double-labeled metabolic maps of memory. Science 233:1167–1175

John ER, Prichep LS, Easton P (1987) Normative data banks and neurometrics. Basic concepts, methods and results of norm constructions. In: Gevins AS, Rémond A (eds) Methods of analysis of brain electrical and magnetic signals. EEG Handbook. Elsevier, Amsterdam (revised series, vol 1)

Neuroanatomical Contributions to Individual Differences in P300 Morphology

J. Polich

1 Overview of P300

The P300 or P3 component of the event-related brain protential (ERP) is produced when subjects attend and discriminate stimulus events which differ from one another on some dimension. It is a large (ca. 10–20 μV), positive-going potential with a latency of approximately 300 ms when elicited with a simple auditory discrimination task, and is of maximal amplitude over the midline central and parietal scalp areas (Donchin et al. 1986; Pritchard 1981). Although the neural orgins of the P3 are still being sought, depth electrode recordings and magnetic field studies in human suggest that at least some aspect of the P3 is generated in the medial temporal lobe areas, most likely including portions of the hippocampus (Halgren et al. 1980; McCarthy et al. 1982; Okada et al. 1983). Since these areas have been associated with learning and memory, the P3 has been linked to fundamental brain events involved with memory-updating processes. This theoretical interpretation has been supported by a variety of studies which have demonstrated changes in the P3 that reflect differential memory performance between experimental conditions (Fabiani et al. 1985; Johnson et al. 1985; Karis et al. 1984; Neville et al. 1986), as well as individual differences for normal and impaired cognitive function (Brown et al. 1982; Howard and Polich 1985; Polich et al. 1983, 1986; Squires et al. 1979). However, because the P3 is sensitive to these cognitive brain events, wide individual variation exists in its overall morphology and timing. This chapter will review some of the initial studies which have attempted to determine the sources of this individual variability in the P3 component of the ERP.

2 Normal P300 Variation

The P3 was first reported in 1965 and investigated subsequently in a wide variety of paradigms which have attempted to delineate the underlying causes and theoretical underpinnings for its generation. However, basic studies concerning the effects of stimulus factors (Magliero et al. 1984; Papanicalaou et al. 1985; Polich et al. 1985) and response task parameters (Pfefferbaum et al. 1985; Polich 1986a, 1987a, b; Ragot 1984) have been reported only recently. Hence, the role of experimental variables which contribute to variations in the P3 across conditions and individuals are still being explored.

Springer Series in Brain Dynamics 2
Edited by Erol Başar and T. H. Bullock

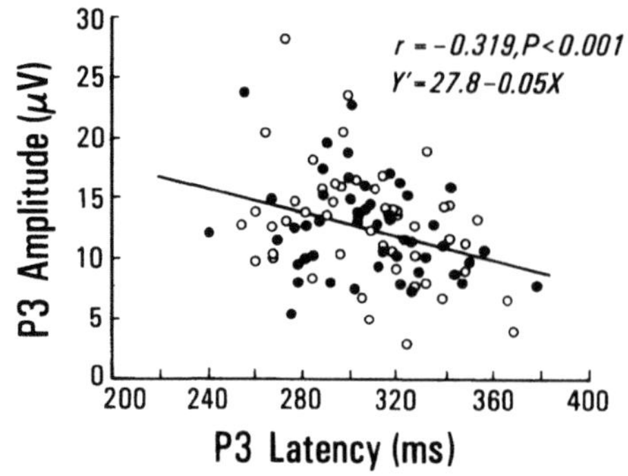

Fig. 1. Scattergram of P3 amplitudes and latencies from 50 male (*closed circles*) and 50 female (*open circles*) undergraduate university students (after Polich 1986 b)

In addition to investigations of paradigmatic influences on the P3, several reports also have begun to describe how individuals may vary in their P3 responses from trial to trial. Studies of individual waveform replicability have suggested that the P3 component is reasonably reliable between measurement times for an individual, with correlation coefficients for P3 amplitude ranging between 0.60 and 0.80 and P3 latency measured at 0.40–0.70 (Fabiani et al. 1987; Polich 1986 b; Sklare and Lynn 1984). Despite this relative statistical consistency within an individual, a wide variety of P3 values are observed between individuals. When the P3 ERP was elicited with a simple auditory discrimination task in a large ($n=100$) highly homogeneous sample of university undergraduates, P3 amplitude was found to vary from about 2–3 μV to 30–40 μV and P3 latency from 240 to 380 ms. Moreover, P3 amplitude and latency variation were associated, since individuals with relatively large amplitudes have relatively shorter latencies and vice versa (Polich 1986 b). This relationship is illustrated in Fig. 1 and indicates that the amplitude and peak latency of the P3 component were negatively correlated across individuals. Thus, the specific neurons which produce the P3 appear to respond in larger numbers and more rapidly in some individuals, but in fewer numbers and more slowly in others.

It should be noted also that no statistically reliable effects for subject sex were obtained from these data, although a tendency was demonstrated by females to produce somewhat larger P3 amplitudes than male subjects, especially at the Pz electrode site. In addition, no effects of P3 amplitude or latency from a similar paradigm have been observed between the onset of menses and the time of ovulation, with the same pattern of results observed for both spontaneous cyclers and birth control pill users (Fleck and Polich 1988). Even though these findings seem to imply that there are no sex differences for the P3 component as have been found with sensory evoked potentials, very few studies have examined this possibility explicitly using a number of different processing tasks. However, given the lack of gender effects from the large sample described in Fig. 1, and no apparent influences from the menstrual cycle, it seems reasonable to conclude that any such sex differences would be relatively weak, at least in similar young adult subjects.

3 P300 from Identical Twins

Previous evoked potential studies have found marked waveform similarity for monozygotic twins using repetitive stimuli in the absence of a performance task, with the comparison of mono- and dizygotic twins suggesting a strong genetic basis for this similarity (Buchsbaum 1974; Lewis et al. 1972; Rust 1975). Hence, some of the between-subject variability observed for the P3 ERP may be attributed to the variation in the physical structure of the neural mechanism which generates this component if similar waveforms are obtained from individuals who share the same physical makeup. This hypothesis was examined by comparing the P3 components from ten pairs of identical twins with those of controls who were matched on age, sex, and educational level (Polich and Burns 1987). ERPs were elicited with a simple two-tone discrimination task and are presented Fig. 2 for both groups. The similarity between ERPs from the twin pairs compared to the relative lack of agreement for the control subjects is striking. The degree of association between individual pair members was quantified by measuring P3 amplitudes and latencies for each pair and computing correlation coefficients for the two samples on each measure. These results are illustrated in Fig. 3 and confirm the impressions gleaned from the waveform data: the twin pairs demonstrated virtually identical P3 values, while there was no statistical relationship between the data obtained from the control pairs. Similar findings were obtained between the twin pairs and their controls for most of the other ERP components (N1, P2, and N2).

An additional pair of twins not included in the original sample also were measured, with their ERP waveforms illustrated in Fig. 4. The twin represented by the solid line was normal throughout her life, while the twin represented by the dashed line had epilepsy as a young child. Although the early and latter portions of the waveforms from each subject are quite similar, suggesting identical neural structures, the P3 component is remarkably changed in the affected twin. Thus, despite a genetic identity, the P3 ERP generating system was altered in one of the subjects relative to the other because of the disease process.

The morphological similarity in the twin waveforms relative to the control pair ERPs implies that individual differences in the P3 component are determined by the underlying neurophysiological structures associated with its generation. While it can be argued that the twin pairs are mentally alike and that this variable could account for the observed ERP consistency, the simple two-tone discrimination paradigm employed required relatively low-level cognitive processes, in contrast to stimulus evaluation, recognition, or memory retrieval tasks. Because a wide variety of P3 values can be obtained from a cognitively homogeneous sample (see Fig. 1), the similarity of electrophysiological responses between the twin pairs compared to the matched controls most likely stems from their identical neuroanatomical structure rather than cognitive backgrounds or strategies. The ERPs from the additional twin pair illustrating the effects of a brain disorder in one but not the other P3 also support this view. In addition, the significant correlations for the exogenous components from the twin pairs compared to the control subjects implies an underlying structural similarity, since these potentials are more sensitive to stimulus than to cognitive factors. Thus, a large proportion of the

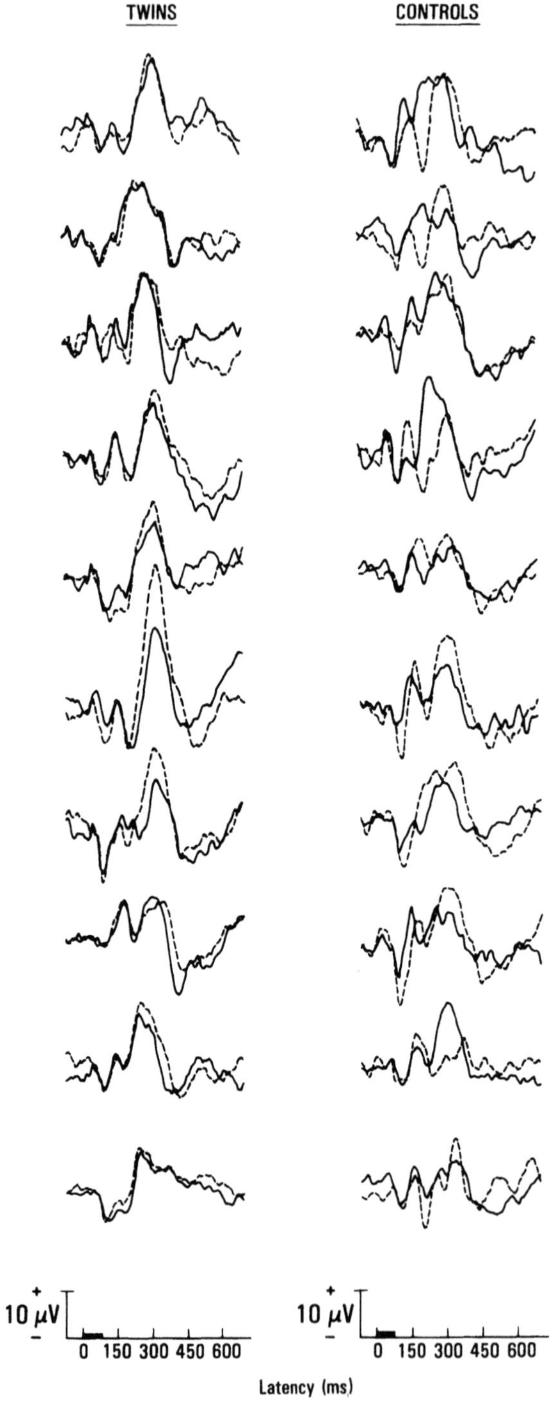

Fig. 2. Event-related brain potential waveforms from pairs of identical twins and matched control subjects (Polich and Burns 1987)

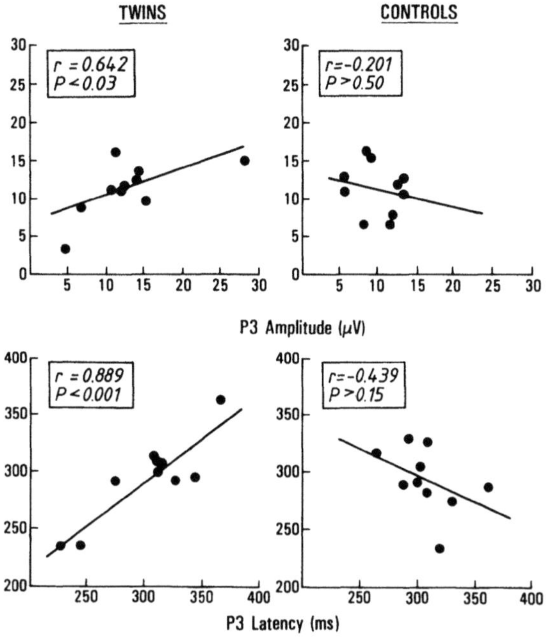

Fig. 3. Scattergrams of P3 amplitudes and latencies from identical twin and control pairs ($n = 10$ pairs/group)

variability observed for P3 morphology can be attributed to individual differences in the neural mechanisms responsible for the generation of this ERP component.

4 P300 and Skull Thickness

If individual differences for ERPs are determined by variation in the neuroanatomical mechanism responsible for its production, as suggested by the twin wave-

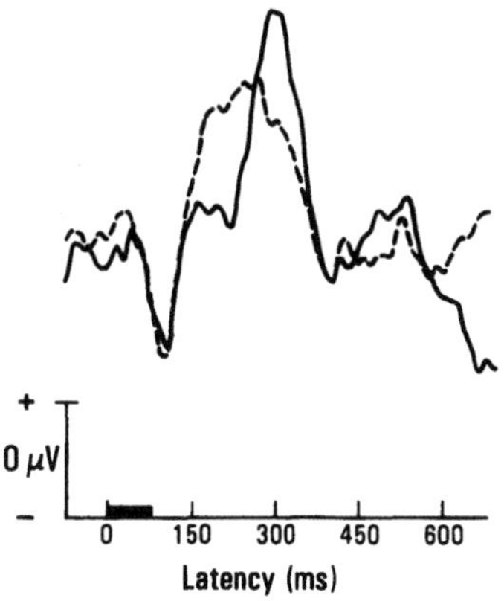

Fig. 4. Event-related brain protential waveforms from a pair of identical twins not used in the original study. The *solid line* was from the normal twin, while the *dashed line* was from the twin who had epilepsy as a child

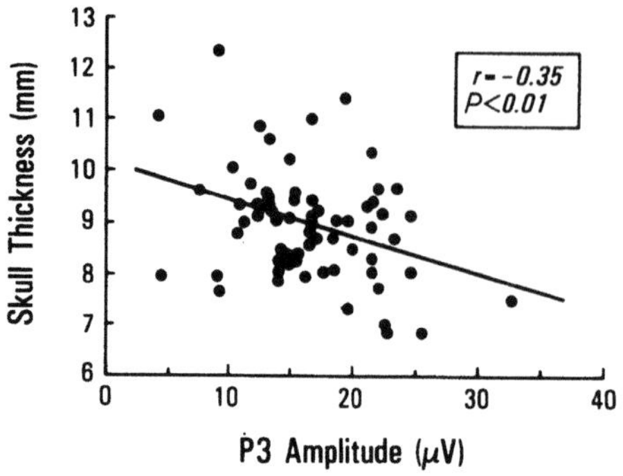

Fig. 5. Scattergram of P3 amplitude and skull thickness assessed from CAT scans in a sample ($n=67$) of adult subjects (after Pfefferbaum and Rosenbloom 1987)

form data, then differences in the morphological structure of the waveform after it is produced should be affected also by skeletal factors (Allison et al. 1986). However, these sorts of relationships are difficult to obtain since such factors as head size or distances from hypothetical generators are hard to quantify in live subjects. Some progress has been made by employing computerized imaging techniques to provide information about the thickness of the skull in the same individuals from whom P3 components have been collected with a visual stimulus discrimination task (Pfefferbaum and Rosenbloom 1987). A scattergram illustrating the relationship between P3 amplitude taken over all stimulus conditions and mean skull thickness as assessed from several image slices for a large sample of adults is presented in Fig. 5. As skull thickness increases, P3 amplitude decreases, with a statistically reliable relationship obtained between the two variables. Hence, variation in skeletal structure also contributes to P3 differences between individuals, although wide variability for individuals still can be observed.

5 Summary

The P3 ERP has been associated with memory performance and individual differences in cognitive capability. Moreover, even within a relatively homogeneous population, P3 amplitude and latency display wide variation despite their mutual association with one another. Comparison of waveforms from identical twins with matched controls suggests that the variation in P3 values stems in part from individual differences in the neurophysiological structure of the ERP-generating mechanism and extracerebral factors such as head shape, since twin waveforms were highly similar to one another compared to the nontwin data. This assertion is supported also by the inverse relationship between measures of skull thickness and P3 amplitude. Thus, the role of possible neuroanatomical and cranial differences between subjects should be considered when comparisons between populations are made with the P3 ERP component.

References

Allison T, Wood CC, McCarthy G (1986) The central nervous system. In: Coles MGH, Donchin E, Porges SW (eds) Psychophysiology: systems, processes, and applications. Guilford, New York, pp 5–25

Brown WS, Marsh JT, LaRue A (1982) Event-related potentials in psychiatry: differentiating depression and dementia in the elderly. Bull Los Angeles Neurol Soc 47:91–107

Buchsbaum MS (1974) Average evoked response and stimulus intensity in identical and fraternal twins. Physiol Psychol 2:365–370

Donchin E, Karis D, Bashore TR, Coles MGH, Gratton G (1986) Cognitive psychophysiology and human information processing. In: Coles MGH, Donchin E, Porges SW (eds) Psychophysiology: systems, processes, and applications. Guilford, New York, pp 244–267

Fabiani M, Karis D, Donchin E (1986) P300 and recall in an incidental memory paradigm. Psychophysiology 23:298–308

Fabiani M, Gratton G, Karis D, Donchin E (1987) The definition, identification, and reliability of measurement of the P300 component of the event-related brain potential. In: Ackles PK, Jennings JR, Coles MGH (eds) Advances in psychophysiology, vol 2. JAI, Greenwich, pp 1–78

Fleck K, Polich J (1988) P300 and the menstrual cycle. Electroencephalogr Clin Neurophysiol 71:157–160

Halgren E, Squires N, Wilson C, Rohrbaugh J, Babb T, Crandall P (1980) Endogenous potentials generated in the human hippocampal formation and amygdala by infrequent events. Science 210:803–805

Howard L, Polich J (1985) P300 latency and memory span development. Dev Psychol 21:293–289

Johnson R, Pfefferbaum A, Kopell BS (1985) P300 and long-term memory: latency predicts recognition performance. Psychophysiology 22:497–507

Karis D, Fabiani M, Donchin E (1984) "P300" and memory: individual differences in the von Restorff effect. Cognit Psychol 16:177–216

Lewis EG, Dustman RE, Beck EC (1972) Evoked response similarity in monozygotic, dizygotic and unrelated individuals: a comprehensive study. Electroencephalogr Clin Neurophysiol 23:309–316

Magliero A, Bashore TR, Coles MGH, Donchin E (1984) On the dependence of P300 latency on stimulus evaluation processes. Psychophysiology 21:121–186

McCarthy G, Wood CC, Allison T, Goff WR, Williamson PD, Spencer DD (1982) Intracranial recordings of event-related potentials in humans engaged in cognitive tasks. Neurosci Abstr 8:976

Neville HJ, Kutas M, Chesney G, Schmidt AL (1986) Event-related brain potentials during initial encoding and recognition memory of congruous and incongruous words. J Mem Lang 25:75–92

Okada YC, Kaufman L, Williamson SJ (1983) The hippocampal formation as a source of the slow endogenous potentials. Electroencephalogr Clin Neurophysiol 55:417–426

Papanicolaou AC, Loring DW, Raz N, Eisenberg HM (1985) Relationship between stimulus intensity and the P300. Psychophysiology 22:326–329

Pfefferbaum A, Rosenbloom M (1987) Skull thickness influences P3 amplitude. Psychopharmacol Bull 23:493–496

Pfefferbaum A, Ford JM, Weller BJ, Kopell BS (1985) ERPs to response production and inhibition. Electroencephalogr Clin Neurophysiol 60:423–434

Polich J (1986a) Attention, probability, and task demands as determinants of P300 latency from auditory stimuli. Electroencephalogr Clin Neurophysiol 63:251–259

Polich J (1986b) Normal variation of P300 from auditory stimuli. Electroencephalogr Clin Neurophysiol 65:236–240

Polich J (1987a) Task difficulty, probability, and inter-stimulus interval as determinants of P300 from auditory stimuli. Electroencephalogr Clin Neurophysiol 68:311–320

Polich J (1987b) Response mode and P300 from auditory stimuli. Biol Psychol 25:61–71

Polich J, Burns T (1987) P300 from identical twins. Neuropsychologia 25:299–304

Polich J, Howard L, Starr A (1983) P300 latency correlates with digit span. Psychophysiology 20:665–669
Polich J, Howard L, Starr A (1985) Stimulus frequency and masking as determinants of P300 latency in event-related potentials from auditory stimuli. Biol Psychol 21:309–318
Polich J, Ehlers CL, Otis S, Mandell AJ, Bloom FE (1986) P300 latency reflects the degree of cognitive decline in dementing illness. Electroencephalogr Clin Neurophysiol 63:138–144
Pritchard WS (1981) Psychophysiology of P300. Psychol Bull 89:506–540
Ragot R (1984) Perceptual and motor space representation: an event-related potential study. Psychophysiology 21:159–170
Rust J (1975) Genetic effects in the cortical auditory evoked potential: a twin study. Electroencephalogr Clin Neurophysiol 39:321–327
Sklare DA, Lynn GE (1984) Latency of the P3 event-related potential: normative aspects and within-subject variability. Electroencephalogr Clin Neurophysiol 59:420–424
Squires N, Galbraith G, Aine C (1979) Event-related potential assessment of sensory and cognitive deficits in the mentally retarded. In: Lehmann D, Callaway E (eds) Human evoked potentials. Plenum, New York, pp 397–413

Sensory and Cognitive 40-Hz Event-Related Potentials: Behavioral Correlates, Brain Function, and Clinical Application

D. E. SHEER

1 Focused Arousal

The main thesis of this chapter is that two measurement operations of the 40-Hz event-related potentials (ERPs) provide a significant systems analysis technique for a developing body of interdisciplinary knowledge labeled as Cognitive Neuroscience – the information processing between sensory input and motor ouptut. Hierarchically organized, increasing neural connectivity forms the basis for more complex cognitive processes. In the hierarchical progression, first-order functional components operate as simple neural systems. These, in turn, interact to produce higher-order functional components, more complex neural systems. The final interactive resultant, operative in cognition and expressed in behavior, must consist of many critical zones throughout the neuraxis.

The first-order functional components are not observed directly in behavior. If we are to more definitively understand information processing, both in cognition and in the brain, it is essential to dissociate by converging operations the functional components and analyze how they work together in cognition and as neural systems. We need to begin with the isolation of simple systems, first-order functional components, which clearly account for significant variances in information processing. These need definitive measurement operations which can provide connecting links for both cognitive and neural analyses, and for implementation in the clinic. Posner and Rothbart (1985) have similarily proposed "a program of sustained research to trace energetic and computational events in terms of their component facilitations and inhibitions in information processing and then to observe the origin of these components in the neural systems that support them."

This emphasis requires some comment regarding the level of reductionism in the physical measurement operations that are most congruent with higher-order information processing. Cognitively, attention, perception, and thinking are dynamic, continually changing processes, with both momentary and sustained fixations of different degrees and durations. The time constants of dynamic perturbations of organized electricity, such as periodicities, coherences, and spatial distributions of amplitudes, are more congruent with the time constants of such information processing. These perturbations are a reflection not only of the specificities of single units, but also of changing ensembles of cells and the time constants in their recurrent neural networks.

A difficulty in dealing with inferential abstractions, such as first-order functional components, has been the surplus and overlapping meanings of common-

Springer Series in Brain Dynamics 2
Edited by Erol Başar and T. H. Bullock

sense language used to denote such inferences. A striking illustration of this difficulty has been the long enduring (Yerkes and Dodson 1908) and many-faceted (Tolman 1948; Moruzzi and Magoun 1949; Lashley 1950; Hebb 1955; Sharpless and Jasper 1956; Broadbent 1957; Duffy 1957; Easterbrook 1959; Malmo 1959; Berlyne 1960; Sokolov 1960; Bindra 1961; Pfaff 1969; Killeen et al. 1978; Eysenck 1982) concept of arousal. The problem has been that arousal is a very basic output property of low-order, facilitatory neural processing, which, in its ascending interactions at different levels of the neuraxis, may be quite differently expressed, electrocortically (Montplaisir 1975), through peripheral autonomic effectors (Lacy 1967) and in behavior (Kahneman 1973), depending upon the contingent operations of its elicitation.

A dissociation can be made between tonic arousal and phasic arousal as anatomically distinct components of the brain stem reticular system (Sharpless and Jasper 1956). A lower brain stem system is concerned with tonic arousal, general wakefulness, and alertness, and a more rostral system with a shorter acting, more restricted phasic arousal. Reports of differential brain stem lesions have provided anatomical support for the dissociation between tonic and phasic arousal (Chatrian et al. 1964; Kaada et al. 1961; Kimura 1964; Landsell and Mirsky 1964; Loeb and Paggio 1953; McDonald and Burns 1964).

A dissociation between these two arousal systems has also been shown in the pattern of cortical EEG desynchronization. The dissociation was demonstrated in monkeys with CS-UCS conditioning by Morrell and Jasper (1956). Novel auditory, tactile, and visual stimuli produce diffuse desynchronization over the entire cortex. These same stimuli as CSs produce desynchronization restricted to the relevant cortical areas when the CS-UCS pairings are just being learned in the acquisition phase. Similar findings with conditioning have also been reported by Gastaut et al. (1957) and Jouvet et al. (1956).

Lindsley and associates (Anchel and Lindsley 1972; Coleman and Lindsley 1975; Macadar et al. 1974) have differentiated two distinct behavioral arousal patterns in cats with electrical stimulation of two different brain stem–hypothalamic zones. Pattern I was characterized as "alerting, orienting, scanning, and investigatory activity." Pattern II was characterized as "activation of attentive processes, but in a more focused and restrictive manner ... to narrow the field of view, whether external or internal ... attention is temporarily 'locked on' to specific aspects of the environment." The stimulation zones for Pattern I were medial hypothalamus and specific sites in the pons, while stimulation zones for Pattern II were lateral hypothalamus and different sites in the rostral pons.

There are converging sets of data which bear on the question of cholinergic neural circuitry as a substrate for phasic arousal. Cholinergic agonists increase and antagonists decrease both cortical acetylcholine and EEG desynchronization (Celesia and Jasper 1966; Montplaisir 1975). With acetylcholine blocking and increased slow wave EEG, there is still a wakeful, alert behavioral state. This dissociation between electrocortical and behavioral arousal is in agreement with a cholinergic rostral arousal system, sparing the tonic arousal system concerned with general behavioral wakefulness. With acetylcholine blocking and the slow wave EEG, the effect behaviorally is impairment of learning and memory and improvement with agonists of acetylcholine (Drachman 1977; Karczmar 1979;

Squire and Davis 1981). Carbachol injections into the pontine reticular formation of cats, increasing acetylcholine, produce a cortical desynchronized state resembling rapid eye movement (REM) sleep, which does not occur with either midbrain or medullary injections (Baghdoyam et al. 1981). Jasper and Tessier (1971) found acetylcholine liberation from cortex during REM sleep, and Sitarem et al. (1976) were able to induce REM during sleep with physostigmine infusion. There is a clear dissociation between a lower brain stem tonic arousal system and a rostral brain stem cholinergic phasic arousal system.

Focused arousal, a first-order functional component of information processing, is still different. It represents an interactive circuitry between rostral brain stem reticular and sensory cues, modulated by limbic and cortical circuitry. This strong significant relationship between phasic arousal and a sensory cue function has been repeatedly demonstrated in animals by means of direct brain stem reticular stimulation (Block 1970; Block and Hebb 1956; Brown and Cohen 1959; Fuster and Uyeda 1962; Hoyman and Kelsey 1977; Lindsley 1958; Mahut 1964; Proctor et al. 1957), and in humans using peripheral autonomic and EEG desynchronization measures of arousal (Beatty 1982; Beatty and Wagoner 1978; Lansing et al. 1959; Payne et al. 1968; Stennett 1957).

Dissociations can be made between focused arousal and phasic and tonic arousal. The contingent facilitation of the phasic arousal, rostral brain stem reticular system is necessary to activate focused arousal in specific sensory circuitry. But the measurement operations of focused arousal – coherent resonance at an optimal periodicity – are restricted to the specific sensory circuitry. The neural substrate for focused arousal is specific sensory circuitry, interactivity activated through sensory cues contingent with phasic arousal facilitation. By facilitating coherent sensory cell assemblies and, at the same time, not becoming a coherent "locked-in" part of such assemblies, the phasic arousal can serve a more adaptive, flexible processing function, on the ready to facilitate different sensory circuitry, as in shifts of attention.

Tonic arousal, diffuse cortical EEG desynchronization, represents initial responding to novel stimuli within the complex matrix of irrelevant environmental stimuli. It is an oscillatory, unstable state of the organism, in which many different subassemblies of the intrinsic electrical activity are firing in different patterns, i.e., nonsynchronously. Not being focused during this orientation to novelty actually represents an adaptive ready state for significant occurrences. With focusing behavioral operations, such as reinforcement contingencies, subassemblies of the electrical activity now fire in coherent organizations restricted to the relevant circuitry. The desynchronized EEG is no longer diffuse, but it still may appear as desynchronized without finer grained analysis because the coherent subassemblies, restricted to limited cortical areas, are now submerged within the total ongoing electrical activity.

Behavior measurement operations for focused arousal need directing, warning, focusing, or reinforcement contingencies (Killeen 1975; Killeen et al. 1978; Pfaff 1969) to dissociate it from tonic arousal. A case in point is a study by Lansing et al. (1959). They examined EEG desynchronization – alpha blocking – in relation to latency of reaction time to a visual stimulus. A brief auditory stimulus presented from 0.3 to 1 s before the visual stimulus as a forewarning, inducing the desynchronization pattern, reduced the reaction time from 280 to 206 ms. However, when reaction times were obtained on spontaneously occurring EEG desynchronization trials, there was no effect; the reaction times were not reduced as compared with alpha trials. This is exactly the point. Spontaneously occurring EEG desynchronization reflects tonic arousal; EEG desynchronization with a warning signal contingency reflects focused arousal. EEG desynchronization, as a measurement operation without finer grained analysis, does not distinguish between tonic arousal and focused arousal. With unitary sensory inputs, the measurement operation of focused arousal in specific sensory circuitry is the sensory 40-Hz ERPs.

Focused arousal, in higher-order information processing, interacts with more complex sensory cues, both antecedent and concurrent, in multisensory circuitry. Much variance in information processing is associated with the content of antecedent experience in long-term store as interactive with the content of immediate experience in short-term store – stimulus set analysis. Out of such experience, distinctive unitary features must be extracted and meaningful organizations constructed. Considerable research has concentrated on isolating distinctive features in visual and auditory modalities (Posner and McLeod 1982). It is clear that there are hierarchical groupings of orthographic, phonological, syntactic, semantic, and certainly other distinctive features that operate in information processing. These features may be organized into structures identified as perceptual and semantic codes (LaBerge 1976), or extracted and grouped into structural codes by primary mechanisms of "unitization" and "categorization" (Shiffrin 1976); alternatively, multiple unitary features may be detected and integrated into a unified percept (Posner 1978).

It is important to understand how structures or codes are assembled from unitary features, and how they operate in a more or less unified form. One source of organization is, certainly, the number and internal characteristics of the associations themselves, e.g., orthographic and phonological rules, and semantic and meaning contexts. In addition, there must be some energizing function that binds and temporarily holds in place unitary features and ties together isolable codes of the same stimulus set. This view is central to cognitive theories developed from behavioral data. For LaBerge (1976), activation added to a code and associations between codes facilitates the processing of stimulus sets. For Shiffrin and Schneider (1977), activation is required to assemble structures necessary for performing a task. For Posner and Rothbart (1985), specific pathway activation provides orienting efficiency and self-regulatory selection. For the parallel distributed processing group, activation forms the energetic component in computational models of information process (McClelland et al. 1986; Rumelhart et al. 1986).

Focused arousal interactive with stimulus set analysis serves such an energizing function – the function of selective specificity. This selective specificity is expressed in the physical measurement operation as distinctive spatial amplitude patterns of coherent resonance at an optimal periodicity in relevant multisensory circuitry. Using the olfactory bulb as a model, Freeman and associates (Freeman and Skarda 1985; Freeman and Viana Di Prisco 1985, 1986; Freeman and Baird 1987; Freeman and Grajski 1987) have detailed the neural mechanisms for selective specificity in the relatively simple neural substrate of the olfactory bulb. Distinctive spatial amplitude patterns of 40-Hz oscillations replicated across the bulb convey specific odor information.

We can generalize the neural mechanisms beyond the olfactory bulb without, of course, the detail that Freeman has worked out for the bulb. Spatial amplitude patterns of coherent oscillations are determined by unique templates of synaptic connections, forming, with focusing behavioral operations such as reinforcement contingencies, distinctive nerve cell assemblies for discriminative information. The specificity of the information conveyed is determined by the distinctive spatial pattern of unit cell specifications, defined by the modulatory controlled content of the multisensory input: stimulus set analysis. The energy that momentarily fixes the unique spatial pattern at a significantly high amplitude level is the coherent resonance at an optimal periodicity: focused arousal. It is the configurational property of the unit specificities – ensembles of interconnected neurons replicated across relevant circuitry – that is distinctive, and not individual units which are involved in multiple patterns. Also, any individual unit may or may not enter into a specific organized pattern from time to time. For this higher-order functional component – focused arousal interactive with stimulus set analysis – the cognitive 40-Hz ERPs act as a carrier frequency.

2 Measurement Operations

The designation "40-Hz EEG" has been used as shorthand notation for different narrow frequency bands across species in the high-frequency gamma range (35–85 Hz), which center on specific coherent frequencies with focusing behavioral operations. As a matter of fact, the bands and specific coherent frequencies may also vary for different neuronal networks in the same species, e.g., 38 Hz in the olfactory system of the cat (Bressler and Freeman 1980) and 46 Hz in the visual system (Kaufmann 1985). It is understandable that these narrow bands of oscillations and their specific coherent frequencies do vary across species and neural systems because they are dependent upon the different time constants of their relevant recurrent neuronal networks. Whatever the specific frequency number may be for different species and different neural circuitry, it is clear that oscillations in the EEG at maximal power and entrainment within the limits of stability for optimal synaptic efficiency occur in a narrow high-frequency band in the gamma range.

The 40-Hz EEG occurs in large amplitude, highly synchronous bursts recorded from the olfactory bulb and other rhinencephalic structures, through a range of species from catfish to man [see references in Bressler and Freeman (1980) and Sheer and Granstaff (1970)]. The amplitude bursts, synchronized with sniffing, are particularly large in quadruped animals, where olfaction is an important distance receptor and sniffing is a highly adaptive orienting response for exploratory, food, and sex behavior. As a matter of fact, in such animals the amplitudes are so large that they are volume conducted to even closely spaced bipolar electrodes at distant sites, so that the 40-Hz EEGs from nonolfactory areas have to be recorded in animals with olfactory bulbs removed or with nostrils occluded with foam rubber (Sheer 1970; Sheer and Grandstaff 1970). At the neocortical level, where laminar structure is far more complex than olfactory bulbs, the 40-Hz EEG is at a much lower amplitude in a more complicated electrical background, but it can still be observed visually on the oscillograph from epidural leads with high-pass filtering at 20 Hz and fast paper speeds.

The 40-Hz EEG, when recorded as an index of focused arousal, is actually loosely coupled ERPs. They are event related because the coherent 40-Hz potentials occur consistently only with behavioral conditions which focus the phasic arousal. They are loosely coupled because – with more complex focusing behavioral contingencies such as stimulus sets and, in humans, instructional sets and problem-solving – they cannot be response averaged, tightly coupled to a specific stimulus.

The operations of the recording procedure are as important in specifying the obtained measure as the measurement technique. This is demonstrated in Fig. 1, where reinforced light flashes were presented to olfactory bulbectomized cats at 4/s for 10 s on a 1-min variable interval schedule. The animals were required to press a bar during the 10-s flash period in order to obtain a milk reinforcement. Depth and epidural recording were made from the lateral geniculate nucleus (LGN) and primary visual cortex (VC) during baseline and flash periods.

In Fig. 1 are shown the recordings from the right and left LGNs and VC of cat LG3 after the animal was responding to the flashes with a bar press during the acquisition phase of learning, defined as responding just beyond chance expectancy. In the acquisition phase of learning the reinforced response to the light flash, 40-Hz oscillations occurred in the LGNs of these animals with each 4/s light flash. In this learning situation, with the 40-Hz oscillations tightly coupled to the stimulus, they can be response averaged for 40 samples over the 10-s period. These oscillations can also be measured as the number of rectified oscillations (80 peaks per second) above a preset amplitude threshold over the 10-s period.

The response averaging technique provides information on the different latencies and amplitudes of aperiodic oscillations, but it requires tight coupling with the event which triggers the response averaging. The second technique can be used when the electrical responses are loosely coupled to a contingent event which does not allow for response averaging, particularly, when the responses are uniform oscillations. In Fig. 1, the 40-Hz oscillations in the LGN, tightly coupled to the 4/s reinforced flashes, are clearly evident in the graphic records. In this case, both techniques, the response averaging and the number of above threshold peaks, are equally valid indices of these ERPs. Where more complex contingent events are loosely coupled with the 40-Hz oscillations, only the peak counts can be used as an index of the ERPs. It should also be noted in Fig. 1 that the concurrent 40-Hz oscillations from the primary visual cortical leads, barely visible in the graphic records, can also be extracted with computer analysis, using the peak counts technique.

The two measurement operations of focused arousal are both 40-Hz ERPs – one, tightly coupled with a contingent specific sensory stimulus, and the other, loosely coupled with more complex stimulus sets. Both measures index focused arousal processing. The one, tightly coupled with a sensory stimulus, indexes the

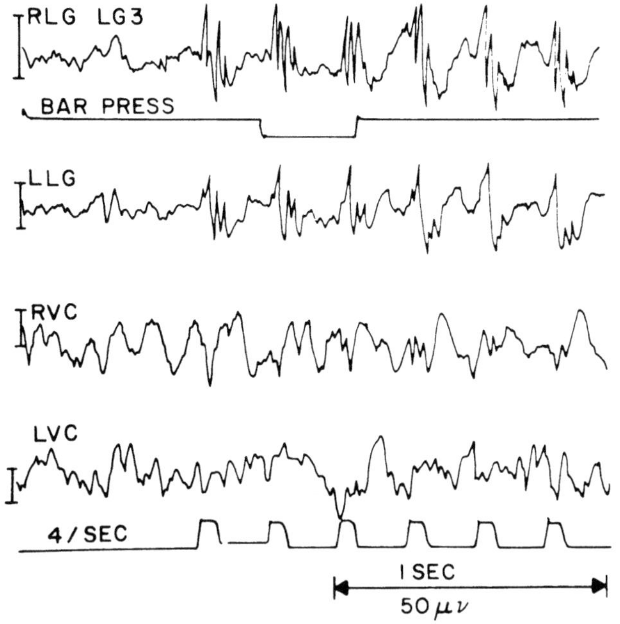

Fig. 1. EEG records from cat LG3 during the acquisition phase (reponses just beyond chance expectancy) of operant conditioning, in which animals were required to press a bar during a 10-s SD period (4/s light flashes) to obtain a milk inforcement. *RLG*, right lateral geniculate nucleus; *LLG*, left lateral geniculate nucleus; *RVC*, right visual cortex; *LVC*, left visual cortex

operation of focused arousal in a specific sensory system. To distinguish between the two measures, this one is labeled the sensory 40-Hz ERP. The other, loosely coupled with more complex multisensory stimuli, indexes the operation of focused arousal in a system convergent at parietal–temporal–occipital (PTO) cortex. Recorded from the PTO cortical zone, this measure, labeled the cognitive 40-Hz ERP, reflects the operation of focused arousal in higher-order information processing.

2.1 Sensory 40-Hz ERP

Galambos et al. (1981) demonstrated that, at stimulus repetition rates of 40/s (monaural clicks or tone bursts), the middle latency response of the auditory evoked potential, normally occurring at 8–80 ms immediately following the stimulus, was significantly augmented in the form of a series of 40-Hz oscillations. This "forcing function" of the 40/s auditory stimulus rate generated a 40-Hz ERP, recordable also from other sensory modalities (Galambos 1982). The 40-Hz ERP is significantly reduced in amplitude under anesthesia and with sleep (Galambos et al. 1981).

In a standard procedure for recording the sensory 40-Hz ERP, electrodes are applied at Cz of the Ten-Twenty System for auditory ERPs, at Oz for visual ERPs, and on the ipsilateral ear. A midline forehead electrode serves as ground, and the interelectrode impedances are maintained below 3 kΩ. For auditory stimuli, clicks are delivered through standard audiometric headphones, 50 dB above individually obtained thresholds, at rates of 40/s and 20/s. For visual stimuli, flashes are delivered about 5 ft in front of the subject with a photic stimulator at

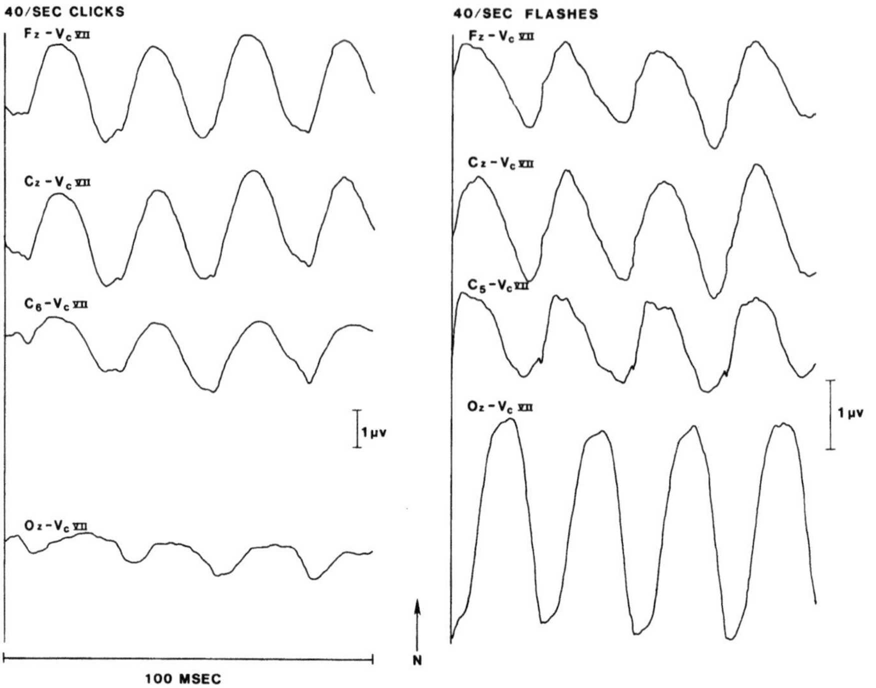

Fig. 2. Sensory 40-Hz ERPs recorded from a normal subject to 40/s clicks and flashes, with an averaging window of 100 ms for 1024 samples. The successive traces are from the Fz midline, Cz midline, C6 (clicks) and C5 (flashes), and Oz midline active electrode sites, referenced to VCVII, the seventh cervical spinous process

rates of 40/s and 20/s. Activity from the Cz or Oz electrodes and the ear electrode is differentially amplified by Grass 10A5 amplifiers, with Cz or Oz fed into grid two of the amplifier so that negative events at the active leads produce an upward deflection. Activity is amplified times 50000 and bandpass filtered from 10 to 100 Hz with 60-Hz filters switched out. The amplified and filtered activity is then fed to a Grass Model 10 Bio-response averager. A total of 1024 sweeps, 100 ms in duration, are averaged for each 40-Hz ERP. Two ERPs are obtained successively to demonstrate reproducibility and then combined. The ERPs are plotted on an X-Y plotter at a vertical scale of X16; they can also be stored in digital form on a microcomputer diskett for off-line analysis.

A topographical mapping of the 40-Hz sensory ERPs on the scalp, comparing auditory clicks and visual flashes, both at 20/s and 40/s stimulus rates in the same subject, was carried out with ten normal adults (Judd 1987). The active electrode sites compared, both with clicks and flashes, were at Fz, Cz, Oz, and C5 with left ear auditory input and C6 with right ear input. The standard procedure for recording the 40-Hz ERP was used, except that all active leads were referenced to an indifferent electrode at the seventh cervival spinous process.

A representative record for the 40/s stimulus rate with clicks and flashes is shown in Fig. 2, and for the 20/s stimulus rate in Fig. 3. The results to be described

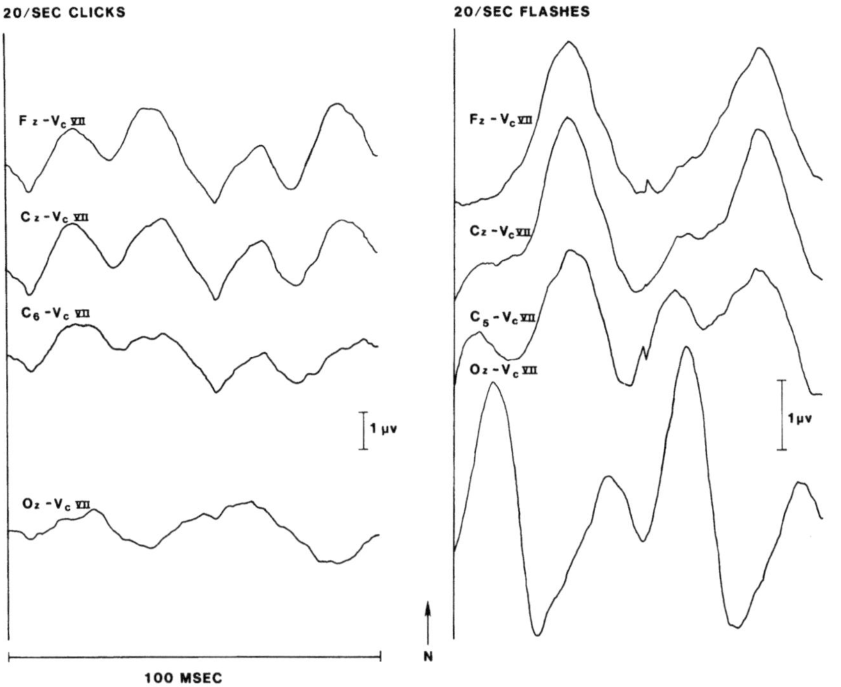

Fig. 3. Sensory 40-Hz ERPs recorded from a normal subject to 20/s clicks and flashes, with an averaging window of 100 ms for 1024 samples. The successive traces are from the Fz midline, Cz midline, C6 (clicks) and C5 (flashes), and Oz midline active electrode sites, referenced to VCVII, the seventh cervical spinous process

from these representative records were statistically significant for the ten subjects. In Fig. 2 with the 40/s stimulus rate, for the auditory clicks the maximum amplitudes were at the active Fz and Cz leads, decreasing at C5 and C6 and lowest at Oz. The Fz and Cz leads were not significantly different from each other, but they were significantly different from C5 and C6, which were also significantly different from Oz.

For the visual flashes, the maximum amplitudes were at Oz, being significantly different from the other three leads. The next highest were at Fz and Cz, which were not different from each other but were significantly different from C5 and C6.

This consistent distribution of amplitudes across the scalp, with both auditory and visual inputs, provides some inferential information on the signal sources. With this clear dissociation, visual input amplitudes being higher at Oz, over the visual cortex, and auditory amplitudes higher at C5 or C6, over the auditory cortex, it is difficult to rule out a cortical source. Somesthetic input amplitudes are also comparatively higher over somatosensory cortex (Galambos 1982).

However, amplitudes are generally higher at vertex electrode sites from subcortical sources, and for auditory inputs, the auditory amplitudes are significantly higher at Cz and Fz than at C5 or C6. This raises the question of an additional subcortical source along with the cortical source, volume conducted to the surface electrode sites. The hypothesis presented here, developed further in the section on neural mechanisms, is that the sensory 40-Hz ERP reflects coherent resonance of specific thalamic relay nucleus and sensory cortex, volume conducted to the active electrode sites.

For the 20/s stimulus rate with auditory inputs (Fig. 3), four waves occur at Fz, Cz, and C5 or C6, and are smoothed out to two at Oz. With visual inputs, four waves appear at Oz and are gradually smoothed out to two at Fz. For the 100 ms time period over which the responses are averaged with a 20/s stimulus rate, only two stimulus inputs are presented. The appearance of four waves at the more active sites is due to endogenous resonance. The appearance of two waves instead of four at the less active sites, with the same signal, is due to the smoothing effects of the volume conduction. The resonance effect of two stimulus inputs producing four waves may be smoothed to two waves, presumably depending upon the relationship of the dipoles to the active recording sites.

The sensory 40-Hz ERP, recorded with 20/s stimulus rates, is less stable and more variable, and in about one-third of normal subjects, even with recording sites at Cz for clicks and Oz for flashes, two waves may occur instead of four. It becomes a trade-off between stability and sensitivity. With the 40/s stimulus rate, the "forcing function" of the 40-Hz ERP is more stable and consistent. With the 20/s stimulus rate, the 40-Hz ERP is more variable, but, at the same time, more sensitive to changes in focused arousal, as affected by relevant state conditions, behavioral contingencies, and brain impairments.

2.2 Cognitive 40-Hz ERP

Since the response-averaging technique is inappropriate with loosely coupled, multistimulus inputs, serious problems are encountered in trying to record reliably short, aperiodic bursts of 40-Hz potentials from the human scalp. One is the greater attenuation of the higher frequency part of the EEG spectrum as it is volume conducted to the electrodes (Pfurtscheller and Cooper 1975). This reduces the absolute amplitude of the 40 Hz generated from a restricted neuronal assembly to a still lower level. Another is the frequency overlap of the 40 Hz with the higher amplitude of the EMG from the scalp muscles, which actually goes down to 14 Hz but shows a considerable increase in power above 30 Hz across a broad spectrum to 100 Hz and above. What is needed to overcome these difficulties are critical scalp electrode placements, considerable signal preprocessing, and "state-of-the-art" special purpose computer technology. Techniques for carrying out these objectives have been evolving (Sheer 1976, 1977, 1984; Spydell and Sheer 1982, 1983). The present procedure for recording cognitive 40-Hz ERPs concurrently with focusing behavioral operations consists of a special-purpose computer with analog filter, digital filter, master control, and user interface sections (Raghavan et al. 1986).

The EEG is recorded from two critical electrode placements, one in the middle of the triangle formed by placements 01-P3-T5 of the Ten-Twenty System on the left hemisphere, and one in a corresponding position relative to 02-P4-T6 on the right hemisphere, both with reference to Cz. Signal preprocessing takes the form of a high-pass cutoff at 20 Hz and a low-pass cutoff at 90 Hz with a sharp 60-Hz notch filter. The control for muscle artifcat depends on the broad frequency spectrum of muscle, and an algorithm based on covariance analysis. EEG activity in the one-third octave range of 36–44 Hz with a center at 40-Hz is specified as uncorrected 40 Hz ($\Sigma y2$), and activity in the one-third octave window of 62–78 Hz with a center frequency at 70 Hz is specified as muscle ($\Sigma \times 2$). The corrected power function $\Sigma y'2 = \Sigma y2 - \Sigma(xy)2$ represents the power of the 40-Hz EEG independent of muscle. The two electrode placements go first to two commercial bioamplifiers with high common mode rejection specifications. Dual outputs from the amplifiers go to a dual-beam oscilloscope for monitoring and to the digital filter section of the computer.

The digital filter section consists of two identically programmed Intel 2920 analog microprocessors, one for each channel. The 2920s include A/D converters on the chip and are designed for high-speed signal processing applications. The 2920s first digitize the two input channels, and then implement bandpass filters and other detection algorithms to determine the presence of 40-Hz and 70-Hz bursts which are then rectified to provide peaks (2x cycles). The two 2920s work independently and give their results to the master control section.

The master control section consists of a single Intel 8751 microprocessor. The 8751 is a fast single-chip microprocessor with RAM, EPROM, I/O ports, timers, and a serial port all on the chip. The 8751 receives the results from the digital filter section for each channel (left and right), calculates the muscle algorithm, and presents the final results through the user interface section.

The user interface section consists of an eight-channel thermal printer, a keyboard, and an eight-character display. The user is able to specify operating parameters, such as threshold levels, number of cycles, duration of epochs, and start and stop computer analysis manually and by remote control. The thermal printer provides a printout of the final results. The entire system is packaged together with its own power supply.

The eight-channel thermal printer, for each designated epoch, prints out the following for both the left (L) and right (R) hemispheres: uncorrected 40-Hz peak power ($\Sigma y2$); 70-Hz peak power ($\Sigma x2$); number of 70-Hz peaks; corrected 40-Hz peak power ($\Sigma y'2$); and corrected number of 40-Hz peaks (N). Mean peak powers $\Sigma y'2/\mathrm{N}$ and normalized asymmetries, (L-R)/(L+R), for both corrected 40-Hz peak powers and number of peaks can be calculated for each epoch.

In the 1960s, Freeman and associates published a series of studies on evoked potentials from the prepyriform cortex of cats (Biedenbach and Freeman 1965; Freeman 1962a, b, c, d, e). They studied characteristics of these evoked potentials with changes in stimulation parameters, and with arousal, motivational, and learning conditions. A general finding of these studies was the association of high-amplitude 40-Hz resonant oscillations in the evoked potentials with maximally effective stimulation parameters, performance efficiency in learning, and the behavioral conditions of high arousal and motivation. Freeman (1975) demonstrated empirically the close relationship between these evoked potentials and the 40-Hz EEG oscillations in the olfactory system of the cat; they are both measures of the same neural mechanisms. Başar et al. (1987), using stop-band filters to analyze the relationship between 40-Hz EEG and the middle latency components of the auditory evoked potentials, reported that they are "... complementary related activities ... the wave complex P25, N50, N60 seems to represent modulation of the 40-Hz mechanisms. The wavelets of this complex are minima and maxima of the same oscillations, namely of the resonating 40-hz wave packet."

3 Neural Substrate

The neural substrate for focused arousal as distinguished from phasic arousal will be detailed. The neural circuitry for phasic arousal involves sets of ascending cholinergic reticular pathways, originating in the parabrachial nucleus of the rostral pons, at the border of midbrain. One set of pathways, passing through the lateral hypothalamus, synapse in the substantia innominata and pallidal areas of the basal forebrain, and then go on to cortex (Coyle et al. 1983). Another set of pathways ascend directly to the cortex, and a third set synapse on nucleus reticularis of the thalamic reticular system (Kimura et al. 1981; Saper 1981, 1982; Saper and Lowey 1980; Singer 1979, 1980; Singer et al. 1976). Cholinergic facilitation has been directly demonstrated by the excitatory effects of rostral brain stem reticular stimulation upon cortical cells (Singer 1979, 1980; Spehlman 1971; Spehlman and Downes 1974; Steriade 1980) and by disinhibition of the nucleus reticularis (Ben-Ari et al. 1976; Singer 1977).

In support of the rostral pons as a focal zone, Mirsky and Pragay (1988), recording units from monkeys in the mesopontine brain stem (a junction areas between the mesencephalic and pontine reticular system), have identified what they termed Type II "attention units." Recorded simultaneously with performance on a modified continuous performance task, these units responded not to changes in the specific stimuli, but to the onset and offset of the stimulus cues, and particularly to reinforcement contingencies.

In the studies by Lindsley and associates (Anchel and Lindsley 1972; Coleman and Lindsley 1975; Macadar et al. 1974), in which they elicited what they termed a Pattern II arousal – "activation ... in a more focused and restrictive manner" –, effective stimulation sites were at the rostral pons level. They used the cat stereotaxic atlases of Snider and Niemer (1961) and Berman (1968) in which the parabrachial nucleus, as defined by Kimura et al. (1981), was not identified. In a follow-up study with cats (Kaufmann 1985; Sheer 1988a) the Kimura et al. (1981) sections were collated with the Berman (1968) atlas to place electrodes in the parabrachial nucleus, consistent with the sites for the Pattern II behavior. The effective stimulation sites in these cats were at P 0.9, L 4.0, H 0.5. The projection of this site on parasagittal section 2.5 from the Berman atlas (Macadar et al. 1974) shows a significant overlap with the effective sites plotted there for eliciting pattern II arousal.

The proposed neural substrate for the phasic arousal circuitry is shown in Fig. 4. The major source of ascending cholinergic pathways from the brain stem is the parabrachial nucleus, extending from the level of the genu of the facial nerve caudally to the level of the trochlear nucleus rostrally. From here a prominent dis-

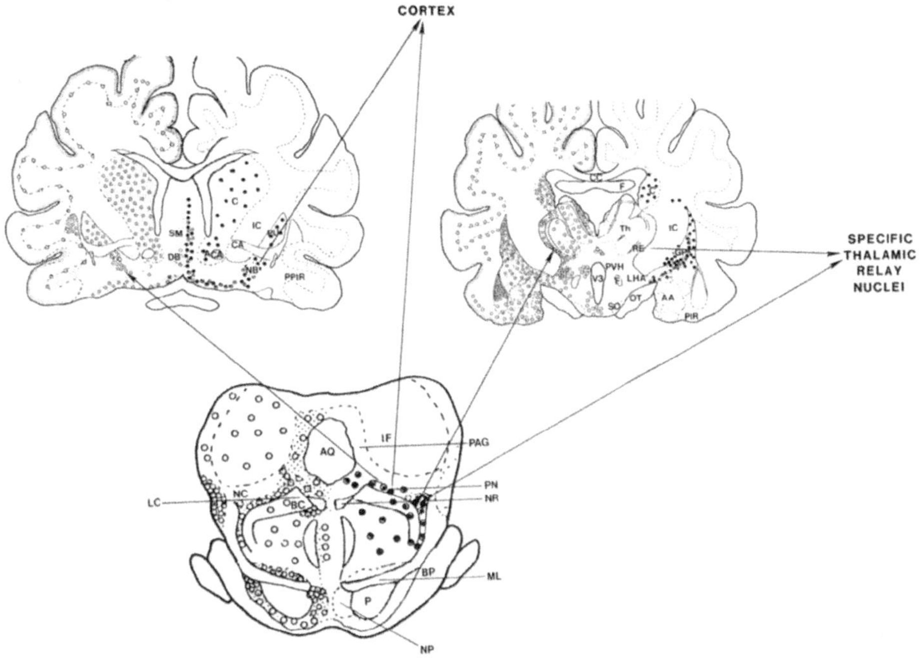

Fig. 4. Neuronal connectivity of the cholinergic phasic arousal system facilitating the focused arousal in specific sensorimotor circuitry. The origin of this system is the parabrachial nucleus of the rostral pons, extending from the level of the genu of the facial nerve caudally to the level of the trochlear nucleus rostrally. Cholinoceptive cells are indicated by *open circles* on the left side and cholinergic cells by *filled circles* on the right side of each section. *A*, accumbens nucleus; *AA*, amygdala; *AQ*, cerebral aqueduct; *BC*, brachium conjunctivum; *BP*, brachium pontis; *C*, caudate; *CA*, anterior commissure; *CC*, corpus callosum; *DB*, diagonal band of Broca; *GP*, globus pallidus; *F*, fornix; *IC*, internal capsule; *IF*, inferior colliculus; *LC*, locus ceruleus; *LHA*, lateral hypothalamic area; *NB*, basal nucleus of Meynert; *NC*, cuneiform nucleus; *ML*, medial lemniscus; *NP*, nuclei pontis; *NR*, raphe nuclei; *OT*, optic tract; *P*, pyramidal tracts; *PAG*, periaqueductal gray; *PIR*, pyriform cortex; *PN*, parabrachial nucleus; *PPIR*, prepyriform area; *PU*, putamen; *RE*, nucleus reticularis; *SM*, medial septal; *SO*, supraoptic nucleus; *Th*, thalamus; *V3*, third ventricle. (Adapted from Kimura et al. 1981)

tribution, through the lateral hypothalamus, goes to the basal forebrain, including the substantia innominata, caudate putamen, and accumbens, which, in turn, distribute to neocortex. Another pathway from the brain stem also distributes to the cortex, where dense terminal fields are found in layers V and VI. Another prominent distribution goes to the nucleus reticularis of the thalamic reticular system and other pathways go to specific thalamic relay nuclei, including the lateral geniculate and medial geniculate bodies. With efferent and modulatory descending pathways from limbic structures and cortex, particularly frontal cortex, and limbic interconnections, not shown in Fig. 4, this is the phasic arousal system, the neural substrate for the facilitation of focused arousal in specific and multisensory circuitry. The neural substrate for focused arousal itself is the facilitated momentarily coherent sensory circuitry, as demonstrated in the following study (Kaufmann 1985; Sheer 1988a).

In a series of five cats, bipolar electrodes were sterotaxically implanted in the lateral geniculate nucleus (LGN) and the parabrachial nucleus (PN), and disk electrodes on the occipital cortex (OC). The experiment, conducted in a standard sound-resistant test chamber, consisted of CS-UCS pairings, with the CS as a 7/s visual stimulus followed, in 12 s, by an electric shock across the chest. EEG and response-averaged 40-Hz evoked potentials were recorded during baseline trials, orientation to the novel CS and habituation trials, and CS-UCS conditioning and extinction trials.

In the analysis, coherences between the different combinations of OC, LGN, and PN were plotted across frequencies up to 50 Hz, for the different trials, with a 0.977 frequency resolution. Spectral distributions of coherences between OC and LGN for two representative animals will be detailed here. The coherence measures for cat B3 were plotted for the different trials after this animal had previously been habituated to the 7/s photic stimulation as a novel stimulus. In this series of trials, the significantly high coherences, 0.67 and 0.70 at 46 Hz, occurred only on trials two and three with the CS-UCS pairings in this one-trial learning situation. The coherence no longer occurred with trial four, nor with baseline or habituated 7/s stimulus trials.

For cat G4 the 7/s flashing light was a novel stimulus. For the first three 7/s novel stimulus trials, there was a variable, broad number of significantly high frequency coherences at 30, 32, 36, 37, 38, and 47 Hz, together with the significant coherences at 46 Hz, indicative of a desynchronization pattern with tonic arousal. Only on trial two of the CS-UCS pairings, after the first CS-UCS pairing, was there a high coherence of 0.51 at 46 Hz, as in cat B3, fairly well limited to this frequency and not present on trial three. For focused arousal, as defined by the acquisition phase in learning, there are high coherences at a very specific frequency, 46 Hz in the cat. In this one-trial learning situation, the high coherences at this specific frequency occur consistently only for one to two trials, during the acquisition phase of learning.

The coherences are confined to the specific sensory circuitry, OC and LGN, and not the PN. The coherences between the PN and both the OC and LGN on all trials were consistently of a low order and nonsignificant. But the parabrachial pathways do interact significantly in the processing. The number of peaks in a 40-Hz EEG band were also computed for PN, OC, and LGN on ten trials through baseline, stimulus novelty, habituation, CS-UCS pairings, and extinction trials, for both cats. Rank order correlations, not coherences, between PN and OC and between PN and LGN for cat B3 were 0.72 and 0.87. For cat G4, they were 0.61 and 0.59. The ascending cholinergic parabrachial excitation does not coherently "lock-in" with the specific afferent input during the focused arousal state. The coherence is restricted to the specific sensory circuitry, but the PN does significantly bias this circuitry, as indicated by these high correlations with components of the relevant sensory circuitry.

These central findings on coherence need replication. After the experiments reported here were completed, a study by Bouyer et al. (1981), on thalamocortical coherences at 40-Hz, was found. In this study, 24 cats were implanted with 15–25 cortical electrodes and also with depth electrodes in the posterior thalamus. EEG recordings were made from these sites during observations of the animals' behaviors in a freely moving situation. Exploratory and general orienting behavior was accompanied by diffuse desynchronization of neocortex. However, as the animal became immobile in an attitude of focused attention, with attention apparently oriented toward a given target of the surrounding space – visual, auditory, or olfactory – EEG activity at 35–45 Hz developed only over two specific cortical foci, while the rest of the cortex remained desynchronized. The two foci were sigmoid cortex, localized in motor area 4, and posterior parietal cortex, centered in area 5. During this focused behavior, the 35- to 45-Hz activity also occurred from a site in the posterior thalamus identified in the posterior nuclear group, as described by Poggio and Mountcastle (1960). Electrolytic lesions in a restricted locus in the thalamus abolished the 35- to 45-Hz rhythm from the thalamus and also from posterior parietal area 5 but not sigmoid area 4.

This posterior nuclear group as described by Poggio and Mountcastle was associated with pain sensibility, but it is now known to have little to do with pain (Jones 1985). The posterior thalamic area in the cat is customarily said to have a lateral posterior nucleus divided into three or more subnuclei and a single pulvinar nucleus. The monkey's posterior thalamic area has a single lateral posterior nucleus and at least four pulvinar nuclei. The four major nuclei of the pulvinar in the monkey can be considered equivalent to the lateral posterior complex and pulvinar of the cat (Jones 1985). The lateral posterior–pulvinar system has prominent interconnections with the posterior parietal lobe (Burton and Jones 1976; Weber and Yin 1984). The pulvinar in the cat, next to the nucleus reticularis, has the second most prominent cluster of cholinoceptive cells in the thalamus (Kimura et al. 1981). This pulvinar–posterior parietal area 5 circuitry has been significantly associated with directed attention, focused on spatial location (Posner et al. 1984b).

Bouyer et al. (1981) computed coherences within the 35- to 45-Hz range between parietal area 5 and sigmoid area 4 and between each of the cortical areas and the thalamic site during the observed focused behavior. Coherences between the two cortical zones were very low and not significant, and coherences obtained on the 35- to 45-Hz activity recorded from the olfactory bulb and cortex and olfactory bulb and thalamus were also very low and not significantly different from background noise. The coherences computed between posterior parietal area 5 and the thalamic locus, during the focused behavior, were significant, with values as high as 60%, 67%, and 69%.

In a behaviorally free situation, the focusing of attention is associated with a high coherence at 40 Hz over a limited circuitry of parietal area 5 and pulvinar in the thalamus. In our study, a specific visual stimulus, focused by reinforcement contingencies, produced a limited coherent circuitry restricted to the specific sensory system focused on vision. In the free behavioral situation with no sensory stimulus to specifically direct attention, the coherent 40-Hz activity occurred in the limited circuitry apparently concerned with a focus on spatial location – parietal area 5 and pulvinar. Whether the coherent 40-Hz EEG occurs in both the parietal area 5 and pulvinar circuitry and in specific sensory circuitry with relevant focusing sensory stimuli, and whether there is a coherent "lock-in" between these two circuits, are open experimental questions.

4 Neural Mechanisms

The olfactory bulb can serve as a model for the mechanisms operative in specific and multisensory circuitry as focused arousal. It provides a convenient model because laminar structure is first encountered here in a much simpler form than in neocortex. Freeman (1975, 1986) has demonstrated, both with equations derived from a theoretical network and with empirical recordings from the bulb, that an optimal level of sensitivity compatible with stability gives rise to rhythmic oscillations at gamma frequencies in the EEG. Rall and Shepherd (1968) and Shepherd (1978) have made a detailed analysis of dendrodendritic synapses as the fine structure for this rhythmic activity in the olfactory bulb. Air flow (steady background input) excites bipolar receptor cells in the olfactory mucosa, whose axons form the olfactory nerve which synapses within encapsulated glomeruli with dendrites of tufted and mitral cells. Depolarization in mitral cells results, through basal dendrites, in synaptic excitation of granule cells; these granule cells than deliver graded inhibition to mitral cells. This inhibition cuts off the source of synaptic excitation of granule cells. As the granule cell activity subsides, the amount of inhibition delivered to the mitral cells is reduced, which permits the mitral cells to respond again to excitatory input from the glomeruli. In this way, a sustained excitatory input to the mitral cells is converted into a rhythmic sequence of excitation followed by inhibition, locked in timing to rhythmic activation of the granule cell pool. The timing of this action is dependent upon the time constants of the local neural network. In the olfactory bulb of the cat, it is in the order of 4×6 ms a time constant of 24 ms, giving rise to rhythmic oscillations in the order of 40 Hz.

This property of recurrent inhibition, functionally contributing to the phasing of rhythmic discharges, has also been demonstrated in the LGN (Famiglietti 1970; Fertziger and Pupura 1971), hippocampus (Andersen et al. 1963), and motor cortex (Stefanis and Jasper 1964; Jasper and Stefanis 1965), and in thalamocortical interactions (Andersen and Andersson 1968). At the neocortex there exists a much more complicated situation for which details on specific neuronal geometry are lacking. However, analogous to airflow in the olfactory bulb, at neocortex the ascending cholinergic brain stem reticular system provides the steady background input to bias the level of depolarization in subsets of functionally connected cell assemblies. In the olfactory bulb, concerned with one sense modality, airflow is the background input, while at the multisensory neocortex the biasing input of the ascending brain stem reticular system is multisensory, activated by collaterals into the brain stem reticular system from the specific sensory systems.

Sensitivity in the form of neural gain depends on the level of depolarization bias as a consequence of steady background input. As cell assemblies become excited by background input, the neural gain increases, driving the neurons closer to their firing threshold. Hence, the probability of a subsequent signal input firing a neural response is increased. However, as sensitivity increases, there is a concomitant increase in instability. If numerous cells with excitatory connections were driven into a state of increased gain, an uninterrupted series of excitation would exist. One neuron would deliver an excitatory pulse to adjacent cells where it would stimulate neurons with heightened gain. This, in turn, would then feed back to the first cell, further increasing its excitatory gain. To prevent runaway excitation, inhibition of approximately equal magnitude interrupts the cycle of mutual excitation. This balance between excitatory and inhibitory influences provides system stability. The feedback loops resulting from mutually excitatory cell connections, mutually inhibitory connections, and excitatory–inhibitory interactions, i.e., recurrent inhibition, give rise to periodic oscillations recorded in the EEG.

The coherent ERPs reflect repetitive stimulation at a constant frequency for a limited time over a finite specific circuitry. The circuitry is defined behaviorally by focusing operations – the spatial–temporal patterning of sensory inputs, motor outputs, and reinforcement contingencies. Repetitive periodic excitation of cells maximizes the efficiency of synaptic transmission over the limited circuitry. Eccles (1964) has well documented this property of frequency potentiation by demonstrating the maximal size of excitatory postsynaptic potentials that develop with repetitive activation at different optimal frequencies for different neural networks.

The duration of repetitive periodic discharges at an optimal frequency is probably as significant in the transfer of information as the number and specificities of cells involved. Granit (1963), in his quantitative studies on the control of tonic motor neurons of the spinal cord, concluded that frequency of firing was the main code determining rate of continuous discharge. Brookhart and Kubota (1963) found a clear relationship between frequency of stimulation of the dorsal root and firing efficiency of motor cells in the spinal cord of frogs. At a frequency of 40/s the neurons behaved in a "driven" fashion at a firing rate of unity, while below and above this frequency the firing pattern was more variable. For cortical pyramidal cells in cats, Stefanis and Jasper (1964) found that their axon collaterals became particularly effective at relatively high frequencies of repetitive excitation. They also noted that the negative feedback of recurrent inhibition provides an automatic control on level of excitation: the greater the excitation, the more the feedback of inhibition. Stimulating epidurally the human somatosensory cortex, Libet et al. (1967) showed that subthreshold stimulus pulses could elicit conscious sensory experience only if they were delivered repetitively at 20–60 pulses per second.

A significant outcome of this loop – wherein recurrent inhibition sets up periodicity which sets up frequency potentiation – is the property of contrast. The negative feedback of recurrent inhibition particularly depresses those synapses that are weakly excited in the "surround" and so serves to sharpen further the focus of excitation. Behaviorally it is reflected, on the input side, in focusing of attention on relevant stimuli, and, on the output side, in greater precision of relevant movements. The operation of contrast as a function of surround inhibition has been detailed for spinal cord motor neurons (Granit 1963), thalamocortical relay cells (Eccles 1964), cortical motor cells (Stefanis and Jasper 1964), somesthetic sensory system (Mountcastle 1961), visual system (Hubel and Wiesel 1963), and audition (Whitfield 1965).

A final point concerns the resonant property of the facilitated coherent circuitry. Singer (1979) demonstrated that cellular networks within the visual cortex display properties of resonance ("ringing," a rhythmic discharge which persists after an input is removed) and hysteresis (a property of a system wherein it fails to return to baseline after an input is removed). These properties in cortical networks are facilitated by a central core of ascending cholinergic brain stem pathways. The firing patterns within their networks are analogous to distinct functional "states," each of which is associated with a balance between excitatory and inhibitory influences. "Any particular functional state tends to be stabilized by reverberation in positive feedback loops, aided by long time constants of inhibition" (Singer 1979, p 1105).

Singer et al. (1976) have also demonstrated the importance of spatial temporal contingencies in the cholinergic facilitation of visual cortex cells. "An interesting phenomenon was the hysteresis in the facilitation of light responses. When the brainstem reticular formation stimulus was applied at the appropriate interval, it facilitated the light response over as long as one sec and more. When, however, the conditioning stimulus was shifted by only 100 msec or less, it completely lost its facilitatory effect over the whole duration of the light stimulus. This suggests, first, that a precise timing between reticular activation and the initial transient component of the light response is necessary to obtain facilitation and, second, that once subliminal excitation has reached threshold, it becomes self-sustaining" (Singer et al. 1976, p 87).

In order to study this apparent resonance with the sensory 40-Hz ERP, stimulus trains of repetitive auditory clicks were presented at 10/s, 20/s, and 40/s rates for train durations of 120 ms and 200 ms, with an averaging window of either 200 ms or 500 ms. Subjects listened to click trains of 120 ms and 200 ms, alternating with periods of silence. Two samples of 512 click trains each were obtained and the two average waveforms combined to form a composite average for each stimulus rate (Borda 1984). A sample record with a stimulation rate at 40/s for a 120-ms train is shown in Fig. 5. As can be seen in this figure, the oscillations reach a maximum amplitude after the fourth to fifth click and continue at a high amplitude for some oscillations beyond the stimulus train.

There is a resonance effect in the 40-Hz ERP, seen also with subharmonics of the 40/s stimulus rate, such as 20/s. This resonant response could occur if the

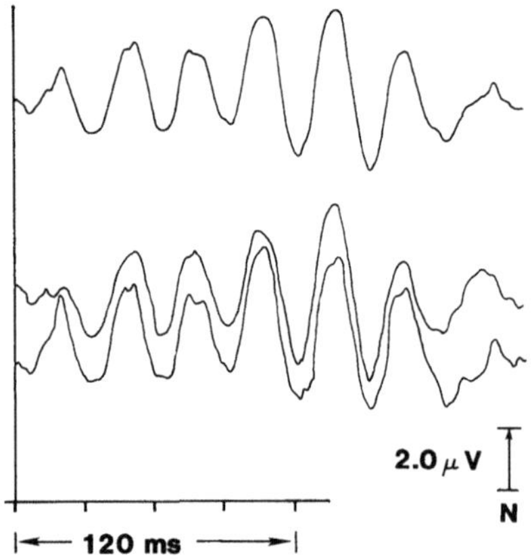

Fig. 5. Sensory 40-Hz ERPs recorded from a normal subject to a 120-ms train of 40/s clicks, with a response-averaging window of 200 ms. The *top trace* is the composite average of the *second* and *third traces*, each of which is an average of 512 samples. The *bottom line* is the stimulus marker with the stimulus indicated by vertical pips

neural circuitry generating the response has an optimal inherent rhythmicity near this frequency. The output of the circuitry would be greatly augmented if a series of repetitive stimuli were to enter the circuit at a rate near the resonant frequency, approaching the augmentation with subharmonics, and declining with rates which produce smoothing and cancellation effects. In physical models this phenomenon is known as "tuned resonance."

5 Behavioral Correlates

There have been an extensive series of studies carried out relating the 40-Hz ERPs to behaviors with significant focused arousal components. In eight normal adults, sensory 40-Hz ERPs were recorded concurrently from Cz and Oz to both clicks and flashes occurring simultaneously at rates of 40/s (Sheer and Schrock 1986). For each subject, the clicks came in one ear and a patch covered the eye on that side, so that the flashes were seen only with the contralateral eye. With the response averaging triggered by the 40/s auditory clicks, in one situation the subject was instructed to focus all of his energy, concentration, and attention only on the auditory clicks. With the averaging still triggered by the clicks, the subject was then instructed to concentrate and focus all of his energy on the flashes. In situation two, the response averaging was triggered by the 40/s light flashes and the subject was instructed to focus all of his energy on the flashes. With the averaging still triggered on the light flashes, he was instructed to concentrate and focus all of his energy on the clicks. Focused arousal, as defined by these instructional sets, had a significant effect on the amplitude of the 40-Hz ERP across sensory modalities. With both auditory and visual stimuli simultaneously presented at 40/s, actively focusing on one stimulus reduced the amplitude of the 40-Hz ERP to the other stimulus and vice versa.

A dissociation between the sensory and cognitive 40-Hz ERPs was demonstrated with a vigilance task, using signal detection measures of d′ (perceptual sensitivity) and Beta (response bias) with seven normal subjects (Appleby 1986). The stimuli were 0.2-ms clicks at 72 dB for the targets and 70 dB for the nontargets, presented one per second for 64 min at a ratio of nontarget to target stimuli at 6.5 : 1. This was a difficult vigilance task, near the sensory difference threshold, as evidenced by the sensitivity d′ measures.

The cortical 40-Hz ERP measures were not related to the signal detection measures, d′ and beta, but they were not random. There were significant, high correlations between the left and right hemispheres for every subject, with the level of 40-Hz activity greater in the left than in the right hemisphere. There was a strong inverse correlation between 40-Hz activity and time on task for each subject. The 40-Hz activity decreased during the 64 min of task performance in both the left and the right hemisphere.

With 16 target stimuli occurring in each 2-min block of the 64-min sessions, 512 responses were averaged for the target stimuli and a random 512 responses for the nontarget stimuli for each session. Records for replicated sessions for subject #1 are shown in Fig. 6 and records for subjects #2 and #3 are shown in Fig. 7.

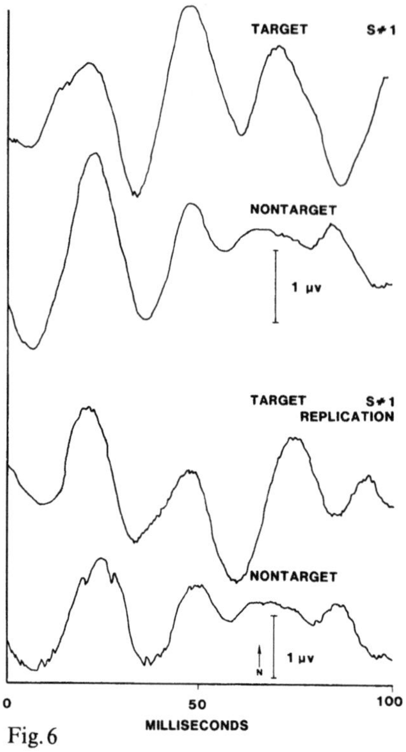

Fig. 6

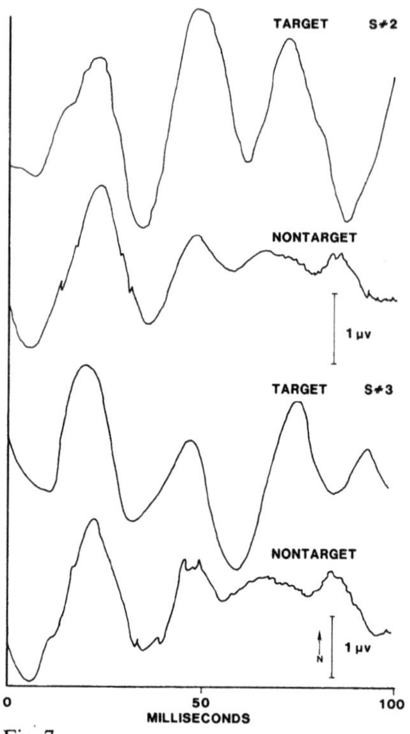

Fig. 7

Fig. 6. Sensory 40-Hz ERPs recorded from a normal subject (S #1) to 72-dB targets and 70-dB nontargets in two 64-min signal detection sessions. The auditory ERP is recorded from Cz, referenced to the ear, for 512 targets during the 64-min sessions and 512 random nontargets, with a response-averaging window of 100 ms

Fig. 7. Sensory 40-Hz ERPs recorded from two normal subjects (S #2 and S #3) to 72-dB targets and 70-dB nontargets in 64-min signal detection sessions. The auditory ERP is recorded from Cz, referenced to the ear, for 512 targets during the 64-min session and 512 random nontargets, with a response-averaging window of 100 ms

At the sensory threshold level of processing in this signal detection task, there was a significantly more pronounced resonant effect for the target stimuli as compared with the nontarget stimuli in the sensory 40-Hz ERPs.

With the cognitive 40-Hz ERPs, a series of studies in both animals (Bauer and Jones 1976) and humans (Bird et al. 1978 a, b; Ford et al. 1980; Sheer 1977) were carried out with operant conditioning techniques to increase and decrease patterns of the 40-Hz activity and observe the concomitant behaviors. In the animal study, cats trained to increase 40 Hz in the visual cortex were associated with the behaviors of arousal and "fixed staring." In the hippocampus, the increases in 40 Hz were associated with immobility (Bauer and Jones 1976). In the human studies, 40-Hz ERPs could be both increased and suppressed with operant conditioning. Throughout the conditioning and both pre- and postconditioning problem-solving sessions, the 40-Hz ERPs showed no correlations with EMG recordings from the scalp and only very low correlations with beta EEG, measured as 21–31 Hz. The subjective state consistently reported with 40-Hz conditioning included, in a Q-sort procedure, such words as attentive, concentrating, vigilant, and effortful, as contrasted with such beta words as active, energetic, excited, and restless. In control tests carried out up to 1 year after successful conditioning, subjects were able to control voluntarily the "on" and "off" of 40-Hz ERPs in ten alternative 2-min periods. The subjects could not turn off the 40-Hz ERPs in the alternate "off" periods when required, at the same time, to solve problems.

The meditation state provides a situation in which subjects are relatively immobile. Das and Gastaut (1955), recording from occipital leads in seven trained yogis, reported high amplitude levels of 40-Hz activity during the samadhi state, the final most intense concentration in this form of meditation. Banquet (1973), studying 12 subjects practicing transcendental meditation with recordings from left occipital and frontal leads, also observed 40 Hz during the third deep stage of meditation in subjects with advanced training. The 40-Hz activity has also been reported during REM sleep from the occipital cortex, in both cats (Parmeggiani and Zanocco 1963) and humans (Itil 1970). Depth electrographic studies in humans have demonstrated 40-Hz rhythms from the occipital lobe evoked by light stimuli (Chatrian et al. 1960) and by looking at pictures and objects only when interested (Sem-Jacobsen et al. 1956), and from the posterior temporal lobe in response to a range of sensory stimuli, which habituated rapidly (Perez-Borja et al. 1961).

In another series of studies (Loring and Sheer 1983; Spydell and Sheer 1982; Spydell et al. 1979), the task-dependent lateralization of 40-Hz activity was demonstrated in left and right cerebral hemispheres, concomitant with verbal and visual–spatial problem-solving. Such asymmetries as expressed in laterality scores $(L-R)/L+R)$ for number of 40-Hz peaks, were not seen with 40-Hz EMGs recorded from temporal and splenius muscles. Alpha and beta (21–31 Hz) EEG activity both decreased during problem-solving, alpha significantly more than beta. There were no significant correlations of the 40-Hz ERPs with EMG activity, or with alpha or beta.

Focused arousal has also been shown to operate in discrimination learning and conditioning. In a series of studies with cats in a successive visual discrimination task (Sheer 1970, Sheer and Grandstaff 1970), consistent relationships were obtained between 40-Hz activity and the acquisition phase of learning, defined by correct bar pressing just beyond chance expectancy. In the 10-s epoch of a 7/s SD flickering light, a burst of 40 Hz occurred in visual and motor cortex about 0.5 s before a correct bar press response, and continued for about 1 s thereafter. There was significant coherence in the sensorimotor circuitry involved in the discrimination task, visual and motor cortex.

Other reports of the association of 40-Hz activity with learning and conditioning include: the caudate nucleus and globus pallidus of cats with learning that the last in a series of 11 clicks leads to an unavoidable electric shock (Galambos 1958); the ectosylvian and lateral gyri and medial geniculate nucleus of cats with acquisition of an auditory CS paired with an electric shock UCS (Rowland 1958); the auditory, somesthetic, and motor cortex of dogs with conditioned limb responses with tone as the CS and excitation of the skin as the UCS (Dumenko 1961); the sensorimotor cortex of dogs with conditioned flexion of the contralateral leg after pairings with different conditioned stimuli (Sakhiulina 1961); and the lateral geniculate nucleus of cats with discrimination of a correct visual pattern in a group of three patterns (Killam and Killam 1967).

Schneider and Shiffrin (1977) investigated the differential effects of automatic and effortful processing in a visual search paradigm with memory load as a co-variable. Subjects were required to detect targets among distractors in rapidly presented stimulus sets. The dependent measures were response times in a single-frame task and accuracy scores in a multiple-frame task. The type of processing was experimentally manipulated, varying the relationship between target and distractor stimuli across trials. Automatic detection occurred where there was a consistent mapping of targets and distractors; that is, the target and distractor sets never changed over trials. Effortful search occurred in varied mapping conditions where target and distractor stimuli were thoroughly mixed across trials. When automatic processing was used, detection was rapid and accurate, and subjects reported minimal levels of attention and effort. Memory load, number of different targets to remember, and number of distractor items in the display set had little effect on performance with automatic processing. When effortful search was used, performance was slower and less accurate, particularly as memory load was increased, and subjects reported high levels of attention and effort.

This experimental paradigm was adapted in a computerized behavioral task, with concurrent cognitive 40-Hz ERPs recorded during each successive 12-s search trial for four subjects (Rogers 1984). A memory stimulus, the target to be detected, was presented for 3 s on the monitor at the beginning of each trial, digits for the automatic processing and letters for the effortful processing. Subjects were instructed to fixate, at the beginning of the trial, on a fixation point in the center of the screen and search for the target. At the end of each trial, they reported how many targets they had detected. Each trial consisted of 20 frames, each one on for 80 ms, followed by a masking frame. In each display frame, the four corners of the screen had either memory items, distractors, or a masking stimulus – a dot matrix the same size as the other characters.

In automatic processing, subjects searched for a digit among two (load 2) or four (load 4) letter distractors. The digit remained the same from trial to trial in the same block. In effortful processing, subjects searched for a letter among two (load 2) or four (load 4) letter distractors. The target letters were different from trial to trial. There were 12 blocks of trials run. Each block consisted of 25 trials with load 2 and 25 with load 4 counterbalanced.

The results for two of the four subjects are shown in Figs. 8 and 9. For both subjects, there is a sharp, consistent decline in percent errors during automatic processing, with a concurrent drop in 40 Hz of the left hemisphere. The correlations between left hemisphere 40-Hz activity and percent errors in search behavior are 0.95 for subject #2 and 0.92 for subject #3. For the other two subjects, they are 0.77 and 0.93. In both subjects in Fig. 8 and 9, as the percent errors approached zero with automatic processing, the 40-Hz activity in the right hemisphere increased. The correlations between right hemisphere 40 Hz and percent errors in search behaviors are −0.62 for subject #2 and −0.060 for subject #3, and for the other two subjects they are −0.77 and −0.60. With effortful processing the 40-Hz activity remains the same throughout the blocks of trials, at a consistently high level in the left hemisphere and at a low level in the right hemi-

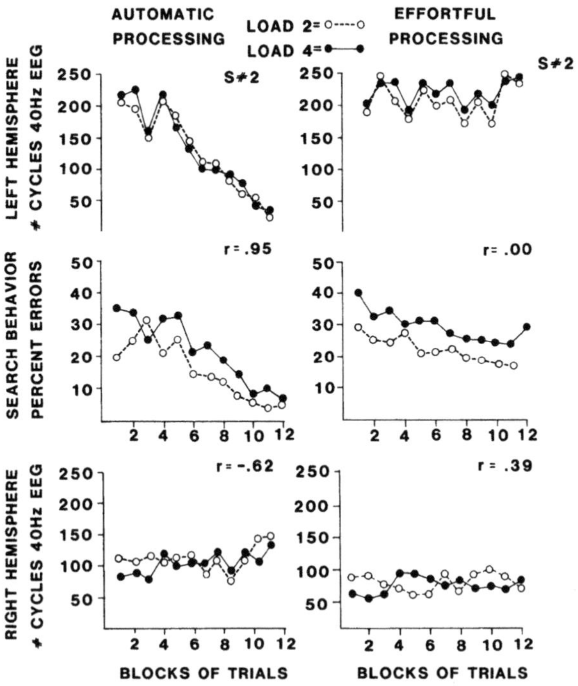

Fig. 8. Concurrent behavioral performance and the cognitive 40-Hz ERPs with automatic and effortful processing on a visual search task. Results for subject #2 are shown for the different measures as indicated in the figure. The correlations (*r*) are between the left and right hemisphere ERP and the behavior measure as indicated

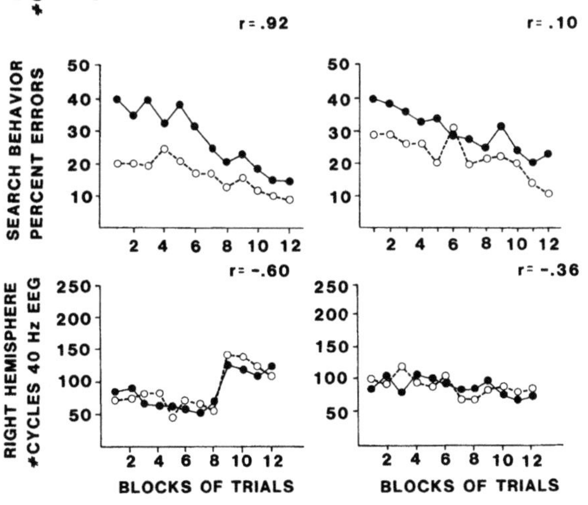

Fig. 9. Concurrent behavioral performance and the cognitive 40-Hz ERPs with automatic and effortful processing on a visual search task. Results for subject #3 are shown for the different measures as indicated in the figure. The correlations (*r*) are between the left and right hemisphere ERP and the behavior measure as indicated

sphere. The percent errors in search behavior, along with left hemisphere 40-Hz activity, are significantly different between automatic and effortful processing for all four subjects.

There is a highly significant, consistent relationship between the cognitive 40-Hz ERPs and attention processing in visual search behavior. With effortful processing, the 40-Hz activity is consistently at a high level in the left hemisphere. It drops markedly with automatic processing and the right hemisphere 40-Hz activity increases significantly. For exactly the same visual search task, the left hemisphere is dominant in effortful processing and the right hemisphere in automatic processing; Luria and Simernitskaya (1977) have made this same distinction between left and right hemispheres in volitional versus automatic memory function, using behavioral operations.

In these extensive series of behavioral studies, where focusing behavioral operations of reinforcement contingencies and instructional sets were used in specific sensory discriminations, operant conditioning, discrimination learning, problem-solving, and visual search and memory with both animals and humans, there were highly significant effects on two measures of focused arousal – the sensory and cognitive 40-Hz ERPs. In the visual search task, where the functional component of focused arousal was maximally involved, the rank order correlations between a behavioral measure of percent errors and a brain electricity measure, the cognitive 40-Hz ERP, where 0.77, 0.95, 0.92, and 0.93 for four subjects. A lateralization of the cognitive 40-Hz ERP was also demonstrated: left hemisphere with a verbal task and right hemisphere with a visual–spatial task. The operation of focused arousal has been further extended to motor programming by DeFrance and Sheer (1988) and emotional behaviors by DePascalis et al. (1987), and integrated in a cognitive model of attention by Sheer (1988a).

6 Clinical Application

The effectiveness of the sensory 40-Hz ERP as a diagnostic technique in brain damage was investigated by Schrock (1985). Two clinical subgroups in which there was verification of brain damage in the brain stem ($n=8$) or the frontal lobes ($n=9$) were compared with each other and with a matched normal group ($n=10$). The standard procedure for recording the sensory 40-Hz ERP was used, with stimulus rates of 40/s and 20/s. These stimuli were presented to one ear both with and without a cafeteria noise distraction from the G-F-W Auditory Skills Test (Goldman et al. 1974) to the contralateral ear. Four patients in the brain stem group had abnormal auditory brain stem evoked potentials.

The combination of 40/s stimulus rates, 20/s stimulus rates, and the distraction provided an analytical tool for dissecting out the differential brain damage in these patients. With the 40/s stimulus rate, the amplitudes of the 40-Hz ERPs were significantly greater in eight of the nine frontal lobe patients as compared with all patients in the brain stem group and eight of the ten subjects in the normal group. With the 20/s stimulus rate, there were zero amplitudes on peaks two and four in all eight brain stem patients, and zero amplitudes on peaks one and three in seven of the eight patients. No frontal patients and no normals had zero peak amplitudes with the 20/s stimulus rate.

The 40-Hz ERP recorded with the distraction in the contralateral ear showed consistent amplitude decreases in both the normal and the frontal lobe group, but consistent increases in the brain stem group. Four patients in the brain stem group also had some frontal lobe impairment and four did not. Using a ratio of distraction minus no distraction divided by distraction plus no distraction, these two subgroups could be identified. The four patients in the frontal lobe impaired subgroup showed a negative ratio on all peak amplitudes and the four patients without frontal lobe impairment showed a positive ratio in all peak amplitudes.

With damage to the rostral brain stem and/or ascending pathways, the sensory 40-Hz ERPs were markedly disrupted, apparently because a critical neural link in the generation of the ERP was impaired. The increase in peak amplitudes with the concomitant distraction condition may be due to increased concentration on the part of the patients under this condition or perhaps to more activation of residual circuitry with more sensory input. The consistently increased peak amplitudes with frontal lobe damage may indicate the disruption of a cortical descending modulatory link in the generation of the ERP. However these effects are finally worked out, empirically the sensory 40-Hz ERP provides an effective diagnostic technique in the differential assessment of brain damage. It was able to identify all eight brain stem cases, while the auditory brain stem evoked potential was abnormal in four cases. Other studies have confirmed abnormal 40-Hz ERPs with midbrain and thalamic lesions (Spydell et al. 1985), and with brain stem lesions in comatose patients (Firsching et al. 1987).

For the cognitive 40-Hz ERP in clinical application, a standard automated testing procedure was developed, presenting verbal and visual–spatial trials synchronized with the start and stop of the computer analysis of the ERP. The verbal trials consist of a sentence repetition task (modified from Baker and Leland 1967 and Spreen and Benton 1969) at three levels of difficulty. The subject listens to a sentence, during which the ERP is recorded, then repeats it upon a signal. The visual–spatial trials have been revised a number of times, but now consist of an unfamiliar facial recognition task (modified from Benton et al. 1975) also at three levels of difficulty. On each trial, the subject observes a face for 3 s and then two faces (difficulty level I), three faces (difficulty level II), or four faces (difficulty level III), each presented for a few seconds. The ERP is recorded during the presentation of the faces, and upon a signal the subject reports how many faces are the same as the first face. This testing procedure, contingent with the cognitive 40-Hz ERP, has been used with learning disability groups and Alzheimer patients (Sheer 1988 b).

Learning disability is a complex, multidetermined problem which, at the psychological functioning level, certainly includes a number of subtypes (Fletcher and Morris 1986; Rourke 1985). There is no question that at the molar level of language, reading, and arithmetic skills there are subgroupings of learning disabilities. Whether these subgroups are significantly determined on a biological basis or through ontogenetic learning is an open question; both may very well be involved. Clinical management and treatment must obviously be focused on the individual child, his social setting, specific functional skills, and major assets. But all this does not do away with the necessity to search for etiological abnormalities. In the initial learning of functional skills, focusing arousal on relevant associations and sequencing plays an important part in the development of cognitive processing.

The cognitive 40-Hz ERP has been shown to be significantly decreased during cognitive processing for subgroups of learning disabled children as compared with matched children at normal grade level in a series of studies (Auerbach and Sheer 1988; Sheer 1974, 1976). Matson (1987) has also been able to differentiate a general language impaired subgroup from a specific arithmetic disability subgroup. The language disability subgroup showed a significantly decreased 40-Hz ERP in the left hemisphere during the verbal trials as compared with the right during visual–spatial trials. The specific arithmetic subgroup showed a corresponding decrease of the 40-Hz ERP in the right hemisphere during visual-spatial trials as compared with the left during verbal trials. The cognitive 40-Hz ERP does more than offer a technique for the diagnosis of learning disabilities. By its close ties to neuronal circuitry and neural mechanisms, as well as to a functional component significantly involved in molar cognitive functions, it provides a way to gain some understanding of the defective cognitive processing and its underlying substrate.

The cognitive 40-Hz ERP, with the clinical testing procedure, was used in a series of studies to differentially diagnose Alzheimer patients. In the first study (Spydell and Sheer 1983), with an early Alzheimer and a matched geriatric control group, the task-dependent lateralization for the geriatric controls was essentially the same as for young adults (Sheer 1984). There was a significant difference in the laterality ratio $(L-R)/(L+R)$ for the verbal task $(+0.44)$ as compared with that for the visual–spatial task (-0.24). In the Alzheimer group, there was no task-dependent lateralization, and for all conditions (baseline, verbal, and visual–spatial) the absolute level of 40-Hz ERP activity was markedly lower and significantly different from that for the geriatric controls.

In a second study (Loring et al. 1985), Alzheimer patients $(n=12)$ were compared with both normal geriatric $(n=12)$ and multi-infarct groups $(n=12)$. The mean absolute levels of 40 Hz in the Alzheimer group under all conditions (baseline, verbal, and visual–spatial tasks) were significantly different from the levels for both the geriatric control and the multi-infarct group, and there was no task-dependent lateralization. Further, individual scores for all groups indicated predictive efficiency for individual cases. Under the verbal task condition, not one Alzheimer case was higher than the entire distribution of geriatric controls and only two cases overlapped with the distribution of multi-infarcts. Under the visual–spatial task condition, only two Alzheimer cases overlapped in the distribution of both the geriatric control and the multi-infarct group. Also, under the visual–spatial task condition, one multi-infarct case had the same 40-Hz level as four Alzheimer cases.

In a third study (Keller 1985), Alzheimer patients $(n=10)$ were compared with both normal geriatric controls $(n=10)$ and a retarded depressed group $(n=10)$. The control and depressed groups demonstrated task-dependent hemispheric lateralization on the cognitive 40-Hz ERP, while the Alzheimer group did not. The Alzheimer group also showed a uniformly lowered level of ERP activity under all conditions (baseline, verbal, and visual–spatial). The retarded depressed group were not significantly different from the controls on the cognitive 40-Hz ERP, while clinically they appeared as impaired as the Alzheimer patients and scored as low on neuropsychological tests.

The cognitive 40-Hz ERP is an effective diagnostic technique in differentiating Alzheimer patients from the normal elderly, multi-infarct dementia patients, and the retarded depressed. Primary sensory and motor cortex have been shown to be relatively spared in patients with early Alzheimer's dementia (Benson et al. 1983). The fact that the sensory 40-Hz ERP is not different in Alzheimer patients as compared with normals (Borda 1984) may have significant implications for cognitive processing. There are two ways in which the circuitry for phasic arousal interacts with cortex. From Fig. 4 in the neural substrate section, an evolutionary newer and, in humans, the major ascending pathway goes first to the basal forebrain, which then projects to association cortex. Another pathway, older and more significant phylogenetically, projects directly to primary cortex. The circuits to cortex, interactive through basal forebrain connectivity, provide for higher-order information processing. The direct connections with cortex provide for simpler, S-R, processing. These two processing modes may be expressed behaviorally in the distinctions that have been demonstrated clinically with mediobasal cortical lesions, where memory and sequential processing are severely impaired but simple habits and S-R connections are retained. Squire (1982) has termed these two processing modes declarative and procedural memory, and Mishkin et al. (1984) have termed them information memory and habit formation.

7 Summary

A developing body of interdisciplinary knowledge labeled Cognitive Neuroscience is concerned with information processing in the brain between stimulus inputs and motor outputs. This processing is expressed behaviorally as correlates of such molar cognitive processes as attention, perception, and memory. It is essential to partition these processes into first-order functional components so that the different brain systems which subserve these separable functions can be delineated and their relationships studied. Most important in such a program is the development of independent measurement operations for these component functions.

Focused arousal is a first-order energetic functional component, distinguished from both tonic and phasic arousal. It represents interactive neural circuitry between the phasic arousal, rostral brain stem reticular ascending pathways and specific sensory and multisensory cues. To function in such molar cognitive processes as attention, perception, and memory, focused arousal requires directing, warning, focusing, or reinforcing behavioral operations, contingent with the sensory inputs. At the neural systems level of analysis, the time constants of dynamic perturbations of organized electricity, such as periodicities, coherences, and spatial distributions of amplitudes, are more congruent with the time constants of this information processing.

Two measurement operations – sensory and cognitive 40-Hz ERPs – have been detailed. The first, tightly coupled with a contingent specific sensory stimulus and using a response-averaging technique, indexes the operation of focused arousal in a specific sensory system. The second, loosely coupled with multisensory stimuli, both antecedent and concurrent, and using peak counts above a preset threshold, indexes the operation of focused arousal in circuitry convergent at parietal-temporal-occipital cortex. The designation "40 Hz" is shorthand notation for different narrow frequency bands across species in the high-frequency gamma range (35–85 Hz), which center on specific coherent frequencies with focused arousal. These measurement operations have been used to work out the linked neural substrate, neural mechanisms, behavioral correlates, and clinical applications of focused arousal.

The contingent facilitation of the ascending cholinergic pathways originating in the parabrachial nucleus of the rostral pons is necessary to activate focused arousal in specific sensory and multisensory circuitry. But the measurement operations of focused arousal – reflecting coherent resonance at an optimal periodicity – are restricted to the specific sensory and multisensory circuitry. The neural substrate for focused arousal is finite sensory and multisensory circuitry, interactivity activated through sensory cues contingent with phasic arousal facilitation. By facilitating coherent sensory cell assemblies and, at the same time, not becoming a coherent "locked-in" part of such assemblies, the phasic arousal can serve a more adaptive, flexible processing function, on the ready to facilitate different sensory circuitry, as in shifts of attention.

The coherent and resonant properties of these finite cell assemblies were demonstrated in a series of experiments. Spatial amplitude patterns of coherent oscillations are determined by unique templates of synaptic connections, forming – with focusing behavioral operations – distinctive nerve cell assemblies for discriminative information. The specificity of the information conveyed is determined by the distinctive spatial pattern of unit cell specificities, defined by the modulatory controlled content of the specific sensory and multisensory inputs. The energy that momentarily fixes the unique spatial pattern at a significantly high amplitude level is the coherent resonance at an optimal periodicity. The coherent 40-Hz ERPs reflect repetitive stimulation at a constant frequency for a limited time over finite cell assemblies. Repetitive periodic excitation of cells maximizes the efficiency of synaptic transmission over the limited circuitry – the property of frequency potentiation.

In an extensive series of behavioral studies, where focusing operations of instructional sets and reinforcement contingencies were used in specific sensory discriminations, operant conditioning, discrimination learning, problem-solving, and visual search and memory with both humans and animals, there were highly significant effects on the two measures of focused arousal. A dissociation was demonstrated between the sensory and the cognitive 40-Hz ERPs on an auditory signal detection task at the sensory threshold level of processing. The sensory 40-Hz ERPs showed resonant oscillations on the averaged responses focused on the target stimuli and no oscillations with the nontarget stimuli. There were no relationships between behavioral measures of this sensory processing and the cognitive 40-Hz ERPs. A lateralization of the cognitive 40-Hz ERPs was demonstrated in problem-solving – left hemisphere over right with a verbal task, and right hemisphere over left with a visual–spatial task. In the visual search task, where the functional component of focused arousal was maximally involved, the rank order correlations between a behavioral measure of percent errors and a brain electricity measure – the cognitive 40-Hz ERPs – were 0.77, 0.95, 0.92, and 0.93 for four subjects.

The combination of a 40/s stimulus rate, a 20/s stimulus rate, and distraction in the contralateral ear with the sensory 40-Hz ERPs, provided a diagnostic technique for assessment in brain damage. With the 40/s stimulus rate, the amplitudes of the 40-Hz ERPs were significantly greater in eight of nine frontal lobe patients as compared with all eight patients in the brain stem group and eight of the ten subjects in a normal group. With the 20/s stimulus rate, there were zero amplitudes on peaks two and four in all eight brain stem patients, and zero amplitudes on peaks one and three in seven of the eight patients. No frontal patients and no normals had zero peak amplitudes with the 20/s stimulus rate. Four patients in the brain stem group also had some frontal lobe impairment and four did not. Using a ratio of distraction minus no distraction divided by distraction plus no distraction, these two subgroups could be identified. The four patients in the frontal lobe impaired subgroup showed a negative ratio on all peak amplitudes and the four patients without frontal lobe impairment showed a positive ratio.

The cognitive 40-Hz ERP, with a standard automated testing procedure presenting synchronized verbal and visual–spatial trials, was used in the differential diagnosis of learning disabilities and Alzheimer's dementia. In a series of three studies, the 40-Hz ERP was significantly decreased during cognitive processing in learning disability children as compared with matched children at normal grade level. In a fourth study, a general language impaired subgroup was differentiated from a specific arithmetic disability subgroup by significantly decreased 40-Hz ERP in the left hemisphere during verbal trials for the former and decreased 40-Hz ERP during visual–spatial trials for the latter. In a series of three studies, Alzheimer patients were significantly differentiated from a matched geriatric control group, from multi-infarct dementia patients and normal geriatrics, and from a retarded depression group and geriatric controls. The geriatric control, infarct, and depressed groups showed task-dependent hemispheric lateralization, while the Alzheimer patients did not. The Alzheimer patients also had markedly lower levels of the 40-Hz ERP with little overlap of individual cases. The retarded depressed group were not significantly different from the controls on the cognitive 40-Hz ERP, while clinically they appeared as impaired as the Alzheimer patients and scored as low on neuropsychological tests.

References

Anchel H, Lindsley DB (1972) Differentiation of two reticulo-hypothalamic systems regulating hippocampal activity. Electroencephalogr Clin Neurophysiol 32:209–226

Andersen PO, Anderson SA (1968) Physiological basis of alpha rhythm. Appleton Century Crofts, New York

Andersen PO, Eccles JC, Loyning Y (1963) Recurrent inhibition in the hippocampus with identification of the inhibitory cell and its synapses. Nature 198:540–542
Appleby MS (1986) The relationship between theory of signal detection statistics d′ and Beta and 40-Hz EEG during performance of an auditory vigilance task. Unpublished Doctoral dissertation, University of Houston
Auerbach V, Sheer DE (1988) Forty Hertz EEG in learning disabled and normal boys. J Learn Disabil (in press)
Baghdoyam RA, McCarley RW, Hobson JA (1981) Cholinergic generation of desynchronized sleep signs: localization to the pontine reticular formation. Paper presented at Neurosciences Society meeting, Los Angeles
Baker H, Leland B (1967) Detroit tests of learning aptitude. Bobbs-Merrill, Indianopolis
Banquet JP (1973) Spectral analysis of the EEG in meditation. Electroencephalogr Clin Neurophysiol 35:143–151
Başar E, Rosen B, Başar-Eroglu C, Grietschur F (1987) The associations between 40 Hz EEG and the middle latency response of the auditory evoked potential. Int J Neurosci 33:103–117
Bauer RH, Jones CN (1976) Feedback training of 36–44 EEG activity in visual cortex and hippocampus of cats: evidence for sensory and motor involvement. Physiol Behav 17:885–890
Beatty J (1982) Task-evoked pupillary responses, processing load, and the structure of processing resources. Psychol Bull 91:276–292
Beatty J, Wagoner BL (1978) Pupillometric signs of brain activation vary with levels of cognitive processing. Science 199:1216–1218
Ben-Ari YP, Dinglederi J, Konazawa I, Kelley JS (1976) Inhibitory effects of acetylcholine on neurons in the feline nucleus reticularis thalami. J Physiol 261:647–671
Benson DF, Kuhl DE, Hawkins RA (1983) The fluorodeoxyglucose 18F scan in Alzheimer's disease and multi-infarct dementia. Arch Neurol 40:711–714
Benton AL, Van Allen MW, Hamsher K deS, Levin HS (1975) Test of facial recognition; Form SL. University of Iowa Hospital, Iowa City
Berlyne DE (1960) Conflict, arousal, and curiosity. McGraw-Hill, New York
Berman AL (1968) The brain stem of the cat. University of Wisconsin Press, Madison
Biedenbach MA, Freeman WJ (1965) Linear domain of potentials from the prepyriform cortex with respect to stimulus parameters. Exp Neurol 11:400–417
Bindra D (1961) Components of general activity and the analysis of behavior. Psychol Rev 68:205–215
Bird BL, Newton FA, Sheer DE, Ford M (1978a) Biofeedback training of 40 Hz EEG in humans. Biofeedback Self Regul 3:1–11
Bird BL, Newton FA, Sheer DE, Ford M (1978b) Behavioral and electroencephalographic correlates of 40-Hz EEG biofeedback training in humans. Biofeedback Self Regul 3:13–28
Block V (1970) Facts and hypotheses concerning memory consolidation processes. Brain Res 24:561–575
Block V, Hebb DO (1956) Étude des phénomènes d'enrayerment et d'activation du comportement per stimulation thalamique et reticulaire chez le ret non anesthesie. Psychol Fr 1:8–9
Borda RP (1984) The 40/sec middle latency auditory response in Alzheimer disease and age-matched normals. Unpublished Doctoral dissertation, University of Houston
Bouyer JJ, Montaron MF, Rougeul A (1981) Fast frontoparietal rhythms during combined focused attentive behavior and immobility in the cat: cortical and thalamic localizations
Bressler SJ, Freeman WJ (1980) Frequency analysis of olfactory system EEG in cat, rabbit and rat. Electroenceph Clin Neurophysiol 50:19–24
Broadbent DE (1957) A mechanical model for human attention and immediate memory. Psychol Rev 64:205–215
Brookhart JM, Kubota K (1963) Studies of the integrative function of the motor neurone. In: Moruzzi G, Fessard A, Jasper HH (eds) brain mechanisms. Vol I: Progress in brain research. Elsevier, New York, pp 38–49
Brown GW, Cowen BD (1959) Avoidance and approach learning motivated by stimulation of identical hypothalamic loci. Am J Physiol 197:153–157
Burton H, Jones EG (1976) The posterior thalamic region and its cortical projections in New World and Old World monkey. J Comp Neurol 195:249–301

Celesia GC, Jasper HH (1966) Acetylcholine released from cerebral cortex in relation to state of activation. Neurology 16:1053–1083
Chatrian GE, Bickford RG, Uihlein A (1960) Depth electrographic study of a fast rhythm evoked from the human calcarine region by steady illumination. Electroencephalogr Clin Neurophysiol 12:167–176
Chatrian GE, Lowell E, White JR, Cheng-Mei S (1964) EEG pattern resembling wakefulness in unresponsive decerebrate state following brain-stem infarct. Electroencephalogr Clin Neurophysiol 16:285–299
Chedru F, Le Blanc M, Lhermitte F (1973) Visual searching in normal and brain-demaged subjects (contributions to the study of unilateral attention) Cortex 9:94–111
Coleman JR, Lindsley DB (1975) Hippocampal electrical correlates of free behavior and behavior induced by stimulation of two hypothalamic-hippocampal systems in the cat. Exp Neurol 49:506–528
Coyle JT, Price DL, De Long MR (1983) Alhzeimer's disease: a disorder of cortical cholinergic innervation. Science 219:1184–1190
Das NN, Gastaut H (1955) Variations de l'activité électrique du cerveau, du coeur et des muscles squellettiques au cours de la méditation et de l'extase yogique. Electroencephalogr Clin Neurophysiol 6:211–219
DeFrance J, Sheer DE (1988) Focused arousal, 40-Hz EEG and motor programming. In: Giannitrapani D, Murri L (eds) The EEG of mental activities. Karger, Basel, pp 153–168
DePascalis V, Marucci FS, Penna PM, Pessa E (1987) Hemispheric activity of 40-Hz EEG during recall of emotional events: differences between low and high hypnotizables. Int J Psychophysiol 5:167–180
Drachman D (1977) Memory and cognitive functions in man: does the cholinergic system have a specific role? Neurology 22:783–790
Duffy E (1957) The psychological significance of the concept of "arousal" or "activation." Psychol Rev 64:265–274
Dumenko VN (1961) Changes in the electroencephalogram of the dog during the formation of a motor conditioned reflex stereotype. Pavlov J Higher Nerv Act 11:64–68
Easterbrook JA (1959) The effect of emotion on cue utilization and the organization of behavior. Psychol Rev 66:183–201
Eccles JE (1964) The physiology of synapses. Springer, Berlin Göttingen Heidelberg New York
Eysenck MW (1982) Attention and arousal. Springer, Berlin Heidelberg New York
Famiglietti EV (1970) Dendro-dendritic synapses in thalamic neurons during synchronizing inhibition. Brain Res 20:181–190
Fertziger AP, Pupura DP (1971) Diphasic-PSP's during maintained activity of cat lateral geniculate neurons. Brain Res 33:463–467
Firsching R, Luther J, Erdelberg E, Brown WE, Story JC, Boop FA (1987) 40 Hz-middle latency auditory evoked response in comatose patients. Electroencephalogr Clin Neurophysiol 67:213–216
Fletcher JM, Morris R (1986) Classification of disabled learners: beyond exclusionary definitions. In: Ceci S (ed) Handbook of cognitive, social, and neuropsychological correlates of learning disabilities. Erlbaum, New York, pp 55–80
Ford M, Bird BC, Newton HA, Sheer DE (1980) Maintenance and generalization of 40 Hz EEG biofeedback effects. Biofeedback Self Regul 5:193–205
Freeman WJ (1962a) Alterations in prepyriform evoked potential in relation to stimulus intensity. Exp Neurol 6:70–84
Freeman WJ (1962b) Changes in prepyriform evoked potential with food deprivation and consumption. Exp Neurol 6:12–29
Freeman WJ (1962c) Comparison of thresholds for behavioral and electrical responses to cortical electrical stimulation in cats. Exp Neurol 6:315–331
Freeman WJ (1962d) Linear approximation of prepyriform evoked potential in cats. Exp Neurol 5:477–499
Freeman WJ (1962e) Phasic and long-term excitability changes in prepyriform cortex of cats. Exp Neurol 5:500–518
Freeman WJ (1975) Mass action in the nervous system. Academic, New York

Freeman WJ (1986) Analytic techniques used in the search for the physiological basis of the EEG. In: Gevins AS, Remond A (eds) Methods of analysis of brain electrical and magnetic signals. Elsevier, Amsterdam, pp 583–664 (EEG Handbook Revised Series, vol 1)
Freeman WJ, Baird B (1987) Relation of olfactory EEG to behavior: spatial analysis. Behav Neurosci 101:393–408
Freeman WJ, Grajski G (1987) Relation of olfactory EEG to behavior: factor analysis. Behav Neurosci 101:766–777
Freeman WJ, Skarda CA (1985) Spatial EEG analysis, nonlinear dynamics, and perception: the neo-Sherringtonian view. Brain Res Rev 10:147–175
Freeman WJ, Viana Di Prisco G (1985) EEG spatial pattern differences with discriminated odors manifest chaotic and limit cycle attractors in olfactory bulb of rabbits. Proceedings of the conference on brain theory, Trieste 1984
Freeman WJ, Viana Di Prisco G (1986) Relation of olfactory EEG to behavior: time series analysis. Behav Neurosci 100:753–763
Fuster JM, Uyeda AA (1962) Facilitation of tachistoscopic performance by stimulation of midbrain tegmental points in the monkey. Exp Neurol 6:384–406
Galambos R (1982) Tactile and auditory stimuli repeated at high rates (30–50 per sec) produce similar event-related potentials. Ann NY Acad Sci 388:722–728
Galambos R (1958) Electrical correlates of conditioned learning. In: Brazier M (ed) The central nervous system and behavior. Macy, New York, pp 375–415
Galambos R, Makeig S, Talmachoff P (1981) A 40-Hz auditory potential recorded from the human scalp. Natl Acad Sci USA 78:2643–2647
Gastaut H, Jus C, Morrell F, Storm van Leeowen W, Doniger S, Naquet R, Regis H, Roger A, Berkerng D, Kamp A, Werre J (1957) Étude topographique des réactions electroencephalographiques conditionnes chez l'homme. Electroencephalogr Clin Neurophysiol 9:1–34
Goldman R, Fristoe M, Woodcock RW (1974) Goldman-Fristoe-Woodcock auditory skills test battery. American Guidance Service
Granit R (1963) Recurrent inhibition as a mechanism of control. In. Moruzzi G, Fessard A, Jasper HH (eds) Brain mechanisms. Vol I: Progress in brain research. Elsevier, New York, pp 23–31
Hebb DO (1955) Drives and the C.N.S. (conceptual nervous system) Psychol Rev 62:243–254
Hoyman L, Kelsey JE (1977) Facilitation of flicker discriminability by electrical stimulation of mesencephalic reticular formation of the rat. J Comp Physiol Psychol 91:951–961
Hubel DH, Wiesel TN (1962) Receptive fields, binocular interaction and functional architecture in the cat's visual cortex. J Physiol 160:160–154
Itil TM (1970) Digital computer analysis of the electroencephalogram during rapid eye movement state in man. J Nerv Ment Dis 130:201–208
Japser H, Stefanis C (1965) Intracellular oscillatory rhythms in pyramidal tract neurons in the cat. Electroencephalogr Clin Neurophysiol 18:541–553
Jasper H, Tessier J (1971) Acetylcholine liberation from cerebral cortex during paradoxical (REM) sleep. Science 172:601–602
Jones EG (1985) The thalamus. Plenum, New York
Jouvet M, Benoit O, Courjon J (1956) Aspects EEG de la formation de liasons temporaires dans le cerveau. Abstract of the XXth International Congress of Physiological Sciences, Excerpta Medica, Amsterdam, p 475
Judd BW (1987) Topography of the 40-Hz evoked-potential in the auditory and visual modalities. Unpublished Masters thesis. University of Houston
Kaada BR, Harmark WY, Stokke O (1961) Deep coma associated with desynchronization in EEG. Electroencephalogr Clin Neurophysiol 13:785–789
Kahneman D (1973) Attention and effort. Prentice Hall, Englewood Cliffs
Karczmar AG (1979) Brain acetylcholine and animal electrophysiology. In: Davis KL, Berger FA (eds) Brain acetylcholine and neuropsychiatric disease. Plenum, New York, pp 265–310
Kaufmann P (1985) 40-Hz EEG and focused arousal behavior in the cat. Unpublished masters thesis. University of Houston
Keller WJ (1985) Forty Hertz EEG activity in the elderly: dementia, depression, and normal aging. Unpublished Doctoral dissertation, University of Houston

Killam KF, Killam EK (1967) Rhinencephalic activity during acquisition and performance of conditional behavior and its modification by pharmacological agents. In: Adey WR, Tokizane T (eds) Progress in brain research, vol 27. Structure and function of the limbic system. Elsevier, New York, pp 388–399

Killeen P (1975) On the temporal control of behavior. Psychol Rev 82:89–115

Killeen P, Hanson SJ, Osborne SR (1978) Arousal: its genesis and manifestation as response rate. Psychol Rev 85:578–581

Kimura D (1964) Cognitive deficit related to seizure pattern in centrencephalic epilepsy. J Neurol Neurosurg Psychiatry 27:291–296

Kimura H, McGeer PL, Peng JH, McGeer EG (1981) The central cholinergic system studied by choline acetyltransferase immunohistochemistry in the cat. J Comp Neurol 200:151–201

LaBerge D (1976) Perceptual learning and attention. In: Estes WK (ed) Handbook of learning and cognitive processes. Vol 4:Attention and memory. Wiley, Hillsdale, pp 268–276

Lacey JL (1967) Somatic response patterning and stress. Some revisions of activation theory. In: Appley MH, Trumbull R (eds) Psychological stress. Appleton Century Crofts, New York, pp 14–42

Landsell H, Mirsky AF (1964) Attention in focal and centrencephalic epilepsy. Exp Neurol 9:463–469

Lansing RW, Schwartz E, Lindsley D (1959) Reaction time and EEG activation under alerted and non-alerted condition. J Exp Psychol 58:1–7

Lashley KS (1950) In search of the engram. Symp Soc Exp Biol 4:454–482

Libet B, Alberts WW, Wright EW Jr, Feinstein B (1967) Responses of human somatosensory cortex to stimuli below threshold for conscious sensation. Science 158:1597–1599

Lindsley DB (1958) The reticular system and perceptual discrimination. In: Jasper HH, Proctor LD, Knighton RS, Noskay WC, Costello RT (eds) Reticular formation of the brain. Little Brown, Boston, pp 513–528

Loeb C, Paggio G (1953) Electroencephalograms in a case with pontomesencephalic hemorrhage. Electroencephalogr Clin Neurophysiol 5:295–296

Loring DW, Sheer DE (1983) Laterality of 40-Hz EEG and EMG during cognitive performance. Psychophysiology 21:34–38

Loring DW, Sheer DE, Largen JW (1985) Forty Hertz EEG activity in dementia of the Alzheimer type and multi-infarct dementia. Psychophysiology 22(1):166–121

Luria AR, Simermitskaya EG (1977) Interhemispheric relations and the functions of the minor hemisphere. Neuropsychologica 15:175–178

Macadar AW, Chalupe LM, Lindsley DB (1974) Differentiation of brainstem loci which affect hippocampal and neocortical electrical activity. Exp Neurol 43:499–515

Mahut H (1964) Effects of subcortical electrical stimulation on discrimination learning in cats. J Comp Physiol Psychol 58:390–395

Malmo RB (1959) Activation: a neuropsychological dimension. Psychol Rev 66:367–386

Matson A (1987) The occurrence and lateralization of 40-Hz EEG activity in LD and normal children. Unpublished Masters thesis, University of Houston

McClelland JC, Rumelhart DE, PDP Research Group (1986) Parallel distributed processing: explorations in the microstructure of cognition. Vol 2: Psychological and biological models. MIT, Cambridge

McDonald RD, Burns SB (1964) Visual vigilance and brain damage: an empricial study. J Neurol Neurosurg Psychiatry 27:206–209

Mirsky AF, Pragay EB (1988) Brainstem reticular system. In: Sheer DE, Pribram KH (eds) Attention: cognition, brain function, and clinical application. Academic, New York

Mishkin M, Malamut B, Bechevolier JB (1984) Memories and habits: two neural systems. In: Lynch G, McGaugh JC, Wemberger NM (eds) Neurobiology of learning and memory. Guilford, New York, pp 65–83

Montplaisir JY (1975) Cholinergic mechanisms involved in cortical activation during arousal. Electroencephalogr Clin Neurophysiol 38:263–272

Morrell F, Jasper HH (1956) Electrographic studies of the formation of temporary connections in the brain. Electroencephalogr Clin Neurophysiol 8:201–215

Moruzzi G, Magoun (1949) Brainstem reticular formation and activation of the EEG. Electroencephalogr Clin Neurophysiol 1:455–473

Mountcastle VB (1961) Duality of function in the somatic afferent system. In. Brazier M (ed) Brain and behavior, vol I. American Institute of Biological Sciences, Washington, p 67
Parmeggiani PL, Zanocco GA (1963) Study of the bioelectrical rhythms of cortical and subcortical structures during activated sleep. Arch Ital Biol 101:385–412
Payne DT, Parby ME, Harasymin SJ (1968) Percentage of pupillary dilation as a measure of item difficulty. Percept Psychophys 4:139–143
Perez-Borja C, Tyce FA, McDonald C, Uihlein A (1961) Depth electrographic studies of a focal fast response to sensory stimulation in the human. Electroencephalogr Clin Neurophysiol 13:695–702
Pfaff D (1969) Parsimonious biological models of memory and reinforcement. Psychol Rev 76:70–81
Pfurtscheller G, Cooper R (1975) Frequency dependence of the transmission of the EEG from cortex to scalp. Electroencephalogr Clin Neurophysiol 38:93–96
Poggio GF, Mountcastle VB (1960) A study of the functional contributions of the lemniscal and spinothalamic systems to somatic sensibility. Central nervous mechanisms in pain. Johns Hopkins Hospital Bull 106:266–316
Posner MI (1978) Chronometric explorations of mind. Erlbaum, Hillsdale
Posner MI, McLeod P (1982) Information processing models – in search of elementary operations. Annu Rev Psychol 33:477–514
Posner MI, Rothbart MK (1985) The concept of energy in psychological theory. Paper presented at the NATO advanced research workshop on: energetical aspects of human information processing, Les Ares, August 23–28
Posner MI, Cohen Y, Choate L, Hockey R, Maylor E (1984 a) Sustained concentration: passive filtering or active orienting. In: Kornblum S, Requin (eds) Preparatory states and processes. Erlbaum, Hillsdale, pp 49–65
Posner MI, Walker JA, Friedrick AJ, Rafal RD (1984 b) Effects of parietal lobe injury on covert orienting of visual attention. J Neurosci 4:1863–1874
Proctor LD, Knighton RS, Churchill JA (1957) Variations in consciousness, produced by stimulating reticular formation of the monkey. Neurology 7:193–203
Raghavan N, Glover JR, Sheer DE (1986) A microprocessor-based system for diagnosis of cognitive dysfunction. IEEE Trans Biomed Eng 33:942–948
Rall W, Shepherd GM (1968) Theoretical construction of field potentials and dendodentritic synaptic interactions of olfactory bulb. J Neurophysiol 8:884–915
Rogers RC (1984) 40-Hz EEG correlates of automatic and effortful information processing. Unpublished Doctoral dissertation, University of Houston
Rourke BP (ed) (1985) Learning disabilities: advances in subtypal analysis. Guilford, New York
Rowland V (1958) Discussion in Morrell, F. Electroencephalographic studies of conditioned learning. In. Brazier J (ed) The central nervous system and behavior. Macy, New York, pp 347–357
Rumelhart DE, McClelland JC, PDP Research Group (1986) Parallel distributed processing: explorations in the microstructure of cognition. Vol 1: Foundations. MIT Cambridge
Sakhiulina BT (1961) EEG manifestations of tonic cortical activity accompanying conditioned reflexes. Pavlov J Higher Nerv Act 11:48–57
Saper CL (1981) Brainstem and basal forebrain-projections to the cerebral cortex in the rat. Paper presented at the Neurosciences Society Meetings, Los Angeles
Saper CL (1982) Reciprocal parabrachial-cortical connections in the rat. Brain Res 242:33–40
Saper CL, Loewy AD (1980) Efferent connections of the parabrachial nucleus in the rat. Brain Res 197:291–317
Schneider W, Shiffrin RM (1977) Controlled and automatic human information processing. I: Detection, search, attention. Psychol Rev 84:1–66
Schrock BJ (1985) The 40-Hz event-related potential in brain damage: effects of lesion site, distraction, and stimulus rate. Unpublished Doctoral dissertation, University of Houston
Sem-Jacobsen CW, Peterson MC, Dodge HW, Jacks QD, Lazarte JA, Holman CB (1956) Electric activity of the occipital lobe in man. Am Med Sci 232:243–251
Sharpless S, Jasper H (1956) Habituation of the arousal response. Brain 70:655–680
Sheer DE (1970) Electrophysiological correlates of memory consolidation. In: Ungar G (ed) Molecular mechanisms in memory and learning. Plenum, New York, pp 177–211

Sheer DE (1974) Electrophysiolographic studies in learning disabilities. In: Eichenwald H, Talbot A (eds) The learning disabled child. University of Texas Press, Austin, pp 47–58
Sheer DE (1976) Focused arousal and 40-Hz EEG. In: Knight RM, Bakker DJ (eds) The neuropsychology of learning disorders. University Park Press, Baltimore, pp 71–87
Sheer DE (1977) Biofeedback training of 40 Hz EEG and behavior. In: Kamiya J et al. (eds) Biofeedback and self-control, 1976/77. Aldine, Chicago, pp 435–472
Sheer DE (1984) Focused arousal, 40 Hz EEG and dysfunction. In: Elbert T, Rockstroh B, Lutzenberger W, Birbaumer N (eds) Self-regulation of the brain and behavior. Springer, Berlin Heidelberg New York, pp 63–84
Sheer DE (1988 a) A working cognitive model of attention – to fit in the brain and in the clinic. In: Sheer DE, Pribram KH (eds) Attention: cognition, brain function, and clinical applications. Academic, New York
Sheer DE (1988 b) Brain damage and dysfunctions. In: Sheer DE, Pribram KH (eds) Attention: cognition, brain function, and clinical application. Academic, New York
Sheer DE, Grandstaff N (1970) Computer-analysis of electrical activity in the brain and its relation to behavior. In: Wycis HT (ed) Current reseach in neurosciences: topical problems in psychiatry and neurology, vol 10. Karger, Basel, pp 160–172
Sheer DE, Schrock B (1986) Attention. In: Hannay J (ed) Experimental techniques in human neuropsychology. Oxford University Press, New York, pp 95–137
Shepherd GM (1978) The synaptic organization of the brain. Oxford University Press, New York
Shiffrin RM (1976) Capacity limitations in information processing, attention and memory. In: Estes WK (ed) Handbook of learning and cognitive processes. Wiley, Hillsdale, pp 177–236
Shiffrin RM, Schneider W (1977) Controlled and automatic human information processing. II: Perceptual learning, automatic attending and a general theory. Psychol Rev 84:127–190
Singer W (1977) Control of thalamic transmission by cortifugal and ascending reticular pathways in the visual system. Physiol Rev 57:386–420
Singer W (1979) Center-core control of visual cortex function. In: Schmitt FO, Worden FG (eds) The neurosciences: fourth study program. MIT, Cambridge, pp 1093–1110
Singer W (1980) State dependent changes in ascending pathways affecting visual cortex functions. Neurosci Res Program Bull 18:92–200
Singer W, Tretter F, Cynader M (1976) The effect of reticular stimulation on spontaneous evoked activity in the cat. Visual cortex. Brain Res 102:71–90
Sitarem N, Wyatt N, Dawson SP, Gillin J (1976) REM sleep induction by physostigmine infusion during sleep. Science 191:1281–1283
Snider RS, Niemer (1961) A stereotoxic atlas of the cat brain. University of Chicago Press, Chicago
Sokolov EN (1960) Neuronal model of the orienting reflex. In: Brazier M (ed) The central nervous system and behavior. Macy, New York, pp 187–276
Spehlman R (1971) Acetylcholine and the synaptic release of nonspecific impulses to the visual cortex. Brain 94:139–150
Spehlman R, Downes K (1974) The effects of acetylcholine and synaptic stimulation on the sensorimotor cortex. I: Neuronal responses to stimulation of the reticular formation. Brain Res 74:139–150
Spreen O, Benton AL (1969) Sentence repetition test. Manual of instructions and norms. Neuropsychology Laboratory, University of Victoria, Victoria
Spydell JD, Sheer DE (1982) Effect of problem solving on right and left hemisphere 40 Hz EEG activity. Psychophysiology 19:420–425
Spydell JD, Sheer DE (1983) Forty Hertz EEG activity in Alzheimer's type dementia. Psychophysiology 20:313–319
Spydell JD, Ford MR, Sheer DE (1979) Task dependent cerebral lateralization of the 40-Hz EEG rhythm. Psychophysiology 16:347–350
Spydell JD, Pattee G, Golde WD (1985) The 40 Hz event-related potential: normal values and effects of lesions. Electroencephalogr Clin Neurophysiol 62:193–202
Squire LR (1982) The neuropsychology of human memory. Annu Rev Neurosci 5:241–273
Squire LR, Davis HP (1981) The pharmacology of memory. A neurobiological perspective. Annu Rev Pharmacol Toxicol 21:323–356

Stefanis C, Jasper H (1964) Recurrent collateral inhibition in pyramidal tract neurons. J Neurophysiol 27:855–877
Stennett RG (1957) The relationship of performance level to level of arousal. J Exp Psychol 54:54–61
Steriade M (1980) State-dependent changes in the activity of rostral reticular and thalamocortical elements. Neurosci Res Program Bull 18:82–91
Tolman EC (1948) Cognitive maps in rats and men. Psychol Rev 55:189–208
Weber JT, Yin TCT (1984) Subcortical projections of the inferior pariental cortex (area 7) in the stump-tailed monkey. J Comp Neurol 224:206–230
Whitfield EC (1965) Edges in auditory information processing. Abstract of the XXIII International Congress of Physiological Sciences, Excerpta Medica, Amsterdam, p 245
Yerkes RM, Dodson JD (1908) The relation of strength of stimulus to rapidity of habit formation. J Comp Neurol Psychol 18:459–482

The CERP: Event-Related Perturbations in Steady-State Responses

S. Makeig and R. Galambos

1 Introduction

1.1 Steady-State Responses

In brain research, the term "steady-state response" (SSR) refers to electrophysiological activity driven by a train of stimuli delivered at a high enough rate so that responses evoked by successive stimuli overlap (Regan 1982). Visual SSRs are evoked by probe stimuli delivered at rates between a few and 60 Hz; these can be sinusoidally modulated light, shifts in checkerboard patterns, or trains of brief flashes (Mundy-Castle 1953; Regan 1982). The auditory SSR reaches a response amplitude maximum at rates of stimulation near 40 Hz (Galambos et al. 1981); auditory probe stimuli are often trains of clicks or brief tone pips, but sinusoidally amplitude-modulated (Green at al. 1986; Kuwada et al. 1986; Rickards and Clark 1984) or frequency-modulated (Rodriguez et al. 1986) continuous tones can also be used. Recent research on auditory SSRs has been encouraged by their possible application in audiology (Kankkunen and Rosenhall 1985; Klein 1983; Makeig and Galambos 1983; Picton et al. 1987; Sammeth and Barry 1985; Shallop 1983; Stapells et al. 1987; 1988; Sturzebecher et al. 1985).

1.2 SSR Modulators

The term "steady-state response" suggests that once initiated, SSRs do not vary, but this is often not the case. In an earlier paper in this series (Galambos and Makeig 1988), we identified and discussed three types of physiological modulation of the auditory SSR and categorized these as spontaneous, sleep-related, and stimulus-linked. A major class of spontaneous modulations are circa minute-scale fluctuations in amplitude and phase (minute-rhythms). Sleep-related modulations are more or less abrupt, reversible changes in SSR amplitude of 50% or more when subjects fall asleep (amplitude drops) or wake up. Both spontaneous and sleep-related modulations appear to be initiated and controlled by intrinsic brain mechanisms. Stimulus-linked modulations, by contrast, refer to alterations in auditory SSR amplitude and phase produced by, for example, presenting moderate levels of noise to the contralateral ear (see experiment 4 below). Our earlier report gives details on these and other aspects of SSR dynamics, and includes a full description of the methods to obtain and measure them.

Springer Series in Brain Dynamics 2
Edited by Erol Başar and T. H. Bullock

1.3 The Complex Event-Related Potential

This study defines and describes another class of stimulus-linked SSR modulations. We define the complex event-related potential (CERP) as a time series of complex numbers describing perturbations in amplitude and phase of the EEG at a single analysis frequency, time-locked to a discrete event of experimental interest. Our CERP examples appear when a subject is presented with numerous repetitions of some experimental event while concurrently receiving repetitive probe stimuli that create an SSR. The CERP analysis frequency is the SSR stimulus repetition rate, or a harmonic of it. Where possible, we phase-lock the experimental events to the probe stimulus train.

To extract a CERP from the EEG, we first order the data into time-locked response epochs, reject any containing gross artifacts, and average the remainder. Next we separate the portion of the evoked response containing the frequency of interest from lower and higher frequency portions by digital filtering, and then, using a moving-average discrete Fourier transform (DFT; Oppenheim and Schafer 1975), convert the single-frequency, coherently averaged response into a complex frequency-domain time series that describes perturbations in ongoing EEG activity at the analysis frequency time-locked to the experimental event.

Mathematically, the CERP can be represented in polar coordinates as a complex time-series in the phase-plane, or equivalently, in Cartesian coordinates as a pair of real time-series revealing the time course of SSR amplitude and phase perturbations surrounding the time-locking event. These two representations are interconvertible; below we display CERPs as sets of two real time-series that separably estimate amplitude and phase perturbations at the analysis frequency. Phase perturbations are further exposed by subtracting from the CERP phase record the circularly increasing phase advance which would be associated with an absence of perturbations at the analysis frequency ("phase unwrapping").

The examples of CERPs we present are mostly perturbations in auditory SSRs produced by probe stimuli delivered at a rate near 40 Hz. The experimental events we introduce consist mostly of altering some feature of the probe-stimulus train – but theoretically CERPs in any stimulus modality, extracted at any repetition rate, and time-locked to any perturbing event (or series of events) can also be produced and examined.

The CERP may be considered to be a narrow-band frequency-domain variant of the conventional wide-band time-domain event-related potential (ERP). CERP experiments can also be thought of as steady-state extensions of ERP paradigms in which pairs or short trains of stimuli are used to study adaptation, habituation, and recovery cycles (Freedman et al. 1987; Woods et al. 1980). The unique feature of CERP experiments is that CERPs and ERPs are evoked concurrently, and can be separated and compared. These 40-Hz CERP and ERP measures give independent information, and so the 40-Hz CERP technique opens a new, high-frequency window into event-related brain dynamics.

2 Experiments

Six experiments are presented below. The first two are mainly descriptive; they illustrate the CERP analysis technique, and describe new CERP data that await physiological explanation. The last four take first steps toward providing such a physiological analysis.

We then summarize the physiological observations, point out limitations and possible applications of the CERP concept and methodology, and offer suggestions about the physiological framework into which explanations of the origins of the phenomena may fit.

2.1 Methods Common to the Experiments

The experiments all involve steady-state auditory stimulation presented to the right ear of a normal-hearing adult subject through a headphone. Probe stimuli (commonly 40–60 dBSL, 0.1 ms square-wave clicks) are delivered at a rate near but not exactly at 40 Hz (to avoid phase-locking to harmonics of the 60-Hz line frequency). The moment of occurrence of the experimental event normally is phase-locked to the moment of delivery of a probe stimulus.

The EEG is collected from one or more sites of the International 10–20 System referred to a balanced noncephalic lead (Woods and Clayworth 1986), and digitized at a rate exactly eight times the probe rate (under common computer clock control). Electrode impedances are adjusted to less than 5 kΩ. EEG is amplified 50 000 times, analog filtered (using a 0.1- to 100-Hz bandpass), converted to 12-bit digital format, and stored on magnetic disk for off-line analysis.

Subjects recline in an easy chair in a quiet room, with eyes closed, and silently count the experimental events they perceive during runs lasting 6–8 min. After each run they report the number of events counted. The counting task is used in order to reduce subject boredom, monitor alertness, and minimize drowsness. To obtain sufficiently high signal-to-noise ratio at the stimulus rate to produce useful CERPs, runs are pooled to assemble the number of averages needed (normally 350–1200).

2.2 Experiment 1: A Frequency-Shift CERP

In experiment 1 the probe stimuli were brief (ca 5 ms, 2-1-2) tone pips. The first vertical line in Fig. 1 marks where the frequency of these pip probes changed from 1000 Hz to 1200 Hz ("pitch up"), and the second where it changed back again ("pitch down"); these events were separated ed by 1.5 s, and occurred every 3–3.5 s. The intensities of the 1000- and 1200-Hz probe trains were both set at 50 dBSL. The EEG electrode was placed at the vertex (Cz), where the auditory 40-Hz SSR is largest.

Figures 1 and 2 illustrate the method used to extract CERPs from such data. In Fig. 1, the top trace (ALL) is the grand average of 729 shift-pair responses. In

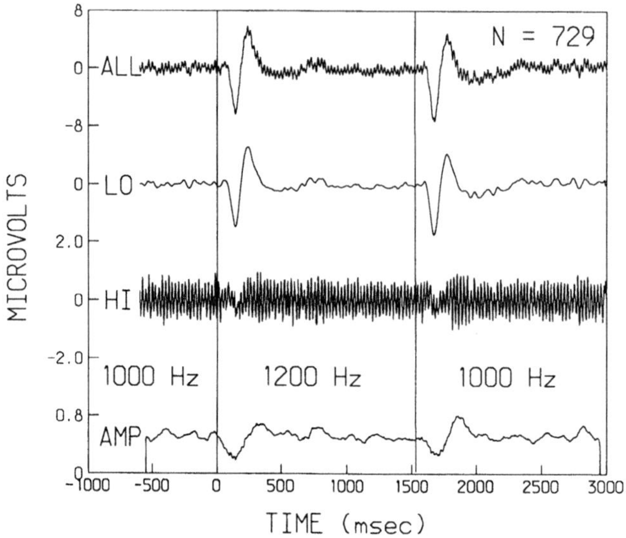

Fig. 1. Extraction of ERPs (evented-related potentials) and CERPs (complex event-related potentials) from the EEG of a person listening through an earphone to short tone bursts delivered at a rate of 40 per s, which are shifted up and then down in pitch every few seconds. The *top trace* contains both the rapid sinusoidal potentials evoked by the individual tone pips (the 40-Hz SSR) and the conventional ERPs produced by the pitch changes. Filtering the EEG at exactly the repetition rate of the tone bursts (40 Hz) extracts the sinusoidal SSR (steady-state response) shown in the *third trace* and leaves behind the ERPs shown in the *second trace*. The *bottom trace* plots the amplitude fluctuations visible in the *third trace* and represents one of two components of the CERP. The second CERP component, the SSR phase changes invisible here, is pictured in Fig. 2

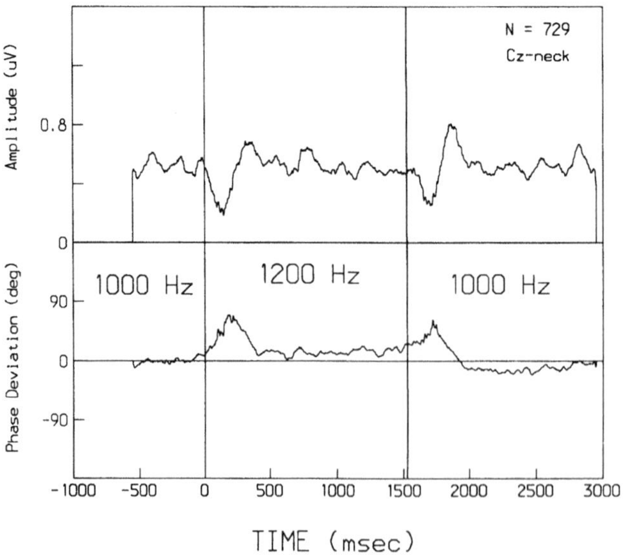

Fig. 2. Continuation of the CERP/ERP demonstration of Fig. 1. The SSR amplitude fluctuations shown in the bottom line of Fig. 1 are reproduced here in the *top line* for comparison with the SSR phase deviations (*bottom line*) that accompany the shifts in pitch of the tones to which the subject was listening. (The procedure used to compute these phase deviations is described in the text)

it one sees two large ERPs, the first after the pitch-up event, the second after the pitch-down, and the 40-Hz sinusoid-like SSR running throughout. The second trace (LO) displays the low-pass or ERP component isolated by passing a moving-average ("boxcar") digital filter with a null at exactly the probe stimulus frequency through the original record (ALL). It contains all the EEG frequencies except the stimulus rate (40 Hz) and any of its harmonics not canceled by the 0.1- to 100-Hz analog filter. The third trace (HI) displays what the filter removed, that is, the SSR portion of the original record. The bottom trace, finally, shows the CERP amplitude record, computed by moving a two-cycle (50 ms) DFT through the SSR data.

As far as we are aware, combined ERP and CERP displays such as these have not been produced and examined previously. Both ERPs and CERPs are the result of the changes in probe frequency. The CERP wave shapes are basically diphasic in morphology – SSR amplitude first drops below the baseline, then rises above it, each deviation lasting about 200 ms. It is clear that the ERP and CERP time-courses differ, which suggests that the mechanisms responsible for them are not identical.

Figure 2 completes the CERP analysis. It adds the CERP phase perturbations (bottom trace) to the CERP amplitude perturbations (top line, same as in Fig. 1), both computed using a 50-ms moving window. Following each stimulus event, phase immediately rises above the baseline (implying earlier SSR peaks) to reach about 60 degrees of deviation (1/6 of the probe stimulus interval, or about 4 ms advance) at about 200 ms, then falls back toward the baseline over another 200 ms. It does not, however, reach its original baseline, since SSR phase advances slightly with rise in tone-stimulus frequency (Galambos et al. 1981; Stapells et al. 1987). The phase changes following the pitch-down event are similar, except that phase falls below initial baseline and takes more than a full second to reach the original 1000-Hz response phase.

The latencies of the amplitude and phase deflections are short and similar, but the phase deflection is more nearly monophasic, not diphasic. These wave-shape differences suggest that complicated physiological processes may be involved in CERP generation.

2.3 Experiment 2: An Intensity-Shift CERP

The subject of experiment 1 also served in experiment 2. The probes were clicks delivered under two conditions; a 6-dB change in probe intensity was the experimental event in both. In condition 1 probe intensity first increased then decreased; in condition 2 this sequence was reversed. In each case, the more intense stimulus was set at 50 dBSL. Intensity-shift number and duration, and the inter-shift-pair intervals, were like those in the previous experiment.

The resulting ERPs and CERPs closely resembled those of Fig. 1. We will examine here only the CERP phase deviations. Figure 3 pictures four of these produced in experiment 2 along with the corresponding ones produced by the pitch-shifts of experiment 1. All show a circa 60 degree phase advance peaking near 200 ms, with return to or near baseline at 400–500 ms. Phase advances following

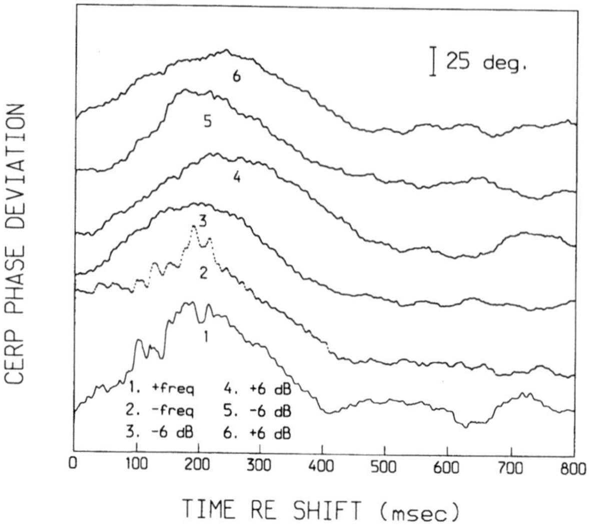

Fig. 3. CERP phase deviations are stable and replicable. Whether elicited by the pitch shifts in experiment 1 (*traces 1 and 2*) or by the small changes in intensity in experiment 2 (*traces 3–6*), they closely resemble each other

the intensity increases (traces 4 and 6) peak slightly later than in the other conditions.

2.3.1 Summary of Experiments 1 and 2

- Alterations in 40-Hz auditory SSR probe stimuli produce both conventional ERPs and novel CERPs.
- Auditory 40-Hz CERP amplitude is diphasic – amplitude drops below baseline for about 200 ms, then rises above baseline to a peak at about 300 ms. We refer to this augmentation in SSR amplitude seen in so many auditory CERP experiments as A300 (meaning an amplitude augmentation peaking near 300 ms; Makeig and Galambos 1987).
- ERP and CERP amplitude deflections do not follow the same time courses, which suggests that their underlying mechanisms are not identical.
- While CERP amplitude perturbations are visible in the raw SSR record, the phase perturbations are invisible (see Fig. 1, third trace) until extracted by the analysis method.
- Auditory 40 Hz CERP phase rises for 200 ms, then falls for about the same period to slightly below the original baseline (or a new one, when probe stimulus frequency changes); and then slowly returns to baseline phase over 1 s or more.
- Phase shifts produced by increase or decrease of either frequency or intensity resemble each other closely (Fig. 3); furthermore the small differences found can be reproduced in the same subject.
- The peak latencies of the CERP amplitude and phase perturbations are not identical (Fig. 2).

2.4 Experiment 3: A Comparison of Auditory and Visual CERPs

Visual SSRs have been extensively studied (Regan 1982). We have shown that alterations in a circa 40-Hz flash or pattern-stimulus probe train can also produce a CERP (Makeig 1985). In this experiment we explore whether a CERP in one modality initiates a CERP in another. To do this we deliver visual and auditory probes concurrently at slightly different rates. We then ask: will an auditory event that produces an auditory CERP also produce a visual CERP? And conversely, does a visual event producing a visual CERP simultaneously produce an auditory CERP?

The auditory probe was a (39 Hz) click train at 44 dBSL. The visual probe was a train of square-wave light pulses (duration 6 ms) in a checkerboard pattern, of comfortable intensity, delivered to the subject's left eye through LED goggles. The rate used (44.6 Hz) is near the 50-Hz "high frequency" peak in the flash-evoked SSR initially reported by Regan (1968). For ease of analysis, the probe stimulus rates were locked in exact 8 : 7 ratio to the EEG sampling rate. SSRs to two such probes presented at different repetition rates are readily separated using Fourier measures (see Galambos and Makeig 1988, for details of the procedure).

In the first condition, time-locking events were like those in experiment 2: 6-dB intensity increases followed 1 s later by 6-dB decreases. In the second condition, time-locking events were decreases in visual probe stimulus intensity (produced by changing the pulse width from 6 to 4 ms), also followed after 1 s by intensity increases. Two channels of EEG were collected: at Cz, where the auditory response is largest, and at Oz, where the visual response is largest.

In the auditory event condition, the auditory CERP was extracted from the SSR as already described. To extract the visual CERPs, the same EEG epochs used to create the auditory CERP were first individually shifted in time to phase-align the data to the flash onset nearest the onset of the intensity shift; then these phase-shifted epochs were averaged. The phase-shift amounted to at most ½ a visual probe stimulus interval (± 11.2 ms, a minor percentage of total CERP duration), and was performed to allow coherent phase-aligned averaging of the visual SSR, and subsequent computation of the visual CERP. In the visual event condition, the auditory epochs were similarly phase-aligned before averaging.

In Fig. 4, the top trace in each box shows CERP amplitude, and the bottom trace, CERP phase deviation (both computed using 100-ms moving windows). Plots A and C show CERPs produced by auditory intensity shifts; plots B and D, CERPs produced by visual intensity shifts. The upper plots (A and B) show CERPs for events of the same modality as that of the probes (auditory and visual respectively), while the lower plots (C and D), show CERPs for the cross-modality probes (visual and auditory). To simplify Fig. 4, we show visual CERPs at Oz only, and auditory CERPs at Cz only.

In Fig. 4 A typical auditory CERPs (both amplitude and phase perturbations) can be seen to follow each click intensity change, but *no* CERP is extracted from the visual probe train concurrently being presented (Fig. 4 C). Similarly, in the reverse experiment, visual CERPs appear at Oz after the visual intensity change (Fig. 4 B), but *not* in the auditory SSR being recorded at Cz (Fig. 4 D).

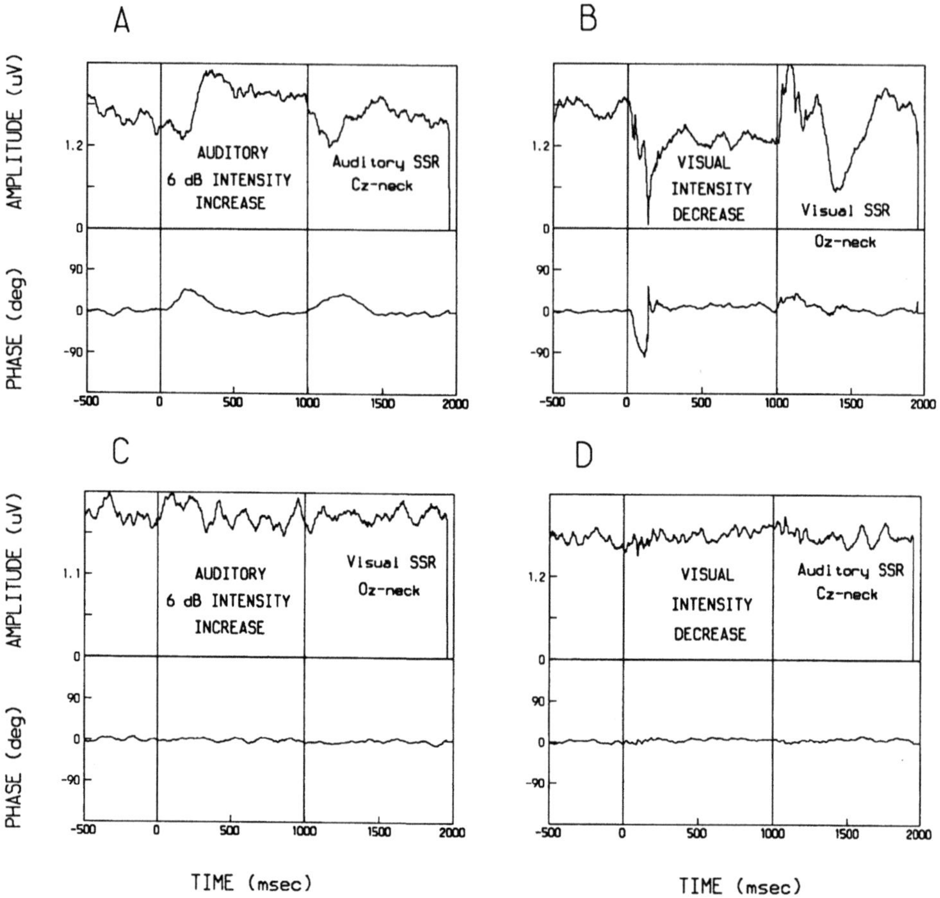

Fig. 4A–D. CERPs produced by a subject listening to clicks and watching light flashes at the same time. The stimuli were presented at slightly different rates near 40 Hz; the two SSRs produced in this way were separately extracted fro the EEG and processed as in Fig. 1 to extract the CERPs which are displayed in each box (*top trace,* CERP amplitude; *bottom trace,* CERP phase, as in Fig. 2). (A) and (C) replicate experiment 2: click intensity was raised and lowered where indicated by the *vertical lines.* In (A), the click intensity changes produce the expected amplitude and phase perturbations, but these do not appear in (C), the CERPs extracted from the concurrent visual SSRs. (B) and (D) show the results of the reverse experiment: when flash intensity was raised and lowered, large CERP perturbations were extracted from the flash SSRs (B), but not from the click SSRs (D)

From these results we conclude:

- Both visual and auditory CERPs can be separated efficiently by the analysis procedure.
- Each of these CERPs is modality-specific; the analysis that extracts CERPs from SSRs in one modality yields a flat line when applied to the SSRs generated concurrently in another modality.
- Despite that fact that the overall durations of ERPs and CERPs are similar, their origins must be different. In this experiment large ERPs were produced

at Cz in both conditions (not shown), but a concurrent auditory CERP appeared there only when the time-locking events were auditory.

Our information on the scalp topography and inter- and intrasubject variabilities of visual CERPs is meager, and we are uncertain about the possible contributions of small eye blinks, pupil changes, and/or retinal adaptation to visual CERP data such as these. We cannot be sure, therefore, that the features of the visual CERP of Fig. 4 B will prove to be typical.

These uncertainties cannot, however, invalidate the following two claims:

- It is feasible to examine circa 40-Hz CERPs in two (or more) modalities, separately and/or concurrently, and using various probe-types and time-locking events.
- No auditory CERP is produced by the visual events in this experiment.

2.5 Control Experiments

Auditory physiologists in search of possible artifacts always look for middle ear muscle (MEM) contractions at stimulus onset (and relaxations at offset) since these mechanical events change the acoustic energy delivered to the cochlea and so can introduce unwanted distortions into records obtained from the scalp. With

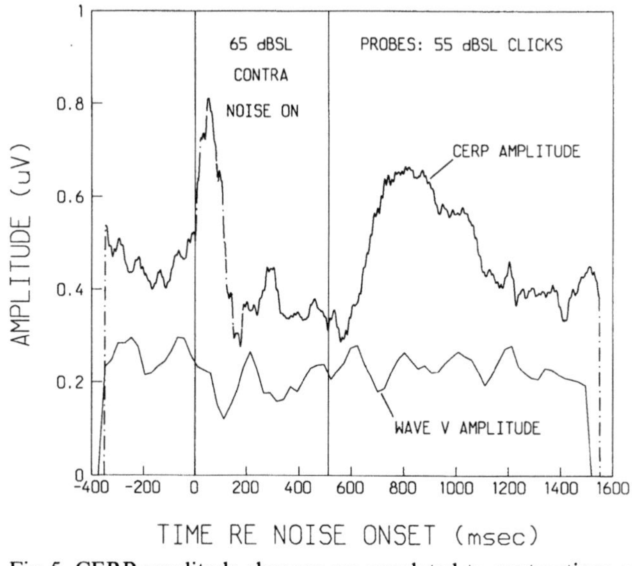

Fig. 5. CERP amplitude changes are unrelated to contractions of the middle-ear muscles. The CERP amplitude (*top trace*) extracted from the click SSR, as previously described, shows prominent amplitude increases following both onset and offset of a burst of white noise presented to the opposite ear through an earphone. The *lower trace* shows no corresponding changes in the amplitude of wave V, a prominent component of the auditory ERP (the auditory brainstem response, ABR) generated by neurons in the brainstem. If the noise had initiated middle-ear muscle contractions, wave V amplitude would have been decreased at noise onset and increased at noise offset

this in mind we have performed a number of control studies to establish whether MEMs contribute to or possibly even produce the auditory CERPs under discussion here. The first of these, summarized in Fig. 5, is an experiment similar to those already discussed; click probes were delivered continuously to the left ear, and every few seconds a (65 dBSL, 0.5 s) wide-band noise burst was introduced into the right ear. Subject and electrode placement were the same as in experiments 1 and 2. In order to record both the SSR and the higher-frequency auditory brainstem response (ABR), the EEG sampling rate was increased (5 kHz) and the analog bandpass widened (0.3–1000 Hz).

2.5.1 Experiment 4: A CERP to a Contralateral Noise Burst

In Fig. 5, vertical lines mark the events producing CERP perturbations (at noise-on and noise-off). The upper trace represents CERP amplitude, and the lower, wave V amplitude.

Consider first the CERP amplitude trace: the initial portion of the transient rise following time zero is evidently created by 40-Hz components of the noise-onset ERP. The small constant amplitude depression thereafter is the contralateral noise-induced modulation of the SSR described briefly in the Introduction and in detail elsewhere (Galambos and Makeig 1988). Following noise-offset a very large CERP amplitude enhancement peaking at 300 ms appears. (Similar data suggest that the small enhancement at 300 ms after noise onset is also an A300.)

The first point to make is that typical CERPs can be created not only by altering the characteristics of the probe stimulus train, as demonstrated in experiments 1–3, but also by introducing a different stimulus in the same modality (in this case noise bursts applied to the ear opposite the one generating the SSRs). Note that it was, in fact, *removal* of the noise stimulus – a decline in the total amount of stimulus energy delivered – that caused be brain to generate the larger of the two CERPs in the figure.

2.5.2 MEM Controls

With respect to the question of MEM artifacts, examine now the trace in Fig. 5 labeled "wave V amplitude." It is the result of an analysis on the data left behind after further filtering the high-pass SSR portion of the response to remove the 40- and 80-Hz SSR components, leaving the higher-frequency (100–1000 Hz) ABR. After computing a moving-average of these ABRs with the same 100-ms window used to construct the CERP amplitude record, another measure was obtained – a moving-average time-domain measure of ABR wave V amplitude (defined as the potential difference between the largest positive peak in the interval 5–10 ms and the succeeding potential minimum). The lower trace of Fig. 5 presents these estimates of ABR wave V amplitude.

The experimental question was whether the CERP amplitude perturbations correlate with wave V amplitude changes. If MEM contractions and relaxations did alter the amount of energy delivered to the cochlea, wave V amplitudes would presumably decrease at noise onset and increase again at noise offset. The data do not in fact suggest any dramatic changes in wave V amplitude at either of these junctures, and in particular there is no indication whatever of an MEM relaxation at noise offset to correlate with the 300-ms CERP amplitude maximum.

We add here three other facts that reduce the likelihood of MEM contributions to our auditory CERP recordings:

- Sound-induced MEM contractions are normally absent at intensities below 80 dB SPL (Jepsen 1951); the sounds that produced the CERPs in this report did not reach this level.
- CERPs can be readily extracted from SSRs produced when the probes are delivered through a bone conduction transducer applied to the forehead. Since bone-conducted stimuli, unlike air-conducted ones, bypass the middle ear to stimulate the cochlea directly, MEM contractions and relaxations seem an unlikely explanation for the CERPs extracted in such an experiment.
- In experiments 1 and 2 the small changes in stimulus intensity and frequency that produced the CERPs were not far from the subject's discrimination threshold. We think it unlikely that MEM activity accompanies such small stimulus changes. To test this supposition the pitch-shift experiment 1 (Fig. 1 and 2) was repeated on a second subject who also produced CERPs to these events. While acoustic impedance measurements made in the contralateral ear of this subject showed the expected impedance changes (i.e., the MEMs contracted and relaxed) to intense noise and tone bursts, no such impedance changes accompanied the pitch shifts that produced the CERPs.

We conclude from these observations that MEM contraction and relaxation contributes little if anything to the auditory CERP recordings in experiments such as these. However, this is not to say that sudden large changes in stimulus intensity could not produce MEM contractions which might contaminate the CERP, and this possibility should be kept in mind whenever such stimuli are employed.

2.6 Relationship of Auditory CERPs to MLR Modulations

Since CERPs are extracted from SSRs, the search for their physiological basis properly begins with an examination of the origins of SSRs. There is widespread agreement that auditory SSRs are made up largely if not exclusively out of the electrical events that follow a sound stimulus with latencies of 10–100 ms. These events, forming the so-called middle latency response (MLR), come after the ABR in the auditory ERP sequence, and can be approximately modeled as a series of three or four sinusoids with periods of 25 ms (Galambos et al. 1981). Causing these electrical events to overlap by delivering a probe train at 40 Hz would theoretically yield the largest summed response amplitude, and this approximates what actually happens. We recently demonstrated that the amplitude and phase of a subject's actual 40-Hz SSR are both reasonably closely predicted by the amplitude and phase of a response synthesized by summing four consecutive 25-ms segments of the same subject's 10-Hz MLR (Stapells et al. 1988).

These arguments indicate that the key to an adequate explanation of auditory CERPs will be the description of how the MLRs making up the SSR are altered by the time-locking event. The two experiments that follow were designed in part to shed light on this problem.

2.7 Experiment 5: An Omitted-Click CERP

Figure 6 shows the response in 16 channels to the omission of single clicks from a 45-dBSL 40-Hz click train. To test the ability of the auditory CERP to rapidly recover its original size, the clicks were omitted in pairs separated by 512 ms. The omitted-clicks pairs occurred every 3.1–3.6 s during the experimental runs of 7–8 min. At the end of the experiment, 368 response-pairs were selected by the computer and averaged. The result, in Fig. 6, superimposes the 16 channels of raw data in which one can recognize the by-now familiar sinusoidal 40-Hz SSRs superimposed on two slow-wave ERPs, each time-locked to one of the omitted clicks. Note that the ERP to the second omitted-click is smaller than the first, an example of the well-known failure of ERP late waves to recover at short interstimulus intervals.

Figure 7 and 8 show the end-result of subjecting each channel to the filter routines that divide the raw data into two parts, slow-wave ERPs and single-frequency SSRs. Figure 7 displays the ERPs in topographic order; though to our knowledge 40-Hz omitted-click ERPs have not previously been examined, they resemble in wave shape and distribution ERPs in the literature, and will receive no further treatment here.

Figure 8 shows the 16 SSR channels extracted by the narrow band-pass filter at the probe rate. (The data have been slightly smoothed for presentation using

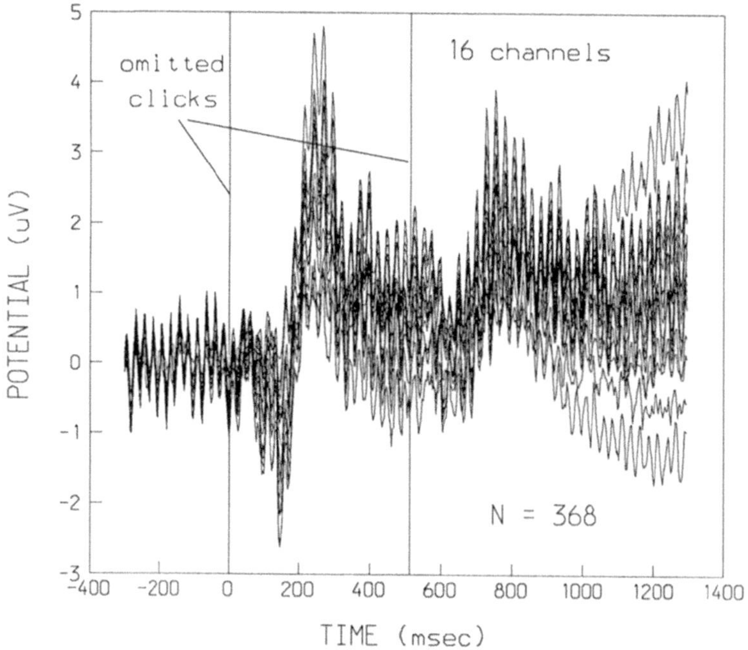

Fig. 6. Omitting pairs of single clicks (512 ms separation) from the train producing an auditory SSR initiates both ERPs and CERPs. The raw data from 16 scalp electrodes is shown superimposed

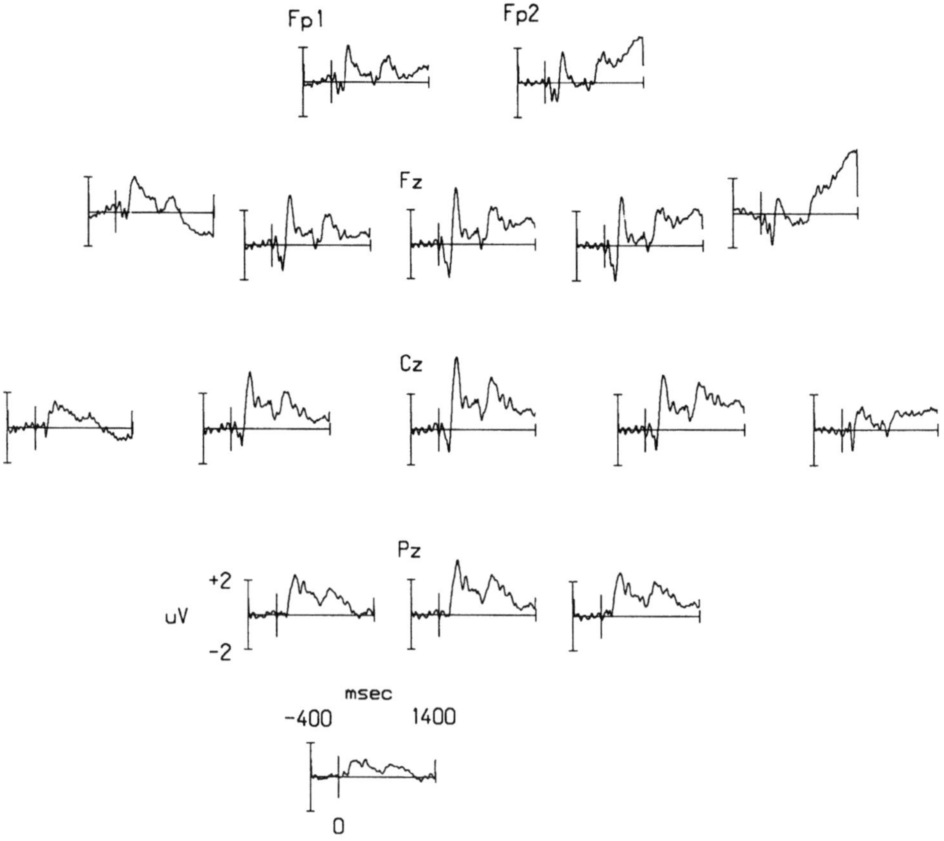

Fig. 7. Topographic map of the ERPs extracted from the raw data of Fig. 6. The *vertical lines* at time 0 mark the position of the first omitted-click of each pair

a 50-ms moving-average filter.) Note that in every channel, a stable pre-event SSR baseline (ms −400 to 0) exists, and this is first reduced then augmented following each omitted-click. The two asterisks at 200–300 ms poststimulus mark approximately where SSR amplitudes reach their maximum.

To obtain the data shown in Fig. 9 we first normalized each SSR for amplitude by dividing by its mean baseline (−400 to 0 ms) channel amplitude. Then we ran a (two-cycle, 50 ms) moving DFT filter through the normalized data to extract CERP amplitude. Figure 9 gives the resulting SSR amplitude perturbations for the 13 channels with largest SSR amplitudes. Note that the CERP amplitude perturbations to the second omitted-click are larger than those to the first, whereas the corresponding ERPs (Figs. 6 and 7) are smaller.

Figure 10 differs from Fig. 9 only in that it narrows the time base and plots the mean normalized amplitudes (of the 13 channels) for the two CERP responses superimposed. (Here, a 100-ms moving window was used to compute CERP amplitudes, which explains why the response peaks appear to be broader and smoother than those in Fig. 9, in which a 50-ms window was used. In this, as in many signal-processing situations, there is no unique "best" estimate of instanta-

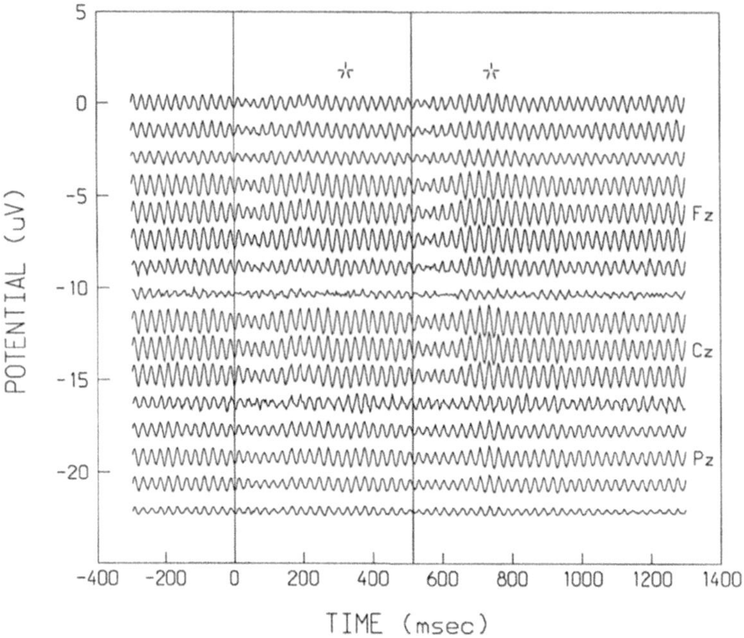

Fig. 8. 40-Hz SSR portion of the evoked responses at 16 scalp channels shown in Fig. 6. *Vertical lines* mark the omitted-clicks; asterisks mark moments in which SSR amplitude rises above baseline

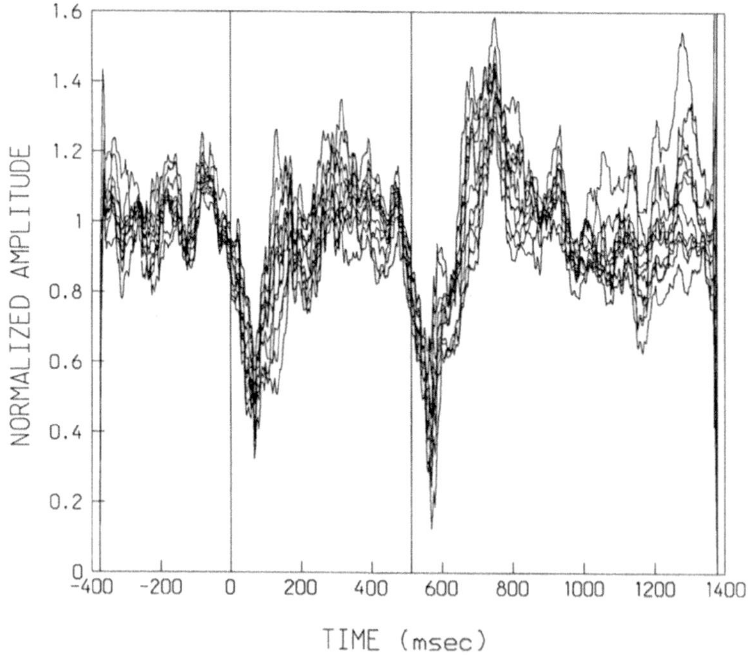

Fig. 9. SSR amplitude perturbations produced by the omitted-clicks (*vertical lines*) all have the form of a decrease followed by an increase. Responses at 13 central scalp sites are shown, normalized to equal baseline amplitude

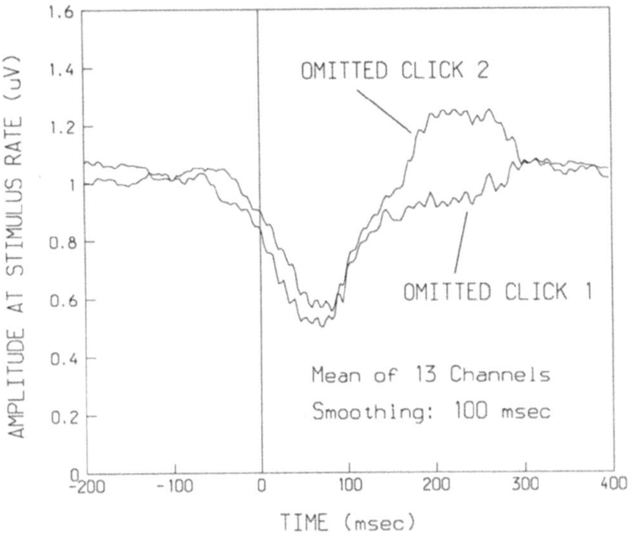

Fig. 10. Comparison of the mean CERP amplitude changes of the data in Fig. 9 produced in this subject by the first and second omitted-clicks shows them to differ in the size of the enhancement near 300 ms

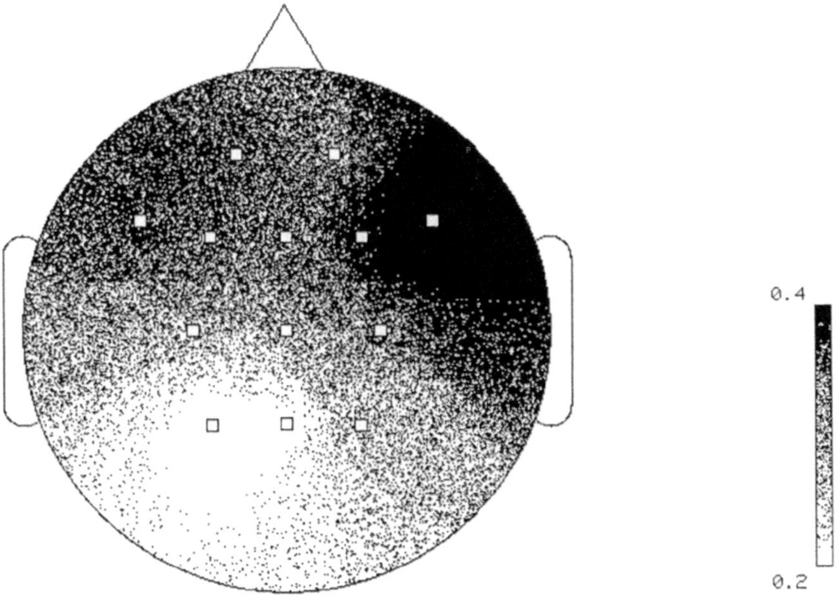

Fig. 11. Map of the distribution across the head of the difference in normalized amplitude enhancement (pictured in Fig. 10) at 216 ms following the omitted-clicks (electrode locations shown by *small squares*)

neous amplitude. The estimator of choice depends on the phenomena to be revealed, and the experimental questions being asked.)

In Fig. 10, CERP amplitudes to first and second omitted-clicks return to baseline only after about 300 ms, and in the interval 150–290 ms the second-click CERP is considerably larger than the first. Figure 11 maps the topographic dis-

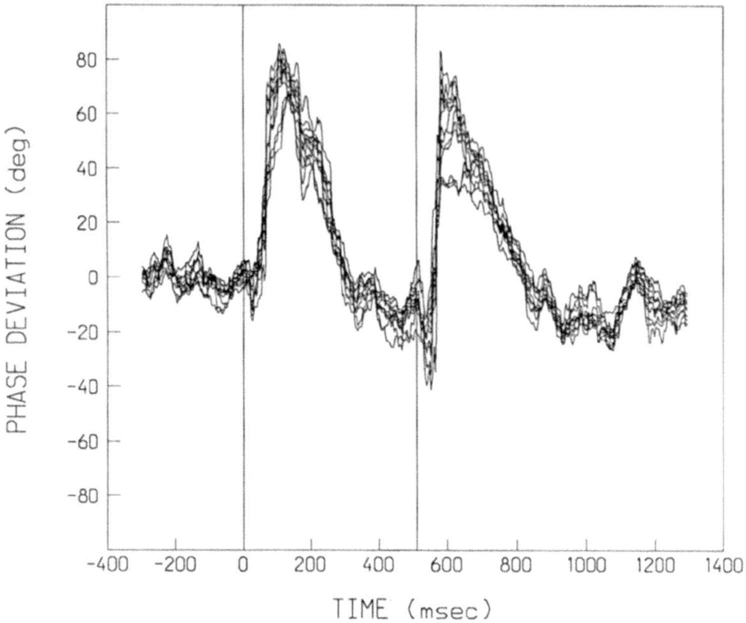

Fig. 12. CERP phase deviations following omitted-clicks (*vertical lines*) in the 13 central scalp channels shown in Fig. 11

tribution of this mean amplitude difference at 216 ms, near where it is maximum. The topographic map shows strong right temporal and weak left temporal foci, suggesting the presence of at least three SSR generators – one in each auditory cortex and one or more elsewhere – each of them apparently somewhat differently perturbed by the first and second omitted-clicks.

Figure 12 plots the CERP phase deviations in the same 13 channels. These phase deviations to omitted-clicks resemble those in experiments 1–3, where the stimulus events were tone-pip frequency and click-intensity shifts. Their initial latency is short; they return to baseline in 300–400 ms, and then drop below it for several hundred ms or more. Like the CERP amplitude perturbations they accompany (Figs. 9 and 10), these phase deviations demonstrate a rapid recovery of the brain events that produce them: both deviations are similar in size and shape despite the short interevent interval (512 ms).

2.7.1 The "Missing-Click MLR"

If the SSRs were nothing but the overlapping sum of stable MLRs to successive probe clicks, then the SSR changes due to the omission of a single click from the probe train would be limited to the first 100 ms or so. Figure 13 shows that the CERP perturbations in this experiment cannot be accounted for in this way. It compares the Cz first omitted-click response with the MLR response recorded from the same subject. The top trace is a fragment taken from Fig. 9 showing a sample of SSR before and after an omitted-click; the second trace plots a "pro-

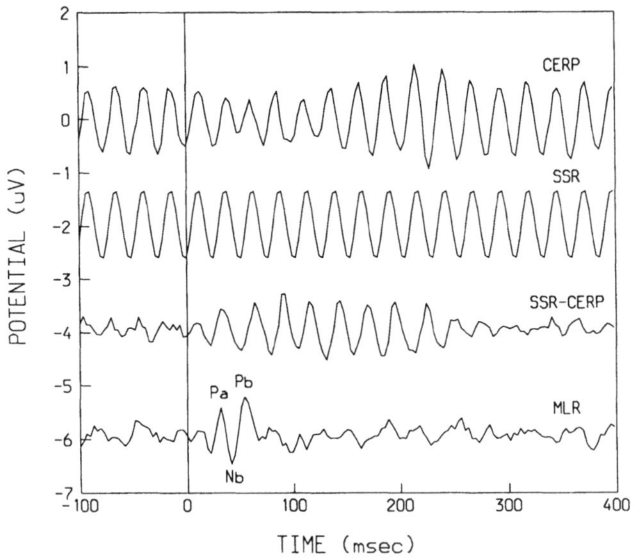

Fig. 13. A demonstration that CERP amplitude perturbations following a single omitted click (*top trace*) cannot be due to the absence of the ERP it would have produced. The 40-Hz component of an ERP to a single click is shown in the *bottom trace*, and an approximation of the perturbation a single omitted-click actually produces is in the *trace above it*. See text for further details

jected" SSR, a computer-synthesized 40-Hz sinusoid adjusted in amplitude and phase to match that of the prestimulus baseline of the top trace.

The next two traces demonstrate that the difference betwen the real and projected SSRs amounts to more than the mere omission of a single MLR. The third trace shows the result of subtracting the actual high-pass filtered data from the projected SSR; this result, a kind of "missing-click MLR," is a time-domain SSR representation of the CERP amplitude and phase perturbations. The lowest trace shows an actual (50 dBSL, 0.2 ms) click-evoked response recorded in another experiment on this subject and then band-pass filtered at 40 Hz like the data in the top trace. Because only the 10–100 ms MLR features (Pa, Nb, Pb, etc.) appear, it is clear that the auditory ERP contains little if any energy at 40 Hz beyond the first 100 ms. Clearly, omitting a single one of these MLRs from an SSR-producing train could not create an event with the duration and wave-shape of that SSR-CERP difference wave shown in the third trace.

2.7.2 Summary of Experiment 5

Two pieces of evidence suggest that CERPs are the result of long-lasting modulator processes that operate on the production of a considerable number of MLRs over several hundred ms.

- First, MLRs must play a key role in CERP generation because auditory ERPs contain appreciable 40-Hz energy only during the MLR portion (the first 100 ms). This means MLRs make the major if not the only contribution to the SSRs from which the CERPs are extracted. Changes in SSR amplitude and phase must therefore reflect changes in the MLRs that sum to produce the SSRs.
- Second, all the CERP amplitude perturbations we have shown, including those following the omission of a single click from the 40-Hz probe train, per-

sist for at least 300 ms. Removal of a single MLR from the SSR trace thus would yield a far smaller perturbation than that actually observed.

Three other observations are important:

- The biphasic shape of the amplitude perturbation, a decline followed by a rise (A300), implies that the modulator process involved is also biphasic; the nearly monophasic shape of the phase perturbation, by contrast, suggests that auditory CERP phase and amplitude modulations may have different origins.
- Although auditory CERPs display similar form and time-course at all scalp locations, suggesting common control, first and second omitted-click CERP amplitude responses may differ across scalp locations. This indicates the amplitude modulations are unlikely to be produced by a single brain procees.
- At short interstimulus intervals (here 512 ms), the phase and amplitude features of the second CERP response show full recovery (see Fig. 9); the second ERP, however, does not (Figs. 7 and 8). These dissimilar recovery properties make it highly unlikely that CERP and ERP processes are the same.

2.8 Experiment 6: The MLR and the A300

The previous experiment rejected the hypothesis that the amplitude and phase features of 40-Hz auditory CERPs can be accounted for by an alteration in a single MLR, and replaced it with the idea that many MLRs must be modulated over a time-course of several hundred ms. The following experiment attempts to define these MLR changes.

Clicks at 50 dBSL were delivered to the right ear at a rate near 10 Hz (not 40 Hz). At random intervals of 3.4–6.2 s, a single click was omitted from the 10-Hz probe train. A total of 624 omitted-click epochs were averaged; Fig. 14 shows the mean MLR (0–100 ms) responses to clicks occurring in four time periods surrounding the omitted-click, labeled I–IV in the figure. Thus, the MLR trace labeled III is the mean of responses to the second and third clicks following the omitted-click; during this period, the MLR features Pa (30 ms) and especially Pb (50 ms) are augmented.

Figure 15 (square trace) compares amplitude at 40 Hz in the MLRs from the four time periods in Fig. 14, with actual moving-averaged 40-Hz CERP amplitude for the same subject (dotted trace), recorded at Cz to second omitted-clicks in experiment 5 (cf. Fig. 9). To create the square trace, four 25-ms segments of each MLR in Fig. 14 were summed, mimicking the overlapping of MLRs by successive probe stimuli presented at 40 Hz instead of 10 Hz. Then the 40-Hz amplitudes in the four SSRs so synthesized were measured using a 25-ms DFT, and plotted.

The (square trace) pattern of diminution and augmentation of MLR 40-Hz amplitude in Fig. 15 fits the actual (dotted trace) pattern of 40-Hz CERP amplitude perturbations well. While the experiment does not indicate whether the augmented MLR following the omitted-click represents responses to the immediately preceding probe clicks, or also includes responses to still earlier probe stimuli, since in single-click ERP recordings 40 Hz activity lasts at most 100 ms, this

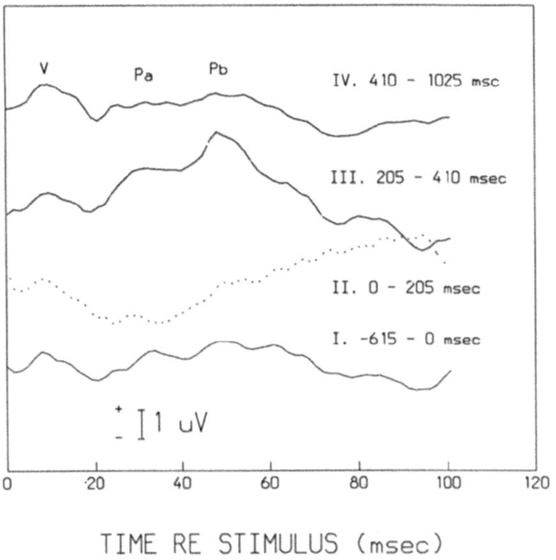

Fig. 14. MLR responses to clicks in a 10-Hz probe train are shown in the *top* and *bottom traces*. Omitting single clicks from the probe train results in alterations in subsequent MLRs (*middle traces*)

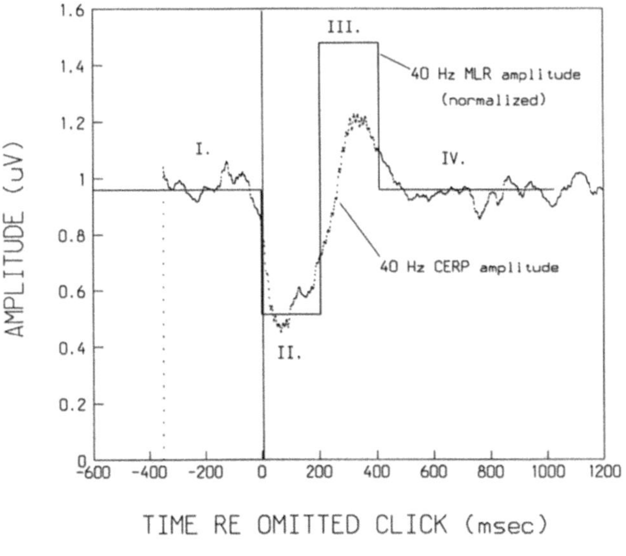

Fig. 15. Comparison of 40-Hz amplitudes in the MLRs shown in Fig. 14 (*square trace*, scaled to equate baseline amplitudes) with an actual CERP amplitude record obtained from the same subject in a separate experiment (*dotted trace*)

seems unlikely. We conclude, therefore, that the A300 is predominately composed of enhanced MLR responses to stimuli occurring 200–300 ms following the time-locking event.

Note that the square trace is a simple example of CERP amplitude analysis at a multiple of the stimulus rate (here four). Here, block-averaged amplitude was used instead of continuous moving-average amplitude in order to improve the stability of the 10-Hz experiment data, though at a cost of latency resolution. (To facilitate comparisons of the two traces, in Fig. 15, MLR amplitudes have been normalized to equate MLR and CERP baseline amplitudes.)

2.8.1 Summary of Experiment 6

- CERP experiments can employ various rates of stimulation, and can be analyzed at a multiple of the stimulus rate, if activity at the higher frequency is present in the evoked response.
- The characteristic pattern of 40-Hz auditory CERP diminution and succeeding augmentation (A300) following an omitted-click probe stimulus seems to be produced by changes in circa 50 ms MLR responses to probe stimuli following the omitted-click.

3 General Discussion

3.1 The 40-Hz Auditory CERP: Experimental Definition

The experiments reported here first create a continuous physiological response, the SSR, using periodic stimulation at a high rate. Next, measurements are made on the ways certain arbitrarily introduced second stimulus events perturb the SSRs. We have dealt mostly with auditory SSRs created by clicks delivered at a rate near 40 Hz, and with two types of perturbing stimulus events: changes in the characteristics of the probe train (omitting one of the clicks; or abruptly changing their intensity; or, using tone pips as probes, changing their frequency), and delivering a second stimulus (noise bursts to the contralateral ear). In one experiment, the SSR perturbations were demonstrated to be modality-specific: visual stimulation fails to perturb auditory SSRs and vice versa.

CERP analysis, performed in the frequency domain, yields separate amplitude and phase estimates of how the baseline SSR is perturbed by the experimental event. These perturbations have been found to be reliable and replicable within and across subjects (Makeig and Galambos 1987). There is an amplitude change in the form of a drop followed by a rise, each approximately 200 ms in duration. The phase change occurring at the same time is more nearly monophasic, a phase advance of about 60 degrees lasting about 400 ms, often followed by a longer-lasting rebound to slightly below baseline phase. These concurrent SSR amplitude and phase changes are referred to as the 40-Hz auditory complex event-related potential (CERP).

3.2 Physiological Origins: CERP Generators/SSR Modulators

First of all, CERPs differ from conventional evoked or event-related potentials (ERPs) recorded during these experiments. ERPs, which also represent perturbations produced by the stimulus event, are potential shifts taking place in the wide-band EEG at frequencies either lower (ERP) or higher (ABR) than the CERP analysis frequency. When the concurrently generated ERPs and CERPs are compared for latency, wave shape, and duration, their differences strongly suggest that the physiological events responsible for the one are unlikely to be those responsible for the other. As is the case with ERPs, topographic maps of CERPs and of functions derived from them indicate that their generators are probably multiple and widespread.

Second, various control studies exclude involvement of middle-ear muscle contractions and relaxations. Auditory 40-Hz CERPs appear to be generated within the brain, not by extracranial events.

We have cited several reasons for our claim that changes in the physiological events composing the MLR create the auditory CERP: MLRs contain most of the 40-Hz energy in the auditory ERP, and, when overlapped at stimulus rates near 40 Hz, they produce the SSR that is perturbed. The simplest possible explanation for the CERP, namely that an event such as an omitted click will subtract a single MLR from the SSR, is unsatisfactory: CERPs last several hundred ms, while the omitted MLR would perturb the SSR for less than 100 ms. Further, we have demonstrated altered MLRs in response to click probes presented some 300 ms following omission of a click from a 10-Hz probe train.

As for CERPs in other modalities, the visual example in Fig. 4 B resembles auditory CERPs in duration, but otherwise the two share few details. Such dissimilarities must be expected if MLRs do indeed give rise to auditory CERPs, as proposed. Visual CERPs, like the auditory ones, are extracted from SSRs containing the ERP energy at the probe stimulus rate. Hence a visual equivalent of the MLR should exist, though it has not been described, but there is no a priori reason to suppose that event-related modulations of the SSR it produces will closely resemble those that from the auditory CERP.

Finally, since a somesthetic SSR exists (Galambos 1982), so too must a somesthetic analogue of the MLR; where in the ERP this analogue is to be found, however, and whether stimulus events perturb somesthetic SSRs to produce CERPs, remain to be established.

As for the CERP-producing modulators, the outcome of experiment 3, which shows that CERPs are modality-specific, invites speculation about where they might be located. If CERP processes are to act independently of each other they must somehow be isolated functionally; perhaps the simplest arrangement would be to locate them within the sensory pathways upon which they operate. Any progress in this analysis, however, clearly awaits answers to questions about the MLR and its analogues in the visual and somesthetic modalities. All SSR modulations, including those that yield CERPs, can be traced back to the brain events that produce ERP energy at the SSR frequency; what these brain events are, and where they take place have not been fully resolved. Whether or how CERP perturbations are related to one or more of the wide variety of other known modulations of evoked responses (Bourbon et al. 1987; Freedman et al. 1987; Moruzzi and Magoun 1949; Picton et al. 1977; Woods and Courschene 1986); or ongoing EEG activity (Van Winsum et al. 1984) are other open questions. Also CERPs may have already been observed in different contexts (for instance, Steriade and Demetrescu 1962).

3.3 CERP Limitations and Advantages

3.3.1 Limited Modulator Bandpass

The CERP measures perturbations in amplitude and phase of driven or spontaneous EEG at a single analysis frequency. However, there is no unique or natural definition of instantaneous amplitude or phase (Oppenheim and Schafer 1975); these can only be estimated within a time window whose length determines the bandwidth of the perturbations that can appear in the CERP being derived. In practical terms, the length of the estimation window must be at least one probe stimulus interval. If the signal-to-noise ratio of the SSR data is poor, a longer window is necessary to produce a stable baseline SSR against which CERP perturbations can be evaluated. Thus, the length of the moving estimation window used to compute a 40-Hz CERP must be at least 25 ms. Modulatory processes acting more rapidly than this will not be represented in the CERP.

Fortunately, peak driving frequencies near 40 Hz exist in auditory, visual, and somesthetic modalities (Galambos 1982; Galambos and Makeig 1988). These are several times higher than the frequencies prominent in the slow-wave ERP (1–10 Hz). Therefore 40-Hz CERPs can measure CERP perturbations occurring in the same frequency range as ERP activity – if these exist, and the experiments above imply that they do.

The 40-Hz CERPs do in fact reveal actions of perturbations whose speed and latencies parallel those of the well-known ERP components (N200, P300, etc.). To observe the action of more rapid modulators, CERP analysis could also be applied to even higher-frequency periodic responses, for example, in the auditory modality, the circa 500-Hz frequency-following response (FFR; Hius in't Veld et al. 1977); in this case it might be possible to observe perturbations as brief as a few ms.

3.3.2 Limited Response Bandpass

The CERP extracts information about modulations of the evoked response bandpassed at a single frequency only, and ignores other frequencies in the response. CERP analysis, therefore, is a specialized tool for evoked-response analysis; to make complete use of wide-band data, CERP analysis should be accompanied by wider-band ERP analyses. One choice is Başar's combined EEG-EP analysis method, in which the broad-band evoked response is divided into a number of frequency bands whose limits are chosen according to peaks in the spectrum of the EEG or evoked response (Başar 1980). Such band-limited methods may be considered to be midway between broad-band time-domain ERP analysis and narrow-band frequency-domain CERP analysis.

A practical advantage of 40-Hz CERP analysis is its ability to record effects and processes occurring over relatively long periods. Since it uses only a high-frequency portion of the EEG spectrum, CERP experiments do not require special DC recording equipment to record very long-lasting CERP perturbations. The first experiment based on the CERP concept in our laboratory used as a time-locking stimulus a piece of recorded music lasting over 10 min; in that experiment SSR phase remained slightly below its baseline value for several *minutes* following music onset (Galambos and Makeig 1988; Makeig 1985).

3.3.3 Non-uniqueness

Fixed-frequency analysis is not be the only method available to model the complexities of perturbed SSRs. For example, CERP amplitude and phase perturbations could in some cases also be modeled as frequency and phase perturbations. When an externally applied driving frequency is not present, CERP analysis is still possible, but in this case there may be no adequate reason to represent any portion of the evoked response as a fixed-frequency CERP. For example, in an alpha-blocking experiment, straightforward CERP analysis would model changes in the evoked response at the alpha EEG frequency as perturbations at that frequency, not allowing for the possibility that the alpha frequency itself might have been perturbed by the time-locking event. In some cases, this limitation could be mitigated by modeling and then removing frequency-modulation trends in the data, as Freeman has done in recent studies (Freeman and Di Prisco 1986), before performing CERP analysis.

3.3.4 Latency Ambiguity

Relatively late auditory CERP modulations represent changes in relatively early MLRs; thus, CERP "latency" has a double meaning. The A300 feature of the auditory CERP, then, arises from modulation of circa 50-ms responses to probe stimuli delivered at 275, 250, and 225 ms following the time-locking event. However, in return for this added complexity, CERP analysis produces a reliable, objective, and easy-to-compute continuous time-series capable of revealing the size, distribution, latency, and dynamic character of concurrent response-generating and -modulating processes.

3.3.5 Peripheral Gating

CERPs do not distinguish between SSR modulations of peripheral and central origin; CERPs to very bright flashes embedded in a steady-state flash (flicker) train could be expected to include perturbations arising from pupillary or retinal adaptation; CERPs to very loud sounds could similarly include components arising from MEM contractions (see Section 2.6.2). Peripheral effects can, however, be distinguished from central ones by several tests: Do accompanying ERPs show any signs of efferent control (eye blinks, postauricular muscle potentials)? Are the suspect CERP features present at all scalp locations proportionately? Do they appear in subjects whose efferent gating systems are missing or not functional? Do they remain even when the time-locking event is decreased in intensity or otherwise made unlikely to produce peripheral gating? Finally, can concurrent direct measurements (of ear impedance, pupil size, etc.) demonstrate that no changes in stimulus input occurred?

3.3.6 Data Requirements

An acceptable signal-to-noise ratio in CERP experiments, as in nearly all ERP experiments, requires averaging a sufficient number of responses to the time-locking event. The 40-Hz auditory SSR is small (normally 0.1–1.0 μV peak-to-baseline); for most adult subjects a stable response requires averaging of at least 10–20 s of data containing 400–800 probe stimulus responses (Stapells et al. 1987). To obtain a comparable signal-to-noise ratio, a similar number of time-locked CERP responses needs to be averaged. Because the time-locking events generating the CERP must be repeated at rates much slower than the SSR stimulus presentation rate (for example, once every few seconds), CERP experiments are time-consuming, and may exceed practical and/or physiological limits. (It is probable that insufficient data contributed to the failure of Linden et al. (1987) to obtain reliable CERP perturbations in their reanalysis of an experiment designed to test the effects of selective attention on the SSR. They recorded only 100 sums per subject, and their barely perceptible experimental events may have produced small CERP perturbations.)

3.4 What More?

If an organism is to survive and evolve, it must be able to detect environmental events, interpret their implications, integrate these with ongoing knowledge and behavior, and respond to them appropriately. Nearly all modern brain and behavioral models for the processes involved in these activities include multiple interacting stages and pathways. The SSR technique produces measurements that could be useful in testing these models; it has already revealed the existence of novel intrinsic modulatory events and processes (the spontaneous minute-rhythms and the sleep-related changes), and the auditory CERP adds a new class of stimulus-linked ones.

Since brain events responsible for CERPs appear not to be identical with those reponsible for generating the concurrent auditory late-wave ERPs, a range of cognitive experiments in which a CERP analysis runs in parallel with an ERP analysis are conceptually feasible, and their outcomes might shed light on cognitive brain events and the neural pathways and levels at which they take place. Such experiments would at the very least complement and expand the ERP data, perhaps in unexpected and useful ways.

In sum, the robustness and evident complexity of the auditory 40-Hz CERPs, even in the straightforward experiments we have presented here, raise hopes that further applications of the CERP technique may open new windows on the ways sensory systems interact and collaborate in carrying out the business of the brain perception, interpretation, integration, and response selection.

4 Summary

We describe CERPs (complex event-related potentials) produced by abruptly changing intensity or frequency of brief stimuli used to produce a steady-state response (SSR). We demonstrated that these events all perturb the auditory, 40-Hz SSR in much the same way: its amplitude drops then rises over a period lasting 400–500 ms, and its phase advances and then falls during a period lasting 300–

400 ms, with a slight phase-retard often persisting for several hundred ms more. The time-locking events produce conventional event-related potentials (ERPs) as well as CERPs, but these differ sufficiently in latency, wave-shape, and duration to justify concluding that the processes responsible for ERPs are not the same as those that produce CERPs. The auditory SSR is composed of overlapping middle-latency responses (MLRs). CERPs are produced by changes in MLRs produced during several hundred ms following a novel event. CERPs recover rapidly; a second CERP may be elicited as soon as 500 ms after a first. If two SSRs are generated at the same time, one by clicks to an ear and another by light flashes delivered at a slightly different rate to the eye, stimulus events in either modality perturb the SSR of that modality only, which suggests a remarkable specificity for the processes responsible for the phenomenon. The brain activities CERPs identify begin less than 100 ms after a novel stimulus arrives at a sense organ and continue for up to 1 s. This latency and duration are appropriate for the brain processes involved in attending to, orienting toward, and preparing to respond in environments where appreciation of novelty is important. It is therefore possible that CERPs will provide useful information about brain processes related to such psychological phenomena.

Acknowledgments. This research has been supported by NIH research grant NS17490 (R.G.), NIMH grant MH43967 (SM), and training and dissertation grants from NIH and UCSD (S.M.). The technical assistance of Phyllis Galambos and Anne Makeig, and the support of Dr. Donald Krebs and his staff at the Children's Hospital Research Center are gratefully acknowledged

References

Başar E (1980) Brain dynamics. Elsevier, Amsterdam

Bourbon WT, Will KW, Gary HE, Papanicolaou AC (1987) Habituation of auditory event-related potentials: a comparison of self-initiated and automated stimulus trains. Electroencephalogr Clin Neurophysiol 66(2):160–166

Freedman R, Adler LE, Gerhardt GA, Waldo M, Baker N, Rose GM, Drebing C, Nagamoto H, Bickford-Wimer P, Franks R (1987) Neurobiological studies of sensory gating in schizophrenia. Schizophrenia Bull 13(4):669–678

Freeman WJ, Di Prisco GV (1986) EEG spatial pattern differences with discriminated odors manifest chaotic and limit cycle attractors in olfactory bulb of rabbits. In. Palm G, Aertsen A (eds) Brain theory. Springer, Berlin Heidelberg New York, pp 97–119

Galambos R (1982) Tactile and auditory stimuli repeated at high rates (30–50 per s) produce similar event related potentials. Ann NY Acad Sci 388:722–728

Galambos R, Makeig S (1988) Dynamic changes in steady-state potentials. In. Başar E (ed) Dynamics of sensory and cognitive processing of the brain. Springer, Berlin Heidelberg New York, pp 102–122 (Springer series in brain dynamcis, vol 1)

Galambos R, Makeig S, Talmachoff P (1981) A 40 Hz auditory potential recorded from the human scalp. Proc Natl Acad Sci USA 78(4):2643–2647

Green GGR, Rees A, Kay R (1986) Steady-state evoked responses to sinusoidally amplitude-modulated sounds recorded in man. Hear Res 23:123–133

Huis in't Veld F, Osterhammel P, Terkildsen K (1977) Frequency-following auditory brainstem responses in man. Scand Audiol 6:27–34

Jepsen O (1951) The threshold of the reflexes of the intratympanic muscles in a normal material examined by means of the impedance method. Acta Otolaryngol 39:406–408

Kankkunen A, Rosenhall U (1985) Comparison between thresholds obtained with pure-tone audiometry and the 40-Hz middle latency response. Scand Audiol 14:99–104

Klein AJ (1983) Properties of the brainstem response slow-wave component. II. Frequency specificity. Arch Otolaryngol 109:74–78

Kuwada S, Batra R, Maher VL (1986) Scalp potentials of normal and hearing-impaired subjects in response to sinusoidally amplitude-modulated tones. Hear Res 21:179–192

Linden RD, Picton TW, Hamel G, Campbell KB (1987) Human auditory steady-state evoked potentials during selective attention. Electroencephalogr Clin Neurophysiol 66(2):137–144

Makeig S (1985) Studies in musical psychobiology. Doctoral dissertation, UCSD, San Diego, University Microfilms, Ann Arbor

Makeig S, Galambos R (1983) Phase consistency of evoked responses to auditory stimuli delivered at high rates. J Acoust Soc Am 74 (Suppl 1):S65

Makeig S, Galambos R (1987) Event-related perturbations in auditory steady-state potentials. Soc Neurosci Abstr 13:331

Moruzzi G, Magoun HW (1949) Brainstem reticular formation and activation of the EEG. Electroencephalogr Clin Neurophysiol 1:455–473

Mundy-Castle AC (1953) An analysis of central responses to photic stimulation in normal adults. Electroencephalogr Clin Neurophysiol 5:1

Oppenheim AV, Schafer RW (1975) Digital signal processing. Prentice-Hall, London

Picton TS, Hillyard S, Galambos R (1974) Habituation and attention in the auditory system. In: Keidel WD, Neff WD (eds) The auditory system: clinical and special topics. Springer, Berlin Heidelberg New York, pp 343–387 (Handbook of sensory physiology, vol V/3)

Picton TW, Vajsar J, Rodriguez R, Campbell KB (1987) Reliability estimates for steady-state evoked potentials. Electroencephalogr Clin Neurophysiol 68(2):119–131

Regan D (1968) A high frequency mechanism which underlies visual evoked potentials. Electroencephalogr Clin Neurophysiol 25:231–237

Regan D (1982) Comparison of transient and steady-state methods. In: Bodis-Wollner I (ed) Evoked potentials. NY Academy of Science, New York, pp 45–71 (Annals of the New York Academy of Science, vol 388)

Rickards FW, Clark GM (1984) Steady-state evoked potentials to amplitude modulated tones. In: Nodar RH, Barber C (eds) Evoked potentials. II. Butterworth, Boston, pp 163–169

Rodriguez R, Picton T, Linden D, Hammel G, Laframboise G (1986) Human auditory steady state responses: effects of intensity and frequency. Ear Hear 7:300–313

Sammeth CA, Barry SJ (1985) The 40-Hz event-related potential as a measure of auditory sensitivity in normals. Scand Audiol 14:51–55

Shallop JK (1983) Electric response audiometry: the morphology of normal responses. Adv Otorhinolaryngol 29:124–139

Stapells DR, Makeig S, Galambos R (1987) Auditory steady-state responses: threshold prediction using phase coherence. Electroencephalogr Clin Neurophysiol 67:260–270

Stapells DR, Galambos R, Costello JA, Makeig S (1988) Inconsistency of auditory middle latency and steady-state responses in infants. Electroencephalogr Clin Neurophysiol 71:289–295

Steriade M, Demetrescu M (1962) Reticular facilitation of responses to acoustic stimuli. Electroencephalogr Clin Neurophysiol 14:21–36

Sturzebecher E, Kuhne W, Berndt H (1985) Detectability of the acoustically evoked composite response (40 Hz potential) near threshold. Scand Audiol 14:23–25

Van Winsom W, Sergeant J, Genze R (1984) The functional significance of event-related desynchronization of alpha rhythm in attentional and activating tasks. Electroencephalogr Clin Neurophysiol 58:519–524

Woods DL, Clayworth CC (1986) Age-related changes in human middle latency evoked potentials. Electroencephalogr Clin Neurophysiol 65(4):297–303

Woods DL, Courchesne E (1986) The recovery functions of auditory event-related potentials during split-second discriminations. Electroencephalogr Clin Neurophysiol 65(4):304–315

Woods DL, Courchesne E, Hillyard SA, Galambos R (1980) Recovery cycles of event-related potentials in multiple detection tasks. Electroencephalogr Clin Neurophysiol 50:335–347

Noninvasive Functional Investigations into Human Cortical Motor Physiology, Motor Learning, and Motor Imagery

L. Deecke, W. Lang, G. Goldenberg, and I. Podreka

1 Introduction

In humans, registration of regional cerebral blood flow (rCBF) and electrophysiological recordings offer the possibility to (a) study directly parameters of brain activity and (b) establish functional–anatomical relations in behavioral tasks. Current trends in imaging of brain functions include the use of scalp-recorded negative DC potential shifts which are time-locked to the performance of cognitive or motor tasks. A reason for utilizing this phenomenon is the notion of the physiological significance of DC potential shifts as recordable from the scalp surface; it has clearly been established that excitatory synaptic currents of cortical neurons generate electrical dipoles that are negative on the surface and positive at depth. The spatial resolution of scalp-recorded DC potential shifts is limited but is, for example, sufficient for separating neuronal activity of circumscribed cortical areas involved in the execution of movements of different parts of the body (Boschert and Deecke 1986). Spatial transformations of EEG recordings (e.g., Laplacian) can offer additional possibilities for estimating current sources contributing to surface-recorded electrical shifts and may help to distinguish different sources underlying these shifts.

Another possibility for localizing task-specific cortical activity is the use of other techniques such as magnetoencephalography und rCBF. For instance, the role of the supplementary motor area for voluntary movements has been partly elucidated by these techniques (Deecke and Kornhuber 1978; Lassen et al. 1978; Deecke et al. 1985). In the present studies on sensory-guided tracking, two parameters, performance (DC-) negativity and rCBF, were combined in order to describe task-related cortical activity.

2 Methods and Results

2.1 Experiment I: Movement-Related Potentials in Sensory-Guided Skills

Sixteen right-handed students participated in the experiment. The subjects fixed their gaze on a fixation point straight in front of them. As they lowered a stylus with their right hand to a plate, a light spot in the left hemifield of vision started moving for 1 s in a random direction, and subsequently for 1 s in another random direction. By this means, sensory information was projected to the sensory areas

Springer Series in Brain Dynamics 2
Edited by Erol Başar and T. H. Bullock

of the right hemisphere. The major events of the scenario, such as the start of the stimulus program and its change in direction were, thus, predictable in time but not in space. Subjects had to track the light spot, which moved at a constant speed. Similarly, tactile stimuli were applied to the left palm by a modified XY-plotter and had to be tracked in a similar manner (for details cf. Deecke et al. 1984; Lang et al. 1984). Sensory events such as the onset of stimulus, its change in direction, and the end of the stimulus program were all preceded by slow negative potential shifts and followed by positive peaks.

Modality-specific effects of brain potentials were demonstrated by (a) comparing corresponding recordings of the two hemispheres and (b) comparing results of tactile and visual tracking selectively for each recording. In the visual task, significant hemispheric differences in the stimulus-related negative potential shifts were obtained in the parietal and occipital recordings, with the higher amplitudes occurring over the right hemisphere. In the tactile task, however, hemispheric asymmetry had a clear maximum in central and parietal recordings. Thus, the right hemisphere, receiving the respective sensory input, generated higher amplitudes of performance-related slow negative potential shifts than did the left hemisphere, which did not receive the stimuli but controlled the tracking movement of the right hand. Interestingly, as mentioned above, this hemispheric asymmetry differed in topography: While tactile experiments were characterized by a centroparietal maximum, the visual task exhibited a parieto-occipital distribution of the hemispheric asymmetry. When performing the comparison in individual leads, stimulus-related slow negative potential shifts differed in C4 (overlying the right somatosensory cortex), with higher amplitudes in the tactile task, and in O2 (overlying the right occipital cortex), with higher amplitudes in the visual task.

2.2 Experiment II: Joint Investigation of Movement-Related Potentials and Tc-99m HMPAO SPECT in Visuomotor Learning

In continuation of similar evoked potential experiments on visuomotor learning (cf. Lang et al. 1983, 1986), a joint study was carried out by measuring both performance-related DC shifts and rCBF in the same group of subjects under similar scenarios (Lang et al. 1988). rCBF was measured using single photon emission computer tomography (SPECT) and a new technetium-labeled compound (hexa-methyl-propylene-amine-oxime,HMPAO) which crosses the blood–brain barrier and is trapped in brain cells within 1–2 min. Tracer concentration shows excellent correlation with rCBF (Podreka et al. 1987). In a learning and a control task, subjects had to track a visual stimulus by moving a stylus with their right hand. The visual target was displayed on a TV screen and moved in three random trajectories for 1.5 s each. The experimental design was similar to that in experiment I, except that now visual information was projected into both hemifields of vision and the stylus position was coupled back to the TV screen in order to provide continuous feedback. In a special learning task (inverted tracking, IT), subjects had to track the target in a horizontally inverted manner, i.e., target movements to the right required hand tracking to the left and vice versa. Movements up and down were not inverted. In the tracking task (T task), the same visual stimulus program was applied but subjects tracked the target in the normal noninverted manner.

The SPECT study showed significant differences between the two tasks; there was increased rCBF during IT as compared to T in the following brain regions: middle frontal gyrus on either side, frontomedial cortex, basal ganglia, and cerebellum (Lang et al. 1988).

In the DC potential study; subjects likewise voluntarily initiated the onset of the stimulus program of 4.5 s duration, which also consisted of three random trajectories of 1.5 s each. After the 4.5 s of tracking, subjects waited for an irregular interval before again voluntarily initiating the next task, and so on. The voluntary initiation of the stimulus program was preceded by a readiness potential, while the skilled tracking movement was accompanied by a sustained negative potential shift (performance negativity). Amplitudes of the performance negativity were significantly higher over frontal and central areas (F3, Fz, F4, FCz, Cz, and C3) in the IT task than in the T task. There were almost no differences between the two tasks in parietal (P3, P4) and occipital (Oz) recordings.

2.3 Experiment III: DC Shifts when Mentally Imagining vs Actually Performing the Tracking

The experimental arrangement was the same as in the DC potential study of experiment II. A new group of subjects ($n = 18$) participated in this experiment. In one task, subjects started the stimulus program but did not track it (no tracking, NT). In the other they had to track the stimulus in an inverted manner (IT, as in experiment II). The order of these tasks was balanced in runs of eight trials along with a third condition (mental tracking, MT), which required the subjects to mentally track the stimulus program in an inverted manner without actually moving the hand.

There were three main results (see Fig. 1):

1. In NT, stimulus presentation was accompanied by a very small DC performance negativity in parietal recordings only – in other recordings DC shifts were absent. In MT, DC shift increased signficantly in parietal recordings; shifts were also present at occipital and central sites.
2. Only in IT (real tracking) did frontolateral recordings show significant negative DC shifts.
3. In IT, DC shifts were lateralized in central recordings, with the higher amplitudes occurring contralateral to the performing hand. In MT, a slight contralateral preponderance of DC shifts was still present despite the absence of any muscular activity in the EMG.

2.4 Experiment IV: Tc-99m HMPAO SPECT and Motor Imagery

Experiment IV was devoted to measuring rCBF changes when a movement is imagined but not actually executed. In order to investigate motor imagery, the following four tasks were selected and performed by 28 right-handed paid stu-

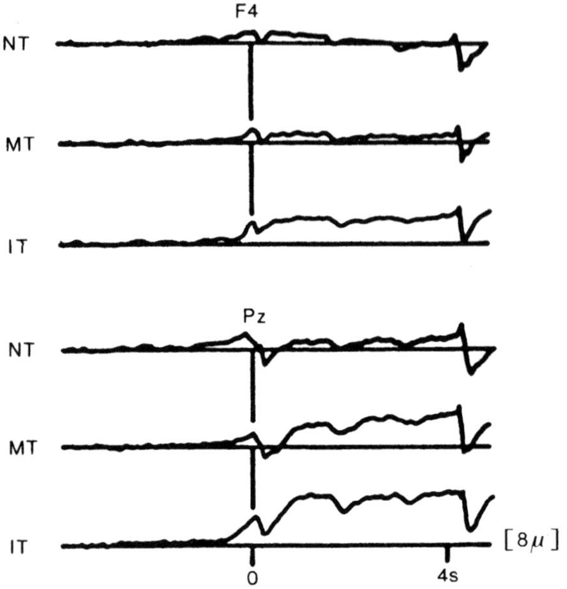

Fig. 1. Grand averages across all subjects of two typical recording sites (F4 and Pz). The *vertical line* indicates the onset of movement ($t=0$). Three conditions are compared: NT (no tracking), MT (mental tracking), and IT (inverted tracking). Note that Pz shows a marked increase of negative DC shifts in MT as compared to NT, but that DC shifts in F4 are almost identical for these two tasks. In Pz (representative for the parieto-occipital recordings) as well as in F4 (as a typical frontolateral recording), negative DC shifts become larger in IT as compared to MT, the percent increase being larger in frontal recordings

dents: (1) yes–no, (2) low imagery, (3) visual imagery, and (4) motor imagery. The experiments were done with eyes closed under a blindfold. An audio tape supplied the subjects via headphones with (a) instructions and (b) the stimuli for the respective task. Subjects held a flashlight in one hand, a cubital i.v. catheter being in place in the other arm. Subjects were divided into an imagery group and a control group. Within a group, every subject performed two experiments at an interval of 1 week so that the order of tasks and hands (holding the flashlight) was balanced within each group.

1. Yes–no. Subjects heard the words "yes" or "no" in a random order followed by a peeping tone after 5 s. They were instructed to ignore "yes" but to actuate the flashlight upon "no."

2. Low Imagery. Subjects heard 50 sentences, each with a pause of 5 s followed by the peeping tone. They had to judge the correctness of the sentences and switch on the flashlight for incorrect sentences. The sentences tested abstract school and general knowledge thought to be retrievable without imagery.

3. Visual Imagery. The procedure was similar to (2); however, the evaluation of the sentences (flashlight for incorrect) required visual imagery.

4. Motor Imagery. The procedure was similar to (2) and (3); however, the evaluation of the sentences required the imagination of body movements. Subjects were strictly requested to avoid executing any movements with the single exception of the gentle press on the button of the flashlight to indicate incorrect. Obedience to this instruction was controlled by EMG in some subjects.

The most salient finding of the experiment concerned the visual imagery condition. Compared to the control conditions, it showed significant elevations of local blood flow in the left inferior occipital region (Fig. 2) and in inferior temporal regions of both hemispheres. The activation of the left inferior occipital region

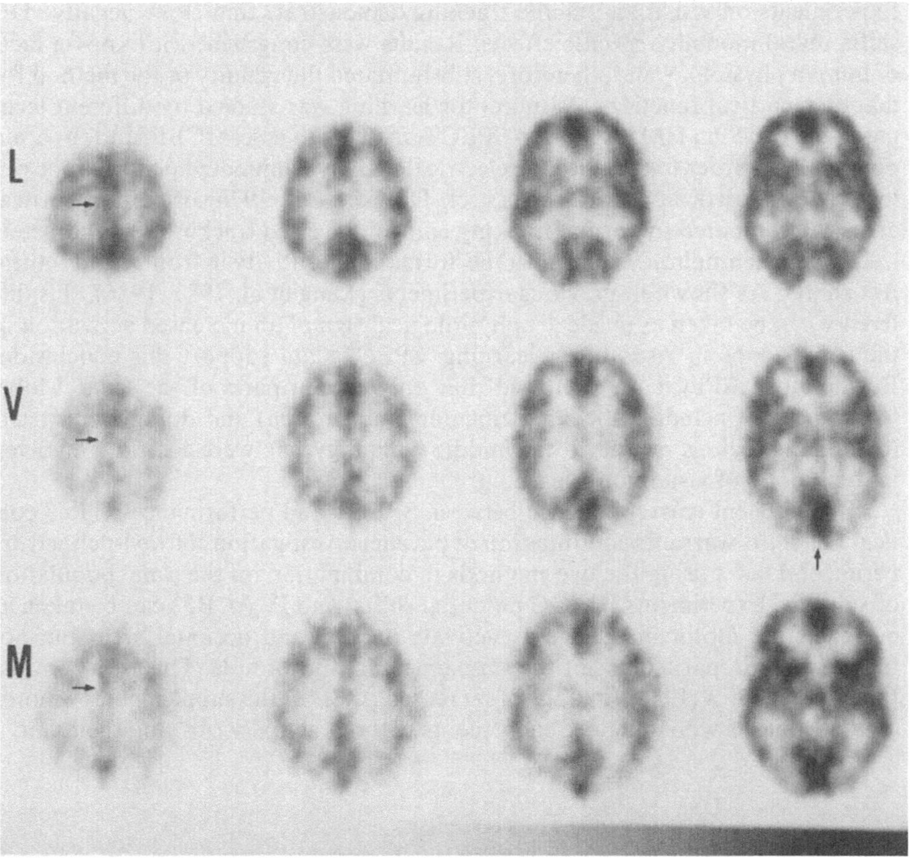

Fig. 2. *Top row:* SPECT study of a subject who judged the correctness of low imagery sentences (*L*); *middle row:* study of a subject who judged visual imagery sentences (*V*); *bottom row:* the same subject as with visual imagery sentences, judging motor imagery sentences (*M*). The right side of each image corresponds to the right side of the brain. Slices in the *leftmost column* show the medial superior frontal region presumed to contain the supplementary motor area (*horizontal arrow*). In these particular studies flow rates in this area are somewhat lower in both high imagery studies than in the low imagery study, but statistically there was no difference across groups. The slices in the *rightmost columns* show the inferior occipital region. Note the activation in the inferior occipital region (*vertical arrow*) in the visual imagery study

in particular confirms previous findings (Farah 1984; Goldenberg 1987; Goldenberg et al. 1987) of a prominent role of this region in visual imagery. Results of the motor imagery condition, however, were less spectacular. Higher local flow rates were generalized over many brain regions. Only in the central regions were the flow rates significantly higher than for visual imagery, but inspection of the data suggested that this was due to low values for visual imagery rather than to high ones for motor imagery.

3 Conclusions

Experiments on visual and tactile tracking demonstrate that slow, negative DC shifts reveal modality-specific effects. Results were congruent with known facts of human physiology and, therefore, substantiated the validity of the method for mapping cortical functions. Visuomotor learning was studied by different techniques: (a) Tc-99m HMPAO brain SPECT, in order to assess rCBF, and (b) brain potentials, in order to measure the electrical activity (summated postsynaptic potentials of superficial cortical layers, cf. Deecke et al. 1976) of certain cortical areas. As compared to normal tracking control, inverted tracking was characterized by higher amplitudes of the DC performance negativity in frontal and central recordings. As shown in previous experiments (Lang et al. 1983, 1986), this difference can be taken as an electrophysiological sign of an increased activation of the frontal lobe in visuomotor learning. SPECT data support this conclusion. Furthermore, SPECT data showed that only certain parts of the frontal lobe, frontomedial (including the supplementary motor area) and dorsolateral parts (containing regions related to the middle frontal gyrus), were critically involved in the process of visuomotor learning.

The excellent correspondence between SPECT and performance-related cortical DC shifts warrants continuation of parallel investigations of well-defined experimental tasks using the two methods in combination on the same population of subjects. Experiments III (DC potential shifts) and IV (rCBF) can be taken as evidence that motor imagery may activate parietal and occipital areas but not frontal ones, in particular not the supplementary motor area. This is in contrast to Roland et al.'s (1980) findings of increased rCBF in the supplementary motor area of subjects who imagined movements without actually carrying them out.

4 Summary

The human cortex integrates perceptions and motivations, i.e., external needs from the environment and internal needs of the organism, with the result of goal-directed behavior. Patterns of task-related cortical activation can be measured by changes of electric and neuromagnetic fields or of rCBF and metabolism. Applications of these methods in man have yielded insights into human voluntary movement physiology. In this contribution, recent experiments are reviewed, in which sensory-guided manual tracking movements were investigated with regard to certain aspects such as (a) the effect of stimulus modality, (b) the effect of learning, and (c) differences between actually performing the task and merely imagining it. In these experiments, two different physiological parameters were assessed: (a) cortical DC shifts preceding (Bereitschaftspotential) and accompanying (performance negativity) motor acts, and (b) rCBF. The following results were obtained: (a) performance-related DC shifts over the primary and secondary sensory areas vary with stimulus modality and depend on the subject's directed attention towards important events; (b) the necessity to develop new skills for sensorimotor integration causes an additional activation of the frontal lobes, and (c) motor imagery activates primary and secondary visual areas but not the supplementary motor area.

References

Boschert J, Deecke L (1986) Cerebral potentials preceding voluntary toe, knee and hip movements and their vectors in human precentral gyrus. Brain Res 376:175–179

Deecke L, Kornhuber HH (1978) An electrical sign of participation of the mesial "supplementary" motor cortex in human voluntary finger movements. Brain Res 159:473–476

Deecke L, Grözinger B, Kornhuber HH (1976) Voluntary finger movement in man: cerebral potentials and theory. Biol Cybern 23:99–119

Deecke L, Heise B, Kornhuber HH, Lang M, Lang W (1984) Brain potentials associated with voluntary manual tracking: Bereitschaftspotential, conditioned pre-motion positivity, directed attention potential, and relaxation potential. Anticipatory activity of the limbic and frontal cortex. Ann NY Acad Sci 425:450–464

Deecke L, Boschert J, Brickett P, Weinberg H (1985) Magnetoencephalographic evidence for possible supplementary motor area participation in human voluntary movement. In: Weinberg H, Stroink G, Katila T (eds) Biomagnetism: applications and theory. Pergamon, New York, pp 369–372

Farah MJ (1984) The neurological basis of mental imagery: A componential analysis. Cognition 18:245–272

Goldenberg G (1987) Neurologische Grundlagen bildlicher Vorstellungen. Springer, Vienna New York

Goldenberg G, Podreka I, Steiner M, Willmes K (1987) Patterns of regional cerebral blood flow related to memorizing of high and low imagery words – an emission computer tomography study. Neuropsychologia 25:473–485

Lang W, Lang M, Kornhuber A, Deecke L, Kornhuber HH (1983) Human cerebral potentials and visuomotor learning. Pflugers Arch 399:342–344

Lang W, Lang M, Kornhuber A, Deecke L, Kornhuber HH (1984) Brain potentials related to voluntary hand tracking: motivation and attention. Hum Neurobiol 3:235–240

Lang W, Lang M, Kornhuber A, Kornhuber HH (1986) Electrophysiological evidence for right frontal lobe dominance in spatial visuomotor learning. Arch Ital Biol 124:1–13

Lang W, Lang M, Podreka I, Steiner M, Uhl F, Suess E, Müller C, Deecke L (1988) DC-potential shifts and regional cerebral blood flow reveal frontal cortex involvement in human visuomotor learning. Exp Brain Res 71:353–364

Lassen NA, Ingvar DH, Skinhoj E (1978) Brain function and blood flow. Sci Am 239:62–71

Podreka I, Suess E, Goldenberg G, Steiner M, Brücke T, Müller C, Lang W, Neininckx RD, Deecke L (1987) Initial experience with technetium-99m-HM-PAO brain SPECT. J Nucl Med 28:1657–1666

Roland PE, Larsen B, Lassen NA, Skinhoj E (1980) Supplementary motor area and other cortical areas in the organization of voluntary movements in man. J Neurophysiol 43:118–136

Signs of Model Making by the Human Brain

A. S. GEVINS

1 Introduction

The brain is a type of Rorschach test for scientists. Each scientist describes the brain's structure or function with the language and imagery of his or her own particular trade, be it mathematics, physics, computer science, neurophysiology, or psychology. But these descriptions inevitably fall short because the brain itself is its own metaphor. With this in mind, I am going to speculate that recent neuroelectric measurements we made from people performing simple cognitive tasks may be interpreted as signs of the internal models that our brains use to interpret stimuli and take action. But first, I should review the methods and results.

2 Signal Processing Methods

2.1 New Methods for Measuring Neurocognitive Neuroelectric Patterns

During the last 15 years, my colleagues and I have devoted a full-time effort to developing mathematical procedures and computer systems for extracting more specific information from EEGs (reviews in Gevins and Yeager 1972; Gevins et al. 1975; Gevins 1980, 1984, 1987a, b, c, d; Gevins and Morgan 1986, 1988; Gevins and Bressler 1988). So far, we have applied these methods to six experiments which attempted to measure human higher cognitive functions with increasing specificity (Gevins et al. 1979a, b, c, 1981, 1985, 1987, 1988, 1989a, b). For each set of two experiments, we developed a novel set of experimental stimuli, analytic procedures, and software systems to overcome the shortcomings of the last set.

The increasing capability of laboratory computers has been the engine of our progress. Fifteen years ago, I naively thought that if I could only apply a trillion arithmetic operations to a billion bytes of neuroelectric data, I could find out something basic about how human brains worked. Now that this capability is routine, I am eagerly awaiting the 100-trillion operation, 10-billion-byte experiment.

The need for this much computation results from the increasing number of scalp EEG channels, as well as the increasingly complicated analyses required to measure subtle spatiotemporal neurocognitive signals from large sets of single-trial data. Currently, we are routinely recording 64 EEG channels, and analyzing

Springer Series in Brain Dynamics 2
Edited by Erol Başar and T. H. Bullock

about 25 200–400 ms intervals spanning each 4- to 6-s task trial. Each trial consists of a cue, stimulus, response, and feedback about performance accuracy; the intervals are time registered to the evoked potential peaks elicited by these events. (In February, 1988, we completed the expansion to 128 channels).

Much of our effort has dealt with developing various spatial and temporal signal enhancement techniques, of which the method of event-related covariance (ERC) is the latest. This method provides a window, albeit limited, on the dynamic functional neural networks which produce simple goal-directed behaviors. ERCs measure two features of short segments of event-related potentials (ERPs): the similarity of waveform and the timing between electrodes. ERC analysis is based on animal studies which showed that when a brain region becomes involved in stimulus-related processing, synchronization of a subset of neurons in that region is manifested as a change in the form of the region's extracellularly recorded, low-frequency macropotentials (reviewed in Gevins and Bressler). Covariance and correlation can measure waveform similarity and timing of macropotentials from different areas of the brain, so these measurement techniques may characterize the spatial organization of coordinated functional activity of the areas involved in a goal-directed behavior. Of course, recording at the scalp, rather than within the brain, introduces some complications into this simple idea. Although for the specific cases we have measured ERCs seem to reflect the coactivation of underlying cortical areas, the net effect of these complications is that we cannot yet definitively say whether this is true in general. Our current research is focused on resolving this issue in some important paradigmatic cases.

2.2 Computing Event-Related Covariances (ERCs)

There are ten steps in computing ERCs (Gevins et al. 1987; Gevins and Bressler): the first four reduce spatial smearing and select intervals and trials that contain task-related information in order to enhance the signal-to-noise ratio and reduce the amount of data. The next six steps measure ERCs on bandpass-filtered, enhanced averages computed from the reduced data set.

The ten steps involve:

1. Recording at least 100 trials of each task condition of the experiment, using at least 24 electrodes
2. Applying the Laplacian operator to the potential distribution of each electrode not located at the edge of the scalp, in order to remove the effect of choice of reference electrode and reduce spatial blur from volume conduction through the skull and scalp
3. Removing data that were contaminated with artifact
4. Finding trials with consistent event-related signals, and computing enhanced averages from these trials (optional)
5. Selecting digital bandpass filters and intervals for analysis by examining ERPs, amplitude distribution maps, and (optional) Wigner time-frequency distributions
6. Computing the multilag cross-covariance functions between all pairwise (by channel) combinations of the interval's enhanced averages

7. Using the magnitude of the maximum cross-covariance function and its lag time to characterize the ERC
8. Estimating significance of ERCs by the standard deviation of the "noise" ERC estimated from randomized data
9. Graphing the most significant ERCs in each interval
10. Statistically comparing ERC distributions between experimental conditions.

2.3 Validating ERC Measurements

The current ERC analysis has been applied to data recorded from several experiments. The validity of the method is demonstrated in analysis of visual stimulus processing and response execution intervals of a visuomotor task (Gevins et al. 1987). During the period of visual processing, ERC patterns involved posterior sites followed by anterior parietal and premotor sites. This finding was consistent with neuroanatomical considerations and clinical neuropsychological studies. Likewise, ERC patterns for movement-related intervals corresponded to prior functional neuroanatomical studies of primates and patients with intracerebral

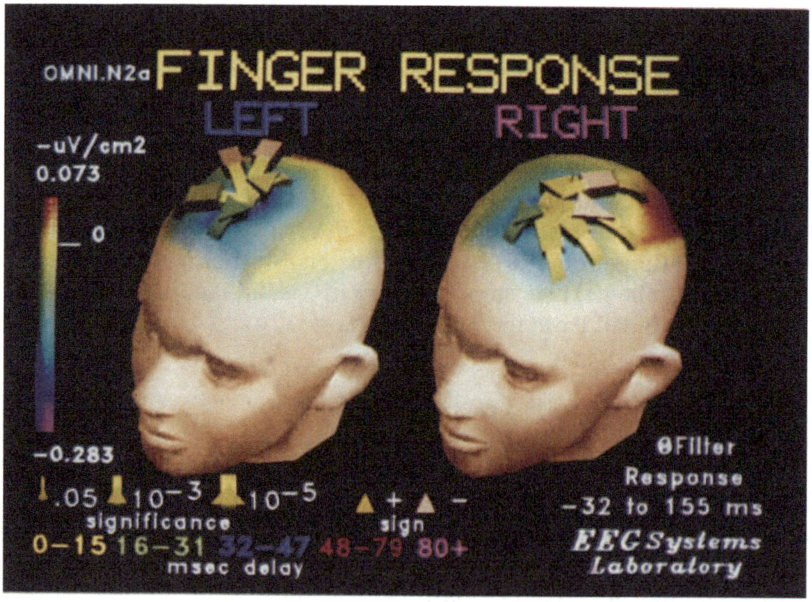

Fig. 1. Most significant (top one standard deviation) ERC patterns at the peak of the finger pressure for right- and left-hand trials from seven right-handed subjects, superimposed on maps of Laplacian response potential amplitude. The thickness of an ERC line is proportional to the negative log of its significance (from 0.05 to 0.00005). The color of a line shows the lag time (ms delay) between sites. Note that the anterior midline precentral (aCz) electrode is the focus of all ERCs. The patterns are distinctly lateralized according to responding hand. The sign of the aCz ERCs is positive (yellow) for lateral frontal, and negative (tan) for lateral central and anterior parietal electrodes

electrodes (Fig. 1): the midline precentral electrode that overlays the premotor and supplementary motor cortices was the focus of all movement-related ERC patterns, and the other most significant ERCs involved pre- and postcentral sites that were appropriatly contralateral to the responding hand. Moreover, the positive sign of the ÉRCs between midline precentral and lateralized precentral sites and the negative sign between midline precentral and postcentral sites are consistent with the sharply focused current sources and sinks spanning the hand areas of somatosensory and motor cortices.

While these ERC patterns appear to reflect neatly the functional coordination of the immediately underlying cortical areas, we must again emphasize that their actual neural sources are, in fact, not yet known. Some caution should therefore be exercised in interpreting the patterns too literally.

3 Preparation and Feedback During Simple Visuomotor Judgment Tasks

3.1 The First Two Experiments

In a series of four experiments, we attempted to characterize distributed networks of cortical function during simple tasks. In the first experiment (Gevins et al. 1981), we measured event-related cross-correlations between 91 pairwise combinations of recordings from 15 electrodes during performance of two simple tasks, one requiring a spatial judgment and the other a numeric judgment. A cue before each trial indicated the type of task to be performed. The cross-correlation patterns of the two tasks differed during the last 300 ms of the cue interval. These differences, which included midline frontal, parietal, and occipital electrode sites, suggested that neural networks associated with preparation were both task specific and spatially detailed. These "preparatory networks" were better visualized in a second experiment, which extracted event-related cross-correlations that distinguished two similar spatial tasks (Gevins et al. 1983, 1985). Spatially distributed, task-specific "preparatory networks" were found again. This time, the midline frontal (Fz) electrode was the only site that showed a task-dependent difference during the last 300 ms of the cue interval.

3.2 The Third Experiment

We designed a third experiment based on these encouraging results. We wished to manipulate preparatory sets more explicitly, and to investigate feedback about performance accuracy (Gevins et al. 1987, 1989 b). Along with a new experimental design, we had to develop new signal processing procedures that more robustly measured event-related covariance between electrode sites.

Seven healthy, right-handed, male adults participated in this study. A visual cue, slanted to the right or to the left, prompted the subject to prepare to make a response pressure with the right or left index finger. One second later, the cue was followed by a visual numeric stimulus (number 1–9) indicating that a pressure of 0.1 to 0.9 kg should be made with the index finger of the previously indicated hand. A two-digit number, presented 1 s after the peak of the response pressure, provided feedback that indicated the subject's exact pressure. On a random 20% of the trials, the stimulus number was slanted in the opposite direction to the cue; subjects were to withhold their responses on these "catch trials." The next trial followed 1 s after disappearance of the feedback. Each subject performed several hundred trials, with rest breaks as needed.

Twenty-six channels of EEG data, as well as vertical and horizontal eye movements and flexor digitori muscle activity from both arms, were recorded. All single-trial EEG data were screened for eye movement, muscle potential, and other artifacts. Contaminated data were discarded.

Intervals used for ERC analysis centered on major event-related potential peaks. ERCs were computed between each of the 120 pairwise combinations of the 16 nonperipheral channels. Intervals were set from 500 ms before the cue to 500 ms after the feedback.

We first calculated the mean error (deviation from the required finger pressure) over all trials from the recording session. Individual trials were then classified as accurate (trial error less than mean error) or inaccurate (error greater than mean error).

ERC patterns during a 375-ms interval centered 687 ms postcue (spanning the late contingent negative variation–CNV), regardless of subsequent accuracy, involved left prefrontal sites, as well as appropriately lateralized central and parietal sites. Inaccurate performance by the right hand was preceded by a very simple pattern (Fig. 2a), while inaccurate performance by the left hand was preceded by a complex, spatially diffuse pattern (Fig. 2b). The relative lack of ERCs preceding inaccurate right-hand performance may simply reflect inattention on those trials, while the strong and complex patterns preceding inaccurate performance with the left hand may reflect effortful, but inappropriate, preparation by the right-handed subjects.

ERC patterns related to feedback about accurate and inaccurate performances were similar immediately after the onset of feedback, but began to differ in an interval centered at 281 ms, that spanned the early P3 peak (P3E) (Fig. 3). The ERC patterns for feedback to accurate performance by the two bands were very similar (bootstrap correlation = 0.91 ± 0.01), involving midline anterocentral, central, anteroparietal, parietal, and antero-occipital sites, left anteroparietal and anterocentral sites, and right parietal, anteroparietal, anterocentral, and frontal sites. These accurate patterns involved many long-delay (32–79 ms) ERCs. The waveforms of the frontal and anterocentral sites consistently lagged those of more posterior sites. For feedback to inaccurate performance, patterns for both hands were also very similar (bootstrap correlation = 0.90 ± 0.02), and involved most of the same sites as the accurate patterns, with the striking exception of the left and midline frontal sites. Again, frontal waveforms lagged those of the more posterior sites with which they covaried. There were even more long-delay ERCs than in the accurate patterns.

We suggest that our prestimulus ERC patterns characterize a distributed preparatory neural set that is related to the accuracy of subsequent task per-

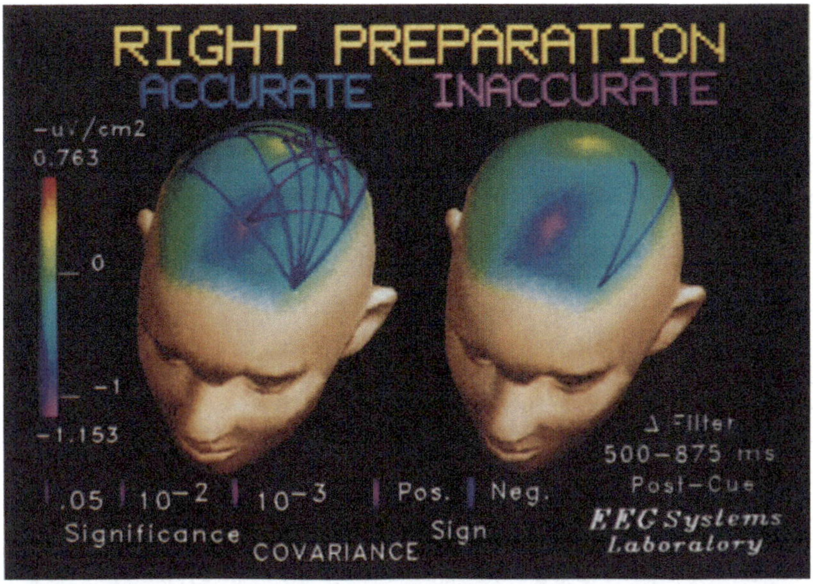

a

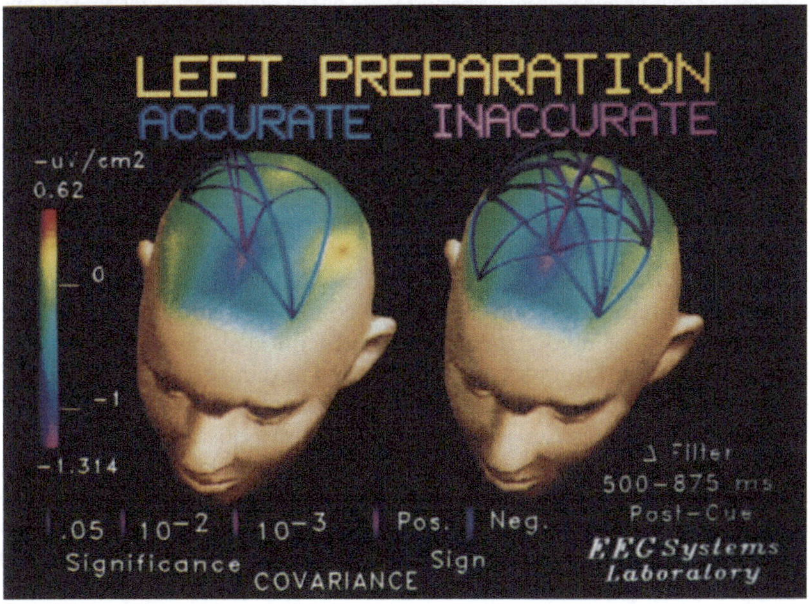

b

Fig. 2a, b. Significant ($P<0.05$) prestimulus ERC patterns superimposed on maps of Laplacian CNV amplitude. Same subjects and task as Fig. 1. Measurements are from an interval of 500–875 ms after the cue for subsequently accurate and inaccurate right- (a) and left-hand (b) visuomotor task performance by seven right-handed men. The thickness of a covariance line is proportional to its significance (from 0.05 to 0.005). A violet line indicates the covariance is positive, while a blue line is negative. ERCs involving left frontal and appropriately contralateral central and parietal electrode sites are prominent in patterns for subsequently accurate performance of both hands. The magnitude and number of ERCs are greater preceding subsequently inaccurate left-hand performance by these right-handed subjects, and are more widely distributed compared with the left-hand accurate pattern. For the right-hand, fewer and weaker ERCs characterize subsequently inaccurate performance. The amplitude maps are very similar for the four conditions, and do not indicate any of the specific differences evident in the covariance patterns

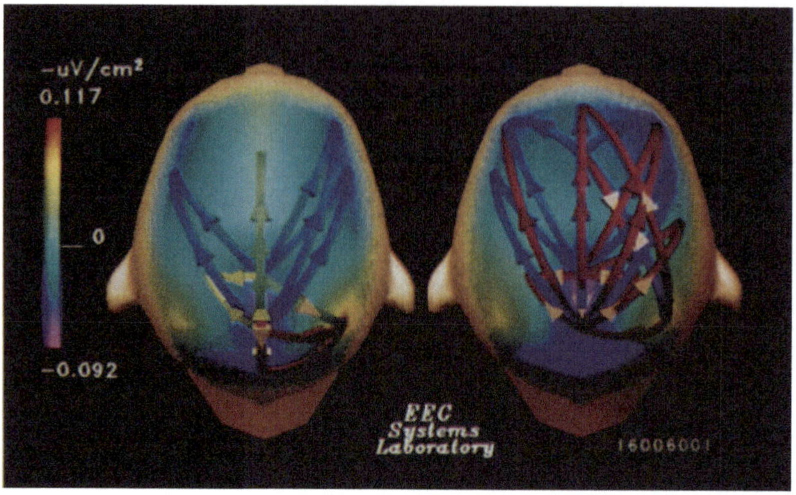

Fig. 3. Most significant (top two largest standard deviations) early P3 (P3E) feedback ERC patterns for feedback to accurate and inaccurate performance by the right hand. Same subjects and task as Fig. 1 and 2. Same scale as Fig. 1. Both the ERCs and amplitude maps are derived from a 187-ms-wide interval, centered at 281 ms after feedback onset, on seven-subject averaged waveforms. A major difference between accurate and inaccurate patterns occurs at the left and midline frontal sites, which are only involved in the inaccurate patterns. The involvement of these sites may reflect the fact that greater processing is required following inaccurate performance in order to update the preparatory set for the next trial

formance. This network appears to involve distinctive cognitive (frontal), integrative-motor (midline precentral), and lateralized somesthetic-motor (central and parietal) components. The involvement of the left frontal site is consistent with clinical findings that preparatory sets are synthesized and integrated in prefrontal cortical areas, and with experimental and clinical evidence indicating involvement of the left dorsolateral prefrontal cortex in delayed response tasks. A midline anterocentral integrative motor component is consistent with known involvement of premotor and supplementary motor areas in initiating motor responses. The finding of an appropriately lateralized central and parietal component is consistent with evidence from primates and humans for neuronal firing in motor and somatosensory cortices prior to motor responses.

Since ERC feedback patterns of accurate or inaccurate performance (involving either hand) were more similar than those between accurate and inaccurate patterns for one hand, we infer that the feedback patterns were related more to performance accuracy than to the hand used. The fact that ERC patterns following disconfirming feddback involved more frontal sites than did patterns following confirming feedback is consistent with the idea that greater resetting of performance-related neural systems is required following disconfirming feedback. Likewise, the front focus of these differences is consistent with the importance of the frontal lobes in the integration of sensory and motor activities (Fuster 1984; Stuss and Benson 1986).

3.3 The Fourth Experiment (Gevins et al. 1988)

After learning and practicing a difficult visuomotor–memory task (VMMT) until their performance was stable, each of five healthy, right-handed, male subjects returned to the laboratory the next morning and performed several hundred trials of the task. This task required subjects to remember two continuously changing numbers, in the presence of numeric distractors, and produce precise finger pressures. Each trial consisted of a warning symbol, followed by a single-digit visual stimulus to be remembered, followed by the subject's finger-pressure response to the stimulus number presented two trials ago, followed by a two-digit feedback number indicating the accuracy of the response. For example, if the stimulus numbers in five successive trials were 8, 6, 1, 9, 4, the correct response would be a pressure of 0.8 kg when seeing the 1, 0.6 kg for the 9, and 0.1 kg for the 4. To increase the task difficulty, subjects were required to withhold their response on a random 20% of the trials. These "no-response catch trials" were trials in which the current stimulus number was identical to the stimulus two trials ago. After the experiment, we divided each subject's responses into "relatively accurate" and "relatively inaccurate" trials, depending on whether the error of an individual trials was less than or greater than that subject's mean accuracy over the session. Accurate and inaccurate trial sets were randomly distributed across the session.

EEGs were recorded with either 33 or 51 channels set in a nylon mesh cap. Vertical and horizontal eye movements were also recorded, as were the respond-

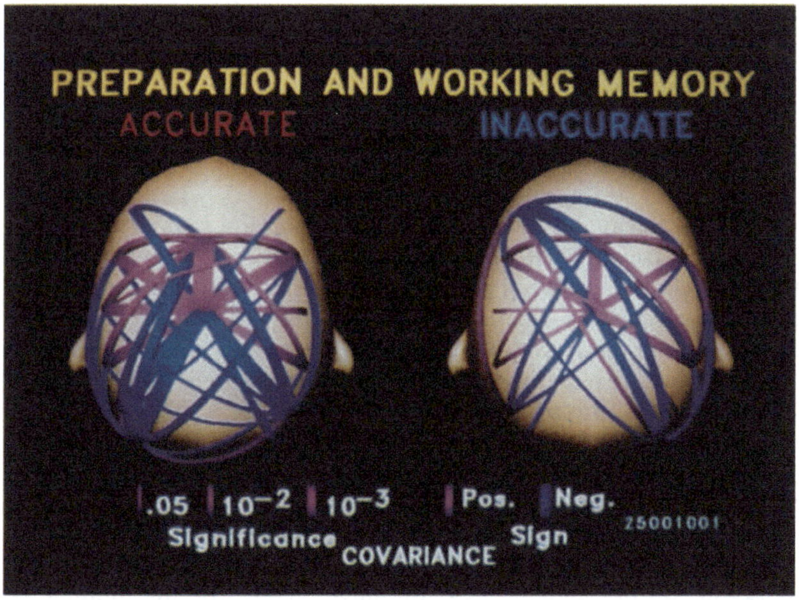

Fig. 4. ERCs preceding accurate and inaccurate performance of a task that required subjects to produce precise finger pressures proportional to numbers held in memory. Significant ($P<0.05$) ERC patterns during a 500-ms interval when five subjects were remembering two numbers and awaiting a new number to be remembered. The ERC pattern preceding accurate performance is stronger, and differs from the inaccurate pattern principally in the posterior channels

ing flexor digitori muscle potentials, electrocardiogram, and respiration. Three-axis magnetic resonance image scans were made of three of the five subjects.

Grand-average (over the five subjects) ERPs were time-locked to the presentation of the numeric stimulus. Spatiotemporal neuroelectric patterns were then quantified by measuring ERCs between all 153 pairwise combinations of the 18 nonperipheral electrodes in the common subset of 33 channels. ERCs were measured on the grand-average ERPs across a 500-ms interval, centered 312 ms before the numeric stimulus. During this interval, subjects were maintaining the last two visually presented numbers in working memory, and preparing for the next stimulus.

Complex, but distinct ERC patterns characterized the subsequently accurate and inaccurate data sets (Fig. 4). The patterns differed both in overall magnitude ($t=3.3$, $df=48$, $P<0.01$) and in topographic distribution (bootstrap correlation $= 0.37 \pm 0.14$), the accurate preparation having stronger ERCs overall. Midline central (607), midline precentral (377), left posteroparietal (257), and right parietal (204) electrodes had the largest summated ERCs in the accurate data set. For the inaccurate data set, the four largest summated ERCs were at midline central (569), midline precentral (392), left frontal (322), and right anterocentral (316). Deferring discussion of the implications of the shift from posterior to anterior ERC foci, we note here that the ERC differences preceding accurate and inaccurate performance of each trial of the visuomotor memory task provide another example of the importance of task-specific, preparatory cortical networks in correct action.

4 Conclusions

4.1 Methodological

Event-related covariance analysis seems to represent a useful improvement over topographic brain electrical activity maps by showing what may prove to be signs of functional interrelations between areas of the cerebral cortex. Furthermore, differences between conditions, which are evident in ERC patterns of preparatory sets, are not apparent on the corresponding topographic maps (Fig. 2) (Gevins et al. 1989a). In our current research we are trying to determine the distributed sources in the brain that produce the ERC patterns at the scalp.

4.2 Brain Model

Although it is convenient to describe brains using the language of computer science, our brains are not actually like computers. They are not stimulus-response machines that passively await stimuli, process them according to fixed algorithms, and then make responses. Rather, we know from introspection, observation, and inference that our brains continuously maintain and update detailed representations (models) of what we imagine our self- and world-states to be at

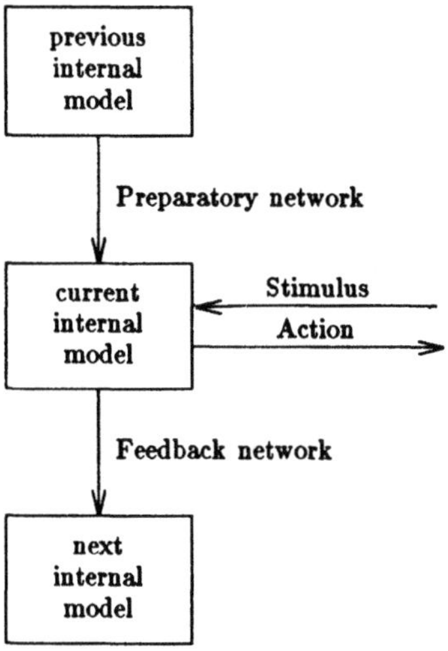

Fig. 5. Schematic diagram of how the previous moment's internal model influences the current moment's perception, decision, and action, and how the feedback from the present moment influences the next moment's internal model. Spatiotemporal neuroelectric signs of preparatory and feedback networks have recently been measured

the current moment. Based on these models, we use our body's effector and sensory systems to actively probe the environment for stimuli relevant to our needs (Fig. 5). When the model for a particular task is incorrect, as with incomplete or inappropriate preparation, subsequent performance is usually incorrect. When feedback follows action, the brain model updates.

I interpret the preparatory and feedback "networks," aspects of which we have measured, as signs of these situation-specific internal models. With further development, we can expect to have increasingly detailed direct measurements of how our experience of reality is synthesized in the brain.

5 Summary

Using state-of-the-art recording and signal processing procedures, we measured spatiotemporal neuroelectric patterns as subjects prepared to produce a precise finger pressure to a visual stimulus, as they made the response, and as they received feedback information about the accuracy of their response. For each of these three situations, we calculated ERC patterns between pairs of electrodes.

ERCs that preceded accurate left- or right-hand performances overlay the left frontal cortex, and the motor and parietal cortices contralateral to the performing hand. In addition, strong ERCs overlying midline motor and premotor cortices preceded left-hand performance. This "preparatory network" comprised distinct cognitive, somesthetic–motor, and integrative motor cortical areas. We used its presence or absence to predict the accuracy of responses that began 0.5–1 s later.

Similarly, we measured a "feedback network" in an interval spanning an early P300 peak. Signals from bilateral and midline frontal sites were involved in patterns related to inaccurate feedback, whereas the left and midline frontal sites were relatively silent following accurate feedback. ERC patterns of accurate performances by either hand, or of inaccurate performances by either hand, were more similar to each other than were accurate and inaccurate performances by a single hand. We concluded that the "feedback network" related more to performance accuracy than to the hand used.

These results suggest that brief, spatially specific patterns of neural activity (referred to as "functional networks") may be signs of the internal models of self and environment which are a hallmark of human consciousness.

Acknowledgments. This research was supported by grants and contracts from the following agencies of the United States Federal Government: National Institutes of Neurological and Communicative Diseases and Strokes, Air Force Office of Scientific Research, and National Science Foundation. Thanks are due to Judith Gumbiner for manuscript preparation.

References

Fuster JM (1984) Behavioral electrophysiology of the prefontal cortex, Trends Neurosci 7(11):408–414
Gevins AS (1980) Pattern recognition of brain electrical potentials. IEEE Trans Patt Anal & Mach Intell 2(5):383–404
Gevins AS (1984) Analysis of the electromagnetic signals of the human brain: milestones, obstacles and goals. IEEE Trans Biomed Eng 31(12):833–850
Gevins AS (1987a) Overview of computer analysis. In: Gevins AS, Remond A (eds) Handbook of electroencephalography and clinical neurophysiology. Vol 1: Methods of analysis of brain electrical and magnetic signals. Elsevier, Amsterdam, pp 31–83
Gevins AS (1987b) Recent advances in neurocognitive pattern analysis. In: Başar E (ed) Dynamics of sensory and cognitive processing of the brain. Springer, Berlin Heidelberg New York, pp 88–102
Gevins AS (1987c) Statistical pattern recognition. In: Gevins AS, Remond A (eds) Handbook of electroencephalography and clinical neurophysiology. Vol 1: Methods of analysis of brain electrical and magnetic signals. Elsevier, Amsterdam, pp 541–582
Gevins AS (1987d) Analysis of multiple lead data. In: Rohrbaugh J, Johnson R, Parasuraman R (eds) Event-related potentials of the brain. Oxford University Press, New York
Gevins AS, Bressler SL (1988) Functional topography of the human brain. In: Pfurtscheller G (ed) Functional brain imaging. Hans Huber, Bern
Gevins AS, Morgan NH (1986) Classifier-directed signal processing in brain research. IEEE Trans Biomed Eng 33:1058–1064
Gevins AS, Morgan NH (1988) Applications of neural-network (NN) signal processing in brain research. Proc IEEE Acoustics, Speech, and Signal Processing 36(7):1152–1161
Gevins AS, Yeager CL (1972) EEG spectral analysis in real time. Digital Corporation Users Society Proc Comp Med 3(4):71–74
Gevins AS, Yeager CL, Diamond SL, Spire JP, Zeitlin GM, Gevins AH (1975) Automated analysis of the electrical activity of the human brain (EEG): a progress report. Proc IEEE 63(10):1382–1399
Gevins AS, Zeitlin GM, Doyle JC, Yingling CD, Schaffer RE, Callaway E, Yeager CL (1979a) Electroencephalogram correlates of higher cortical functions. Science 203:665–668
Gevins AS, Zeitlin GM, Yingling CD, Doyle JC, Dedon MF, Schaffer RE, Roumasset JT, Yeager CL (1979b) EEG patterns during "cognitive" tasks. I. Methodology and analysis of complex behaviors. Electroencephalogr Clin Neurophysiol 47:693–703
Gevins AS, Zeitlin GM, Doyle JC, Schaffer RE, Callaway E (1979c) EEG patterns during "cognitive" tasks. II. Analysis of controlled tasks. Electroencephalogr Clin Neurophysiol 47:704–710

Gevins AS, Doyle JC, Cutillo BA, Schaffer RE, Tannehill RS, Ghannam JH, Gilcrease VA, Yeager CL (1981) Electrical potentials in human brain during cognition: new method reveals dynamic patterns of correlation. Science 213:918–922
Gevins AS, Schaffer RE, Doyle JC, Cutillo BA, Tannehill RL, Bressler SL (1983) Shadows of thought: shifting lateralization of human brain electrical patterns during brief visuomotor task. Science 220:97–99
Gevins AS, Doyle JC, Cutillo BA, Schaffer RE, Tannehill RS, Bressler SL (1985) Neurocognitive pattern analysis of a visuospatial task: rapidly-shifting foci of evoked correlations between electrodes. Psychophysiology 22(1):32–43
Gevins AS, Morgan NH, Bressler SL, Cutillo BA, White RM, Illes J, Greer DS, Doyle JC, Zeitlin GM (1987) Human neuroelectric patterns predict performance accuracy. Science 235:580–585
Gevins AS, Cutillo BA, Fowler-White RM, Illes J, Bressler SL (1988) Neurophysiological patterns of operational fatigue: Preliminary results. NATO/AGARD conference Proceedings 432:22 –22–7
Gevins AS, Bressler SL, Morgan NH, Cutillo BA, White RM, Greer DS, Illes J (1989 a) Event-related covariances during a bimanual visuomotor task. Part I: Methods and analysis of stimulus- and response-locked data. Electroencephalogr Clin Neurophysiol 74(1):58–75
Gevins AS, Cutillo BA, Bressler SL, Morgan NH, White RM, Illes J, Greer DS (1989 b) Event-related covariances during a bimanual visuomotor task. Part II: Preparation and feedback. Electroencephalogr Clin Neurophysiol 74(2):147–160
Stuss DT, Benson DF (1986) The frontal lobes. Raven, New York

The Clinical Use of P300 Cartography in Diseases with Disturbed Cognitive Processing of the Brain

K. Maurer

1 Introduction

Until recently evoked potentials (EPs) were recorded mainly with few electrodes. For evaluation of information processing in normals and patients, latency measurements were usually performed. This procedure proved very valuable when early auditory evoked potentials (EAEPs) were applied to evaluate hearing function and to localize lesions in the lower auditory pathway (Starr and Achor 1975; Stockard and Rossiter 1977; Maurer et al. 1979, 1982). Topodiagnosis in neurootology became possible by virtue of the fact that EAEPs originate in well defined structures along the auditory nerve and in the brain stem (Möller et al. 1981; Maurer and Mika 1983). It is acknowledged that the generator site of wave I is the auditory nerve while that of waves II to V is the proximal part of the auditory nerve (II) and adjacent portions of the brain stem such as the lower pons (wave III), upper pons (wave IV), and midbrain (wave V). The generator sites of later waves, such as AEPs of middle and late latency, could not be determined as accurately as in the case of EAEPs, with the consequence that the diagnostic value of the late waves has remained limited until now.

The AEPs of early, middle, and late latency are also known as "exogenous potentials" due to the fact that their amplitude, latency, and shape are determined by physical characteristics of the stimulus, such as intensity, frequency, and duration. Since 1965 (Sutton et al.) it has also been known that potentials can be elicited under psychological conditions. Since the generation of these waves is closely linked to cooperation, concentration, and ability to count and discriminate between different kinds of stimuli, these waves have been classified as "endogenous potentials."

The approximately 15 components which can be elicited after acoustic stimulation are shown in Fig. 1. Only late waves in the latency range of 200–400 ms show the ability to respond to a target stimulus and to develop cognitive components or endogenous potentials. The best known "endogenous wave" is the auditory elicited P300, which was first described by Sutton et al. (1965). Since then the P300 has become very popular and enables the evaluation of cognitive processing of the brain in normals and patients with a cognitive deficit.

Clinical P300 studies with few electrodes were performed in patients suffering from dementia, schizophrenia, and depression (Pfefferbaum et al. 1984). It was very soon realized, however, that the measurement of latencies and amplitudes alone was insufficient for evaluation of cognitive processing in diseases. The reason was the lack of specifity of the P300 test, which exhibited similar latency or

Springer Series in Brain Dynamics 2
Edited by Erol Başar and T. H. Bullock

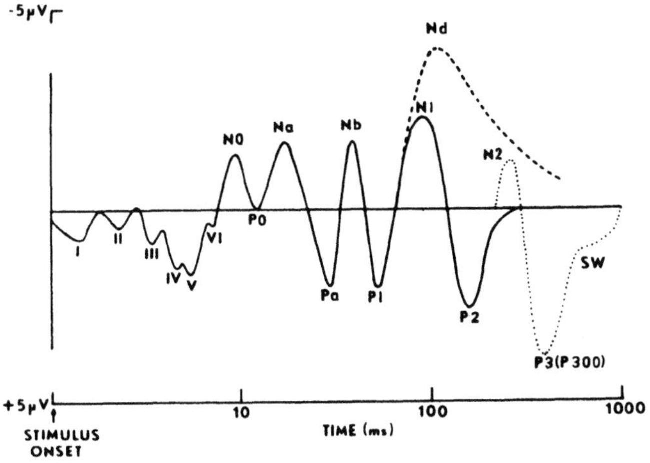

Fig. 1. Sequence of approximately 15 waves which can be evoked after acoustic stimulation

amplitude changes in diseases of different etiology. The amplitude reduction of P300, for example, occurred not only in dementia but also in schizophrenia and depression (Roth et al. 1981). In 1979 Duffy et al. described a method of topographic display of EEG and EPs and opened up a new field in electrodiagnosis. In the case of EPs the application of 16–20 electrodes to the scalp provided the clinician with new information from areas where no data could previously be obtained. Thus for frontal and temporal regions, Maurer and Dierks (1987) found evidence of cognitive dysfunction in schizophrenia and dementia.

The main aim of the following investigation was to delineate the topography of the cognitive endogenous wave P300 in normals and in patients suffering from diseases with a cognitive deficit, such as dementia and schizophrenia. For cartography we used a quasi three-dimensional head model which allowed the evaluation of amplitude behavior in a more obvious way than the two-dimensional head formats formerly used, especially when phenomena such as hypo- and hyperfrontality occurred.

2 Methods

The electrical activity of the brain was recorded by 20 electrodes which were attached to the head according to the 10–20 system. The ears served as points of reference. After analog-digital conversion, P300 cartography was calculated by linear interpolation.

The P300 was elicited by using frequent 1000-Hz and infrequent 2000-Hz tones which were randomly presented to both ears via earphones. The intensity was 80 dB. The ratio of frequent to infrequent tones was 5:1. Volunteers and patients were asked to pay attention to the rare high tones and to count them. If the task was performed properly a well defined positive wave appeared with a peak at approximately 300 ms (P300). For display of P300 cartography only the three-dimensional head models will be used, which show the P300 topography by using peak values of the P300 waves.

3 Results

3.1 P300 Cartography in Normals

Normal cartography was obtained in a group of ten students (mean age: 27.7 years). The P300 distribution showed a characteristic pattern with maximal amplitudes in the parietal region at points P3, Pz, and P4 (Fig. 2). The amplitudes became smaller towards temporal and frontal structures, where values of about 5 μV were measured bifrontally.

3.2 P300 Cartography and Aging

The P300 amplitude behavior was no different in a population of ten persons aged 43.8 years. In geriatric controls, however, we saw a generalized amplitude reduction in the order of about 5 μV (Fig. 3). The amplitude reduction was not linked to a change in cartography, which still showed the typical pattern with maximal positive fields at points P3, Pz, and P4 and a slight decrement towards frontal structures.

3.3 P300 Cartography in Senile Dementia of the Alzheimer Type

Ten patients suffering from senile dementia of the Alzheimer type (SDAT) were chosen for P300 cartography (mean age: 70.1 years). All patients were able to perform the P300 task properly. The score on the Brief Cognitive Rating Scale was between 24 and 32, which corresponds to a moderate degree of dementia. The P300 cartography was different to that found in age-matched controls, with an evident increase of amplitudes bifrontally (hyperfrontality) and a decrease in temporo-parieto-occipital areas (Fig. 4). The bifrontal increase was statistically significant ($P < 0.01$).

3.4 P300 Cartography in a Reversible Psycho-organic Syndrome Caused by Wilsons's Disease

We were able to perform P300 cartography in a patient with a disturbance of the copper metabolism. P300 was measured before and during therapy with penicillamine. At admission the patient showed a moderate degree of dementia (Short syndrome test $>9<18$).

The corresponding P300 cartography revealed augmented amplitudes bifrontally as a sign of hyperfrontality. The pattern was similar to that obtained in

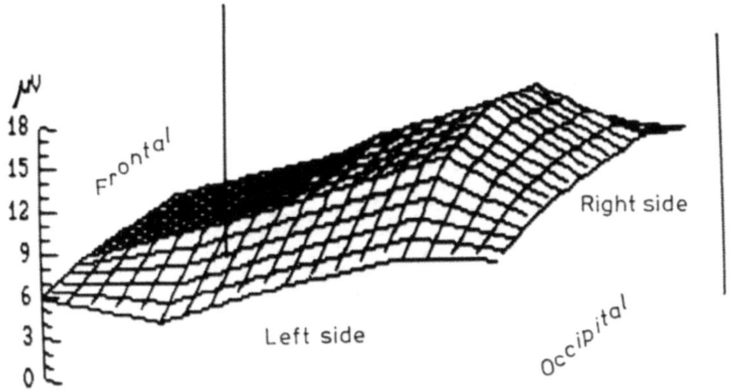

Fig. 2. Three-dimensional display of P300 peak amplitudes; left side view. P300 amplitudes are maximal at P3, Pz, and P4 and show a decrement towards frontal structures

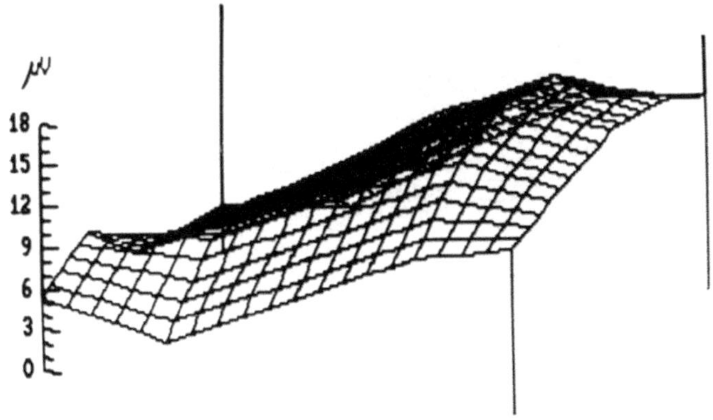

Fig. 3. P300 topography in a geriatric control population: generalized amplitude reduction without a change in P300 topographic pattern

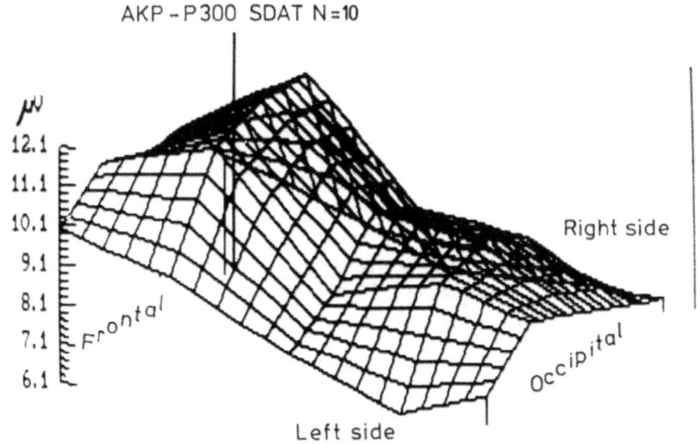

Fig. 4. P300 topography in SDAT patients with increased bifrontal amplitudes and a decrement towards parietal structures

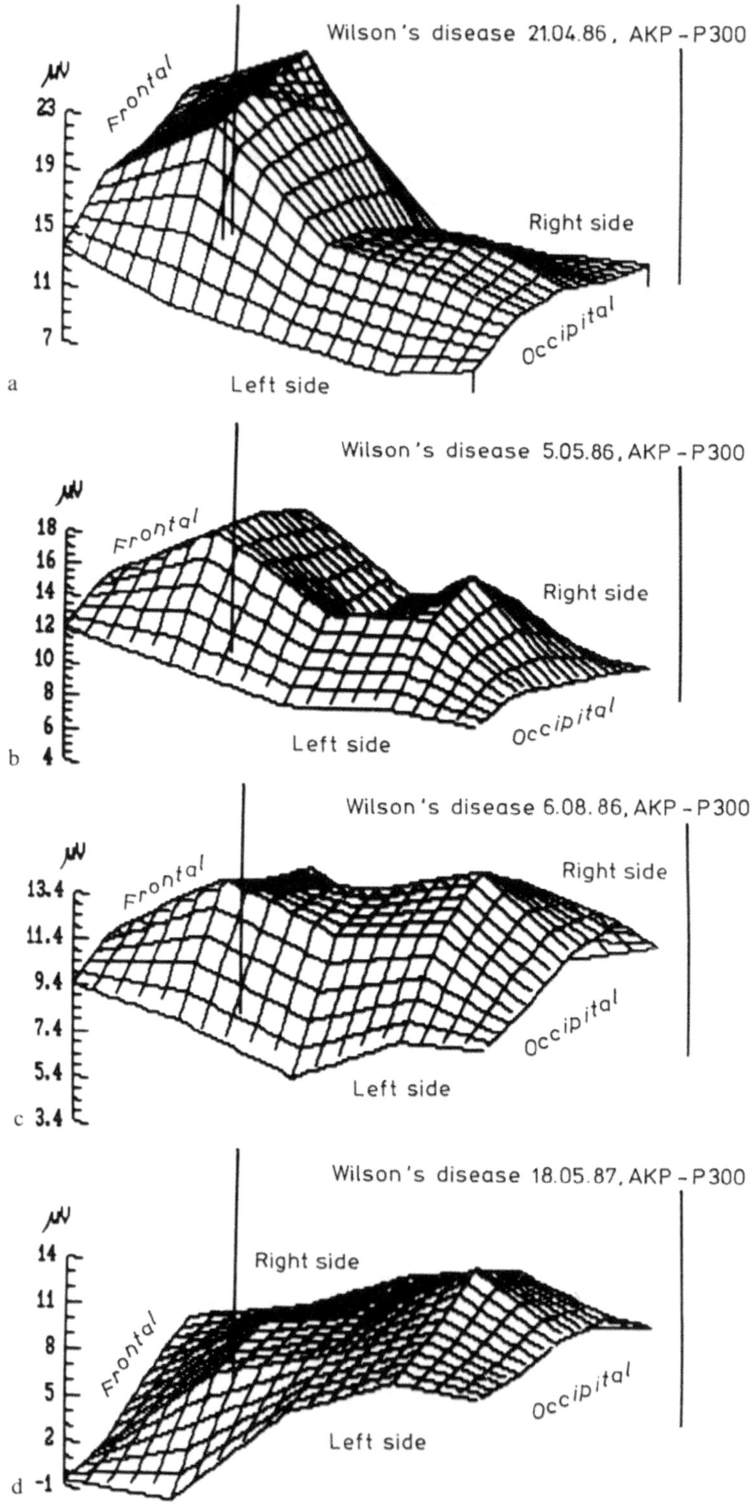

Wilson's disease 21.04.86, AKP - P300
μV
23
19
15
11
7
Frontal
Right side
Occipital
Left side
a
Wilson's disease 5.05.86, AKP - P300
μV
18
16
14
12
10
8
6
4
Frontal
Right side
Occipital
Left side
b
Wilson's disease 6.08.86, AKP - P300
μV
13.4
11.4
9.4
7.4
5.4
3.4
Frontal
Right side
Occipital
Left side
c
Wilson's disease 18.05.87, AKP - P300
μV
14
11
8
5
2
-1
Right side
Frontal
Occipital
Left side
d

SDAT patients (Fig. 5). During therapy with penicillamine a change in cognitive field pattern occurred, with signs of an anterior–posterior displacement of the P300 peak within 12–14 months. In the further course of the therapy a nearly normal P300 pattern could be observed.

3.5 P300 Cartography and Schizophrenia

In schizophrenic patients amplitudes of P300 were generally lower. In the hebephrenic and paranoid subgroups, however, a different cartography could be detected.

In contrast to SDAT patients, the hebephrenics (ICD-9; 295.5) showed a very low frontal field (Fig. 6a). *t*-Values were maximal at points Fp1, F3, Fz, and F4, exhibiting values between −1.6 and −2.4. The hypofrontality was significant

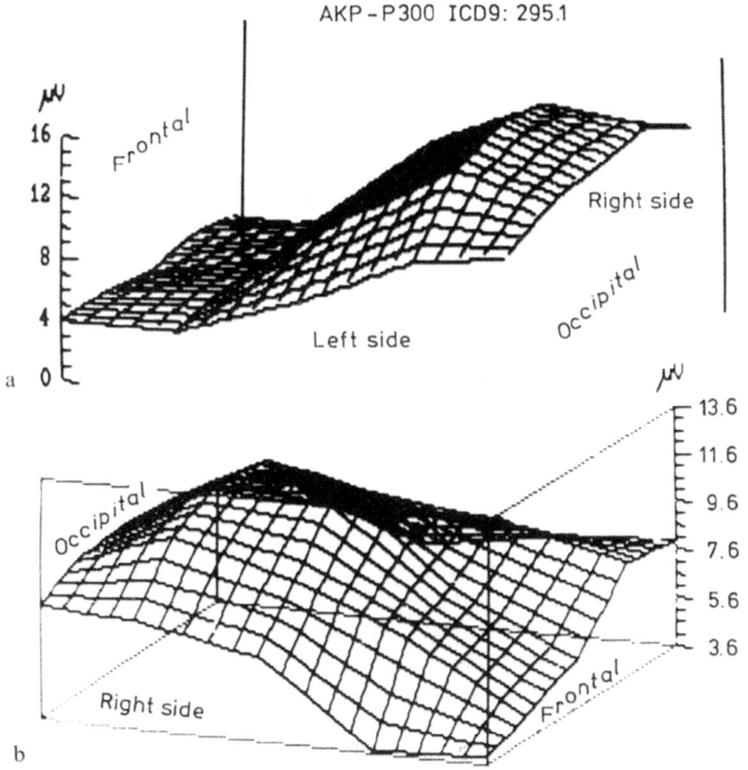

Fig. 6a, b. P300 topography in schizophrenia. (a) P300 cognitive field in the hebephrenic subgroup with low frontal amplitudes (hypofrontality). (b) P300 cognitive field in the paranoid subgroup with an amplitude decrement in the right-sided frontotemporal area; right side view

Fig. 5a–d. P300 topography in a patient with Wilson's disease. (a) First investigation (21.4.1986) with increased bifrontal amplitudes. (b) Second investigation (5.5.1986) with a decrease in frontal and an increase in parietal amplitudes. (c) Third investigation (6.8.1986) with amplitudes of similar magnitude at frontal and parietal sites. (d) Fourth investigation (18.5.1987) with a parietal maximum of P300 amplitudes

($P<0.01$). In the paranoid subgroup (ICD-9; 295.3) a lateralized P300 pattern could be observed, with lower amplitudes on the right side in the frontotemporal area (Fig. 6b). *t*-Value maps revealed that the differences between normals and schizophrenics were highest at points P4 and T6.

4 Discussion

The terms "hypofrontality" and "hyperfrontality" have been used to describe phenomena of changed frontal function in the course of schizophrenia and demential processes (Ingwar and Franzenk 1974; Itil 1978; Stigsby et al. 1981; Farkas et al. 1984). In schizophrenia, for example, the reduced blood flow bifrontally measured by rCBF and the diminished glucose utilization revealed by PET led to the term "hypofrontality." The reverse occurred in dementia, where Stigsby et al. (1981) found a hyperfrontal flow distribution. The rCBF findings showed a good correlation with the neuroanatomical findings of Brun and Gustafson (1976), namely a relative lack of degeneration at frontal lobes and the anterior cingulate gyrus. The main lesion pattern was observed in temporo-parieto-occipital areas, where cortical degeneration took place.

By means of EEG measurements and analysis it was difficult to reproduce phenomena like hypo- and hyperfrontality detected by functional imaging procedures such as rCBF and PET. It was assumed by some authors that bifrontal slowing would coincide with low frontal function and therefore hypofrontality (Morstyn et al. 1983). Later, however, it emerged that frontal EEG slowing was due to artifacts caused by blinks and eye movements (Karson et al. 1987). Another difficulty was the absence of any clear correlation between function and frequency in the EEG which would enable us to define hypo- and hyperfunction in the brain. The alpha frequency band did not contribute toward detection of hyper-/hypofunction in the frontal lobe because of its occipital prevalence, with the consequence that the concept of decreased or augmented vigilance could not be applied to the frontal lobe.

The work of Maurer and Dierks (1987) and others revealed that the topography of P300 shows not only a very reliable parietal field in normals but also a change of topography in psychoses and dementia. Whereas in the hebephrenic subtype signs of hypofunction were found (hypofrontality), in dementia an increased function was observed bifrontally (hyperfrontality). In normals, drugs such as cholinergic substances (physostigmine) and nootropics were able to augment the P300 fields. These findings in normals and patients indicated that P300 cartography is as efficient as other functional imaging procedures in defining signs of frontal hypo- or hyperfunction.

Regarding P300, however, the two assumed sites of origin – cortical and subcortical – have to be considered when further discussing the correlation between brain function and P300 amplitude fields. In respect of the subcortical origins, there is extensive evidence that most P300 activity comes from the hippocampus and nucleus amygdalae. This was shown by Halgren using depth recording techniques to establish the limbic sources and by Okada et al. (1978), who, employing magnetoencephalographic techniques, found evidence of a hippocampal dipole responsible for the positive parietal P300 fields in normals. Consequently only the cortical part of P300 can be compared with findings of rCBF and PET studies because both these methods describe the function of cortical structures. Concerning the limbic system sources there is still the question how the P300 generator affects hypo- and/or hyperfrontality. In the case of schizophrenia we can only speculate and we have to take into account our understanding of basal ganglia and limbic system pathology (Bogerts et al. 1985). When schizophrenia is caused by a lesion in those areas, a bifrontal decrement of P300 can result, which could then cause signs of hypofrontality.

In the case of dementia, hyperfrontality is easier to interpret because a deep hippocampal dipole is assumed to be responsible for the positive P300 fields. It is known from the work of Ball et al. (1987) that in Alzheimer-type dementia a lesion occurs in the hippocampus. Accordingly it may be speculated that the dipole changes its direction, which leads ultimately to a displacement of the P300 field. This assumption is supported by the fact that the total positivity of P300 is unchanged in dementia, solely the local distribution being maximal in the frontal lobes.

Summarizing, P300 topography made the P300 findings more specific and enabled the determination of well defined cognitive fields in diseases such as dementia and psychoses. In contrast to EEG topography, the P300 fields delineated phenomena such as hypo- and hyperfrontality. P300 cartography makes possible the visualization of functioning of the brain, which until now could only be achieved by imaging procedures such as rCBF and PET. Advantages of P300 cartography over the latter imaging procedures are its noninvasiveness and its short analysis times, which allow the analysis of cognitive processing of the brain in the millisecond range.

5 Summary

After acoustic stimulation approximately 15 AEP components can be elicited. Waves which are determined by physical characteristics of the stimulus are named "exogenous potentials." In addition there are late waves in the latency range of 200–400 ms which are generated due to a target stimulus. Of these so-called endogenous potentials the best known wave is the auditory elicited P300. With the aid of computer techniques a topographic display of P300 was performed using three-dimensional head models. By this means P300 cartography was carried out in an aging population and in patients with diseases such as Alzheimer-type dementia, Wilson's disease, and schizophrenia. It emerged that with P300 cartography differentiation between dementia and psychosis was possible.

References

Ball MJ, Hachinski V, Fox A, Kirshen AJ, Fisman M, Blume W, Kral VA, Fox H (1987) A new definition of Alzheimer's disease: a hippocampal dementia. Lancet 5:14–16

Bogerts B, Meertz E, Schönfeldt-Bausch R (1985) Basal ganglia and limbic system pathology in schizophrenia. Arch Gen Psychiatry 42:784–791

Brun A, Gustafson L (1976) Distribution of cerebral degeneration in Alzheimer's disease. Arch Psychiatr Neurol 223:15–33

Duffy FH, Burchfiel JL, Lombroso CT (1979) Brain electrical activity mapping (BEAM): a method for extending the clinical utility of EEG and evoked potentials data. Ann Neurol 5:309–321

Farkas T, Wolf AP, Jaeger J, Brodie JD, Christman DR, Fowler JS (1984) Regional brain glucose metabolism in chronic schizophrenia. Arch Gen Psychiatry 41:293–300

Halgren E, Squires NK, Wilson CL, Crandell PH (1982) Brain generators of evoked potentials: the late (endogenous) components. Bull Los Angeles Neurol Soc 47:108–123

Ingwar DH, Franzen G (1974) Abnormalities of cerebral blood flow distribution in patients with chronic schizophrenia. Acta Psychiatr Scand 40:425–462

Itil TM (1978) Qualitative und quantitative EEG-Befunde bei Schizophrenien. Z EEG EMG 9:1–13

Karson CN, Coppola R, Morihisa JM, Weinberger DR (1987) Computed electroencephalographic activity mapping in schizophrenia. Arch Gen Psychiatry 44:514–517

Maurer K, Dierks T (1987) Brain Mapping – topographische Darstellung des EEG und der evozierten Potentiale in Psychiatrie und Neurologie. Z EEG EMG 18:1–9

Maurer K, Mika H (1983) Early auditory evoked potentials (EAEPs) in the rabbit. Normative data and effects of lesions in the cerebello-pontine angle. Electroencephalogr Clin Neurophysiol 55:586–593

Maurer K, Leitner H, Schäfer E, Hopf HC (1979) Frühe akustisch evozierte Potentiale, ausgelöst durch einen sinusförmigen Reiz. Dtsch Med Wochenschr 104:546–550

Maurer K, Leitner H, Schäfer E (1982) Akustisch evozierte Potentiale. Enke, Stuttgart

Möller AR, Jannetta PJ, Möller BM (1981) Neural generators of brainstem evoked potentials. Results from human intracranial recordings. Ann Otol 90:591–599

Morstyn R, Duffy FH, McCarley RW (1983) Altered topography of EEG spectral content in schizophrenia. Electroencephalogr Clin Neurophysiol 56:263–271

Okada YC, Kaufman L, Williamson SJ (1983) The hippocampal formation as a source of slow endogenous potentials. Electroencephalogr Clin Neurophysiol 55:417–426

Pfefferbaum A, Wenegrat BG, Ford JW, Roth WT, Kopell BS (1984) Clinical application of the P3 component of event-related potentials. II. Dementia, depression and schizophrenia. Electroencephalogr Clin Neurophysiol 59:104–124

Roth WT, Pfefferbaum A, Kelly AF, Berger PA, Kopell BS (1981) Auditory event-related potentials in schizophrenia and depression. Psychiatr Res 4:199–212

Starr A, Achor LJ (1975) Auditory brainstem responses in neurological disease. Arch Neurol 32:761–768

Stigsby B, Johannesson G, Ingwar D (1981) Regional EEG-analysis and regional cerebral blood flow in Alzheimer's disease. Electroencephalogr Clin Neurophysiol 51:537–547

Stockard JJ, Rossiter VS (1977) Clinical and pathologic correlates of brainstem auditory response abnormalities. Neurology (Minneap) 27:316–325

Sutton S, Braren M, Zubin J, John ER (1965) Evoked potential correlates of stimulus uncertainty. Science 150:1187–1188

Influence of Physostigmine on Cognitive Processing of the Brain

R. IHL, L. FRÖLICH, T. DIERKS, and K. MAURER

1 Introduction

The ascending cholinergic system of the central nervous system plays an important role in cognitive processing and memory formation of the brain. This was suggested by pharmacological studies of the CNS effects of anticholinergic drugs (Drachman et Leavitt 1974; Peters and Levin 1979; Drachman and Sahakian 1980; Summers et al. 1981). Another line of evidence indicating the involvement of cholinergic mechanisms in cognitive processing arises from the cholinergic hypothesis of geriatric memory dysfunction (Bowen et al. 1976; Bartus et al. 1982). In patients with dementia of the Alzheimer type (DAT), various markers of central cholinergic activity are reduced and there is a correlation between the degree of dementia and the reduction of the cholinergic marker cholinacetyltransferase. Pharmacological manipulations to increase acetylcholine availability may ameliorate certain cognitive deficits. However, the clinical relevance of cholinomimetics is limited. Most cholinergic drugs have a plasma half-life of only a few minutes, lack CNS specificity, cause considerable side-effects, and have a narrow therapeutic range. Physostigmine, a reversible inhibitor of acetylcholinesterase, is capable of improving cognitive processing in a dose-dependent manner (Christie et al. 1981; Davis and Mohs 1982). Physostigmine may be a valuable reference substance for evaluating the effects of drugs for the treatment of senile DAT (Mohs and Davis 1987). Brain mapping has been proven to be a good method for evaluating the actions of drugs on the brain (Maurer and Dierks 1987; Ihl et al. 1989). The topographic distribution of electrical brain activity in electroencephalograms (EEGs) and evoked potentials (EPs) contains information about the main localization of drug effects. P300 is supposed to be generated in the hippocampal formation (Halgren et al. 1982; Okada et al. 1983), an important area for learning and memory. The amplitude, latency, and topography of P300 describe information-processing capacity. Topographic changes in EEG frequency bands firstly construct a picture of drug activity and secondly include information about cognitive processing. The present study was performed to evaluate the effects of physostigmine on the parameters described above.

2 Methods

A 25-year-old right-handed female volunteer from the staff of the University Hospital, Würzburg, lay relaxed with closed eyes for recording of electrical brain ac-

Springer Series in Brain Dynamics 2
Edited by Erol Başar and T. H. Bullock

tivity. Twenty electrodes placed according to the international 10–20 system led activity to a Biologic III Brain atlas, which calculated maps with the method for four nearest neighbors. The method is described in detail in Maurer and Dierks (1987) and Duffy et al. (1979). The stimulus rate for target of acoustic P300 was 0.2 while the interstimulus interval was 2 s. Samples of 30 targets were averaged. The stimuli were produced and applied binaurally by a Telephonics 39 TDH headphone. Before evaluation the EEG was transformed with Fourier transformation. Four frequency bands were analyzed (delta 0.0–3.5, theta 4.0–7.5, alpha 8–11.5, and beta 12–31.5 Hz).

To minimize peripheral side-effects of the cholinomimetic drug physostigmine, 0.2 mg glycopyrrolate was applied subcutaneously after recording preapplication values. Half an hour later 0.25 mg physostigmine was given intravenously. Measures were repeated 5 times, every 30 min after administration.

3 Results

The subject reported neither effects nor side-effects of the drug. EEG and EP records were altered simultaneously. After 30 min P300 activity increased all over the scalp except for occipital electrodes. The highest increases were recorded at F3, Fz, F4, F8, T4, and Cz (Fig. 1 a, b). After 60 min frontal areas showed lower values compared to the 30-min values. However, activity under 01, Oz, and O2 increased (Fig. 1 c). After 90 min maximum of alteration was reached (Fig. 1 d). At 120 min after application (Fig. 1 e) activity diminished under all electrodes except for the prefrontal ones, which showed diminished activity after 150 min (Fig. 1 f).

EEG changes were observed in all frequency bands (Fig. 2) Alpha activity was altered most obviously. After 30 min it was increased under all electrodes. After 60 min the increase was concentrated on parieto-occipital areas. After 90 min higher values were also recorded in the posterior temporal regions. After 120 min activity began to decrease again.

Delta activity decreased in the frontotemporal areas, whereas other electrodes showed no change. Theta activity increased slightly after 30 min. Beta activity increased under occipital electrodes after 30 min and decreased thereafter.

4 Conclusions

In this study physostigmine increased brain electrical activity over the whole recorded period. In EEG the most important changes were increased alpha activity and reduced delta activity. Alpha activity was increased under all electrodes. Highest increase was recorded in parieto-occipito-temporal areas. The increase can be attributed to activation of the cholinergic system.

The increase in P300 activity after 30 min is most pronounced in frontal and temporal areas. It can be assumed that the hippocampally generated P300 pro-

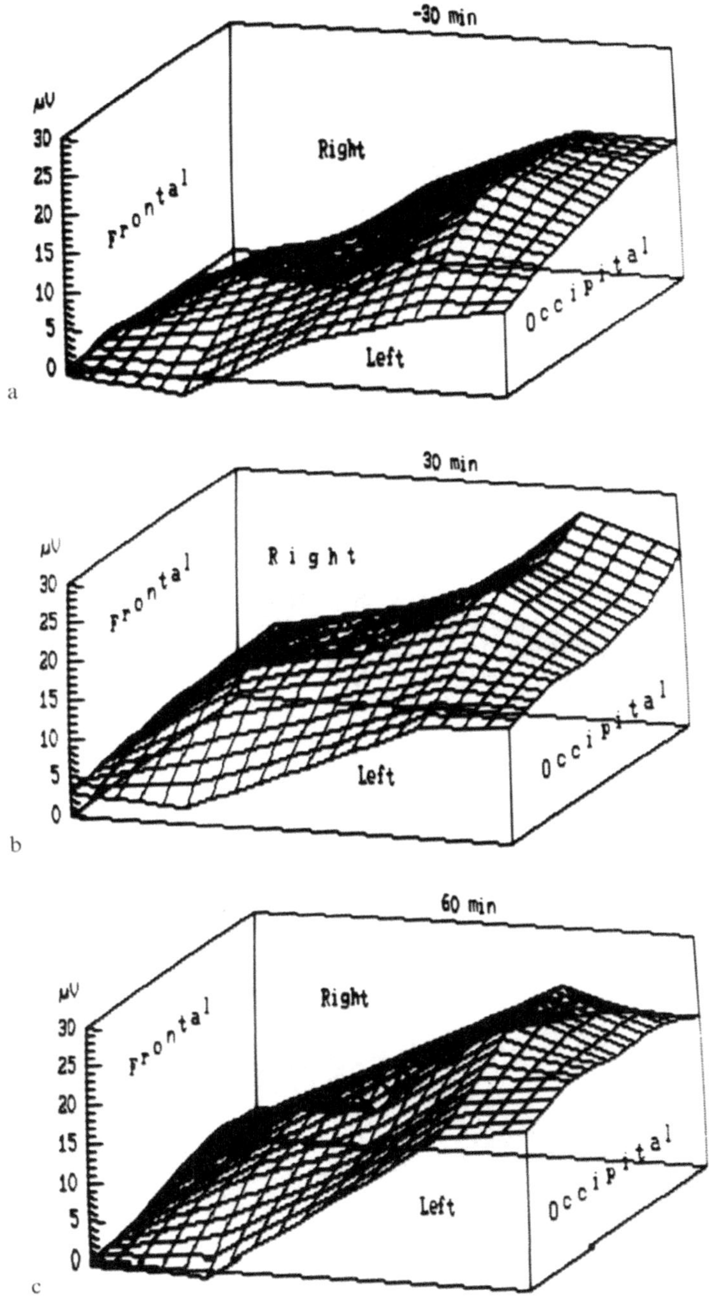

Fig. 1 a–f. P300 activity. (a) baseline; (b) after 30 min; (c) after 60 min; (d) after 90 min; (e) after 120 min; (f) after 150 min

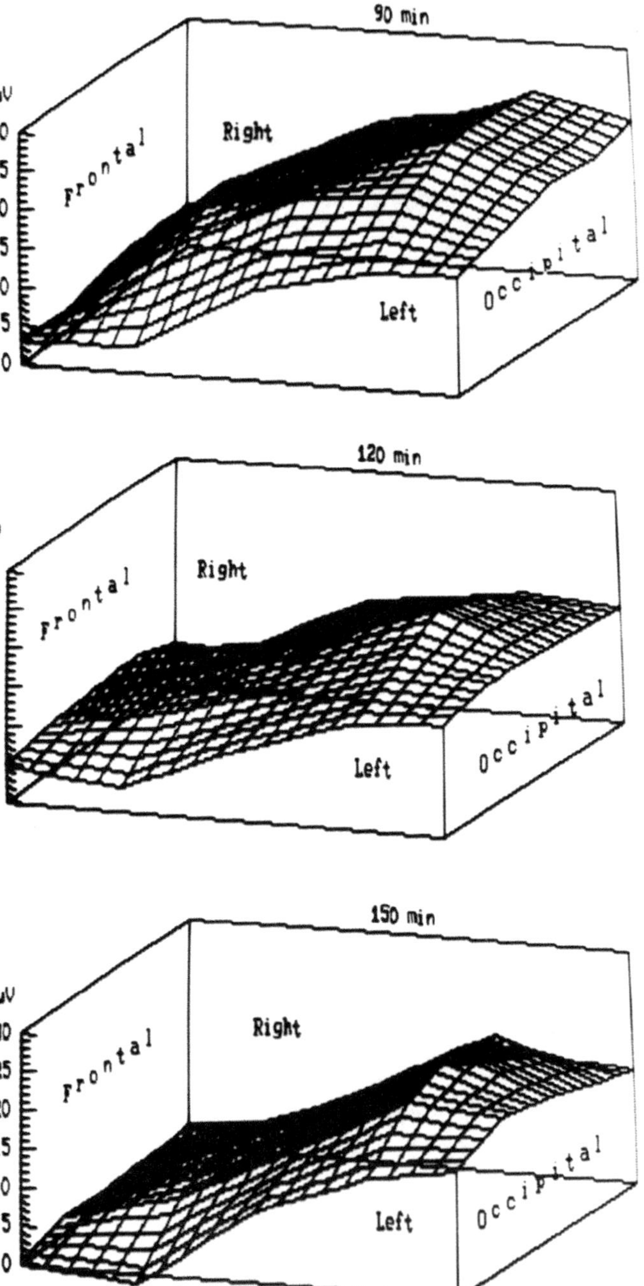

d

e

f

Abb. 1 d–f. Legend see page 431

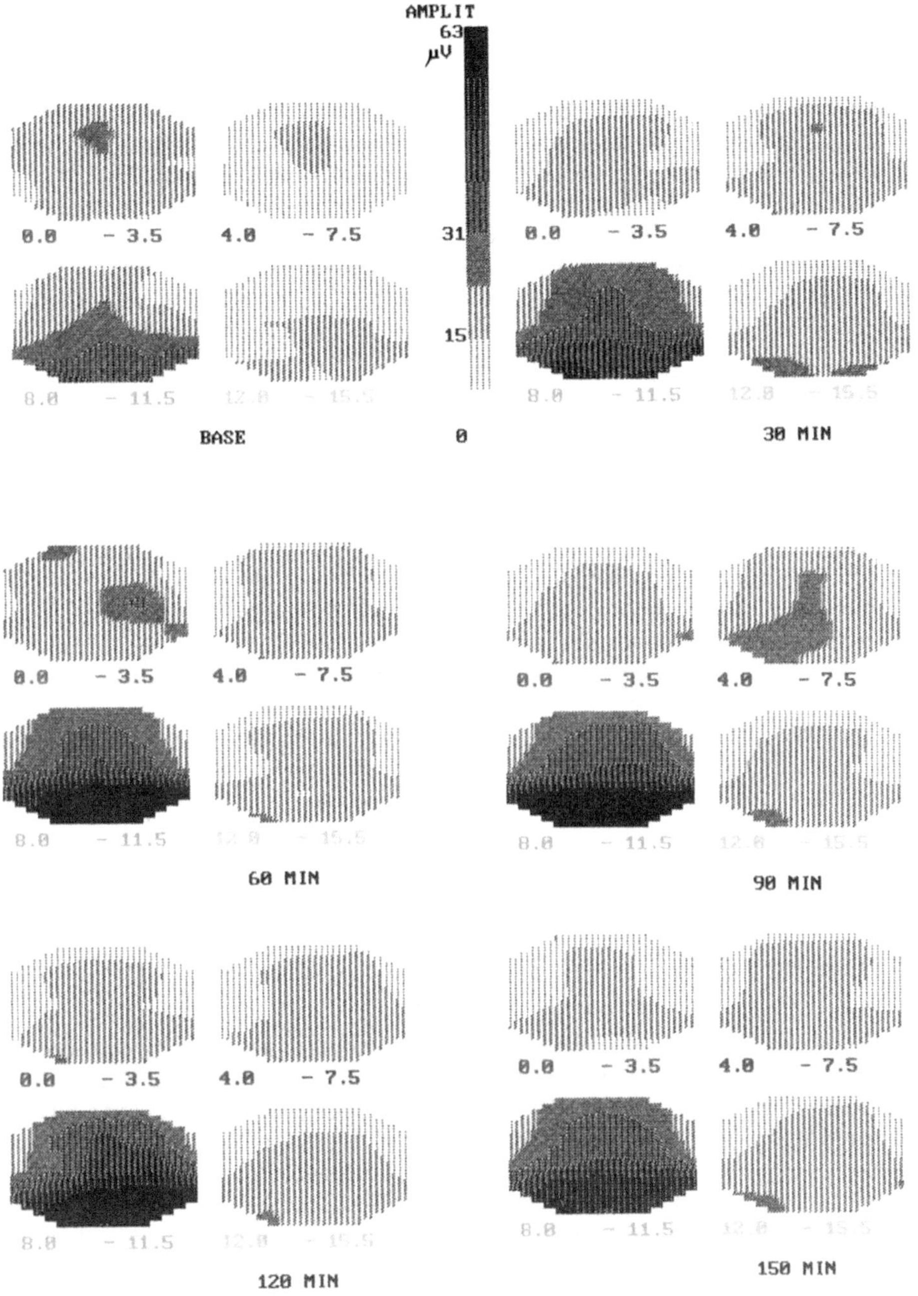

Fig. 2. The four frequency bands before and five times after drug administration. Each shade of black represents 7.5 µV for all 24 maps

duces this increase in cortical activity. After 60 min high activity is also found over parietal and occipital areas. The increase is most prominent in areas with low preapplication values. Frontal neurons seem to react most strongly to acetylcholine activation. After this first effect increase centers to the usual point of maximal activity (Pz). It appears clear that cholinergic modulation affects not only brain alpha rhythm but also the topography of cognitively evoked potential P300.

The half-life of physostigmine is between 10 and 30 min. However, the main influence in the reported subject was seen after more than two half-lives. Nevertheless the delayed alteration is in accordance with the findings of Muramoto et al. (1984), who reported an improved ability to copy geometric figures after 30–60 min. These effects could be due to a compartmentation of acetylcholine concentration at the receptor after the application of the acetylcholinesterase inhibitor physostigmine.

Two findings show connections to DAT. The alpha activation is principally parieto-occipito-temporal. It is assumed to be due to cholinergic activation. In DAT the same area shows low alpha values (Ihl et al. 1989) and is most involved in the disease (Brun and Gustafson 1976). The reduced alpha values can be due to cholinergic inhibition or a loss of cholinergic neurons. So both findings underline the cholinergic hypothesis of DAT.

Secondly, P300 increase was observed most prominently in frontal areas. These areas are also important in DAT. They are supposed to be less vulnerable (Brun and Gustafson 1976). These findings were underlined by Stigsby et al. (1981), who reported signs of "hyperfrontality" in DAT, and by Maurer et al. (1988, 1989), who reported P300 to be higher in frontal areas. The same pathways which keep their function for a long time in DAT may be activated intensively in normals after administration of physostigmine, producing the selectively augmented P300 field.

5 Summary

Brain electrical activity mapping of EEG and EP data is a noninvasive procedure for studying cognitive processing of the brain. Here it was used to evaluate the influence of physostigmine on these parameters. A female right-handed proband received 0.25 mg physostigmine intravenously. To minimize peripheral side-effects 0.2 mg glycopyrrolate was applied subcutaneously 30 min earlier. EEG and EP data were recorded before, 30, 60, 90, and 120 min after drug application. P300 activity increased in frontotemporal regions after 30 min and additionally, in occipital regions after 60 min. In the EEG, alpha activity increased in parieto-temporal areas up to 90 min after application. Delta activity decreased. Physostigmine influenced electrical brain activity in the EEG and EPs. Response was delayed compared to the short half-life of the drug. This can be regarded as an effect of compartmentation. The sensitivity of temporoparietal areas for increased alpha activity may be seen as a connecting link to findings in dementia of the Alzheimer type, where the same regions show a reduced activity.

References

Bartus RT, Dean RL, Beer B, Lippa AS (1982) The cholinergic hypothesis of geriatric memory dysfunction. A critical review. Science 217:408–417

Bowen DM, Smith CB, White P, Davison AN (1976) Neurotransmitter related enzymes and indices of hypoxia in senile dementia and other abiotrophies. Brain 990:459–496

Brun A, Gustafson L (1976) Distribution of cerebral degeneration in Alzheimer's disease. Arch Psychiatr Neurol 223:15–33

Christie JE, Shering A, Ferguson J, Glen AIM (1982) Physostigmine and arecoline: effects of intravenous infusions in Alzheimer presenile dementia. Br J Psychiatry 138:46–50

Davis MD, Mohs RC (1982) Enhancement of memory processes in Alzheimer's disease with multiple dose intravenous physostigmine. Am J Psychiatry 139:1421–1424

Drachman DA, Leavitt J (1974) Human memory and cholinergic system: a relationship to aging? Arch Neurol 30:113–121

Drachman DA, Sahakian BJ (1980) Memory, aging and pharmacosystems. In: Stein D (ed) The psychobiology of aging. Problems and perspectives. Elsevier, Amsterdam, pp 347–368

Duffy FH, Burchfield JL, Lombroso CT (1979) Brain electrical activity mapping (BEAM): a method for extending the clinical utility of EEG and evoked potential data. Ann Neurol 5:309–321

Ebenezer IS (1986) The effects of atropine on event related slow potentials in the rat. Eur J Pharmacol 120:371–374

Halgren E, Squires NK, Wilson CL, Crandall PH (1982) Brain generators of evoked potential: the late (endogenous) components. Bull Los Angeles Neurol Soc 47:108–123

Ihl R, Dierks T, Maurer K, Wannenmacher U (1988) EEG and EP mapping in patients with senile dementia of the Alzheimer type. In: Maurer K (ed) Topographic brain mapping of EEG and evoked potentials. Springer, Berlin Heidelberg New York, pp 241–248

Maurer K, Dierks T (1987) Brain mapping – topographische Darstellung des EEG und der evozierten Potentiale in Psychiatrie und Neurologie. Z EEG EMG 18:4–12

Maurer K, Ihl R, Dierks T (1988) Topographie der P300 in der Psychiatrie – II. Kognitive P300 Felder bei der Demenz. Z EEG EMG 19:26–29

Maurer K, Dierks T, Ihl R, Laux G (1989) Mapping of evoked potentials in psychiatry. In: Maurer K (ed) Topographic brain mapping of EEG and evoked potentials. Springer, Berlin Heidelberg New York, pp 458–473

Mohs RC, Davis KL (1987) The experimental pharmacology of Alzheimer's disease and related dementias. In: Melzer HY (ed) Psychopharmacology: the third generation of progress. Raven, New York

Muramoto O, Sugishita M, Ando K (1984) Cholinergic system and constructional praxis: a further study of physostigmine in Alzheimer's disease. J Neurol Neurosurg Psychiatry 47:485–491

Okada YC, Kaufman L, Williamson SJ (1983) The hippocampal formation as a source of the slow endogenous potentials. Electroencephalogr Clin Neurophysiol 55:417–426

Peters BH, Levin HS (1979) Effects of physostigmine and lecithin on memory in Alzheimer disease. Ann Neurol 6:219–221

Stigsby B, Johanneson G, Ingvar DH (1981) Regional EEG analysis and regional cerebral blood flow in Alzheimer's and Pick's disease. Electroencephalogr Clin Neurophysiol 51:537–547

Summers WK, Viesselman JO, Marsch GM, Candelora K (1981) Use of THA in treatment of Alzheimer-like dementia: pilot study in twelve patients. Biol Psychiatry 16(2):145–153

Semantic Distance and the Electrophysiological Priming Effect

T. F. MÜNTE, H. KÜNKEL, and H. J. HEINZE

1 Introduction

Since the first reports on event-related potentials (ERPs) of the human brain and their putative relations to psychological processes in the mid-1960s much has been learned about ERPs and behavior. A rapidly growing literature is concerned with ERP changes and language processes. An area especially accesible with the ERP approach is the facilitation of semantic processing of stimuli by contextual cues, so-called semantic priming. An adequate description of this phenomenon was given by one of the founders of experimental psychology, William James, as early as 1890: "Each word is doubly awakened; once from without by the lips of the talker, but already from within by the premonitory processes irradiating from the previous words ..."

A vigorous investigation of semantic priming, using a simple "one-word context," followed a report by Meyer and Schvanefeldt (1971), who found shorter reaction times to words which were preceded by semantically related items. One of the paradigms used in this kind of research is the lexical decision experiment, where the subject has to indicate as quickly as possible whether a given stimulus is a "word" or a "nonword." With this lexical decision task it was found that significant semantic priming occurs between antonyms (Becker 1980), super- and subordinate categories (Becker 1980), items of similar physical appearance (Flores d'Arcais et al. 1985; Schreuder et al. 1984), words of the same semantic category (e.g., Schreuder et al. 1984), etc.

In the ERP domain most of the experiments conducted so far have used the sentence-reading paradigm: Kutas and Hillyard (1980 a, b, c 1983, 1984), in a series of such experiments, demonstrated a negative component – labeled the N400 – which varied as a function of the semantic relatedness of the terminal word of a seven-word sentence and the preceding six words. This effect, first seen for visually presented written words, could also be demonstrated for the auditory modality (McCallum et al. 1984) and for American sign language (Kutas et al. 1987), thus qualifying as an endogenous potential. Polich (1985) found an identical ERP pattern to semantically incongruous words terminating a short list of words of one semantic category (e.g., flowers).

Using lists composed of words and nonwords in conjunction with a lexical decision task similar to those of the behavioral literature, Bentin and co-workers (1985) obtained an ERP difference between "prime" and "target" stimuli beginning approximately 200 ms after stimulus onset, the ERPs to the primes being

Springer Series in Brain Dynamics 2
Edited by Erol Başar and T. H. Bullock

considerably more negative in the 200–600 ms range. In their cautious interpretation of the findings, Bentin et al. (1985) noted that the electrophysiological priming effect reached its maximum near the peak of a negative going deflection at 400 ms, which in their data was present for all four stimulus categories (targets, primes, fillers, nonwords). While the prime–target difference waves of their experiment revealed a strong similarity of morphology, latency, and scalp distribution of this electrophysiological priming effect to the N400 of Kutas and Hillyard, Bentin et al. (1985) found that there was not enough evidence to interpret their data in terms of an N400 effect.

The goal of the present experiment was to explore further the relationship between ERPs and semantic priming by varying the task of Bentin et al. (1985). It was reasoned that – if indeed the electrophysiological priming effect is related or identical to N400 – an increase of the cloze probability (CP, see Kutas and Hillyard 1984) of the target stimuli should result in a disappearance of the negative deflection peaking at 400 ms for the targets and in an increase in the prime–target difference. To cover different levels of strength of semantic relationship between primes and targets, three conditions were used: a condition employing the immediate superordinates (= category names) as primes and members of the category as targets was designed to give a rather low CP for the targets. A higher CP was assumed in a condition were the subordinates (= members of category) served as primes and the superordinates as targets. Finally, a very high CP was expected in a condition were antonym pairs served as primes and targets. Thus, the largest difference between primes and targets was expected for this last condition.

2 Method

2.1 Subjects

Twenty healthy subjects (12 males, 8 females) were recruited from the student population and hospital personnel. Vision was normal or corrected to normal. All except two were right-handed as assessed by the Edinburgh Handedness Inventory (Oldfield 1971). The age range was 20–30 years (median 24.8). Subjects were naive with respect to the hypotheses of the experiment.

2.2 Stimuli

Common German words with word length ranging from 4 to 11 letters (mean 6.4) were selected by the experimenters to generate word lists. For each condition six stimulus lists were constructed. For the first condition 96 pairs of superordinate/subordinate nouns were chosen, where the subordinate represented a common member of the category denoted by the superordinate (e.g., flower–rose). This condition will henceforth be called Super/Sub. The second priming condition used the same 96 word pairs but with reversed order. This part of the study will be named Sub/Super. For the third condition 78 pairs of adjectives with an antonymic relation (e.g., hot–could) were constructed (Antonym). For each condition six stimulus lists were generated, consisting of 96 (Super/Sub and Sub/Super) or 78 stimuli (Antonym) each. Half to the stimuli were words; the other half were made up of pronounceable nonwords generated by changing one or two letters of the words. The words were further divided into "primes," i.e., the first words of the word pairs, "targets," i.e., the second word of a pair, and fillers, that is words used to create an unpredictable list, which were selected from the same word pool as the primes and targets. Stimulus lists were presented one word at a time on a videomonitor under the control of a microcomputer. Stimulus duration was 200 ms. The presentation rate was fixed at one word every 2 s. Words were presented in white against black background at the center of the screen and subtended a vertical visual angle of 0.4°.

2.3 Design

Subjects were divided into two groups of ten. The first group received the Super/Sub condition and was subsequently tested in the Antonym condition. The second group was tested in the Sub/Super and Antonym condition. This was done because pilot studies had indicated a substantial carry-over effect between the two categorical conditions.

Subjects were seated in a comfortable chair with a touch-sensitive button beneath each index finger. The instructions were to press the right button as fast as possible whenever a word stimulus appeared on the screen, and the left button whenever a nonword appeared. For half of the subjects the side of the buttons were reversed. In addition, subjects were reminded to sit still during a run and to minimize eye movements. One experimental run lasted 3.2 min (Antonym 2.6 min). Between blocks subjects were given a 3-min rest period. After a practice run subjects first received six lists in the Super/Sub (or Sub/Super) condition followed by six runs in the Antonym condition (all subjects).

2.4 ERP Recording

Ag/AgCl electrodes were affixed to Fz, Cz, Pz, C3, and C4 sites of the 10/20 system after carefully cleansing and abrading the skin. The scalp channels were referenced to the left mastoid. In addition two electrodes were placed at the lateral outer canthus and the inferior orbital ridge of the left eye to monitor the EOG. The EEG signals were amplified by a Van Gogh 50000 polygraph with time constants of 10 s and low pass filter settings at 35 Hz. A/D conversion was done online at a rate of 167 cps. After rejection of trials contaminated by artifacts the ERPs were averaged separately for each stimulus type and electrode. Only trials with correct answers were included. Averaging epochs began 200 ms before stimulus onset and were 1400 ms in duration. In addition to the averages a number of difference waves were computed to allow a better visualization of the differential processing of the different word types.

Table 1. Parameters

Name	Time window (ms)	Baseline (ms)	
Averages:			
1. N100/PK	50– 120	−200–0	Peak
2. N400/AR	200– 600	−200–0	Area
3. N400/PK	300– 500	−200–0	Peak
4. N400/LT	300– 500	−200–0	Latency
5. AR600-1000	600–1000	−200–0	Area
Difference waves:			
6. D-N400/AR	200– 600	−200–0	Area
7. D-N400/PK	300– 500	−200–0	Peak
8. D-N400/LT	300– 500	−200–0	Latency
9. D-N400/QPL	300– 500	−200–0	Quarter-peak latency

2.5 Quantification and Statistical Analysis

Several peak amplitude, peak latency, and area measures were determined for the averages and the difference waves by a computer program. All of the parameters were calculated using a 200-ms prestimulus baseline as reference. To obtain a measure for the onset of the negative difference component the "quarter-peak latency" procedure of Hansen and Hillyard (1980) was used. The measures are summarized in Table 1. The labels of the measures (e.g., N400/PK) used in this paper were chosen for ease of communication. It was not intended to equate these measures with known ERP components.

For statistical evaluation of the present data only nonparametric tests were applied, since these procedures make fewer assumptions about the normality of the data. The Friedman test, if significant, was supplemented by the Wilcoxon–Wilcox test. The Wilcoxon test was used to compare the different conditions. In these tests one of the categorical prime conditions (Super/Sub, Sub/Super) was compared with the Antonym condition (using the 10 subjects serving in both conditions).

3 Results

3.1 Performance

The mean reaction times (RTs) with standard deviations and error rates for all three conditions are shown in Table 2. The statistical results, summarized in Table 3, indicated that words in general were answered faster than nonwords. Furthermore, targets were answered significantly faster than fillers and in two instances the prime–target difference reached significance. The RT to primes and

Table 2. Mean RTs and error rates

Stimulus type	Super/Sub				Sub/Super				Antonym			
	T	P	F	N	T	P	F	N	T	P	F	N
RT (ms)	556	641	653	741	652	713	741	858	560	672	719	774
SD (ms)	81	63	70	90	72	62	59	95	108	87	96	112
Error %	3.2	4.0	7.1	6.0	2.6	3.2	6.1	5.0	3.0	6.1	8.8	5.1

T, target; *P*, prime; *F*, filler; *N*, nonword

Table 3. Statistical analysis of RTs

Condition	Friedman	Wilcoxon–Wilcox					
		T/P	T/F	T/N	P/F	P/N	F/N
Super/Sub	0.01	0.1	0.05	0.05		0.05	0.1
Sub/Super	0.01		0.05	0.05		0.05	
Antonym	0.01	0.1	0.05	0.05		0.05	

fillers did not differ significantly. Error rates were lowest for targets (not significant).

3.2 Event-Related Potentials

To illustrate the basic ERP effects of this experiment the grand average over 20 subjects for the Antonym condition is depicted in Fig. 1. After an initial negative deflection with a peak latency of approximately 80 ms, a sharp positive drop is seen for all four stimulus types. Whereas for the target ERPs a broad positivity is seen with a centroparietal maximum and a peak at about 530 ms, a negative going deflection can be observed for the primes and fillers in the 200–600 ms range, peaking at about 400 ms.

For the nonwords this negative deflection is seen with a larger amplitude but with a similar scalp distribution. Note that for the targets – except for the lateral sites and Pz – no negative deflection in the 200–600 ms range is seen.

In Fig. 2 grand average potentials of the three conditions are shown. The difference between targets and primes is smallest for the Super/Sub condition (upper panel) and largest for the Antonym condition. Furthermore, for the Super/Sub and less so for the Sub/Super conditions the target ERPs displayed a negative deflection comparable to the primes and fillers, though with an attenuated amplitude.

The statistical tests on the measures obtained on the waveforms revealed no significant differences between word types and conditions for the N100/PK measure.

The results for the N400/AR area measure are summarized in Table 4 for the midline electrodes. The lateral sites were comparable. While the Friedman test

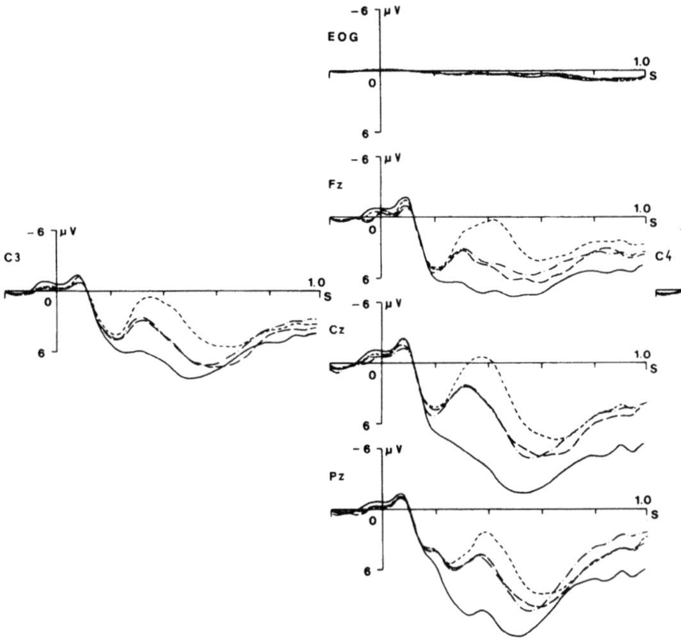

Fig. 1. Scalp distribution of grand average potentials ($n=20$) recorded in the Antonym condition. Stimulus presentation at $t=0$. For discussion of waveforms see text

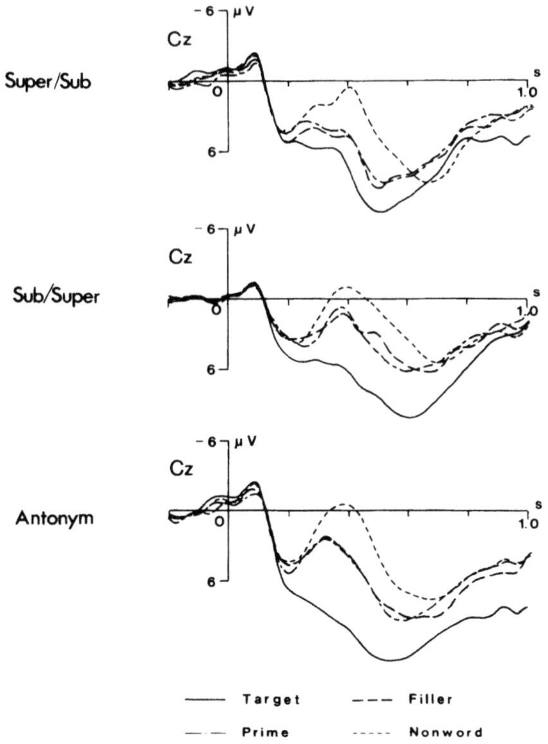

Fig. 2. Comparison of grand average potentials recorded in the three experimental conditions. Only recordings from the Cz electrode are presented

Table 4. Statistical analysis of N400/AR measure

	Friedman	Wilcoxon–Wilcox					
		T/P	T/F	T/N	P/F	P/N	F/N
Super/Sub							
Fz	0.01			0.05		0.05	
Cz	0.01			0.05			
Pz	0.01			0.05			
Sub/Super							
Fz	0.01		0.05	0.05		0.05	
Cz	0.01			0.05		0.05	
Pz	0.01			0.05		0.05	
Antonym							
Fz	0.01	0.05	0.05	0.05			
Cz	0.01	0.05	0.05	0.05		0.05	
Pz	0.01		0.05	0.05		0.05	

Table 5. Statistical analysis of N400/PK measure

	Friedman	Wilcoxon–Wilcox					
		T/P	T/F	T/N	P/F	P/N	F/N
Super/Sub							
Fz	0.01	0.05		0.05			
Cz	0.05			0.05			
Pz	0.01		0.05	0.05			
Sub/Super							
Fz	0.01			0.05		0.05	
Cz	0.01			0.05		0.05	
Pz	0.01			0.05		0.05	0.05
Antonym							
Fz	0.01	0.05	0.05	0.05		0.05	0.05
Cz	0.01	0.05	0.05	0.05		0.05	0.05
Pz	0.01		0.05	0.05		0.05	0.05

was significant ($P<0.01$) for all electrodes, significant and consistent differences (Wilcoxon–Wilcox) between targets and primes (and fillers) were attained only in the Antonym condition.

In all three conditions significant differences between targets and nonwords could be shown (Table 4). Comparing the Super/Sub with Antonym condition this measure (Wilcoxon test) showed significant differences for the targets (all $P<0.01$ except Pz $P<0.05$) and for the primes (Fz, C3, C4 $P<0.05$). The comparison between Antonym and Sub/Super condition was significant only in two instances (fillers Cz, C4).

The N400/PK maximal amplitude measure exhibited a comparable pattern of results, shown in Table 5. The differences between conditions were similar to the N400/AR measure.

For the N400/LT latency measure no significant differences for word types or conditions were found.

The AR600–1000 area measure displayed consistent significant differences in the Friedman test only for the Antonym condition, primes and targets differing significantly from nonwords in most channels. Comparing Super/Sub and Antonym conditions revealed differences between targets (Fz, Cz $P<0.05$; others $P<0.01$), primes (all $P<0.05$ except Pz), and fillers (Fz, Cz: $P<0.05$). No differences were found between Sub/Super and Antonym conditions.

In addition a series of analyses was carried out to assess topographical differences, which did not show a consistent pattern. In particular no differences were found for the lateral electrodes.

3.3 Difference Waves

Three different sets of difference waves were computed to show differential processing of the word types: The prime − target difference waves (P − T) were of primary interest since they were assumed to represent the electrophysiological concomitants of semantic priming. To assess for differences in processing of words and nonwords, filler − nonword difference waves (F − N) were calculated. Finally, to show the equality of fillers and primes, prime − filler difference waves (P − F) were generated.

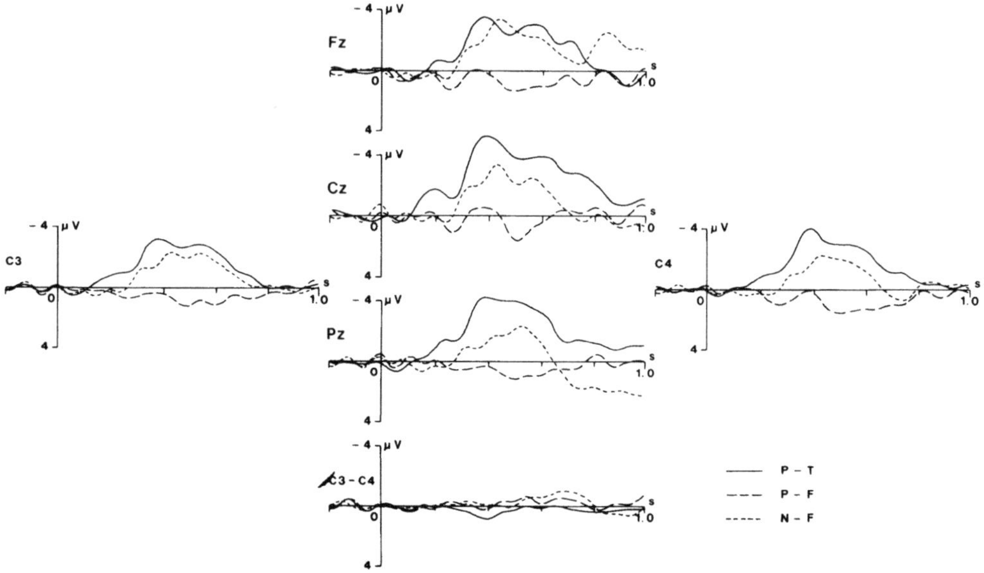

Fig. 3. Scalp distributions of the three classes of difference waves recorded in the Sub/Super condition. *P − T*, prime minus target; *N − F*, nonword minus filler; *P − F*, prime minus filler)

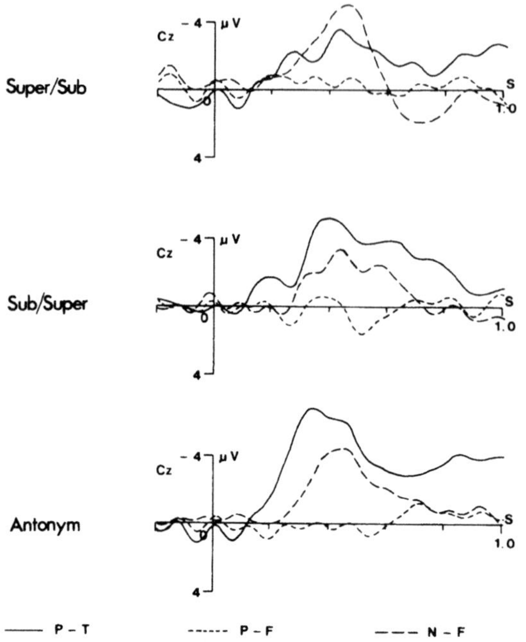

Fig. 4. Comparison of difference waves recorded in the three experimental conditions. Only recordings from the Cz electrode are presented

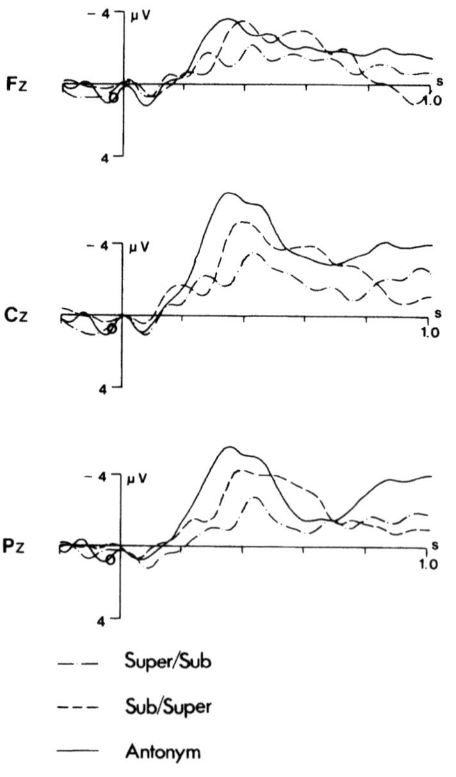

Fig. 5. Comparison of prime minus target difference waves from the three experimental conditions for the midline electrodes

Table 6. P–T difference waves (Cz, average of subjects)

Condition	Quarter-peak (ms)	Peak latency (ms)	Maximum (μV)
Super/Sub	350	420	−3.6
Sub/Super	270	380	−5.3
Antonym	200	320	−7.0

In Fig. 3 the difference potentials for the Sub/Super condition are represented, showing an onset of the negative difference in P − T of about 180–200 ms (grand average). The distribution of the negative difference component peaking at about 400 ms is widespread, with a slight preponderance at Cz.

In the F − T potentials a similar negative going difference with a somewhat later onset and a comparable scalp distribution was present. The P − F differences, as was to be expected, showed no observable effect. In Fig. 4 the difference waves of the three conditions (Cz) are depicted. It is clearly visible that the P − T difference is most pronounced for the Antonym condition with the least effect in the Super/Sub condition. The F − N waves were rather similar in morphology and amplitude for the different conditions.

In Fig. 5 the P − T differences for the different conditions and electrodes are shown, confirming a larger amplitude and earlier peak latency for the Antonym condition (see also Table 6).

Statistical evaluation of the difference waves was done with the Wilcoxon test comparing between the Super/Sub (or Sub/Super) and Antonym conditions. For P − T the D-N400/AR measure showed a significant difference ($P < 0.01$) between Super/Sub and Antonym for Cz, Pz, and C3. No differences were seen for Sub/Super and Antonym.

For the quarter-peak latency measure, differences were found between P − T waves of the Super/Sub condition and Antonym (C3) and the Sub/Super condition and Antonym (Cz). The D-N400/PK measure was comparable to D-N400/AR for P − T. For the Filler–nonword difference waves all tests were nonsignificant.

4 Discussion

The behavioral indices clearly show a semantic priming effect in all conditions of the present study, ranging from 61 ms (Sub/Super) to 112 ms (Antronym). Surprisingly a small, statistically nonsignificant difference was found between the reaction times of primes and fillers, the RTs to the fillers being somewhat slower. One explanation for this effect, also found by Bentin et al. (1985), would be a difference between prime and filler word pools. This speculation is corroborated by the error rates.

A related problem is that the design of the present study precluded targets and primes/fillers from coming from the same word pool (in the categorical conditions), which might also have caused some uncontrolled RT differences. However, since substantial priming was observed in both conditions, this effect, if present, seems to play a minor role. The ERPs recorded in the three conditions of the present study were remarkably similar in scalp distribution and nature of the prime target difference to those presented by Bentin and co-workers (1985), indicating a high stability of the observed effects across subject populations and different languages (English and German).

In all three conditions of the present experiment a negative broad deflection could be demonstrated in the P – T difference waves, which had a widespread scalp distribution similar to the N400 of Kutas and Hillyard. The F – T difference waves, presumably reflecting the differential processing of words and nonwords, were characterized by a similar negative going potential in the 200–600 ms range, suggesting that processing of the nonword stimuli is differentiated from word processing only quantitatively.

One of the current theories on semantic priming, the verification model (Becker 1976, 1979, 1980), envisions a word detector field receiving sensory (of the presumed target) as well as semantic (of the preceding stimuli) input. If a mismatch between sensory and semantic input occurs, in the next stage all items of the word field activated by the previous stimuli are checked. When no match between an item of the activated semantic field and the sensory input information is detected at this stage, a time-consuming verification of the sensory information is initiated.

While any interpretation of the electrophysiological concomitants of semantic priming has to be speculative at this point, it is intriguing to think about the magnitude of the ERP priming effect in terms of the extent to which a search operation in the semantic lexicon has to be carried out. Further support for this tentative interpretation is provided by the comparison of the ERPs to the target words of the three conditions (Fig. 2): while for the Antonym condition virtually no negative peak in the 200–600 ms range is visible, the targets of the categorical conditions, in particular the Super/Sub condition, were associated with a negativity in this latency range.

The primary difference between the three conditions of the present experiment is the different predictability (or CP) of the target stimuli. Where an almost complete determination of the target by the prime is seen in the Antonym condition, a large group of targets is primed in the Super/Sub condition. Thus, as a rather large group of possible targets has to be looked up, a larger negative peak would be expected in the latter condition.

In light of the foregoing considerations, how can the P – T difference waves (Fig. 5) be best understood? The negativity reflecting the differential processing of targets and primes is clearly largest for the Antonym condition and smallest for Super/Sub. Of course, both less negativity to the targets and increased negativity to the primes would result in a larger negativity in the P – T difference wave. As the primes themselves are presumably unprimed and equally unexpected across the conditions, it seems appropriate to consider the smaller negativity in the difference wave of the Super/Sub condition primarily as an effect of an increased negativity in the same latency range to the target potentials. Another interesting feature revealed by the difference waves, however, is the latency difference of the negative component. At the Cz site the peak latency between Antonym and Super/Sub differed by as much as 100 ms, the onset latency estimated by the quarter-peak latency method by 150 ms. In line with the previous considerations, this effect might be due to an earlier termination of the semantic look-up process in the targets of the Antonym condition.

Before accepting the results and their implications discussed above as a sound basis for further studies, a number of possible flaws in the design of the present study have to be considered. First, the different pattern of effects observed in the categorical and antonymial conditions might be due to an order effect, since the Antonym condition always followed either the Super/Sub or Sub/Super condition. While an order effect cannot be fully ruled out, the ERPs in the Antonym condition (preserved negativity for the primes, less negativity for the targets) render it unlikely. A second note of caution pertains to the fact that different groups of subjects participated in Super/Sub and Sub/Super conditions, thus impairing the comparability of these two conditions.

In summary, the present experiments favor an interpretation of the electrophysiological priming effect in terms of an N400 component. The lexical decision paradigm might thus constitute an alternative to the sentence reading paradigm in the study of structure and access of the human mental lexicon.

Studies eliminating the possible procedural shortcomings of the present experiment are currently underway in our laboratory.

Acknowledgments. Without the computational assistance of Michael Scholz and the laborious efforts of Knut Rahmel in data acquisition, this experiment would not have been possible.

References

Becker CA (1976) Allocation of attention during visual word recognition. J Exp Psychol [Hum Percept] 2:556–566

Becker CA (1979) Semantic context and word frequency effects in visual word recognition. J Exp Psychol [Hum Percept] 5:252–259

Becker CA (1980) Semantic context effects in visual word recognition: an analysis of semantic strategies. Mem Cognit 8:493–512

Bentin S, McCarthy G, Wood CC (1985) Event-related potentials, lexical decision and semantic priming. Electroencephalogr Clin Neurophysiol 60:343–355

Flores d'Arcais GB, Schreuder R, Glazenborg G (1985) Semantic activation during recognition of referential words. Psychol Res 47:39–49

Hansen JC, Hillyard SA (1980) Endogenous brain potentials associated with selective auditory attention. Electroencephalogr Clin Neurophysiol 49:277–290

James W (1980) The principles of psychology. Holt, New York

Kutas M, Hillyard SA (1980a) Reading between the lines: event-related brain potentials during natural sentence processing. Brain Lang 11:408–420

Kutas M, Hillyard SA (1980b) Reading senseless sentences: brain potentials reflect semantic incongruity. Science 207:203–205

Kutas M, Hillyard SA (1980c) Event-related brain potential to semantically inappropriate and surprisingly large words. Biol Psychol 11:99–115

Kutas M, Hillyard SA (1983) Event-related brain potentials to grammatical errors and semantic anomalies. Mem Cognit 11:539–550

Kutas M, Hillyard SA (1984) Brain potentials during reading reflect expectany and semantic association. Nature 307:161–163

Kutas M, Neville HJ, Holcomb PJ (1987) A preliminary comparison of the N400 response to semantic anomalies during reading, listening and signing. In: Ellingson RJ, Murray NMF, Halliday AM (eds) The London symposia. Elsevier, Amsterdam
McCallum WC, Farmer SF, Pocock PV (1984) The effects of physical and semantic incongruities on auditory event-related potentials. Electroencephalogr Clin Neurophysiol 59:477–488
Meyer DE, Schvanefeldt RW (1971) Facilitation in recognizing pairs of words: evidence of a dependence between retrieval operations. J Exp Psychol 90:224–234
Oldfield RC (1971) The assessment and analysis of handedness: the Edinburgh inventory. Neuropsychologia 9:97–113
Polich J (1985) Semantic categorization and event-related potentials. Brain Lang 26:304–321
Schreuder R, Flores d'Arcais GB, Glazenborg G (1984) Effects of perceptual and conceptual similarity in semantic priming. Psychol Res 45:339–354

The Search for Face-Specific Evoked Potentials

K. Bötzel, O.-J. Grüsser, B. Häussler, and A. Naumann

1 Introduction

The visually evoked potentials (EPs) recorded from the skull of man in the waking state depend on elementary properties of the stimulus, such as intensity, size, color, and retinal location, as well as on internal factors of the subjects, such as general and spatially directed attention, vigilance, expectation, etc. (for review see Regan 1972). The spatial frequency content of the visual stimulus, its geometric configuration, the contextual meaning, and the tasks subjects have to perform during stimulus presentation may also change the EPs (for reviews see John et al. 1967; Johnston and Chesney 1974; MacKay and Jeffreys 1973; Neville et al. 1982). These EP-influencing factors appear to interact in a more or less complex way.

It seems rather questionable whether the search for EPs "specific" to a selected pattern or class of patterns is meaningful; it may even be hopeless. Nevertheless, during the last few years we have analyzed EPs for face-responsive components. In recording EPs from different points of the skull, a new stimulus technique was applied by which the early, elementary components of the EPs were minimal. The motivation to commence this research came from results obtained by extensive neuropsychological studies on the ability to perceive and recognize faces, facial expressions, and gestures in normal subjects, brain lesioned patients, and schizophrenics (e.g., Berndl et al. 1986a–c; Grüsser and Kirchhoff 1988; Grüsser and Kremer 1988). Since the face is not only the source of expressive signals for nonverbal social communication, but also the most important and obvious characteristic structure of man's individuality, it does not seem totally hopeless to search for face-specific potentials. Moreover, some evidence exists that special structures of the brain control the execution and understanding of facial expression and gestures:

1. Circumscribed lesions on both sides of the basal occipitotemporal region in the cortical hemispheres of man lead to the symptoms of *prosopagnosia,* i.e., the inability to recognize previously known persons by the face alone (Bodamer 1948; Hécaen and Angelergues 1962; Hécaen 1981; Meadows 1974; Damasio et al. 1982; Grüsser 1984). Landis et al. (1986) and Bliestle et al. (1986) recently provided some evidence that extended unilateral lesions of this region in the right hemisphere might also lead to at least a transitory prosopagnosia.

2. Microelectrode recordings performed in awake monkeys indicate that face-specific neuronal networks exist in the primate brain located in cortical areas

Springer Series in Brain Dynamics 2
Edited by Erol Başar and T. H. Bullock

bordering on the central region of the superior temporal sulcus of rhesus macaques (Perrett et al. 1982, 1984; Rolls 1984; Desimone et al. 1984; Baylis et al. 1985; Rolls and Baylis 1986). The neuronal network responsible for face-specific responses in the temporal lobe most likely has connections with parts of the *nucleus amygdalae,* where neurons responding selectively to face stimuli have also been found (Rolls 1984; Leonard et al. 1985). Creutzfeldt (1987) described responses of neurons in the "rostral area 19" of rhesus macaques to threatening expressions of a human face but not to friendly or neutral human faces. More recently, face-specific neurons were also demonstrated in the temporal lobe of sheep (Kendrick and Baldwin 1987). These findings in animals suggested that face-specific neurons are located in different regions of the brain, including limbic structures.

In the following we will review the results from different series of experiments in which we tried to evaluate EPs obtained with different classes of face and non-face stimuli.

2 Methods

About 70 subjects, mainly unpaid medical students (19–32 years of age, both sexes) whose right-handedness had been confirmed by the Edinburgh Inventory (Oldfield 1971), served as subjects in the different series of experiments. All subjects had normal binocular vision and no signs of neurological, visual, or oculomotor impairment. Myopia, when present, was compensated by eye glasses. Except in the experiments with known or famous faces, the test persons were not informed until completion of the series that faces were the crucial stimuli.

2.1 Stimulation

The subject sat in a moderately comfortable chair and rested his/her head against the back. The experimental room was separated by a wall from the room containing the recording and stimulation equipment (Fig. 1). The subject was instructed to relax and fixate either a small red light spot (0.05 degrees in diameter, 2 m away at eye level on a vertical reflecting white screen) or the center of the continuously visible projection field (3.9 × 5.9 degrees visual angle). The stimuli consisted of a sequence of slides projected onto this field for 2.5–4 s. The average luminance of the stimulus pattern was between 5 and 15 $cd \cdot m^{-2}$ for the set of "positive stimuli" (black line drawings on a white background, black and white photos) and 0.5 $cd \cdot m^{-2}$ for the "negative stimuli" (white line drawings on a black background). The background luminance was about 0.01 $cd \cdot m^{-2}$. The slides were projected alternately from one of two carousal slide projectors and directed precisely onto the same field. A combined shutter, driven by a pneumatic system and controlled by means of electromagnetic valves, alternately opened and closed the beams of the two projectors (Fig. 1). The shutter was so constructed that the moment one beam was turned off, the other was turned on. Switching from one stimulus pattern to the next took less than 6 ms. Shutter movement and slide

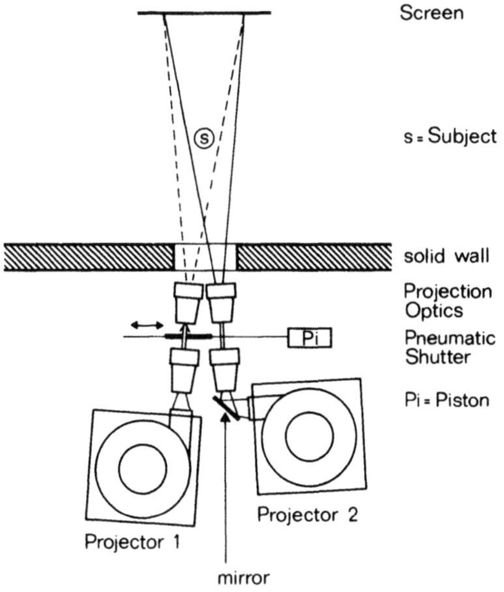

Fig. 1. Scheme of the stimulus set-up. Two carousel projectors were used to project the stimuli alternately, each for 2.5–4 s, onto a reflecting screen. By an automatically driven "pneumatic" shutter one projector was turned off and the other turned on simultaneously within less than 6 ms

change in the two carousel projectors were controlled by computer signals. By means of the signals generated by a small light source, two holes in the slide frame and two photocells, each slide could be classed by the computer into one of four stimulus categories.

2.2 EEG and EOG Recordings

The electroencephalogram (EEG) was recorded by means of Beckmann AG/AgCl electrodes attached to the skull of the subject with collodium. The skin was prepared beforehand with a cleaning paste. The electrode cups were filled with Beckmann EEG electrode paste. After a few minutes, the electrode impedance was measured; it never exceeded 2 kΩ. The reference electrode consisted of "linked earlobes" or "linked mastoids." We could record from six sites at the most simultaneously. In the different studies, responses were evaluated from the following EEG electrodes: Oz, Pz, Cz, Fz, T3, T5, and T7 (left) and T4, T6, and T8 (right) (international 10–20 system, Jasper 1958). The EEG signals were amplified by Grass P511a amplifiers (bandpass 0.1–100 Hz). At the same time vertical and horizontal eye movements were recorded by a DC electronystagmogram with amplifiers and optical insulation units operating at a bandpass frequency of 0–100 Hz. The EEG and electro-oculogram (EOG) signals were digitized (1 : 2048 for the amplitude) with a sampling rate of 400 Hz per channel, then stored on magnetic discs and evaluated off-line with an HP1000 computer.

2.3 Data Evaluation

A program written by G. Frey (Berlin) and extended by one of us (K.B.) allowed backward and forward averaging of the EEG signals from the moment of slide change. Sweeps containing large eye movements or blinks were rejected from the averaging. As a rule the EEG responses were averaged separately according to the stimulus categories for a period beginning 100 ms before the stimulus change and continuing over 800 ms. In addition to the average values, the time course for the standard deviations was measured. Further data processing revealed subtraction signals between electrodes or between responses to different stimulus patterns and finally the *grand averages* of responses obtained in a larger group of subjects. From individual or grand averages, autocorrelation and cross-correlation functions were computed. Negative and positive peaks were determined from the average responses of each subject and the algebraic mean of these values was computed for the group of subjects participating in a given task. These values were compared with the latency and amplitude of the negative and positive peaks found in the grand averages of the respective responses.

3 Results

3.1 Responses to Highly Schematic Faces

In a first study *(paradigm 1)* a highly schematic face was constructed consisting of a circle, two dots, a hook (nose), and a small curve (mouth, Fig. 2). The four structures inside the circle were changed to achieve a continuous degradation from a face to a nonface. Thirty-one male and female medical students were asked to put the set of faces in a row of continuous transition from the "best" face to the most evident nonface. Figure 2 gives the results of this study. For the EP studies we selected three of these face schemes (nos. 1, 6, and 12, paradigm 1). Figure 3 exhibits grand averages obtained with six different recording electrodes from 14 subjects (seven females and seven males). Category-specific differences were only obtained at electrodes T5, Cz, and Oz, while the differences from other electrodes

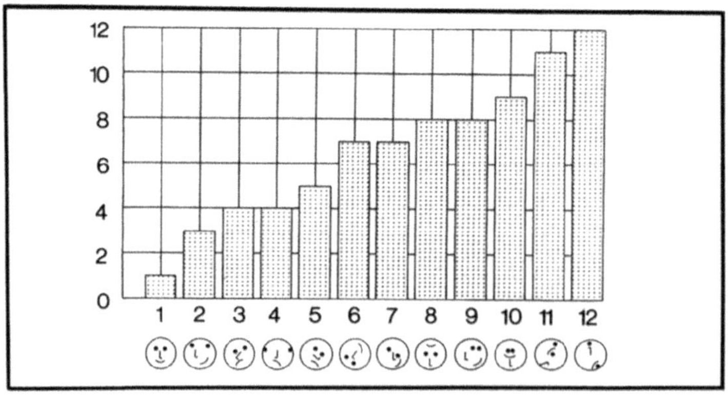

Fig. 2. A set of 12 highly schematic faces and nonfaces was constructed and the subjects had to rank these 12 items from the "best" face to the most distinct nonface. Average rank order obtained in 31 medical students

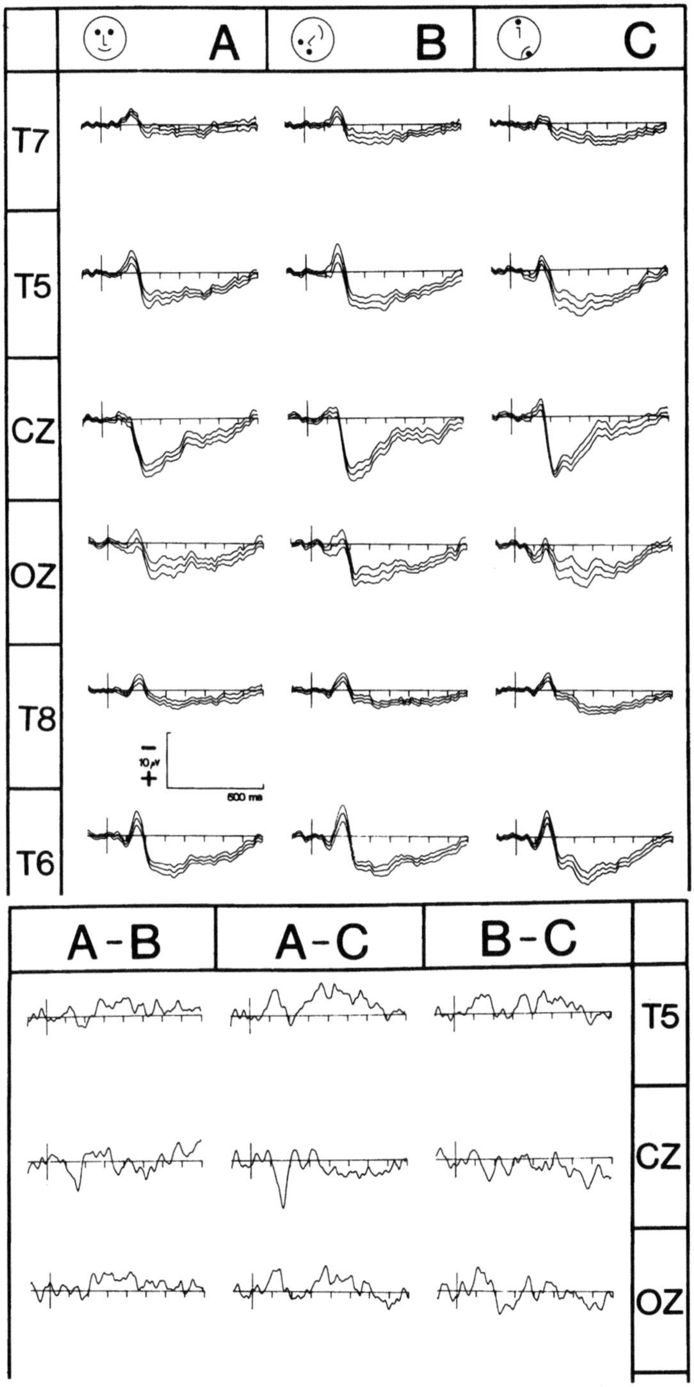

Fig. 3. EPs obtained with patterns 1, 6, and 12 of Fig. 2. EP grand averages (± standard error) of 14 subjects. Recording electrodes as indicated. In the *lower part of the figure*, difference curves according to stimulus categories are shown for three electrodes as indicated

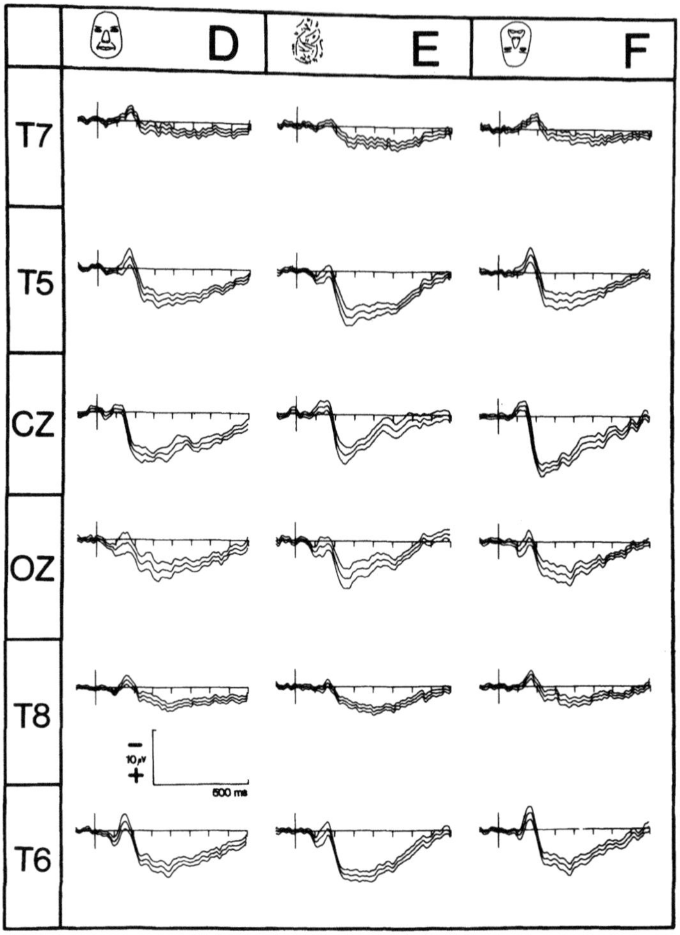

Fig. 4. Grand average of the responses to a schematic face, the same stimulus "scrambled" (increasing average spatial frequency content), and the stimulus turned upside-down. Grand averages of 14 subjects. Recording electrodes as indicated. The schematic faces were taken from Thornton and Pilowski (1982)

remained within the noise level. Some evidence was obtained that a slight difference existed between the average responses recorded from male and female subjects (Grüsser and Häussler, in preparation). When "bipolar recordings" (referred to Cz) were computed, the latency of the peak N1 (about 100 ms) was significantly shorter for the "good" face (1) than for the "poor" face (6) and the nonface (12) of Fig. 2. Over the temporal region, these differences were seen more clearly at electrodes T5 and T7 on the left side than at electrodes T6 and T8 on the right.

In *paradigm 2* with not so simple schematic faces (Fig. 4), the most significant category-specific differences (schematic face/scrambled stimulus or schematic face/upside-down face) were also obtained at the electrode Cz (Fig. 4a, b). As with the first stimulus category, the differences between the grand averages of EPs re-

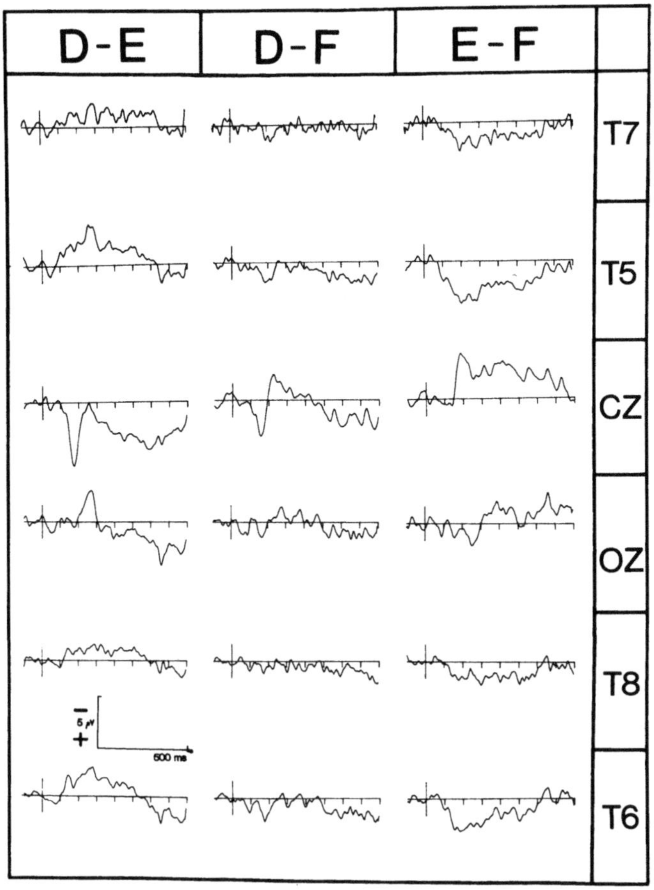

Fig. 5. Category-specific difference curves of the recordings shown in Fig. 4

corded from male and female subjects by the second type of schematic face stimulus were also most pronounced at the electrode Cz, in both the normal and the upside-down position (Fig. 5).

3.2 Responses to "Realistic" Drawings of a Face, a Tree, and a Chair

The stimulus sequence of *paradigm 3* consisted of 160 slides of three different simple drawings (54 × face, 53 × tree, and 53 × chair, Fig. 6), presented in a semi-random order. Test subjects were five female medical students. The slide series was shown twice to each subject. In the first series, black drawings on a white background were used (P-slides) and in the second, negatives of the same drawings (N-slides). Figure 7 exhibits grand averages of the responses obtained at electrode Cz. In contrast to the EPs aroused by the highly schematic faces applied in paradigms 1 and 2, a more "structured" EP was obtained. Similarly, category-de-

Fig. 6. Stimuli used in paradigm 3 (P stimuli). The same drawings white on black were the set of N stimuli

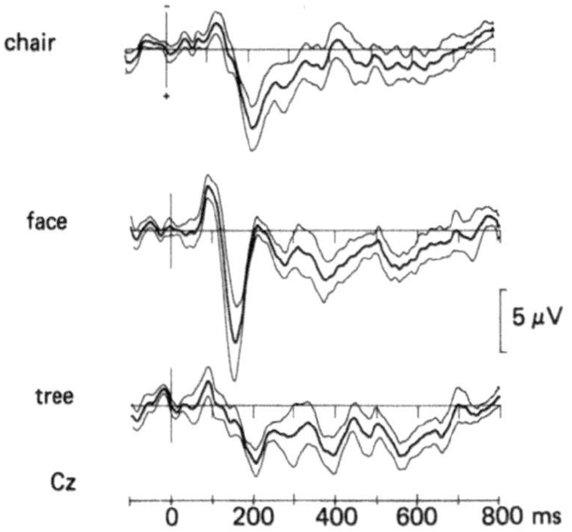

Fig. 7. Grand average (± standard error) of EPs recorded through electrode Cz in five female subjects. Stimuli were black line drawings of a chair, a face, and a tree, as shown in Fig. 6; binocular stimulation (Bötzel and Grüsser 1988)

pendent differences from the EPs found at electrode Cz were also recorded from electrode Pz, while the stimulus category-induced differences were small at the temporal electrodes. The differences between EPs obtained with N and with P stimuli were minimal. Figure 8 displays mean latency and mean amplitude of the peaks relative to a prestimulus base line measured for all three stimulus categories in the average of the curves of the individual subjects and in the grand average curves of all subjects. It is evident that a close correlation exists between these two types of data, but the category-dependent differences were remarkable. This is shown once more by the category-related difference curves for the grand average EPs obtained at electrode Cz (Fig. 9). A detailed analysis of autocorrelation and cross-correlation functions of the grand averages or the category-dependent difference curves emphasized that with these simple drawings from three categories, the responses to faces "popped out." For further detail, the reader is referred to Bötzel and Grüsser (1988).

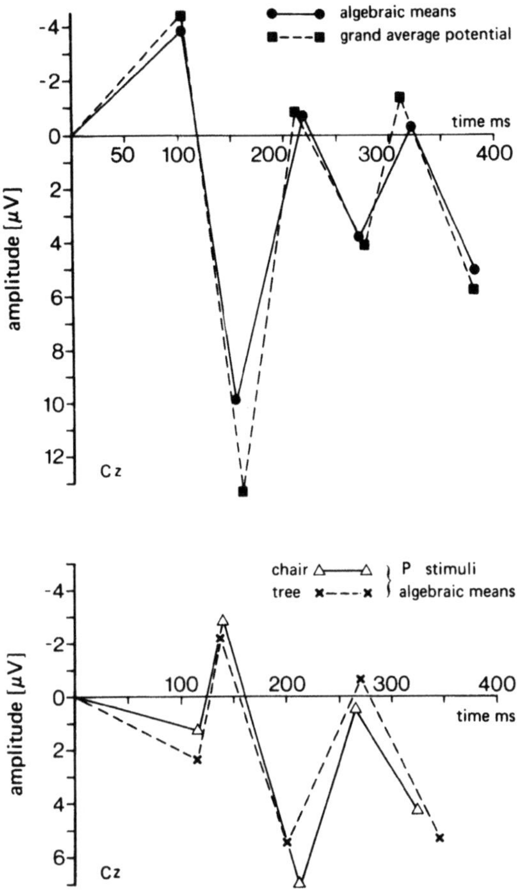

Fig. 8. Amplitudes of negative and positive EP peaks (*ordinate*) are plotted as a function of the respective peak latencies (*abscissa*). Data from five female subjects; stimulus pattern as in Fig. 6. *Above:* Mean values of the responses to face stimuli (*dots*) and peak values of the corresponding grand average curve (*squares*); electrode Cz. Note the similarity in both diagrams. *Below:* Mean values of the evoked responses (electrode Cz) from different subjects to chair and tree stimuli are similar to each other, but deviate considerably from the corresponding diagram obtained from the responses to face stimuli, as shown above. (Bötzel and Grüsser 1988)

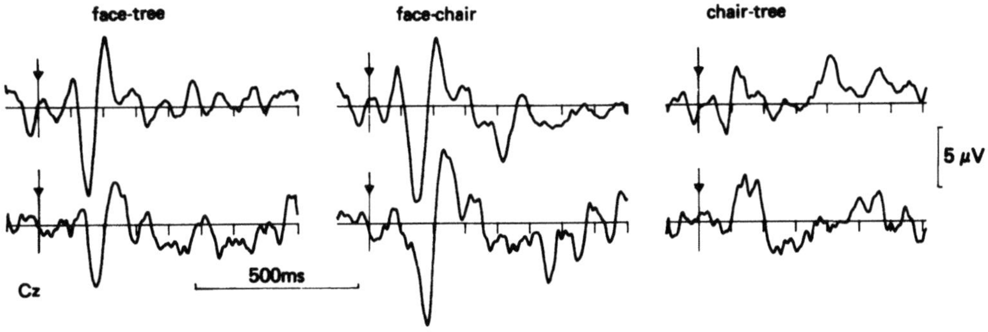

Fig. 9. Subtraction curves of the EP grand averages shown in Fig. 7. Note the considerably larger difference when face stimuli were involved for the responses obtained between 100 and 350 ms after stimulus change (*arrow*). Data obtained with P stimuli are shown in the *upper row*, those obtained with N stimuli (white line drawings on black background) in the *lower row*

3.3 Responses Evoked by Black and White Photographs of Faces and Nonfaces

In *paradigm 4,* the stimuli consisted of 54 black and white photos of different *human faces* of both sexes (frontal view) unknown to the subjects, 53 photographs of different ornamented *vases,* and 53 photographs of different *pairs of worn shoes* photographed from above. Shoes and vases were chosen as comparison stimuli, because they share with faces the features of partial axial symmetry, complexity, and individuality. Most of the vases were of the art nouveau style. The expressions on the faces ranged from neutral to mildly friendly. To raise the attention of the subjects they were told to watch the slides carefully in order to be able to recognize them later. Figures 10 and 11 exhibit the grand averages obtained, in which the prominent face-specific responses were again found at electrode Cz.

An early negative peak, occurring about 140–160 ms after the change in stimulus, exhibited the shortest latency and highest amplitude at Cz when the

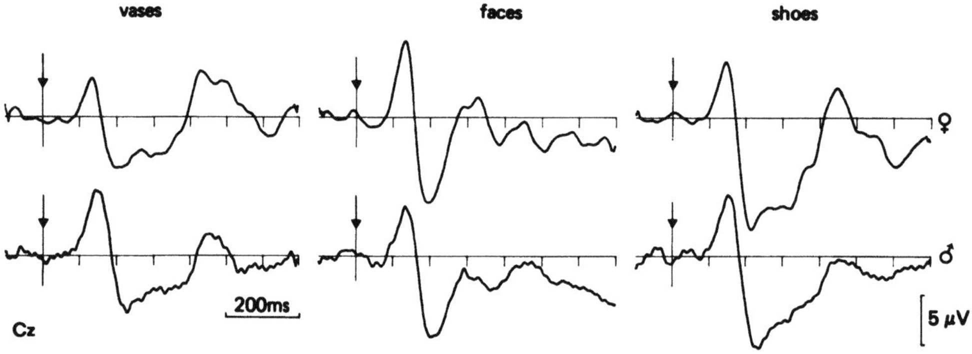

Fig. 10. Grand averages of evoked responses recorded at electrodes Cz in five female (*upper row*) and six male (*lower row*) subjects. Stimuli were different black and white photographs from slide to slide of vases (*left*), faces (*middle*), and pairs of worn shoes (*right*)

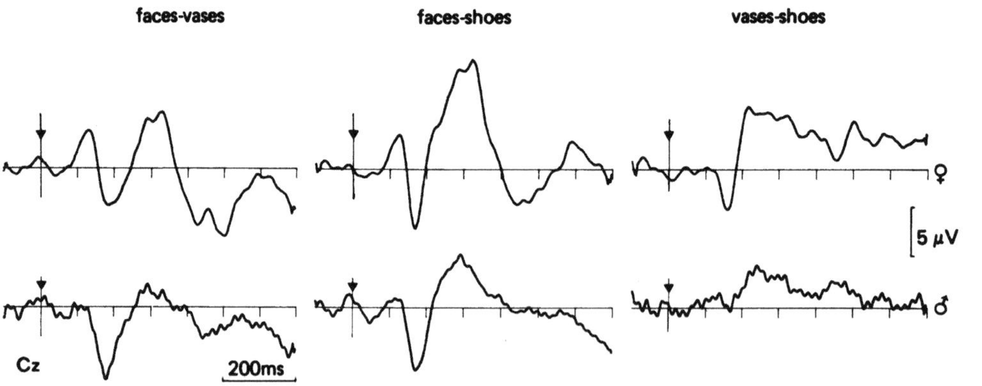

Fig. 11. Subtraction curves of the grand averages of evoked responses shown in Fig. 10. (Bötzel and Grüsser 1988). Separate data from female and male subjects

stimulus was a face (average values for women/men: face 132/140, vase 145/153, shoes 146/154 ms, Fig. 10). The first large *positive peak* was identified at 210–240 ms. The latency of this peak measured in the recordings from the temporal electrodes T5 and T6 was about 12 ms longer than that from electrode Cz. At all three electrodes significantly shorter latencies were found for the first positive peak in the curves obtained with face stimuli. The latency of this positive peak was about the same for men and women (214 ms for faces, 230 for vases, and 234 for shoes).

In most subjects a negative peak appeared at about 300 ms with face stimuli, while with the nonface stimuli this peak was absent or small. It was most pronounced at the electrodes T5 and Cz and was higher in women than in men. Neither the latency nor the amplitude of two later peaks (N460 and P600) was dependent on stimulus category or subject gender. The difference curves of the EP grand averages, as shown in Fig. 11, clearly demonstrate the face-responsive EP components.

These findings were confirmed in general in another set of experiments in which a fourth category of stimuli (flowers) was introduced and the subject had to perform a recognition task in addition whenever a flower stimulus appeared (for details see Bötzel and Grüsser 1988).

3.4 Responses to Known and Unknown Faces

In 15 subjects (seven female and eight male medical students) EEG responses were obtained during a face recognition task *(paradigm 5)*. The right-handed subjects had to press a key with the right hand whenever the face projected was known to them. The subjects inspected 11 photographs (slides) twice (6 and 4 s) about 10 min before the recording experiments started. They had to judge the faces during the first inspection series as "friendly," "unfriendly," or "neutral." In the second inspection series they had to signal that they recognized the faces previously evaluated as "neutral" by pressing a bar. In the set of black and white photographs projected during the recording session about half of the faces were known, i.e., the known faces appeared several times. The unknown faces, however, were not repeated in the course of the recording experiment. Figure 12 shows grand averages of the responses obtained at electrodes T3, T4 and Cz for the two face categories. It is evident that under these stimulus conditions a left–right difference was present for electrodes T3–T4 but the most prominent responses were again recorded at electrode Cz. With known faces the maintained electrical negativity began about 400 ms after the stimulus change and was of a longer duration than in the response to unknown faces (Grüsser and Naumann 1988).

3.5 Responses Evoked by Famous Faces as Compared to Unknown Faces

In the same group of 15 subjects, the EPs to 45 "famous" faces intermingled with unknown faces (all slides of black and white photographs) were measured *(paradigm 6)*. The subjects were asked to press a key during the recording session when

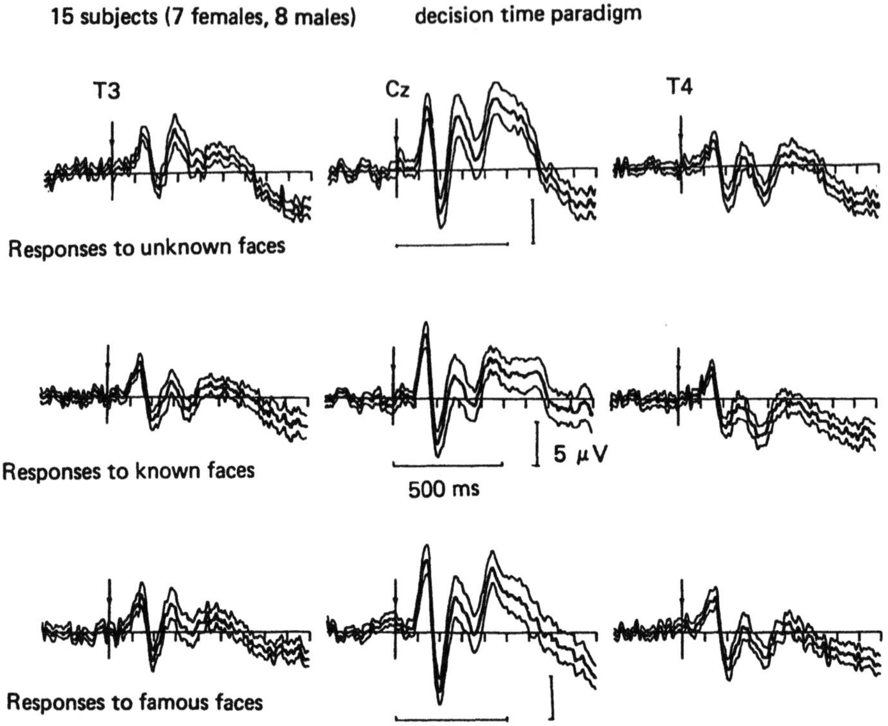

Fig. 12. Grand averages (±standard error) of EPs. Stimuli were known, unknown, and famous faces (15 subjects). For further explanation see text

a famous face known to them appeared. The average response time for this task was 730 ms and the correct-response average was about 75%. Less than 6% "false-positive" responses were made, i.e., the subjects pressed the bar when unknown faces were projected. The grand averages of the EPs are shown in Fig. 12. Here again, the most prominent responses were obtained at electrode Cz, but the differences between the EEG responses evoked by the two categories of faces were minimal and within the noise level at all electrodes. EPs were included in the average only when the subject pressed the bar on the appearance of a famous face and did not press the bar when an unknown face was presented.

3.6 Responses to Known and Unknown Flowers

To investigate the importance of known versus unknown complex visual stimuli, we repeated paradigm 5, as described in Sect. 3.5, in 12 of 15 subjects (five females, seven males), but instead of faces, black and white photographs of *flowers* were used. Figure 13 demonstrates grand averages obtained for the EPs at electrodes T3, T4, and Cz. The positive waves occurring between 200 and 400 ms were significantly more pronounced and lasted longer with known stimuli. Comparing the recordings obtained in this and the preceding paradigm, however, one recog-

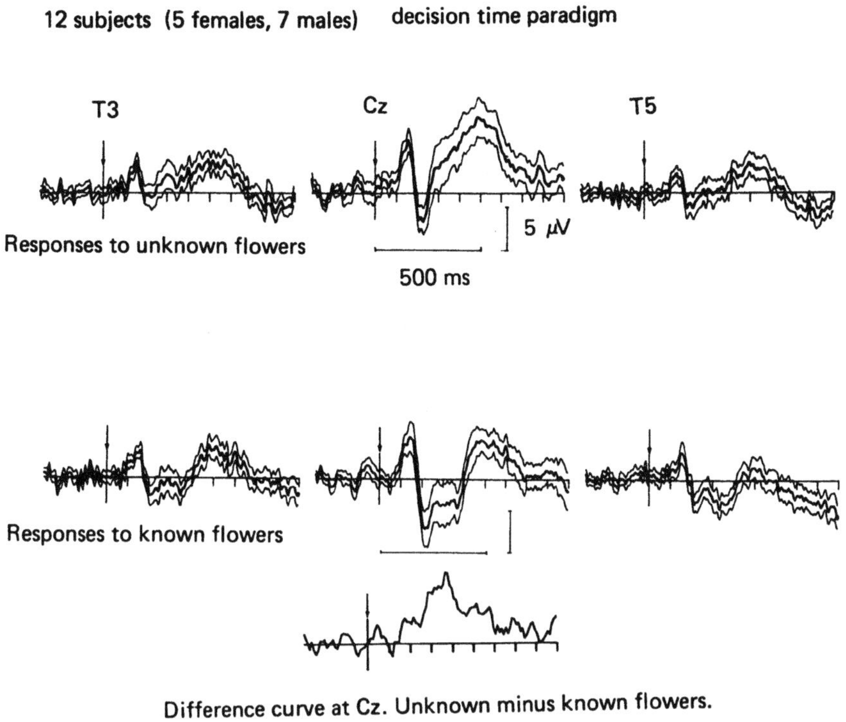

Fig. 13. Grand averages (±standard error) of EPs. Stimuli were known and unknown flowers (slides of black and white photographs). For further explanation see text

nizes that especially the face-responsive EP components recorded at electrode Cz differed from those evoked by flowers, despite the formal structural complexity of the stimuli being about the same for the two categories. The most pronounced amplitude-difference was the negative peak occurring at about 280 ms. A comparison of Figs. 12 and 13 supports the idea of face-responsive or even face-specific components in the EPs.

4 Conclusions

The perception of a face by a human observer presumably evokes a complex set of processes in his brain which differ in part from those occurring when he perceives other complex visual objects. In the EP recordings described in the present study, the elementary visual components of the evoked responses were small, since luminance and average spatial frequency content of the stimulus patterns varied little from stimulus to stimulus. One can therefore assume that the main components evoked by the stimulus change are caused by *structural differences* between the stimuli or by a higher order precognitive or cognitive process. The data presented indicate that in comparison to other complex visual stimuli (tree, chair, shoes, vases, flowers), the EEG responses evoked by face stimuli contain components indicating that face-specific responses may occur in the human brain. The most prominent face-responsive components were recorded, however, not through the temporal or temporo-occipital electrodes but with the midline electrode Cz. This finding could mean that the face-responsive EP components are *unspecific* and are caused, for example, by an increased degree of interest or attention, leading to EPs differing from those obtained with other, less fascinating stimulus categories. It is well known that meaningful and interesting stimuli evoke other EPs than stimuli which do not engage the attention. Only fairly extensive studies with a large set of stimulus categories could settle this question. The data described in the Sects. 3.4 and 3.5 indicate that it is not the recognition or the memory process which leads to the EP components considered to be face-responsive. Furthermore, our hypothesis that the EP components appearing are perhaps face-specific is supported by the increase in EP complexity when the responses to highly schematic faces, a drawn "realistic" face, and face photographs are compared. The most prominent face-responsive EPs were not evoked in the temporal or temporo-occipital region, as one would expect on the basis of clinical findings on prosopagnosia, but consistently at the midline electrode Cz. Difference recordings from electrodes Cz/Pz and Cz/Fz revealed that it is indeed the region around Cz which yields the most prominent EEG responses. Two explanations come to mind:

1. The face-responsive components recorded through Cz do not originate in the temporo-occipital cortical "face region" but in fairly symmetrically organized deep temporal structures. There the cross electrical dipoles may change during the evoked responses in such a way that a *passive* summation occurs most strongly at electrode Cz.

2. A more physiological explanation would be the assumption that responses occurring in deep limbic structures and/or responses in the gyrus cinguli were evoked more strongly by face than by nonface stimuli. Since the gyrus cinguli receives inputs from the nucleus amygdalae, which, according to the data published by Rolls (1984) and Baylis et al. (1985), contains a considerable number of face-specific neurons, this physiological projection hypothesis seems feasible.

Assuming that face-responsive components do indeed exist in the EPs described in the preceding sections, another conclusion, in our opinion, is inevitable: face-responsive neuronal mechanisms most certainly exist in the human brain outside of the "prosopagnosia region" located in the basal surface of the temporo-occipital lobe. This hypothesis is corroborated by findings obtained in preceding neuropsychological studies: face recognition is impaired in patients with brain lesions in all parts of the neocortex outside of the occipital lobe and without any signs of prosopagnosia, i.e., without involvement of the primary visual cortex or the basal occipitotemporal border zone (Grüsser and Kirchhoff 1988).

Several laboratories have investigated EPs of the human brain when photographs (or slides) of faces were presented. Small (1983) compared EEG potentials evoked by pictures of known and unknown faces and by geometric figures. She reported a higher amplitude of the P300 wave recorded over the right hemisphere with face stimuli than with nonface stimuli. As in the present study, she found no differences in the EPs when known and unknown faces were used as stimuli. Neville et al. (1982), however, reported a higher P300 amplitude in the EEG responses evoked by pictures of persons, paintings, or places known to the subjects as opposed to unknown items. Srebro (1985a, b) tried to localize face-specific electrical brain activity in the temporal lobe of man by means of a vector-analysis technique, computing the spatial "Laplacian" vector response. He also evaluated the differences between the EPs evoked by stimulus patterns of faces and nonfaces (noise-masked patterns). In this study only slight differences were found between EPs aroused by face and by nonface stimulation. More recently, however, Jeffreys and Musselwhite (1987) reported on face-responsive visual EPs and described a midline positive response maximum at the electrodes Cz and Pz, peaking at 140–160 ms latency. Inversion or rotation through 90° increased this latency by 20–25 ms. These findings correspond well to our responses obtained with schematic face drawings. The findings of Sobotka et al. (1984), who described some hemispheric differences in EPs to face stimuli, are difficult to compare with the present data, since these authors used Cz as the reference electrode.

The essential neural processes for face recognition occur within the first 400 ms after stimulus onset. In recent studies performed in our laboratory on the psychomotor response time in normal young adults, it was found to be in the range of 400 ms for the decision face/nonface. When the subjects had to decide whether the face had been seen or not before the experiments, the decision time increased by 120–150 ms. The cerebral process of simple "structural" face recognition is evidently accomplished within the average time occurring between normal inspection saccades. Human subjects perform such saccades at somewhat irregular intervals every 300–600 ms. Should a more specific response analysis be performed in the future in the continuing search for face-responsive EPs, technically more complex equipment than in the present study would be required, e.g., simultaneous recordings through many more electrodes and the computation of the spatiotemporal distribution of EPs across the human skull. Even under these conditions, one could restrict the data analysis to the first half-second after the stimulus change. This limit would make our goal realistic, namely to test experimentally our hypothesis that some of the face-responsive components described in the present study are indeed expressions of the operation of a face-specific neuronal network in the human brain.

5 Summary

In six different experimental series EEG responses evoked by changes in pictorial patterns were recorded in about 70 adult human subjects. Quantitative data evaluation was performed for the responses from electrodes T4–T8, Cz, Pz, and Oz. The stimuli were projected to a 4×6 degree binocularly viewed field on a vertical reflecting tangent screen. The patterns changed within 6 ms every 2.5–4.5 s.

1. Average evoked responses (EPs) obtained with a highly schematic face, a "half-face," and a "nonface" stimulus were compared. Slight category-dependent differences in the EPs at electrodes Cz and T5 were found. Similar data were obtained with another less schematic face, whereby the responses to the schematic

face were compared to those evoked by the upside-down schematic face and by a "scrambled" stimulus.

2. Simple realistic drawings of a face, a tree, and a chair were used as stimuli, either black on white or white on black (altogether 160 slides in a semirandom order), in the third series of experiments. At electrodes Cz and Pz, EPs characteristic for faces were found, with three prominent peaks that were absent in the EPs aroused by the chair or tree stimuli. Category-difference curves (face–chair, face–tree, chair–tree) and cross-correlation functions between these difference curves supported the hypothesis of the existence of face-responsive components in the EPs.

3. In another series of experiments black and white photographs (slides) of different faces, ornamented vases, and worn pairs of shoes were used as stimuli. "Face-responsive" EP components were more pronounced in female than in male subjects and were most prominent at electrode Cz. When a *recognition task* was included, essentially the same EPs were recorded, but a late positive potential appeared in addition (450–600 ms) in the responses to all stimulus categories.

4. The face-responsive components in the EPs, most prominent at electrode Cz, were seen in responses both to *known* and *unknown* faces. Compared to each other, these two categories gave slightly different EPs. No significant differences were seen between the EPs to unknown and to famous faces.

5. When the same paradigm was applied for known and unknown *flowers,* a clear difference was present between the averaged EPs recorded at electrode Cz. The EPs recorded with flower stimuli, however, did not contain the components regarded as "face specific" and recorded in the preceding series of experiments with face stimuli.

It is concluded that the search for face-responsive EP components may not be totally futile. Such components, however, appear most prominently at the midline electrodes and not at electrodes above the temporo-occipital regions.

Acknowledgments. The work was supported in part by a grant of the Deutsche Forschungsgemeinschaft (Gr 161). We thank Mr. J. Lerch, Dipl. Ing. H. Nitert, and Dipl. Ing J. Petsch for valuable assistance in the construction of the stimulation equipment, Dipl. Ing. L.-R. Weiß for assistance in computer programming, Mr. P. Holzner for photographic work, Mr. M. Winzer for the drawings, and Mrs. J. Dames for her valuable help in the English translation of the manuscript. A. Naumann was supported by a "Doktoranden-Stipendium" of the NAFÖG system. The senior author (O.-J. G.) was supported by an "Akademie-Stipendium" of the Volkswagen Foundation.

References

Baylis GC, Rolls ET, Leonard CM (1985) Selectivity between faces in the response of a population of neurons in the cortex in the superior temporal sulcus of the monkey. Brain Res 342:91–102

Berndl K, Dewitz W, Grüsser O-J, Kiefer RH (1986a) A test movie to study elementary abilities in perception and recognition of mimic and gestural expression. Eur Arch Psychiatry Neurol Sci 235:276–281

Berndl K, von Cranach M, Grüsser O-J (1986b) Impairment of perception and recognition of faces, mimic expression and gestures in schizophrenic patients. Eur Arch Psychiatry Neurol Sci 235:282–291

Berndl K, Grüsser O-J, Martin M, Remschmidt H (1986c) Comparative studies on recognition of faces, mimic and gestures in adolescent and middle-aged schizophrenic patients. Eur Arch Psychiatry Neurol Sci 236:123–130
Bliestle A, Regard M, Landis T (1986) Prosopagnosia: an autopsied case. J Clin Exp Neuropsychol 8:127
Bodamer J (1948) Die Prosop-Agnosie. Arch Psychiatr Nervenkr 179:6–53
Bötzel K, Grüsser O-J (1988) Electrical brain potentials evoked by pictures of faces and nonfaces. A search for "face-specific" EEG-potentials. Exp Brain Res (in press)
Creutzfeldt O (1987) Gibt es eine Mechanik des Denkens? Aus Forschung und Medizin 2(1):7–19
Damasio AR, Damasio H, van Hoesen GW (1982) Prosopagnosia; anatomical basis and neurobehavioral mechanisms. Neurology (NY) 32:331–341
Desimone R, Albright TD, Gross C, Bruce C (1984) Stimulus-selective properties of inferior temporal neurons in the macaque. J Neurosci 4:2051–2062
Grüsser O-J (1984) Face recognition within the reach of neurobiology and beyond it. Hum Neurobiol 3:183–190
Grüsser O-J, Kirchhoff N (1988) Face recognition in brain-lesioned patients. Eur Arch Psychiatry Neurol, (in preparation)
Grüsser O-J, Kremer H (1988) Delayed face recognition in schizophrenic patients. Eur Arch Psychiatry Neurol, (in preparation)
Grüsser O-J, Naumann A (1988) Brain mechanisms for recognition of faces, facial expression and gestures: Neurophysiological and electrophysiological studies in normals, brain-lesioned patients and schizophrenics. Proceedings of the ARNMD-Meeting, New York 1987, (in press)
Hecaen H (1981) The neuropsychology of face recognition. In: Davies G, Ellis H, Shepherd J (eds) Perceiving and remembering faces. Academic, London, pp 39–54
Hecaen H, Angelergues R (1962) Agnosia for faces. Arch Neurol 7:92–100
Jasper HH (1958) The ten-twenty electrode system of the international federation. Electroencephalogr Clin Neurophysiol 10:371–375
Jeffreys DA, Musselwhite MJ (1987) A face-responsive visual evoked potential in man. J Physiol (Lond) 390:26P
John ER, Herrington RN, Sutton S (1967) Effects of visual form on the evoked response. Science 155:1439–1442
Johnston VS, Chesney GL (1974) Electrophysiological correlates of meaning. Science 186:944–946
Kendrick KM, Baldwin BA (1987) Cells in temporal cortex of conscious sheep can respond preferentially to the sight of faces. Science 236:448–450
Landis T, Lehmann D, Mita T, Skrandies W (1984) Evoked potential correlates of figure and ground. Int J Psychophysiol 1:345–348
Landis T, Cummings JL, Christen L, Bogen JE, Imhof HG (1986) Are unilateral right posterior cerebral lesions sufficient to cause prosopagnosia? Clinical and radiological findings in six additional patients. Cortex 22:243–252
Leonard CM, Rolls ET, Wilson FAW, Baylis GC (1985) Neurons in the amygdala of the monkey with responses selective for faces. Behav Brain Res 15:159–176
MacKay DM, Jeffreys DA (1973) Visually evoked potentials and visual perception in man. In: Jung R (ed) Handbook of sensory physiology, vol 7/3. Springer, Berlin Heidelberg New York, pp 647–678
Meadows JC (1974) The anatomical basis of prosopagnosia. J Neurol Neurosurg Psychiatry 37:485–501
Neville H, Snyder E, Woodas D, Galambos R (1982) Recognition and surprise alter the human visual evoked response. Proc Natl Acad Sci USA 79:2121–2123
Oldfield RC (1971) The assessment and analysis of handedness: The Edinburgh Inventory. Neuropsychologia 9:97–113
Perret DI, Rolls ET, Caan W (1982) Visual neurons responsive to faces in the monkey temporal cortex. Exp Brain Res 47:329–342
Perret DI, Smith PAJ, Potter DD, Mistlin AJ, Head AS, Milner AD, Jeeves MA (1984) Neurons responsive to faces in the temporal cortex: studies of functional organization, sensitivity to identity and relation to perception. Human Neurobiol 3:197–208

Regan D (1972) Evoked potentials in psychology, sensory physiology and clinical medicine. Chapman and Hall, London

Rolls ET (1984) Neurons in the cortex of the temporal lobe and in the amygdala of the monkey with responses selective for faces. Human Neurobiol 3:209–222

Rolls ET, Baylis GC (1986) Size and contrast have only small effects on the responses to faces of neurons in the cortex of the superior temporal sulcus of the monkey. Exp Brain Res 65:38–48

Schulman-Galambos C, Galambos R (1978) Cortical responses from adults and infants to complex visual stimuli. Electroencephalogr Clin Neurophysiol 45:425–435

Small M (1983) Asymmetrically evoked potentials in response to face stimuli. Cortex 19:441–450

Sobotka S, Pizlo Z, Budohoska W (1984) Hemispheric differences in evoked potentials to pictures of faces in the left and right visual field. Electroencephalogr Clin Neurophysiol 58:441–453

Srebro R (1985a) Localization of visually evoked cortical activity in humans. J Physiol 360:233–246

Srebro R (1985b) Localization of cortical activity associated with visual recognition in humans. J Physiol (Lond) 360:247–259

Thornton M, Pilowski I (1982) Facial expression can be modelled mathematically. Br J Psychiatry 140:61–63

Area-Specific Regulation of Slow Cortical Potentials

B. Rockstroh

1 Introduction

A negative dc potential shift typically develops prior to an anticipated event or performance. These slow potentials (SPs, a typical example of which is the contingent negative variation, CNV) have been related to behavior within the frameworks of attention (e.g., Tecce 1972), preparation (e.g., Gaillard 1978; Loveless 1979), and controlled processing (e.g., Cooper et al. 1979). However, it seems still to be topic of discussion to what extent SPs represent nonspecific states of activation/arousal or preparation, or whether they rather indicate preparatory processes in a *specific* manner, i.e., within distinct brain regions for distinct behavioral responses. In an attempt to integrate the various findings and interpretations reported in the literature we have conceptualized negative SPs as an indicator of "cerebral potentiality" that facilitates anticipated cerebral performance (Rockstroh et al. 1982). With regard to specificity of potentiality supply it has been assumed that potentiality or processing resources are allocated to specific cortical regions or neuronal networks depending on actual as well as anticipated processing requirements (Elbert and Rockstroh 1987). From this model it follows that SPs should be generated and regulated in a *specific* manner, i.e., specific with respect to temporal aspects of SP generation, in that "the timing of the E-wave is related to when in time the process it reflects is needed" (Ruchkin et al. 1986), as well as area specific, in that resources are allocated to those cortical neuronal networks that will be involved in the anticipated cerebral performance.

The present chapter discusses this hypothesis of area-specific SP regulation related to lateralized performance as investigated by means of different experimental approaches, the two-stimulus paradigm, biofeedback-induced SP shifts, and application of transcephalic dc currents. Lateralized effects, i.e., asymmetry of scalp-recorded SPs as well as asymmetry in performance measures related to the hemisphere-specific processing, would argue against a nonspecific functional significance of SPs.

2 Methods and Results

2.1 Two-Stimulus Paradigm

In the two-stimulus paradigm hemispheric asymmetry in SPs has been demonstrated under conditions of lateralized responding: Hemispheric asymmetry of

Springer Series in Brain Dynamics 2
Edited by Erol Başar and T. H. Bullock

the CNV with more pronounced negativity over the central location contralateral to the responding hand was demonstrated under reaction time conditions (Gaillard 1978). Similarly, CNV asymmetries were reported for sensorimotor tasks (e.g., by Syndulko and Lindsley 1977), but also in anticipation of cognitive tasks such as arithmetics, pattern matching (Butler and Glass 1974; Birbaumer et al. 1981), reading, writing, and listening to music (Altenmüller 1986; Grözinger et al. 1980; for a review see also Rugg 1983). A straightforward evaluation of lateralized processing and SP generation should include somatosensory processes which take place predominantly in the lateral hemisphere of the brain. We therefore performed an experiment in which subjects anticipated sensorimotor stimulation of and performance with the two hands. The primary cortical somatosensory and motor areas for each hand are known to be located in the contralateral hemisphere next to the central sulcus. They should be involved in preparation for and processing of lateralized responses with the hands, which should show up in an asymmetry of SP patterns. In order to maximize lateralized preparation and processing the tasks were arithmetic (left hemispheric) or spatial (right hemispheric) in nature.

Twenty-one right-handed subjects participated in the experiment. Subjects sat in a reclining chair and placed the index and middle fingers of each hand on a pad from which pins could be protruded for tactile stimulation. The protrusion of all four pins constituted the warning stimulus (S1). The pins protruded for 1 s at the hand that was to be stimulated again (S2) after 5 s (i.e., the entire anticipatory interval lasted for 6 s). The task-related S2 lasted for 2 s. The right-hand task was a decision according to the number of pins (either 1–2 or 3–4). The direction in

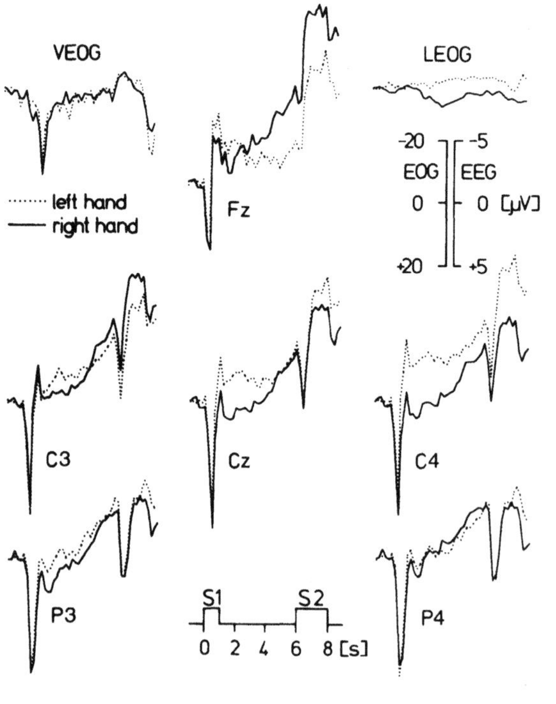

Fig. 1. Monopolar SPs and eye movements (vertical, *VEOG*, and lateral, *LEOG*, electrooculogram) during 1 s prestimulus baseline, 6-s anticipation interval, and 2-s task presentation, averaged across subjects. (Lutzenberger et al. 1985)

which a lever was moved with the right thumb indicated the solution. The left-hand task was to compare patterns of pins between the two fingers and to decide whether they were equal or different; in this case a response with the left thumb was required. Thirty-two trials with right-hand stimulation/task and 32 trials with left-hand stimulation/task were presented in pseudorandom order. If a wrong response or no response within a 4-s interval following task onset was recorded, an error was counted. The intertrial interval varied between 10 and 25 s. The EEG was recorded from both central and parietal leads (C3, C4, P3, P4) as well as along the midline (Fz, Cz, for details see also Lutzenberger et al. 1985).

A task-dependent asymmetry showed up for components of the event-related potential (N1, P2, P3; late positive complex, LPC) as well as for the CNV (recorded bipolarly from C3 and C4; see Fig. 1). For all components the amplitudes were larger over the hemisphere contralateral to the stimulated hand, the asymmetry being most pronounced at the recording site at which the particular component exhibited the predominant amplitude (e.g., at the central location for N1, P2, and CNV, at the parietal location for P3). (This was confirmed by significant interactions of Task × Laterality × Electrode Location with $F(1,20)=66.4$ for N1a, $F=23.4$ for N1b, $F=7.1$ for P2, $F=10.0$ for P3, $F=92.9$ for slow wave (SW) and $F=23.0$ for LPC, all $P<0.05$. For the bipolar SPs the interaction Task × Electrode Location provided $F(1.20)=38.1$, $P<0.01$). Lateral eye movements were small and showed no significant effects that might have pointed to an influence on the reported ERP asymmetries.

No overall asymmetry was found for response latency (RL) and error rates. Subjects reponded faster with the right than with the left hand and less errors were made when tasks were presented to the right as compared to the left hand.

Three results of this study seem to be of interest: Firstly, preparatory negative SPs develop in two-stimulus paradigms with entirely somatosensory stimulation. Secondly, a tactile stimulation associated with different (hemisphere-specific) types of task for the left and the right hand induces hemisphere-specific preparation and processing, as indicated by EP components and SPs. This supports the hypothesis that the preparation of a hemisphere-specific sensorimotor task is lateralized in a similar way as the input processing itself. Finally, it seems interesting to note that negativity was maintained and even increased *during* task presentation and performance. The relatively long task performance interval might have allowed additional negativity to be generated after S2 onset, negativity that might represent ongoing allocation of processing resources as well as anticipation of feedback (Kakigi et al. 1978). As will be outlined later, lasting negativity might also suggest that thresholds for neuronal excitation should be set low in distinct neuronal networks as long as information processing or performance-related processes are required.

2.2 Biofeedback-Induced Shifts in Slow Potentials

A complementary approach to the functional significance of area-specific SP generation includes the modification of SPs by means of biofeedback and instrumental learning. It is well established that human subjects can learn to increase or decrease surface-negative SPs upon the presentation of discriminative stimuli if adequate feedback is provided (Elbert et al. 1980; Ueda et al. 1985; Trimmel 1986; for summary of results see also Birbaumer et al. 1981; Rockstroh et al. 1982). The next step was to investigate whether SP regulation within this paradigm can be achieved in an area-specific manner and to evaluate effects of such area-specific self-induced SPs on behavioral responses in tasks that are assumed to involve the respective cortical areas. Proceeding along the line of experimental approach started with the first study described above, the precentral areas (C3–C4) were chosen for the training of SP regulation in two studies. Behavioral effects of the self-induced SPs were tested by means of the performance in lateralized sensorimotor tasks as described above, because the underlying somatosensory projection areas of fingers and hands project radially to the precentral surface.

In the first study, continuous visual feedback of SPs was provided by the outline of a small rocket ship that moved from bottom to top on a TV screen in front of the subject, taking 6 s to do so. Two "goals" (bars, one blue, the other yellow) were marked at the upper frame of the screen and subjects were asked to direct the rocket into one of the goals depending upon which of two discriminative stimuli (S^D: a yellow or a blue floodlight) was presented during the 6-s feedback interval. Two types of trial could occur in pseudorandom order. In *L-trials* the required goal was reached whenever a shift toward increased left – but decreased right – central negativity was produced. In *R-trials* the reverse asymmetry would direct the rocket into the goal indicated by the S^D.[1] The deviation of the rocket from its vertical axis at any time point *t* during its flight was a linear function of the intergrated bipolar EEG from the onset of the S^D to the actual time *t*, referred to a prestimulus baseline of 4 s. Various control procedures were implemented to prevent artifactual influences of vertical and lateral eye movements, electrode displacements, or head movements on the required SP shifts (for further experimental details see Rockstroh et al. 1989). The same lateralized sensorimotor tasks that were described before for the two-stimulus paradigm were used to assess the properties of feedback-derived asymmetry. These tasks were performed at the end of "transfer" trials on which the S^Ds indicating the required hemisphere-specific SP polarization were presented, but feedback was removed. Within each of two identical experimental sessions 80 feedback trials were followed by 64 transfer/task trials. In the second experiment, which used the same feedback algorithm as the first one[2], a forced choice task was introduced on transfer/task trials. Subjects held a button in each hand and were asked to press one of the buttons immediately with the termination of each 6 s S^D. Within one experimental session 120 feedback and 60 transfer trials were given.

In both experiments about 50% of the subjects (21 out of 45 subjects in experiment I, 20 out of 48 subjects in experiment II) produced reliable hemispheric asymmetries of more than 2 μV under feedback conditions. For experiment I learning was confirmed by a significant difference in SP asymmetry between L- and R-trials by analysis of variance [for the area between the SP curves during

[1] Assignment of goal, required, shift, and S^D was counterbalanced across subjects.

[2] The design differed with regard to the feedback mode: No continuous visual feedback was provided. Subjects were informed that they should induce different "brain states" and that they would receive feedback and reinforcement at the end of each S^D interval.

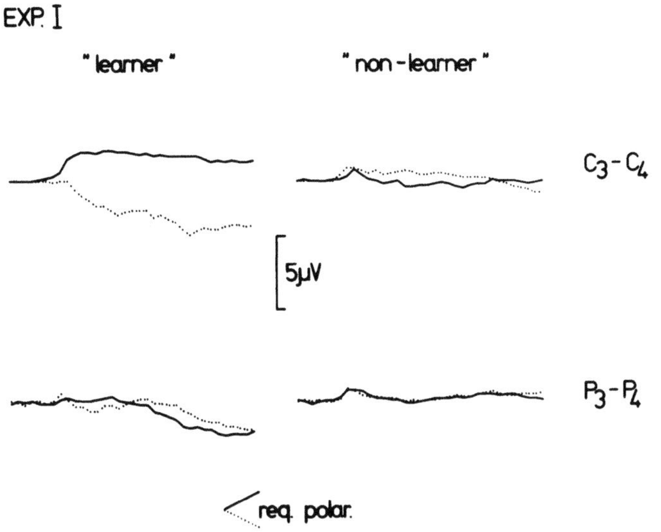

Fig. 2. Bipolar central and parietal SPs averaged separately for "learners," i.e., subjects achieving the criterion of reliable 2 μV differentiation between the hemispheres, and for "nonlearners" and the different asymmetry requirements (*solid lines*: averages across L-trials; *dashed lines*: averages for R-trials) under feedback conditions. (Rockstroh et al. 1989)

L- and R-trials $F(1,44)=4.6$, $P<0.05$]. (The poorer performance of subjects in experiment II was attributed to the lack of immediate and continuous reinforcement; Rockstroh et al. 1989.) While this SP control remained restricted to feedback trials in the first session of experiment I, significant differentiation was also achieved under transfer conditions without feedback in the second session. In both experiments the feedback-induced SP asymmetry was restricted to the central – the feedback – location and did not spread to the parietal locations (Fig. 2).

In experiment I those subjects who achieved the required SP polarization demonstrated more efficient responding in the lateralized sensorimotor tasks (i.e., smaller response latency and less percentage of errors[3] with the hand contralateral to the hemisphere in which they had induced increased negativity. In experiment II a forced choice of hand was required at the end of each S^D interval. Those subjects who reliably induced SP polarization between the hemispheres of more than 2 μV on the average in feedback trials and who did not make their choice of hand contingent upon a signal light in an arbitrary manner (thereby reducing variability to zero) showed a more pronounced right-hand preference on L-trials, in which they had induced more left–central negativity [$t(10)=2.5$, $P<0.05$]. This asymmetry did not show up on R-trials, in which subjects chose the right hand almost as often as the left hand, nor was it evident in subjects who failed to produce SP asymmetry (see Fig. 3).

[3] RL difference between L- and R-trials -46 ± 39 ms for the right and 69 ± 25 ms for the left hand; interaction 115 ms for the first, 101.6 ms for the second session ($P<0.05$). The comparable interaction in error rates was $9.5 \pm 3.2\%$ ($P<0.01$), again for "learners" in the first session.

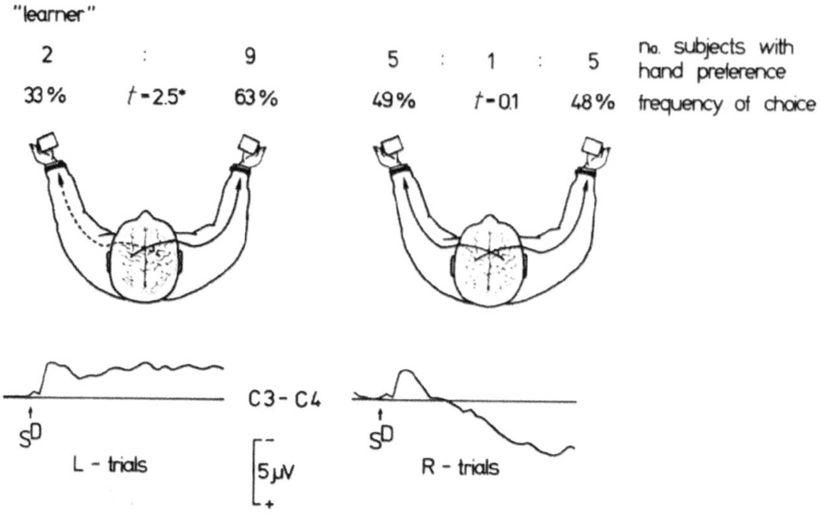

Fig. 3. Schematic drawing of the relationship between SP asymmetry and choice of hand. Of the 11 "learners," 9 preferred to respond with the right hand on 63% of the L-trials, but responded only on 33% of L-trials with the left hand. On R-trials no comparable asymmetry in the choice of hands became evident. If it is assumed that the left hemisphere controls responding primarily of the right hand, an increase in left-central negativity on L-trials should facilitate right-hand responding. If the right hemisphere is assumed to control responding with both hands, increased right-central negativity on R-trials should affect responses with both hands equally. (Rockstroh et al. 1989)

Two results of these experiments are noteworthy with regard to the hypothesis of area-specific SP regulation. Firstly, it was demonstrated that SPs can be modified in an area-specific manner using instrumental feedback procedures. The effects reported in the present article for one experimental session may seem insufficient (only about 50% of the subjects achieved small but reliable SP differentiation). However, these results are strengthened by the increasing learning success in the second session as well as by another study using the same feedback algorithm in which five experimental sessions were scheduled (Birbaumer et al. 1988). Significant hemisphere-specific SP differentiation was obtained after the fourth experimental session. We may conclude that continuous feedback and prolonged training may be important experimental variables that enable area-specific self-regulation of SPs.

Secondly, a relationship between area-specific SPs and performance in lateralized tasks was found only in subjects who achieved area-specific SP control. In experiment I subjects responded more efficiently with the hand contralateral to the hemisphere in which they had increased negativity. In experiment II hemisphere-specific SP modulation enhanced (on L-trials) or counteracted (on R-trials) the right-hand preference normally observed in right-handed subjects. Thus, the "bias" of right-handedness was reduced consequent upon the self-induced SP asymmetry – in particular during trials with increased negativity over the left hemisphere –, while it prevailed in "nonlearning" subjects.

2.3 Application of Transcephalic dc Currents

The third experimental approach to the investigation of area-specific SP generation and its behavioral significance involves the application of transcephalic weak electric currents. We have provided evidence that a current source favors the generation of SP negativity in the underlying cortical tissue, whereas a sink tends to inhibit a negative shift (Elbert et al. 1981; Lutzenberger and Elbert 1987). Shorter response latencies were obtained when positive transcephalic currents were applied to the vertex electrode (Elbert et al. 1981). The studies reported here were an attempt to elaborate whether currents applied laterally across the centroparietal cortex would differentially interfere with tasks that require lateralized processing. In two studies currents were applied via the electrodes placed over C3 and C4, which should result in a polarization of the centroparietal cortex. In the third study the currents were applied through the electrodes placed over F3 and F4, which should result in the lateralized polarization of the frontal cortex. Effects on the sensorimotor performance and the forced choice of handedness as described above were investigated.

In experiment I subjects ($n=28$) received sensorimotor stimulation and tasks as in the two-stimulus experiment described above. Subjects of one group were exposed to a 0.3 mA constant direct current applied between the central electrodes, with 64 trials with dc+ at C3 and dc− at C4 and 64 trials with the reverse polarization occurring in pseudorandom order (for details see also Jäger et al. 1987). Current was switched on together with the presentation of the tactile S1 and lasted until subjects had responded to the S2 onset. Subjects of the control group were submitted to the same procedure with the exception that currents were never switched on. In the second experiment with another 17 subjects a forced choice of hand as described above was required in response to tone pips that were presented in regular intervals of 2 s. Three trial series with 60 tones of either polarity (C3+/(C4− vs C3−/C4+) and 60 tones without current were presented. In the third experiment the same forced choice task was introduced but currents were applied between the frontal electrodes for intervals of 6 s each.

In these experiments the behavioral responses provided evidence for a differential effect of current-induced polarization. In the first experiment, the experimental group showed faster responses with the right hand when the current source was applied to C3+ as compared to the opposite current flow (mean difference in RL, 21.4 ms), and faster responses with the left hand when the current source was at C4 and the current sink was ipsilateral to the responding hand (mean RL difference 24.3 ms; for the interaction within the experimental group $P<0.05$). In the control group a right-hand predominance prevailed (subjects being faster with the right hand than with the left hand by 53 ms, $P<0.01$).

Similar results for RL were obtained in the second experiment, in which a forced choice of hand was required. As before, the left hand was faster when a C4+ polarity was met, while the right hand tended to be faster under C3+ conditions (for the interaction between dc polarity and RL for right and left hands, $P<0.05$).

It is interesting to note that the effect was largely due to a profit in response speed by anodal polarization over the primarily involved hemisphere compared

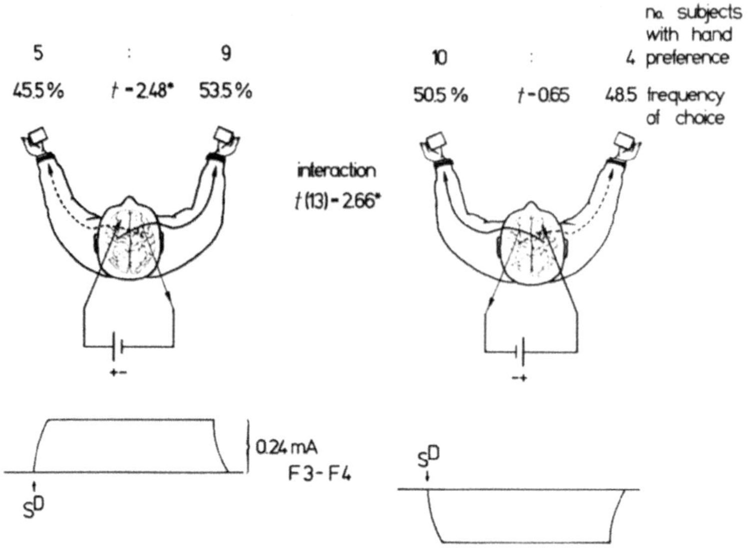

Fig. 4. A similar illustration to that in Fig. 3. Distribution of percentage of responses with the left or the right hand depending upon current application through the frontal electrodes. *Left graph*: current source over F3; *right graph*: current source over F4

to the no current condition. The disappearance of the effect of handedness in the experimental group was a surprising result. It is remarkable that the left hand particularly was speeded up under both polarity conditions compared to the control group. A possible explanation of this effect is that especially the left hand profits from current application. Another possible explanation is that activity with the left hand is more variable and dc effects are therefore more pronounced.

On the other hand, no effect of current application on the choice of which button to press was found in experiment II, in which the currents were applied through the central electrodes. In order to evaluate whether a forced choice task would be effected rather by polarization of *frontal* areas and whether a 2-s polarization might not be long enough to exert its effects, the currents were applied for 6-s intervals through the frontal electrodes (F3–F4) in the third experiment.

As illustrated in Fig. 4, the application of the current source to the left frontal cortex produced significantly more responses with the right than with the left hand, while a tendency toward a preference for the opposite hand occurred for a left-frontal current sink and a right-frontal current source.

3 Discussion

These results obtained from different experimental approaches clearly indicate a relationship between lateralized asymmetry in SPs and lateralized preparation and performance of responses with fingers or hands. These specific effects are not likely to result from changes in nonspecific arousal associated with, for example, feedback or discriminative stimuli. This supports the hypothesis that SPs may be generated and regulated in an area-specific manner in those cortical regions that are involved in the anticipated information processing and response performance. On the neurophysiological level, cerebral potentiality or negative SPs represent depolarization of apical dendrites in cortical pyramidal cells. This depolarization or allocation of "potentiality" is assumed to be regulated depending upon task demands and according to a "minimax principle": "the time during which reinforcement-oriented performance is required and during which simultaneous potentiality is available, is maximized, while the overall potentiality is reduced to a minimum" (Rockstroh et al. 1982, p 96; see also Ruchkin et al. 1986). This should also be true for area-specific SP generation.

We have more recently proposed (Elbert and Rockstroh 1987) that negative SPs might represent changes in thresholds of cellular excitation in localized cortical regions, i.e., neuronal networks. Based on biocybernetic ideas about cell assemblies and threshold regulation in neuronal networks (Braitenberg 1984; Palm 1982), it is assumed that neuronal networks with mutual excitatory connections, forming the substrate of information processing in the cerebral cortex, might activate each other beyond control. This possibility should be avoided by a threshold control of cortical excitability. The tuning of cortical excitability is realized via control of the depolarization in the apical dendrites, which gives rise to surface negative SPs. It is likely that information about the amount of ongoing activity is projected via the basal ganglia to the thalamus. This in turn projects back to the apical dendrites. The thalamocortical afferent system is controlled by the mediothalamic–frontocortical system (according to Skinner and Yingling 1977). Threshold control should enable the interruption of ongoing activity when relevant information is received, since thresholds would be set high, so that ongoing activity will drop to a low level instantaneously. Furthermore, thresholds can be set in advance, so that attention and preparation can be directed to anticipated information processing and response performance. Such a cortical tuning would improve processing of expected events. Support for such a model is provided by the relationship between negative SPs, e.g., in the two-stimulus paradigm, and efficient responding to the S2, but also by the relationship between feedback-induced negative SPs and facilitated responding (for a review of results see Rockstroh et al. 1982, 1984).

The results presented here may also be taken as first evidence supporting the hypothesis that the proposed threshold regulation loops function in a rather specific manner, i.e., for specific neuronal networks that are activated consequent upon task demands or stimulus conditions. This should show up in the scalp distribution of SPs, behavioral effects of area-specific SP self-regulation, and behavioral effects of cortical tissue polarization through transcephalic dc currents, as they were reported in the present article.

4 Summary

Slow potentials (SPs) of the brain that develop prior to an anticipated event or performance are considered to represent preparatory processes in the sense of cerebral resources, or, in neurophysiological terms, depolarization of apical dendrites. It is assumed that resources are allocated, hence, SPs are generated in those cortical neuronal networks that will be involved in the anticipated cerebral performance. This hypothesis of *area-specific* SP regulation was investigated in different experimental approaches:

1. The anticipation of lateralized tactile tasks presented to the fingers of each hand within a two-stimulus RT paradigm induced lateral asymmetry of event-related potential components (LPC, late CNV) recorded from the primary projection areas of fingers and hands (C3, C4).
2. A potential gradient between left- and right-central (C3–C4) SPs could be self-induced within an instrumental feedback paradigm; after acquiring SP self-regulation, subjects responded more efficiently in various tasks with the hand contralateral to the hemisphere in which they increased negativity "upon command."
3. When weak direct currents were applied externally through the electrodes placed over C3 and C4, the resulting lateralized polarization of centroparietal cortex led to facilitated motor responding if the current source was contralateral and the current sink ipsilateral to the responding hand.

These results, obtained from different experimental paradigms, support the hypotheses (a) that surface negative SP shifts facilitate performance, representing cerebral resources, and (b) that these resources are allocated in an area-specific manner.

Acknowledgments. Research was supported by the Deutsche Forschungsgemeinschaft (SFB 307). The reported studies were performed together with Drs. Thomas Elbert, Werner Lutzenberger, and Niels Birbaumer.

References

Altenmüller E (1986) Hirnelektrische Korrelate der cerebralen Musikverarbeitung beim Menschen. Eur Arch Psychiatry Neurol Sci 235:342–354

Birbaumer N, Lang P, Cook E, Elbert T, Lutzenberger W, Rockstroh B (1988) Imagery effects on slow cortical potentials. Int J Neurosci 39:101–116

Birbaumer N, Elbert T, Rockstroh B, Lutzenberger W (1988) Biofeedback produced hemispheric asymmetry of slow cortical potentials and its behavioral effects. Int J Psychophysiol (in press)

Braitenberg V (1984) Vehicles. MIT, Cambridge

Butler S, Glass A (1974) Asymmetries in the CNV over left and right hemisphere while subjects await numerical information. Biol Psychol 2:1–16

Cooper R, McCallum WC, Papakostopoulos D (1979) Bimodal slow potential theory of cerebral processing. In: Desmedt J (ed) Cognitive components in cerebral event-related potentials and selective attention. Karger, Basel, pp 182–186

Elbert T, Rockstroh B (1987) Threshold regulation – a key for the understanding of the combined dynamics of EEG and event-related potentials. J Psychophysiol 1/4:317–333

Elbert T, Rockstroh B, Lutzenberger W, Birbaumer N (1980) Biofeedback of slow cortical potentials. Electroenc Clin Neurophysiol 48:293–301

Elbert T, Lutzenberger W, Rockstroh B, Birbaumer N (1981) The influence of low-level transcortical DC-currents on response speed. Int J Neurosci 14:101–114

Gaillard AWK (1978) Slow brain potentials preceding task performance. Academic Pers BV, Amsterdam
Grözinger B, Kornhuber H, Kriebel J, Szirtes J, Westphal K (1980) The Bereitschaftspotential preceding the act of speaking. In: Kornhuber H, Deecke L (eds) Motivation, motor and sensory processes of the brain. Elsevier, Amsterdam, pp 798–804
Jäger D, Elbert T, Lutzenberger W, Birbaumer N (1987) The effects of externally applied transcephalic weak direct currents on lateralization in choice reaction tasks. J Psychophysiol 1:127–133
Kakigi S, Matsuda T, Hagino G (1978) Factors affecting the resolution of contingent negative variation (CNV) Jpn Psychol Res 20:187–190
Loveless N (1979) Event-related slow potentials of the brain as expression of orienting function. In: Kimmel HD, Van Olst EH, Orlebeke J (eds) The orienting reflex in humans. Erlbaum, Hillsdale, pp 77–100
Lutzenberger W, Elbert T (1987) Assessment of the effects of weak electric currents on brain and behavior. In: Schmidt A (ed) Safety assessment of clinical NMR techniques. Thieme, Stuttgart, pp 36–45
Lutzenberger W, Elbert T, Rockstroh B, Birbaumer N (1985) Asymmetry of brain potentials related to sensorimotor tasks. Int J Psychophysiol 2:281–291
Palm G (1982) Neural assemblies. An alternative approach to artificial intelligence. Springer, Berlin Heidelberg New York
Rockstroh B, Elbert T, Birbaumer N, Lutzenberger W (1982) Slow brain potentials and behavior. Urban and Schwarzenberg, Baltimore
Rockstroh B, Elbert T, Lutzenberger W, Birbaumer N (1984) Operant control of slow brain potentials: a tool in the investigation of the potential's meaning and its relation to attentional dysfunction. In: Elbert T, Rockstroh B, Lutzenberger W, Birbaumer N (eds) Self-regulation of the brain and behavior. Springer, Berlin Heidelberg New York, pp 227–242
Rockstroh B, Elbert T, Birbaumer N, Lutzenberger W (1988) Biofeedback produced hemispheric asymmetry of slow cortical potentials and its behavioral effects. Int J Psychophysiol (in press)
Ruchkin D, Sutton S, Mahaffey D, Glaser J (1986) Terminal CNV in the absence of motor response. Electroencephalogr Clin Neurophysiol 63:445–463
Rugg M (1983) The relationship between evoked potentials and lateral asymmetries of processing. In: Gaillard A, Ritter W, Kok A (eds) Tutorials in event related potential research: endogenous components. Elsevier, Amsterdam, pp 369–383
Skinner J, Yingling C (1977) Central gating mechanisms that regulate event-related potentials and behavior. In: Desmedt J (ed) Attention, voluntary contraction and event-related cerebral potentials. Karger, Basel, pp 30–69
Syndulko K, Lindsley D (1977) Motor and sensory determinants of cortical slow potential shifts in man. In: Desmedt J (ed) Attention, voluntary contraction and event related cerebral potentials in man. Karger, Basel, pp 97–131
Tecce JJ (1972) Contingent negative variation (CNV) and psychological processes in man. Psychol Bull 77:73–108
Trimmel M (1986) DC potentials of the brain. In: Papakostopoulos D, Butler S, Martin I (eds) Clinical and experimental neuropsychophysiology. Croom Helm, London, pp 312–338
Ueda M, Furumitsu I, Kakigi S (1985) Self-regulation of contingent negative variation (CNV) using immediate feedback. Jpn J Physiol Psychol Psychophysiol 3:1–9

V How Brains May Work

Introductory Remarks

T. H. Bullock

An unusual feature of the 1985 Conference[1] was a session entitled "How brains may work," organized and edited by E. Başar and T. Melnechuk, in which several scientists were invited to speculate or summarize their personal view of the principles of operations of this most complex system known. For the 1987 Conference some of the same people and some new people were invited to do the same thing, or to state how their views may have changed in the past few years, or to make a statement about directions and needs in research on how brains work.

The following chapter is Hilton Stowell's lightly edited transcript of some of the statements, reviewed and corrected by the speakers. They as well as other members of the Conference were offered the opportunity to submit expanded, written remarks and the following chapters represent such contributions, edited and revised. As the organizers had hoped, the result is a highly diversified series of statements that underline the major role in science of the subjective element in guiding what scientists do and how they react to their hard-won data. Well aware that there are vehement advocates of expunging verbatim discussion from the printed literature, we are more influenced by those who have enthusiastically welcomed the taste of scientists reacting to each other without premeditation. Some of the spontaneous discussion is therefore preserved following the more of less prepared "Remarks." We believe both forms of revealed thought have more than historical value; they may help in understanding contemporary controversies and differences in approach, interpretation, and emphasis.

[1] Included in E. Başar (ed) (1988) Dynamics of sensory and cognitive processing by the brain. Springer series in brain dynamics, vol 1. Springer, Berlin Heidelberg New York

Springer Series in Brain Dynamics 2
Edited by Erol Başar and T. H. Bullock

Panel Discussion: How Brains May Work

EDITED BY H. STOWELL, T. H. BULLOCK, and E. BAŞAR

Dr. Bullock chaired a panel consisting of Drs. W. R. Adey, A. Babloyantz, E. Başar, A. de Callataÿ, L. Deecke, W. J. Freeman, R. Galambos, O.-J. Grüsser, E. R. John, and A. Starr. He invited them in turn to make remarks along the lines indicated in the Introduction and following each one called on anyone in the panel or in the audience to comment. With some omissions and alterations, and separate chapters for Drs. Adey, de Callataÿ, Mpitsos, and Freeman, there follows here a digest of the remarks and discussion.

The discussion following the remarks by Dr. de Callataÿ and the discussion contribution by Dr. Babloyantz and others are appended following the chapter of Dr. de Callataÿ in Part II of this volume.

Remarks by R. GALAMBOS:

The Glia-Neural Approach

My speculations about how brains might work will expand further on some of the problems and possibilities a combined glia-neural point of view offers the interested neuroscientist.

The usual answer to the question "How are EPs generated" calls upon physiological events at neuronal membrane surfaces to initiate electrical currents that then move through brain fluids to reach distant electrodes on the scalp surface. These neuronal membrane events are, of course, either propogated axon depolarizations or the local de-and hyperpolarizations continuously going on at cell-body surface and between axon and dendrite terminals in the neuropile. The possibility that only a small percentage of the current generated in this way could appear at the scalp is rarely if ever entertained.

With the introduction of astrocytes, as the retinal experiments demand, this possibility becomes real, for at least two new factors enter the analysis of the problem. One is the extracellular space, variously estimated at 15%–25% of total brain volume, another the astrocyte itself, with its special membrane and metabolic properties. We need much new information about both of these brain compartments. Changes in size and composition of the extracellular space (where neuronal events are continuously changing $[K^+]_0$) are under study, as is the astrocyte membrane-cytoplasm system which transport K^+ to and from the endfeet on blood vessels and brain surface where the exchanges take place. Very little is known, however, about the formation and removal of the gap junctions that in-

Springer Series in Brain Dynamics 2
Edited by Erol Başar and T. H. Bullock

terconnect the astrocytes, and this knowledge may be critical for the development of any comprehensive theory of cortical electrogenesis.

Since scalp electrodes are in fact separated from neuronal activities by both a large fluid space and a system of glial cells unique in their geometry and physiology, the EPs and EEGs these electrodes visualize must at best be only distantly related to the transmembrane voltage differences microelectrodes measure. EPs must be viewed as *derived*, not *direct* measurements of neuronal events, and derived data inevitably incorporate distortions of several sorts. *Temporal* distortions enter because the movement of ions through fluids and across neuronal and astrocyte membranes takes time. *Spatial* distortions can in theory at least be impressive: imagine, for instance, the extreme case where very large K^+ fluxes into and out of a small region of extracellular space exactly cancel; net $[K^+]_0$ change would be zero and nearby astrocyte membrane would sense nothing.

If current through astrocyte cytoplasm does in fact contribute to the potentials recorded at the scalp surface, some theoreticians attempting to specify the biological sources of EPs will have to reexamine their assumptions. Many now model the brain as a homogeneous isotropic conduction medium, and assume that the potential field mapped at the surface approximates that generated by a dipole source. If K^+ current were to flow radially through astrocyte cytoplasm and be returned by different ions through narrow and tortuous clefts of extracellular space, these assumptions would appear to be invalid.

One advantage EP measurements share with no other method is that EPs deliver a voltage-time display which *integrates* for the viewer the elaborate event-sequences under way in the neuropile. I illustrate and discuss the synthesis and dissection of such event-sequences in my main paper, "conditioning" a monkey with puffs of air adds particular voltage wavelets to the EP sequence, "habituation" removes them but "dishabituation" brings them back again. I interpret these changes as evidence that particular neuronal inputs become added to and subtracted from the neuropile region being sampled, a hypothesis brilliantly confirmed by Pirch and his colleagues (1986) for a 50 microvolt steady potential that is no longer added to the rat cortical EP after the nucleus basalis (cholinergic) input to the region under the electrodes has been destroyed.

My paper also states that similarities in the EPs recorded from widely spaced electrode pairs (the E. Roy John cat experiment) may result from structural changes in the system of glial cells responsible for conducting the K^+ currents away from neuropile regions. Put simply, *the number and location of gap junctions interconnecting the astrocytes is conceived to vary over time.*

Confirmation of this hypothesis, if it ever comes, would open wide the question whether glial cells do indeed actively participate in organizing neuronal activities throughout life, not merely during embryology, when radial astrocytes so clearly guide the migrating neurons.

In human experiments, where microelectrode measurement upon brain neurons is unlikely to become routine, EP methods offer the only feasible approach to the study of brain-behavior correlates, and are being actively pursued. In the context of the present discussion how does one interpret the following pair of cognitive waves recordable through scalp electrodes from normal subjects? One is the P300 (a scalp-positive wave measuring 5 μv or more having a latency near 300 msec), the other the N400, a uniquely human brain wave generated whenever subjects encounter a semantic error in the sentence they are smelling. (Note to the proofreader: Do not correct that semantic error!).

Surely this information about human brain activities which P300 and N400 waves deliver cannot be wholly useless to the neuroscientist. To me they provide unequivocal evidence that two very interesting brain events have surfaced, thanks to the psychologists. It is still an empirical question, however, whether the first one, the P300, represents a K^+ surge into extracellular space from neuron terminals lying deep within the cortex, followed promptly by a K^+ surge out of astrocyte endfeet into the subarachnoid space beneath the recording electrodes on the scalp. The N400 process could similarly originate in cortical neuropile, but this time the initial K^+ flux would have to be into, not out of, the neuronal elements, and the corresponding timelocked K^+ ion flow at the brain surface would have to be out of the subarachnoid space and into the endfeet. (NB: it may be that I reverse the direction of flow, but the idea is, I hope, clear.) The *neural* processes that initiate the P300 and N400 K^+ surges at the neuropile level are different, of course, and both remain to be described.

A final thought: If interactions between astrocytes and the neuropile are capable of generating the spectrum of electrical frequencies seen in the EPs, one must wonder whether the similar spectrum encountered in the spontaneous EEG (from which the EPs are extracted) has the same or a different electrogenesis. The principle of parsimony suggests that a general theory of the origins of all scalp-recorded electrical events may be ready for formulation and testing.

Who is prepared to test this proposition: *llinked movements of K^+ into and out of astrocytes, the extracellular space and neuron terminals make important contributions to all scalp-recorded electrical events, including the EEG.*

References

Pirch JH, Cozbus MJ, Rigdon GC, Lyness WH (1986) Generation of cortical event-related slow potentials in the rat involves nucleus basalis cholinergic innervation. Electroencephalogr Clin Neurophysiol 63:464–475

Remarks by O.-J. Grüsser:

Neurons and Glia in the Retina

Grüsser: I want to continue Dr. Galambos' thoughts by demonstrating the limitations of both macroelectrodes and microelectrodes in determining electrogenesis in tissue, using as an example the fairly well-organized structure of the retina. There are about 60 morphological types of cells, classifiable into seven large classes – pigment epithelium, glial (Müller) cells, receptors, horizontal cells, bipolars (on or off), amacrines (feedforward or feedback), and on-center or off-center ganglion cells. In the cat one can compare extracellular and intracellular recordings from the different layers with the overall difference potential, the electroretinogram (ERG) recorded between macroelectrodes placed at the inner and outer parts of the retina. The ERG to a ganzfeld transient has the a-wave, a slower b-wave, and a much slower c-wave. If one destroys all cells except the receptors and pigment epithelium by interrupting the blood flow through the central retinal arteries, the b-wave of the ERG disappears and we are left with a large a-wave and c-wave. Microelectrode recordings show that pigment epithelium cells respond to light flashes with a very slow potential lasting several seconds; this behavior corresponds to the c-wave of the ERG. The a-wave, however, is mainly caused by the photoreceptor potentials.

Photoreceptors respond to a short light flash with a short hyperpolarization. A similar potential is recordable by microelectrodes from the horizontal cells. Bipolar cells respond with a transient depolarization or hyperpolarization, without generating action potentials. Amacrines depolarize and discharge small (presumably Ca^{2+}) spikes. The on- and off-ganglion cells, however, behave quite differently. They fire with three to four rhythmical bursts of action potentials, which are correlatable with the psychophysically observed short afterimages. The oscillatory activation periods of the on-center ganglion cells are correlated with the inhibitory phases of the off-center ganglion cell responses. Such oscillatory reponses occurring within 1–2 s after the short flash at the output level of the retina cannot be seen at all in the mass recordings (ERG). In the normal ERG the a-wave corresponds to the receptor potential and perhaps to the horizontal cell potentials. The b-wave is very similar to the glial cell potential. Accordingly, there is a glial cell response visible in the ERG, while at the same time the oscillatory activity of about a million ganglion cells of the retina is not easily represented by the macroelectrode recordings.

This observation indicates that a temporal and spatial microstructure of the response exists, realized in a local network, the operation of which is hidden from the macroelectrode recordings. The antagonistic functional organization of the ganglion cells cannot be discovered by mass recordings. There are limits of prediction from b-wave to neuron activity, but also from single units to the mass activity, because intracellular recordings will not predict the absence of oscillation in the b-wave.

Switching now from retina to the visual cortex and the work of Dr. Mitzdorf, presented at this conference, I want to recall work on the cat visual cortex and to mention Chang's early work with potentials evoked by electrical shocks applied to the optic nerve. The EP of the primary visual cortex to a single optic nerve shock consists of four to five fast early waves and a slow potential lasting about 20 ms. The first two waves recorded at the surface of the cat visual cortex are not caused by cortical discharges but by afferent optic radiation fiber action potentials, highly synchronized due to this "unnatural" stimulation. The next two fast waves are related to postsynaptic potentials and action potentials mainly in layer IV cells, possibly also in layer III and V pyramidal cells, and occur with a delay of 0.4–0.6 ms. Hence, the first fast waves of the EP represent input signals from afferent visual radiation fibers and postsynaptic potentials from first and second order cortical cells. But what about the later slow wave? In the 1950s this wave was described by Chang as rather variable from shock to shock. We have found very few single unit discharges correlatable with this large late slow wave. As Dr. Mitzdorf has convincingly demonstrated, such slow potentials might reflect the spread of activity into the dendritic branches. In addition, this slow potential might also represent a rather slow glial cell response.

We can support the latter idea with data from experiments in which direct electrical shocks were applied to the cortical surface and intracellular recordings were obtained from either nerve cells or glial cells. While the neurons respond to a single surface electrical shock with 1–3 high-frequency discharges and return thereafter to spontaneous activity, the glial cells have very slow potentials and respond with a delayed depolarization. With repeated stimulation, depolarization increases up to a plateau. By means of DC recordings, one finds a DC shift at the surface parallel to the glial cell depolarization.

In summary, depending on the stimulus conditions, all possible neural and glial structures can contribute to the surface EEG potential: afferent and efferent axons, single nerve cells, postsynaptic potentials, and action potentials including slower spread of membrane potential changes to the dendrites. Under certain conditions, even glial cells – or at least those oriented predominantly perpendicularly to the cortical surface – might contribute to the EP. One cannot say that the EEG or the EP are caused by *either* PSPs *or* action potentials *or* glial cell potentials. These mass recordings seem to be a mixture and what type of cortical cell activity predominates seems to depend on the stimulus condition.

Anonymous question from the floor: I'm not sure if the EEG has any behavioral relevance. Does the EEG really explain how the brain works?

Grüsser: This is a question I have not addressed so far, but I can shift my comments to this topic. I think that if you select the right type of experimental paradigm and recording, you can learn a lot about how the brain works. You can also be misled, however, unless you choose the stimulus and your hypothesis very carefully. The point I have been trying to make is that neither microelectrode nor macroelectrode studies *alone* are sufficient. At the microelectrode level one can discover more, but you can also get lost in the manifold possible responses. The local EEG tells you something about the "democratic vote" of a restricted neural population, but if these neurons are antagonistically organized in their temporal responses, as is the case for on-center and off-center ganglion cells, you are immediately misled. You must have some idea about the local organization of a neuronal network and whether it is antagonistically or cooperatively connected, before drawing from the EEG response any conclusion about the single unit level. For example, compare the visual and auditory cortex: the latter has a tonotopic organization and very few auditory "off-neurons" are found, while the visual cortex has as many units inhibited as excited by a local light spot or contrast border. So one must be familiar with the main principles of cortical organization to be able to interpret EEG responses and evoked potentials. Under these conditions I think EPs are a very useful tool and can with discretion be correlated to observations at the perceptual or behavioral level.

Deecke: It might also be that we see *only* excitation from the surface?

Grüsser: Absolutely! Take Dr. Mitzdorf's model of inward/outward current, a Na^+ current and an opposing K^+ current. If you have an IPSP, K^+ and Cl^- conductivity of the membrane increases and K^+ ions move outward, whereas for an EPSP the reverse is the case and Na^+ ions move into the nerve cell. Under these conditions, the local sink/source mechanisms have changed. How these conditions change depends entirely on the activity level. If the neurons are highly activated, the difference between the K^+-equilibrium potential and the membrane potential is high, then you have a considerable K^+ current; but if you are near the equilibrium potential, you have a small K^+ current and you might have only a Cl^- current which plays a certain local role in the currents of inhibitory synapses. As a rule, however, there is a large difference between the membrane potential of a normally activated nerve cell and the Na^+-equilibrium potential. Therefore, on average, one tends to see more excitation at the surface of the cortex than inhibition. In cortical nerve cells, the depolarization can reach higher values than the hyperpolarization because the membrane potential of a neuron is generally nearer the K^+-equilibrium potential and very far away from the Na^+-equilibrium potential.

Roy John: True, but if you use a baseline – some kind of resting state – that's not a silent nervous system, but a random level of firing, diminutions *below* that, which are inhibition, can then be seen relative to that baseline; so I don't think we should give up trying to see inhibitory processes ...

Grüsser: No, we should not give up this goal. I think inhibition can be seen in the surface EEG, but the chance of seeing it is about ten times less that of seeing excitation.

Remarks by E. R. JOHN:

Comments on Integrative Function in the CNS

My views have not really changed since 1985; they have been reinforced, despite the impressive instrumental armamentarium I have shown you at this conference. There are two possible explanations for this: (a) I am inflexibly imposing my preconceptions in the data; (b) that preconception is compatible with the data that emerge.

My problem is the same as it was two years ago: If central neural functions are as distributed as the data indicate, then how does the system integrate those functions? That is the central problem as I see it. Whether you call this consciousness, or the nonrandom activity in neural ensembles, the problem is still the integrative function. There is no neuron, or ensemble of neurons, that can sample the nonrandomness in all the other ensembles. Further, when you look at the information processing of the neuron in one of those processing ensembles, you find that its capacity is really distributed; it's a weak signal; it's noisy. Yet despite the apparent tenuousness of this superpuzzle, it's not difficult to find very lawful things when you observe how this distributed system operates.

I'm thinking of something called "cognitive style" – the stability of intraindividual approach to task performance – as reflected in cerebral electrophysiology, if you believe that such electrophysiology reflects process. Interindividual approaches may be quite different, but there is remarkable stability in the individual processing. My experience comes from psychiatric patient groups; I argue, perhaps naively, that a common clinical classification could only emerge if there were something similar about those patients. This is where one finds a lot of stability, within groups. Now these groups, consisting of the clinically classified mildly cognitively impaired, demented, alcoholic, or with mild head injuries showing cognitive complaints a year later, unipolar depressed with negative affect, and bipolar tending to be euphoric most of the time, differ sufficiently in their behaviors outside the laboratory, and in their general *Weltanschauung,* to be classified as belonging more to one group than another. Now, our laboratory topographic maps of these people's EEG and EP show that all groups, averaged together, have a significant shift away from the independent normal baseline, but also that intergroup discriminant functions can be constructed to classify individuals within a particular clinical group at 80%–90% accuracy, depending on the group. So consistent differences can be found in the "housekeeping" operations of the nervous system, between members of different groups. Let's say the background EEG is housekeeping, then it's as if the janitors have forgotten certain corridors and rooms of the house, and those rooms are characteristically the same for each disease. I find it remarkable that people with thought disorders or cognitive impairment differ so much on these measures from normal people. Maybe there's something in here that reflects the disease, and the disease is a *mental* disease – something peculiar about the way a person interacts with himself and his environment.

Still more remarkable are our recent data from the noncontingent probe EP paradigm in groups of normals, elderly depressed and schizophrenics: The correlation between the changes of each group, relative to each group's baseline, on this task is about 0.8. So, it is as if, when you impose a mental task on a "crazy" person and he performs it adequately, he makes a change relative to his "crazy" baseline which is the same change as you see in a normal person relative to his normal baseline. In other words, there's enough similarity in the mental machinery so that when the same task is performed in three different groups, two of which are classified as having psychiatric diseases, the correlation between the map changes, relative to their own baselines, is 0.8. There's some robustness in this fragility. That kind of correlation suggests that these are not epiphenomena, and that they may be related to how the brain works, and how it processes information. But the problem remains: Look at those maps and the contours; where is the process that can compute all the hilliness of those maps and integrate it?

Grüsser: There is now good evidence from both neuropsychological and evoked-potential studies for a difference between the function of male and female brains. What would happen to your normal, reference group of some 50 subjects if you divided them into male and female groups? What would happen to your z scores if you did this, forming to subgroups of about the same size each as your experimental groups? Did you try this?

John: No, but others have, and there are significant male/female differences. But we have deliberately pooled sex differences in order to identify what is common;

so we have inflated our variance and you see mixed M/F groups in all our samples.

Grüsser: Yes, but what would happen if you divided your two groups or took 25 out at random and of mixed sex and compared this smaller group with your reference group, normal group with normal? Still no difference? The question is: How variable are these data if you just repeat with normals or take out randomly selected normals?

John: The normal or reference group was an *independent* group not used to construct the norms. All the individuals in that group were excluded from the norm formation process, which is an independent replication where the incidence of positive (i.e., abnormal – Ed.) findings is at the 5% chance level. Now I don't know whether that 5% chance level is differently distributed among males and females; no striking differences appear for outliers, which we have studied. We inflated our variances deliberately, in order not to be distracted by differences which we considered unimportant.

Galambos: I'm still fixated on my glial business. These voltage differences which you are measuring, and which we see as your red, white, and blue spots and contours, are upwellings of potassium through the endfeet of astrocytes – not current flowing around individual cells – at least that's what I'm proposing. The power for the current flow is in the neuropil and must be defined in some way by the volume of astrocyte cytoplasm connected through gap junctions. We may assume that the current flows freely from one collection of neurons to another – from below one of your white spots to below one of your red spots, let's say – but without too much crosstalk. That's the kind of primitive conceptualization I now have, in contrast to the more traditional view I once had. How do you see the mechanisms for the origins of these colored spots which you present on your maps?

John: Those spots are certainly not generated by neurons! I think they are generated by charge systems. Some years ago I used the word "hyperneuron" to suggest a charge system distributed nonrandomly in space, produced by glial cells and neural cells, but which was a much more diverse and much more information-carrying thing than a neuron. The vehicle for distribution of that charge system may very well be an interlacing system of glial cells. I have difficulty holding those charge distributions on neuronal bodies for as *long* as I have observed these effects. Radio potassium studies show that at least 25% of the potassium in the brain is not labile; so it's being held by something, and my guess is that it's being held by an extended glial structure; and that structure may be some near relative to what I think of as a "hyperneuron" – that distributed charge inhomogeneity which somehow doesn't dissipate.

Starr: I feel like one of your glial cells sitting down here. But I have two comments: (a) We see a number of patients having abnormal glial structure and I often wondered about the relationship between their abnormalities and their slow-wave EEG – dominant delta band for instance; your glial hypothesis may have something to say here. (b) On quite another matter, I may not have looked at quite as many abnormal EEGs as you have, but my strong impression is one of a lack of EEG difference between different patient classes. I think the reason for this is our different criteria for calling something different, and this is a statistical problem. Your *z* transforms have a ± 1.5, but what if you raise that to the kind of level that I use, say ± 2.5 or ± 3? If you call someone "abnormal" when they are normal, they get very nervous! Would your patients be "abnormal" using my criteria?

John: Yes, many of our patients had at least one routine EEG in the past and are almost always "normal" by that criterion; but a *z* transform relative to age-appropriate population norms from healthy people, whom we examine, sharpens your ability like a zoom lens. You're quite right; if you raise my scale and so decrease the sensitivity of the system, all those heads can look alike. But our purpose is to find the differences.

Starr: But what does ± 1.5 mean?

John: 1.5 standard deviations from the mean!

Starr: By that measure I could find lots of people "abnormal" in this room!

John: We find 5%, at the $P < 0.5$ level, which is 1.96.

Starr: Yes, and we don't like that, because we could call many of the 70 people in this room "abnormal" by that criterion! If you were to raise your *z* transform level, how "abnormal" would your abnormals be?

John: Yes, well taken, but the fact is that at ± 2.5 a significant proportion of our population is still "abnormal." Our maps show you univariate, not multivariate features; we can show you the incidence of abnormality at the $P < 0.01$ and 0.001 levels in those populations; they are still interesting. Regional cerebral bloodflow (rCBF) and our *z* transform neurometric measures were applied to groups of patients in a joint study in Amsterdam; some of these were asymptomatic at time of examination, while some had persistent symptoms; the asymptomatic patients had 32% abnormalities by rCBF, but by our neurometric criteria they had 88% abnormalities, with 3% in an age-matched group of 64 healthy Dutch controls. Now, when you examine the data matrices for these asymptomatic patients, very few indeed reach the ± 1.5 standard deviation level on any univariate measure, while the multivariate measure of *relationship* still showed Malhanobis distances across that set of measures with *z* values of 10. We don't do our diagnostic classification by maps, but by discriminant functions using univariate and multivariate features constructed from the matrices. One of those maps is simply a row in a matrix, that's all it is. When you consider the whole matrix, and particularly the covariance matrix between those elements, I don't think you'll find a high percentage of the population "abnormal" under those circumstances!

Remarks by E. BAŞAR:

Brain State Matrix – A Preliminary Proposal to Approach Brain Function by Using EEG-EP Feynman Diagrams

My tentative, but important assumption is that we may approach the understanding and prediction of brain function by using a large number of indicators, linear and nonlinear (with an adequate computer) and by taking the appropriate, related brain states into consideration. Always finding a great danger in distinguishing sensory from cognitive processes, I insisted upon the expression "dynamics of sensory and cognitive processing" already several times in my own publications and conferences (as in the current volume). The EEG, as an active signal, shows processing in a manner which combines sensory and cognitive information.

My proposal is to develop a system of rules like the physicist's Feynman diagrams for brain function. This should be considered as a tentative step to combine several simultaneous measurements of brain processes and approach the question of "How brains may work" globally. The rules will be completely different from physical Feynman diagrams. I think, nevertheless, we may use a similar way of thinking. I also hope that this step can be enhanced by developing a "brain dynamics study program" including the characterization of the brain's pathological states, the brains of lower vertebrates and invertebrates, and also mathematical models.

In our earlier papers and also a few days ago during the present conference, my coworkers and I showed that in the human brain ordered and coherent EEG states can be observed without the application of physical stimulation. This is an extension of our earlier results which have shown that physical sensory stimulation brings the brain to a temporarily coherent state (Başar 1980). During a learning paradigm the human brain can create replicable phase-ordered patterns in the 4, 10, and 40 Hz ranges (Başar et al., this volume). In other words, if one designs an experiment properly, one can find in the EEG of an experimental subject templates of EEG fragments which repeat themselves if the subject under study is in replicable initial experimental conditions. I agree with the statement of Dr. Grüsser that experimental design and concept are primarily important in brain research. My concept here is the fact that, for designing an experiment, I do not consider the EEG as a constant or passive background in a cognitive process. According to the experiments which I described 3 days ago, and according to my work over the past 15 years or so, I consider the EEG as an active ground signal. The present experiments demonstrate that, as with evoked potentials, the EEG can be treated deterministically.

The evidence is twofold: The replicable EEG patterns are almost congruent in time space and this congruence can be demonstrated by using correlation coefficients (Başar et al., this volume). These findings demonstrate that during mental tasks the EEG depicts a transition from a higher to a lower dimension with the properties of a strange attractor. (For the definition of a strange attractor see papers by Röschke and Başar , this volume; Babloyantz et al. 1985; Başar 1988). In this place I also want to mention the important work of Walter Freeman on replicable 40-Hz patterns in the olfactory bulb of the rabbit brain during defined experimental conditions (Freeman and Skarda 1985; Freeman 1988, and this volume; see also Skinner et al., this volume).

According to this new type of experimental design, it seems to me that for comprehending the event-related potentials we have a new set of parameters to consider in our work on the paramount EEG. I tentatively name the set of useful parameters as "brain indicators." Here is a provisional list of indicators: the nonlinear correlation dimension (influences the degree of order), phase angle of the brain waves and their amplitude modulaton envelopes (as Dr. Bullock described), Rms values of various EEG frequencies, degree of order, coherence in space and coherence in time for each frequency. Using these indicators we can go to a new kind of file, which I call the "Brain State Matrix."

What is the Brain State Matrix? In my earlier publications I have tried to make a picture of this matrix by stating that the brain state could be described by several measures of instantaneously defined EEG properties, outlined above as indicators, during a given short period, say of about 0.05–1 s (Başar 1983). The knowledge of parameters in such a matrix enables the experimenter to predict roughly the shape, amplitude, and frequency content of an event-related potential. As earlier described, the amplitude of an evoked response depends strongly on the amplitude of the ongoing activity (Başar 1980). In the experiments I described on a previous day we could show that, if the subject's EEG is already in a coherent state, the physical sensory stimulation does not create a new, more coherent state. There is no evoked potential to a physical stimulus in such a coherent state of the brain (see Figs. 16 and 19 in Başar et al., this volume). One can extend the results with nonlinear descriptors by stating that, if the dimension of the EEG is low, the transition of the EEG to a lower dimension is not possible or difficult. In other words, as I tried to explain earlier, if the brain's electrical activity shows a low entropy state (high-order), the transition to a lower entropy state is difficult (Başar 1980). The amplitude and the shape of the outgoing response or of the outgoing activity (evoked activity) are inversely correlated. The outgoing response is a function of the ongoing activity. I use here the expression of ongoing and outgoing activities to come to an important analogy in elementary particle physics. The *S*-matrix introduced by Heisenberg to elementary particle physics described the nuclear reactions by considering *ongoing* and *outgoing* waves.

The Feynman diagrams which are used in particle physics have been developed in order to describe and predict the electromagnetic processes, whereby electrons and photons interact. This can give rise to drawings that look complicated, such as the one in Fig. 1. In this Feynman diagram, a single electron enters on the left at A, and then a single electron emerges on the right at B. To a nonphysicist it looks as though one electron has directly moved from A to B. The interactions are, indeed, complicated; there is a sort of "grammar" to these diagrams, which allows only certain pictures to be realized in nature. This grammar is a result of basic laws of physics, such as conservation of energy and conservation of electric charge. Particle physicists have found that this complexity should be handled in a reduced form, and in order to understand the behavior of electrons and photons they use approximations which neglect all but simple Feynman diagrams. By considering roughly the simplest hundred diagrams for certain processes, physicists have been able to predict important relations precisely.

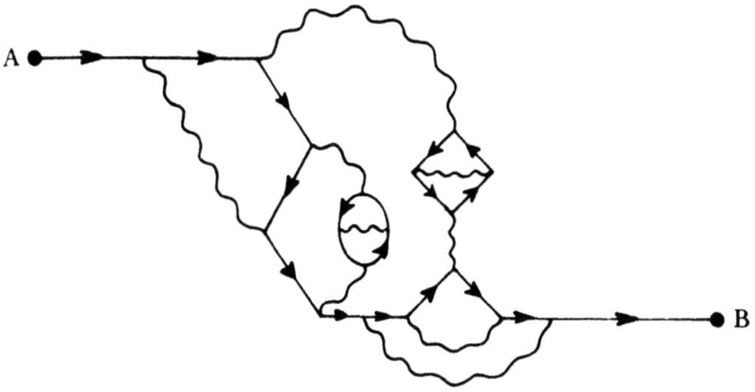

Fig. 1. A Feynman diagram showing the propagation of a renormalized electron from A to B. In this diagram, time increases to the right

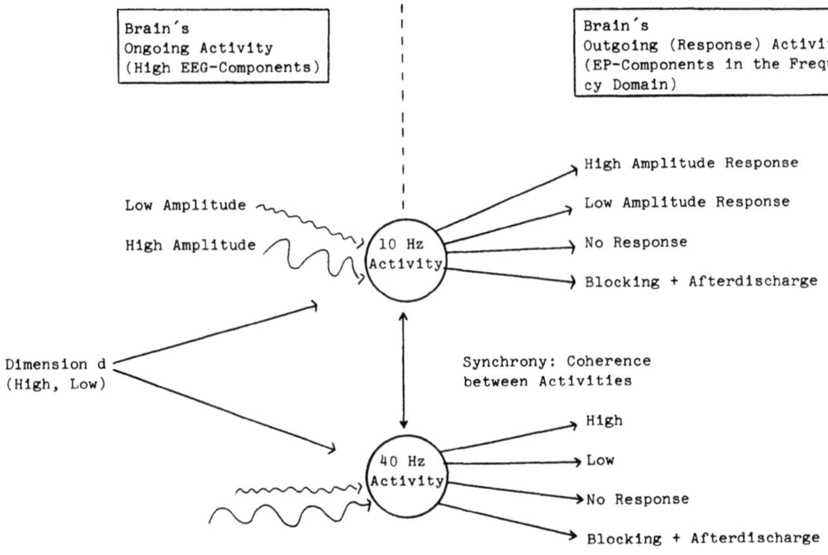

Fig. 2. A rough schematical explanation for an EEG-EP equivalent of Feynman diagrams. Transitions from ongoing (EEG) to outgoing (EP) activity in the frequency domain can be roughly predicted. Evoked potential components in various frequency channels are inversely proportional to EEG amplitudes in the frequency domains. Couplings between frequency components also influence the EP components. Dimension or degree of chaos in the unfiltered EEG is also an additional measure to predict the reaction. These drawings should be extended to other frequency ranges

My proposal is that by introducing the Brain State Matrix which is composed of several EEG indicators, it would be possible to predict several brain reactions which we analyze as sensory evoked and event-related potentials. In brain research we often describe the evoked potentials as a transition from an unexcited brain state to an excited one. However, the brain waves are often excited due to as yet hidden sources. Thus, the EEG can often be considered as internally evoked potentials stemming from unknown sources within the CNS (see Başar et al., this volume). Accordingly, transitions of the brain waves which occur without defined external stimulation should also be considered as an active transition and approached with the method of Feynman diagrams. For the time being, however, I choose the simpler way to consider only transitions following defined sensory stimulation.

As my coworkers and I have shown (Başar 1980, 1983; Başar et al. 1987, 1988) there are several allowed and unallowed transitions of the EEG following a stimulation. For example, if a subject emits abundant high amplitude alpha waves prior to application of a sensory stimulation, usually no enhancement of that frequency is seen in the encountered response. On the contrary, we then observe an alpha blocking. The same rule is true also for 40-Hz activity (Başar et al. 1987 b). Further, if the overall coherency between various structures of the brain is high, again the enhancement of EEG activity is low or vanishes. Furthermore, there exist couplings between frequency components and also between amplitudes of various EEG components among different brain structures – for example, there is important coupling or similarity between 10-Hz activities of formatio reticularis and thalamus as has been previously described (Başar 1983; see also Fig. 2). By starting with a Brain State Matrix and developing new rules step by step (which should be experimentally evident and would allow to facilitate or prohibit several transitions of the brain rhythms) it could be possible to predict a large number of brain reactions that we analyze as the brain's compound response potentials.

To my way of thinking the EEG is not only an activity that reflects some brain state, but also an activity that anticipates reactive mechanisms and controls the output to a stimulation. Accordingly, the introduction of this new type of "grammar" may also serve to design experiments for the understanding of a large number of cognitive processes.

References

Babloyantz A, Nicolis C, Salazar M (1985) Evidence of chaotic dynamics of brain activity during the sleep cycle. Phys Lett [A] 111:152–156

Başar E (1980) EEG-brain dynamics. Relation between EEG and brain evoked potentials. Elsevier/North-Holland, Amsterdam

Başar E (1983) Toward a physical approach to integrative physiology. I. Brain dynamics and physical causality. Am J Physiol 245(4):R510–R533

Başar E (1988) (ed) Dynamics of sensory and cognitive processing by the brain. Springer, Berlin Heidelberg New York Tokyo (Springer series in brain dynamics, vol 1)

Başar E, Başar-Eroglu C, Greitschus F, Rosen B (1987) The associations between 40 Hz-EEG and the middle latency response of the auditory evoked potential. Int J Neurosci 33:103–117

Başar E, Başar-Eroglu C, Röschke J (1988) Do coherent patterns of the strange attractor EEG reflect deterministic sensory-cognitive states of the brain? In: Markus M, Müller S, Nicolis G (eds) From chemical to biological organization. Springer, Berlin Heidelberg New York, pp 297–306

Freeman WJ (1988) Nonlinear neural dynamics in olfaction as a model for cognition. In: Başar E (ed) Dynamics of sensory and cognitive processing by the brain. Springer, Berlin Heidelberg New York Tokyo, pp 19–29 (Springer series in brain dynamics, vol 1)

Freeman WJ, Skarda CA (1985) Spatial EEG patterns, non-linear dynamics and perception: the neo-Sherringtonian view. Brain Res Rev 10:147–175

Bullock: Your list of indicators is very intriguing, because I've said we need more descriptors, more ways of distinguishing brain voltages as signs of brain activity. Do you think your list of indicators would distinguish the stream of voltages going to a radio loudspeaker, when a Chinese was talking, or a German speaker, or when a mocking bird was singing? I want to be able to distinguish the ongoing activity of the brain between, say, a mammal and a fish. Differences in state are wonderful, as between a psychiatric patient and a waking or sleeping normal, but we've got to be able to distinguish also between advanced brains and simple brains, brains with a cortex and without, brains with little differentiation versus brains with high differentiation. Right now I can't do that; will your list of indicators help me to distinguish people speaking different languages or people from birds?

Başar: If I could answer this question, then I'd be able to say how the brain works! I cannot answer this question. But I'll say this: To make understanding the brain less complicated, we must first try to simplify the questions. What are the effects of *un*complicated inputs to brains? As you know, using simply the naked eye it is almost impossible to tell the difference between the EEG of cat cortex and of some brain structures of a ray. But I think that step-by-step application of indicators such as I have mentioned may lead to some quantification. For example, *Aplysia* ganglion has a type of 10-Hz activity, but you don't call it "alpha" because qualitatively it looks different. Why does *Aplysia* 10 Hz have smaller magnitudes, and why does the movement of recording electrodes produce sudden differences in coherence? As I have shown, one can generate states of EEG in humans which are predictable, but it's very difficult to do it in cat brain. By using these indicators we can get more deterministic, step by step, but we are not going to solve the problem of Chinese or Russian. Perhaps we shall reach more predictable reactions, but certainly not 100% predictability.

Freeman: I want to ask Dr. Başar if he would consider expanding his picture from characterization to states of *transitions* between states, especially sequences of states which might be called a trajectory, or an evolution from one state to another resulting in some quasi-endstate. This might be more valuable to your goal, Dr. Bullock, of describing a hierarchy of animals achieving states of increasing complexity.

Başar: Thank you for going further than I wanted to! Yes, the bifurcations are more important. I had a letter recently which advised me to look at the transitions of bifurcations between brain states, rather than at the states themselves, if I wanted to understand cognitive processing in humans. But I am still trying to define states, so, if you want to go to bifurcations, please do those experiments yourself! (Note added after conference: See transition of the EEG activity of a cat from sleep to hippocampal theta state in Röschke and Başar, this volume.)

John: I like this idea of a "Brain State Matrix"; we call our machine (for neurometric diagnostics – Ed.) a "Brain State Analyzer." We have absolute power, relative power, coherence, and symmetry measures in each of the frequency bands that factor analysis shows are independent, and that is a multivariate space. An individual subject or patient is a vector extending into that space, and one can summarize all I've said about discriminant functions like this: The brain state vectors in brain state space point in different directions for different diseases. Interestingly, the brain state vector is age dependent; so you can ask, for any individual, is there any age at which this brain state vector would have been appropriate for a human being? This vector has been checked in six different countries and is certainly culture-, language-, and class-free, as long as the subject has been adequately nourished. That vector changes drastically from age 6 to 90, because we've characterized it along that life span. With regard to Dr. Bullock's question, it would be interesting to test every animal with a decent EEG to see if any species has any maturational lag which would make that vector appropriate for a human being at some stage of development. Maybe some of these animal EEGs are congruent with the brain state vector of early stages of human development. If one took the EEG of animals in the phylogenetic scale and determined the dimensionality of each brain state vector, one might be able to make a brain scale that would reflect complexity of development. We know now how many dimensions we are dealing with in the human brain. It would be interesting, like many other things, but would occupy about three lifetimes!

Başar: Seven years ago we published a few data on the EP of children, and last year we did some more on babies and small children, using this kind of brain state matrix. By looking at the EEG I said that a particular child should have larger slow-wave or intermediate latencies, etc., because as the EEG changed the oscillatory responses (of the EP – Ed.) got faster also; as the ongoing EEG became faster during the child's development, so the oscillatory responses contained higher frequencies. Some children developed 10 Hz by 3–4 years, as expected, some did not, also as expected. The maturer responses are really predictable in these cases; I was almost prepared to bet a bottle of champagne on some of them! I am unfamiliar with evolution, but maybe one could try to make some phylogenetic predictions from such brain state matrices, from invertebrates to lower vertebrates, etc. It seems to work in children.

Bullock: Why I chose languages like Portuguese and Russian – ones I cannot understand at all but which I can distinguish on the radio in a second or two, and quasi-white noises too – is the fact that such distinctions can be made so rapidly. The time dimension is *the* important thing for speech, for bird calls, for distinguishing a waterfall from air bubbling. So, could we add something to your list of indicators or descriptors, by looking for these very rapid changes in the fraction of a second domain? Or, could we add some kind of segmentation of the EEG, such as Creutzfeldt and colleagues, among others, have done, looking for the number and sequencing and the character of switching from one segment to another?

Başar: Yes! We should consider more matrix elements; I originally included envelopes in my list of descriptors, but forgot to bring that list with me. Especially if I am recording from cerebellum, it is important to control not just stimulation frequency but also the envelope duration, which is the "Bullock envelope" of the 300-Hz packages; this is a very important parameter; also the number of envelopes, e.g., a 40-Hz oscillation usually persists for three or four wave packets and stops, so first there is a state transition and then a certain *number* of wave packets.

Freeman: One comment, from Dietrich Lehmann who isn't here: In a paper on the topic of EEG segmentation he found segments of 200 ms, which I think are exactly what you are talking about.

John: How large a sample do you have to take for the statistics to become stationary – that would be very different for, e.g., a waterfall and somebody speaking – and another question, what is the dimensionality of the system? The brain dimensionality decreases with increasing hypothermia and decreasing blood flow; in intensive care units the comatose patient has a brain dimensionality of one! Differentiation occurs with dementia, with disease, and with hypothermia during heart surgery, while redifferentiation of EEG is prognostic of recovery. So dimensionality of the signal space and duration needed for EEG stationarity are very important indicators too.

Başar: Yes, and yet another dimension could be the superposition of various states, because the duration of a 40-Hz wave packet is different from that of a 10-Hz packet; and the superposition of these two wave packets would make the situation even more complicated. We shall see these things on the faster computers. Dr. Lehmann may have something to say about this.

Remarks by D. LEHMANN (added after the conference):

Temporal Characteristics in the Brain's Electrical Records

This comment brings up the basic conceptual question whether brain work can meaningfully be partitioned into different portions which are electrically manifest as different temporal frequency bands. A consideration of brain global functional states leads to the concept that there is at each moment in time a particular global functional state in correspondence with the present higher-order function; this state consists of a mosaic of numerous different local functional substates, associated with local functional suboperations. Even though brain mass electrical activity can numerically be decomposed into different temporal and spatial frequency bands, a given global functional state is characterized by all bands. The basic units of higher order functions (conscious experiences, percepts, recall) are hypothesized to have some finite durations in time, but the momentary spatial distribution of potential (the instantaneous map) characterizes the global state. The slowest temporal frequency necessary to define electrically the functional brain state associated with a given higher order function indicates its basic package duration. These considerations lead to a strategy to identify states in multichannel brain electric records. Grossly different behavioral states, such as, for instance, sleep, drowsiness and wakefulness are associated with slower of faster dominant temporal frequencies of the changes of the spatial standard deviation of the brain potential field data (Global Field Power, see Lehmann, this volume): this parameter can be used for an adaptive segmentation into gross states of stationary temporal characteristics, which, for recognition of the basic homogeneous building blocks of higher functions is followed by space-oriented adaptive segmentation into microstates of stationary spatial characteristics, i.e., constant landscapes.

Remarks by W. R. ADEY:

Bioinstrumentation – Cutting Edge and Limiting Factor in the Future of Brain Research

The question, "How does the brain work?" is a perennial one heard for many years at every meeting presuming to summarize aspects of brain organization. Why are we asking this question now? Is it that we believe that we have moved perceptibly closer to offering significant answers? If so, do we wish to address the brain-mind question, or do we merely wish to understand the machinery of the brain? It must be emphasized that there is still and may always be a substantial and even a vast gulf fixed between an understanding of the brain's machinery and an understanding of mental and cognitive processes. Will an historical perspective help us to understand the relative importance of aspects of the current research presented here? In this connection, I raise some issues on which I would welcome your thoughts.

First, it is clear that there has been a serendipity in the application of instrumentation to brain research. If we look back to Caton in 1875 and his first observations in the alpha wave using a string galvanometer, we will recall that he did not devise this galvanometer specifically for the purpose of studying brain function. This typifies much that has happened in the evolution of brain research. Berger was in the same situation. Adrian used the most primitive of three-valve amplifiers for his first studies of brain function. They were perceived as marvals of ingenuity; but neither he nor the engineers who built these amplifiers were pioneers in the field of instrumentation. The same can be said of the digital data and analysis systems that we have all applied. They came from elsewhere with no prime awareness on the part of their designers of our intention to use them in brain research.

In the same context, it seems that there has been an existentialism in planning our research projects. We do the easy and the obvious. I am mindful of the thoughts of my mentor in Oxford when I was a postdoctoral fellow in 1950. Professor Sir Wilfrid LeGros Clark in the Department of Human Anatomy was deeply involved in questions of relations between the limbic system and the neocortex. He remarked that if the brain had been placed in the skull upside down, we would know vastly more about the limbic system than the neocortex, simply because it would be easily accessible for research. I am not facetious in observing that we take a very adventitious approach to what we look at and why we look at it. But I also believe that winds of change are upon us, in great measure from our discontent with the lack of instrumentation that will give a complete three-dimensional transform of brain state in structural and functional terms. Thereby, as we have already discussed here today, we may hope to understand both the organization of cerebral domains as well as the question of segmentation in time, since domains will be dynamics that change with time.

I cite here some prospects that make me mindful of the poet's remark, "westward look, the land is bright." A quiet revolution has taken place in studies by those not always directly involved in the brain problem, where we regard ourselves as aficionados. There is the prospect of making functional and spatial transforms of the brain using dielectric dispersion techniques. This is neither a figment of the imagination nor is it something extremely complex to understand.

Key basic studies were done by Barnett Rosenberg, a biophysicist at the University of Michigan, working for the US Office of Naval Research. The work has been independently replicated and extended by James Kelley, a psychiatrist not steeped in biophysics but a wondrous tinkerer with devices and medical instrumentation. In simple but very revealing experiments, Rosenberg (Abey and Rosenberg 1982) showed that if a weak microwave signal is passed through the head, the emerging scattered beam has amplitude information of an EEG-like character. It is an essential aspect of scattering transforms that they also carry exquisitely detailed position information. Thus, in a single approach, there appears to be an opportunity to achieve a transform that will precisely define its changing spatial and electrical characteristics. Kelley's studies with similar weak microwave fields scattered through the heads of rats showed rapid evoked transients closely resembling electric evoked potentials in response to light flashes (Kelley et al. 1984).

This dielectric behavior of tissue is a topic also studied in detail by those not involved in major aspects of brain research. Herman Schwan (1957, 1977) and his colleagues at the Moore School of Engineering in the University of Pennsylvania have pioneered the field of dielectrometry across the spectrum from ELF to microwaves, with many significant observations on the frequency dependency of dielectric dispersion in vitro and more recently in the whole brain.

By examining the scattering transform of a microwave signal passing through an isolated dog kidney, Larsen and Jacobi (1979) successfully reconstructed the internal configuration of the organ. Lawrence Larsen, an MD with a strong mathematical background, was first attracted to studies of EM field interactions with tissues while working in my laboratory at the UCLA Brain Research Institute. His studies with an engineering colleague, John Jacobi, at the Walter Reed Army Institute of Research have opened the door to yet another tissue imaging technique that may have special importance in defining tubular structures and tissue discontinuities. Perhaps the pursuit of these and other new opportunities may lead us away from the continuing tunnel vision that characterizes so much biological research, with strong tendencies to continue doing what one has always done in the belief that therein there may be some solution to a basic problem.

Interdisciplinary collaboration has led to new applications of imposed EM fields at the cellular and molecular levels. Some of these are described in my paper. One of the great pioneers in the physics of superconductivity, Herbert Fröhlich of the Oliver Lodge Laboratory of Physics in the University of Liverpool, began writing to biologists in 1968 in an almost evangelical way that they were missing a great deal about the organization of tissue by not searching for a wide range of resonant interactions at cell surfaces, based on coherent states of fixed charges on surface macromolecules (Fröhlich 1968, 1975, 1986). His model predicted a major resonance in the millimeter microwave region at 10^{12} Hz. Fröhlich and a biophysicist colleague, the late Herbert Pohl of Oklahoma State University, also noted that if one imposes fields at frequencies between 10 Hz and 1.0 MHz on cells in culture, there are some quite unexpected effects (Pohl 1984). Many cell types spin rapidly and may then take on a linear arrangement, and Pohl has inferred that these sensitivities relate to intrinsic oscillations. These findings have taken a significant place in the immunology of monoclonal antibodies. Since cells exposed to these fields line up in string formations, Zimmermann at Marburg has developed devices that are used worldwide to join cells together in hybridoma formations – an essential technique in the production of monoclonal antibodies.

May I remark, perhaps too polemically, that we do not see the horizon because we have not looked. Nonetheless, it has so often been the case in science that the tinkerer, the researcher who is willing to go the extra mile, often outside established ways of doing things, has come upon the door to broad new knowledge of the utmost significance.

Hierarchical organization in intercellular communication has been extensively discussed at this conference. Concepts of hierarchies have carried us in a century of biological research from organs to tissues to cells and finally to molecules. Now we stand at a new frontier where events at the atomic level are the focus of interest. Stuart Hameroff (Hameroff et al. 1984), an anesthesiologist and molecular biologist at the University of Arizona, has modeled the cytoskeletal lattice of microtubules and microfilaments, considering implications of information processing through coupling of Davydov solitons, Fröhlich coherent oscillations, and other nonlinear electrodynamic phenomena of conformational states of these grid-like subunits of microtubules. Phosphate nucleotides such as guanosine triphosphate (GTP) bind rapidly and reversibly at specific tubulin monomer sites. Thus, solitons may be introduced or supported at specific "preprogrammed" GTP binding subunits, and may propagate between alpha- and beta-tubulin dimer regions, altering conformational states and binding of Ca^{2+}. I have pointed out (Adey 1987) that in the breakdown of molecules like ATP and GTP to cAMP and cGMP, thermal energy so released may be as a coherent excitation. This coherent energy would be like a narrow-band radio signal with both direction and signal content and confined to a very narrow field of molecules in its immediate environment. We may speculate that millimeter microwave/far infrared fields may play a physiological role within the cell, providing highly specific communication over short distances.

We do not possess instrumentation that would enable us to determine the existence of this field as a coherent radiation. However, it does indicate the need to reconsider our traditional preoccupation with the low-frequency portions of the electromagnetic spectrum, specifically at EEG frequencies, as the prime region, and even as the sole region involved in biological interactions. We should be aware that these low-frequency oscillations may be merely the reflection of events occurring at far higher frequencies, in the very same way that physicists concerned with the most fundamental aspects of the physics of matter point to the role of quasiparticles (Lawrence and Adey 1982).

In an historical perspective, in the sequence of studies from Boltzmann to Bohr to Einstein, there has evolved the concept of the statistical mechanics of matter, which says in essence that there is a total randomness to the organization of its constituent particles at atomic and subatomic levels at any temperature above absolute zero. This has been thrown into grave doubt in recent years by work showing that in its ultimate organization, matter may exhibit a granularity in the occurrence of quasiparticles which we have discussed here cursorily in terms of solitons. It is a humbling thought that as we have approached and passed a succession of frontiers in biological research over the past century, we have believed with a fervor amounting to zealotry at each succeeding phase that we have been confronting an ultimate organization, only to find that we have been dealing with a fugue on the reality of matter as we now see it in a biological context. Instead of revealing an ultimate organization, each new technology has removed but one more veil, more fundamental than the last, but still not identifiable as a final essence.

And so it is with equal fervor that we now launch ourselves into nonlinear electrodynamics of biomolecular systems, in the hope of discerning the domain properties that characterize this "whispering together" of brain cells. We face a new journey hand in hand with physicists working at the frontiers of the physics of matter. With a not uncharacteristic bluntness that may well reveal my ethnic origins in the southern hemisphere, I now ask, "Are we as neurobiologists the most appropriate people to pursue the future study of the brain?" I humbly suggest that we examine ourselves and our associations and the way in which we do experiments. If not, there will be an increasing tendency for us to form yet another masonic lodge, or at least to fall victims to ecclesiasticisms that tell us that the ways of our forefathers are ways that we should continue to follow.

At the same time, we should be aware that, for better or worse, there are others, who, in the fashion of "Corporate America," will plan to take over our little corporation of brain research. In the 1960s, this was the fashion of engineers, some of whom wrote books on brain function that all too frequently trivialized the problem in the feckless simplicity of their analogies and models. As I mentioned in my paper, Maddox (1986) wrote a fascinating critique in *Nature* asking, "Are physicists about to hijack DNA, and when are molecular biologists going to buy a klystron and some waveguide" to perform experiments based in physical, rather than chemical percepts of biological organization. In any event, there are those who would come among us who share our interests and our goals. We would do well to view them with interest and compassion, rather than with a xenophobia that would brand them as savages bent on an evil purpose.

I conclude this brief evaluation of what we have heard here with a personal insight. I have been changed by my associations in recent years with molecular biologists, cell biologists, and physicists to the point that in considerable degree I should admit to being "born again." Yet it is being born again into the very environment in which my early interests began, namely, what is the function of this wave activity in brain tissue?

References

Abey T, Rosenberg B (1982) Microprobe for monitoring biophysical phenomena associated with cardiac and neural activity. US Patent 136, 317, Aug, 1982

Adey WR (1987) Biological models of electromagnetic interactions with tissues: a review and synthesis of recent findings. In: Anderson LE, Kelmar BJ, Weigel RJ (eds) Interactions of biological systems with static and ELF electric and magnetic fields. US Department of Energy, Washington DC, p 237 (Symposium series DOE60)

Fröhlich H (1968) Long-range coherence and energy storage in biological systems. Int J Quantum Chem 2:641–649

Fröhlich H (1975) The extraordinary dielectric properties of biological materials and the action of enzymes. Proc Natl Acad Sci USA 72:4211–4215

Fröhlich H (1986) Coherent excitation in active biological systems. In: Gutmann F, Keyzer H (eds) Modern bioelectrochemistry. Plenum, New York, p 241

Hameroff S, Smith SA, Watt WC (1984) Nonlinear electrodynamics in cytoskeletal protein lattices. In: Adey WR, Lawrence AF (eds) Nonlinear electrodynamics in biological systems. Plenum, New York, p 567

Kelley JT, Everett R, Reilly EL, Colton GS (1984) The relationship between flash evoked potentials and evoked amplitude-modulation patterns of an applied UHF electromagnetic field in the rat. Bioelectromagnetics 5:365–376

Larsen LE, Jacobi JH (1979) Microwave scattering parameter imagery of an isolated canine kidney. Med Phys 6(5):394–403

Lawrence AF, Adey WR (1982) Nonlinear wave mechanisms in interactions between excitable tissue and electromagnetic fields. Neurol Res 4:115–153

Maddox J (1986) Physicists about to hi-jack DNA? Nature 324:11

Pohl HA (1984) Natural AC electric fields in and about cells. In: Adey WR, Lawrence AF (eds) Nonlinear electrodynamics in biological systems. Plenum, New York, p 87

Schwan HP (1957) Electrical properties of tissue and cell suspensions. Adv Biol Med Phys 5:147–209

Schwan HP (1977) Some tissue determinants of interactions with electric fields. Neurosci Res Program Bull 15:88–98

Remarks by W. J. Freeman:

Brain Dynamics and Differential Equations

Dr. Adey, you're a tough act to follow – remarkable eloquence, and our model from early days! Having had this ambition of dealing with how brains work for about 35 years, I take it very seriously. I contend that even now we know, in some sense, how brains work. We have a conception, by no means complete and possibly never to be completed, that brain is a dynamic system, or, as Dr. Bullock would put it, that each species of brain is a dynamic system.

For me dynamics is process or exchange and transformation of energy in matter, and I prefer to use the language of classical physics, which is differential equations, to describe dynamics. When I'm able to write a set of differential equations which suffices to predict the performance of a brain region with respect to a certain set of initial conditions and an output, this tells me *how* the brain is working. We don't seem to hear that old dictum so often nowadays, that the business of science is to explain how, not why; not why does the apple fall, but in what manner does it fall? Now, as Dr. Adey mentioned, what do we mean when we say we can explain something? The traditional dualism of explanation says there are two approaches, supposedly antagonistic, called reductionism and holism. We can work downwards to progressively smaller parts, or upwards to greater complexity; downwards from neurons to organelles or upwards from nuclei to cell assemblies and to cortex. Each direction has meaning in terms of explanation at multiple hierarchies, going from one to another. But the common interpretation of this duality – the trees versus the forest – misses the point, because the important thing is what satisfies us, what gives this internal sense of "Yes, now we *understand!*" I think this satisfaction in "understanding" comes in two forms: One is the causal chain, as from imposed current to depolarization to conductance change, and so on and on, from A to B to C; at each step there is something that makes sense to us. Another is establishing predictive relationships among variables, so we can write a set of equations by specifying the initial conditions and then we can solve them in order to have a prediction of what a system will do, but without necessarily understanding what we mean by causal connections within the system. Prediction is not necessarily explanation, but since the work of Kenneth Craik it has become commonplace to interpret explanation as the building of a model which will represent what we know about the system.

Both of these senses of explanation are valid, but I suggest that both are in deep trouble, when it comes to understanding brain function. Taking our first, our causal chain, in elemental steps: essentially they begin to fall apart when we introduce feedback. Until the first third of this century the concept of feedback in nervous systems was formally disallowed by our intellectual leaders, such as Ramon y Cajál, who said it could not exist with reference to brain functions, because it led to indeterminacy and a nervous system could not work in that manner. This may seem faintly absurd to us now, yet is was clearly documented. Feedback does introduce indeterminacy and to that extent it weakens our notion of causality. This becomes more acute when we go to distributed parallel processing (PDP), where we have large numbers of elements and no longer have a clear distinction between cause and effect. Can we have any concept of causality at all in larger systems? I recall an old concept of clinical practice that I was taught, and one especially relevant to the autopsy and pathology: You cannot say that X died *of* Y, only that X died with it. This difficulty with causality is compounded, of course, when we introduce nonlinearity in the elements, which are engaged in this kind of "whispering," as Ross Adey so elegantly put it.

So we have a breakdown in our ability to reason about cause, even if it does exist out there in the "Ding an sich" – the Noumenon – but we cannot know it. This breakdown is paralleled now by an imminent collapse of our ability to make predictive relationships, which used to be our alternative. The latter can often be nonintuitive, which is a disadvantage; certainly noncausal. But now we have to rely for our predictions increasingly on numerical simulation with digital computers which will shortly become too small and too slow, so that we can no longer handle the large numbers of relationships among the numbers of variables that we require for a solution within a reasonable time. The first problem, of causality, has been with us since at least the time of David Hume, obviously; it will continue to obtrude as we go to higher complexities – particularly when we go to the problems of basin and attractor theory, which we have just touched on in our workshop.

The second difficulty I think does have solutions, and here is how I hope we shall go in the next few years: My own approach in neurophysiology for the past 30 years has been to record the waveforms of multiple events simultaneously and then to construct differential equations, starting with linear equations based in the time-honored principle of superposition, then going to coupled nonlinear, ordinary differential equations; so that now I am able to have expanded solution sets which include not only equilibrium and limit-cycle solutions but also chaotic solutions. This is clearly a significant advance, though by no means a terminal position.

Where do we go next? The state variables are now too various for for digital simulation, so I must develop hardware simulations allowing a reversion to analog computation and large-scale integration with dedicated chips; this may permit simulation of neural elements equivalent to those of the olfactory bulb, prepyriform cortex, entorhinal cortex, hippocampus, etc. by devices which have the speed, flexibility, and adaptiveness that we find in the nervous system. We are no longer limited by the number of connections involved, a difficulty which has restricted us to nonphysiological simulations until now. This kind of very large-scale integration (VLSI) raises another apparent major concern – connections. Connections are the present problem for engineers who have taken to neural networks like ducks to water and who, as Ross Adey has pointed out, may well swamp us if we fail to keep our heads above water. However, engineers do have inherent limitations in that they show innate distaste for oscillations or chaos. So, if we can handle oscillations and chaos we shall keep several steps ahead, to our benefit. But how to make the connections and then verify or describe the proof that these connections are appropriate? This is the problem of describing attractors.

One motive for this conference was how to describe the geometry of attractors. In any set of differential equations description of the equilibrium points is the main point – description of those states to which the system will go and stay until you boost it out again, or the limit cycle. Description of these chaotic attractors will challenge the best of us, especially when it goes beyond our current limitations of 3D representation. That's why I think Otto Rössler's concept of the 4D flight simulator is so potent, and, as he has pointed out, Poincaré first suggested kinesthetic, muscular involvement in the control of the perceptions laid onto the retina as "higher dimensional." As Otto puts it, experiencing a 4D cube is like going into a house with eight rooms; if you put a bit of furniture into each room, as you go from one to another, so you know where you've been, then on your return you'll know how you got there. Keyboard manipulation of the self in relation to this structure can, within 30 min or so, make the self at home with a 4D structure, whereas a visitor looking over your shoulder will have no idea what's going on. In our laboratory we now have some personal experience with this kind of modelling, using an Evans and Sutherland PS300 – as our programmer puts it, we have "kluged" high-speed 3D Fortran microcode to simulate 4D. So we do now have this technology, and we should take full advantage of these new windows, as Dr. Adey has said. Moreover, I feel that we should continue this work directly with human recordings, since our focus is our own brain, and that is why I am looking forward to collaboration with Alan Gevins and his group in San

Francisco. Here the problem is how to deal with many electrodes and the resulting mass of data, for one trial may involve 1–3 million bits in a few seconds, while an afternoon may produce 100 million bits. Immense computing power will be needed if we are to complete anything useful within the lifetime of one research grant! Yet these multiplying complexities are within reach of our hands and minds. We may be optimistic about what we can hope to learn and literally understand about brain function during the coming decade.

Remarks by L. DEECKE:

From Readiness Potential Onwards

Well, I know how the brain functions: very economically! There are estimates that only half a watt is taken up when the living brain is taken into account. I want to make a few additions to the discussion by Dr. Galambos and others of yesterday, from our viewpoint of the readiness potential *(Bereitschaftspotential)*. We see from surface electrodes many different generators, as Dr. Grüsser said – incoming fiber activity, glial cell activity – but we hardly see neuronal spike activity, and this has also been shown by Dr. Creutzfeldt; the action potential itself cannot be seen from the surface. I think the participation of glial cells is the most important thing in *propagation* of the neural electrogenesis, because the glia form a syncytium with all kinds of tight junctions, which can distribute electrical phenomena very easily in the brain. It we ask what our readiness potentials are, we cannot explain them, but we can go a step further in description by calling them a negativation of the cortical steady potential. Although this steady potential and its negativation are epiphenomena, they can tell us where in the brain there is more and less activity.

Does this help us to understand how the brain functions? Unfortunately, no! As I've put into print, we can see only "where the wires are hot." We cannot read the messages going through these wires. We just have an indicator of activity or no activity. If you ask me about the future of learning how the brain works, I think the information in the wires is the thing, and this can be recorded only from single cells.

What Dr. Eckhorn is doing with multielectrode correlations is the way of the future, but it's immensely complicated because it must be done with *identified* neurons. Identification of the neurons from which you record is paramount in cortical electrophysiology; and the classical way of identifying pyramidal neurons is another method by which we can move stepwise forward, but it's complicated.

Now, what is this computer-brain? It's easy, but probably incorrect, to say that the brain is a huge computer, but in calling it an information processor we often fail to recall that the human brain is the *end result* of a very successful optimization. Given enough time, selection must produce an instrument optimally adapted to habitation on this planet. And I think this is what our brain is. It has been called the most complicated 1.5 kg of matter in the universe, but that's still an open question as long as we are ignorant about extraterrestrial intelligence. To take the extreme zoological view, one can say that the end product is a brain which has this ability of coping with conditions on this planet Earth. But this does not mean that we have either "pure mathematics" or "pure logic" in our brains, as Kant, among several others, pointed out: There is no pure logic, and we don't have a rational apparatus, only a "ratiomorph" which looks like rational logic but has actually been conditioned by trial and error mechanisms.

We should keep that in mind, and I am reminded of it by recent records from single units in the temporal lobe of conscious man, work done by Creutzfeldt in Vienna; now of course this was done with all the ethical and practical constraints of human neurosurgery in patients with brain pathology. He tested these single neurons with a variety of stimuli, both psychophysical and natural, and found it impossible to find out what these single neurons did. I mention this only as an illustration of how difficult it is to identify single units in cortex. Perhaps we shall never be able to find the final secrets of the brain.

Finally I want to ask about consciousness: Why did nature invent it? Is it necessary? One could think of living without it. An oversimplified answer might be that in phylogenesis consciousness was just a necessity of data reduction. A hierarchy of selection processes was necessary in order to cope with the abundance of data coming in, and only the most important thing, where the subject focuses attention, comes from antechamber to antechamber to consciousness. Whatever metaphors we use – Jung's spotlight on the stage or whatever – I think conscious experience is simply a mechanism enabling us to focus, for a moment, on the things we are interested in, and to disregard all the others. But it takes this high-selection machine to know what can be discarded without danger; and this is the purpose of a higher brain. This should make us also modest, because we should always remember that brains are intended to adapt us to this planet. This aspect must be taken into account when we think about how the brain works.

Chen: I want to recall what Dr. Bullock said about rapid recognition of languages which we don't understand semantically. I think that the past 40 years of EEG study, and our own past 20 years of ERP study have given us a lot of ideas, about the context, but not the content. But we are searching for the content. So the question ahead of us is how are we going to propose some kind of microtheory or macrotheory for this content. We need help from communications experts. Could we invite some experts from that area to one of our future meetings?

Another point is our confounding of micro- and macrotheory; it is difficult to expand a mathematical theory into a big model to explain such a variety of things.

Finally, we haven't paid much attention to one major information area, and that is imagery. We analyze a lot, but we haven't much analysis of imagery: How do we mentally create our imagery as part of our information, and how do we study this kind of domain? This looks like another challenge.

Deecke: Thank you very much for making this point, because we do now have tools for investigating silent thinking, for instance with positron emission tomography (PET). Silent thinking is coming close to what we do when we make discoveries or are creative. We now have a chance to observe the topography of such higher functions. But I don't think that consciousness will show in the PET.

Sheer: Since I come from a simplistic discipline, let me try to present a simplistic notion. We don't really know what the problem is. We have two sides of an equation where each side is unknown – one side is the physical workings of the brain, expressed in the physical operations of structure, electricity, and chemistry; the other is mind, cognition, and behavior. On the left-hand side we have a tremendous armamentarium, in terms of technology and data, which we don't know how to put together; on the right we have terminology about which we know nothing in terms of brain, just molar rubrics developed from behavioral observations and which are chapter headings in psychology textbooks, such as "memory, perception, and learning." All of the latter have to do with the whole brain working, and there is no way to make definitive relationships between those kinds of words and some of the organizations of the physical measurement operations on the left-hand side. We must break these molar rubrics down to words which have meaningful relevance to the physical functioning of the brain, words like coding, energizing, and storage; there could be other words, these are just successive approximations. For example, the word "storage" does not seem to be resolvable at the level of electrophysiology, because it is essentially a structural chemistry problem. But I think words like "energizing" which suggest dynamic processing, are resolvable at the electrophysiological level. Why don't we agree at this meeting on a problem, a first-order interaction of a simple system that has to do with the energizing function and define that; then decide that we shall all use our armamentarium to work on this one problem for two years and return with our progress?

Deecke: I couldn't agree more. An interdisciplinary approach, using many different tools on the same question, should be encouraged. For example, EEG, magnetoencephalography (MEG), and PET, perhaps with rCBF, could be so used.

John: It's an excellent idea and I've been sitting here recalling some animal preparations that have fallen by the wayside. Dr. Galambos did some interesting lesion studies, so did others, transecting the optic pathways so that prechiasmically about two-thirds of the optic nerve was cut and postchiasmically another two-thirds was cut. So from retina to cortex it was hard to find an uninterrupted pathway. Those animals retained, and could acquire new, pattern discriminations. There's an enormous literature on the absent or minimal effect of multiple-stage lesions, where the same lesions imposed in one step would have had catastrophic behavioral results. Similar effects have been shown for multiple-stage, bilateral lesions of the reticular formation, where the multiple-stage animals retained their sleep-wake, arousal, retention, and learning almost unimpaired, whereas coma would have been predicted from similar single-stage lesions. There are ways to challenge the distribution of function so that you take away from the brain what it usually uses and the function goes some place else. "Nature's experiments" in the human animal – strokes, injuries, neurosurgery – provide examples too; a location in the brain isn't there any more, yet recovery of behavior occurs. Are we sufficiently using opportunities given by neuropathology in man and by new technology for studying those pathological defects, in detail and analytically, in other animals? We now have high-quality imaging techniques designed for humans which we should be applying experimentally, in a sophisticated way, to animal paradigms. These tools were built for man, and we've lost sight of the fact that they now give us a whole program of new studies in other animals. I can't recall seeing one picture of a PET scan of an animal performing a task, and then of the same animal performing the same task after a multiple-stage lesion. This is a big loss of an opportunity!

Remarks by A. STARR:

View of a Neurologist

Roy John has opened up the area I want to talk about, and that is what we learn about how the nervous system works, from ourselves. There are two approaches: (a) That there is localization of function; and (b) that it's distributed. I think both approaches are correct, and there may be many other approaches. Perhaps the Rashomon movie metaphor would best describe my view: An event occurs while we are traveling on the road; a movie-maker interviews each of us separately and objectively; we all are trying to tell the truth but we all have different stories. This may be the continuing theme of brain research, that there are multiple nervous systems and multiple ways it can function.

For the clinician, the only unifying theme is that the stories patients can tell following a lesion of a particular part of the brain have a remarkable internal coherence. After a lesion of the left occipital lobe they lose the ability to see to the right. People with lesions in the right frontal lobe will say there's nothing wrong with them. So we tend so say there's localization of function in the part of the brain that's been lost. It's logically incorrect, since what we see is how the rest of the nervous system functions in the absence of that structure. I retain a romantic reference for saying that the lost function resides in that structure! By looking at patients and listening to patients one gets a sense that there is a localization of function to particular parts of the nervous system. I'm thinking of descriptions of how the temporal lobe functions in epilepsy, for instance by Dostoevsky. Such descriptions are in many ways akin to normal sensations that we have, though grandiose.

I use tools like the EEG and EP to tell me more about the nervous system than the patient can tell me. Our visual evoked potential records (VEP) from patients with multiple sclerosis have revealed very abnormal function of the visual pathway, while those same patients thought they had normal vision and told us so; at some point in their life they had suffered a lesion of the optic nerve. This can only mean that we haven't been testing visual function correctly. One possible interpretation is that optic nerve function is distributed, and after the lesion other parts of the brain take over. But I don't think this is the answer, because many stroke patients who have "recovered" may take a sleeping pill and wake up in the morning paralyzed again; the barbiturate has done something. So I think that there's an incorrect assessment underlying this concept of distributed function. I like Roy John's approach; yes, we must go back to patients; but I use these tools to see how the rest of the nervous system functions, thinking that the loss of function is *in* the temporal lobe, or *in* the optic nerve.

John: I'm thinking about Sprague's experiments when he removed visual cortex unilaterally and had contralaterally hemianopic monkeys. Then he lesioned ipsilateral superior colliculus and showed disappearance of their hemianopia. A subsequent explanation said that the visual cortex normally inhibits the superior colliculus, while the latter inhibits the lateral geniculate, so the intact cortex allows information through the lateral geniculate for distribution to at least 11 retinotopic mappings; but when the cortex is damaged the superior colliculus is released from the cortical inhibition and can inhibit the lateral geniculate, so you become hemianopic. Visual function did not reside in the visual cortex, but inhibition of the superior colliculus resided there. I disagree with Dr. Starr. The message is not *in* the node which you cut and through which it passes.

Starr: I repeat that I'm a romantic and I *want* the function back in the occipital lobe! I'm thinking of those patients who lose the ability to use certain classes of names, such as flowers, while retaining those of birds. There must be a brain *structure* underlying these specific losses and retentions, and similarly for our storage and classification of parts of speech in language, for instance, where analogous human losses and retentions can be shown. I don't know whether such specific functions reside in nodes or cell groups, but I feel that individual words are coded in particular groups of cells. Perhaps the concept of chaos reflects the ability of a mathematician to look inside his own nervous system and say how it's working.

Başar: I sense an interesting dynamical change in our discussion, from oscillations at the atomic level with Poincaré to patients with cortical lesions! I thank our two neurologists for showing us how to understand something about brain function by observing patients. I'd like to remind you that Dr. Starr began the study of P300 waves in the event-related brain potential (ERBP) with patients having brain pathologies; this may be an interesting new direction. Now, in thinking about so-called lower levels (if indeed superior colliculus is functionally lower), I wonder about the study of animals which have no cortex ("Nature's lesions") or have different sensory systems from ours, such as electric fishes or lampreys.

Deecke: To return briefly to your tectal vision, Drs. John and Starr. You said that the hemianopia disappeared, but we both agree that this recovery reflected a different kind of vision. I think Dr. Grüsser would agree, for it is his story, that this is not really "conscious vision" which is restored.

Starr: I'm sorry, but how do you know that? He was still performing pattern discrimination.

Deecke: Yes, that's right! My point is that the recovery was shown only behaviorally.

Starr: The concepts of "conscious and notconscious vision" are too artificial. I have two models: (a) The child born with one occipital lobe lost; these people do not show detectable hemianopia until they are adult, when they go to an ophthalmologist for eyeglasses. He finds half the optic nerve white. Only then, under controlled conditions, do they reveal their hemianopia. There had been no evidence of a behavioral problem until then, because the child had developed head movements or other maneuvers for pulling in the visual information. Only if you restrict their movements can you pick up the hemianopia. The nonneurologists among you would fail to notice the important details about a patient, and I could say that you are all blind. I'd have to teach you to look for certain things. Vision is a dynamic process, and Roy John in a way is right, that the cat is behaving and doing the task, but in other ways it is blind. The person who is blind can *behave* perfectly normally; vision is dynamic and clinical data tell us that.

John: Experiments by Cameron and Chase (those are the names I associate with the studies, anyway) showed that multiple-stage occipital-pole resections in rats, bilaterally, prevented them from performing or learning pattern discrimination. Other groups had a unilateral destruction and were kept in the dark for 2 weeks; then the second lobe was destroyed. They too could neither perform previously learned nor new pattern discriminations. A third group was allowed to live in light for 2 weeks and it didn't show pattern discrimination ability when the destruction was made bilateral. A fourth group was carried through the pattern-discrimination maze, by being pulled in a cart by another rat who knew how to solve the maze; this fourth group had the visual experience, but not the sensorimotor experience, and they too could not perform the pattern discrimination, or learn it. The most interesting group was allowed to perform pattern discrimination for 2 weeks before the resection was made bilateral, by running through the maze; they alone could subsequently perform pattern discriminations and learn new ones. Now I don't care if the fifth group was blind or what you wish to call the *quality* of the vision. What I want to debate is whether it is reasonable to say that an animal, looking at patterns and learning associations which enable him to search for and find food, has something sufficiently analogous to vision, sufficient for us to say that energy reaching his retina, within a certain range, has information that is useful to him. The suggestion is that the association of visuomotor simultaneous processes enabled a new system which could do something that couldn't be done before because it wasn't necessary. As long ago as 1884 experimenters said that when things work together in the brain a system gets established so that *part* of that system can do the whole job. I could cite human studies – the woman with childhood temporal lobe damage who developed uncontrollable seizures and at age 34, when she was both hemiparetic and hemianopic, has a radical hemispherectomy because she could not be managed by drugs. This case got published in *Neurology* because the authors thought it was interesting that she was neither hemianopic nor hemiparetic after that hemisphere had been removed. Their tests may not have been the kind we would perform, but their clinical impression was that she was no longer hemianopic or hemiparetic. This was not a cat but a human.

Starr: I didn't want us to get embroiled in who is right or wrong; these patients have a controversial phenomenon and our approach is to do the electrophysiology, biochemistry, or whatever, because we have a definite function and definite behavior. I just wanted to present our strategy towards understanding how brains may work, and this is the kind of dilemma that I like to get into.

Sheer: The interpretation of those multiple-stage lesion studies, and also of these controversial human data, is not by a sensorimotor connection but by the interdiction of an energizing function – an effortful locomotion which produces the intergroup differences. The rehabilitation-from-brain-damage literature suggests the importance of effortful, conscious training for any kind of restitution of function.

Another comment: I think it's unfortunate that we haven't got into any violent arguments at this meeting. A lot of good research emerged from the bitter disputes between Golgi and Cajál or Volta and Galvani! Now, I don't want to start anything ... [laughter]. I don't think that diagnosing schizophrenia with molar electricity will lead to much understanding of the brain; nor will the discrimination of exquisite interrelationships between the identified neurons of *Aplysia;* these two approaches are too far removed, conceptually and phylogenetically. We must start somewhere about the middle of the phylogenetic scale, and in the middle of our conceptual ladder, rather in the manner of Sherrington. Our successive approximations should start somewhere in the middle, with low-order concepts, make connecting links, and then proceed only with small inferential steps. But if we continue to use these tremendous computer programs and simulations, and these complicated color displays (I don't understand them!), and with them try to understand mechanisms, it's going to take us a long, long time.

Mpitsos: I often wonder if what I do is really useful and you may too in terms of understanding brain function, I mean. I once heard a humanities undergraduate telling a science major that real science is literature. Over the years I've wondered if that may not be the case. If we want to understand how the brain functions, maybe we ought to deal with it in those terms, which deal with whole brain function. It may be nice to know what neurotransmitters are, what the cytoarchitecture is, what synapses can do; but those are questions in cell biology. Is there really anything in neurocircuit analysis? I don't necessarily agree with this attitude. I'm just asking another question.

Bullock: I think it's appropriate that Dr. Mpitsos should have the last word, as the token minority here representing the poor sea slugs. He has just used a nice word for what Dr. Babloyantz called "peddling." She said she is peddling her science, and much of science, of course, is trying to sell one's point of view to others. Dr. Mpitsos has called it literature!

On behalf of all present, panelists and listeners, I thank you, Dr. Başar and all your staff. We look forward to the next meeting – and the new understandings which we may possibly attain – as this one is now adjourned.

Analysis of Strange Attractors in EEGs with Kinesthetic Experience and 4-D Computer Graphics

W. J. FREEMAN

1 Introduction

Newly developed techniques for numerical analysis of the time series derived from physical and chemical systems displaying turbulence and other seemingly random behavior have recently been applied to electroencephalographic potentials (EEGs). Estimates for the lower bounds of measures of the fractal dimension, Hausdorff dimension, Lyapunov exponents, Kolmogorov entropy, and other related coefficients have indicated that by suitable means of observation the background "noise" generated by brains may be nonrandom, and may be constrained by as yet undefined aspects of neural mechanisms generating it. In a word, brain activity, at least in some states and in some brain parts, appears to be chaotic and not stochastic (Rapp et al. 1985; Freeman and Viana Di Prisco 1986; Babloyantz and Destexhe 1986; Havstadt and Ehlers 1987; Meyer-Kress 1987; Freeman 1987a, b, 1988).

This inference may have profound consequences for theories of brain function, leading to better understanding both of the neural mechanisms generating chaos, and the purposes that it may serve in the genesis of goal-directed behavior (Skarda and Freeman 1987). In order to exploit this new opportunity we need to know much more about the dynamics of neural masses that underlies the appearance of the EEGs recorded in behaving animals and man.

2 Methods

One way to study the dynamics is to plot the set of points in a time series (an electroencephalographic or EEG trace as exemplified in Fig. 1) in a coordinate space, in which one axis is the amplitude at each time t, and each other axis (Fig. 2) is the amplitude at a time lag $t+nT$, where n is an integer 1, ... up to a desired number of axes (a Ruelle plot) and T is a time increment usually expressed as a multiple of a digitizing interval. The figure will appear as a cloud of points that may be rotated for projection or section into any plane in the space for display. It is convenient to connect successive points with a line segment for display as a continuous trajectory. When it is plotted in 3-D or a higher dimension the figure may be rotated and translated to inspect its shape.

A variant on this approach is to record EEGs at two or more sites and use these rather than one trace repeatedly lagged. Yet another is to rotate the coor-

Springer Series in Brain Dynamics 2
Edited by Erol Başar and T. H. Bullock

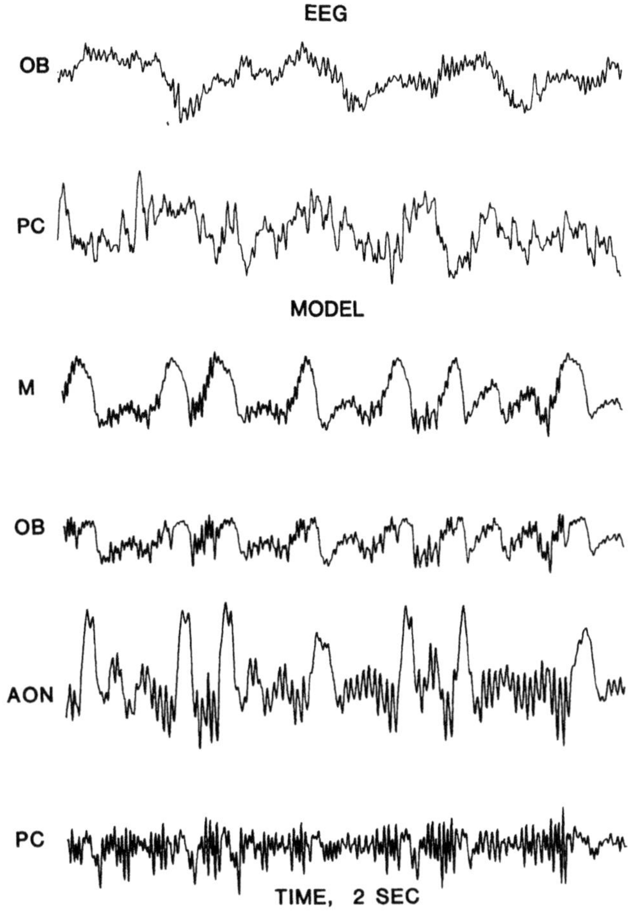

Fig. 1. Examples of EEG activity recorded from a waking rat from the olfactory bulb (*OB*) and prepyriform cortex (*PC*), and of simulated EEGs from a model of the olfactory system including outputs from the mitral cell population (*M*) and the anterior olfactory nucleus (*AON*). (From Freeman 1987b)

dinate axes by singular value decomposition or factor analysis, so that the axes are orthogonal for plotting the n-dimensional distribution of the variance (Mees et al. 1987).

A successful outcome of these procedures is the detection of spatial structure in the projections into one or more planes of section. The cloud of points or line segments manifests "flow" of "activity" in the system, that may expand in some directions and contract in others, and that may be folded back into itself to form loops and whorls. These structural features are manifestations to be expected of the constraints or invariants of the dynamics that generates the neural activity.

If the duration of the time series, the temporal digitizing interval, and the time lags and recording locations are all varied and ultimately optimized to enable the extraction of these visual features, then rotation and translation of the cloud or of the position of observation should convey to an observer a perceptual experience that is reproducible over repeated sample traces of the EEGs from the same or similar subjects in the same or similar states. The "object" is then communicable by a photograph, sketch, or cinema.

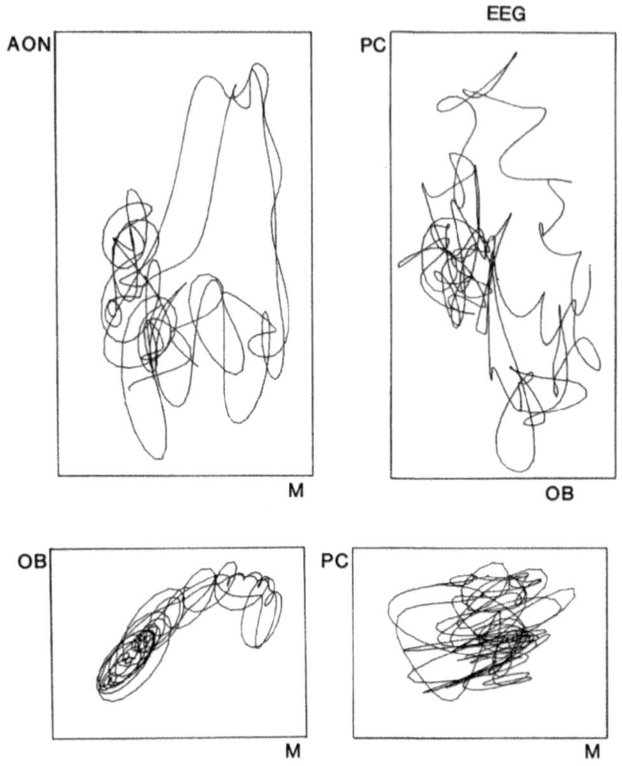

Fig. 2. Examples of tangles generated by plotting one variable against another for the EEG (*upper right frame*) and the outputs of the model. The dimension estimated by the Grassberger and Procaccia (1983) and Guckenheimer (1984) methods averaged 5.46 for the model and 5.92 for the EEG (Freeman 1988). (From Freeman 1987 b)

3 Modes and Models

These prescriptions are of little use with the conventionally recorded scalp EEGs, which tend to consist of potential fields from unknown numbers of neural "generators" all summing in the volume conductor of the brain and scalp tissues, and over time epochs that in the waking or conscious condition will include indefinite numbers of state transitions subserving the ongoing evolution of goal-directed behavior. Contributions from electrode noise, muscle potentials and movement artifacts may serve further to inflate the overall variability of the data. The variability may be arbitrarily reduced by improper use of smoothing and detrending filters, and by use of an excessively small digitizing interval or an overly brief duration of recording epoch (Albano et al. 1986). Examples of Ruelle plots of EEG data typically lack interesting and reproducible structure and resemble bowls of spaghetti.

What is needed is some prescription on what look for by way of structure. In our experience this requirement is best met by constructing models of the dynamics of neural masses, which consist of sets of coupled nonlinear ordinary differential equations (ODEs), for which the solutions generate time series and spatial patterns that simulate the corresponding features of EEGs. The most useful models are those that closely hew to the known anatomical connections and open-

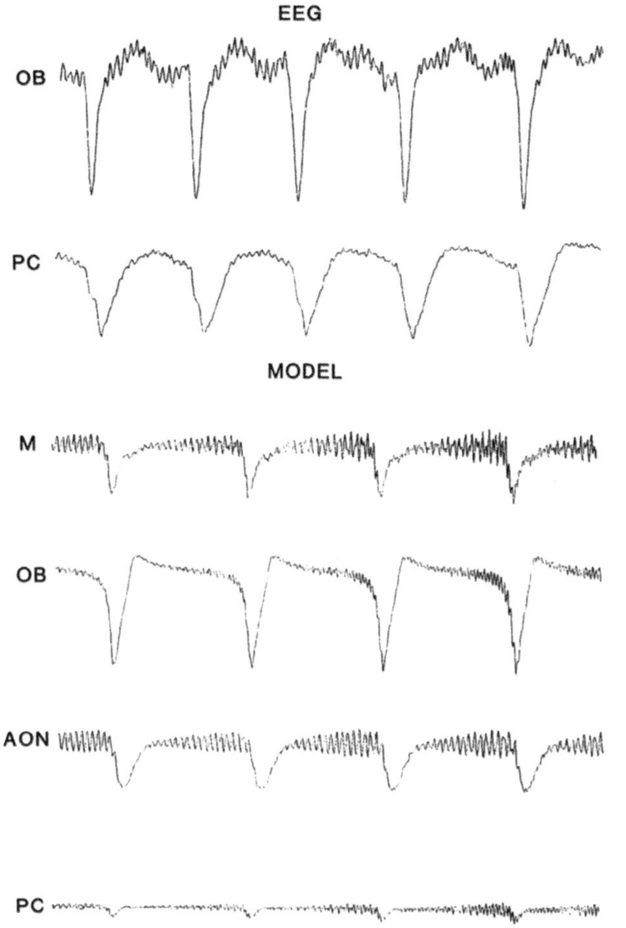

Fig. 3. Examples of seizure activity from the rat EEG and from the model. The correlation dimension of the output of the model averaged 3.76 and that of the EEG was 2.52 (Freeman 1988). (From Freeman 1987 b)

loop properties of component neurons (Freeman 1975, 1987 a, b; Freeman and Skarda 1985; Baird 1986).

Our best example is a model of the olfactory system that suffices to generate chaotic activity that with appropriate parameter settings is not only indistinguishable from low-pass filtered random activity (so-called colored noise as distinct from white noise) by all conventional statistical tests but is also indistinguishable from olfactory EEGs at various sites, whether seen in time series or Ruelle plots (Figs. 1 and 2). The model consists of 30 first-order ODEs, but the dynamic elements representing local subsets of neurons are formed into four main groups with delays between them. This leads to the hypothesis that the dynamics modes, degrees of freedom, or in some sense "dimension" of this system is on the order of 4, not 30, and that the number of modes in the corresponding olfactory system is less than the numbers of nuclear clusters of cells or glomeruli (which are not readily defined) or the numbers of neurons (which may well exceed 100 million).

This hypothesis was evaluated by applying several numerical algorithms to the real and simulated EEGs from the "normal" basal state. The results have shown that the two sets of estimates of correlation dimensions are equal to each other within the standard error of measurement, and have suggested that they lie between 5 and 6 (Freeman 1988).

This part of the brain has the capability for generating an abnormal EEG pattern of partial seizure that is accompanied in animals by a form of epilepsy (Fig. 3). Experimental analysis shows that its onset is occasioned by a catastrophic "structural" failure, in that a synaptic connection fails between two of the main parts of the system. This "disconnection" can be construed to reduce the number of modes to three, with the expectation that the fractal dimension is reduced to between two and four. The same abnormal pattern is generated by the model when it is similarly disconnected (Freeman 1986). Our estimates of fractal dimensions by the Grassberger and Procaccia (1983) and the Guckenheimer (1984) methods show that our prediction of four dimensions was too low (Grajski and Freeman 1988) but that there is agreement between values for the EEG and the model, both giving values of five to six dimensions.

4 3-D and 4-D Displays

This condition of the "disconnected" state yielding seizure activity gives a highly structured cloud of points and connected line segments that, from various perspectives in Euclidean space, can be seen to conform to a filled-in 2-torus (Freeman 1987b). Analysis of the model shows that this pattern of activity can be reached by the Ruelle-Takens quasiperiodic route to chaos, starting from a point attractor and passing through two Hopf bifurcations before the phase transition to chaos. The structure can be construed to be a chaotic attractor, because in the model its size and shape is independent fo the path by which it is reached from an arbitrary starting point, although the form of the time series of the activity is strongly dependent on the initial conditions (Freeman 1986). The same conclusion holds for the animals because the seizure can be repeatedly induced with substantially the same time course to recovery. Within the seizure state the time course is not altered by transient perturbations induced by superimposed electrical stimulation, but no two time series are identical. This example is felicitous, because it has a chaotic attractor that can readily be embedded and examined in normal 3-D space (Fig. 4). It is a good test object for improving our methods of data procesing and display. However, a major problem emerges when the animal recovers from the seizure. Typically there is an abrupt return to normal behavior and a concomitant shift of the EEG back to its normal pattern. The Ruelle plots of the EEGs in this normal condition reveal a structure that is both interesting and reproducible, but it cannot yet be identified with or encompassed within any known geometric form. The simulated EEG as well yields structure in its phase portraits that is beautiful to behold but baffling. Both portraits manifest chaotic attractors, because they are stable under perturbation and recur from differing starting conditions at the time of transition into these reproducible states.

Our analysis of the model and our measurements on the fractal dimensions of the real and artificial normal EEGs lead us to believe that the attractor and

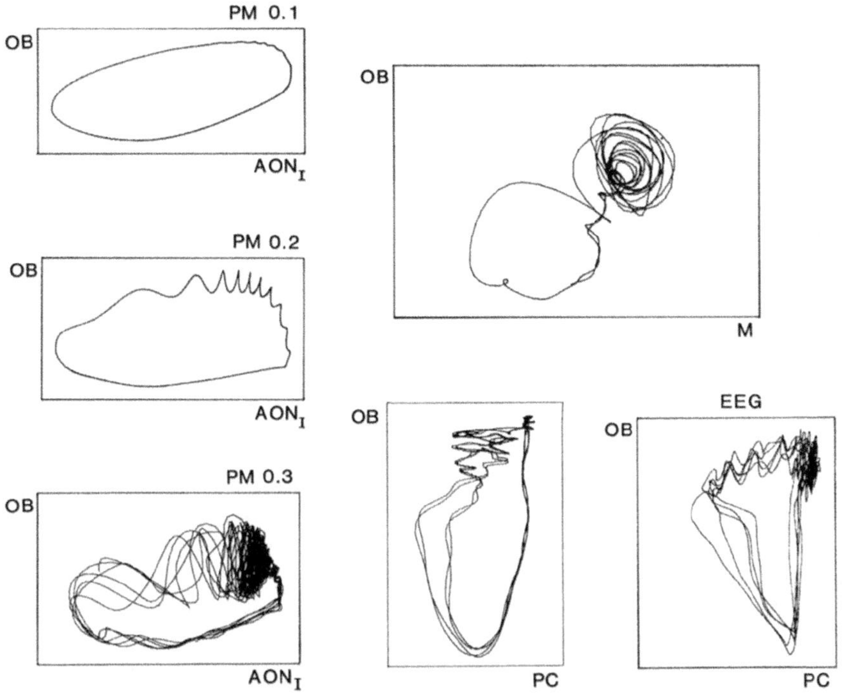

Fig. 4. Examples of phase plots of the EEG (*lower right*) and of the model. The series at *left* shows the sequence of two Hopf bifurcations in the model starting from a point attractor, culminating in the onset of chaos with increase in a bifurcation parameter *PM*, a synaptic connection strength in the model. (From Freeman 1987 b)

its simulacrum exist in 4-D or yet higher space. It is incumbent on us, therefore, to postulate a 4-D structure and deduce what 3-D geometric forms will appear when planar sections are taken through it. This procedure has been followed in constructing images of the 4-D hypercube, which in 3-D sections give cubes, octahedrons, and other regular solids (Banschoff 1984, 1986), and the 4-D hypersphere, which yields spheres and 2-tori in various projections (Banschoff 1986). A promising candidate for the normal basal EEG attractor is the filled-in 3-torus (hypertorus), because the model contains three coupled oscillators with incommensurate characteristic frequencies (Freeman 1986). One of the geometric forms that is predicted for a planar section through a 3-torus is a 2-torus, and some of the "views" we have of the attractor in three dimensions suggest the appearance of a 2-torus. Kaneko (1986) has shown that this structure also provides a route into chaos. However, there is not yet available an intensive study of this structure on which to base our expectations on what to seek, or to tell us how to define and characterize the features that will support or exclude the 3-torus as a hypothesis.

5 Kinesthetics and the EEG

The major experimental and conceptual problem we face now is how to explore the geometry of 4-D structures with the aid of computer graphics. A solution proposed by Rössler (1978, 1981) is to "fly the attractor" with the aid of a flight simulator. An example is given by H.-O. Peitgen's film (1985), "Fly Lorenz," for this famous strange attractor that can be visualized in 3-D. Experience has shown that an observer can rapidly become familiar with a 4-D object, when he is given the ability to move his position with respect to the object in the same way that an aircraft pilot moves his point of observation with respect to his visual world.

This experience is in close conformity to our understanding of how our visual nervous system operates. Our retinal input is restricted to 2-D images. From these we construct 3-D conceptions of our world, and the essential means is to move ourselves through the world and to move objects in that world by use of our limbs (Poincare 1958, 1963). Locomotion by voluntary goal-directed muscular coordination is the key to 3-D visual perception. There is good reason to suppose that our nervous systems can adapt to 4-D in much the way that they do to 3-D. As O. E. Rössler (personal communication) points out, experiencing a hypercube is like entering a house with eight rooms. Even in our 3-D world we must experience the rooms one at a time and build our conception by walking through them serially. The operator at a graphics terminal, if the visual feedback from his movements is fast enough to be "realistic," can quickly become at home in a hypercube, but someone watching the video screen over his shoulder will not know what is going on. Numerous studies in visual psychophysics have anticipated this result and strongly support this interpretation.

6 Implications

This convergence of neurophysiology, mathematics, and computer science holds great promise for all three fields. For the physiologist the immediate application is in the exploration of the hypergeometry of neural strange attractors revealed by EEGs, with the hope of deducing how they both shape and express the underlying neural dynamics, the ways in which they are stable, how they change with behavioral state, how pathological attractors can develop, and what remedial actions can be taken to reestablish normal dynamics. The use of computer graphics in 4-D or possibly in higher spaces may be the most important technical innovation in electroencephalography since the invention of the ink-writing polygraph as a clinical diagnostic tool.

The insights derived from our experience with 4-D perception may provide new ways to reformulate our ideas about visual perception and concept formation (Rucker 1984). The scientists most likely to benefit may be those who are intent on building systems and devices that can "see" in the real world and control their behaviors in 3-D space. The physiologist and psychophysicist already understand that perception is the incorporation of external information into the nervous system by movements into the world. Cognitive scientists place too great emphasis on the response to stimuli as distinct from the grasping of stimuli. If one desires a machine to read, one must first make it to write and to read what it has written. If one wishes to fly a plane, one must fling one's body about the sky with one's hands on the controls, until the machine becomes a part of the self and one's body image extends to the wingtips and the tail; reading about it will not suffice.

Mathematicians have made substantial progress in the past decade in their understanding of systems with three equations, but still little has been done with sets of four, which are the threshold for the domain of "hyperchaos" (Rössler 1979, 1983). The barrier seems to be in the inuitive apprehension of ("grasping at") the hypergeometric structures from the phase portraits of these systems. The ready access to these experiences through kinesthetics and computer graphics may open a field of new human endeavor in hypergeometry as broad as that of the clinical applications for the EEG.

Finally, the technology of computer graphics is already rich with new art forms (Schwenk 1976; Peitgen and Richter 1986). As cinema put new dimension into photography, and sound recording added another to that, the fusion of visual, reafferent and proprioceptive images by use of a graphics terminal and a mouse in each hand (and perhaps under each foot as well – this arrangement would be familiar to an experienced glider pilot) may open wide a door that has been sought by scientists and philosophers for a century or more (Abbott 1952). The cost of the necessary equipment at present is far too high for general availability. Those machines that are fast enough and have the capacity for handling the large numbers of data points needed have their microcode written in firmware for 3-D. Various tricks and strategems to circumvent this design limitation have their drawbacks, but the realization of full 4-D or even 5-D capability is a matter of cost and demand. We can expect within a short time to witness an explosive development with many as yet unforeseen new applications in "applied hypergeometry." Certainly our bodies are limited in dimensions but our minds are not.

Acknowledgment. This work was supported by a grant MHO6686 from the National Institute of Mental Health, United States Public Health Service.

References

Abbott EA (1952) Flatland, 6th edn. Dover, New York

Albano AM, Mees AI, deGuzman GC, Rapp PE (1986) Data requirements for reliable estimation of correlation dimensions. In: Holden AV (ed) Chaotic biological systems. Pergamon, New York

Babloyantz A, Destexhe A (1986) An instance of low-dimensional chaos in epilepsy. Proc Natl Acad Sci USA 83:3513–3521

Baird B (1986) Nonlinear dynamics of pattern formation and pattern recognition in the rabbit olfactory bulb. Physica 22D:150–175

Banschoff TF (1984) The hypercube (16 mm color film) International Film Bureau, Chicago

Banschoff TF (1986) Visualizing two-dimensional phenomena in four-dimensional space: a computer graphics approach. In: Wegman EJ, DePriest DJ (eds) Statistical image processing and graphics. Dekker, New York, pp 187–202, 18 plates

Freeman WJ (1975) Mass action in the nervous system. Academic, New York

Freeman WJ (1986) Petit mal seizure spikes in olfactory bulb and cortex caused by runaway inhibition after exhaustion of excitation. Brain Res Rev 11:259–284
Freeman WJ (1987a) Techniques used in the search for the physiological basis of the EEG. In: Gevins A, Remond A (eds) Handbook of EEG and clinical neurophysiology, vol 3 part 2. Elsevier, Amsterdam, chap 18
Freeman WJ (1987b) Simulation of chaotic EEG patterns with a dynamic model of the olfactory system. Biol Cybern 56:139–150
Freeman WJ (1988) Strange attractors that govern mammalian brain dynamics shown by trajectories of electroencephalographie (EEG) potential. IEEE Trans Circuits and Systems 35:781–784
Freeman WJ, Skarda CA (1985) Spatial EEG patterns, nonlinear dynamics and perception: the neo-Sherringtonian view. Brain Res Rev 10:147–175
Freeman WJ, Viana Di Prisco G (1986) EEG spatial pattern differences with discriminated odors manifest chaotic and limit cycle attractors in olfactory bulb of rabbits. In: Palm G, Aertsen A (eds) Brain theory. Springer, Berlin Heidelberg New York, pp 97–119
Gleick J (1987) Chaos. Making a new science. Viking, New York
Grassberger P, Procaccia I (1983) Measuring the strangeness of strange attractors. Physica 9D:189–208
Guckenheimer J (1984) Dimension estimates for attractors. Contemp Math 28:357–367
Havstadt JW, Ehlers CL (1987) Attractor dimension for small data sets. Preprint, La Jolla CA
Kaneko K (1986) Collapse of tori and genesis of chaos in dissipative systems. World Scientific, Singapore
Mees AI, Rapp PE, Jennings LS (1987) Singular value decomposition and embedding dimension. Physical Rev A36:340–346
Meyer-Kress G (1987) Application of dimension algorithms to experimental chaos. In: Hao Bai-Lin (ed) Directions in chaos, World Scientific, Singapore, pp 122–147
Peitgen H-O (1985) Fly Lorenz (16 mm color film). International Film Bureau, Chicago
Peitgen H-O, Richter P (1986) The beauty of fractals. Images of complex dynamical systems. Springer, Berlin Heidelberg New York
Poincare H (1958) The value of science. Dover, New York, pp 37–74
Poincare H (1963) Last essays. Dover, New York, pp 25–44
Rapp PE, Zimmerman ID, Albano AM, deGuzman GC, Greenbaum NN, Bashore TR (1985) Experimental studies of chaotic neural behavior: cellular activity and electroencephalographic signals. In: Othmer HG (ed) Nonlinear oscillations in biology and chemistry. Springer, Berlin Heidelberg New York, pp 175–205
Rössler OE (1978) Deductive biology – some cautious steps. Bull Math Biol 40:45–58
Rössler OE (1979) Chaos. In: Guttinger W, Eikemeier H (eds) Structural stability in physics. Springer, Berlin Heidelberg New York, pp 290–309
Rössler OE (1981) An artificial cognitive-plus-motivational system. Prog Theor Biol 6:147–160
Rössler OE (1983) The chaotic hierarchy. Z Naturforsch 38a:788–801
Rucker R (1984) The fourth dimension. Houghton-Mifflin, Boston
Schwenk T (1976) Sensitive chaos. Schocken Books, New York; Steiner, Dornach
Skarda CA, Freeman WJ (1987) How brains make chaos in order to make sense of the world. Brain Behav Sci 10:161–195

Chaos in Brain Function and the Problem of Nonstationarity: A Commentary

G. J. Mpitsos

1 Introduction and Historical Perspective

Over 16 years ago, the work of two of the participants in the present conference clearly pointed to the need to consider the brain as a noisy processor (Adey 1972) in which statistical mechanisms (John 1972) lead to the production of organized behavior. The difficulty in pursuing this problem further in terms of the individual neurons that comprise the network has been in the lack of conceptual tools with which to understand the informational language arising from simple neuroanatomic structures such as the convergence of two neurons onto a third (Mpitsos et al. 1978), or, as Sperry (1981) has stated it, in our inability to handle the "three-bodies problem." Bullock (1984) has observed that "circuit analysis," the once great hope of neurobiology, will not by itself "provide the major insights necessary to understand the emergent mechanisms present in complex systems," even if these systems are as simple as the stomatogastric ganglion of the lobster (Selverston 1980). The realization of this fact itself is perhaps a major achievement in invertebrate neurobiology. Nonetheless, while major strides have been made in understanding the self-organizing processes in distributed, parallel networks (e.g., Amit et al. 1985; Grossberg 1980; Hopfield 1982; Jeffrey and Rosner 1986; Kleinfeld and Sompolinsky 1987; Pellionisz and Llinas 1983; Rumhelhart et al. 1986; Sejnowski and Rosenberg 1987; Werbos 1974), the subject of noise has not been broadly addressed, and only recently has it received rekindled interest.

This interest has come about through the analysis of the dynamics in spontaneous and stimulus-evoked neural responses, and through observations of behaviors and of the neural patterns that generate them. From the application of dynamical techniques, it seems now that chaos may be involved in sleep and seizure EEGs (Babloyantz and Destexhe 1986; Babloyantz et al. 1985), baseline olfactory bulb EEGs (Freeman and Skarda 1985; Skarda and Freeman 1987), bulbar EEGs in response to odors (Skinner this volume, the responses of electrically driven neurons (Aihara and Matsumoto 1986; Hayashi and Ishizuka 1986), spontaneous firing of cortical neurons (Rapp et al. 1985), and in computer simulation of single neuron and network properties (Chay 1985; Labos 1986).

From studies of behavior, interest in variability has arisen in human studies which show that certain types of fluctuations in hand and finger movements may provide a mechanism for phase transitions to occur (Kelso et al. 1986). In our own work, observations of extensive variability in behaviors, motor patterns, responses of single neurons, and in their interrelated responses has led us to propose

Springer Series in Brain Dynamics 2
Edited by Erol Başar and T. H. Bullock

that variability may not be simply a product of supposed inexactness of biological systems, but, rather, that it may be an essential feature by which neural activity self-organizes (Cohen and Mpitsos 1983 a, b; Mpitsos and Cohan 1986 a, b). State space analyses of the dynamics in such motor patterns have supported this assertion by showing that the responses of key neurons in the generation of motor patterns may be governed by chaotic attractors (Mpitsos et al. 1988 a, c). Moreover, as has been proposed for the transition between chaos and limit cycles (Freeman and Skarda 1985; Freeman and Viana Di Prisco 1986; Skarda and Freeman 1987), we have suggested that chaos and other forms of variability may provide a means by which activity can shift from one motor pattern to another. Sources of variations may also be useful in optimization of network activity (Burton and Mpitsos 1988 a, b; Kirkpatrick et al. 1983; Jeffrey and Rosner 1986).

The above findings permit us to answer in the affirmative Adey's (1972) question: "Is it possible to envisage an information-processing system in which the very presence of on an ongoing noise-like activity produced no degradation in information-handling ability, and might even enhance it?" We can go further and build on Adey's insight to say that while there may be many forms of noise in brain function, among which we include variations having white spectra, or even invariant perturbations, variations arising from chaos cannot be separated from the informational content of the signal because they themselves are an integral part of the information and are generated by mechanisms that need not contain explicit elements of noise.

In the remainder of this chapter, I shall address the subjects of chaos, difficulties in assessing it in behavioral systems, the problem of nonstationarity, and the ability of neural networks to read and transmit chaos. Inasmuch as the intent of this chapter is to provide a commentary on the implications of chaos, and on difficulties in obtaining evidence for it, rather than an exposition of the findings themselves, I refer readers to the publications listed above for details on the specifics of the experimental findings and for the application of dynamical theory to their analysis. Further introductions and applications of dynamical theory may be found in Abraham and Shaw (1983); Babloyantz (this volume); Freeman (this volume; Garfinkel and Freeman (this volume); Gleick (1987); Shaw (1986); and in the volumes edited by Holden (1986) and Degn et al. (1986).

2 Duration Multifunctionality, and Nonstationarity of Coordinated Activity

2.1 Experimental Preparation

The experimental foundation, which we use as a point of departure to address broader issues in the following discussions, stems from observations of whole animal behavior, from motor patterns recorded in relatively intact preparations (e.g., McClellan 1982 a; Mpitsos and Cohan 1986 a; Mpitsos et al. 1978), and from the responses of individual neurons that take part in generating integrated output of activity in isolated (deafferented) nervous systems (Cohan and Mpitsos 1983 a, b; McClellan 1982 b; Mpitsos and Cohan 1986 b; Mpitsos et al. 1988 c).

Our experimental animal is the carnivorous marine mollusc *Pleurobranchaea californica,* a snail-like animal having no shell. As in humans, the most complicated behaviors of this animal involve its mouth, lips, jaws, and tongue. These behaviors include several components of feeding and regurgitation, rejection, which resembles a reverse of the sequential bite-ingestion movements of feeding, defensive biting, and self- and interanimal gill-grooming. (For photographs of feeding behavior see Mpitsos et al. 1978.) Although these behaviors produce obviously different effects (consider the difference between ingestion and regurgitation), their repetitive movements are quite similar to one another.

The nervous system of *Pleurobranchaea* contains about 10 000 neurons, several thousand of which may become active in a given behavior, but the crucial information for all coordinated buccal-oral behaviors must pass through and is generated by only about 30 neurons. We have examined the activity of individual neurons of this group and of the motoneurons that they drive, but the goal that our work is directed to is to determine the role of variability in the self-organization of coordinated activity within the group. Stating this in scale-independent terms, the experimental question has to do with the process by which individuals influence the activity of the group, and, in turn, how the group affects the function of the individual. Although our experimental animal may be idiosyncratic in many respects, we believe that the neurointegrative strategies found in it relating to variability and their role in individual and group action may have broad applicability. Moreover, we believe that the problems we have encountered in addressing variability in our experimental animal may also apply to other biological systems.

2.2 Duration and Variability of Biological Responses

Unlike mathematical or physical and chemical systems, such as the Lorenz equations (1963) and the Belousov-Zhabotinskii (Roux et al. 1983) reaction that can be generated indefinitely, biological responses are often short-lived. Moreover, the same behavioral effect may be generated by different movements on different occasions, first, because the external stimuli are probably not the same in those occasions, and, second, because motor patterns encoded within the central nervous system may be inherently variable. Thus, while it is possible to generate thousands of cycles in mathematical attractors, a biological response may be equally well established as an attractor but may last for only several cycles and, thereby, yield only marginally adequate information for characterization of the attractor. For example, bite-ingestion behavior in *Pleurobranchaea* and eating behavior in humans may involve so small a number of repetitive movements of the buccal-oral system that the available analytical tools for chaos may not be applicable, but the behavior may be adaptive and generated by a robust chaotic attractor despite the fact that we may not be able to characterize it as bona fide chaos.

2.3 Multifunctionality, Blending, and Nonstationarity

By multifunctionality we imply, first, the ability to produce different behaviors with the same motor system, and, second, the ability to produce the same behavior in different ways. Blending implies the ability to prodce a behavior that involves a smoothed combination of different behaviors (Bellman 1979; Mpitsos and Cohan 1986a). Blending can happen progressively; for example, the animal may begin feeding with a particular structure of its motor pattern, and, then as the response progresses, the same effect of feeding may occur but the motor pattern may change into one that has components resembling certin phases of regurgitation. Similarly, when the animal is turned upside down, it may begin to twist its foot in order to right itself, but when inverted and presented food, the foot may attempt to execute two behaviors simultaneously so as to twist in order to right itself, and fold around the food in order to grasp it. In this case, blending happens because of the demands of the stimulus environment, and, therefore, the behavior appears nonstationary because of external forces. However, we have found that deafferented nervous systems themselves can also produce nonstationary motor activity in which an apparently homogeneous, though variable, motor pattern can abruptly or gradually shift into a structurally different pattern (e.g., Mpitsos et al. 1988c). As in the case of variability in homogeneous responses, the shifts of activity in nonstationary patterns appear to be unpredictable, but further analyses are needed to determine whether they are statistically stationary.

Because of blending, it has not been possible in controlled and blind experiments to identify which behavior the animal was executing solely by examining the electrically recorded motor patterns (Mpitsos and Cohan 1986a). A similar situation occurs in the olfactory EEGs of the rabbit (Freeman and Viana Di Prisco 1986; Freeman and Skarda 1985; Skarda and Freeman 1987). In a remarkable series of experiments, animals were trained to lick to one of two different odors, yet the structure of the EEGs occurring in response to both odors was similar. Having evidence that the animals had learned to distinguish between the two stimuli, Freeman and coworkers reasoned and then demonstrated that the information conveying the different odors must reside within the dynamical aspects of the odor-evoked EEGs, despite the fact that such information could not be distinguished solely from the overall structure of the unprocessed EEG patterns. In our experiments, we recorded the motor patterns from muscles at the same time that the animals repeatedly performed particular behaviors. The electrically recorded motor patterns for similar and different behaviors were quite variable and usually indistinguishable from one another. From this similarity we concluded that the information for the different behaviors probably resided within the dynamics of the motor patterns recorded from the muscles rather than in the appearance of their temporal structure. As we shall discuss below, the process by which this information is generated does not reside exclusively within the nervous system but arises from a greater context resulting from interactions between the animal and the environment.

Because of similar drifts and blends in the activity of single neurons and in the pattern of activity arising from efferent roots, it has not been possible to determine the functional role of individual neurons in establishing motor patterns. This led us to propose:

1. From multifunctionality: That the same neuroanatomic framework consisting to a given set of neurons, their synaptic weights, and background of neuromodulators can generate different behaviors. That is, connectivity does not determine a behavior but the potentiality of a large set of activities.

To be sure, changes in the neuroanatomic framework, such as selective neuromodulation and synaptic changes produced by learning, can also lead to the production of different behaviors, but it is necessary to distinguish these possibilities from our proposal which focuses on the emergence of different motor patterns solely through the dynamics occurring within a particular neuroanatomic framework. Harmon's (1964) neuromime studies were the first to show that the same network can produce different patterns of activity. More recent computational studies have confirmed this (e.g., Amit et al. 1983; Kleinfeld and Sompolinsky 1987), and we shall discuss it below regarding chaos in simulation networks. However, because of their uncontrollability, even the simplest biological systems pose difficult experimental problems for studies aimed at proving emergent multifunctionality. Therefore, many questions we wish to address may be answerable only in computer studies that simulate the known characteristics of the biological systems.

2. From context: That knowledge of the functional identity of a given neuron depends on the context of its activity as it takes part in firing with other neurons.

3. From the fluid quality of contextual activity: That the interrelationship of firing in pools of neurons is always shifting because of the inherent variability and chaos in the firing of individual neurons. As it relates to chaos, what comes out of the pool of neurons is deterministic but has long-term unpredictability.

4. From context of function in the activity of sets of synapses: That, given the multifunctional capabilities of the neuronal network, it is conceivable that synaptic changes resulting from training to one adaptive situation may also be used in many different responses such that the notion of context applies not only to groups of neurons but also to groups of synapses. One important implication is that a synaptic change resulting from learning of one behavior does not exclude the possibility that the same synaptic change can code for another learned or unlearned behavior. Whether chaos can act as an informational signal in brain function shall be discussed in the section dealing with simulation networks.

5. From observations of blending, and from findings that motor patterns may not remain stationary nor have a homogeneous flow of variable activity (i.e., the activity can abruptly or smoothly shift from one structure of variable activity into another), it may be useful to consider that:

a. Some attractors may be nonhomogeneous, with each topologically different area representing structurally different neural activity. There may be attractors having long cycles (in the order of minutes) representing the recurrence of blended motor patterns, as well as the shorter cycles (in the order of tens of seconds) representing subcomponents of buccal-oral behavior.

b. The trajectory of the activity may transiently visit different attractors representing different expressions of the same behavior.
c. The different attractors may not be stationary; i.e., there may be a continual, fluid shift in the form of attractors, their position in state space, in their interrelated position to one another, and attractors may blend into one another.
d. A particularly interesting idea to examine is the possibility that metastable chaos may form between the borders of two stable attractors (see discussion of "fuzzy borders" in Decroly 1986).

2.4 The "Right" Attractor

One of the most obvious qualities of animal behavior is that it is prone to error. As we indicated above, a set of neural connections may be able to generate many and variable attractors. Some of these may be adaptive for certain environmental demands, while others may have little adaptability. What determines the appropriate response at any given time, we believe, is not the selection of a particular centrally programmed neural structure or attractor, but rather the process arising from the continual interaction between the animal and the environment. In this respect, the possible fluidity of attractors and the margin left open for error allow for the greatest possible use of the capabilities of a set of connections. When viewed in this way, the concept of attractor is scale-independent: it takes into account a variety of different types of activity that occur within single neurons, among groups of neurons, between areas of the brain, and through the interactions between the animal and the environment.

3 Chaos

3.1 Evidence for Chaos

In order to determine the natural tendency for the central nervous system (CNS) to generate variable activity, we recorded motor patterns in isolated nervous systems so as to remove the effects of sensory inputs (Mpitsos et al. 1988 a, c). Two patterns were examined, one relating to the bite-swallow phase of feeding and one to the active phase of regurgitation. In each case, we examined the activity of a small set of multifunctional neurons in the buccal ganglion and their motoneuron targets in the cerebral ganglion. The buccal-ganglion neurons are essential for coordinating all buccal-oral behaviors and are the major source of the central pattern generator. The recordings were made intracellularly and the spike trains were analyzed using both an interpolation method and an equal-interval sampling method to generate equal-interval time series. The findings of the following analyses provided evidence that the motor patterns were generated by chaotic attractors:

1. The autocorrelation functions of the time series rapidly drop to zero within about four cycles of activity representing the opening and closing of the jaws; i.e.,

the probability for predicting the future evolution of the activity decreases with time with respect to starting conditions.

2. The value of the principal Lyapunov exponent ($\lambda 1$) was positive for all cells we examined; the lowest value in bits/s was 0.15 and the highest value was 0.55; Lyapunov exponents were calculated using the algorithm of Wolf et al. (1985). In accordance with the autocorrelation functions, this indicates that there was a gain of information per unit time with respect to initial conditions as the neural activity evolved (and an equivalent loss in the precision of initial measurements). The positive value of the exponent indicates that the trajectories exponentially diverge from one another in state space, causing the surface of the attractor to grow as $2\lambda 1$ in its principal axis; positive growth must be accompanied by folding in order for the activity to remain bounded.

3. The 1-D maps of successive Poincaré sections through the attractors show that the trajectories lie in folded sheets that appear to stretch and fold, which is consistent with the finding of positive Lyapunov exponents. In addition, because the trajectories lie in sheets rather than at random in the 1-D map, the activity appears to be deterministic.

4. The correlation dimension of the attractor is low and non-integer (fractal), and the attractor is embedded in low-dimensional phase space. Values calculated so far for the correlation dimension range between 1.75 and 2.5 for different neurons, and the embedding dimensions range from 2 to 6; somewhat higher values were obtained in calculation of correlation and embedding dimensions of time series consisting of the unequal spike intervals themselves (unpublished). The fractal value of the attractors is consistent with chaos (Mandelbrot 1985; Grassberger and Procaccia 1983). The low-dimensional value of the embedding space provides some indication that the mechanisms generating the repetitive motor patterns may be simple enough to be analyzable.

3.2 Difficulties in Assessing the Results of Chaos Analysis

Chaos analyses of repetitive activity rely primarily on the number of orbits that the activity takes through the attractor. A large number of observations may be useful, but if the number of orbits is too small, high samples do little good. Because the characteristics of biological attractors are usually unknown, experimenters are hard-pressed to determine what constitutes appropriate data lengths (Albano et al. 1987; and Wolf et al. 1985 discuss some requirements). The short duration of many biological responses poses difficulties in obtaining sufficient numbers of orbits. In our experience, homogeneous bite-swallow activity usually lasts no longer than 10 to 20 cycles; if the attractor is simple enough, there is some hope of characterizing it. Longer patterns of activity run the risk of nonstationarity. Difficulties arising from nonstationarity have to do with the selection of appropriate homogeneous sections of data for analysis. Thus, while activity having short periods may appear to yield enough orbits for analysis, other aspects of the activity may not yield sufficient numbers of orbits for analysis but may represent an important feature of the same attractor.

The form of the data is also important. Although it is possible to analyze unequally spaced samples (Shaw 1986), more tools are available for data sampled at equal intervals. However, some forms of continuous measures often generate so much data that it may be difficult to analyze in even large computers. For example, a typical bout of 10 feeding cycles in *Pleurobranchaea* may last 200 s. Sampling the intracellular membrane potential of a neuron at a modest rate of 1 kHz generates 200000 points. Analysis of the correlation dimensions in 10-dimensional state space for this data would require months of CPU time in even the fastest laboratory computer. To avoid long calculation times, the data may be compressed by examining the intervals between action potentials rather then membrane potentials. But in doing so, one now has to handle unequal time intervals. Such data series may be converted to equal-time series (Mpitsos et al. 1988 a, c), but all manipulations in such conversions, and in subsequent handling, run the risk of inducing spurious correlations or removing real ones.

Yet another difficulty arises from the fact that application of the available methods of analysis requires some subjective judgements. For example, there is considerable leeway in the selection of the lag for constructing state space and in selection of slopes in the analysis correlation dimensions, both of which can influence the results (Albano et al. 1987, 1988; Mpitsos et al. 1988 c). Similarly, Lyapunov exponents are sensitive to scaling factors (Wolf et al. 1985; Mpitsos et al. 1988 c). Because of such subjectivity, the exact dimensions of an attractor may be quite difficult to determine. Moreover, given the possible nonstationarities discussed above, the same attractor may not occur twice or similar behavior may be produced by somewhat different attractors. Because nonstationarities are difficult to characterize, their contribution to the measures obtained in a time series may not be obvious to experimenters. Therefore, the validity of similarities or differences in the numbers one obtains from various calculations may be difficult to assess. Because of these problems, evidence of chaos may rely on circumstantial incrimination rather than definitive numerical proof.

3.3 Why is Chaos Important?

Chaos may aid in the generation of organized neural activity. For example, Freeman and Viana Di Prisco (1986), Freeman and Skarda (1985), Skarda and Freeman (1987) found that the rabbit olfactory bulb generates a baseline chaotic EEG that is attributable to chaos. As the animal inhales odorants, the chaotic activity gives way to one of a number of limit cycles that encode specific odors. Our own studies of motor behavior indicate that chaotic attractors themselves may represent the information relating to specific behaviors. Unlike externally added or amplified noise that tends to obscure the neural signal underlying behavior, variability arising from chaos is not separate from this signal, it is the signal. Therefore, by gaining an understanding of the variability arising from chaos, one immediately gains some understanding of the mechanism generating that aspect of the behavior.

Chaos may also have a role in the generation of rapid adaptation to changing environments. Nonvarying signals, such as limit cycles, carry no new information into the future than what they contain at some previous time; activity at one time can be used to predict the evolution at future times: By contrast, the long-term evolution of chaos is not predictable, information at one time does not correlate well with information obtained later on, despite the fact that the activities at the two times are equally deterministic and generated by the same mechanism. Such unpredictability represents a gain of information (e.g., see Abraham and Shaw 1983, vol. 2, p. 107) by which the brain naturally creates new response possibilities.

The sensitivity to initial conditions in chaotic activity may provide a more efficient means for dissipating perturbations than might occur through more predictable mechanisms (Conrad 1986; Mpitsos et al. 1988a). In chaotic systems, there are as many ways to express a behavior as there are countless trajectories that fill the surface of the attractor that generates it. A small perturbation in the activity at one time leads to a large difference later on, but although the movements of the animal may appear different, their overall effect will be the same; that is, the energy introduced into the system is dissipated through the subsequent deviation in the movements of the animal that produce the same behavior in different ways. By contrast, repetitive activity generated by limit cycles must return asymptotically to the same course of action that occurred before the perturbation. Here, the energy introduced into the system must be dissipated as heat loss to the environment through the tissues of the animal. In either case, sensory feedback and control mechanisms are undoubtedly involved, but the feature that sets chaos apart is that through it the central nervous system itself has a natural tendendy to accommodate unexpected inputs.

4 Can the Nervous System Read and Repeat Chaotic Information?

4.1 Simple Networks Learn to Read the Dynamical Features of Chaotic Attractors

A repeating signal embedded in noise has a finite probability of being read (Shannon 1963). In this respect, it is easy to see that limit cycles can provide a readable signal in brain function. Because of their inherent variability and long-term unpredictability, chaotic signals are more difficult to envisage as information sources. Consequently, having obtained evidence that brain function in our experimental animal is chaotic, it was necessary to determine whether such information arising in one part of the nervous system could be read by another.

In order to avoid the complexity of even "simple" neuro/behavioral preparations, we chose to answer this question through computer simulation of simple connectionist networks (Rumelhart et al. 1986; Werbos 1974). Although these networks do not simulate realistic properties of neurons, they nonetheless can be used to answer questions aimed at the flow of information. The network we constructed consisted of one input unit, one output unit, and four hidden (intermediate) units, with each input and output unit being connected to all hidden units (Mpitsos et al. 1988a). Most studies on such circuits have used binary, nonchaotic signals. Recently, Lapedes and Farber (1988) used chaotic signals in order to determine whether the network could predict the future flow of the input signal. Our study examined the question of whether the network could learn to repeat the 3.60, 3.95 logistic (May 1976) and Rössler (1979) equations, all of which are chaotic.

The results showed that the network can learn to distinguish one chaotic signal from another. Interestingly, the output was always distorted by comparison to the input, and, rather than recognizing specific aspects of the fine details of the input, the network learned the fundamental qualities of the input from which it could generalize to other signals. For example, training the network with random noise proved sufficient for the network to respond immediately to other chaotic attractors as well as it would have following tens of thousands of training trials in the absence of preconditioning with random noise. The one feature that all chaotic attractors have in common is variability. Once the network had learned this quality of the input stimulus through preconditioning, it could immediately characterize other variable inputs. However, the fine parametric details about the input in state space were only approximated by the network, whereas the fundamental aspects of the dynamics that characterized the attractors were reproduced accurately. For example, the Lyapunov exponent $\lambda 1$, which indicates the rate of loss or gain of information in the principal direction of growth in the attractor, was similar in both the input and output attractors. In the same way one might say concerning human perception, that the sensory environment may be perceived first through abstractions of its fundamental qualities. The details of the visual field may be inexact by comparison to the input, and in need of continual updating, even in cases of familiar objects, as, for example, the face of a friend.

4.2 Multifunctionality of Networks

The ability of such networks to learn the fundamental qualities of one chaotic signal and then to generalize from these qualities so as to distinguish rapidly two unlearned chaotic signals was not accompanied by changes in the synaptic weights and thresholds; that is, the network could distinguish several types of signals with the same parametric set of its connections. To be sure, neurophysiological systems can change the type of pattern they produce by using different subsets of the network, different values of the connections, and different neurohumoral modulation (Getting and Dekin 1985; Marder and Hooper 1985). But the network simulations indicate, and agree with our previous proposal (Mpitsos and Cohan 1986a, b), that different types of chaotic motor patterns can also be produced by the same set of connections.

5 Conclusion

In a recent review, Bullock (1984) discussed the need for a revolution in comparative neuroscience. Our own views have arisen from watching animals, and it has seemed to us that a better understanding of behavior requires the merging of two paths that could force such a "quiet revolution." One path holds that self-organized neural output arises from a reciprocal interaction between the animal and the environment. What is not clear to us is whether this dialectic involves simple reductionist cause and effect or whether it leads to higher levels of emergence. Scale-independent concepts such as "attractor" may be useful tools in the study of such a dialectic between animal and environment. What I have attempted to illustrate here is that the description of neural connectivity (including connection weights and neuromodulators) does not determine a behavior, but rather the possibility of many behaviors, some representing adaptive responses and some representing error or nonsense. What determines the functional "neurocircuit" is the relationship of the animal to the environment. Without the dynamic between animal and environment, one can examine such a network in detail and not know what it can do. One is reminded here of the analogy drawn by Gould and Lewontin (1979) between the capability of the brain and paintings placed in the spandrels arising from intersection of cathedral arches: structure provides the bounds, but what becomes expressed through structure is determined by external input.

The second path holds that variability is an integral and inseparable part of such neurointegrative mechanisms, not just an expression of sloppy biological function. Chaos is interesting because of its deterministic character, but nondeterministic processes such as thermal or random noise may also be important (Kirkpatrick et al. 1983; Jeffrey and Rosner 1986; Burton and Mpitsos 1988 a, b). We have discussed some of the major difficulties in assessing whether chaos occurs in behavioral processes. Given the weight of some of these problems, it may not be possible to prove beyond circumstantial evidence whether chaos actually exists in a particular behavior. But even as a heuristic tool, chaos may provide access to new formulations about brain function because unlike added or amplified noise, it does not have a separate existence from the neural code underlying behavior: to the extent that chaotic variation exists in motor processes, it is the code.

6 Summary

Our studies of variability of buccal-oral behaviors in the sea slug *Pleurobrancheae californica,* and simulation of network activity, has led us to propose that it may not be possible to assign specific functional identities to neurons, their synapses, and even to groups of neurons, first, because they can produce qualitatively different motor patterns (they are multifunctional), and second, because the interrelated firing of neurons can change even during the expression of a given motor pattern. While reflexes undoubtedly occur in neural function, the emergent properties that arise from variability in parallel processing may go beyond the reflex in the same way that quantum physics superceded classical physics. Whereas classical reflex neurophysiology may propose that a learned behavior is ascribable to a particular synaptic change, the newer perspective cautions that the same change may be used in other learned and even unlearned behaviors, and that variability in the responses of individual neurons, and in the interrelated responses of groups of them, plays a central role in generating behaviors. Therefore, we define three integrative levels: anatomical, dynamical, and self-organizing, representing, respectively, the structural framework, the temporal characteristic of linear and nonlinear processes, and emergent qualities. Self-organization arises from the first two, but, through the quality of variability, it also supercedes them.

Although there may be many sources of behavioral variability, as may arise from random noise in neural processing or from perturbations produced by sensory inputs, our analyses of motor patterns in deafferented nervous systems indicate that chaos in the responses of individual neurons may be one contributing factor. The frequency of firing in these neurons was examined by means of state-space analyzes and was found to be unpredictable over the long-term, but, nonetheless, to be generated by deterministic mechanisms: 1) Autocorrelation functions of spike-frequency time series rapidly decreased, indicating that the activity was sensitive to initial conditions. 2) Accordingly, phase portraits exhibited state-space trajectories that diverged and mixed, and 3) the value of the principle Lyapunov exponent was positive, providing a quantitative measure of the exponential divergence of nearby trajectories which would cause the surface of the phase portrait to stretch. 4) Poincaré sections taken through the phase portraits provided graphic evidence of folding that should accompany stretching. 5) From calculation of integral autocorrelation functions, phase portraits were found to have fractal correlation dimensions (ranging from about 1.75 to 2.5) embedded in low-dimensional state space (from 2 to 6 dimensions). Together, these findings indicate that the activity was generated by chaotic attractors, and the low-dimensionality of the activity suggests that the underlying processes may be simple enough for further experimental study.

Whatever the source of variability, the aspect of animal behavior, and probably of other biological systems, that most taxes the application of dynamical theories is that it can be nonstationary. By nonstationary, we imply the ability of patterns of activity to change abruptly or blend gradually into other patterns that appear qualitatively different from the preceding though variable activity, and the pattern overall may not be statistically stationary. Although such changes may represent shifts into different behaviors, they can also occur during the execution of the same ongoing behavior. In addition, significant differences may occur between repetitions of either short- or long-duration behaviors. Such nonstationarities may occur over the long-term, but they also may occur within several cycles in behaviors that involve many repetitions. Because of graded blending, it is often difficult to determine whether the pattern changes constitute nonstationarities or are simply part of the long-term variability in a stationary pattern. Finally, even when robust, stationary attractors generate adaptive biological responses, the duration of the responses is often too short for application of the available mathematical tools. Together, these aspects of biological systems compromise the applicability of extant dynamical theory which has been developed primarily through well-defined mathematical and physical systems. Consequently, we believe that numbers arising from application of the available chaos theory must be viewed cautiously, more as indicators of the possible qualities rather than as absolute measurements of particular dimensions or characteristics of the underlying dynamics.

Given that adaptive behaviors may be generated by chaotic attractors, the unpredictability of such attractors required us to ask whether activity arising in one part of the nervous system could be read and reproduced by another. Because there are no neuro/behavioral systems that are sufficiently simple and controllable for inquiring into this question quantitatively, we studied the informational aspects of the problem in computer simulations of a simple connectionist network consisting of one input, one output, and four hidden units. As determined by quantitative comparisons of the input chaotic signal with the output of the network, recognition and repetition of such signals appear to be accompanied by distortion. What is learned or remembered has more to do with fundamental qualities in the dynamics of the input signal, such as the rate of loss of information or the structure of folding in the attractor surface, rather than the specifics of its parametric details such as the value of the input at each successive interval. It seems that after learning some fundamental quality of a signal, the same underlying network of connection weights and thresholds can repeat other, previously unlearned signals as well as if it had learned them first; i.e., the ability to respond to different signals arises from generalization through the characteristics that are shared by these signals. For example, after being trained to a random signal, the network can recognize and repeat different chaotic signals because they share in the quality of appearing random, but whether generalized or learned, the distortion in the output is the same.

Thus, our findings on physiological and computational systems indicate that connectivity does not determine a behavior, but rather the potentiality of many behaviors arising as discrete and separate from one another or as blends, some of which may be adaptive and some may represent errors. We have proposed that variability arising from chaos, random noise, and feedback from the environment are essential parts of such self-organizing processes. The aspect of chaos that makes it particularly interesting is that through its tendency to gain information with respect to initial conditions i.e, the nervous system itself has a natural capacity to generate new informational space for adaptation to an unpredictable environment.

Acknowledgments. I thank the Air Force Office of Scientific Research for support (grant 86-0076), W. Ross Adey, Robert M. Burton, Theodore H. Bullock for critical reading of a previous version of this manuscript, and Lavern Weber for making space available for me at the Hatfield Marine Science Center.

References

Abraham RH, Shaw CD (1983) Dynamics. The geômetry of behavior, vols 1–3. Aerial, Santa Cruz

Adey RW (1972) Organization of brain tissue: is the brain a noisy processor? Int J Neurosci 3:271–284

Aihara K, Matsumoto G (1986) Chaotic oscillations and bifurcations in squid giant axons. In: Holden AV (ed) Chaos. Princeton University Press, Princeton, pp 257–269

Albano AM, Mees AI, de Guzman GC, Rapp PE (1987) Data requirements for reliable estimation of correlation dimensions. In: Degn H, Holden AV, Olsen LF (eds) (1986) Chaos in biological systems. Plenum, New York

Albano AM, Muench J, Schwartz C, Mees AI, Rapp PE (1988) Singular-value decomposition and the Grassberger-Procaccia algorithm. Physical Rev 38A:3017–3026
Amit DJ, Gutfreund H, Sompolinsky H (1985) Storing infinite numbers of patterns in a spin-glass model of neural networks. Phys Rev Lett 55:1530–1533
Babloyantz A, Destexhe A (1986) Low dimensional chaos in an instance of epilepsy. Proc Natl Acad Sci USA 83:3513–3517
Babloyantz A, Salazar JM, Nicolis C (1985) Evidence of chaotic dynamics of brain activity during the sleep cycle. Phys Lett 111A:152–155
Bellman KL (1979) The conflict behavior of lizard, *Sceloporus occidentalis*, and its implication for the organization of motor behavior. PhD dissentation, University of California, San Diego
Bullock TH (1984) Comparative neuroscience holds promise for quiet revolution. Science 222:473–478
Burton RM, Mpitsos GJ (1988 a) A constant-noise mechanism for enhancing learning in neural networks and its comparison to simulated annealing. Soc Neurosci Abstr 14:104–114
Burton RM, Mpitsos GJ (1988 b) Noise enhanced learning in neural networks: comparison to simulated annealing, and applicability to biological systems. (To be puldished)
Chay TR (1985) Chaos in a three-variable model of an excitable cell. Physica 16D:233–242
Cohan CS, Mpitsos GJ (1983 a) The generation of rhythmic activity in a distributed motor system. J Exp Biol 102:25–42
Cohan CS, Mpitsos GJ (1983 b) Selective recruitment of interganglionic interneurons during different motor patterns in *Pleurobranchaea*. J Exp Biol 102:43–58
Conrad M (1986) What is the use of chaos? In: Holden AV (ed) Chaos, Princeton University, Press, Princeton, pp 3–14
Decroly O (1986) Interplay between two periodic enzyme reactions as a source for complex oscillatory behavior. In: Degn H, Holden AV, Olsen LF (eds) Chaos in biological systems. Plenum, New York, pp 49–58
Degn H, Holden AV, Olsen LF (eds) (1986) Chaos in biological systems. Plenum, New York
Freeman WJ, Skarda CA (1985) A perspective on brain theory: nonlinear dynamics of neural masses. Brain Res Rev 10:147–175
Freeman WJ, Viana Di Prisco GV (1986) EEG spatial pattern differences with discriminated odors manifest chaotic and limit cycle attractors in olfactory bulb of rabbits. In: Palm G, Aertsen A (eds) Brain theory. Proceedings the first Trieste meeting on brain theory, 1–4 October, 1984. Springer, Berlin Heidelberg New York
Getting PA, Dekin MS (1985) *Tritonia* swimming: a model system for integration within rhythmic motor systems. In: Selverston AI (ed) Model neural networks and behavior. Plenum, New York, pp 3–20
Gleick J (1987) Chaos. The making of a new science. Viking-Penguin, New York
Gould SJ, Lewontin RC (1979) Spandrels of San-Marco and the panglossian paradigm: a critique of the adaptationist program. Proc R Soc Lond [Biol] 205–581–588
Grassberger P, Procaccia I (1983) Characterization of strange attractors. Phys Rev Lett 50:346–349
Grossberg S (1980) Studies of mind and brain. Reidel, Boston
Harrnon LD (1964) Neuromimes: action of a reciprocally inhibitory pair. Science 146:1323–1325
Hayashi H, Ishizuka S (1986) Chaos in molluscan neuron. In: Degn H, Holden AV, Olsen LF (eds) Chaos in biological systems. Plenum, New York, pp 157–166
Holden AV (ed) (1985) Chaos. Princeton University Press, Princeton
Hopfield JJ (1982) Neural networks and physical systems with emergent computational abilities. Proc Natl Acad Sci USA 79:2554–2558
Jeffrey W, Rosner R (1986) Optimization algorithms – simulated annealing and neural network processing. Astrophys J 310:473–481
John ER (1972) Switchboard versus statistical theories of learning and memory. Science 177:850–864
Kelso JAS, Scholz JP, Schöner G (1986) Nonequilibrium phase transitions in coordinated biological motion: critical fluctuations. Phys Lett 18A:279–284
Kirkpatrick S, Gelatt CD, Vecchi MP (1983) Optimization by simulated annealing. Science 220:671–680

Kleinfeld D, Sompolinsky H (1988) Associative neural network model for the generation of temporal patterns: theory and application to central pattern generators. Biophysical J 54:1039–1051
Labos E (1986) Chaos and neural networks. In: Degn H, Holden AV, Olsen LF (eds) Chaos in biological systems. Plenum, New York, pp 195–206
Lapedes A, Farber R (1987) Nonlinear signal processing using neural networks: predictive and system modeling. Los Alamos Laboratory Technical Report, LA-UR-87-2662
Lorenz EN (1963) Deterministic non-periodic flows. J Atmos Sci 20:130–141
Mandelbrot BB (1985) The fractal geometry of nature. Freeman, New York
Marder E, Hooper SL (1985) Neurotransmitter modulation of the stomatogastric ganglion of decapod crustaceans. In: Selverston A I (ed) Model neural networks and behavior. Plenum, New York, pp 319–338
May RM (1976) Simple mathematical models with complicated dynamics. Nature 261:459–467
McClellan AD (1982a) Movements and motor patterns of the buccal mass of *Pleurobranchaea* during feeding, regurgitation, and rejection. J Exp Biol 98:195–211
McClellan AD (1982b) Re-examination of presumed feeding motor activity in the isolated nervous system of *Pleurobranchaea.* J Exp Biol 98:212–228
Mpitsos GJ, Cohan CS (1986a) Comparison of differential Pavlovian conditioning in whole animals and physiological preparations of *Pleurobranchaea:* implications of motor pattern variability. J Neurobiol 7:499–516
Mpitsos GJ, Cohan CS (1986b) Convergence in a distributed motor system: parallel processing and self-organization. J Neurobiol 7:517–545
Mpitsos GJ, Collins SD, McClellan AD (1978) Learning: a model system for physiological studies. Science 199:497–506
Mpitsos GJ, Burton RM, Creech HC, Soinila SO (1988a) Evidence for chaos in spike trains of neurons that generate rhythmic motor patterns. Brain Res Bull 21:529–538
Mpitsos GJ, Burton RM, Creech HC (1988b) Connectionist networks learn to transmit chaos. Brain Res Bull 21:539–546
Mpitsos GJ, Creech HC, Cohan CS, Mendelson M (1988c) Variability and chaos: neurointegrative principles in self-organization of motor patterns. In: Kelso JAS, Mandell AJ, Shlesinger MF (eds) Dynamic patterns in complex systems. World Scientific, Singapore, pp 162–190
Pellionisz A, Llinas R (1983) Space-time representation in the brain: the cerebellum as a predictive space-time metric tensor. Neuroscience 7:2949
Rapp PE, Zimmerman ID, Albano AM, Deguzman GC, Greenbaum NN (1985) Dynamics of spontaneous neural activity in the simian motor cortex: the dimension of chaotic neurons. Phys Lett 10A:335–338
Rössler OE (1979) An equation for continuous chaos. Phys Lett 57A:397–398
Roux J-C, Simoyi RH, Swinney HL (1983) Observation of a strange attractor. Physica 8D:257–266
Rumelhart DE, McClelland JL, PDP Group (eds) (1986) Parallel distributed processing: explorations in the microstructure of cognition, vol 1. Foundations. MIT Press, Cambridge
Sejnowski TP, Rosenberg CR (1987) Parallel networks that learn to pronounce English text. Complex Syst 1:145–168
Selverston AI (1980) Are central pattern generators understandable? Behav Brain Sci 3:335–371
Shannon CE (1963) The mathematical theory of communication. In: Shannon CE, Weaver W (eds) The mathematical theory of communication. University of Illinois Press: Urbana (Reprinted from Bell Syst Tech J 1948, July and October)
Shaw R (1986) The dripping faucet. Aerial, Santa Cruz
Skarda CA, Freeman WJ (1987) How brains make chaos in order to make sense of the world. Behav Brain Sci 10:161–195
Sperry RW (1981) Changing priorities. Ann Rev Neurosci 4:1–15
Werbos PJ (1974) Beyond regression. New tools for prediction and analysis in the behavioral Sciences. PhD thesis. Harvard University, Cambridge
Wolf A, Swift JB, Swinney HL, Vastano JA (1985) Determining Lyapunov exponents from a time series. Physica 16D:285–317

Epilogue

Signs of Dynamic Processes in Organized Neural Tissue: Extracting Order from Chaotic Data

T. H. Bullock

Introduction

It used to be said that anatomy has all the new techniques and physiology has fallen behind. This collection of papers, broad surveys and particular examples, like other recent samplings of a moving front, illustrates some of the range and ingenuity of methods now available for attempting to decipher the signs of dynamic, distributed, parallel processing in the brain. Besides the battery of constantly expanding tricks used on cells and parts of cells, especially the single-unit spike form of their signalling, and the armamentarium of methods for visualizing the glucose metabolism, local blood flow, local surface temperature with temperature-sensitive dyes (Shevalev et al. 1985, 1986a, b), optical changes, with and without voltage-sensitive dyes, movement of positron emitters, and impedance changes (methods represented in other literature), in this volume we see evidence of the new hardware and software for recording and analyzing electric and magnetic fields, for exploiting single and multiple electrodes, for arranging stimulus paradigms to distinguish sensory and cognitive events, for reducing volumes of data and discriminating randomness from deterministic dimensions in chaotic data.

This volume reflects an historical moment when a relatively neglected field of basic research upon signs of the dynamic processes ongoing in organized neural tissue is almost explosively expanding, to complement the other approaches, just indicated in a partial list. The expansion is not simply a further use of the EEG and EP in the classical sense of widely spaced macroelectrodes at some distance from the active cells. We look forward to developments not represented in this sample; for instance, further detailed studies of the distribution and dynamics of impedance of neural tissue on a fractional millimeter and fractional second scale, as well as subsurface optical and local temperature changes.

While the pressure is higher than ever for additional tools with better resolution, both spatial and temporal, to make still finer measurements, the ancient question still confronts us daily, what conclusions are really permitted from such data?

We Need More Windows Through Which to See the Brain in Operation

Fortunately, we are not limited to the two approaches, of studying the composite signs of mass activity, on the one hand, and the microelectrode record of unit

Springer Series in Brain Dynamics 2
Edited by Erol Başar and T. H. Bullock

spikes, on the other – only two blind men with a large, hairy mammoth. To be sure, these two approaches, both of which record dynamic processes in the brain "on line," are improvements in that respect over the purely indirect methods when we had to make inferences about the brain from its input/output relations, its reflexology, psychophysics, behavioral physiology, and experiments in learning, with and without lesions or brain stimulation. Powerful and suggestive as these inferential methods are, we need to see what is going on directly.

The two classical windows onto brain operation named above are both severely limited: the first is like a record from a microphone suspended above a city; the second is not so much like recording from an element in a computer as from one individual among many, not identical but quasiequivalent in the whole human species, or more exactly, to recording that individual's speech, without knowing the expressions, gestures, touches, odors, and other outputs. No wonder each method makes discriminations the other cannot and predicts the other extremely poorly. No wonder investigators differ drastically in their informal brain models and their views about what is hopeless, ridiculous, dead-end, soft and hard – and cannot even understand the jargon of a closely related field.

This conference and book were planned as efforts against these formidable barriers. A shortfall from the high objectives is obvious; nevertheless there is here a broad sample, incomplete of course, of the state of the art and of recognized issues, as well as a rich assortment of heuristic propositions about how the brain might work, and how we might gain further insights into this most complex of known systems.

Fortunately, there are more than two windows. Yet, the strongest impression, was that we urgently need new and wider windows (see especially the chapters by Adey and Freeman), not only for simultaneous glimpses at many cells or loci, (exemplified here by Eckhorn et al., John et al., Gevins, Freeman and others), but for imaging qualitatively different sets of elements.

Conferences Afford Glimpses Through Stained Glass Windows

Galambos wants to see the array of glia that are connected by high potassium conductance; are they primarily radial? Grüsser makes me want to see all the cells aroused by familiar faces and those aroused by familiar flowers or shoes, both intrinsic and projection neurons, in the temporal lobe and elsewhere. Freeman wants to *feel* the brain attractors in 4-D by kinesthetic sensations, over time. Adey calls for new instrumentation to visualize high radio-frequency fields within the nerve cell.

The contributions by Babloyantz, Rössler, Başar and Röschke, Skinner, Mpitsos, and others make me hopeful that we will soon see the dimensionality of many parts of the brain at the same time, second by second, in cats, catfish, and octopus, as rest changes into arousal, directed attention, and recognition. Taking these authors at face value, the computer programs are at hand and we have only to put the data samples through them, with proper precautions and controls. Section 5, below, raises some questions, however.

Başar proposes the physiological equivalent of the physicist's Feynman diagrams, specifying a series of indicators or descriptors of the ongoing compound field potentials for each place in the brain and behavioral state, including nonlinear dimensionality, phase angle, amplitude, amplitude modulation envelope for different frequency bands, degree of order, coherence in space, and coherence in time for each frequency. Such a battery should include the sophisticated measures of John, Gevins, Lehmann and others, and with microelectrodes might permit those of Eckhorn and the multiunit spike analysts.

A major message from the contributions in this volume is that a whole domain of important dynamic events in a working brain will not be explained or predicted by an extensive knowledge of the dynamics of molecules, channels, spikes, or their linear differential equations. Of course, these are basic in the long run, and are proving in the short run to provide rich insights at their levels. Yet the common tendency to concentrate solely on molecular and cellular levels overlooks the brain and its feats as an organized system.

The brain is the extreme case of a many-body problem. As resistant as even the three-body problem has been over the years, we can expect to require a convergence of approaches, as many as possible, to gain an understanding of the operating principles of higher vertebrate brains as organized information machines. This volume, like Brain Dynamics I (Başar 1988), exhibits a selection, not only of approaches but of attitude, standpoint, and interpretation within a given approach. The organizers should not apologize for the lack of integration and the diversity of styles and stages of analysis; these are inherent in the goal. One can understandably be discouraged, reading such an array of chapters, by the divergence more than convergence in what knowledgeable people see, looking at the same organ, recalling again the blind men and the elephant. Yet this is a reality we must work with, compounded by differences in training, skills, and interests. Confucius – or someone – said "You're down on what you're not up on." It is the inherent problem as well as the merit of such a book, dedicated to bringing together diverse approaches and diverse investigators.

The opportunity of a panel discussion and particularly an epilogue is to comment on various points, without the necessity of a tightly organized theme or logic line. This assignment I undertake humbled by the awareness that Brain Dynamics II will be seen in the perspective of history only a short time hence as representing a primitive and naive stage in the ongoing quest for insight into how the brain works as a system. I commend to readers the epilogue of Başar to Brain Dynamics I (Başar 1988), as well as the chapters of that book, since I will not attempt anything parallel to his, nor claim to discern trends or progress in the interval. Many of the same authors have contributed to both collections, choosing complementary themes that may or may not represent a change in viewpoint.

Evolution Leads to Advances in Levels Plus Divergence Within Levels

A great unifying thread in modern biology has been the outstanding success of simplifying ideas – e.g., the genetic code, the neuron doctrine, and general physiology. Psychology has also felt the impact of "general process theorists," especially in theories of learning, e.g., Thorndike, Hull, and Skinner, who discourage the idea of a variety of mechanisms of learning. The lure of research on cellular and molecular basic mechanisms, and the achievements of general physiology, identified almost automatically with common if not universal mechanisms operate like low-pass filters to channel the thinking of neuroscientists toward simple, general answers. "Complex" is a bad word in many circles; comparative physiology is distorted to mean the use of simple systems to discover general answers; diversity is deliberately glossed over; grades of organization are old-fashioned; "higher" is worse than subjective, it is immoral; evolution is a reality but its only results worthy of attention are adaptive radiation of lower taxa – lateral divergence within the same grade. I am enthusiastically for simplifying ideas, as far as they go; when that is not all the way. I am obliged to recognize diversity.

As I have argued before (Bullock 1984, and in this volume), it is most true of the nervous system that evolution has occasionally and intermittently invented more advanced grades of organization so that the span of complexity from the simplest brains to the highest is very great indeed; this is due to inventions and emergents, novel mechanisms superimposed on older ones, both anatomical and physiological, mainly at the multicellular levels. A prime need and opportunity is to discern, beyond the general and the common mechanisms, the relevant bases for differences in brains and their achievements, differences between higher taxa of clearly distinct grade of organization, between lower taxa adapted to separate habits of life, between sexes, individuals, life-history stages, and pathologic states of the same species.

We Need Further Study of the Generators and the Nature of EEG and EP

Whether the compound field potentials are merely signs of underlying activity or in some degree causal, their basic nature and origin are still poorly known. I have expanded elsewhere (this volume) on some aspects of this theme. Suffice it here to underline the wide variety of known generators, such as plateau potentials, pacemaker potentials, nonclassical hyperpolarization with conductance decrease, and modes of interaction among them, as well as the probability of as yet unappreciated generators. For example, the little-known, standing or slowly changing longitudinal potential difference within a neuron, between the tips of its dendrites and the terminals of its axon, hypothesized by Gesell (1940), might contribute to the extracellular compound field both a DC and an AC component. The loci of generators include dendrites, axons, axon terminals, and somata. Glia, pia, and ependyma might well have contributions, the more so as they are connected by electrotonic junctions.

To underline further the need for basic studies, I make a few assertions about our ignorance of the EEG. (a) First, concerning "synchrony," despite a widespread use of the term, its several possible meanings are usually not distinguished, and no measure is accepted for the degree of synchrony in an assembly of cells.

The most significant sense of the term in our context is synchrony over a wide band, including slow waves, among the generators within a small volume of tissue, hence underlying the compound field of that mass of brain tissue. We still have no method for quantitative evaluation or justification of this descriptor, apart from a few extreme situations such as ictal discharges in epilepsy, unless a recent proposal (Bullock et al. 1984) proves out. We have no means, even by modelling, to specify the degree of coincidence expected by chance in the theoretical case of no synchrony, that is, all generators being independent, so that we cannot say when a small degree of synchrony is above the expectation of the null hypothesis. This is because the known generators, let alone those still unknown, are not simple dipoles and are not oriented in a readily knowable fashion, nor is the volume conductor isotropic. Besides the problem of assessment, we are in need of testable hypotheses for mechanisms of synchronization, of which there are no doubt more than one and probably nonlinear. Old standbys such as field potentials as weak causal influences are still in need of crucial tests. Freeman made a strong case for their ineffectiveness in a particular situation, but one may argue how widely this dismissal applies. Release of effective amounts of transmitter by subthreshold presynaptic events is well established in invertebrates (Graubard et al. 1980; Roberts and Bush 1981) but as yet not evaluated in vertebrate preparations.

(b) The terms and the concepts of "rhythm" and "oscillator" are widely used and familiar. There is no question the concepts are heuristic, and if we would always say "treating the compound field potentials of the brain as though they were a mixture of rhythms arising in oscillators ...," one could not object. We have, however, little justification for proposing that in fact the generators of EEG and EP are oscillators, meaning that their characteristic, synchronized activity is periodic in the 0.1- to 50-Hz range, in the general case. This is true even for the cases of relatively sharp peaks, which are in any event special cases, since most micro-EEG recorded in various species and places in the brain without sharp peaks is a broad spectrum phenomenon. This means that terms such as "oscillation" and "resonance" should for the present be accompanied by qualifying phrases.

(c) The problems just stated are parts of the broader area of relative ignorance, in spite of heroic work by many investigators over many years, namely the semimicrostructure of the ongoing and the evoked compound field potentials, in the tangential as well as the radial dimensions, on a millimeter or finer scale and a fraction of a second time resolution – as is called for in chapter 2, pp 5–12 (Bullock, this volume).

Comment on Chaos Analysis

At least eight chapters in this volume (those by Babloyantz, Başar, Başar and Röschke and Başar, Freeman, Graf and Elbert, Mpitsos, Rössler and Hudson, Saermark et al., and Skinner et al.) make major reference to chaos analysis of brain activity, that is, the application to digitized field potential data (EEG or spike intervals) of computer methods that can reveal chaotic as now distinguished from random activity or predictable periodicity. Chaos is not a rigidly defined but deterministic order predictable only over the short term; it can be numerically estimated by the number of dimensions, including fractional dimensions. Babloyantz has given an introduction to these concepts in her chapters in this volume and in its predecessor (Başar 1988).

It may be timely to raise a few questions from the standpoint of a physiologist who does not have the language of mathematics but – no doubt to a limited degree – appreciates what it can do. These are questions for which I do not find answers in the several chapters here and did not get answers in conversations with knowledgeable cognescenti, yet which seem nontrivial to the naive experimentalist.

The reason for raising questions is that these methods promise to be of value both in research on the brain and in clinical practice by providing potentially sensitive descriptors. They apparently permit fresh insights, exclusions, and quantitative statements from data as simple as single-channel time series, such as an EEG record.

For example, if the curve of correlation dimension (sometimes called D_2) against embedding dimension saturates to a clearly horizontal plateau, one is permitted to say that there is negligible "noise." The latter includes intrinsic random activity that need not be noise in the communication sense of useless interference. This is a strong statement for the electrocorticogram (Röschke and Başar, Babloyantz, Skinner et al., all this volume) or for gastropod neuron spiking (Mpitsos, this volume). We could not heretofore have said any such thing, and would have doubted any claim to this effect. As Mpitsos points out, instead of assuming, as we have, that deterministic signals ride on a background of random variation, it is now possible to show, in given cases, that the latter is insignificantly small and the observed fluctuation is essentially deterministic. It is now possible to distinguish samples quantitatively, by fractional dimensions; these may represent a minimum estimate of the nonlinear interactions as well as the number of distinct parameters like incommensurable periodicities. One might presume therefore that dimensionality grades complexity in a definable sense.

The physiologist evaluating these methods, or results reported from their use, needs to be given some feeling for their sensitivity to several factors. Roeschke (1986) implies that extrinsic white noise has a small effect on the estimate of correlation dimension if the signal-to-noise ratio is >10; it is not clear whether this is true for systems with higher as well as lower D_2. It would help to know how large must be the amplitude of some component in order to exert an effect on D_2; for example, when does an incommensurable periodicity newly appearing, and independent of the rest of the components, reach a threshold amplitude and add its contribution to D_2? How much effect does a single strong rhythm have, presumably suppressing D_2 by pushing some of the weaker signals below threshold as a proportion of the total power?

How sensitive is D_2 to nonstationarity – which comes in all degrees and kinds? Only Skinner, among our authors gives data on stability of D_2 over repeated samples, an elementary requirement for evaluating a biological descriptor.

I skip over questions about the several judgemental aspects of applying the standard algorithms (Albano et al. 1987; Theiler 1987). It should not be overlooked that some of them can be made substantially less subjective, according to new findings by Rapp et al. (1988) and Mees et al. (1987). Clearly the methods are not routine; there are still major differences among the users of these methods in the criteria for selecting values, in the practical compromises and rules of thumb, as well as in their statements on the upper limit of reliable D_2 values. General agreement on such matters will come slowly and unevenly, one imagines.

Beyond these basic robustness questions, one looks forward to help in evaluating the numbers. Is an EEG that goes from D_2=ca. 5 in stage 2 sleep to 6 awake with eyes closed, to 9.5 with eyes open, becoming more complex? Is the complexity comparable to some familiar systems? One wonders, for example, about the dimensionality of a microphone record of a voice that goes from singing to babbling to reading Shakespeare; or of several voices that go from unison recital to two and then three or more speaking independently at once? In my sampling of the literature and conversations with experts it is difficult to find analyses of more or less familiar systems of different degrees of complexity. Turbulence in a certain vessel of water, I am told (by J. Theiler), caused by counter-rotating cylinders, appears to have a correlation dimension of from 3.3 to 5.3, depending on the velocity of stirring. Other systems for which we may have some intuitive feel are, however, almost never examined. Some experts in these methods profess not to trust measurements of actual physical systems because of pitfalls in the application of the algorithms. It is clear that *correlation dimension is not a measure of complexity* of causes, in a common sense meaning, as distinct from elaborateness of consequences; D_2 can change merely by altering some constant in an interaction. D_2 can be an even integer, 3.0, as a result of three independent, incommensurable rhythms, whereas D_2 can be 2.1 as a result of three rhythms plus a nonlinear interaction between them, a more complex system. What does a higher D_2 mean; what does more chaotic mean?

To this observer the currently reported correlation dimensions of 5 to 10 for the mammalian cortex, insofar as this implies about that many variables, must testify to a great smoothing, in space or time or both, in the light of the known diversity of cell types and integrative variables determining their activity. Looking ahead, it is to be hoped that the power of these analyses will be extended to records of the same piece of brain with larger and smaller electrodes that see larger and smaller populations of cells, the latter being more likely to change with a half millimeter shift of locus. The preliminary reports available so far (see refs. in Başar 1988, and the chapters by Röschke and Başar; Babloyantz; Rössler and Hudson; Skinner et al.; Mpitsos; all this volume; Rapp et al. 1985; Hayashi and Ishizuka 1987; Labos 1987) encourage the search for this descriptor in various parts of the nervous system, including intracellular records not only of invertebrates (my old love, the follower cells of the cardiac ganglion of lobsters should be of special interest) but also of cortical, thalamic, limbic, reticular, cerebellar, and spinal neurons and glia, in resting, attentive, and striving states.

Since single-channel records provide such a meager sample of the spatiotemporal distribution of activity in the brain, it is to be hoped that the methods will be applied to recordings with 2, 3, 5, or 50 channels of simultaneous recording from neighboring parts of the brain, even though this increases the already formidable methodological problems. We cannot settle for the equivalent dipole so useful in cardiac physiology, since the imposed simplification would be too great, apart from special cases. In the general case we do not have a dominant rhythm but a broad spectrum. We know that coherence falls with distance in the millimeter range, to a widely varying degree (Bullock et al. 1984; Bullock and McClune 1989), but is appreciable even to centimeters in some frequency bands. Cooperativity among neurons is rife – and ripe for quantitative analysis. Invertebrate ganglia offer prime targets for methods sensitive to nonlinear interactions among a few dozens of cells.

We cannot yet say that all brains at all times have strange attractors, and nonrandom order, of integral or fractal dimensions, or significant coherence in time and space. Albano et al. (1987) and an unreported number of experiments in various laboratories obtain the result quite often that a given data set representing a time series recorded from the brain does not resolve into a stated number of dimensions and cannot be characterized in these terms. But chaotic attractors are at least widespread, and their dimensionality may yet be found to correlate in some degree with higher species, levels of the CNS, and levels of cognitive performance. I, for one, anticipate an exciting period in the near future, while empirical experience accumulates and interpretations are disputed and refined. As one of a set of new descriptors, quite nonintuitive and rooted in the dynamic processes ongoing in the system, it will surely stimulate new experiments, provide surprises, and lend support to some old prejudices.

Summary

On the ancient problem – Can the feeble instrument of the human brain comprehend the incredible complexity of the human brain? – we take the position that a negative answer cannot be justified any more than a positive one, that the goal is so important we must converge every possible tool, open new windows and expect advances on many fronts which will at the same time represent significant insights yet remain far from a basic comprehension. Without attempting to review the diverse contributions in this collection, I see them as reflecting a multipronged advance in the hitherto relatively neglected domain of studying signs of dynamic processes in organized neural tissue, both to explain them and to exploit them for clues to system function. The need is greater than ever for new views into the seething mass of cells, to add to the growing list of technics for visualizing activity, and the local signs of change in the brain.

A selection of specific needs is underlined. Foremost are needs for new data on the fine structure of the compound field potentials of various parts of the brain, ontogenetic stages, working states, and phylogenetic levels of the brain with resolution in the fractional millimeter and fractional second ranges, analyzed with respect to new, multichannel descriptors. Besides the demand for hardware and software to produce data of finer resolution in space and time and to reduce it and extract new measures, the basic questions are highlighted: what conclusions are really permitted from each new measure and display? Often, the first formulation of the findings with a new technic requires cautious restatement. Comments and questions are raised about the use of so-called chaos analysis, a potentially valuable new descriptor for achieving numerical estimates of dimensionality of brain activity in time and space.

References

Albano AM, Mees AI, de Guzman GC, Rapp PE (1987) Data requirements for reliable estimation of correlation dimensions. In: Degn H, Holden AV, Olsen LF (eds) Chaos in biological systems. Plenum, New York, pp 207–220

Başar E (1988) Dynamics of sensory and cognitive processing by the brain. Springer, Berlin Heidelberg New York (Springer series in brain dynamics, vol 1)

Bullock TH (1984) Comparative neuroscience holds promise for quiet revolutions. Science 225:473–477

Bullock TH, McClune MC (1989) Lateral coherence of the electrocorticogram: a new measure of brain synchrony. Electroencephalogr Clin Neurophysiol (in press)

Bullock TH, Lange GD, McClune MC (1984) A measure of synchrony in the cortical EEG: the slow wave drowsy state is slightly more synchronized horizontally than the low voltage fast state. Soc Neurosci Abstr 10:1143

Gesell R (1940) A neurophysiological interpretation of the respiratory act. Ergeb Physiol 43:476–639

Graubard K, Raper JA, Hartline DK (1980) Graded synaptic transmission between spiking neurons. Proc Natl Acad Sci USA 77:3733–3735

Hayashi H, Ishizuka S (1987) Chaos in molluscan neuron. In: Degn H, Holden AV, Olsen LF (eds) Chaos in biological systems. Plenum, New York, pp 157–166

Labos E (1987) Chaos and neural networks. In: Degn H, Holden AV, Olsen LF (eds) Chaos in biological systems. Plenum, New York, pp 195–206

Mees AI, Rapp PE, Jennings LS (1987) Singular-value decomposition and embedding dimension. Phys Rev 36A:340–346

Rapp PE, Zimmerman ID, Albano AM, de Guzman GC, Greenbaum NN, Bashore TR (1985) Experimental studies of chaotic neural dynamics: cellular activity and electroencephalographic signals. In: Othmar HG (ed) Nonlinear oscillations in biology and chemistry. Springer, Berlin Heidelberg New York, pp 175–205

Rapp PE, Albano AM, Mees AI (1988) Calculation of correlation dimensions from experimental data: progress and problems. In: JAS Kelso, AJ Mandell, MF Schlessinger (eds) Dynamic patterns in complex systems. World Scientific Publishers, Singapore, pp 191–205

Roberts A, Bush BMH (1981) Neurones without impulses. Cambridge University Press, Cambridge

Röschke J (1986) Eine Analyse der nichtlinearen EEG-Dynamik. Dissertation. Georg-August University, Göttingen

Shevelev IA, Tsykalov EN, Gorbach AM, Budko KP, Sharaev GA (1985) Thermovision parameters of brain responses to visual stimulation. Hum Physiol 11:227–232

Shevelev IA, Volovik MG, Sharaev GA, Tsykalov EN, Budko KP (1986a) Parameters of photic stimulation and characteristics of thermoreactions in the cerebral cortex of the white rat (in Russian). Neirofiziologiia 18:322–340

Shevelev IA, Kotliar BI, Volovik MA, Martinson IuL, Budko KP (1986b) Detection of the activation of the motor cortex of the cat during movement of an extremity using thermomapping of the brain (in Russian). Neirofiziologiia 18:266–269

Theiler J (1987) Spurious dimensions from correlation algorithms applied to limited time series data. Phys Rev 34A:2427–2433

GPSR Compliance

The European Union's (EU) General Product Safety Regulation (GPSR) is a set of rules that requires consumer products to be safe and our obligations to ensure this.

If you have any concerns about our products, you can contact us on ProductSafety@springernature.com

In case Publisher is established outside the EU, the EU authorized representative is:

Springer Nature Customer Service Center GmbH
Europaplatz 3
69115 Heidelberg, Germany

Batch number: 10838062

Printed by Printforce, the Netherlands